Effective Model-Based Systems Engineering

John M. Borky • Thomas H. Bradley

Effective Model-Based Systems Engineering

John M. Borky
Systems Engineering Program
Colorado State University
Ft. Collins, CO, USA

Thomas H. Bradley
Department of Mechanical & Systems
Engineering
Colorado State University
Ft. Collins, CO, USA

ISBN 978-3-319-95668-8 ISBN 978-3-319-95669-5 (eBook)
https://doi.org/10.1007/978-3-319-95669-5

Library of Congress Control Number: 2018950394

This Springer imprint is published by the registered company Springer Nature Switzerland AG.
The registered company address is: Gewerbestrasse 11, 6330 Cham, Switzerland

Preface

Systems and enterprises (often referred to as systems-of-systems) that employ multiple advanced technologies and achieve ever-higher levels of complexity have become mainstays of the modern world. Managing the complexity of these systems, especially to create solutions to the needs of system users and other stakeholders within acceptable cost and schedule, is a recurring challenge to the Systems Engineering (SE) community. Spectacular progress in mechanical and electronic devices, materials, sensors, energy generation and control, vehicles and propulsion, communications and networking, computing and software, and many other fields creates both new capabilities and new demands for tools and methods to effectively exploit them. Communities as diverse as the sciences, commercial enterprises, education, government, academia, and the military frequently report that programs aimed at delivering new capabilities based on advanced technology have overrun their budgets and schedules, failed to meet their objectives, or simply overwhelmed the capabilities of system developers and users.

The thesis of this book is that there exists a fundamentally better approach to dealing with complex, multi-technology, and information-intensive systems. The approach is based on the concept of using system architecture as a rigorous, objective, quantitative, and measurable discipline for defining, analyzing, and building systems that employ advanced technologies. The current state-of-the-practice in SE is a strategy called Model-Based Systems Engineering (MBSE) that is predicated on this principle.

In this book we describe a methodology called the Model-Based System Architecture Process (MBSAP), which is a specific implementation of the principles of MBSE. Our emphasis is on the practical aspects of translating user needs into operationally effective, affordable, and supportable systems, rather than on formal architecture theory, which is well covered in a number of existing texts, referenced in the Bibliography. With this in mind, throughout the book we illustrate the principles and products of MBSAP with detailed system architecture examples. For one of these, we provide an extensive appendix showing a more complete architecture description than is possible in the main body of the text, intending this as a template for new architecture developments.

System Architecture and Systems Engineering

The notion of architecture is deceptively simple—everyone has looked at a skyscraper or a bridge and thought with admiration that some architect has designed an elegant solution to a human need for space or passage. At the same time, an insightful observer will recognize that the art of beautiful design must be complemented by the science of practical implementation. We adopt the broad notion of architecture as the grand design of complex solutions to complex problems, but we seek to develop deeper insight into the technologies and design methods that make system architectures real and valuable to a system's owners and users. Our examples reflect the areas in which we have amassed much of our professional experience: one from the world of aerospace and another from the electric power industry. Beyond these domains, we, and our students, have successfully applied MBSE in general and MBSAP in particular to a wide range of systems and enterprises. A reader who acquires the architectural viewpoint and masters the skills presented here will be well prepared to function as a system architect and to implement MBSE in most arenas where advanced technology, especially information technology, is the key enabler of capability.

The view that architecture, and specifically the use of an architecture model as the center of an SE process, is the key to success in exploiting advanced technology to meet the needs of today's society is neither new nor original. Many individuals and technical organizations such as the International Council on Systems Engineering (INCOSE) and the Open Group have developed and advocated architecture-centric methods. The INCOSE vision for 2020 stated that the future of SE will be model-based. However, to be effective, such a strategy must be founded on a formal description of architecture, that is to say, on an architecture model built in an appropriate tool and conforming to applicable standards and methods.

It is important to distinguish between *system* architecture and *software* architecture. Most books and articles in this area focus on software, with the implicit assumption that the primary challenges in achieving satisfactory products and systems lie in the software that controls them. In fact, most of the concepts, methods, and standards described in this book have their origins in software science and engineering. The most important of these is object orientation (OO), which is summarized in Chap. 2. The developers of OO theory and practice mostly took the view that hardware such as computers, data storage devices, and user interfaces was a given, typically provided as a set of off-the-shelf commodity products whose characteristics were well understood. Hardware, in this view, exists mostly to give software a place to run and some resources to control.

While conceding this history and acknowledging the importance and difficulty of developing high-quality software, our emphasis in this book is on the entire system, the ensemble of hardware, software, controlled equipment, user interfaces, networks, and other elements that together deliver the solution to a customer's needs. This perspective is especially important in the world of complex, high-performance systems where advanced hardware is commonly developed in parallel with software, and the two must integrate effectively to meet requirements. Even when existing hardware components can be used, there is generally a need for careful performance analysis and trade studies to select the best products. In addition, systems often have stringent process deadlines, safety constraints, harsh operational environments, and many other stressing requirements. An optimized blend of hardware and software components is essential; consequently, system architecture, embracing more than just software, is the essential focus. Even in less demanding situations, the methodology developed in this book applies, and, in fact, it increases the chances that a truly optimized system solution will be achieved because all the factors affecting performance, supportability, upgrading, ease of use, and so on are considered in a balanced and objective fashion.

We place primary emphasis on systems and enterprises with significant information content and processes. This is because the functionality of complex systems today is increasingly dominated by information technology: computing, communications and networking, software and algorithms, human-machine interfaces, and so forth. While conceding this bias, we strongly believe that the principles and methods of MBSAP apply very generally, including to mechanical, chemical, and even biological technologies and systems. The central challenge is to manage the complexity brought about by diverse and rapidly changing technologies and ever more sophisticated products and services. Managing that complexity is what MBSAP is fundamentally about.

Our overall goal, then, is to show system engineers and others concerned with effective SE how MBSE helps them do their jobs and achieve their objectives. We discuss the differences between MBSE and traditional document-based SE and explain the benefits of the former in the processes that make up an end-to-end SE flow, from customer needs to delivered and supported system solutions. We make this specific and practical in the MBSAP methodology. We present detailed examples of system architectures that show how each SE activity is made more efficient and repeatable using architecture modeling as the foundation and how the products of those activities become more clear, unambiguous, and easily communicated compared to document-based approaches.

Organization of the Book

The material of the book falls into three broad sections. Chapters 1 and 2 present the challenge being addressed and the basic concepts of an OO approach to architecture and MBSE. The second section, Chaps. 3–6, spells out the stages of MBSAP and contains the core material of the methodology. After an overview of MBSAP in Chap. 3, Chaps. 4–6 describe in detail the development of Operational, Logical/Functional, and Physical representations of the architecture. Finally, Chaps. 7–15 get into more detail on key topics encountered in many systems such as service orientation, real-time behavior, cybersecurity, networking, architecture prototyping, verification and validation, and architecture governance. In so doing, these chapters explore some of the important variations on the basic MBSAP approach. In the appendices, we provide additional detail in various areas that may be important to some system architects and engineers but would be overwhelming if embodied in the main text. Note on graphics: most of the diagrams in this book conform to the Systems Modeling Language (SysML) standard. In a few cases, alternative notations are used, as is the practice in some architecture modeling tools. These are clearly labelled.

Note to Instructors

This book is a refinement and expansion of material that has been used effectively with hundreds of students over a period of more than 10 years. It can be used with a variety of course structures at both the advanced undergraduate and graduate levels. We assume a typical course is based on three credit hours in a 16-week semester or 12-week quarter. An architecture modeling lab, for which one or more additional credits can be awarded, can be a valuable adjunct to classroom instruction.

Modeling skills must be acquired through actual practice, and so we strongly recommend student projects that apply MBSAP to realistic problems in the form of class projects. These projects can be student-selected, and some of the best have been related to current or projected employment, especially for graduate students who are also practicing professionals. Project development can be incorporated as homework assignments or carried out in a lab setting.

Ideally, students should master a full-featured SysML modeling tool and use it to create their project architectures. However, our experience clearly shows that covering the full body of material in the text while simultaneously dealing with the numerous subtleties of modern architecture tools in a single course can create an unrealistic student workload. We suggest the following as candidate courses that take these facts into account.

- Undergraduate Introduction to MBSE (three credits with the option to add a one credit modeling lab):
  - Structure the course as a follow-on to a general SE introductory course or as a stand-alone by adding an SE survey at the beginning to present the essentials of the discipline.
  - Cover Chaps. 1–6 and 15.
  - Cover selected content from Chaps. 7–14, depending on specific learning objectives and student backgrounds and needs.
  - Require student projects; in a basic three-credit course, these can be created as graphics, text, and tables.
  - With the addition of a modeling lab, make a mainstream SysML modeling tool available, provide suitable tutorials and reference materials, and use the tool to create projects.

- Graduate-Level Introduction to MBSE—similar to the undergraduate course, but with more complete coverage of Chaps. 7–14 and more required content in student architecture projects. The addition of a modeling lab for added credit to allow use of a SysML tool is strongly advised.
- Graduate-Level Course Sequence in MBSE (six or more credits):
  - Cover the full text.
  - Incorporate student projects using a SysML tool with appropriate scope and complexity to apply all of the architecture methods presented in the text, including real-time behavior, enterprise context, cybersecurity, and so forth.
  - Optionally supplement the text with advanced topics such as formal methods, agile SE, and architecture optimization.

Because of the importance of student projects in all of these alternative courses, we include in Chaps. 3–15 summaries of project content to be created as a course or course sequence proceeds.

Summary

The MBSAP methodology has evolved over more than three decades of experience and experimentation, involving a wide variety of systems. It is essentially a selection and synthesis of ideas and proven practices from a very large number of sources. In the end, we hope that this approach and methodology will provide system engineers and architects who are grappling with the task of applying current and rapidly evolving technologies to difficult problems a better set of tools along with confidence that success is possible.

Ft. Collins, CO, USA John M. Borky
Ft. Collins, CO, USA Thomas H. Bradley

Acknowledgments

We are indebted to colleagues and students far too numerous to name for information, insights, reviews, feedback, and encouragement over a period spanning decades without which this book would not have been written. We particularly want to thank the Director, Dr. Ron Sega, and the faculty of the Systems Engineering Program in the College of Engineering, Colorado State University, who helped us in countless ways to refine both the content of the text and the courses in which this material is used, helping to make this topic a core element of an integrated Systems Engineering curriculum.

Contents

Chapter 1
Introduction: Framing the Problem

1.1 The Aesthetic and Scientific Dimensions of Architecture

The concepts of and methods behind the idea of an architecture have been applied in virtually every field of human activity, from the original context of designing buildings and other structures to areas as diverse as human anatomy, business processes, and computer networks. Typical definitions of system architecture address the basic structure and components of a system, its internal and external interactions and behaviors, and the rules and guidelines that govern its initial development and long-term evolution. In this book, we consider the use of architectural principles and formal disciplines to manage the complexity of advanced technology systems and enterprises, especially those that rely on sophisticated information technology. We illustrate these principles with examples from the aerospace and energy domains, representing the authors' backgrounds, but they are applicable to virtually any complex system category.

We use the terms "complex" and "complexity" in the sense of describing something that cannot be adequately dealt with using only simple, intuitive tools and methods. Such complexity in a system or enterprise commonly evolves from large numbers of constituent elements, multiple internal and external interactions, known and unanticipated dependencies among system elements and with external entities, diverse information content, sophisticated algorithms, and other factors. For example, the complexity challenge has reached a point where the Defense Advanced Projects Agency (DARPA) recently launched an initiative called Adaptive Vehicle Make (AVM) that combines approaches such as model-based design, virtual collaboration, and foundries to deal with burgeoning program failures, schedule slips, and cost overruns [1]. The American Institute of Aeronautics and Astronautics (AIAA) has created a Complex Aerospace Systems Exchange (CASE) to spread complex system experience and promote advanced Systems Engineering (SE) methods [2]. Complex systems routinely exhibit behaviors that are difficult to predict or even to understand. The challenge of representing, analyzing, designing, and

J. M. Borky, T. H. Bradley, *Effective Model-Based Systems Engineering*,
https://doi.org/10.1007/978-3-319-95669-5_1

optimizing such systems demands the power and conceptual rigor of mathematical approaches such as modeling.

Maier and Rechtin [3] talk about the "art and science of architecture" to express the concept of separate but related aesthetic and scientific dimensions. Architecture must often concern itself with achieving beauty, elegance, and efficiency in applying resources to achieve results that meet the needs of those who will use its products. Ryschkewitsch et al. [4], who use the phrase "The Art and Science of Systems Engineering," express a similar view. Their perspective derives from long experience in programs of the National Aeronautics and Space Administration (NASA), which has for decades been a leader in the evolution of modern Systems Engineering (SE), and they stress the characteristics of effective system engineers and how they can be developed.

Often, system architecture must embrace both aesthetics (elegant concept) and technology (feasible implementation). The boundary between architecture and engineering, or between conceptualization and design, is necessarily rather vague, but it's helpful to think of the architect as attacking a problem at the outset, when it is often ill-formed and incompletely described, and reducing it to a set of clear and unambiguous tasks and requirements from which engineers can proceed to implement a solution. The architect ideally remains involved throughout the design and building of the product or other solutions to ensure that the integrity and balance of the original vision are maintained and to resolve inevitable conflicts, misunderstandings, and ambiguities that arise in the course of achieving the final result.

A typical illustration of this "art and science" concept lies in an old saying among aircraft designers: "looks good, flies good," meaning that an airplane that pleases the eye sitting on the ramp will usually perform well in the air. From this perspective, the architect is first an artist who brings to a task a sense of order, proportion, relationships, and basic principles, as well as comprehensive knowledge of earlier solutions and successful approaches to similar problems. Following this line of thought, much good architecture practice can be reduced to heuristics, or rules of thumb, that are helpful in the initial stages of assessing a problem and plotting the path to a solution, and we present a number of these in this book.

Architecture becomes real and powerful in dealing with complex challenges through the role of the architect. An architect is often described as a "big picture" person, someone who can understand a multidimensional problem, discover inherent patterns and themes, and cast the problem into a form that allows an orderly approach to its solution. At the same time, the system architect must be a highly skilled technical professional who can apply quantitative tools and methods to analyze, design, optimize, measure, document, communicate, and control the complex entities for which architectures are desired. Under this overarching theme, we explicitly deal in this book with modeling and the use of standard languages to capture and manipulate architectures for complex systems and enterprises. Considerations as diverse as operational economics, safety, and compliance with international law demand a level of rigor and precision that only formal modeling methods can impart. As a result, one of the hallmarks of a good architect is the ability to work in both aesthetic and mathematical realms. To that end, the discussion in this text stresses methods that are model-based and metrics-managed, but along the way, there are a number of heuristics that distill experience from earlier architectures and offer guidance for new ones.

The analogy to architecture in its traditional sense is appropriate. A customer who engages a building architect to design a skyscraper expects that architect to apply knowledge of structures and materials, to account for floor loads, elevator traffic, heating and cooling, lighting, and other factors as well as providing for wind forces, solar heat loads, and other environmental stresses. While doing all this, the same architect, or, more likely, the architecture team, is expected to achieve a result that is beautiful, complements its surroundings, and can be erected by engineers and craftsmen with available tools and products on a predictable schedule and budget. A central tenet of this book is that equivalent expectations should apply to architecting large, high-technology systems.

System, and especially software, architecture has all too often been a rather ad hoc, "make-it-up-as-we-go-along" proposition, sometimes taken as consisting of little more than the documentation of whatever design happens to emerge at the end of a development. We argue for the position, increasingly shared by government policy-makers and by sophisticated system customers in the civil and commercial sectors, that architecture is the critical foundation for developing and acquiring these systems and that it is capable of the same levels of scientific and engineering rigor as are expected in building a skyscraper. Following this premise, we pay attention to applying architectural methods to create systems and enterprises that are well organized, expandable and evolvable, robust under the stresses of real-world use, and affordable to own and operate. In short, the essence of the art and science of architecture is manifested in results that are beautiful in the eyes of their users while satisfying those users' practical needs.

1.2 Model-Based Systems Engineering

Although this book explicitly focuses on architecture, it's important to recognize that this is only one aspect of the broader subject of system science, or what is sometimes called "system thinking" [5]. A system is a set of interacting elements that work together to accomplish some purpose, and it's generally the case that the capability delivered by the ensemble is greater than the sum of the capabilities of the individual parts, or subsystems. The big picture viewpoint of architecture is concerned with maintaining a comprehensive vision and understanding of the whole system, avoiding the temptation to focus too narrowly on the details of its implementation or its functions in any one area.

The importance of such a system-level description is the primary motivation for the emergence of the discipline of system architecture. The power of a formal architecture methodology based on development of a model to account for all the dimensions, behaviors, and components of a system is becoming recognized. This leads to the growing acceptance of MBSE, in which the model is the foundation of the entire

SE process, as further described in Chap. 3, and provides a clear and unambiguous definition of the system. For readers looking to explore the history and underlying theory, Long and Scott provide a concise and insightful summary of the rationale for MBSE [6]. Ryan et al. [7] summarize the advantages of MBSE and present a framework for architecture definition based on an improved trade study methodology that includes variability modeling. More general treatments of SE can be found in books by Kossiakoff and Sweet [8] and by Buede and Miller [9].

Typical definitions of SE run along the lines of "a process for translating customer needs into an effective and affordable solution." Expanding on this basic idea, the International Council on Systems Engineering (INCOSE) defines MBSE in terms of a formal application of modeling to support a wide range of systems engineering activities, including requirements definition, design, analysis, verification, and validation [10]. These activities start in the conceptual design phase and continue throughout development and later life cycle phases.

MBSE formalizes SE using models to:

- Ensure rigor, repeatability, and producibility in SE processes
- Promote quality, completeness, and correctness in system designs
- Reduce risk in requirements analysis, design, integration and test, and other activities
- Enhance communication and synchronization of activities across organizations and disciplines

System engineers are often viewed as the central coordinators of a system development effort, responsible for analyzing requirements, performing trade studies, allocating requirements to specialized hardware and software teams, and coordinating activities such as system integration and test. It will always be essential to deal fully with the hardware, software, data, interfaces, and other details that have traditionally been the primary focus of designers. However, MBSE using an architecture model greatly enhances the ability to deal with overarching issues such as the diverse concerns of a system's stakeholders, stated and derived system requirements, planned system evolution, policy and other constraints, and operational suitability of the delivered system.

This book should be read in the context of MBSE. The architecture methodology it describes, the Model-Based System Architecture Process (MBSAP), is tightly tied to the SE processes that occur over a system's life cycle, which are often portrayed in a "Vee diagram," of which Fig. 1.1 is a simple example. In Chap. 3, we map the MBSAP methodology to a more detailed Vee diagram and show that the architecture modeling methodology maps readily to the various processes associated with SE as it has been practiced for decades while providing better tools for dealing with the complexity of modern systems. Supported by an architecture model, those processes include:

- *Concept development and analysis*—The model supports a dialog with system stakeholders to identify and assess alternative system concepts.

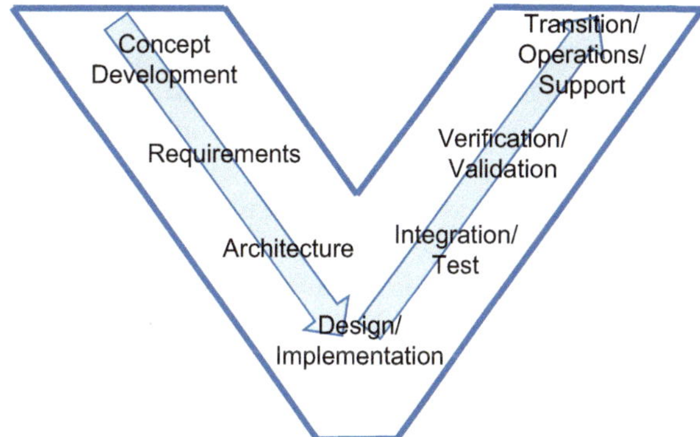

Fig. 1.1 The traditional Systems Engineering Vee diagram shows the primary stages of system development

- *Requirements capture, analysis, allocation, and traceability*—The model allows requirements to be decomposed and allocated to system elements and behaviors and to be documented in artifacts generated from the model.
- *High-level and detailed design*—The model captures and documents the evolving system design baseline and enables supporting analysis and trade studies.
- *Integration and test*—The architecture supports planning and execution of I&T, including identification of stressing conditions that create effective test points.
- *Verification and validation*—The model provides a solid foundation for verifying compliance of the design with requirements and validating that the system meets stakeholder needs.
- *Transition to operations and support*—The model and associated documentation support delivery and installation, personnel training, configuration management, modifications, upgrades, technology refreshment, failure analysis, and other activities associated with long-term system effectiveness and supportability.

1.3 The Evolution of Modern, High-Technology Systems

The origins of concepts that evolved into today's SE practice can be traced to the early years of the twentieth century [11]. Beginning shortly before and especially during World War II, electronics and information technology grew rapidly in importance in both military and commercial systems. This included, first, analog computers that represented numerical values as continuously varying voltages and subsequently digital computers that dealt with quantities as binary numbers represented by strings of bits. A few examples among thousands that could be cited will illustrate the trend. During the war, computers were used to calculate artillery trajectories and to aim the big guns on ships, as well as to support intelligence analysis and other activities. A particularly famous example was the machine used to break the Enigma communications encryption codes of the Axis armed forces. As the Cold War took form, an air defense network called the Semi-Automated Ground Environment (SAGE) became one of the first large and recognizably modern computer-based systems. The designers of SAGE pioneered features that ranged from fully redundant computing machinery (to deal with the failure rate of vacuum tube electronics) to a sophisticated human-machine interface. Mainstream programming languages like COBOL emerged from military software development efforts and were then applied more broadly in business and other arenas. By the 1960s, airplanes like the F-106 and F-111 used general purpose digital computers for such functions as target tracking, and onboard computers made possible the rendezvous, trajectory control, and lunar landing maneuvers of Project Apollo.

During the 1970s and 1980s, embedded (often called "real-time") computers, interconnected by multiplex data busses and networks, became ubiquitous in many kinds of systems. Globally distributed systems entailing enormous amounts of software appeared. Communities as diverse as manufacturing, medicine, energy, and finance became increasingly reliant on information processing and storage, automation, and communications. For example, as electrical power grids grew in size, they implemented centralized monitoring and control, automated protection devices, sophisticated metering, and backup generating capacity, all of which were increasingly reliant on computers and telecommunications.

In the decades since, the amount of information created, exchanged, and used in every part of advanced societies has grown astronomically. Probably the most important technology trend was the development of microelectronics and especially the microprocessor that made powerful computers small enough to be embedded in devices from wrist watches and handheld calculators to guided missiles. Techniques to process and manage this information such as expert systems, adaptive control, artificial neural networks, and fuzzy logic have been tried, admittedly with varying results. When the information needed to make informed and timely decisions is contained in many different, often disparate, forms and locations, a need arises to enable individual systems and units containing the information able to function as nodes in a networked organization or enterprise. In this respect, military, civil, and commercial systems face similar challenges, whether they involve coordinating operations, forecasting tomorrow's weather, or predicting stock prices. In addition to relentless growth in software and information content, systems are required to interact over networks and to undergo continuous evolution, capability upgrades, and technology refreshment. There is no end in sight to this growth in information volume, processing power, and interconnectedness.

Much of this is fueled by the steady improvement of electronic technology often labeled Moore's Law.[1] Military systems like the Minuteman strategic missile system were largely responsible for the early stages of this trend by providing the initial

---

[1]Gordon Moore, a co-founder of Intel Corporation, first articulated the fact that the number of transistors on a chip (called its density) was doubling in a period of about 18–24 months, resulting in exponential growth in chip function and a commensurate decrease in unit function cost. Comparable trends can be seen in complementary technologies such as disk storage capacity and network data rates. The persistence over decades of such rates of progress in a fundamental technology is unprecedented in the history of science and quite astounding, although taken for granted by most users of the resulting products. Gordon Moore, incidentally, has said he wants to be remembered for anything *but* Moore's Law.

mass market for microelectronics. Today, computers as powerful as the large mainframe machines of a generation ago can be built into a smartphone. To go with this tremendous local computing power, advanced satellite communication networks see a future in which global data exchanges at multi-gigabit per second (Gbps) rates become possible. A single data processing center can require petabytes of nonvolatile data storage. All the books ever printed would fit in a small corner of such an information repository. The collective resources of the World Wide Web are now measured in zettabytes, and no end is in sight.

In parallel with the explosion of information technology (IT) and electronics, great advances have been made in other areas. In materials science and manufacturing, the growing use of composite materials, advanced alloys and metallurgy, and additive manufacturing has created new possibilities for structures that are light, strong, faster to design, and less expensive to produce. Renewable energy from wind, solar power, geothermal sources, ocean currents, and other novel generation mechanisms is becoming a significant contributor to meeting humanity's power needs while reducing pollution. Chemistry, biology, computer science, and many other disciplines continue to make great strides. The result is that the architect of a modern system has options for satisfying user requirements that earlier generations could only dream about.

Such progress has also had a downside. With the potential inherent in exploiting multiple advanced technologies come new levels of complexity. Improved system performance has led to increased cost of development and sustainment, and budget and schedule overruns have been and remain commonplace in system development programs. Software is particularly notorious for taking longer to develop and test, delivering less performance, and being fielded with more residual errors than were predicted or could be tolerated. This has resulted in more than a few system development programs being canceled. A little over a decade ago, an advanced tactical aircraft with an early version of an integrated avionics suite had major problems with computer crashes, ultimately corrected by intensive effort and at considerable expense, whose root cause proved to be a lack of adequate tools and discipline for coping with the system's complexity. Continued progress in almost every area of modern society in the information age depends to an important degree on solutions to these problems. Thomas [12] recently published a searching analysis of the importance of SE in the aerospace and defense community and concluded that applying the discipline early and throughout all stages of a program can reduce both cost and time to complete it.

That is the main motivation for this book. In later chapters, we discuss the reality that complex, information-intensive systems have reached a level of complexity where it is sometimes not possible to identify all of the ways a system will perform or respond to inputs based solely on examining the behaviors of its constituent parts [13]. A system's outputs may change drastically, and even unpredictably, in response to relatively small changes in its inputs. Software programs measured in millions or tens of millions of source lines of code (SLOCs) are now commonplace, and new ways to specify, write, integrate, test, modify, and document them are urgently needed. Powerful new modeling and simulation (M&S) technologies are available that have great promise in dealing with these challenges if they can be integrated into architectural and design methodologies. Techniques that have evolved in the world of software engineering can be equally effective in the larger spheres of system and system-of-systems enterprise engineering. In this book, we show how MBSAP weaves these and other techniques together in an architecture-centric methodology for MBSE, embodying best practices that represent the consensus of the systems community.

1.4 Architecture's Tower of Babel

A rigorous architecture discipline demands clear and unambiguous terminology. This has been sadly lacking in the field of system architecture. There is scarcely a word in common use that has as many different and inconsistent meanings as "architecture," and many arguments among developers, users, and policy-makers can be traced to the fact that the disputants are using the same terms for very different things. The first concern, therefore, is a set of definitions that attempt to reconcile competing usages and will, at least, be consistently employed in describing MBSAP.

Among the many senses in which the word architecture has been applied are:

- *Design*—description of the structure, behaviors, and interactions of a complex entity
- *Method*—tools and processes for deriving, optimizing, and documenting the design
- *Standards*—"building codes" that constrain the design for the sake of desired features such as commonality, interoperability, and evolvability (one aspect of "open architecture")
- *Documentation*—drawings, specifications, and models that capture the design
- *Concept*—a grand scheme for that which is to be designed

Architecture has sometimes been treated as if it consisted only of a list of standards or a list of things. Many government agencies and private sector organizations compiled lists of approved or mandated standards, believing the use of these would assure affordable, interoperable, and evolvable systems [14]. It was common to encounter "architectures" that consisted of little more than artist's concepts, with system participants connected in a highly nonspecific way using cartoons connected by lightning bolts. These representations were long on eye appeal but short on technical content and totally inadequate to support actual SE. Other engineering activities have reduced architecture to a hardware block diagram, an entity-relationship diagram, a set of functional flows, or other specific documents and drawings.

In another variation on the theme, the space community commonly takes "architecture" to refer to the early analysis of system concepts, dealing with basic decisions like the number of stages in a booster or of satellites in a constellation. Using this definition, McManus et al. [15] discuss architectures for space in the context of an overall trade space, leading to a "design vector" that defines the number of vehicles, concept of operations, orbital parameters, communications parameters, spacecraft maneuver capability, and so forth.

The problem is that while standards profiles, operational context diagrams, and other examples just cited may be valid architectural artifacts, each only deals with a narrow slice of the subject. Worse, too narrow a focus can provide a convenient excuse to skip over the hard work of real architecture analysis that extends throughout a program's life. More recent government initiatives, including the Federal Enterprise Architecture Program (FEAP) [16] and its equivalents in various departments and agencies, are more complete and rigorous and will be discussed in a later chapter. For now, the goal is a set of terms and definitions adequate to underpin a modern architecture methodology.

The definition of architecture can be expanded to make it more useful in structuring a methodology. Accordingly, throughout this book we define system architecture as the formal representation of a system or other complex entity, using models, in order to elucidate:

- Its structure, interfaces, and internal and external relationships
- The behaviors exhibited by the entity and its elements, both internally and externally
- The global rules to which the entity and its elements must conform in order to meet the requirements allocated to them, initially and over the entity's operational lifetime

This definition makes it very natural to address architecture through structural, behavioral, and contextual perspectives. By way of comparison, Lattanze [17] describes architectural design as addressing:

- Partitioning of a system into parts or elements and interactions among them
- Overall system properties
- A declarative system description

This can be seen as an alternative way to treat the basic factors of structure, behavior, and rules.

In contemporary practice, the various dimensions of system architecture are commonly organized using "Viewpoints" which contain various "Models" or "Views." A Viewpoint typically represents the aspects of the system that are of primary concern to one or more "stakeholders," the individuals or organizations that have a legitimate interest in the system and therefore a voice in how it is defined and implemented. For example, an "Operational Viewpoint" captures the things of greatest importance to the system's end users. Viewpoints and their specified contents are the building blocks of "architecture frameworks" that attempt to provide guidance for architecture development and consistency in the resulting system descriptions. These and related subjects are explored in Chap. 12.

Many architecture categories and varieties exist, and these are so widely used that they too demand clear definitions. Table 1.1 defines a representative set. Each of these arises from the concerns of a particular stakeholder or other interest group, which, naturally, thinks its perspective on architecture is the most important. The

Table 1.1 Architecture categories and variations

Architecture category	Definition
Enterprise architecture	Applies the basic definition of architecture (structure, behaviors, and global rules) to the top level of an ensemble of nodes, systems, elements, or other resources that collaborate to fulfill the functions of an overall organization or business process
System architecture	Applies the basic definition to an ensemble of elements (ultimately hardware and software components) that collaborate to fulfill defined requirements allocated to a node or system (implying that a clear system boundary and user interfaces are defined)
Software architecture	Applies the basic definition to software, focusing on frameworks, software requirements, application programs, infrastructure programs, workflow management, networking and messaging, interfaces, and other aspects of computer programming
Hardware architecture	Applies the basic definition to hardware, focusing on processors, storage, interconnects, operator stations, communications, sensors, effectors, and other hardware elements
Layered or n-tier architecture	Defines functionality in tiers with interfaces that isolate functions in one tier from changes in another; typical layers include: • Client/presentation/visualization layer • Process/business logic layer—further divided into applications and middleware • Repository/resources/networking layer
Service-oriented architecture (SOA)	Describes an architecture pattern in which interactions among elements and with the external environment take the form of behaviors exposed as services, usually invoked in a loosely coupled request/response model using messaging. A service is a callable function or routine that is made available over a network. A service exposes an interface that is controlled by a service-level agreement (SLA) or other contracts and that defines the behavior of the service and the messages it accepts and returns. A service is a unit of work done by a service provider to achieve desired end results for a service consumer. See Chap 7
Technical architecture	Defines a set of allowable standards for components, services, protocols, formats, and other elements
Reference architecture (RA)	A logical/functional abstraction that defines the features and behaviors common to a domain or class of entities. An RA is instantiated by the addition of relevant detail to achieve a physical architecture that satisfies a specific set of requirements within the domain
Executable architecture	Represents the architecture in the form of computer models that can be run to simulate behaviors, perform automatic code generation, verify design correctness, etc. Executables exist at all levels of abstraction used to describe an architecture

architect must maintain a holistic viewpoint, embracing all of these perspectives in a consistent methodology and set of artifacts that capture the full range of architecture characteristics, relationships, and interdependencies. Architecting involves the search for optimum solutions in problem spaces with many dimensions, and the architect must balance competing interests while keeping at the forefront the needs of the ultimate customer. Clear and consistent terminology helps in meeting that challenge.

1.5 The Real Meaning of Open Architecture

One term conspicuous by its absence from Table 1.1 is "open architecture." This subject is so important, and so widely misunderstood, that it warrants a separate discussion as part of the stage setting for our discussion of MBSAP. What might be called the "open architecture vision" evolved over many years of experience with computing systems that were "closed" because their technologies and products were proprietary to a given manufacturer as well as being difficult and expensive to change over time as user needs evolved. Manufacturers took pains to ensure that their products were incompatible with competing products and that a customer had essentially no choice, other than starting over, but to remain with the original supplier for updates, added functions, and infusion of new technologies. By contrast, an open system is characterized by the ability to:

- Select and assemble "best of breed" products from various sources without expensive modifications and integration efforts
- Selectively replace, modify, or upgrade at the level of relatively fine-grained components, such as a processor or software application, with minimal (ideally, no) impacts to the remainder of the system other than using the new or better functionality
- Scale system capabilities by the addition of modular units of hardware and software, achieving performance commensurate with the installed assets; similarly, create new system components that can be reused for equivalent functions in other systems, and reuse such components from other systems
- Interoperate with networks and other systems with minimal (ideally, no) added effort to provide adapters or other special purpose components
- Make extensive use of commercial-off-the-shelf (COTS) products to minimize development effort and enjoy mass market pricing

Following a similar train of logic, the Software Engineering Institute defines an open system as a collection of interacting software, hardware, and human components that is:

- Designed to satisfy stated needs
- Scalable and upgradeable
- Provided with interface specifications of its components that are fully defined, available to the public, and maintained according to group consensus
- Designed so that the implementations of the components conform to the interface specifications

As noted above, early attempts to define open system architecture (OSA) in ways that could be enforced in system development stressed the use of open standards. The emphasis was on interfaces on the principle that standardization here would allow components to be connected and integrated more freely. This is often called the "building codes" approach by analogy to standard interfaces like a 110 V wall outlet that any household appliance can plug into (unless, of course, the appliance finds itself in Europe or Asia where other voltages and plug designs are standard). In practice, primary reliance on this interface strategy has resulted in mostly disappointing and sometimes disastrous results.

An early example that illustrates the problem was the Defense Information Infrastructure Common Operating Environment (DIICOE) [18], which created a standards infrastructure to make networked processors and applications interoperable. It specified a layered structure with application programming interfaces (APIs) and required hardware and software components to be segmented to fit into this structure. DIICOE had some successes in situations where the entire standardized ensemble could be managed to a single COE version and enjoyed stable requirements. However, integration of software products, especially COTS that were not designed for compliance, raised issues with the cost and time involved in segmentation. Another issue was that limited backward compatibility hindered the integration of elements built to different DIICOE versions. Today, DIICOE is generally viewed as a good try that didn't work, and it has been supplanted by a service-oriented architecture (SOA) paradigm. Chapter 7 describes SOA and its attributes in more depth.

The lesson of such experiences is that standards compliance, especially for interfaces, is an essential aspect of OSA, but is not sufficient to realize the intended benefits. No standard can compensate for a system design that is badly partitioned, has tangled control threads, relies too much on critical timing relationships, or has hard-coded workflows. Moreover, standards themselves are often transient and can become obsolete in much less time than the life of a system. The key points are that:

- Effective standards are based on real-world experience and evidence that they work in their intended role; this causes them to evolve more slowly than the systems to which they are applied.
- Application of a standard is a *design decision* made at a point in time on the basis of being the best available solution to a particular system need.
- Effective standardization demands a high-quality underlying architecture.
- The architecture must support an evolving standards profile just as it does changing functions or successive product generations.

These are the principles behind more refined open architecture strategies, and they are central elements of MBSAP.

An example of successful application of open architecture principles to real-world systems is the Radar Open System Architecture (ROSA), now in its second version, ROSA II [19, 20]. This is based on modular components, both hardware and software, and a layered architecture (see Chap. 5) in which application software modules are isolated from the computing platform in such a way that they can be selected and composed to create a tailored version of a radar system. Radar subsystems are loosely coupled and can be heterogeneous. The architecture also supports inclusion of web servers for networking and selective component replacement and upgrading and allows infrastructure products from multiple sources to be incorporated. Recent successes in a variety of system categories have established that open systems are feasible and deliver the desired benefits to systems and their stakeholders. MBSAP and the principles espoused in this book seek to ensure that all the aspects of an effective open architecture are present in the resulting designs.

1.6 An Architecture Taxonomy

One further important task in defining terms is to create a framework or taxonomy within which individual architectures and architecture issues can be consistently treated. Basic architecture tenets are universal, but there are significant differences among the architectures of various categories of systems and enterprises, and an effective methodology must account for these variations. Quite often, arguments and confusion in dealing with architecture trace to the lack of a common frame of reference. Experience has shown that at least three basic dimensions are essential in defining such a taxonomy. These are shown graphically in Fig. 1.2 and defined in the following paragraphs:

- *Level of abstraction.* Architectural issues begin with the operational environment in which a system or enterprise must operate and proceed through the definition

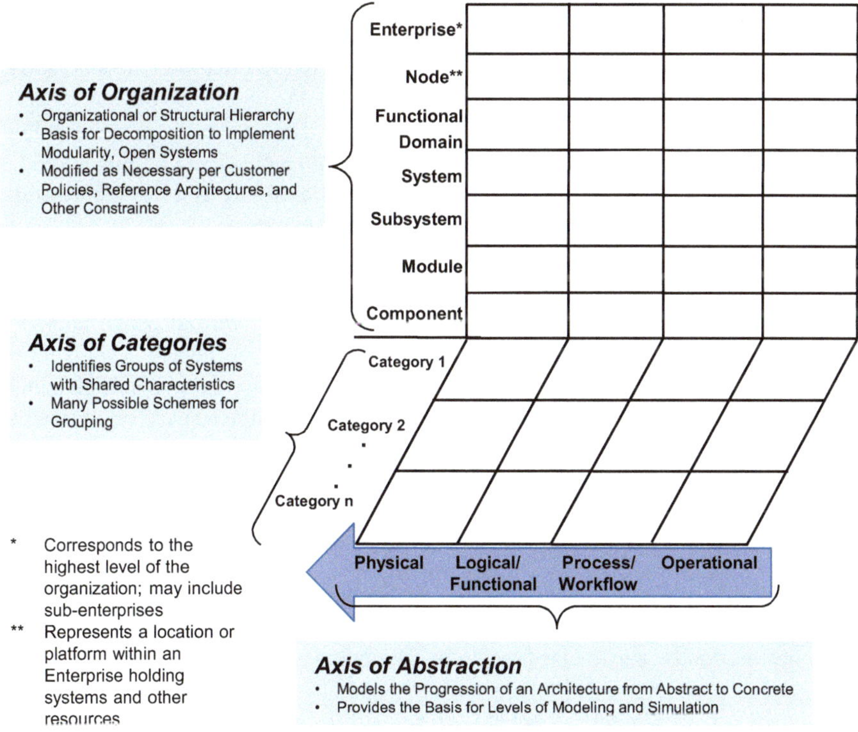

Fig. 1.2 Architecture taxonomy defined by three fundamental dimensions

of required capabilities, structure and behaviors, and, ultimately, physical implementation with quantitative measures of performance. These levels of progressively more concrete system description create an *Axis of Abstraction.* As the figure indicates, the flow of architecture development proceeds from the abstract to the concrete through the creation of Operational, Process/Workflow, Logical/Functional, and ultimately Physical representations. MBSAP is consistent with a process in which increments of capability are realized in successive increments of development, as well as a strategy in which the full set of required capabilities are addressed in a single effort. In either case, architecture development follows this sequence of levels. Importantly, executable representations of an architecture are often needed at each level of abstraction as discussed further in Chaps. 3 and 11, and this taxonomy is useful in dealing with the various types of simulations and associated tools. We also note that the Axis of Abstraction corresponds to the top four rows of the widely used Zachman Framework for Enterprise Architecture™, which is further discussed in Chap. 3 [21].

- *Organizational hierarchy.* Architecture may be needed at any organizational level from that of an entire enterprise or system-of-systems (SoS) down to individual platforms, systems, and subsystems. This organizational hierarchy creates the second dimension, called the *Axis of Organization* to highlight the importance of clearly specifying the organizational level at which an architecture is defined. At any level, a set of OSA principles and standards will apply, and at the higher levels, both standards organizations and many customers who acquire and use systems are defining reference architectures (RAs), reference models (RMs), and policies that govern major aspects of system and enterprise design. Chapter 11 goes into this important topic more deeply. MBSAP transforms an architecture as it evolves along the Axis of Abstraction into a physical implementation at one or more levels of the Axis of Organization, applying a set of architectural tenets and conforming to applicable policies and mandates.
- *Mission and performance category.* Next, a useful taxonomy must account for significant differences in architectural characteristics among systems in various categories that are designed to meet the requirements of various users, each with their own business processes and required levels of performance. This is represented in Fig. 1.2 as an *Axis of Categories.* For example, there is a fundamental distinction between systems whose driving requirements derive from hard real-time processing and those where the emphasis falls on distributed, networked functionality and human-machine interactions. To begin with, it's useful to distinguish the following three architecture categories associated with levels of real-time behavior:

  - *Real-time.* This category includes embedded processing such as a machine controller, guidance system, or software radio that can have deadlines of milliseconds or less. Hard real time also includes distributed configurations of sensors, effectors, and other nodes that can have deadlines of seconds to perhaps a few minutes. Chapter 8 explores this system category, including the distinction between "hard" real time, where deadlines must always be met,

and "soft" real time, where it may be good enough to meet them on average or by some other criterion.
- *Decision/process real-time*. This category includes systems where human action is part of the process. Because deadlines often derive from the need for timely decisions and involve various levels of process automation, ranging from seconds to minutes, the label used is "Decision/Process Real-Time." Examples include manual equipment controls, the management center of an electrical power grid or a communication network, and other time-constrained, human-in-the-loop situations.
- *Management information systems*. Applications in logistics, business administrative functions, and other networked, multiuser systems are often called management information systems. They tend to have large, distributed user communities, large data repositories, and diverse functions, but typically have no, or very relaxed, deadlines.

- Another way to construct the Axis of Categories is based on the business processes or operational uses for which systems are developed. These might include planning, operational management, finance, transportation, manufacturing, logistic support, legal, educational, and a host of others. As a practical matter, a typical enterprise, and sometimes an individual system, needs an architecture that draws characteristics from multiple categories. System categories of several kinds are described and illustrated in later chapters.

- *Time*. Although not shown in Fig. 1.2, there is an important fourth dimension, and that is time. Architecture does not exist in isolation from the requirements laid on systems and enterprises or from the technologies and products available to implement solutions. Systems can remain in use for decades, while technologies turn over every few years. A sound architecture strategy provides for regular technology refreshment and allows for fundamental revision of the architecture, usually at somewhat longer intervals. It is therefore important to clearly declare the epoch in which an architecture is defined and the rules for evolving it in subsequent epochs. This is often expressed by labeling successive versions of an architecture using terms like "as-is," "to-be," "objective," or "final."

As a simple example of the taxonomy in action, Fig. 1.3 maps a computer-controlled manufacturing tool with its embedded electronics. In the hierarchy, this is a system installed in a factory, which is effectively a node in a larger corporate enterprise. The tool falls in a functional domain that might be labeled "Fabrication" or "Tooling." The Axis of Categories in Fig. 1.3 shows a few possible system types that might apply within a factory. Specifically, there is a category for machine control, another for real-time embedded processing, and others for things like communications, planning systems, and, in practice, many more. A machine controller includes sensors, actuators, power control, and other characteristics that are shared with other computer-controlled tools. The tool also incorporates hard real-time embedded processing that responds to external commands and generates signals for the machine controller. Accordingly, this system example spans the two categories

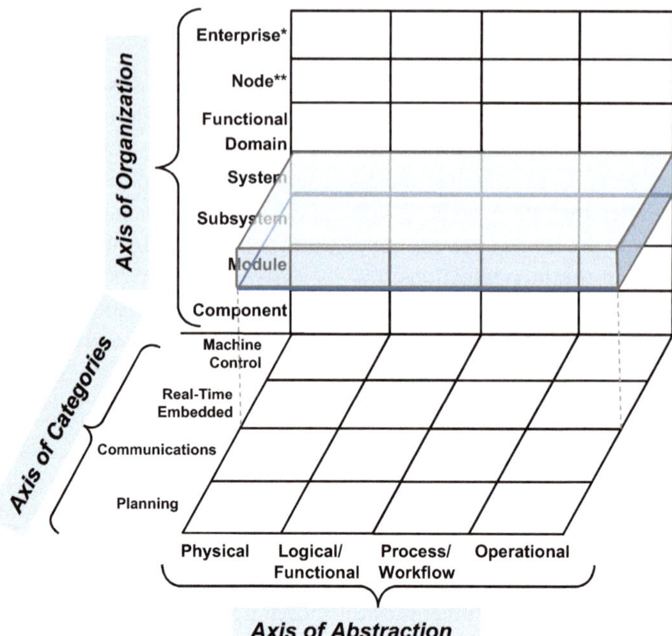

Fig. 1.3 Computer-controlled manufacturing tool mapped to the architecture taxonomy

indicated in the figure. In architecting such a system, all four levels of abstraction will need to be addressed, both in creating the manufacturing tool and in performing modeling and simulation. Following this rationale, the rectangular space shaded blue in the figure represents the placement of this system in the architecture taxonomy.

Even if it's not necessary in a particular development effort to go through the exercise of explicitly mapping the end product and its constituent elements into the taxonomy like this, a shared vocabulary and agreed-upon set of system categories are important to clear communication and consistent effort across teams and technical disciplines. Resolving architecture's Tower of Babel will often eliminate arguments, misunderstandings, mistakes, and wasted time, especially when multiple groups must collaborate in developing complex systems with challenging requirements and advanced technologies.

1.7 Basic Tenets of Good Architecture

Chapter 15 discusses the important subject of assessing the quality of an architecture using objective criteria, especially Quality Attributes. This introductory chapter simply presents a set of general principles and best practices that improve the architect's chances of success. These are distilled from the outcomes, both good and bad, of many real-world architecture efforts. While no two architecture theorists will entirely agree on what these golden rules include, and no set of tenets will perfectly match every situation, the list defined in Table 1.2 is representative and will be useful in most projects. In addition, many of these general attributes of good architecture serve as the basis for a set of objective architecture metrics. With appropriate tailoring to an individual architecture, these can provide valuable insight and a mechanism for achieving and preserving a modular, open, scalable, flexible, and evolvable architecture that will meet customer needs when first delivered and over the long term. The vital subject of formal architecture assessment returns in later chapters, but the ideas in the table are important as general background and are fundamental to virtually everything in the remainder of this book.

1.8 The Role of Architecture in Programs

We conclude this introductory chapter with some observations on the ways in which architecture can and should be used as a fundamental element of successful programs to develop, acquire, and sustain complex systems and enterprises. Architecture is the bedrock on which design and implementation are founded, and in a similar way, it offers an organizing principle for the structure of the program.

The following are some specific examples of effective integration of architecture into a program structure:

- Architecture is central to requirements management and compliance traceability. MBSAP provides a rigorous and traceable mapping from customer-required capabilities to the physical components of the solution, ensuring that all required behaviors are accounted for.
- The architecture process is a vital element in customer dialog. It explicitly addresses increasingly common customer requirements that architecture be a formal deliverable and that it conforms to policies and reference models. Architecture artifacts can visualize and animate a proposed solution in forms that various customer communities can understand and respond to. MBSAP facilitates rapid prototyping, human-in-the-loop simulations, technology demonstrations, and the construction and analysis of trade studies. Architecture helps ensure that the developer and customer have a common understanding and maintain it as requirements are refined and the architectural solution evolves.
- Architecture is at the heart of technical execution. The Chief Architect should be empowered as the "technical conscience" of the program, responsible for ensuring that pragmatic decisions made under the pressures of budgets, schedules, requirements growth, and other factors do not undermine the technical integrity of the solution. Architecture models and other artifacts are the authoritative capture of the evolving design at any point and provide the basic material on which SE processes from risk management to configuration control to test

Table 1.2 General tenets of effective architectures

Architecture tenet	Definition
Modularity	Hierarchically partitioned in a fashion that: • Groups related functions and information, minimizing required interactions across boundaries • Confines the effects of design changes or failures within known boundaries, minimizing the scope of required regression testing • Allows upgrading and technology refreshment • Facilitates the application of component-based design methods
Openness	Designed to realize the characteristics of an open system as described above through: • Modularity and standards compliance (separately defined) • Complete, unambiguous definitions of internal and external interface • Absence of reliance on products and features that are proprietary or otherwise of limited availability • Ability to incorporate proprietary or nonstandard elements when necessary, often by wrapping them to create open interfaces that conform to the architecture. • Isolation of applications and application support services from the underlying operating system and physical platform such that changes to the latter produce small if any impacts to the former (layered architecture; see service orientation)
Scalability	Allows performance or capacity to be increased by the addition of relatively small units of hardware or software with overall performance commensurate with installed resources
Standards compliance	Applies mature, publicly available, consensus-based standards to architecture elements, services, and interfaces following applicable policy
Service orientation	Exposes behaviors as services that can be invoked and delivered through defined interfaces, including layering that provides required services from lower tiers to higher tiers and from the infrastructure to applications
Loose coupling	Minimizes the detailed knowledge that one element requires about another in order to collaborate, exchange information, or provide services; maintains interface definitions independent of the implementation of functions or services accessed through the interfaces (closely related to service orientation)
Error management	Provides robust mechanisms to detect and mitigate failures, exceptions, crashes, data errors, and other anomalous conditions
Accreditability	Possesses the features required to support accreditation for classified information handling, safety, or other critical requirements
Performance	Provides the features needed to comply with requirements such as real-time deadlines, schedulability, preemption, Quality of Service, parallel task execution, information storage, availability, etc. while maintaining adequate margins
Interoperability	Provides the features to support information sharing, common understanding, and collaborative action with other systems and nodes of an enterprise; implements enterprise services and data strategies

(continued)

Table 1.2 (continued)

Architecture tenet	Definition
Composability	Provides the features at each level of the Axis of Organization to allow rapid assembly and integration of subordinate elements to create a capability package tailored to a specific task or situation, especially through interfaces that facilitate vertical and horizontal integration
Adaptability	Extends the composability tenet with features that enable the ensemble and its elements to respond to new or unanticipated tasks and events in ways that make best use of their capabilities to achieve new goals
Clarity	Addresses the important characteristics of an architecture and its model that make them easy to understand and apply; this is central to the aesthetic dimension of architecture

planning can be performed. A shared repository provides a powerful tool for integrated efforts across participating organizations and locations. Architecture metrics provide both technical and program managers with their best insight into the status of the effort and early warning of potential problems.

- Architecture is central to managing and preserving the effectiveness, supportability, and affordability of a system over the long term. The architecture is the best basis for a system Master Evolution Plan that lays out a long-term strategy and process for introducing new or different capabilities and components as well as refreshing technology. Good partitioning and interface control are essential to selective upgrading to deal with technology obsolescence and changing requirements. The object-oriented approach promotes components that are self-contained and have fully defined interfaces, facilitating the task of porting them to a new system platform and even a new architecture when required, e.g., if the enterprise in which the system participates moves to a new model of interaction and governance. The architect of a complex system whose operational lifetime is expected to span multiple technology generations and changing uses or requirements should ensure that Quality Attributes (Chap. 15) are enforced to ensure such evolution is feasible.
- Finally, architecture offers a basis for organizing and managing the program so that the effort is well matched to the challenge. Early mapping of customer requirements to an architectural context highlights the most important technologies and system behaviors, which may, in turn, suggest the best structure of integrated product teams (IPTs), identify high-leverage risk reduction activities and demonstration events, and even offer guidance on the best subcontracting and vendor management approaches for various areas. The key point is that all program structures and activities should be harmonized with the architecture to ensure resource allocations and management practices provide the best possible support to a successful outcome.

Looking at these matters from the perspective of the architect, Muller [22] proposes that his or her role in a program should include:

- Providing feedback for course corrections at every stage of system development
- Owning and enforcing the architecture process
- Developing a set of architecture views (in Muller's work, these include Customer Objectives, Application, Functional, Conceptual, and Realization, the last three of which are very much like the Logical/Functional, Operational, and Physical Viewpoints of MBSAP)
- Major or primary roles in deliverables that include specifications, design, feasibility analysis, program roadmap, and system verification
- A certain amount of creative tension with the program manager, due partly to competing values in the architect's long-term, big picture view and the program manager's near-term, specific problem view

Similarly, Lattanze summarizes the role of a lead architect as:

- Defining the architecture scope and boundaries
- Defining roles and responsibilities of the architecture team
- Defining architecture artifacts
- Coordinating with management

Additional architecture team roles that are essential to fulfilling the overall architecture function include:

- Requirements analyst/operational analyst—responsible for ensuring that the requirements baseline is complete and correct and that requirements are correctly represented in the architecture
- Subject matter experts (SMEs)—responsible for the content of the architecture in specific areas
- Architects, modelers, and analysts—responsible for developing, analyzing, and modeling the architecture under the guidance of the Chief Architect
- Simulation specialists—responsible for developing and employing executable architecture models at various levels of abstraction

1.9 Summary

This chapter has established a view of architecture as the core of a strategy for understanding complex, multidimensional, information-intensive systems and enterprises. It defines essential terminology, summarizes key tenets, and suggests ways in which architecture can inform an entire development and acquisition program. This entire book champions the architect as both visionary and enforcer, seeking the most efficient and balanced design for any set of requirements and then ensuring that the result embodies the critical attributes of good architecture. The next several chapters address the stages of an object-oriented SE methodology, MBSAP, as the engine for translating customer needs into delivered solutions that are both effective and elegant and that remain so over the lifetime of the system or enterprise.

Exercises

1. What distinguishes a system architect from other kinds of engineers?
2. What is the fundamental premise of Model-Based Systems Engineering?
3. Why is a clear and unambiguous language for describing system architecture important?
4. Describe how the definition of system architecture as used in this book provides a foundation for a structured approach and methodology to develop effective solution to customer requirements.
5. Briefly distinguish among Enterprise, System, Service-Oriented, and Executable Architectures, listing the characteristics they have in common and those that are different.
6. List some characteristics of an effective open architecture, and describe how they contribute to the development and use of a system.
7. Sketch an architecture taxonomy graphic similar to Fig. 1.3 for the following systems: (a) a robotic transportation system used in a wholesale distribution center or warehouse, (b) a passenger airliner, and (c) a national telephone network.
8. For at least five of the tenets listed in Table 1.1, explain how these features yield benefits to the developers and users of a system.
9. List some roles architecture can play in a program, and describe how they contribute to successful system development.
10. Considering the functions of a system architect or architecture team leader and the members of such a team, sketch an organization chart including program management, engineering management, and teams responsible for products, integration and test, and system configuration management. Show the relationships of the architect and architecture team to these other organizational entities.

References

1. DARPA (2015) http://www.darpa.mil/program/adaptive-vehicle-make. Accessed 11 Jul 2016
2. DeTurris DJ, D'Urso SJ (2015) Mastering complexity. Aerospace America, Nov 2015, pp 32–36
3. Maier MW, Rechtin E (2009) The art of systems architecting. CRC Press, New York
4. Ryschkewitsch M, Shaible D, Larson W (2009) The art and science of systems engineering. NASA Monograph
5. Senge P (1990) The fifth discipline. Doubleday, New York
6. Long D, Scott Z (2011) A primer for model-based system engineering, 2nd edn. Vitech Corp., Blacksburg, VA
7. Ryan J, Srkani S, Mazzuchi T (2014) Leveraging variability modeling techniques for architecture trade studies and analysis. Syst Eng 17:1
8. Kossiakoff A, Sweet WN (2003) systems engineering principles and practice. Wiley, Hoboken, NJ
9. Buede DM, Miller WD (2016) The engineering design of systems: models and methods, 3rd edn. Wiley, Hoboken, NJ

10. INCOSE SE Vision 2020 (2004) INCOSE-TP-2004-004-02
11. Fagan MD (ed) (1978) A history of engineering and science in the Bell System: National Service in War and Peace, 1925–1975. Bell Telephone Labs, New York
12. Thomas N (2014) Systems engineering: program empowerment for 21st century aerospace projects. IEEE A&E Syst Mag 29(12):18–26
13. Perry DA (1995) Self-organizing systems across scales. Trends Ecol Evol 10:241–244
14. Defense Technical Information Center (2003) Joint Technical Architecture. www.dtic.mil/cgi-bin/GetTRDoc?AD=ADA443892. Accessed 11 Jul 2016
15. McManus H, Hastings D, Warmkessel J (2004) New methods for rapid architecture selection and conceptual design. J Spacecraft Rockets 41(1):10–19
16. FEAC Institute (2014) Federal Enterprise Architecture (FEA) Framework Version 2. https://www.feacinstitute.org/resource/ea-zone/item/federal-enterprise-architecture-fea-framework-version-2. Accessed 11 Jul 2016
17. Lattanze AJ (2009) Architecting software intensive systems: a practitioner's guide. CRC Press, New York
18. DoD Chief Information Officer Council (2001) A practical guide to Federal Enterprise Architecture, version 1.0. http://www.gao.gov/assets/590/588407.pdf. Accessed 11 Jul 2016
19. Reito S (2000) Radar open systems architecture and applications. Paper presented at radar conference 2000
20. Nelson JA (2010) Radar open system architecture provides net centricity. IEEE A&E Syst Mag 25:17–20
21. Zachman JA (2008) A concise definition of the Zachman framework. https://www.zachman.com/about-the-zachman-framework. Accessed 12 May 2017
22. Muller G (2005) System architecting. http://www.gaudisite.nl. Accessed 11 Jul 2016

Chapter 2
Applying Object Orientation to System Architecture

2.1 Motivation for Object-Oriented Architecture

The Model-Based System Architecture Process (MBSAP) applies the concepts of Object Orientation (OO) or, more specifically, Object-Oriented Design (OOD), to the tasks of developing architectures for complex, technology- and information-intensive systems and enterprises. The International Council on Systems Engineering (INCOSE) [1] addresses this through their Object-Oriented Systems Engineering Method (OOSEM), and Chap. 3 discusses this and various other OO methodologies. In this chapter, we provide a summary tutorial on the principles of OO, the Unified Modeling Language (UML), which is the primary and most widely accepted standard for OO modeling and analysis, and the System Modeling Language (SysML) Profile of UML which adapts the basic language to the needs of Systems Engineering (SE). We discuss UML first, both because it is the root language standard on which SysML builds and because MBSAP can be applied to a purely software architecture using UML. Readers familiar with OOD, UML, and SysML can skip this discussion, while those seeking a deeper look into OO architecture principles can consult various texts listed in the bibliography and the references to this chapter.

In Chap. 1, we introduced the background of MBSAP, a bit of the history of SE, the basic principles of good architecture, and the categories and terminology of architecture in general. This book is based on a fundamental assertion that OO represents the best, most effective approach currently known for tackling complex architecture and SE challenges and is a major advance over previous methods like Structured Analysis and Structured Design. Still better methodologies may well arise, and OO itself continues to be refined and extended, but for now it provides the best basis for an architecture strategy. The reasons that this is so include:

- The inherent power of object-based decomposition to break down complex entities and relationships so that the architect can deal with comprehensible and coherent subsets of components and resources, along with their functions and content, without losing the essential perspective of the whole

J. M. Borky, T. H. Bradley, *Effective Model-Based Systems Engineering*,
https://doi.org/10.1007/978-3-319-95669-5_2

- The use of abstraction to discover and exploit underlying patterns and common characteristics across seemingly diverse entities
- The benefits of encapsulation in localizing design and implementation issues, simplifying system integration, minimizing the need for large-scale testing, minimizing the system-level impacts of specific design changes, and promoting reusability of designs and products
- The focus on interfaces among system elements and between a system and its environment that is inherent in OOD
- The fact that these and other aspects of OOD collectively provide a basis for achieving open systems, whose characteristics and benefits are summarized in Chap. 1

2.2 Basic Principles of Object Orientation

Over the past 20 years or so, many authors have spelled out the nature of OO and its underlying rationale [1–6]. The UML and SysML language standards are maintained by the Object Management Group, which publishes a family of related standards and other documents. The released version of UML as of this writing is 2.4.1 with Version 2.5 in beta test [7] . Version 1.4 of SysML has been released. Excellent tutorials are also available from academic institutions and from architecture modeling tool vendors. The remainder of this section is distilled from this enormous body of reference materials and presented from the viewpoint of a practicing system architect who needs to apply OOD effectively but is less interested in refining and extending the language itself.

The foundation of OOD rests on a set of principles that establish an approach to modeling and analyzing complex entities. All of these, in one way or another, contribute to solving the management of complexity challenge. While OOD originated in software architectures, and is still primarily applied to them, it is equally true that beginning with UML Version 2 (UML 2), OO principles have also been applied to hardware and to hardware/software ensembles. SysML largely eliminates the software bias of UML. Because students with a software background are likely to be familiar with UML, and because SysML is a Profile of UML and shares much of the UML semantics, this chapter begins with a summary of UML and then explicitly addresses the changes brought in by SysML. However, in the remainder of this book, we use SysML exclusively since it is the preferred language of SE modeling. By the end, it should be clear why any significant MBSE activity today uses SysML.

Some OO purists maintain that the only really essential principles are encapsulation, generalization, and polymorphism, but a somewhat fuller list is helpful in understanding the nature and power of OO in the kinds of applications addressed by this book.[1] The following is a representative set of OO principles.

[1]Grady Booch, one of the primary authors of UML, lists abstraction, encapsulation, modularity, hierarchy, typing, concurrency, and persistence as the basis of the Object Model.

Abstraction: Abstraction is a way of dealing with complexity by identifying common characteristics shared by a group of entities such as hardware and software components, interactions, or conditions so that they can be represented (modeled) by a more general entity that has these shared characteristics. By far the most common use of this principle in UML is in defining a Class that abstracts the common properties of a set of concrete items in the real world that are referred to as Objects. Indeed, OO gets its name from the thought process of defining such real-world things that are part of a system or enterprise. SysML does much the same thing with Blocks and their Instances. Then a Class or Block captures (abstracts) the essential characteristics that define a set of related real-world Objects or Block Instances[2] and is said to be a Classifier.

An Object in UML is thus a particular instance of a Class, and in SysML, usage of a Block creates an Instance. In this book, a reference to an Object in UML is equivalent to a reference to a Block Instance in SysML. In system architecture, this kind of abstraction can be used to separate the *functional* characteristics of an object or group of objects from the *physical* characteristics associated with implementation that make a given Object or Instance unique. In the original software context of OO, this usually means that a Class is instantiated as an Object by loading code and data into memory in a computer for execution, which can be done multiple times in parallel. In system architecture, it is more appropriate to say that a functional representation of a system element as a Class or Block is realized physically as an Object or Block Instance through point design decisions such as those involved in the development or procurement of a hardware or software component. Many model elements in addition to Classes and Blocks can act as Classifiers by providing mechanisms to describe common structural and behavioral features of things.[3]

Encapsulation: Encapsulation provides a way to partition architecture content into units or sets of information, functions, and resources, each within a defined boundary. This principle descends from early concepts of structured programming where a crucial concept was to hide the details of an Object's implementation ("information hiding") while exposing publicly only those things that must be shared with other Objects. Encapsulation implies well-defined interfaces through which public functions and information are exposed and accessed; such an interface can be viewed as a contract that will be honored by the concealed details of the implementation.

Modularity: Modularity extends the basic idea of encapsulation to the larger process of decomposing a complex entity into a set of simpler entities that can be individually managed and employed. A modular element of a system (more simply called a module) may bundle system elements that are individually encapsulated,

[2] Many authors devote attention to various ways of identifying a good set of Objects or Block Instances within a complex entity, and this is an important aspect of MBSAP, especially in the Logical/Functional Viewpoint (Chap. 5).

[3] Words like Class, Block, Object, System, Dependency, Port, and Flow are capitalized when used in the sense of formal UML/SysML constructs, e.g., when a System is treated as a Classifier; when used generically, these terms are left lowercase.

such as Objects or Block Instances, and must have an interface through which it provides and consumes functions, services, and information or other content. Good partitioning, based on grouping closely related information and processes ("cohesion") within a module, promotes separation of concerns among system elements and supports loose coupling (minimum dependencies among modules), which is a fundamental attribute of a robust, open system.

Generalization and Inheritance: The complementary concepts of Generalization and Inheritance create a kind of *Hierarchy* (rank ordering) involving successive levels of abstraction. A higher-level Classifier (informally called a Superclass) abstracts the common properties of a set of lower-level Subclasses, creating a Generalization. Then Inheritance is the inverse relationship by which the child elements possess (inherit) the characteristics of their parent elements, along with their own peculiar properties. When necessary, inherited properties can be explicitly overridden. A Subclass or specialized Block may inherit from multiple Superclasses ("multiple inheritance"). In this way, Generalization provides an additional tool to manage complexity by defining common characteristics of diverse system elements. It also promotes consistency in an architecture through inheritance from a common core set of properties by many System Classes or Blocks (which inherit from general Superclasses), and it reduces effort and errors in creating multiple detailed Class or Block definitions.

An extension of Inheritance is Typing in which the properties of one Class. Block or other element are defined by another. This can be an enforcement mechanism that constrains implementations to conform to global design decisions. In general an Object or Block Instance of a given Type cannot be substituted by one of a different Type. In system architecture, Typing is generalized somewhat, for example, by using whatever flows through a Port to type the Port.

Aggregation and Composition: Another kind of hierarchy involving successive levels of structural decomposition is created using Aggregation and Composition. In UML, Aggregation creates an "is a part of" relationship between an architecture element and one or more lower-level elements that are contained within it or help to make it up. With an Aggregation relationship, lower-level Classes or Blocks and their Instances can exist independent of the higher level Class. Composition is a stronger kind of Aggregation in which the lower-level Classes or Objects "compose" the higher-level Class or Object, which is responsible for creating or providing them, and they have no independent existence.

In SysML, Aggregation relationships are transformed into Reference Associations in which a Block on one end references the Block on the other end, making the second Block a Reference Property of the first block. This creates a *logical* hierarchy, based on required interactions among architecture elements. On the other hand, Composition relationships in SysML define a *structural* hierarchy. The Block on one end of a Composition Association is declared to be a Part Property (more simply, a Part) of the Block on the other end. We illustrate the nature of these relationships in the examples used throughout the remaining chapters.

Interfaces: Good SE practice maintains a focus on the definition and use of interaction points among the modules or other structural elements of a system or between these elements and the external environment. Although not usually called out as an OO principle in itself, this attention to interfaces is critical to system architecting and works hand-in-hand with encapsulation, modularity, aggregation, and other principles.

Polymorphism: The last and perhaps most esoteric concept on our list provides a means by which a given system element such as an Object or Block Instance can implement a common set of operations in different ways according to the specific way in which the operations are invoked. Most commonly, a module or other resource unit performs variants on common functions such as responding to a service request from a client based on things like client identity or the value of a provided parameter.

These underlying concepts are important in understanding and applying UML and SysML in real system architectures. OO, UML, and SysML are powerful in the tools they offer to the architect but can be daunting in the variety and subtlety of those tools. That reality is what gives full-featured architecture modeling tools their steep learning curves. OO principles come to life in transforming a customer requirement into an effective solution. The next sections of this chapter examine the basics of this modeling tool kit. Appendix A is a summary reference to UML diagrams and semantics. It also includes some more formal aspects of the UML standard, including the metamodel that defines the overall UML syntax, but practicing architects don't usually need to worry about these subtleties as long as mainstream tools that comply with the language standards are used. Appendix B provides a similar concise reference to the SysML Profile.

2.3 Modeling Architecture in the Unified Modeling Language (UML)

2.3.1 Modeling Structure

As discussed in Chap. 1, architecture fundamentally concerns itself with structure, behavior, and rules. It's appropriate to begin this description of architecture modeling in UML with the constructs used to represent structure.

Objects: An Object corresponds to something in the real world, whether it's a physical entity like a computer, an electromechanical actuator, a piece of software, or something virtual or conceptual such as a list of items or an organizational relationship. An Object has three elements: (a) Identity, usually in the form of a unique name; (b) Attributes, representing the content of the Object, usually in the form of parameters or data; and (c) Behavior, representing the Operations the Object can carry out. An Operation is the signature of a Method, which defines the implementing behavior, so that calling an Operation invokes the Method. Only the Identity is absolutely required; some Objects have no Attributes, and some perform no Operations. Optionally, the Object can also have a fourth element, State, which represents a behavioral configuration resulting from previous events or activities

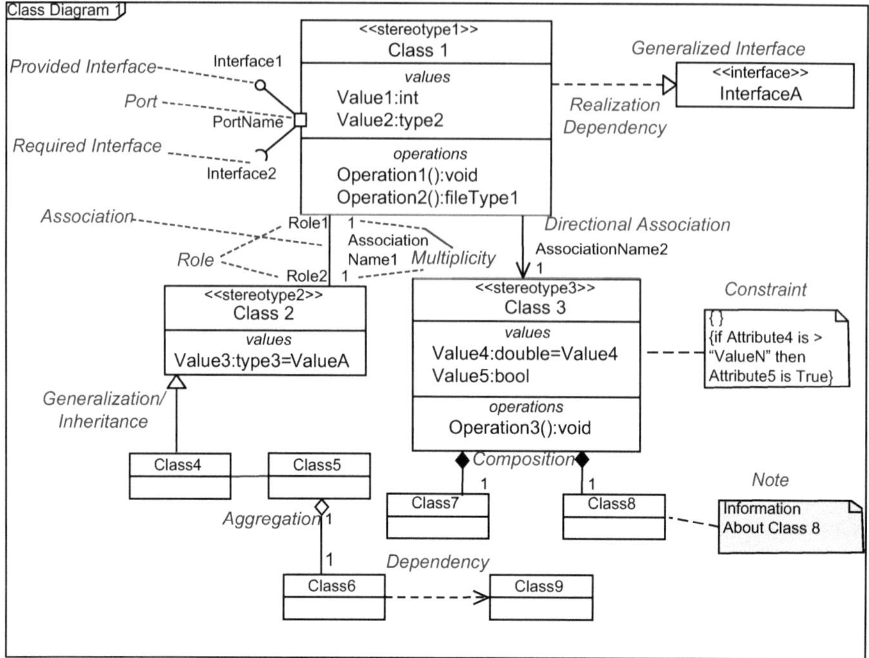

Fig. 2.1 Elements of a Class Diagram

and corresponds to an Object's "memory." Less formally, but very importantly, an Object can have Responsibilities, which collectively describe the purpose the Object fulfills in meeting customer requirements as part of a system or enterprise solution. Objects constitute the building blocks of a system or enterprise and are ultimately associated with the physical system or enterprise.

Classes: A basic Class, which in MBSAP is called a System Class, is the abstraction of one or more Objects and defines their common Attributes and Operations. It's typical for a System Class to inherit its properties through a hierarchy of Generalizations, and Classes higher up in the ranking may be Abstract Classes, meaning that their purpose is to define a category of system elements while they are themselves never directly instantiated as Objects. Abstract Classes can be conveniently designated with the `<<Abstract>>` stereotype (see below). Some authors consider Class and Type to be synonymous in the sense that a Class defines the properties of an Object and thus "types" the object. An Object can be described by more than one Type, and these are not necessarily related. In an Object Diagram, an Object can be shown with a Realization Dependency to the Class that types it, signifying that the Object is the physical realization of the Class abstraction.

Figure 2.1 is an artificial Class Diagram that illustrates some of the more common syntactical elements associated with Classes and showing their graphical representations (icons). Some important features include:

- A Class or Object can be shown as a simple rectangle with a name, a rectangle with horizontal partitioning lines (to distinguish it from any other rectangle), or a rectangle with "compartments" for Attributes and Operations, which is usually preferred because it conveys important information about the nature and purpose of the Class. The compartments can be labeled, or the icons for Attributes and Operations provided by a modeling tool may be enough to distinguish them. An Attribute should normally have a Type to make clear the nature of each element of Class content and can optionally show an initial value and Multiplicity that shows how many copies of the value can exist at any given time. A software Class Attribute will normally have a data Type such as Int (integer), Bool (Boolean), Float (floating point), etc. and can have a complex Type such as Enum (enumerated list). If the Class models hardware, Attributes can denote properties of the system element such as computer throughput, actuator torque, etc.; the actual values may not be specified until the Class is instantiated as a physical Object, but this content is essential to defining a functional specification for this element, which is a key output of the MBSAP methodology. That being said, we emphasize that in this case it's preferable to switch to SysML and model hardware with Blocks, which are much more natural and intuitive.
- Operations model functions or methods and have a return Type to show what, if anything, an Operation produces when invoked. They can also have a list of parameters that are passed to the operation, each of which also has a Type. In architecture modeling, the default return type is void, but a software Class Operation will often need a data type for the returned information. In accordance with the principle of encapsulation or information hiding, Attributes and Operations can have "visibility" to explicitly show how they can be accessed. The standard visibilities and their symbols are:

  + Public—any other Object can access the Attribute or Operation.
  – Private—only the Object itself can access the Attribute or Operation.
  # Protected—only the Object and its descendants have access.
  ~ Package—only Objects in the same Package have access.

- Classes and Objects are commonly annotated with one or more `<<stereotypes>>` to show their nature or general category; in fact, virtually any model element can be stereotyped, and this often adds greatly to the descriptive value of a diagram. A stereotype is essentially a label, and a model element can have multiple stereotypes. UML Version 2 generalized the original construct of a stereotype to that of a keyword, meaning any predefined term used to categorize or describe a model element and notated as `<<import>>` or `{abstract}` depending on the term, but the original stereotype term is still generally used, including in this book. Class and Object names can take the following forms, with an Object name underlined and separated from its Class name by a colon:

Class1	Basic Class name
:Class1	Anonymous Object (instance) of Class1
Object1:Class 1	Named Object (instance) of Class1
Object1:	Named Object of an unspecified Class

An Association shows that two Classes can interact by exchanging messages, information, or even (in terms of hardware) physical quantities. A simple line indicates bidirectional interaction; a "vee" arrowhead on an Association means the exchange is unidirectional. Associations are said to have bidirectional or unidirectional "navigability." It's good modeling practice to provide Roles on one or both ends of an Association to name and summarize the behavior of a Class in the context of its interaction with the other Class.[4] An Association can also have a name to identify the interaction between Classes that it models. Just as a Class is instantiated as an Object, an Association is physically instantiated as a Link.

- Multiplicity (sometimes called Cardinality) on the ends of an Association denotes the number of instances (Objects) that can exist in the physical system. In a software context, Multiplicity indicates the number of instances of a Class that can be active at runtime. In a hardware context, it shows the number of physical Objects a system can contain, e.g., the number of computers, valves, interface devices, or other components typed by a Class that can be present in a system. The most common values are:

1	One and only one
1..*	At least one but up to an unspecified maximum number
m	Exactly m instances
*	Any number, including none
m..n	At least m and no more than n

- A Port provides a more structural way to model the interactions of a Class or Object with other elements and the external environment of the system. It represents a point at which the Class provides or consumes (requires) something and can also be declared on a Component or Subsystem. In general, a Port should be typed and typically has at least one defined Interface that types the Port, which can be Provided or Required, and a Port can have both. Interface definitions spell out the nature of the interactions through the Ports that they type.
- A Generalization/Inheritance relationship, defined as above, is annotated with a solid line having a closed but unfilled triangular arrowhead. The parent Class (or

[4]UML 2 replaces the word Role with Association End Name, but the former is much more intuitive.

other Classifier such as an Actor or a Use Case) is on the arrowhead end, and one or more Subclasses are on the other end(s).

- Aggregation, as defined above, is annotated with a line having an unfilled diamond on the end of the aggregating Class. Composition has a similar annotation except with a filled diamond and is often drawn with a directionality arrowhead on the other end of the line to show that an instance of the target, or subordinate Class, can only be owned by an instance of the source, or superior Class.
- A Dependency is a relationship between model elements in which one requires content, services, interfaces, operations, or something else provided by the other in order to function correctly. The most general notation is simply a dashed line. Usually, a "vee" arrowhead is used to point from the dependent element to the other, independent element. Co-dependent elements have arrows pointing both ways. In general, a Dependency should have a stereotype such as `<<use>>`, `<<call>>`, `<<import>>`, `<<include>>`, or `<<create>>` to indicate the nature of the relationship.
- Realization is a special kind of Dependency showing that the source element creates or "realizes" the destination element. Figure 2.1 shows a common use, where Class1 models a component (or has an internal component) that is responsible for creating InterfaceA. More generally, this dependency can be used to show that an Object realizes (instantiates) a Class.
- A Constraint models a condition or rule that a system element must follow and is annotated as an expression between curly brackets. This is commonly a descriptor such as `{ordered}`, a limit such as `{maxValue=value}`, or an `{if... then...else...}` rule.
- UML allows more or less unlimited informal notation of a diagram using the Note symbol containing explanatory text and anchored with a dashed line to the model element(s) it describes. Liberal use of Notes is encouraged because they can tremendously increase the readability and communicative power of a diagram.

Structured Class: UML 2 introduced a more sophisticated structural construct than simple Classes and Objects. A Structured Class is an alternative way to show composition that allows internal details to be modeled, typically showing an internal structure made up of Parts with Associations. A Composite Class is one that has a Composition relationship with one or more other Classes. This is sometimes used synonymously with Structured Class, but it more properly denotes a Class that incorporates the functionality of other Classes, which is the basis for reusing existing system elements. Parts are properties of the abstract (owning) Class, and an instance of the abstract Class may include instances of the collaboration of Parts it contains. Parts can have multiplicity. Figure 2.2 shows a notional example of a Structured Class that owns three parts, modeled as Objects, and has two Ports, one with a provided Interface (shown as a "lollipop") and the other with a required Interface (shown with a "socket"). Links from the Ports to specific Parts are called "delegations," meaning that the connected Parts are responsible for creating and supporting the corresponding Interfaces. As a practical matter, a system architect

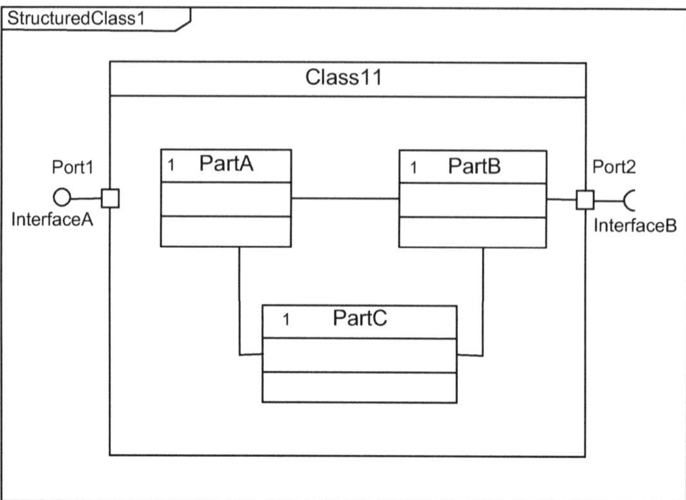

Fig. 2.2 Structured Class showing internal structure with Parts and Ports that have Provided and Required Interfaces

needing to show this level of structural detail should use SysML and employ the power of Block Diagrams.

Packages: The primary way of organizing most UML (and SysML) models involves Packages, which are essentially containers in which related model elements can be placed. Unlike Classes and Objects, they don't correspond to actual hardware and software elements of the system being architected, and so they provide a natural way to model top-level structure in which the system and its architecture are partitioned into what MBSAP calls "Domains" and "Subdomains" that group related resources that interact closely to produce major system functions (see Chap. 4). Also, most UML tools use Packages in a Browser to create the internal structure of a model. Packages have what are sometimes called "lightweight semantics" because they aren't intended to deal with details of structure and behavior, and UML 2 allows primarily Dependencies and Flows between Packages as a way to show their interactions or, more accurately, the relationships among contents of various packages. The icon for a Package is very appropriate, suggesting a file folder into which various related items can be dropped.

Actors: Anything that is outside the boundary of a system but with which the system interacts or that in some way influences system behavior is modeled as an Actor. Actors are most commonly associated with Use Cases, which are part of the behavioral aspect of an architecture that we examine next, but they have an important role in structure as well because they are used to explicitly declare relationships between a system and external entities. The icon, which is a sort of stick figure, as well as the name Actor, strongly suggests that these must be humans. While that is often the case, an Actor can also model an external system or network, a resource that the system uses but does not own or control, an organization, and many others.

One important type of Actor in system architecture modeling is what MBSAP calls a User Role. This is a defined set of actions and characteristics describing a particular kind of interaction between the system and the people or external systems who are the "customers," for whose benefit the system is provided. User Roles often represent system operators or crew members, administrators and other support staff, maintenance personnel, trainers, management authorities, and others who have specific functions, duties, and authorities with respect to a system. When dealing with any system that has significant external interactions and dependencies like these, it's a good practice to develop what might be called an "Actor Diagram" (actually a specialized kind of Use Case diagram) that defines User Roles and perhaps other Actors so that they can be consistently employed in subsequent diagrams. Often, a Generalization hierarchy is useful to show things like overall categories of Actors and their shared characteristics such as training and qualifications. A very important point is that a User Role does not automatically correspond to an individual; many people may share a User Role, and a single person may perform multiple User Roles, depending on the situation. Figure 2.3 illustrates Packages and Actors and includes a notional Actor Diagram.

2.3.2 Modeling Behavior

UML offers a variety of constructs for representing and decomposing system behaviors. The most important of these for system architecture are Use Cases, Activities, Interactions, and States, and each of these has a diagram type. These diagrams are very similar in UML and SysML. They are described in the following paragraphs, and extensions introduced by SysML are discussed later in the chapter. UML and SysML use a set of terms in rather specific ways when prescribing behavioral modeling, and it's important to keep them straight and use them correctly. The key terms include:

- *Behavior*—the generic term for something a system or enterprise does to deal with an input, an internal state change, the expiration of a specified time interval, or other circumstances. A behavior can take the form of an Activity, a State Machine, or an Interaction (usually modeled by a Sequence Diagram); SysML adds a fourth behavioral representation in the form of a Parametric Diagram, which we will introduce later in this chapter. A Classifier such as a Class or Block can have an associated Activity or State Machine Diagram that specifies its Classifier Behavior, which is the main behavior of the Classifier and which is executed at the start of the lifetime of an instance of the Classifier.
- *Activity*—models an increment of behavior involving a series of specific Actions and Object exchanges. Activities commonly focus on internal processes, rather than external events. They are used to model things like algorithms, parallel or simultaneous behaviors, Use Case scenarios, and business processes.

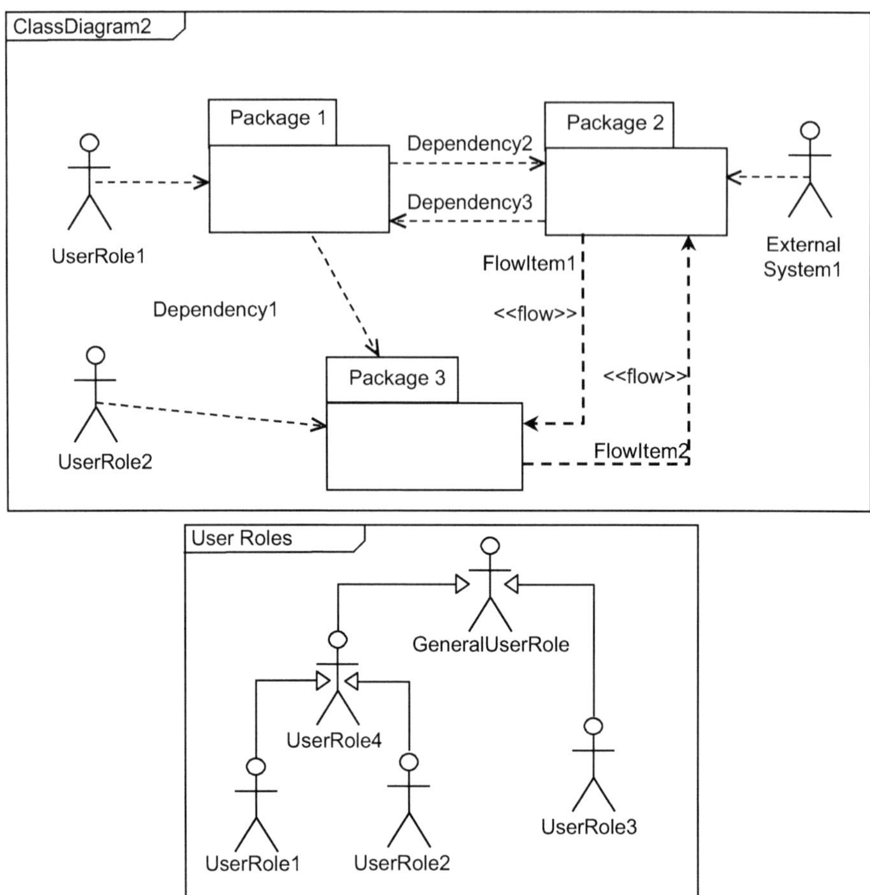

Fig. 2.3 Specialized diagrams showing Packages and User Roles

- *Operation*—a characteristic of a Class, Block, or other element that models an invocable behavior. An Operation serves as the signature for its Method, and the Method is defined by the Behavior associated with the Operation.

In the UML language standard, there are five "Interaction Diagrams" that model the ways Objects (or Blocks) work together to implement behaviors: Sequence, Activity, Communication, Timing, and Interaction Overview Diagrams. The following paragraphs discuss the more important behavioral diagrams; Appendix A gives more detail.

Use Cases: These provide a systematic and intuitive way to identify, organize, and begin to describe system behavior, especially in terms of the observable functions that satisfy customer needs. An exercise in developing Use Cases can be a powerful way for system architects and developers to achieve a common understanding with system users about the purposes and required capabilities of the system. Just as Packages can be used to model the top-level structure of a system, a Use

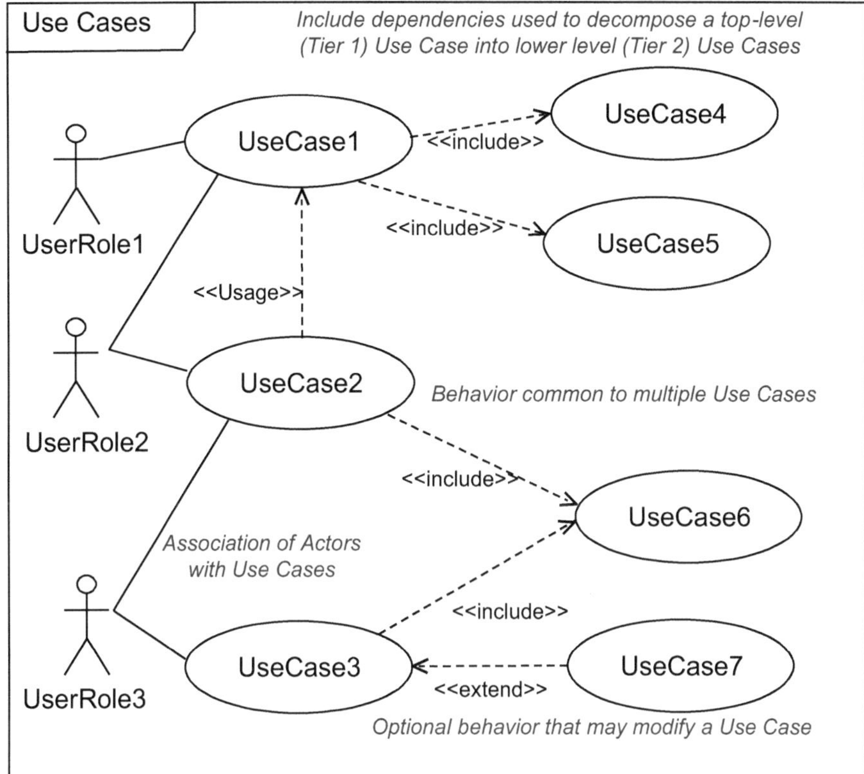

Fig. 2.4 Elements of a Use Case Diagram

Case Diagram can serve as something like an index to its behaviors. All that is really required is to decide on the set of behaviors, give each an appropriate name, and place them in a Use Case Diagram. However, a much richer and more useful diagram can be built by decomposing complex behaviors into simpler ones that are more appropriate for implementation, declaring sub-behaviors that are shared across multiple high-level behaviors, and by declaring associations with Actors. Figure 2.4 shows typical content, and the following paragraphs define some of the key features.

- *Use Case*—Chapter 4 has more to say about creating and specifying Use Cases. For now, it's enough to note that they identify system behaviors and provide a convenient graphical layout that summarizes this fundamental aspect of an architecture in one or a few easy-to-read diagrams. An essential element of a Use Case is a Scenario that spells out the flow of the behavior. Usually, there will be a primary or default Scenario plus one or more alternatives that show what happens under various circumstances such as a User decision or a fault condition.
- *Actor Associations*—A User Role or other Actor may use or interact with one or more behaviors; Associations model these relationships. The Actor Diagram in

Fig. 2.3, including Generalizations, can actually be created on a Use Case Diagram. This may become messy if many Actors are involved, but it can sometimes simplify a Use Case Diagram by allowing a single Association to a high-level Actor to replace numerous Associations to specific Actors.
- *Include Dependency*—In the original software context of UML, `<<include>>` was used to show a shared code object that could be instantiated once and called from multiple points in a main program. In architecture modeling, this idea of a shared behavior is still valid, e.g., a Generate Report Use Case might be invoked to document the final stage of the scenarios of multiple higher-level Use Cases, each of which results in the creation of report content. However, it's also useful in decomposing a large, complex behavior into simpler behaviors that are easier to deal with in designing implementations. In MBSAP, this hierarchy is organized using Tiers, and a major system or enterprise may well have as many as two or three or even more levels of Use Case decomposition. In Fig. 2.4, Use Case 1 in Tier 1 is decomposed into simpler behaviors in Tier 2, modeled by Uses Cases 4 and 5.
- *Extend Dependency*—An `<<extend>>` Dependency models an optional behavior that may modify another behavior, e.g., a Use Case modeling communications might have an optional extension to perform encryption of sensitive information.
- *Other relationships*—Figure 2.4 shows a Usage Dependency among Use Cases modeling a situation where one requires something produced by the other in order to execute correctly. Many other relationships, including Generalizations and Flows, can be used in these diagrams to add clarity and make important details explicit in the model.

Activities: One of the most important elements of the MBSAP methodology is the use of Activity Diagrams (ADs), illustrated in Fig. 2.5, to bring together structure and behavior. An Activity is a primary unit of behavior, and it is therefore normal in architecture modeling to use ADs to graphically depict Use Case Scenarios as well as larger system processes. The MPSAP discussion in this book frequently calls these flows Business Processes or Mission Threads. Also, an Activity is quite often used to define the "main" or "classifier" behavior of a Class or Block.

Anyone familiar with the flowcharts that software and process engineers have used for decades will feel immediately comfortable with Activity Diagrams. An Activity Diagram is a variant of a State Machine Diagram, which is described later, and they share many semantic elements, including the use of the same rounded rectangle icon to represent an Action in an Activity Diagram and a State in a State Machine.

The following paragraphs describe important features, and additional details are in Appendix A and the references to this chapter.

- *Swim Lanes*—Strictly speaking, an Activity is divided by "Activity Partitions," but they are universally called "swim lanes" from their resemblance to a pool. They can be laid out horizontally or vertically; we use the latter. They represent the structural units, Actors (external entities), or other model elements that are

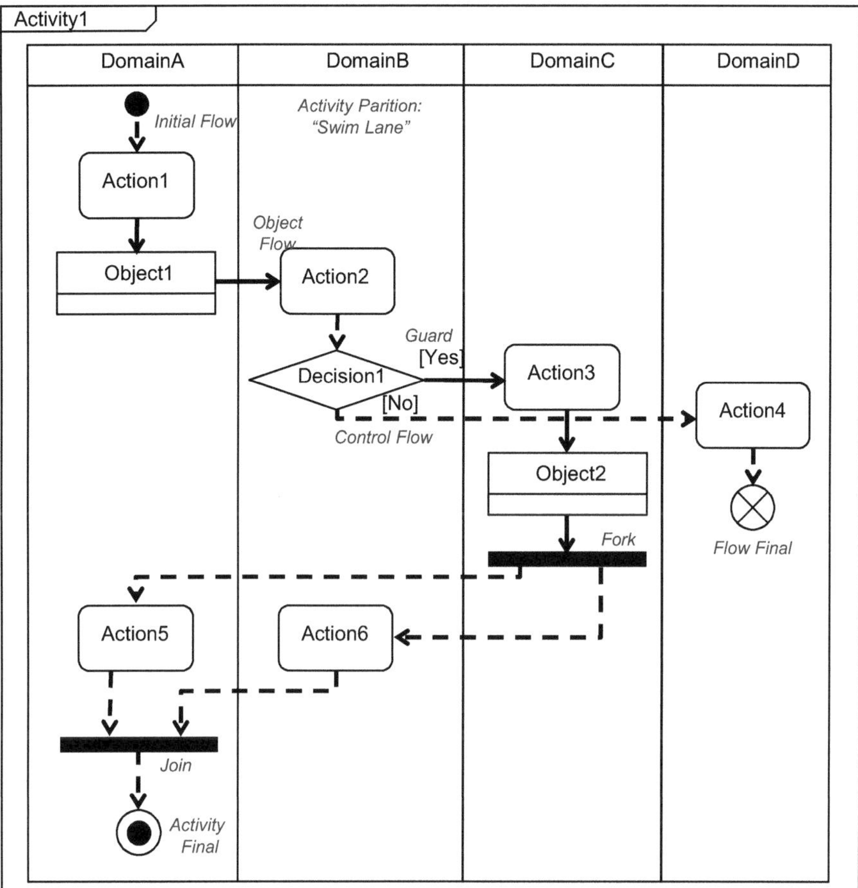

Fig. 2.5 Elements of an Activity Diagram

responsible for one or more Actions involved in a Scenario. An Activity Partition allocates Actions and associated Objects to structural elements of the architecture. This is the key to uniting structure and behavior.

- *Actions*—An Activity is made up of finer-grained behaviors called Actions. An Action is a basic unit of behavior within its activity, but it can be further decomposed into yet more specific behaviors until the desired level of detailed description is obtained. During system execution (in software terms, this means at run time), an Action can model a query to an Object and can invoke other behaviors, including Operations and other Activities. In other words, it's possible to construct Activity hierarchies. UML and SysML define a variety of Actions; the way a particular modeling tool implements these details determines how complicated ADs are developed and evolved. A simple or primitive Action has a name and possibly a text description of the behavior it represents, such as an Object access or a calculation, and may involve properties or variables and specification of values. More commonly, a Call Behavior Action is used, and its specification has a link to the behavior it invokes, usually defined by another AD, a State Machine Diagram (SMD) or a Sequence Diagram (SD) elsewhere in the model. A Call Operation Action, as the name implies, invokes a specific Operation. When Behaviors are allocated to system resources (Classes, Blocks, Interfaces, etc.), Operations are created on these structural elements and linked to the appropriate behaviors. As was described earlier, an Operation can be viewed as the signature of a Method, which is the specification of the associated behavior. Once this is done, a Call Behavior Action can be converted ("refactored") into a Call Operation Action that invokes the behavior as a Method via its Operation signature. This completes the description of the Behavior by combining its definition in a behavioral diagram with the identity of the system element that will execute it. Additional Action types include Opaque Actions, in which a block of software code is embedded to describe the behavior; Send Action (or Send Signal) Actions that create an Event or Signal for other model elements to detect; Accept Action (or Accept Signal) Actions that respond to events; and Accept Time Actions that are linked to some kind of clock or other timing mechanism.

- *Flows*—Although this discussion avoids many of the formal and theoretical aspects of UML, it may add clarity in describing Activities to note that these diagrams, and State Machines as well, use the concepts of Graph Theory, including Token Flows. A graph consists of a set of nodes connected by edges, and tokens can be exchanged between nodes via these edges to show the progression of the process being modeled and to control the activation, or "firing," of successive nodes. Buede and Miller [8] discuss both discrete mathematics (sets, relations, functions, etc.) and graph theory as the formal basis for key aspects of UML and SysML modeling. In an Activity Diagram, the Actions are nodes, and they can exchange two kinds of tokens: Control and Objects. Tokens are exchanged via Pins, which can be shown on the border of an Action.
- *Object and Control Flows*—It's sometimes said that a diagram is "wired" with Object and Control Flows. An Action executes as soon as all input Tokens, including data or possibly other Objects, along with all incoming Controls, are available. A Pin acts as a buffer for tokens, whether entering or leaving the Action, and can have multiplicity to denote the number of tokens either consumed or produced in one execution of the Action's behavior. As a practical matter, Pins are often omitted for Control Flows when each Flow involves the passing of a single Control Token. With Object Flows, the language standards allow Pins to be replaced by Object Nodes between Actions with entering and exiting Object Flows. This explicit depiction of an Object can make a diagram more intuitive and easier to read. This is the convention used in Fig. 2.5, which doesn't contain Pins. However, when multiple and complex Object and Control Flows must be rigorously modeled, the preferred syntax in the language standard uses Input and Output Pins, connected by Control and Object Flows. This helps ensure Pins have compatible types and allows details of the Object exchange to be spelled out in specifications. Pin syntax is generally required if an AD is to be simulated (animated) in the tool. See Appendix A and the references for a summary of this

subject. In either case, the development of Object Flows among Actions is a golden opportunity to discover information and other content of the system by considering the inputs and outputs that are inherent in the Actions making up an Activity.

- *Initial Flow, Activity Final and Flow Final Nodes*—An Activity must have well defined start and stop points. The Initial Flow, which starts on an Initial Flow Node, shows where execution begins when the Activity executes, and this is the point at which any Preconditions associated with the Activity must be satisfied. When the Activity is complete, which may be via one of several alternative flow paths, the Activity Final Node shows this and is the point at which any Postconditions are guaranteed to be satisfied. If there are conditions under which the Activity may terminate without reaching the Activity Final Node, a Flow Final Node is used to model this.
- *Decisions and Merges*—Very often the path of execution through an Activity depends on conditions in the system and its environment. Decision diamonds model this. The flows exiting a Decision have Guards, which are orthogonal Boolean expressions contained in square brackets, the most common being `[Yes]` or `[No]` or `[Condition]` or `[Else]`. The outbound Flow whose Guard evaluates to True is active. A Decision can have multiple exit Flows, each controlled by a Guard. For clear and unambiguous behavior, only one Guard can evaluate to True at any given time, but there can be multiple Decisions in succession to model complex control logic such as If/Else If/Else If and so on. The inverse of a Decision is a Merge, also modeled as a diamond, but with multiple inbound Flows and a single outbound Flow. As soon as any Flow reaches the Merge, it is passed to the next node in the diagram; in other words this is an asynchronous recombination of multiple Flows.
- *Forks and Joins*—Sometimes, the nature of an Activity is such that the flow inherently splits into two or more simultaneous branches. This is modeled with a Fork, as shown in Fig. 2.5. The inverse of a Fork is a Join, which uses the same fat bar icon. All of the inputs to a Join must be available before the flow continues to the next node, so that this is a synchronous recombination.
- *Signals*—Signals provide a mechanism for communication among Activities. A Signal can be generated in one Activity and then detected and acted upon in others. Signals are commonly used to model responses to Events, external control inputs, or other trigger conditions.

Sequence Diagrams: In most systems, groups of Classes, instantiated as Objects, (or groups of Block Instances in SysML) work together to produce required behaviors. Actors, Systems, Subsystems, and anything else that can exhibit behavior can also be participants. In classic UML literature, this principle is expressed as "A Use Case is realized by a collaboration of Classes," and there is a special icon, an oval with a dashed outline, for a Collaboration. Interactions are modeled as Messages between Objects. An SD is a very intuitive way to graphically depict the interactions as a time-sequenced series of messages and other events as well as executions of various Operations of individual participants. UML 2 significantly increased the power of these diagrams to show decomposition of complex behaviors into simpler

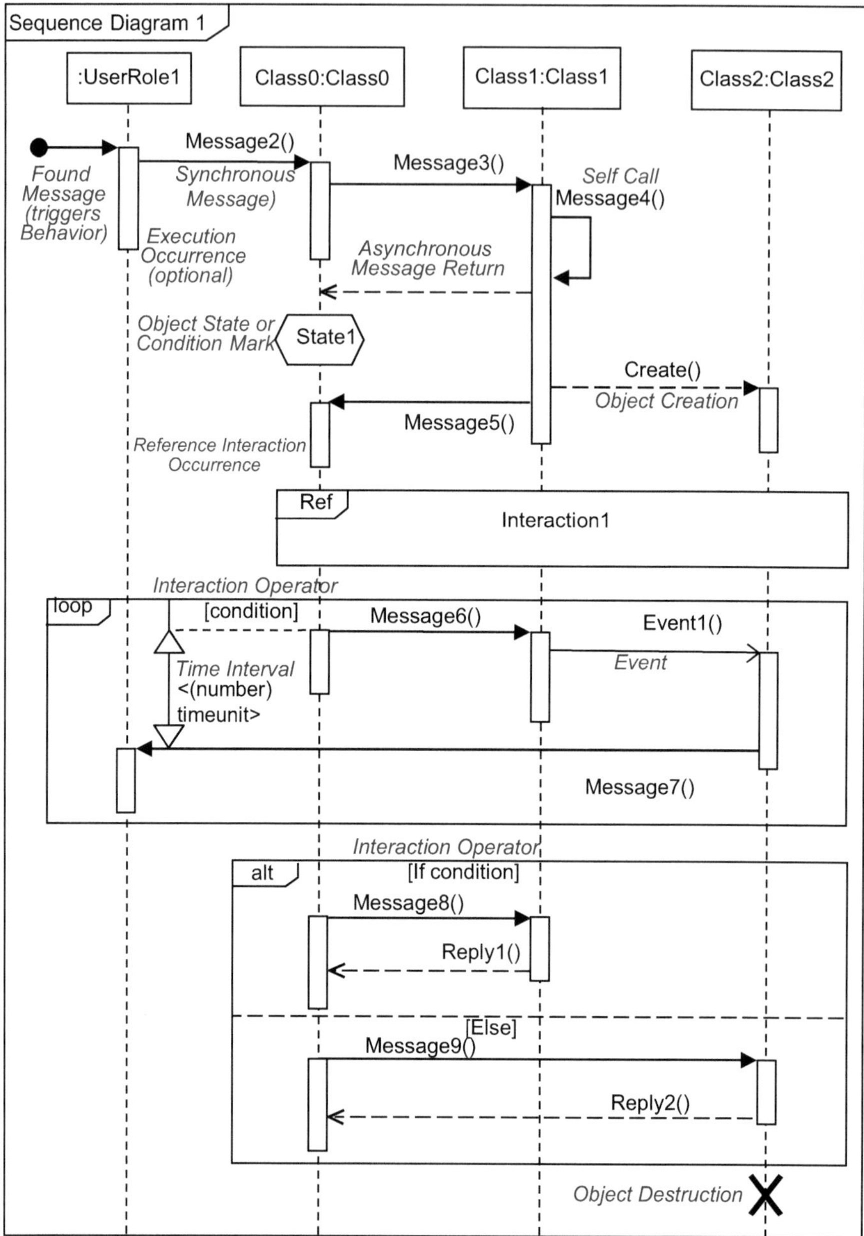

Fig. 2.6 Elements of a Sequence Diagram

ones and to model complex decision logic. Figure 2.6 illustrates some of the main elements of a Sequence Diagram, and the following paragraphs spell out some important points.

- *Interaction Fragments*—These are named elements of an Interaction Diagram that represent units of the interaction. They include:
  - *Occurrences* (strictly, these are Occurrence Specifications) that represent moments in time, usually associated with sending and receiving messages or with the start and end of an execution period
  - *State Invariants*, which represent runtime (during system operation) constraints such as a value of a variable or an internal or external State that must be satisfied at various times
  - *Interaction Uses or Occurrences*, which are discussed below
  - *Combined Fragments*, also discussed below
- *Instance Lines*—The participating Objects, as well as Actors and sometimes other model elements that can exhibit behavior, are shown across the top of the diagram at the heads of a set of dashed lines, strictly speaking called Instance Lines but commonly referred to as "lifelines." Time increases from top to bottom along a lifeline, although the scale is not necessarily linear.
- *Messages*—A variety of message types can be used to model various kinds of interactions. The most common are Synchronous Messages that are sent in sequence at specific points in the execution of the behavior, hence the name of the diagram. When software Objects interact, a Message most often consists of sending some sort of data object, a software signal, or a procedure call, which can include a list of parameters. However, in a hardware context, a "message" might be anything from a pneumatic pressure change to a control voltage that energizes ("picks") a relay. Each Message should have a name, and it may end in parentheses to show that some action is being invoked and to contain any parameters being passed. Dispatching and receiving a Message can be interpreted as a `SendEvent` and a `ReceiveEvent`, respectively, and the latter is often used to trigger a transition between Object States as described under State Machine Diagrams, the next topic in this section. A specialized Message is available to show the sending of other kinds of Events and is notated with an arrow with a "vcc" head. A similar arrow with a dashed line is used when it's important to show a reply such as a return from a procedure call, and can be used for any Asynchronous Message, such as a procedure call, where the sending Object continues executing without waiting for a reply from the Receiving Object.[5] There are special Events for creating and destroying Objects. When applied to software Objects, these generally denote directions to the Operating System (OS) to load (instantiate) and run the Object and then to delete the Object from memory. In a hardware context, these events might be used to activate and deactivate a system resource. Optionally, these Events can be designated with `<<create>>` and `<<destroy>>` or other suitable stereotypes.
- *State/Condition Marks*—Sometimes it's important to show explicitly the State of an Object or some other condition at one or more points in the sequence of

[5] The reader should be aware that various modeling tools take liberties with the details of notating messages and should determine how a given tool shows them.

execution, and a Mark can be placed on the Object lifeline at the appropriate point. Figure 2.6 gives an example. This is one example of a State Invariant, which is a runtime constraint placed on one or more participants in the modeled interaction.

- *Execution Occurrences*—An optional notation in a Sequence Diagram places a narrow rectangle on an Object lifeline to show when the Object is active. This may improve the readability of complex diagrams. Figure 2.6 uses this notation.
- *Frames and Fragments*—It's good modeling practice to create a Frame around any UML diagram (they are mandatory in SysML), but especially so with Sequence Diagrams. A Frame creates a boundary around a unit of model content (the diagram) and provides information like a diagram name and type in a box in the upper left corner. Two kinds of sub-Frames, each of which is an Interaction Fragment, are used in Sequence Diagrams to show shared or optional elements of behavior and to model a variety of control flows such as branching and looping. Instead of a diagram name, the top left corner of such a Fragment contains an Interaction Operator that defines what kind of behavior is being represented.
  - *Combined Fragment*—A Combined Fragment captures a subset of the behavior modeled by a Sequence diagram and expresses two key elements: an Interaction Operator that controls the execution of the fragment and Interaction Operands that include Messages, Events, Execution Occurrences, and other elements that define the behavior. It is drawn within a frame that encompasses the lifelines of the participants involved in the subordinate behavior. Figure 2.6 shows two Combined Fragments, one with a Loop operator and one with an Alternative operator. There are many Interaction Operators, but a few of the more common ones are:

`seq`	Sequential—the default operator, which simply means that the Fragment executes when the flow reaches it
`opt`	Optional—the Fragment executes if a Boolean [guard] evaluates to True
`alt`	Alternative—contains two or more sub-Fragments, each with a [guard], separated by horizontal dashed lines; the sub-Fragment whose guard is True is the one executed
`loop`	Iteration—has integer parameters for a minimum and maximum number of times the Fragment will execute, plus a [guard] that must be True for the loop to continue
`par`	Parallel—contains sub-Fragments that execute concurrently
`ref`	Reference—see next paragraph

  - *Interaction Occurrence (or Use)*. This is an Interaction Fragment with the `ref` Operator. It serves as a pointer to another Sequence Diagram whose behavior is to be inserted into the main diagram at the point where the Fragment is placed. The Interaction Occurrence contains the name of the referenced lower-level diagram. This technique is very frequently used and has two main purposes: to allow decomposition of a complex Sequence Diagram into multiple, more readable diagrams and to allow a specific behavior that occurs in multiple places (e.g., in more than one Use Case) to be modeled once and then inserted by reference wherever it's needed. Diagramming a complex behavior might start with a top-level Sequence Diagram containing nothing but a series of Interaction Occurrences, each separately defined in its own diagram. An Interaction Occurrence can have input and output parameters that tailor its execution to a specific usage. This method of decomposition can be carried on as many times as necessary to achieve readable diagrams showing comprehensible units of behavior.

- *Timing*—Any system that has time-sensitive behavior, meaning that certain things must happen within time constraints such as deadlines, requires an architectural analysis to establish that timing requirements can be met. Very often, a Sequence Diagram provides a useful starting point for this, and with a system that doesn't have complex or rigorous time constraints, laying out timing on a Sequence Diagram may be all the analysis that's required. In more challenging situations, this first step may be useful in assessing overall system timing relationships and in identifying stressing points in system behavior where more thorough analysis will be needed. Time runs vertically from top to bottom of a Sequence Diagram, and the vertical positions of Messages and Events give a sense of where they occur in the overall timeline. This is not precise—there is nothing in the semantics of these diagrams to support accurate placement of content along a "time ruler," although a carefully constructed diagram can have its elements dragged up or down to get an approximate representation of durations and latencies. In addition, a Message arrow can be drawn with a downward slope to represent time consumed in generating, transmitting, and receiving the Message.
- *Time Interval*—A more useful notation is the Time Interval, which allows a numerical time value to be declared between any two points in the diagram; Fig. 2.6 includes an example. A typical timing analysis would begin by identifying an overall Time Interval corresponding to a deadline and using more specific Time Intervals to allocate time to various processes or other elements of behavior such that the sum of their estimated execution times fits within the deadline. It may be possible to calibrate these time estimates with measured data on a prototype or experimental configuration or by comparison to existing system designs that do similar things and to thereby get an assessment of the feasibility of meeting the deadline. Furthermore, it's not uncommon for this sort of timing allocation to become part of the requirements issued to a design team responsible for the system components that must meet a timing constraint. Timing analysis and additional ways to notate timing constraints on Sequence Diagrams are treated more fully in Chap. 8.

State Machines: Objects, Subsystems, and Systems can have States that abstract behavioral modes or status. An Object or Block Instance transitions from one State to another upon receipt of a message that can be either an Event or a Triggered Operation, or the transition may simply occur after completing all the Actions in the prior State. For example, a system might have States for Inactive, Startup, Maintenance, and Active, each of which would have a set of Actions appropriate to these various operational conditions. The system would transition between States in

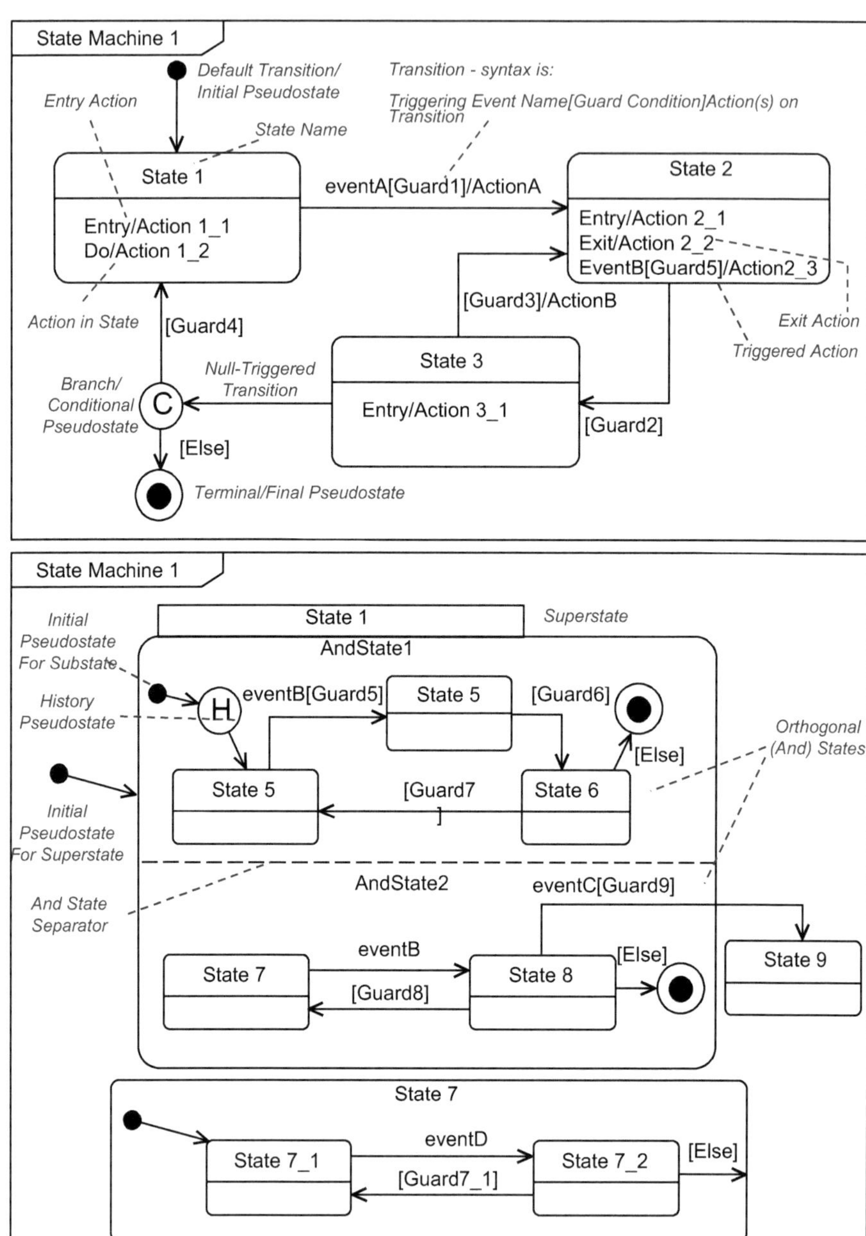

Fig. 2.7 Elements of a State Machine Diagram

response to some triggering event such as applying power (e.g., move from Inactive to Startup) or receiving an operator command to change modes (e.g., move from Active to Maintenance). A processor Object might have States for conditions such as Idle (not currently doing anything), Communicating, Processing, and Storing Data. A State Machine Diagram (SMD), also called a Statechart, can be attached to a Class or Block and will then govern the behavior of the Objects/instances instantiated from the Class/Block. Figure 2.7 illustrates an SMD.

Software engineers have used state machines for decades, and the concept is entirely valid and very useful for hardware and system architectures as well. The developers of UML incorporated a form of SMD called the Harel Statechart, named for its inventor, David Harel [9]. Both SMDs and Activity Diagrams are concerned with portraying the flow of a behavior, and so they share many common elements of syntax, but an SMD deals with a single model element, while an Activity Diagram has a larger scope involving multiple system elements. The following paragraphs give important details.[6]

- *Transitions*—Transitions from State to State are typically triggered by (1) receipt of some signal or event generated outside the Class or (2) completion of the Actions associated with a State under conditions that make it appropriate to transition to a succeeding State. What makes behaviors like this "stateful" is that they depend on the results of previous periods of activity that leave the Object in a particular condition or configuration. Indeed, when a system or Object is activated ("wakes up"), it needs a default State or a History Pseudostate to tell it where to start executing. A State includes the values of Attributes that were valid the last time the Class was instantiated and executed, and default startup values should be defined for the initial instantiation. A State is therefore somewhat analogous to "memory" of an Object or Block Instance created by its recent history. Only Classifiers such as Classes/Blocks, Systems, Components, and Use Cases can define state models, and only their instances can execute state machines.
- *State*—States are notated with rounded rectangles and can have compartments for a text description and for Actions. A State can have one or more Entry Actions that execute as soon as the State is entered, various kinds of Actions that happen while the State is active, and Exit Actions that are performed after all others and just before a Transition out of the State is taken. A State must have a unique name, and attention to naming is important to make diagrams readable and consistent with other parts of an architecture model. States can be decomposed into Substates, in which case the enclosing State is called a Composite State. In a simple SMD, as long as the Object whose behavior is defined by the diagram is active, one and only one State must be active at any given time. This also applies to Substates within a state. Such States are referred to as Or States.

[6] Once again, the reader should be aware that individual UML modeling tools have variations on SMD syntax, some driven by the mechanics of automatic code generation and associated programming languages; these can be annoying but are not of concern in this book.

- *And States*—The upper SMD example in Fig. 2.7 has the features most commonly used in system architectures. However, when a System or Object has two or more concurrent threads or activity flows that must be valid at the same time and can be independent of each other, an SMD uses the form in the lower part of Fig. 2.7. These are And States and should have names that show the specific areas of overall Object behavior they represent. They can have their own histories, can complete their functions at different times, can initiate separate Transitions to other SMDs, and can separately process the same Events when these are received by the parent Object. The State of the Object is essentially the combination of conditions in the various And States. Figure 2.7 shows an overall Initial Pseudostate for the Superstate and individual ones that show how each And State starts execution.
- *Pseudostates*—SMDs have very rich semantics and can model complex behaviors through a combination of decomposition and logic. Pseudostates are not States but can designate a wide variety of behaviors. Appendix A has a more complete description, but the more important Pseudostates illustrated in Fig. 2.7 are:
  - *Initial or Default Pseudostate*—designates the Substate that is entered first in the absence of some other overriding Transition.
  - *Branch or Conditional Pseudostate*—has a single incoming Transition, and two or more outgoing Transitions with Guard conditions to select the Transition to be taken. It can also be notated with a decision diamond like those in Activity Diagrams.
  - *History Pseudostate*—calls for State information from the most recent State of the Object or And State to be retrieved and for execution to continue from that point. Deep History extends this to nested Substates.
  - *Other Pseudostates*—include Pseudostates for junctions, forks, entry and exit points, and other aspects of controlling the flow of execution.

Timing Diagrams. We have already discussed the use of Sequence Diagrams for coarse timing analysis has. A more precise representation of process or execution timing is possible with Timing Diagrams. Figure 2.8 gives a sampling of the kinds of content and formats that can be used. A Reactive Object is one whose behavior responds to Events. The middle diagram in the figure illustrates exchanging events between Objects, and the bottom one shows the kind of detail that might be needed in a highly time-sensitive behavior. Timing Diagrams are deprecated in SysML.

2.3.3 Modeling Implementation

Many of the artifacts that make up the Physical Viewpoint of an architecture consist of familiar things like drawings, specifications, code listings, user interface definitions, and so forth. However, UML provides diagrams that are sometimes useful in

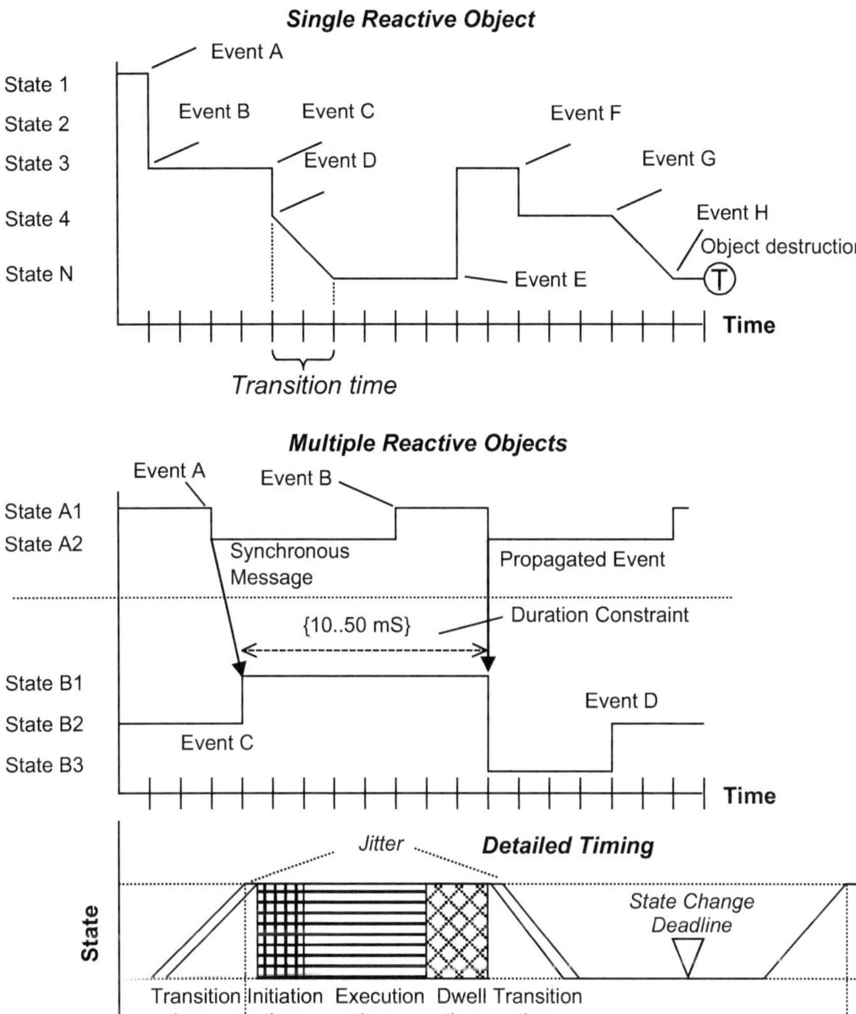

Fig. 2.8 Elements of Timing Diagrams

depicting the physical design of a system. The most commonly used in system architecting are Component and Deployment Diagrams.

Component Diagrams. Applying the principles of encapsulation and loose coupling leads to a design approach that has been called "component based." A Component is a physical entity that encapsulates a set of resources, which may be hardware, software, or both, and exposes only content and functionality that is

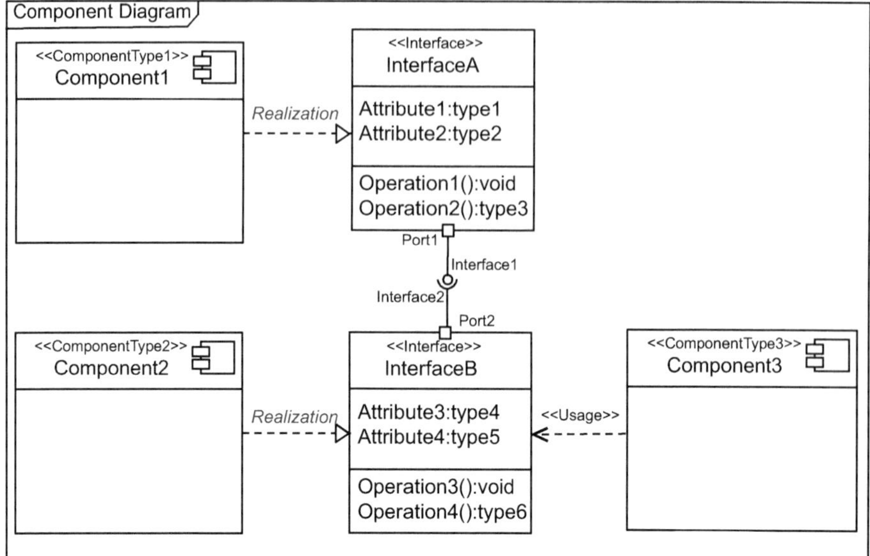

Fig. 2.9 Elements of a Component Diagram

explicitly declared to be public. Because the internal implementation details are hidden, a Component is often called a "black box." However, for the architecture to be complete, the internal design must be defined and documented, and this is commonly done by creating a Structured Class (Fig. 2.2) to describe these details. This information makes the Component a "white box." Obviously, the interface is supremely important and must be fully defined and disclosed to allow the Component to be integrated and used in a system. Figure 2.9 shows typical content for a Component Diagram, and the following paragraphs give some highlights.

- Components are identified by an icon of a square with two small rectangles. A Component can be decomposed into lower-level components, and it can have a stereotype to identify the kind of system element it represents, e.g., a software executable, a processor, a network switch, or a library.
- Figure 2.9 shows explicitly the Interfaces realized by two Components. The Interfaces have ports with associated required and provided interfaces, and there is a `<<usage>>` dependency to show that Interface B gets content from Component 3. The specifications for these Port Interfaces can add detail to the definition of the Component Interface as a whole.
- A Component can also have a direct dependency with an Interface as shown with Component 3.
- Figure 2.9 shows explicitly the Interfaces realized by two Components. An advantage of declaring Component Interfaces this way is that the Interface can show exactly the Attributes (content) and Operations the Component is exposing publicly. The Interface specification can also include any important detail such as applied standards or protocols, application programming interfaces (APIs), quality

of service (QoS) parameters, and many other things. As an alternative, some modeling tools allow required and provided Interfaces to be placed directly on the Component boundary. UML provides a Component Assembly icon in which a provided Interface from one Component is joined to a required Interface from another.

Deployment Diagrams. It is often important in a Physical Viewpoint to show the mapping of software to hardware, especially when they are developed or procured separately. A Deployment Diagram provides a way to model the hardware "boxes" of a system, which are called Nodes in UML, and the connections among them, and then to allocate software Components to the Nodes where they will be loaded and executed. Figure 2.10 shows a simple example, and some additional details are in the following paragraphs.

- *Nodes*—A Node has a unique name and can be stereotyped to show the type of hardware it models (server, network device, workstation, storage, etc.) or even a specific product identifier like a model number. In the model, the specification for a Node provides a convenient place to capture any important information about the associated hardware.
- *Artifacts*—Strictly speaking, UML defines the physical implementation of a Component or Subsystem to be an Artifact, which is notated with a file icon of a small rectangle with a folded-down corner and can optionally have the `<<arti-`

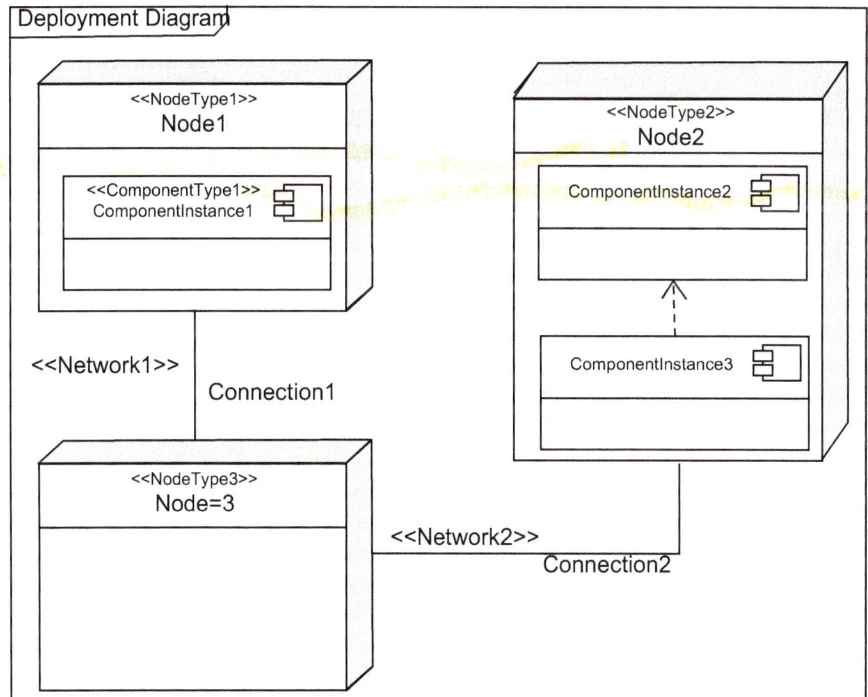

Fig. 2.10 Elements of a Deployment Diagram

fact>> stereotype. An Artifact is said to manifest its defining Component and to be deployed to a Node, shown using Dependencies with <<deploy>> and <<manifest>> stereotypes. An Artifact is sometimes called a Component Instance, but the meaning is the same. Dependencies can be declared between Artifacts on a given Node or across Node boundaries.
- *Flows*—The interconnects among Nodes can be modeled as Flows, with names that identify the connections and stereotypes that show things like the physical connection type, a network or serial link protocol, or anything else that is important to understanding and implementing the "wiring." In Fig. 2.10, Nodes 1 and 2 might be servers hosting applications components, and Node 3 might be a network switch, in which case the connecting flows would be stereotyped with the network type, e.g., Ethernet.

2.4 Using the Systems Modeling Language (SysML) Profile of UML

2.4.1 Summary of the SysML Profile

The software heritage of UML limits its ability to deal with the full dimensionality of complex system architectures. UML Version 1 almost ignored hardware, with only the Deployment Diagram to represent computing resources, and then only as a convenience to specify where software would be loaded and run. With UML 2, a far richer vocabulary for describing physical structure, flows, interfaces (especially Ports), and other matters important to the architect became available. Still more recently, an initiative of the Object Management Group, with extensive participation from industry, tool vendors, academia, Government, and various liaison

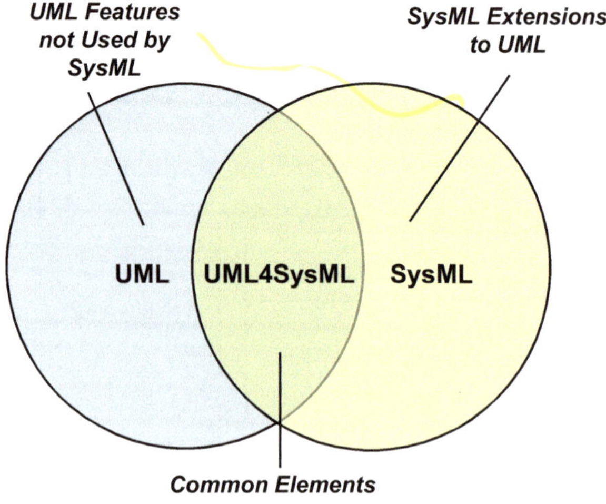

Fig. 2.11 Relationship of SysML to UML

organizations, has led to the specification of the Systems Modeling Language, SysML [10]. Reference material and tutorials can be found at the SysML web site [11].

The explicit intent of SysML is to provide better tools for applying the principles of Object Orientation to the problems faced by System Engineers. SysML supports the specification, analysis, design, verification, and validation of systems that include hardware, software, data, personnel, procedures, and facilities. SysML is a UML 2 Profile and reuses a subset of the basic UML modeling elements, shown as UML4SysML in Fig. 2.11, which is the standard portrayal of their relationship. Unneeded UML elements are excluded from the profile, including Timing, Communication, Interaction Overview, Component, Object, Profile, and Deployment Diagrams on the grounds that these are redundant with constructs made available by SysML. At the same time, two new diagrams are introduced (Requirements and Parametric) and several others are modified. An important practical consideration is that today's mainstream architecture modeling tools fully support the SysML profile.

SysML provides methods to deal with:

- Requirements definition, allocation, derivation, and verification
- System structure using block diagrams with which System Engineers are familiar
- Functional behavior, replacing older modeling methods such as Integrated Definition Function Modeling Method (IDEF0)
- Allocations of various kinds, including functions to structure, requirements to resources, and software to hardware
- Basic testing as a starting point for detailed test planning and execution
- Basic trade studies during specification and design phases

This chapter doesn't attempt a full SysML tutorial; Friedenthal et al. [12], who were key developers of the profile, have done that superlatively, and their text complements this one. Our focus is on the ways SysML provides added descriptive and analytical power to the capabilities of UML. This is especially true in describing complex and decomposed structural details, behaviors involving flows of various kinds among system elements, and relationships among various aspects of a model, such as between requirements and design. In principle, MBSAP can be implemented with UML 2 when software dominates an architecture. However, SysML is greatly preferred, especially in situations requiring:

- Rigorous treatment of requirements in the architecture context, including analysis and allocation to elements of the solution
- Capture of analytical content such as algorithms and functions in the architecture
- Complex configurations where structural details are important
- Attention to the details of physical structure, including not just computers and networks but things like power and cooling, controls, and actuators
- Accounting for information and physical item flows among model elements

One item of SysML syntax that deserves emphasis is that all diagrams must have frames and headers. The header explicitly declares the diagram kind, the model element type, the model element name, and the diagram name, making a diagram easy to identify and place in the context of the overall architecture model. Examples are shown in the figures that follow.

2.4.2 Modeling Requirements

Requirements definition, analysis, allocation, validation, and verification are among the most important SE responsibilities, and the Requirements Diagram explicitly addresses this. The same information can also be expressed in table format. In fact, the Table Diagram can be used to document anything from computed or measured data on system performance to applicable documents such as standards that apply to the system design. References [10–12] give a fuller discussion.

Requirements are modeled using Requirements Blocks, identified by a `<<requirement>>` stereotype, and can have the following properties:

- *Identification*—a unique identifier, which may match the same value in an external requirements data base
- *Text*—a statement of the requirement ("The system shall perform some function with some performance criterion")
- *Category*—an optional user-defined sorting of a requirement into a group
- *Verification*—an optional specification of the verification method to be used

Once requirements have been captured this way, the model can show relationships such as linkages between requirements and the system elements to which they are allocated and to the sources of numerical values that are used to verify them. We note that even when Requirements Diagrams are used, many organizations, when dealing with large and complex requirements sets, prefer to capture them in a mainstream requirements management tool. In addition to being a familiar way of managing requirements, such tools provide analysis and reporting functions. When this is the case, a less detailed representation of requirements in the SysML model is likely to be sufficient to provide the desired traceability to design without incurring a lot of duplicative work to maintain the same requirements in two tools. Most architecture modeling tools support import of requirements data in standard formats such as a spreadsheet or comma-separated variable (CSV) file, and many of them incorporate automated bridges to requirements tools in order to minimize the required effort.

The broader subject of Allocations in SysML is further described below, but this is so important in requirements analysis that it gets special attention in the SysML language standard. In fact, the basic `<<allocate>>` stereotype is not used in conjunction with requirements. Instead, a set of more descriptive relationships is defined, usually shown as Dependencies with appropriate stereotypes. These include:

- *Derivation*—a `<<deriveRqmt>>` stereotype on a Dependency arrow pointing from a derived requirement to the source requirement; a Comment with a `<<Rationale>>` stereotype can be anchored to the Dependency to document the basis for the derivation.
- *Copying*—a `<<copy>>` stereotype on a Dependency pointing from the slave requirement to the master; if the master requirement is updated, all slave copies are automatically updated to match.
- *Satisfaction*—a `<<satisfy>>` stereotype on a Dependency pointing from a Block that has responsibility for meeting a requirement to the Requirement Block defining it; this is commonly shown using a compartment on the Block rather than a Dependency in order to declutter a diagram.
- *Refinement*—a `<<refine>>` stereotype on a Dependency pointing from a Use Case or other model element that adds detail or content to a Requirement Block.

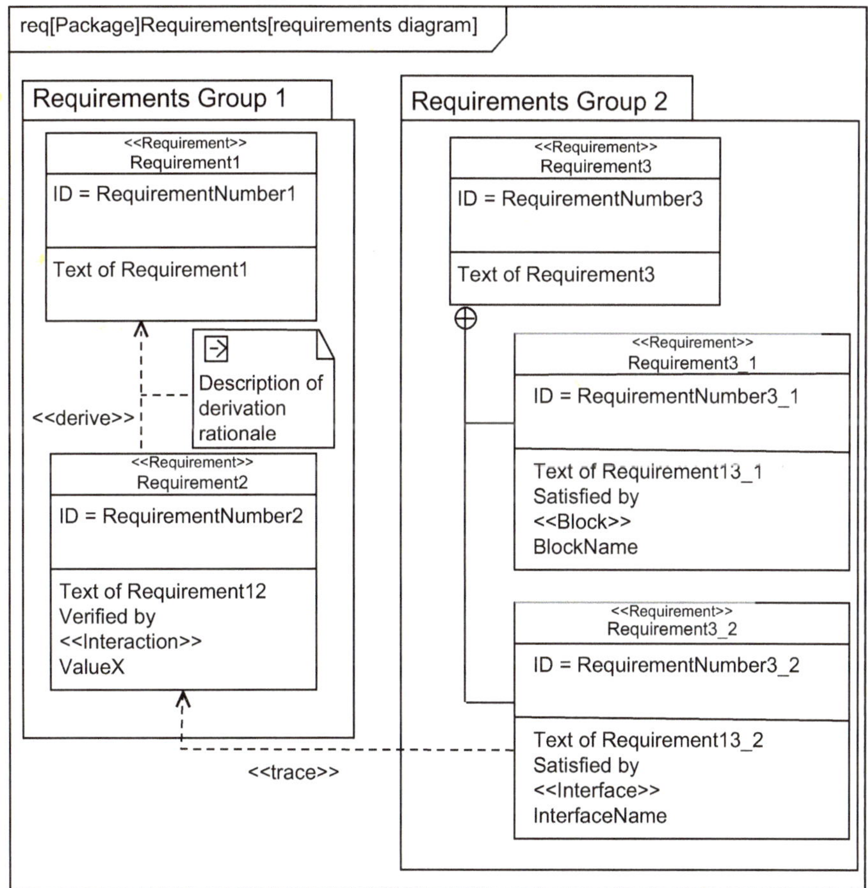

Fig. 2.12 Elements of a Requirements Diagram

- *Tracing*—a <<trace>> stereotype on a Dependency pointing from a Requirement Block to some other model element associated with defining the requirement
- *Verification*—a Block with the <<testCase>> stereotype can be used to model a verification method, and this can then be associated with one or more requirements using a <<verifiedBy>> Dependency; the verification method is the Classifier Behavior of the Block and can be shown in a suitable behavioral diagram.

Figure 2.12 shows some common content found in a Requirements Diagram. The left hand package in the figure shows how a derived requirement is portrayed, complete with a Comment that captures the rationale for the derivation. The specification for the derived requirement block has a `VerifiedBy` entry that tells the reader that Value X, stereotyped as an <<Interaction>>, will be computed or measured and compared to the required value to determine if the requirement is met. The right-hand package shows decomposition of a Composite Requirement into subordinate Requirements, using the containment (or nesting) mechanism shown as a cross in a circle. Each sub-requirement has a `SatisfiedBy` entry showing the architecture elements, in one case a Block and in the other an Interface, that satisfy them. Finally, a <<trace>> Dependency shows that Requirement Group 2 can be traced to Requirement Group 1, for example, the latter might specify a system and the former a subsystem whose requirements can be traced to those of the system. Requirements Diagrams can also include Slave Requirements, Refinement Requirement relationships, Copy relationships, and other subtleties.

2.4.3 Structural Modeling Enhancements

In SysML, the primary structural entity is a Block, which is a modular unit of a system and generally has structural and behavioral features. The replacement of Classes and Objects by Blocks and their Instances is perhaps the largest difference between UML and SysML. When a Block is declared to be included in another Block by a Composite Association, it becomes a Part Property, or simply a Part, of the composite Block that includes it. A Part is typed by the Block of which it is now an instance in the context of the composite Block. Any structural entity should first be defined as a Block and then used where appropriate as a Part.

The main structural representations are the Block Definition Diagram (BDD), which is a modification of the UML Class Diagram, and the Internal Block Diagram (IBD), which is modified from the UML Composite Structure Diagram.[7] A BDD is used to define system elements, including Composition, Reference, Association, and Generalization/Inheritance relationships. An IBD shows how system elements

[7] Since BDD and IBD are acronyms, we prefer to write them in uppercase, even though the SysML standard uses lowercase, and that's how they appear in diagram headers.

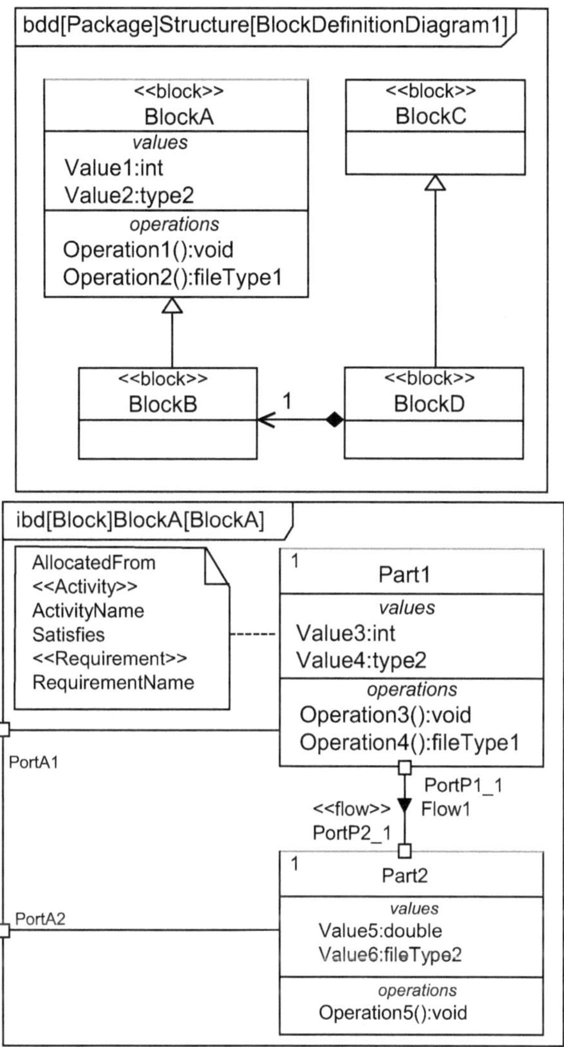

Fig. 2.13 Elements of a Block Definition Diagram and Internal Block Diagram

are used and stresses interactions among elements (typically Parts within a Block), including various kinds of Interfaces and Flows. Figure 2.13 gives a sense of how these diagrams look, including one way of annotating `Allocation` and `Satisfaction` relationships on an IBD. In the IBD, the diagram frame is the boundary of the Block whose structure is being modeled, and Ports for external interactions can be shown on that boundary. This example suggests that Block A gets an input via Port A1, and then Part 2, to which that Port is delegated, operates on the input and sends Flow 1 to Part 1, which does additional processing and makes an output available via Port A2. The diagram identifies a Part by a number in the upper left corner showing how many instances are contained in the Block.

Appendices B and C and the references give many more examples. Some modeling tools show Part multiplicity in other ways.

The two types of Block Diagrams are a good example of a very general principle that can be expressed as "definition vs. usage." In many situations, it's important to first define a model element that can then be used as often as necessary without recreating it each time. A Block, defined in a BDD (or just in the browser of a modeling tool), can be instantiated in one or more IBDs, and the Block becomes the Classifier of these instances, much like the relationship between a UML Class and its Objects. Another way to think about the two types of Block Diagrams from an architectural point of view is that BDDs are excellent for showing taxonomies of architecture elements, while IBDs are the normal way to show structural design patterns. Other examples of model content that should be defined as early as feasible include User Roles and other Actors, Value Properties, Value Types, and Item Flows.

We now need to discuss some important details of a SysML Block that make it more powerful than a UML Class in capturing system structure. Important Block characteristics are often modeled using Properties, which are typed and can be displayed on a Block in a diagram using suitably labeled compartments. The four kinds are Part Properties (Parts), Value Properties (Values), Reference Properties (References), and Constraint Properties (Constraints). A Port can be thought of as a Property but is sufficiently unique to enjoy its own separate syntax and semantics. Properties are summarized in the following paragraphs.

- *Part Property*—A Composition Association creates one or more instances of a Block within the context of another Block. This is the classic "is a part of" relationship between a Part and the Block that contains it. It's the basic SysML mechanism for structural decomposition and can be carried to any desired level of detail. This decomposition can be shown on diagrams by nesting lower-level elements within higher-level ones. The system element modeled by a Part will appear in the browser of an architecture modeling tool once as a Block that defines its characteristics and then one or more additional times as a Part Property of other Blocks. Compositional or structural hierarchies created in this way are central to the Systems Engineering tasks of decomposition and composition.
- *Value Property*—A Value defines a quantitative characteristic of a Block such as the mass or power consumption of a system component. Values can also be used to define data entities much like Attributes are used with a UML Class. Closely related to Value Properties are Value Types, stereotyped as `<<valueType>>`, which can have two optional properties: a unit and a dimension or quantity kind. Value Types are reusable and can be used to type properties anywhere in a model. In particular, a Value Type can be used to type a Block Property. Value Types are among the model elements that should be defined as early as possible in model development to ensure consistent typing. A common approach is to put them in a BDD as part of the data model of the architecture.
- *Reference Property*—A Reference indicates a non-compositional relationship between Blocks that may belong to separate compositional hierarchies. Reference Properties create a logical hierarchy. For example, a container may reference its

contents, but each of these is likely to be classified in a distinct compositional hierarchy. When shown on an IBD as nested within an owning Block, a Reference is indicated as a rectangle with a dashed outline. References are commonly created either using Aggregation Relationships or Associations between Blocks.
- *Constraint Property* –A Constraint is a quantitative or mathematical condition, imposed on properties of a Block, often in the form of an equation, an inequality, or limits. It's typed by a Constraint Block, which is further discussed under Parametric Diagrams below.

Another important consideration in modeling system structure is the proper use of Ports. Ports are adopted from UML 2, but SysML distinguishes among kinds of Ports. The original SysML standard, up through Version 1.2, defined Standard Ports, which typically model points for data or message exchanges via services, much as in UML, and Flow Ports, which can be in, out, or bidirectional and model exchanges of one or more tangible quantities called Flow Items. In these versions, a Flow Port is typed by a Port Specification, shown as a Block with a `<<flowSpecification>>` stereotype. The Flow Specification contains Flow Properties that define the items the Port can flow, and these are the specification's only valid attributes. Flow Properties have a name, direction, type, and multiplicity. An Item Flow defines an item that can flow on a Connector between Flow Ports and is typed by a Flow Property. An Item Property identifies the usage of an item in the context of an enclosing Block and must be typed by a Value Type or Block. A continuous flow is identified with a `<<continuous>>` stereotype.

Beginning with SysML Version 1.3, Flow Ports and Flow Specifications have been deprecated, and the allowed Port types are now `<<Full>>` and `<<Proxy>>`.[8] A Full Port is a Part Property of a Block and is placed on the Block boundary. Like any other Part, it models a set of resources within the Block, and it handles flow, non-flow, and behavioral interactions without exposing any of the Block's internal details. It's typed by a Block, of which it is an instance, and is used to model a physical part of the system. For example, a network interface can be represented by a Full Port that accounts for all the piece parts that make up the interface.

A Proxy Port is not a Part but rather a point at which the Block's inner workings are exposed. Flow, non-flow, and behavioral features are provided or consumed by a Part within the Block, to which the Port is delegated. A Proxy Port is typed by an Interface Block and used when the interaction does not involve an actual separate physical Part such as a cable connector or hose attachment. This new Port scheme is intended to facilitate definition of physical connections. Item Flows and Flow Properties remain with the new Ports and are used as in earlier SysML versions. A Flow Property of a Port describes *possible* Item Flows between a Block and its environments, including data, material, energy, etc. An Item Flow describes an *actual* flow between Blocks or Parts via a Connection or Flow; Item Flows can be allocated from object nodes in activity diagrams or signals sent from state machines.

[8] To maintain backward compatibility with models created in earlier SysML versions, architecture modeling tools commonly support both the previous and the current kinds of Ports.

Blocks can have behaviors that are allocated from other diagrams, making explicit in the architecture which structural elements provide particular required functionalities. This is a leading example of yet another very fundamental and important feature of SysML: the use of Allocations to declare various kinds of relationships among model elements. Allocations are commonly declared early in the development of an architecture and then refined and described in greater detail as the effort proceeds. Allocating behavior to structure honors the "definition vs. usage" dichotomy. For example, an Activity can be allocated to a Block at the definitional stage, while an Action within the Activity can be allocated to a Part in describing the usage of the Part within its owning Block. Another common Allocation is from an Object Flow in an Activity diagram to an Item Flow between Ports on an IBD. The specialized Allocations associated with requirements modeling in SysML were described earlier. References [10, 12] go into this topic in detail. Allocations can be modeled in several ways:

- With a Dependency carrying the `<<allocate>>` or other appropriate stereotype
- Using a Comment anchored to the affected model element with either `AllocatedFrom` or `AllocatedTo` and the name of the other element in the relationship
- Using a compartment on a Block or Part

A UML Class has three basic partitions or compartments: Identifier, Attributes, and Operations. A SysML Block can have many more compartments, and they can be listed in any order. Typical examples include:

- *Properties*, as described above
- *Namespace*—shows Blocks that are defined in the namespace of the Block
- *Structure*—shows connectors and other structural elements of the Block
- *Allocation*—shows model elements that are allocated to the Block
- *Ports*—shows ports that have been created on the Block
- *Other*—additional compartments, often defined by individual modeling tools

2.4.4 Behavioral Modeling Enhancements

Activity Diagrams (ADs) in SysML are modified from UML 2, and the enhanced semantics are particularly useful in modeling detailed behaviors. In particular, SysML incorporates the Enhanced Functional Flow Block Diagram (EFFBD) that is familiar to many System Engineers. The emphasis on sharing and manipulating values in SysML will often call for the use of Activity Parameter Nodes as illustrated in Fig. 2.14. This makes the specific inputs and outputs (Parameters) of an Activity clear and easy to understand. Parameters are usually typed by a Value Type or by a Block and can have multiplicity when multiple Value Tokens are consumed or produced.

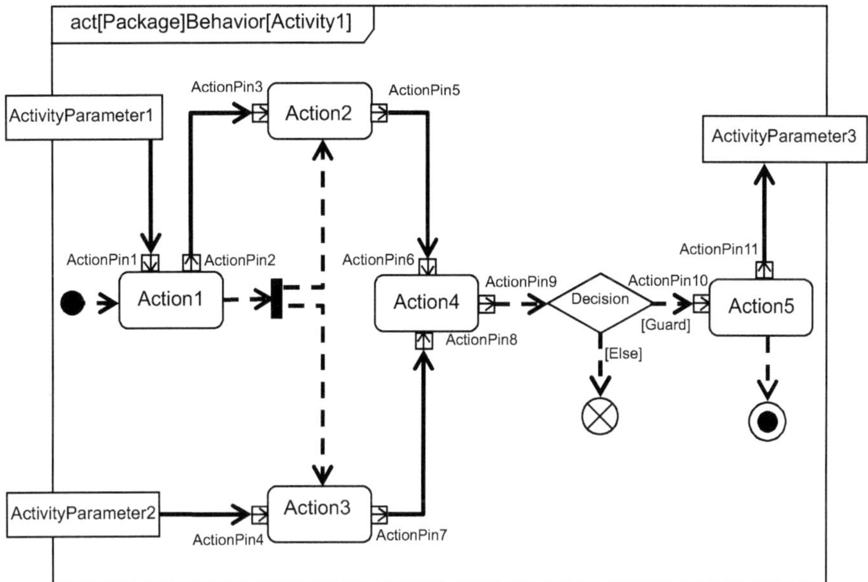

Fig. 2.14 Activity Diagram format example showing the use of Activity Parameter Nodes

As in UML, ADs may use more specific or fine-grained swim lanes (Activity Partitions) so that the behaviors assigned to specific system elements are clearly identified. The use of Pins in ADs was discussed in Sect. 2.3.2. SysML defines several types of Action Pins, drawn as small squares that are graphically reminiscent of Ports in a structural diagram. Pins model flows of Objects and Items, but they are typically omitted for Control Flows. They should be labeled with the object or item unless it is shown explicitly. Pins are also used with Call Actions that can invoke other behaviors and operations, analogous to a subroutine call in software but more general in meaning. A called Operation needs a target that identifies the system element receiving the call. Control Flows are normally shown simply as arrows between actions.

SysML includes a number of advanced techniques for Activity modeling. For example, Activity Parameters can be either Streaming, meaning that input or output can occur during the execution of the Activity, or Non-Streaming if input can only occur before execution and be output after it completes. There are also mechanisms for modeling interrupts, flow rates, probabilistic flows, and many other subtleties of a complex behavior. Reference [10] gives full details.

Use Case, Sequence, and State Machine Diagrams in SysML are essentially identical to those in UML. However, the Parametric Diagram offers an entirely new way to model quantitative behavior by defining mathematical relationships among variables and linking them to the model elements where their values are established and used. Parametrics provide a mechanism to model constraints and relations (typically in the form of equations) among Value Properties. This integrates engineering

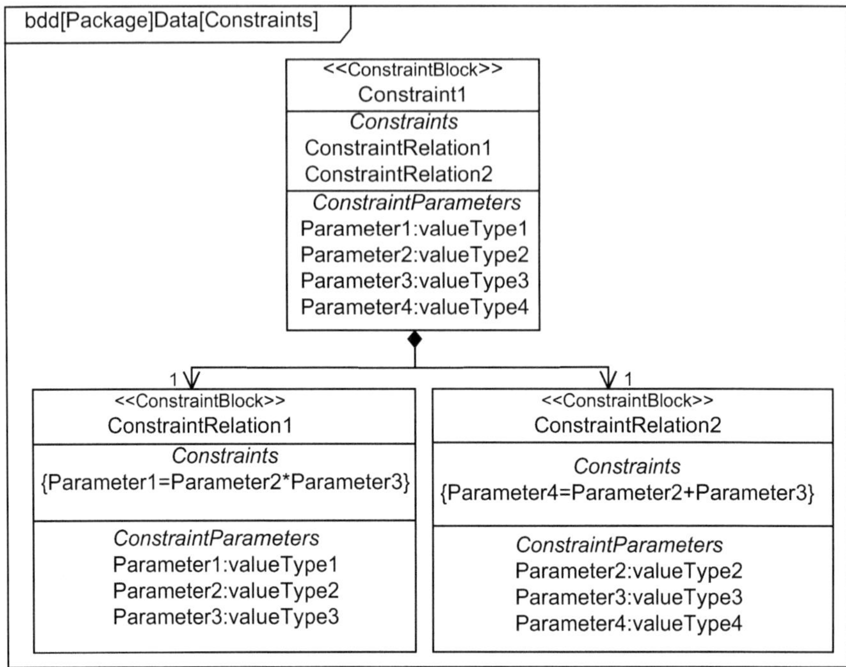

Fig. 2.15 Constraint Blocks defined in a Block Definition Diagram

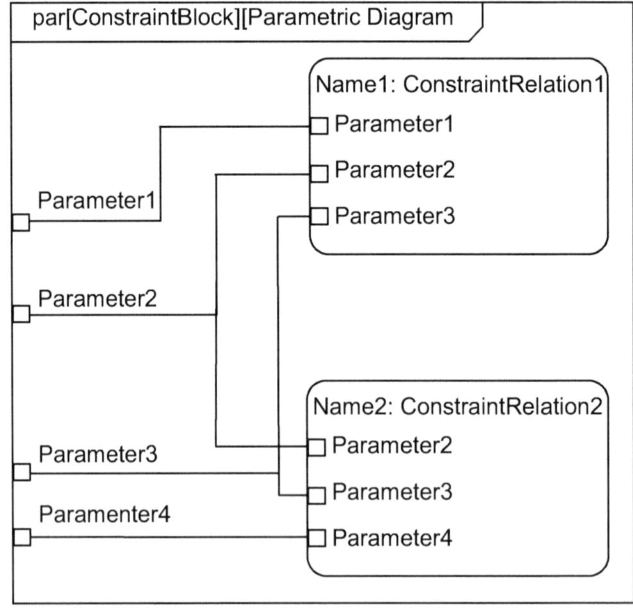

Fig. 2.16 Elements of a Parametric Diagram

analysis, dealing with values for performance, reliability, and other important aspects of a system, with SysML Blocks. A UML Class Diagram can specify mathematical relationships in defining Class methods, but the Parametric Diagram provides an intuitive portrayal of where values come from and go to and of the way various constraints interact to determine system behavior. Figures 2.15 and 2.16 show the format. Some modeling tools further enhance this capability by providing a means to link a Parametric Diagram to external analysis tools such as equation solvers when higher-order mathematics must be evaluated.

In general, a Constraint is a condition or restriction applied to a model element, usually expressed in mathematical form so that it can be evaluated. A Constraint Block packages one or more Constraints in a way that makes them reusable. A Constraint Parameter is a property of a Constraint Block that allows binding to other properties when the Constraint is used, and a Constraint Property is a Block Property typed by one or more Constraint Blocks.

2.4.5 *Implementation Modeling Enhancements*

SysML has major benefits in modeling physical design because of its power to represent structures and their relationships. Blocks that are initially part of a functional design are fleshed out with specifications, product data, and other details. Flows of information and physical quantities are quantified. Ports are described with interface documentation. Anything important about the implementation of the architecture is included either directly on diagrams or in specifications for the various model elements. Detail can be included in the model in various ways, including Notes directly shown on a diagram and links to supporting documentation such as product specifications.

SysML is particularly useful when an architecture model needs to deal with complex design and implementation details of both hardware and software. Block Diagrams can be carried to whatever level of detailed decomposition is needed, and the use of Allocation, Verification, Satisfaction, and other dependencies keeps structural elements tied to requirements, functions, test cases, and other parts of the model. Well-constructed physical diagrams are both rich in content and easy to understand, making the model a powerful communication tool.

2.5 Concluding Thoughts

UML 2 is a rich and complex modeling language, and SysML further enhances an architect's ability to deal with requirements, performance, design, integration, and other SE processes and tasks. This chapter provides a basic introduction that is sufficient to understand how OO principles apply to system architecture and for practicing the MBSAP methodology. The diagrams and general descriptive material in this chapter will be illustrated in more elaborate examples in subsequent chapters.

The System Engineering and Architecture community is overwhelmingly moving to SysML, and we have shown the basis for that trend in terms of enhancements to the constructs of UML 2. A practical reality is the fact that modeling tools from various vendors tend to have (usually minor) variations in the way they implement the UML and SysML standards, which is annoying but not fatal, especially when an organization selects a preferred tool and trains its users accordingly.

It's also true that both the UML and SysML language standards are living documents that are updated with some frequency to correct mistakes or ambiguities or to enhance the descriptive power of the language. For example, SysML Version 1.4 adds the ability to assemble a group of model elements in a BDD using an `<<ElementGroup>>` stereotype without introducing any new definitions. This contrasts with the use of a Package, which brings in namespace semantics. It also adds compartments on a Block for Classifier Behavior and Owned Behavior. Usually, Version updates are of interest chiefly to the developers of architecture modeling tools, and many affect only advanced or esoteric modeling techniques that are beyond the scope of this book. However, as we noted under Ports in describing the changes that came in with SysML Version 1.3, sometimes a change is sufficiently important that all modelers need to take note and adjust their practice appropriately. Even in this case, however, the large number of legacy models built under earlier SysML versions means that the older constructs (e.g., Flow Ports) are likely to be supported by architecture tools for some time to preserve backward compatibility.

A user or group of users whose architecture modeling is focused on a particular application area or system category can tailor the language to make such modeling more efficient and expressive. The mechanism, which is available with both UML and SysML, is the Profile, which defines a set of extensions to language elements targeted to specific usage. Most modeling tools support Profile creation, export, import, and use. As noted earlier, SysML is actually a "heavy-duty" Profile of UML. The Service-Oriented Architecture Modeling Language (SoaML) is a Profile discussed in Chap. 7, and the Modeling and Analysis of Real-Time Embedded Systems (MARTE) Profile is discussed in Chap. 8. Like most Profiles, these are less drastic than SysML in the changes they make to the root language. A Profile typically contains extensions such as stereotypes, tagged values, constraints, customizable icons for diagrams, predefined patterns of elements, and so forth. Specific modeling tools support additional capabilities such as templates and reusable or reference models.

What ultimately matters is exploiting the power of OO and its expression in SysML (or UML for architectures dealing primarily or exclusively with software) as the bedrock of a high-quality Model-Based System Engineering methodology able to cope with the challenges of modern technologies and systems. The remainder of this book describes such an architecture-centric approach and explores some of the variations and subtleties that arise in dealing with the wide range of challenges confronting today's System Engineers and Architects.

Exercises

1. List some reasons for preferring Object Orientation as the approach to defining and developing complex, information-intensive systems.
2. List some principles of OO that are important in system architecting.
3. What is a Classifier? What is the relationship between an Object and a Class? What is the equivalent relationship for SysML Blocks?
4. How are Packages used in a model?
5. What is an Actor in a UML or SysML model?
6. How do Use Cases support modeling of behavior?
7. How do Activity Diagrams relate structure and behavior in a model?
8. Why is a Merge an asynchronous recombination of flows, while a Join is a synchronous recombination?
9. List some ways Interaction Fragments can be used in a Sequence Diagram.
10. What makes the behavior of a Class or Block "stateful"?
11. What are the two primary uses of a Deployment Diagram?
12. List some typical System Engineering tasks that are explicitly supported by SysML.
13. In terms of the MBSAP methodology, list several major differences between SysML and UML 2.
14. Differentiate between the two kinds of Block Diagrams in SysML.
15. How would a normal requirements database and a set of Requirements Diagrams be used in a given project?
16. If a UML architecture model were to be converted to SysML, what would be the primary change in describing system structure?
17. What are some of the SysML enhancements to the UML Activity Diagram?
18. How is a complex behavior modeled in a Sequence Diagram broken down into simpler behaviors?
19. How does a Parametric Diagram contribute to describing a system in a SysML architecture model?
20. List some of the ways in which SysML enhances the modeling of implementation.

References

1. Object Management Group (2011) INCOSE Object-Oriented Systems Engineering Method (OOSEM). www.omgwiki.org/MBSE/doku.php?id=mbse:incoseoosem. Accessed 19 Dec 2016
2. Booch G (1997) Object-oriented analysis and design with applications, 2nd edn. Addison-Wesley, Reading, MA
3. Fowler M (2004) UML distilled: a brief guide to the standard object modeling language, 3rd edn. Addison-Wesley, Boston, MA

4. Muller P-A (1997) Instant UML. Wrox Press, Paris
5. Chonoles M, Schardt J (2003) UML2 for dummies. Wiley, New York
6. Booch G, Rumbaugh J, Jacobson I (1999) The unified modeling language user guide. Addison-Wesley, Boston, MA
7. Object Management Group (2011) UML 2.4.1. http://www.omg.org/spec/UML/2.4.1. Accessed 19 Dec 2016
8. Buede DM, Miller WD (2016) The engineering design of systems: models and methods, 3rd edn. Wiley, Hoboken, NJ
9. Harel D (1987) Statecharts: a visual formalism for complex systems. Sci Comput Program 8(3):231–274
10. Object Management Group (2015) SysML 1.4. http://sysml.org/docs/specs/OMGSysML-v1.4-15-06-03.pdf. Accessed 19 Dec 2016
11. Object Management Group (2016) System modeling language. http://www.omgsysml.org/. Accessed 19 Dec 2016
12. Friedenthal S, Moore A, Steiner R (2015) A practical guide to SysML, 3rd edn. Morgan Kaufman/Elsevier, New York

Chapter 3
MBSAP Methodology Overview

3.1 Basic Principles

This and the next several chapters apply the basics of OO modeling in the MBSAP methodology that is the central element of our approach to system architecture and MBSE. To begin, we introduce the structure and activities of the methodology as a whole and seek to place it in the context of the overall SE discipline. MBSAP has evolved in large part by drawing on best practices of the SE and architecture communities and on lessons learned in a number of earlier architecture development efforts for a variety of systems. Many of the individual elements of MBSAP have been developed, used, and published by various workers, as documented in the chapter references. The contribution to the field that we seek to make in this book is the adaptation and integration of concepts and methods that have been successfully employed in real systems and programs into a coherent and comprehensive architecture strategy. Although the discussion frequently refers to "*system* architecture" and "*systems* engineering," it should be understood that MBSAP applies to any complex, technology- and information-intensive entity, including an enterprise combining multiple nodes and systems.

In tackling complex system challenges, it quickly becomes apparent that architecture and SE are inseparable. Indeed, we argue that SE processes for systems and enterprises like those of interest in this book must be rooted in architecture. SE is concerned with things like requirements analysis and allocation; trade studies involving performance, cost, schedule, and supportability; risk management; design analysis; integration and test; configuration management; security; safety; and other aspects of achieving an optimized and operationally satisfactory solution to a customer's needs. Architecture furnishes the central organizing principle for these SE activities because it is the authoritative representation of the system or enterprise design at any point in time and provides the raw materials used in SE processes.

At the most general level, a model can be thought of as a representation of an entity in a form that abstracts its essential characteristics. It is, in effect, a specification

J. M. Borky, T. H. Bradley, *Effective Model-Based Systems Engineering*,
https://doi.org/10.1007/978-3-319-95669-5_3

for the entity from which a design can be derived. In building a formal model, MBSAP uses syntax defined by SysML, semantics that define the meanings of model elements, and rules for model construction and modification. A well-constructed model allows description and analysis to facilitate understanding of the entity and communication among the parties who are concerned with it, the stakeholders. In engineering and science, a model must be mathematical, and to be useful, it must be based on standards for things like the modeling language employed, the underlying laws and conventions, the units of measure, and the content and format of model artifacts. In system architecture and engineering, a model must adequately define the structure, behavior, rules or universal properties, information content, and other key aspects of the thing being architected. When constructed according to these principles, an architecture can be subjected to modeling, simulation, and analysis (MS&A) as described in Chap. 11 to make quantitative predictions about the physical properties, response to stimuli, human-machine interactions, and other characteristics of the system or other entity it represents.

Although this book does not attempt to provide a comprehensive presentation of the entire subject of SE, it's helpful to consider MBSE and the architecture model that supports it in the context of a traditional SE process. In Fig. 3.1, the left side shows the sequence of activities in a typical SE flow, from initial discovery of customer needs, through design and implementation of a system that meets those needs, and on to fielding, operation, and support of the system. The right side of the figure shows a few representative examples of the support MBSE provides at each stage, once the customer's needs and desires have been defined and documented. We will expand on these and many other particulars in the remainder of the book. The point to be made here is that a fully realized MBSE process offers important benefits to virtually every SE activity across a system's life cycle.

Architecture models provide the subject matter for SE tasks like those listed above. This concept of *architecture-centric* or *model-driven* SE represents a major change from traditional thinking, which today is commonly described as *document-centric* because the basis for describing a system has historically consisted of textual specifications, graphics, tables, and other documents. Moreover, architecture was in the past commonly treated as simply an artifact produced by SE, sometimes consisting of little more than block diagrams of the final design. Today, it is widely realized that architecture is the heart of the system development process and that it is critical to all phases of translating customer requirements into effective solutions.

MBSAP is a specific instantiation of leading industry practices such as The Open Group Architecture Framework (TOGAF™) [1], the Zachman framework™ [2], Model-Driven Architecture (MDA™) [3], the Object-Oriented Systems Engineering Method (OOSEM) [4], and others. It incorporates a number of basic principles that experience has shown to be essential in defining and developing complex systems. The characteristics and tenets of MBSE in general, and of MBSAP in particular, include the following:

- Use of OO, both because it is the mainstream of modern architectural practice and in order to exploit the power of abstraction, encapsulation, generalization,

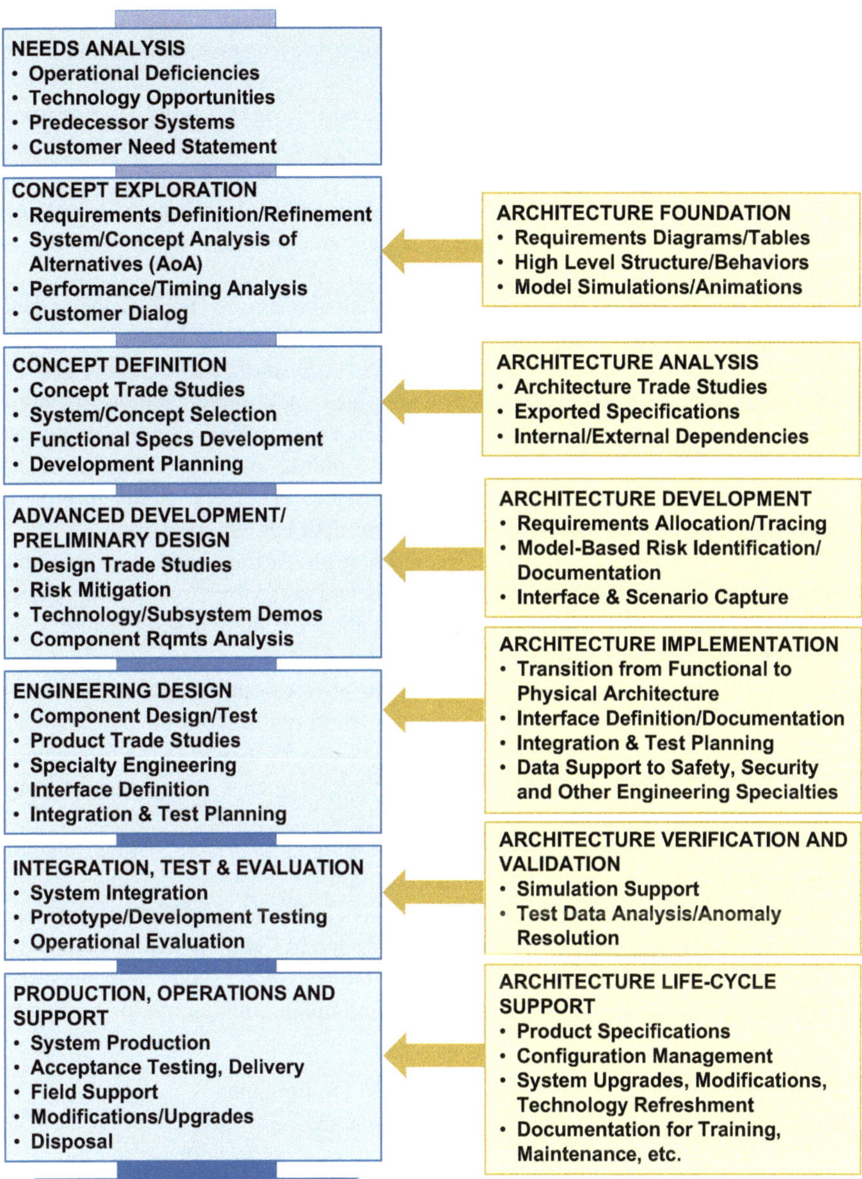

Fig. 3.1 MBSE support to Systems Engineering activities

polymorphism, and other OO principles. Chapter 2 spelled out this rationale in detail. A good model is abstract in the sense of conveying essential content while suppressing inessential details.

- Use of architecture models as the core material for SE processes, as discussed above. A good model facilitates understanding and communication by and among its users.
- Rigorous traceability through the flow of architecture development from customer requirements to the elements of a delivered solution. A good model is accurate, current, and unambiguous.
- Support for implementation of service-oriented architecture (SOA) as described in Chap. 7.
- Enforcement of architecture Quality Attributes as discussed in Chaps. 4 and 15 to maintain the integrity of the architecture and track implementation of nonfunctional requirements. Quality Attributes are based primarily on best practices like modularity, standards compliance, loose coupling, and strong interface management and are tracked using objective measures. They commonly provide a mechanism for evaluating satisfaction of nonfunctional requirements.
- Construction and use of executable representations of the architecture (simulations) at various levels of abstraction when needed to resolve architectural issues. A good model facilitates thorough investigation of alternatives because it is cheaper to build and modify than an actual system.
- Linkage of the evolving architecture to both physical and virtual (simulation) prototypes. A good model supports rigorous prototyping both for analysis and optimization of an existing or emerging design and for assessing the future evolution of that design in response to changes in requirements, operational environments, and other factors.
- Effective means to apply the basic methodology and its principles to a range of architecture categories as addressed in Chap. 1.
- Support to architects in complying with customer policies, especially Government standards and reference architectures as addressed in Chap. 12.
- Improved methods to deal with challenges that have increasingly threatened the success of programs to develop and sustain complex, high-technology systems and enterprises, including:
  - Ambiguous, unstable, and poorly allocated requirements
  - Limited commonality and reuse among systems
  - Problems with integrating system elements, especially from multiple suppliers
  - Lack of interoperability within and among systems
  - Poorly structured or proprietary ("closed") systems that are hard to modify or upgrade and that limit follow-on competition

Ultimately, the motivation for MBSE comes down to giving system engineers better ability to handle technical and programmatic complexity, improved efficiency and quality in activities and products across the SE life cycle, reduced risk, and higher confidence that customer requirements will be transformed into operationally effective and affordable system solutions.

3.2 Expanding the Systems Engineering Legacy

Traditional SE practice has a strong heritage in advanced technology and high-performance hardware. Its central analytical approach is embodied in methods like Structured Analysis and Structured Design (SA/SD) [5] with emphasis on top-down functional decomposition, ultimately resulting in allocation of requirements to hardware and software components, followed by a bottoms-up integration and test

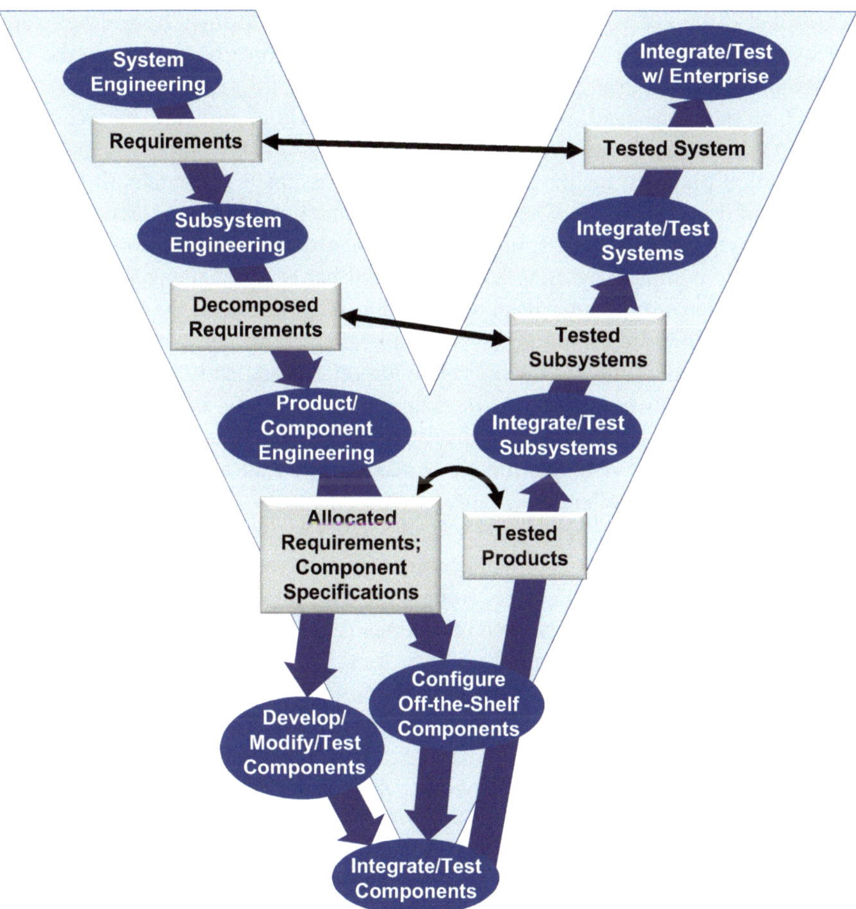

Fig. 3.2 Systems Engineering "Vee" representing requirements analysis and allocation, followed by design, integration, and test to realize the final system

sequence to build up the final system. Figure 3.2 is a more detailed version of the SE "Vee" diagram introduced in Chap. 1. Buede and Miller [6] refer to this as Traditional, Top-Down Systems Engineering (TTDSE). The SE activities in Fig. 3.1 can be mapped to the Vee; for example, Needs Analysis, Concept Exploration, and Concept Definition are associated with System and Subsystem Engineering.

One consequence of methods like SA is a tendency to isolate hardware and software, develop them separately, and trust that they will come together smoothly during integration. Data is usually treated as a distinct problem, separately documented in a data model, data dictionary, database table designs, and perhaps a storage map. Similarly, rules for system behavior are often captured in a separate model. This approach allows engineering specialists to focus on their specialties, but it impedes holistic thinking about the system, and it often delays the discovery of subtle problems until the later stages of the development cycle where they are hardest and most expensive to correct. For these and other reasons, the SA/SD approach has consistently had difficulty when dealing with increasingly information- and software-intensive systems, and serious cost and schedule overruns have become the rule rather than the exception.

OO, on the other hand, is descended from the world of software engineering. It tends to take the hardware as a given and assume that its only purpose is to provide an execution platform for the software, which is assumed to contain all the complexity and do all the work. Chapter 2 discussed the limitations of UML Version 1, and to a lesser extent Version 2, in dealing with hardware and with development and integration of complete systems with significant hardware challenges. In the past, OO methods have sometimes been dismissed as inapplicable to the kind of system developments that are common in situations where both hardware and software must be produced, with their designs evolving interdependently and in parallel. In reality, the inherent power of OO in managing complexity, discovering reusable design patterns, facilitating modularity, abstracting away internal implementation details, and localizing data and functions makes it very attractive for these as well as other modern SE challenges.

Despite their different legacies, OO and SA/SD have many things in common. Both rely on hierarchical functional decomposition to reduce complex entities to a level that makes them manageable. Both emphasize interfaces and threads of control. Both insist on traceability from requirements to design. Both build up the final product by progressive integration of lower level elements. One way to view their basic difference is that SA decomposes functions, while OO decomposes a system into identifiable real-world entities, including the resources making up each entity, the operations it performs, and the data it uses. This bundling creates coherent units of system content. It works this way because the nature of a Block or Block Instance is to encapsulate the Properties (Values, Parts, References, Constraints, etc.) and the Operations (methods or functions) of the entity being modeled and to define an interface in terms of the publicly exposed Operations and Values. Among other advantages, this allows public functions and data to be readily transformed into services as the foundation for SOA.

The challenge, which SysML tackles head-on, is to deal with the limitations resulting from OO's software legacy while exploiting its inherent advantages. This is a recurring theme throughout the book, but the following paragraphs summarize the basics, which form the foundation of MBSAP.

- Treat the elements of an architecture as it is being developed and modeled using Blocks as initially *functional* entities, shown in functional block diagrams, to which responsibilities and relationships are assigned. Defer decisions such as allocating functions to physical hardware and software components and selecting the actual components to be used until a later implementation stage.
- In practicing MBSE, integrate hardware and software engineering in an overarching architectural viewpoint that seeks to create and maintain a balanced and optimized solution to a customer's needs. Perform system-level trade-offs and account for constraints that arise in optimizing the system as a whole. Figure 3.3 illustrates the relationship among Enterprise, System, Hardware, and Software Engineering. An enterprise can be partitioned into the systems that comprise it, which may be hosted by nodes. A system is built by allocating functionality to hardware and software, which are then integrated to create a solution that satisfies the system's requirements. For example, a system might need a specialized processor, and requirements analysis would define the performance and other capabilities that the processor must deliver. The actual processor design, with crucial choices such as implementing its algorithms in software or in custom hardware processor logic, can come later when the Block is physically instantiated. Those details are hidden within the encapsulation boundary because the rest of the system only needs to know how to invoke the processor's functions.
- Supplement a UML/SysML model and its diagrams with additional artifacts. Traditional products like a data dictionary and a classic hardware block diagram can be efficient tools to document and communicate important details of a design. They should be developed from and kept consistent with the architecture model.

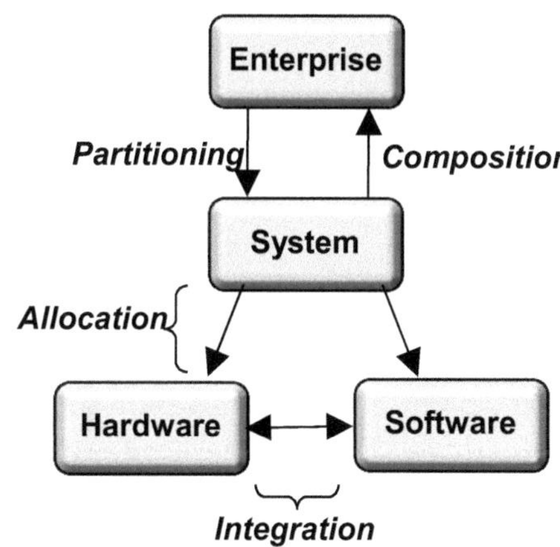

Fig. 3.3 Relationships among enterprise, system, hardware, and software architectures

- Exploit advances like SysML itself and improved modeling and simulation tools and methods, including executable architecture, that enhance the power of OO to cope with the challenges confronted by system engineers dealing with complex, high-technology systems. For example, it's possible that formal methods will be added to this list in the not-too-far distant future, but that subject is beyond the scope of this book.

3.3 The Basic Methodology

The primary motivation for MBSAP is to provide a framework and methodology for applying MBSE and enjoying its benefits in dealing with a wide range of technology-intensive systems and enterprises. However, its applicability extends beyond system design. The MBSAP approach can be applied to challenges such as optimizing organizations and their processes, providing objective analyses and data products for improved technical and programmatic decisions, supporting dialog with diverse stakeholder communities, and creating efficient and economically attractive product lines. In all these areas, an architectural model that unambiguously represents the structure, behavior, and constraints of a complex entity creates the foundation for achieving balanced and effective solutions to the needs of end users and other stakeholders who are the "customers" of an MBSE effort. We discuss and illustrate MBSAP in terms of its primary usage in specifying, developing, producing, and supporting systems, but the fundamentals are the same whether the entity in question is an airplane, a factory, an electric utility, or a government agency in need of more efficient processes.

MBSAP has evolved largely through the incorporation of proven best practices from many sources and through refinement of the methodology based on the outcomes of multiple system development and sustainment programs. It integrates OO and the other principles of MBSE in a tool-supported and metrics-managed process flow with defined work products. The methodology is shown at a high level in Fig. 3.4 and described in detail in the paragraphs that follow. The process begins

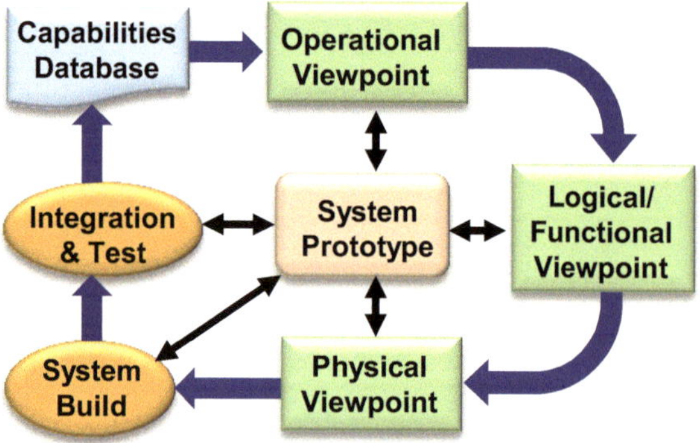

Fig. 3.4 Top-level overview of the MBSAP methodology

with customer needs, progresses through successive phases of architecture development and system realization, and concludes with either delivery of a final product or the starting point for a follow-on increment of development.

Figure 3.4 is drawn as a closed loop, implying that MBSAP is iterative. Many practitioners of OO methods assume that the underlying program strategy will use an incremental approach that is commonly referred to as an "evolutionary spiral" in which subsets of a requirements baseline are implemented in successive rounds of design, implementation, and testing until the entire solution has been achieved. MBSAP supports such an incremental approach but is equally compatible with a unitary program structure (sometimes called a "big bang" approach) in which a single cycle of analysis, design, integration, and test implements the entire required capability set of a system. Spiral development evolved originally in response to the reality that requirements for a complex system are frequently impossible to completely define at the outset of a project.

System users may be unable to describe completely and in detail the behaviors they want their system to exhibit until they can experience and compare alternatives. Hard choices about levels of performance may not be made until their cost, schedule, and risk implications can be quantified. It is therefore commonly the case that requirements analysis and refinement must proceed in parallel with design and that requirements will evolve on the basis of experience gained in the design process. A spiral program "eats the elephant a bite at a time" and decides where to bite next based on how the last morsel tasted. Today, the term "spiral development" has largely given way to various forms of "agile development," but the underlying concepts are very much the same.

Agile or spiral methods can be thought of as implementing the idea of "build a little, test a little" aimed at ultimately achieving the full required capabilities. Each increment of development has allocated to it a subset of those capabilities and results in a prototype that can be tested against the selected requirements and evaluated with system stakeholders. Theorists of the spiral method predict that something like 10–25% of the developed hardware and software in a given prototype is likely to be scrapped or redesigned on the basis of unsatisfactory results. The next increment takes account of this experience along with the next group of capabilities to be implemented and the revisions to system requirements resulting from what has been learned about risk, cost, performance, and supportability. Successive prototypes build up to the final delivered product, whose requirements have also been refined through the spiral experience. Thoughtful involvement of system users and other stakeholders, especially using system prototypes as they evolve, helps ensure that the result will meet their expectations and will avoid unpleasant surprises.

In Fig. 3.4, the starting point is the business case and resulting customer requirements, represented by a Capabilities Database in the upper left corner. Any significant system development is supported by a business case justifying the effort and cost, and each stage of the architecture development must link back to that business case as its fundamental justification. The terms "capabilities" and "requirements" are often used synonymously, but the former term is preferred to make it clear that the desired outcome is a set of operational capabilities that the developer is free to achieve through whatever design proves to be optimal rather than being unduly constrained by a priori design requirements dictated by the customer.

The content of this database can take many forms. In the early stages of program formulation, customer needs may be expressed in terms of an operational concept (CONOPS), white papers and briefings describing the desired operational capability, documentation of business processes and desired enhancements, a high-level requirements document describing the enterprise in which a system is a participant, a capabilities description defining the operational contributions the system is intended to make, a performance specification, or various other contents. In many cases, the customer will impose detailed capability descriptions, applicable policies and reference architectures, sustainment practices, and other requirements that tend to constrain the allowable solutions. Sophisticated customers are increasingly defining specific architecture deliverables and methodology requirements because they recognize the importance of these in achieving a robust, affordable, and evolvable mission solution. Many organizations use a requirements management and analysis tool to codify and manage the customer's required and desired capabilities, including any Threshold (mandatory) and Objective (desired) performance levels, resulting in a Capabilities Database. Alternatively, or in addition, Requirements Diagrams in the model can be used.

The thick blue arrows in Fig. 3.4 represent mappings or transformations from one architecture representation to the next. The first such mapping translates the customer's required capabilities, in whatever form they are expressed, into an architectural context labeled an Operational Viewpoint (OV). This mapping is achieved with Use Cases and other OO constructs as described in Chap. 4. Lattanze [7] is one of many writers who stress the importance of starting with a high-level, coarse-grained view of the architecture before proceeding to detailed design. That is the primary purpose of the OV.

We emphasize again the importance of good model structure to achieving the full benefits of MBSE. In MBSAP, that structure starts with the Viewpoints defined in Fig. 3.4 and then groups the content of each Viewpoint into a set of Perspectives that create a logical and easily searchable content framework. These are described in the next section. When a system development is required to conform to an established architecture framework, the OV can be used, either directly or with appropriate augmentation and reformatting, to create the appropriate operational or business process artifacts.

To be complete, the OV should describe the context within which a system functions. Any external entities, including system operators and customers, other systems and networks, facilities and utility services, data repositories, regulatory bodies, and many other elements of the system environment are normally represented as Actors in structural and behavioral diagrams. This context is generally beyond the control of the system and its owning organization, and so the architecture must have the flexibility to respond to changes and to evaluative their impact.

Another important aspect of the OV is often simulation of the emerging high-level architecture. As the OV takes shape, it may be helpful to represent it in one or more executable models that bring the overall system, its behaviors, and its operational context to life. This is important in ensuring requirements are complete and understood and in supporting dialog with stakeholders. These simulations help identify and resolve issues and ambiguities and support early trade-offs among competing system characteristics and capabilities before significant resources are expended on design and implementation.

The next mapping transforms the OV to the Logical/Functional Viewpoint (LV), which is detailed in Chap. 5. This is where design begins using Block Diagrams to define the details of system elements, services, functions, information exchanges, data entities, and behaviors derived from the OV. The LV continues the Perspectives of the OV and is developed in large part by decomposing and adding detail to the content of the OV. The LV represents a *functional* definition of the system or enterprise that is independent of any particular technologies or products. This helps ensure that MBSAP is consistent with MDA and provides an effective basis for technology refreshment to deal with product obsolescence. No decisions have been made at this point in the architecture development about allocating functionality to hardware or software, and only top-level decisions about choices of standards are in order. Standards selection in the LV is mostly a matter of complying with overarching architectural requirements and policies such as the need to implement enterprise services and to comply with the associated standards.

However, design tenets like modularity, localization of data and functions, and loose coupling are very much in force when developing the LV, and considerations of feasible performance, technical risk, security mechanisms and processes, and other practical details are important influences on the design. For example, it may be very useful at this stage to prototype alternative human-machine interfaces (HMIs) and iterate these with operational users before finalizing design requirements and proceeding to build the system. An initial assessment of the hardware and software resources that will be required, accounting for the capabilities of available products, may also be needed as a feasibility check on the functional design. Continuing the theme of exploiting executable architecture, a variety of tools are available to animate the functional model in order to assess and even quantify behaviors.

Architecture modeling is completed by mapping the LV to the Physical Viewpoint (PV), which is then the basis for actual implementation of the full system or of an increment of system capabilities in a prototype. One way to think about the relationship of the LV to the PV is that the former defines *what* is to be built and the latter *how* it will be realized. One consequence of this, described in Sect. 3.4, is a shift in focus to products and standards whose selection is central to arriving at a physical point design.

Chapter 6 describes the activities involved in the PV. This is the transition from design to implementation and involves the selection at a point in time of specific hardware and software products, software programming language(s), a detailed standards profile, messaging and networking solutions, sensors and actuators, power and cooling, and all the other aspects of a point design. Now, the decisions about allocating specific functionality to hardware and software must finally be made, typically through trade studies based on available technology, cost, risk, development schedule, and other factors. Those decisions may be revisited in later versions of the design as technology advances and operational experience accumulates. The PV must account for system context, now described in quantitative terms. For example, environmental factors such as temperature, electrical power quality, and external network capacity may be included in product trades to ensure a satisfactory system design is achieved.

In MBSAP, the design modeled in the PV is built up in a prototype and goes through integration and test to assess its suitability against the required capabilities that are being addressed, whether the program approach is unitary or incremental. The results of this prototype evaluation help to identify where the implementation is unsatisfactory, resulting in problem reports or other design changes. Additionally, the assessment may turn up problems in the system requirements such as inconsistent or incomplete definitions and requirements that are not feasible or cost-effective to achieve. When that happens, the process provides for updates to the Capabilities Database. In an evolutionary spiral, these results, along with a fresh increment of capabilities, become the basis for the next spiral. In a unitary program, prototyping results support refinements and problem fixes in the system design.

The prototyping environment, drawn in the center of Fig. 3.4, ties the whole MBSAP process together. In most situations, there should be both virtual and physical representations of the system being architected. The virtual prototype is an ensemble of executables, modeling and simulation tools, and artifacts as described in Sect. 3.6. These can include simulations at any of the levels of abstraction defined in Fig. 1.1. One possible approach at the Logical/Functional level, especially when system functions are primarily created by software, is an adaptation of Executable UML (xUML) [8, 9]. Chapter 11 explores prototyping in more detail.

The physical prototype could be a System Integration Laboratory (SIL) or its equivalent where the actual prototype of the system is built up from its physical components, integrated, instrumented, tested, and refined, including problem fixes. Prototyping at the level of an enterprise often involves networking a set of distributed facilities representing various nodes and systems. Once the development is complete, the SIL can be transformed into a sustainment facility that is used over the long-term for modifications and upgrades, diagnosis and correction of latent design errors, training, and other functions.

The arrows in Fig. 3.4 from the prototyping activity to the OV, LV, and PV show that the prototype draws on the results of each stage of architecture definition and that the results of prototyping feed back to each of those stages. For example, the results of an executable architecture model built for the LV might identify erroneous or inconsistent definitions of behaviors in the OV. Similarly, performance modeling for the PV could show that deadlines defined in the OV cannot be met or that the LV fails to account for significant interactions within the system. In fact, although the diagram becomes hopelessly cluttered if drawn this way, there is an implied feedback path from every stage of the spiral to every preceding stage, with the prototyping environment furnishing the most common pathway.

With this initial description of MBSAP in hand, we can complete our discussion of the relationship of MBSE to traditional SE. Figure 3.5 shows the relationship between MBSAP and the traditional SE Vee. Essentially, architecture definition in the OV, LV, and PV maps to the left side of the Vee, while system build, integration,

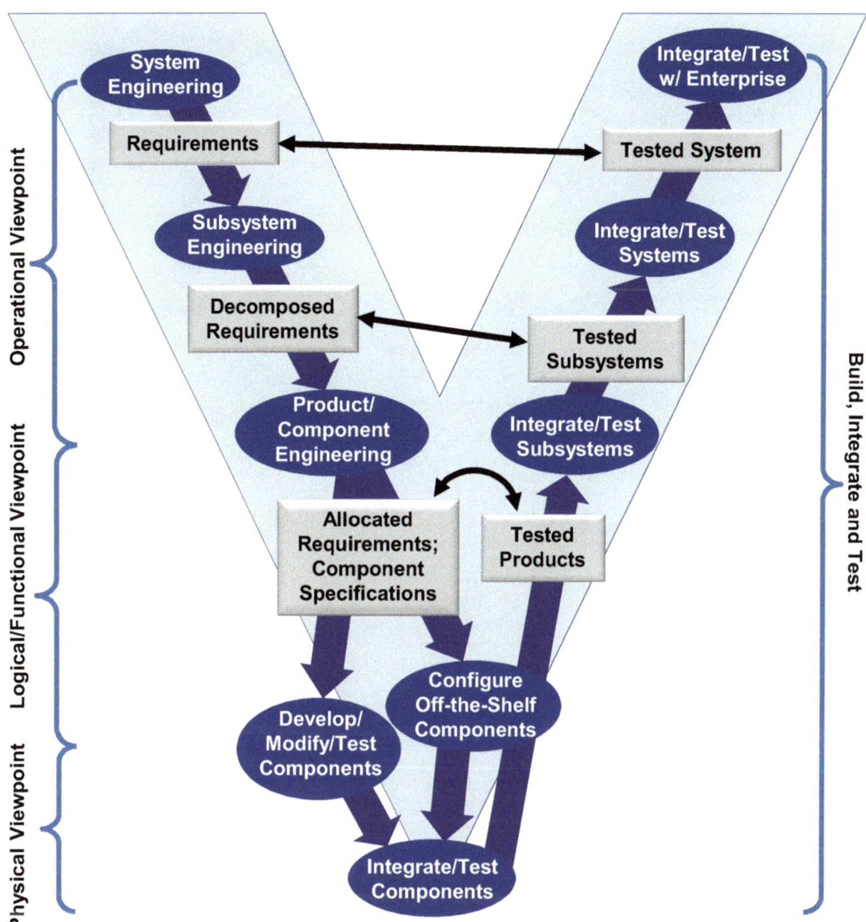

Fig. 3.5 MBSAP Viewpoints and Activities associated with the traditional Systems Engineering Vee diagram

and test correspond to the right side. In Fig. 3.6, the ties between the major SE activities shown in Fig. 3.1 are made more explicit. The yellow arrows show the association of primary SE processes with an architecture model. In some cases, an SE process interacts with the entire model and in others with one or more specific Viewpoints. For example, the transformation of a preliminary or top-level design into a detailed design is mirrored in the architecture as the LV is fleshed out with point design decisions to create the PV. The figure highlights Configuration Management and Governance, including change control, as two especially important SE activities that depend vitally on the architecture.

In the SE process in Fig. 3.6, formal program reviews are normally completed at the end of each major activity. Typical examples include a System Requirements Review (SRR) to approve the requirements baseline, Preliminary and Critical

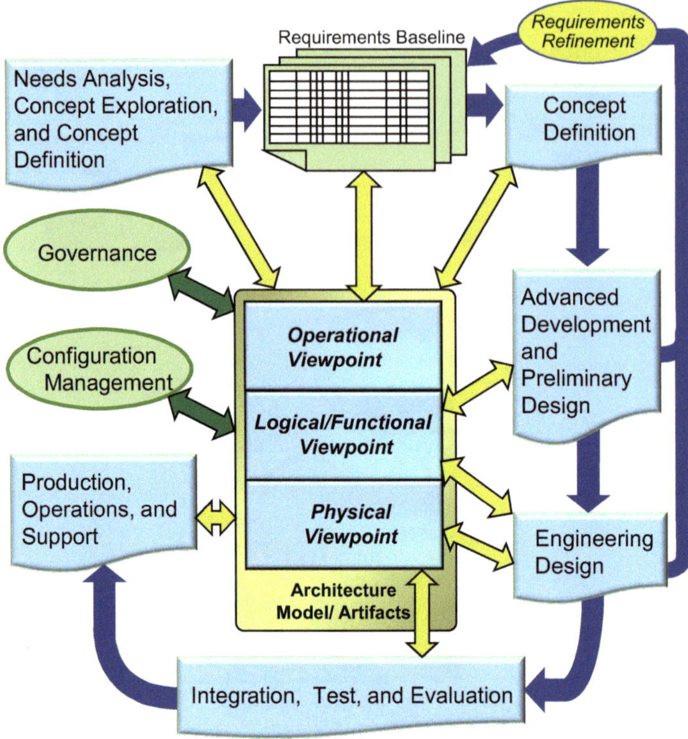

Fig. 3.6 Relationships between the stages of an SE process and the content of a supporting architecture model

Design Reviews (PDR/CDR), Test Readiness Reviews (TRR), and so on. Rigorous evaluations of the evolving architecture are equally important and can be incorporated in the agendas of the larger review events. Depending on the structure and schedule of the program, it may also be useful to hold specific architecture reviews, for example, to approve successive baselines as the OV, LV, and PV are completed. The architecture team should also hold informal reviews to update architecture metrics, coordinate with stakeholders, and ensure the architecture model is being kept current and consistent with the system design. Whatever the scope and timing, an architecture review typically includes:

- Specific artifacts of the architecture model, including model element specifications, as well as any supplemental materials; these are spelled out in detail for the various Views in the next several chapters.
- Requirements traceability, allocation, and compliance, including both functional and nonfunctional requirements.
- Architecture validation using the results of both physical and virtual prototyping.
- Status of the architecture development against the current program plan.

- Architecture issues associated with risk, resources, changes to system requirements or program priorities, and anything else impacting the architecture development.

3.4 Viewpoints and Perspectives

For convenient reference, Table 3.1 summarizes the Perspectives within the Viewpoints of MBSAP. These are largely derived from the fundamental elements of architecture described in Chap. 1: structure, behavior, and rules, with some extensions to facilitate a model structure that is easy to navigate and manage. For the OV and LV, these Perspectives are Structure, Behavior, Data, and Context, plus Services when these are an important part of the architecture. If requirements are included in the model, a Requirements Perspective is also useful. As we noted earlier, in the PV the emphasis is on physical system components that have both structure and behavior, and so we combine those two Perspectives into a Product Perspective. At the same time, it's a good practice to create a Standards Perspective to account for this essential content of an open architecture.

In addition to these primary model elements, it will usually be the case that architecture representations tailored to the special interests of various stakeholders are required. These are called Focused Viewpoints, and they draw material from and serve to complement the OV, LV, and PV. An essential feature of MBSAP is that it supports any set of Focused Viewpoints needed to meet the needs of a particular program and its stakeholders. For example, a system that processes sensitive data will need a Security Viewpoint that collects and portrays all the architecture features associated with achieving a required level of trust. There is no single set of these that meets all situations, but a representative list would include the following, each identified with the chapter where it is described.

- *Runtime Environment Viewpoint*—representation of the software infrastructure and products used by executing mission applications (Chap. 6); of concern to application software developers.
- *Hardware Installation Viewpoint*—traditional drawings and other documents for the hardware configuration of the system (Chap. 6); of concern to system integration and equipment installation engineers.
- *Network Viewpoint*—similar drawings and documents focused on internal networks and interfaces to external networks (Chap. 6); of concern to communications and networking engineers, network administrators, system integrators, and others.
- *Communications Viewpoint*—collection of information on the communications equipment of the system and its configuration and operation (Chap. 6); of concern to communications engineers.
- *Security Viewpoint*—documentation of the mechanisms, processes, and procedures employed to implement a security policy and achieve required protection

Table 3.1 MBSAP viewpoints and perspectives

Perspective	Content	Model artifacts
Operational viewpoint		
• Structural Perspective	Top-level partitioning	Packages/Blocks (Domains)
• Behavioral Perspective	Primary system functions	Use Cases, Activity Diagrams
• Data Perspective	Overall information categories	Conceptual Data Model
• Services Perspective	Domain services	Services taxonomy, Domain specifications
• Contextual Perspective	Supporting information	Documents, graphics, etc.
Logical/functional viewpoint		
• Structural Perspective	Functional components/interfaces; specifications	Block Diagrams
• Behavioral Perspective	Class and Class Collaboration functions	State Machine Diagrams, Sequence Diagrams
• Data Perspective	System information entities; XML schemas	Logical Data Model
• Services Perspective	Specific services (functional)	Services taxonomy, specifications, Sequence Diagrams
• Contextual Perspective	Supporting information	Documents, graphics, etc.
Physical viewpoint		
• Design Perspective	Hardware/software, component specifications	Class/Block Diagrams
• Standards Perspective	Initial standards profile and forecast	Standards tables; model element specifications
• Data Perspective	Database tables	Physical Data Model
• Services Perspective	Physical service definitions	Service-Level Agreements; Service specifications
• Contextual Perspective	Supporting information	Documents, graphics, etc.
Focused viewpoints		
• Runtime Environment	Software infrastructure	Package/Block Diagrams; Component specs
• Hardware Installation	Mounting, cabling, weight/balance, power/cooling, etc.	Diagrams and specifications
• Network	Network topology, design, management, etc.	Block diagrams; Documents and graphics
• Communications	External communications devices	Device specifications
• Security	Information protection features and functions	Diagrams and documents
• Messaging	Information exchange mechanisms	Diagrams and documents
• Services/SOA	Collection of services information	See Services Perspectives
• System Management	Resource monitoring, control, fault recovery, configuration management	Diagrams and documents; product specifications

of sensitive information (Chap. 10); of concern to organizations responsible for approving a system to conduct secure operations.

- *Messaging Viewpoint*—representation of the messaging modes, protocols, policies, and other aspects of message-oriented information exchanges, internally and externally (Chap. 6); of concern not only to networking and communications engineers but to software developers who must correctly employ messaging.
- *Services/Service-Oriented Architecture (SOA) Viewpoint*—definitions of services and the mechanisms through which they are exposed, discovered, and used (Chap. 7); of concern to anyone interested in service-enabled processes, enterprise integration, system access using network services, and many other aspects.
- *System Management Viewpoint*—documentation of the products and procedures used to manage and control system resources, detect and recover from faults, perform information backup, and do other tasks essential to proper system operation (Chap. 6); of concern to network administrators and to personnel responsible for the long-term maintenance and upgrading of the system.

3.5 Foundations for Architecture Development

A number of early activities to prepare for the use of MBSE in a system development can yield major benefits in terms of establishing an effective and well-planned effort and identifying and dealing with issues that may cause problems later on. A premature start on architecture modeling before the essential foundation has been laid is a common mistake in system development programs. The time and effort involved in this preparation will pay significant dividends as the architecture effort proceeds and will reduce the risk of expensive efforts to correct early errors and omissions. We address several of these foundational steps in this section.

3.5.1 Information Gathering

A concentrated effort to compile all available information relevant to the architecture before starting to model it is a valuable way to maximize efficiency and minimize errors and rework. This commonly includes:

- System requirements and associated data, including a Concept of Operations (CONOPS), performance specifications, required system characteristics, etc.
- Data about the system's operational and support environments, including any larger enterprises in which it will function
- Policies, procedures, directives, legal considerations, and other constraints, such as applicable reference models or reference architectures and identification of architecture decision and approval authorities

- Legacy systems and other resources with which the system must interface or from which content is to be reused
- Technologies and products to be used or developed
- Data models or definitions to be employed
- Identification of stakeholders and their interests and concerns
- Any other pertinent data

3.5.2 Stakeholders

Throughout this book, we stress the importance of accounting for and incorporating the needs and priorities of diverse groups who have legitimate interests in the outcome of a system development, collectively referred to as the "stakeholders." They often include:

- The acquisition organization that specifies and procures a new system, including key personnel such as the Program Manager and Chief Engineer
- The ultimate customers who will use or operate the system and consume its products
- Support personnel, ranging from maintenance technicians to system administrators
- The owners and operators of a larger enterprise in which the system will participate, including networking and communications
- Experts in testing, safety, security, and other specialty engineering communities
- The organizations and personnel who will manufacture, deploy, support, and ultimately dispose of the system.

An important element of success is early stakeholder involvement while requirements are still malleable, and trades among competing system characteristics are still possible. It is extremely important that stakeholder concerns be surfaced and fully dealt with in arriving at a requirements baseline and ultimately a system design that is optimized in the sense of achieving the best feasible balance among characteristics desired by various stakeholders. It is all the more vital when those stakeholder interests are in conflict, or at least compete with each other, which is very often the case. Equally important is a process that gives each stakeholder community a sense that their voice has been heard and valued.

One of the best ways to start this is an Architecture Kickoff Workshop, ideally held very early in the architecture development with the fullest possible stakeholder participation. This is spelled out in more detail in a later section. This event may be profitably combined or coordinated with a Requirements Gathering Workshop since the architecture should always reflect stakeholder issues and concerns, which become the source of system requirements. Another key element is to define, with stakeholder involvement, objective measures of Quality Attributes and to regularly display them as management indicators to build confidence that commitments to various stakeholder interests are truly being honored. This is further discussed in Chap. 4.

More generally, an architecture description that gives stakeholders insight into the architecture provides a communications medium that supports a healthy ongoing dialog about architecture evolution and whether it is on course to achieve the desired optimum balance. Rozanski and Wood define the term "architecture description" as a set of products or artifacts that describe the architecture for this purpose. MBSAP, with the contents of the OV, LV, and PV, yields precisely this kind of architecture description and can be used in just this way. Accordingly, one important measure of an architecture developed through this methodology is the degree to which it produces the diagrams, tables, supporting documentation, executable models, and other content that meet the needs of all stakeholders for information and insights, from senior executives and policy authorities down to the system operators and supporters who must often live with the consequences of a development program for many years.

Assessing architecture in this light requires both active, continuing stakeholder involvement and expert judgment on the part of the architecture team. The following are some of the techniques that are available to help with the challenge.

- *Architecture Summary and Overview*—an important early document that sums up the architecture dimension of a program. The process of coordinating this document among stakeholders, especially between the architecture team and program management, can significantly facilitate the discussion over scope, products, resources, methodology, priorities, and other essential aspects of the architecture. If it can be brought to a point where all (or at least most) stakeholders are prepared to sign off that it represents an adequate architecture approach, a solid foundation for the architects to proceed with their work will be achieved. Among other things, this becomes the root of an architecture description as defined by Rozanski and Wood. An Architecture Kickoff Workshop should have as its most important goal to negotiate a solid initial draft of this document with full stakeholder input and discussion. Attachment 1 to Appendix C is an example of an Architecture Summary and Overview document.
- *Quality Attributes*—an objective means of measuring and displaying key system attributes, whose importance has already been mentioned. Quality Attributes (QAts), with their metrics, are the MBSAP mechanism for doing this. Chapters 4 and 15 expand the discussion of nonfunctional requirements that represent essential overall system characteristics and that can be hard to properly evaluate. QAts may evolve over time as experience with the system develops and practical constraints of technology, funding, schedule and performance come into focus. It's essential that these nonfunctional requirements remain traceable to system capabilities and basic architecture principles, that they be assessed with metrics that are objective and as quantitative as possible, and that agreed-upon methods for verifying that they are achieved be defined and enforced. QAts are introduced in Chap. 4 and more fully developed as part of architecture governance in Chap. 15.

- *Ongoing Stakeholder Dialog*—including both formal and informal program reviews and coordination meetings provide opportunities to keep the stakeholder dialog alive. That dialog can provide early indications of whether the evolution of the architecture is being viewed by one or more stakeholders as departing from the agreed balance or otherwise failing to satisfy their interests. These events can be supplemented at a lower level and with greater frequency by discussions between the architects and individual stakeholder representatives.

3.5.3 Conforming to Customer Policies and Mandates

Before beginning an architecture project, it's essential to know about customer policies and mandates so that the effort can be planned and executed to comply with them. A major goal of MBSAP is to help prove that delivered solutions conform to the policies, references, best practices, and standards laid down by customers. Both public and private sector organizations routinely mandate some degree of open, modular, standards-compliant, networked, and service-oriented architecture. They may use checklists, reference architectures, and reference models as tools to enforce and measure compliance. MBSAP includes a number of features aimed at helping architects and system engineers devise solutions that deliver required capabilities at lowest feasible cost and risk while providing evidence that applicable policies and mandates have been followed. The following paragraphs summarize some of the features of MBSAP that support this objective.

- Each Architecture Viewpoint is defined in terms of a set of artifacts that create the products required by customer practices, either directly or by reformatting or other transformation. The basic flow of MBSAP, from OV to LV to PV, is based on mirroring customer architectures while preserving the essentials of good architecture practice.
- Established architecture best practices are embodied in the Architecture Tenets defined in Table 1.2. These are applied, with tailoring as appropriate, at each stage of MBSAP, and objective metrics are defined to allow tracking and visualization of their incorporation in the evolving architecture. The Architecture Tradeoff Analysis Method (ATAM™) [10] is one approach to defining and implementing a set of metrics that yield greatest insight into any particular architecture development challenge. Chapter 15 goes into more detail about measuring architecture quality and integrity.
- MBSAP includes development of a repository of reusable architecture elements, examples, guidance documents, supporting data, and other materials of interest to architects. Among the contents are reference implementations of enterprise services, metadata schemas, sample artifacts, and other materials whose use saves time and money while enhancing a program's ability to demonstrate compliance with customer architecture policy. This is the fundamental principle of reference architecture as discussed in Chap. 12.

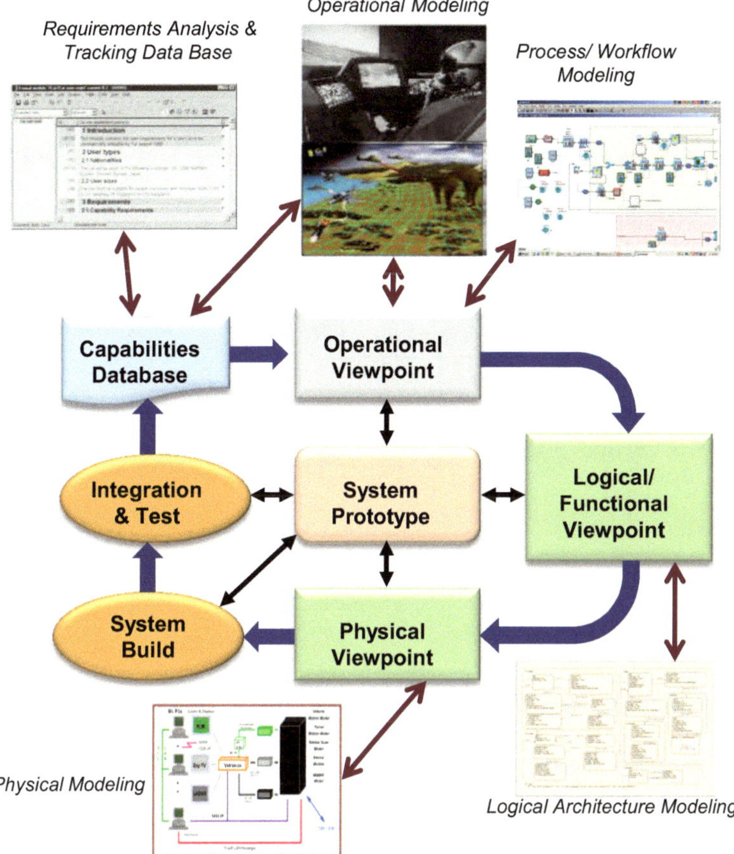

Fig. 3.7 Stages of architecture development supported with visualizations from various tools

3.6 The Role of Modeling and Simulation

Chapter 1 defined an Axis of Abstraction to deal with the reality that architecture development proceeds from an initial representation that is conceptual and abstract to a final one that is completely physical. This is mirrored in the MBSAP process sketched in Fig. 3.4. The next step is to associate a level of modeling, simulation, and analysis (MS&A) with each level of abstraction, as shown in Fig. 3.7. MBSAP incorporates four levels of M&S as summarized in Fig. 3.8, which shows the kinds of questions addressed in each layer, typical information exchanges among layers, and the kinds of tools used. Chapter 11 looks at this aspect of MBSAP in more detail, but the following paragraphs give brief summaries of the MS&A levels.

- *Operational MS&A*—examines architecture from the perspective of system operators and other users who are concerned with accomplishing the tasks for which the system is intended. It deals with the environment in which the system

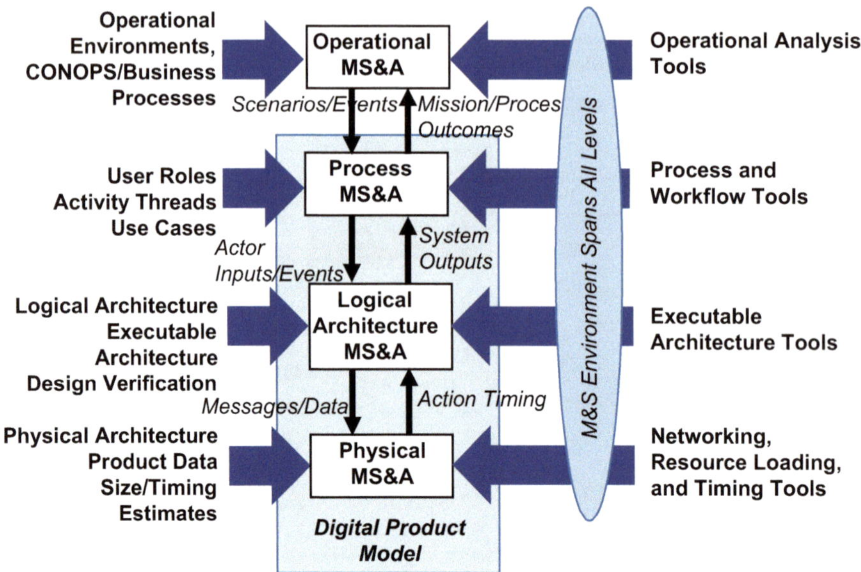

Fig. 3.8 Levels of abstraction in architecture matched with layers of modeling and simulation, with summaries of content, information exchanges, and tools

operates, operational scenarios and interactions of the participants, the outcomes of employing the system in various ways, and measures of operational performance and effectiveness. Accordingly, it can be used to create an overall context for architectural analysis and to visualize the behavior of the system as a whole. An operational simulation can look very much like a video game or computer-generated movie, and modern tools often share technologies with such products. An operational model represents the system at the most general and abstract level but can incorporate inputs from more concrete models, e.g., to establish the coverage provided by a sensor or the error rate in a data network at various points in a scenario.

- *Process and Workflow MS&A*—behavior modeling in the OV using Use Cases and Activity diagrams that represent the primary activities and interactions of system Domains. The corresponding executable representation deals with these behaviors as business processes or organizational missions. Process models include nodes representing resources and connections representing flows of data and control, with parameters such as the time associated with a processing or decision step. They are very helpful in addressing issues such as ensuring that system processes are fully understood, finding bottlenecks and imbalances, identifying the tasks and required skills of operators at various stages of a scenario, characterizing key decision points, and assessing parallel processes and the points where they must be synchronized.
- *Logical Architecture MS&A*—performed as the system design takes shape in the LV. Executable versions of the architecture's elements and diagrams can be used

to check the model for completeness and correctness, to do more detailed timing and performance analysis, to support dialog with customers and developers, and in many other ways. Behavioral diagrams such as Activity, Sequence, and State Machine Diagrams are the primary candidates for simulation. Some UML/SysML modeling tools support automatic code generation, and this software can be run on target hardware as another form of executable architecture.

- *Physical MS&A*—physical models represent the hardware and software with high enough fidelity to reproduce detailed behaviors and to compute processor throughput and loading, network data rates and latencies, antenna patterns, sensor detection ranges, vehicle dynamics, and other physical parameters. Physical modeling is the level with which most engineers are familiar and comfortable. A large number of tools and modeling methods are in use, some commercial products and some created internally by development organizations. This modeling can be expensive and time-consuming, and so it is typically focused on specific design and requirements issues where answers are needed before actual system components can be built and tested. Examples include analysis of whether requirements will be satisfied with adequate margins and whether computing and communications resources are adequate for worst-case loads.

A specific program may not need all of these MS&A levels, but a properly defined virtual prototype that efficiently addresses important system architecture issues can be an invaluable tool in creating and optimizing a system design.

3.7 Architecture Kickoff Workshop

In Sect. 3.5.2, we introduced the idea of an Architecture Kickoff Workshop, to be held during the formative stages of the program with the broadest feasible stakeholder participation. This is an important element of MBSAP that aims to establish the architecture scope, goals, products, and process and create a "contract" between the architecture team and the program's stakeholders, especially program management. To be successful, this kickoff meeting requires careful preparation and coordination, including:

- Designation of the Chief Architect and key members of the architecture team with explicit roles and responsibilities.
- Compilation of information on program scope, objectives, priorities, budget, schedule, constraints, and any other factors that may impact architecture development; in particular, an initial requirements baseline is an essential element, and the kickoff should not be conducted until this is available.
- Consultation with stakeholders to ensure that all the needs and issues with which the architecture must deal are identified; stakeholder involvement in the kickoff event itself and buy-in to the agreements that are reached greatly improves the architectural foundation for the program.

Depending on the scope and complexity of the program, the kickoff meeting may last from a few hours to several days and may be part of a larger overall program initiation and planning event. The key items on which the architecture team needs to achieve stakeholder consensus include:

- Architecture process and methodology, including application of reference architecture (Chap. 12) and choice of tools, data formats, analysis and prototyping approach, modeling and simulation approach, and any other significant aspects of the architecting effort
- Architecture scope, context, and assumptions, including operational concepts and environment
- Schedule and resources
- Architecture purpose and use, including specific artifacts and other products, mapped to the program decisions and SE activities that the architecture supports
- Quality Attributes and associated metrics (Chap. 15)
- Policies, directives, a priori design decisions (e.g., reuse of existing designs and components), business rules and processes, organizational context, and any other constraints on the architecture
- Architecture Governance and provisions for long-term updating and maintenance (Chap. 15)

As a goal, the workshop should produce at least a complete draft of the Architecture Summary and Overview described in Sect. 3.5.2.

3.8 Alternative Methodologies

Not surprisingly, almost as many architecture methodologies have been put forward as there are system architects and organizations that employ them. Many companies have defined formal architecture processes. The International Council on Systems Engineering (INCOSE) maintains a survey of current leading MBSE methodologies [11] and has published a tool-independent Object-Oriented Systems Engineering Method (OOSEM) [4]. In general, these methodologies, including the one described in this book, are variations on a basic set of architecting activities and products, including:

- Choice of Views or Viewpoints and the artifacts each contains; the term Viewpoint emphasizes the fact that each represents aspects of an architecture that are of interest to a particular stakeholder or community.
- Sequence of steps to develop an architecture and the products of each step.
- Choice of tools, formats, metrics, and other aspects of a controlled and measured architecture process.

MBSAP deals with the need for flexibility by encouraging a carefully selected set of Focused Viewpoints as described earlier in this chapter. Chapter 12 looks at architecture frameworks, which are a common way of prescribing the important features of a methodology. TOGAF [1] has already been cited. In this section, we summarize some examples that illustrate alternatives to MBSAP.

The Software Engineering Institute [12] has evolved an approach called Views and Beyond that is based on a very flexible and adaptive methodology in which Views are selected and defined based on the particulars of an architecture project and the needs of its customers. As expected, this is focused on software architectures and is not intended to generalize to systems. In this scheme, views fall into three categories called viewtypes:

- Module viewtype, where the elements are modules, which are units of implementation
- Component-and-connector viewtype, where the elements are components (which are principal units of computation) and connectors (which are the communication vehicles among components)
- Allocation viewtype, where views show the relationship between the software elements and elements in one or more external environments

Another approach is taken in IEEE Standard 1471-2000, which addresses the architectural description of software-intensive systems and provides a reference model and recommended practices. A framework or methodology that satisfies the conformance requirements of the standard will have certain assured capabilities for:

- Expressing the system and its evolution
- Ensuring adequate communication among stakeholders
- Evaluating architectures in a consistent way
- Planning, managing, and executing architecture development
- Verifying compliance

Key points of the standard's recommendations on architecture documentation include:

- An emphasis on data documentation, including details such as date of creation or revision, scope, context, and other aspects of data entities.
- An emphasis on stakeholders and their concerns, very much in-line with the discussion earlier in this chapter; architecture documentation is among the most powerful tools for maintaining stakeholder dialog and proving that the architecture satisfies their concerns.
- Use of Viewpoints that express the perspectives and provide traceability to the concerns of various stakeholders.
- Use of Views that are comprised of models that define the architecture; views should be assessed for compliance with Viewpoints and for consistency among themselves.
- Careful recording of the rationale that supports design decisions and choices among architectural alternatives.

MBSAP implements the key principles of 1471-2000.

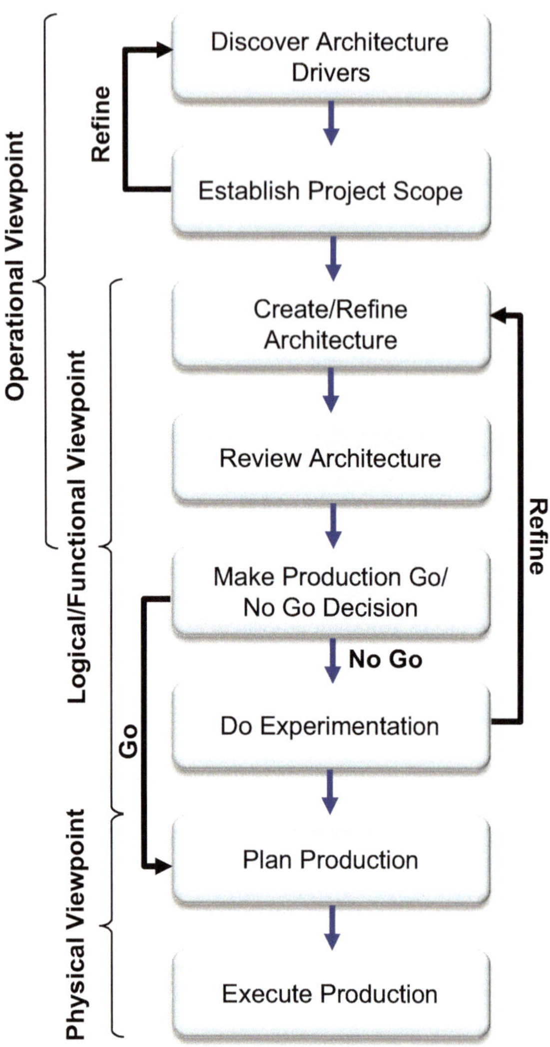

Fig. 3.9 Stages of the Architecture-Centric Design Method (ACDM) mapped to the Viewpoints of MBSAP

As an example of the various ways MBSE principles can be put into practice, Lattanze [7] describes an Architecture Centric Design Method (ACDM). Like Views and Beyond, ACDM explicitly addresses software architectures, while our concern in this book is more general system architecting, but the two approaches have much in common. ACDM is based in part on finding a workable middle ground between lightweight (agile) and heavyweight architecture process frameworks, seeking to achieve the speed and responsiveness of the former while retaining enough of the rigor and repeatability of the latter. Figure 3.9 shows the eight stages of ACDM and maps the MBSAP Operational, Logical, and Physical Viewpoints to them. Stages 3 and 4 of ACDM contain the core architecting activities and are associated with elements of both the Operational and Logical Viewpoints. ACDM is heavily weighted toward the structural dimension of architecture, while MBSAP seeks a balance among structure, behavior, data, and other aspects.

Lattanze stresses a number of key points in applying ACDM:

- Get architecture drivers from stakeholders to ensure all requirements, products, and constraints are accounted for; this is a critical part of the architecture kickoff.
- Iterate the architecture design until it meets the criteria for proceeding into production, supported by experimentation to resolve issues.
- Document the architecture to capture both the design and the underlying rationale.
- Use the architecture as the blueprint for the entire project.

Another standard relevant to the class of architectures of interest in this book is International Standards Organization (ISO)/International Electrotechnical Commission (IEC) Standard 15288, *Systems and Software Engineering—System life cycle processes* [13]. This standard defines a set of processes dealing with organizational, technical, project management, and other aspects of the life cycle of a system, from initial development and acquisition through operations, maintenance, and disposal, organized into Enterprise, Project, and Technical process areas.

Many system architects use the Zachman Framework™ that was cited in Chap. 1 as the basis for an architecture methodology. This approach takes the "what," "how," "where," "who," "when," and "why" of an architecture and captures them in successively more concrete representations, ultimately a detailed design. MBSAP is similar in many important respects (see Chap. 12 for a mapping) but emphasizes object orientation and tools that implement it. Accordingly, the Zachman Framework is not an integral part of our approach. However, it can be helpful, especially in the early stages of a project, in discovering and organizing scope and content, identifying the primary challenges and complexities, and communicating with customers and management.

In their book on software architecture, which despite the subject has important insights into the more general field of system architecture, Rozanski and Woods [14] make stakeholders the central factor in architecture development. They go so far as to say that the whole purpose of architecture is to meet stakeholder needs and that the basic measure of goodness is the degree to which this is achieved. They advocate an approach much like MBSAP in which "Views" or "Viewpoints" that describe the structure and functions of the system being architected and "Perspectives" that describe Quality Properties are developed in a collaborative environment that promotes the identification, prioritization, assessment, and achievement of the interests of various stakeholder groups. What Rozanski and Woods describe as Quality Properties are defined in Chap. 4 as Quality Attributes, but the terms mean much the same thing, especially in the sense that they are a primary source of nonfunctional requirements aimed at defining and measuring overall characteristics of a system such as reliability, supportability, and security.

INCOSE's OOSEM is similar in principle to MBSAP. It brings together functional decomposition, OO principles via SysML, and a very familiar flow of SE activities from stakeholder needs analysis and system requirements development through logical and physical architecture development and on to system verification and validation, all underpinned by an architecture model. OOSEM also stresses requirements traceability throughout the MBSE process. It defines a hierarchy of processes, identified as an SE foundation, things common to any Object-Oriented Systems Engineering (OOSE) methodology, and activities considered to be OOSEM unique. OOSEM is careful to be strictly tool and vendor-neutral.

The Object Management Group (OMG) [3] continues to develop and refine the MDA methodology, mentioned earlier in this chapter. This is specifically a software development approach that depends on automated transformation of an architecture model into code, either source code in a specified language or machine code for direct execution. It follows the UML assumption that the system development task is to build a set of software applications that will run on a computing platform. MDA goes through a succession of three models:

- *Computation-Independent Model (CIM)*—models system requirements and environment; essentially the same as a business model or the MBSAP Operational View
- *Platform-Independent Model (PIM)*—describes the system, but not the implementation details; equivalent to the MBSAP Logical/Functional View
- *Platform-Dependent Model (PDM)*—captures the technology and detailed design used to implement the system; equivalent to the MBSAP Physical View

The success of MDA depends on the quality of the tools that implement it, especially for autocoding from the PIM to the PDM. A basic principle is "maintain the model, not the code," and run the tool whenever a design change is made to update the software. There is an ongoing debate about the feasibility of applying MDA to software in general, but autocoding is used successfully in certain well-structured domains and is likely to grow in popularity as the tools and technology improve. Compatibility with MDA has been an important consideration in our development of the Viewpoints and Perspectives of MBSAP.

Another approach to architecture-centric SE, which is receiving increasing attention as the overall concept of agile processes becomes more and more prominent, is commonly called Agile Systems Engineering (ASE) [15]. In keeping with the overall agile themes of flexibility and reliance, ASE aims to make systems better able to deal with the reality of continuous and unpredictable changes in requirements, operational environments, cybersecurity threats, and other influences on system design and operation. Dove [16] calls this "Design for Transformation." ASE exploits the advantages of modular architecture, including encapsulation of system content in self-sufficient and reusable components with open interfaces. It is based on a deep commitment to open architecture as we discussed in Chap. 1. On that foundation, ASE seeks to achieve flexible system composition using "plug and play" components and systems that can be dynamically reconfigured and even self-reconfiguring. The methodology has much in common with agile software techniques such as

Scrum, which feature a prioritized backlog of customer requirements, short periods of development ("sprints"), and continuous integration and test. MBSAP supports all of the key concepts underlying ASE, especially modular, functionally defined system elements with emphasis on their interfaces.

INCOSE's ASE Life Cycle Model (ASELCM) project [17] is a notable effort to mature this discipline and encourage its adoption. The ASE working group has found that system environments are often characterized by capriciousness, uncertainty, risk, variation, and evolution. In response, a key tenet of ASE is to provide for simultaneous and asynchronous activities across the system project phases of research, concept exploration, development, production, utilization, support, and retirement, embodied in an Agile Life-Cycle Framework. Another central concept is the use of a framework comprised of three nested systems:

- A Target System that is the ultimate objective of development and innovation
- A Target System Life Cycle Domain System that includes the Target System and its external environment
- An Innovation System that includes the first two along with the resources to manage the life cycle

We believe, but cannot yet prove, that an intelligent fusion of MBSAP with ASE can yield an extraordinary MBSE capability [18].

While most of the alternative methodologies in the preceding paragraphs have significant equivalencies to MBSAP, the last one we mention does not. The Object Process Methodology (OPM), developed by Dori [19], takes a fundamentally different approach to that embodied in SysML. While it includes abstraction, generalization, and other OO principles, it emphasizes system functionality, in contrast to the OO integration of function, structure, data, and other characteristics in the system units modeled by Blocks. OPM uses a single type of diagram. OPM has many advocates and has been employed effectively in a variety of architecture analysis situations. However OPM and MBSAP represent fundamentally different approaches to the system architecture challenge, and we regard the former as being outside the scope of this book.

3.9 The Tool Dilemma

One further practical consideration is needed to complete this introduction to MBSE and the MBSAP methodology. Both the readiness with which a new engineering paradigm is accepted and the economic benefits it delivers depend critically on the quality and availability of tools to implement it. The current reality is that, while excellent individual tools exist for the various stages of the MBSE end-to-end process, an efficient and integrated tool environment is still a work in progress. Requirements analysis, architecture modeling, performance and timing analysis, test planning and data analysis, configuration management, risk management, and other processes are commonly supported by stand-alone tools with limited support for automatic information exchanges or workflows among them. A typical example involves linking an architecture model in a UML or SysML tool to an equation solver or simulation engine. An analyst sometimes has to create another version of the architecture using the constructs that are native to the second tool. While this is actually no worse than the equivalent situation in a traditional SE environment, it's clear that the full power and productivity of MBSE will only be realized when robust and flexible tool integration and interoperability are achieved.

Broy [20] and his colleagues have analyzed the challenge of creating a seamless environment for model-based development and have proposed a strategy based on an underlying modeling theory and generic tool framework. However, as they point out, commercial tool development is expensive, and tool vendors have little economic incentive to invest in solving the overall problem. One initiative is a data exchange standard for SE data, AP (for Application Protocol) 233, under the overall ISO 10303 *ST*andard for the *E*xchange of *P*roduct information (STEP). AP 233 uses the popular EXPRESS data modeling language, which is defined in ISO 10303-11, *The EXPRESS Language Reference Manual.* It aims to exchange and integrate information between SE tools. For example, mappings have been developed between AP 233 and SysML.

Denno and his colleagues [21] argue that a robust MBSE discipline must employ both parametric modeling and decision support technology and that these require four basic characteristics:

- Formal specification of a viewpoint
- Associativity, in the sense of specifying correspondence of information, across viewpoints
- Established validity of these associations, especially relationships among properties across viewpoints
- Mechanistic ("built-in") traceability of design commitments ("refinements") from requirements to the current system design (this may be a fruitful opportunity for the application of formal methods)

These and other analyses have their roots in the seminal work of Wymore [22]. Collectively, they establish a reasonable theoretical basis for MBSE and its enabling infrastructure. For example, the four characteristics identified by Denno et al., would be powerfully supported by a common metamodel used by the tools implementing various SE processes.

Despite this promise, the reader should be aware that, at present, any comprehensive SE environment confronts two important limitations. Cole et al. [23] illustrate these very well in an insightful description of an MBSE pilot project at the Jet Propulsion Laboratory. The first is technical and results from the immature state of tool integration just discussed. This causes considerable manual data transfer and reformatting to move model content and other information through a succession of tools, increasing process overhead and cost. Careful selection of the tools that comprise an MBSE environment, along with attention to the ways they will interact, is very important.

The second is cultural. The transition to a new and significantly different methodology with unfamiliar tools, artifacts, formats, and skills can be hard to accept, especially for more senior engineers and architects. Something as simple as a change in the symbols used to represent familiar items in a diagram can evoke rejection and even hostility. The key insight is that careful preparation and an incremental, non-threatening approach to bring model-based tools and techniques into an established organization and its processes are likely to be indispensable. Education of the technical staff; pilot projects to demonstrate the nature and benefits of MBSE; ample support for the transition, especially expert help with new tools; and consistent top-down support from senior management are all required.

While acknowledging these tool difficulties, we end this discussion on an optimistic note. A number of leading architecture modeling tool vendors have made progress in broadening their product portfolios to address all or most of the system architect's tasks and in integrating them in a tool framework that promises to minimize manual data exchanges. Over time, tools and frameworks will evolve that progressively reduce the cost and improve the quality and correctness of system designs that are mathematically derived through modeling. As these product lines mature, a major barrier to the adoption of MBSE will be overcome.

3.10 Summary

This chapter has briefly summarized the scope and content of the MBSAP methodology to provide context for the detailed discussions of the next few chapters. MBSAP draws on a large body of work in the theory and practice of system architecture and seeks to combine a set of best practices that are supported by commercially available tools and that aid programs in conforming to customer policies while delivering effective, affordable systems and enterprises to their ultimate users. The next chapters flesh out this framework by exploring the principles of Object Orientation and then presenting the specifics of developing Operational, Logical/Functional, and Physical Viewpoints.

Exercises

1. Referring to Fig. 3.1, suggest some additional specific examples of ways in which the major SE activities shown in the process flow could be supported by architecture. For example, the flowdown and allocation of requirements in an architecture model supports Initial and Detail Design phases by helping to ensure that all requirements are accounted for.
2. Considering the fundamental principles underlying the MBSAP methodology, suggest some ways in which it could be tailored for greatest effectiveness in a program involving both advanced hardware and complex software vs. one involving only software executing on generic (off-the-shelf) hardware.
3. Describe the nature of the transitions from Operational to Logical/Functional Viewpoints and from Logical/Functional to Physical Viewpoints in MBSAP.

4. Suggest additional Focused Viewpoints (beyond those listed in the text) and the circumstances in which they might be needed to address stakeholder concerns.
5. List some typical stakeholders for a space transportation system used to place satellites in orbit.
6. List some of the ways an *Architecture Summary and Overview* document can be used in the early stages of a program to support an effective architecture development effort.
7. Considering the four levels of modeling and simulation in Fig. 3.8, suggest which levels would be most important in developing architectures for (a) a large, non-real-time information system with multiple users and (b) a real-time control system for a materials handling facility with extensive automation and robotics.
8. Briefly describe the desired outcome of an Architecture Kickoff Workshop and lay out an agenda to achieve that outcome.
9. Referring to Fig. 3.6, suggest some specific architecture content that would be used by the Integration, Test, and Evaluation and the Production, Operations, and Support processes.
10. Describe how an architecture model could be used to implement a Configuration Management process (i.e., a formal way to define, control, and document the design baseline of a system).
11. Consider a local bank in which customers interact with bank personnel (tellers) to accomplish deposits, withdrawals, payments, and so forth. Sketch an Activity Diagram including a customer, a teller, and the bank's information system showing the flow of such a transaction.

Student Projects

As the start of a project, students should be required to prepare a 2–3 page summary of the system for which they propose to create a SysML model. The instructor should review these to ensure they include adequate scope and content, including time-sensitive behaviors for practicing real-time modeling and a larger enterprise context for practicing modeling of enterprise processes and interactions. An optional but valuable supplementary activity is to draft an Architecture Summary and Overview document.

References

1. The Open Group (2017) TOGAF Version 9 Enterprise Edition. http://www.opengroup.org/togaf/. Accessed 18 May 2017
2. Zachman JA (2008) A concise definition of the Zachman framework. https://www.zachman.com/about-the-zachman-framework. Accessed 12 May 2017
3. Object Management Group (2016) Model driven architecture. http://www.omg.org/mda/. Accessed 18 May 2017
4. Object Management Group (2011) INCOSE Object-Oriented Systems Engineering Method (OOSEM). http://www.omgwiki.org/MBSE/doku.php?id=mbse:incoseoosem. Accessed 12 May 2017

5. Yourdon E (1989) Modern structured analysis. Prentice Hall, Englewood Cliffs, NJ
6. Buede DM, Miller WD (2016) The engineering design of systems: models and methods, 3rd edn. Wiley, Hoboken, NJ
7. Lattanze AJ (2009) Architecting software intensive systems: a practitioner's guide. CRC Press, New York
8. Mellor S, Balcer M (2002) Executable UML: a foundation for model-driven architecture. Addison-Wesley, New York
9. Raistrick C et al (2004) Model driven architecture with executable UML. Cambridge University Press, Cambridge
10. Clements P, Kazman R, Klein M (2001) Evaluating software architectures: methods and case studies. Addison-Wesley, Boston, MA
11. Estafan JA (2016) INCOSE MBSE initiative survey of model-based systems engineering methodologies. http://citeseerx.ist.psu.edu/viewdoc/summary?doi=10.1.1.184.9414. Accessed 18 May 2017
12. Clements P, Bachmann F, Bass L, Garlan D, Ivers J, Little R, Nord R, Stafford J (2002) Documenting software architectures: views and beyond. Addison-Wesley, Boston, MA
13. ISO (2008) ISO/IEC 15288:2008 systems and software engineering—system life cycle processes. http://www.iso.org/iso/iso_catalogue/catalogue_tc/catalogue_detail.htm?csnumber=43564. Accessed 18 May 2017
14. Rozanski N, Woods E (2005) Software systems architecture: working with stakeholders using viewpoints and perspectives. Addison-Wesley, New York
15. Douglass B (2015) Agile Systems Engineering, 1st edn. Morgan Kaufmann, New York
16. Dove R (2014) Agile Systems and Systems Engineering: One Day Tutorial. INCOSE Colorado Front Range Chapter, Denver
17. INCOSE (2017) Agility in systems engineering—findings from recent studies. Unpublished working paper. www.parshift.com/s/ASELCM170415-AgilityInSE-Findings.pdf. Accessed 10 May 2017
18. Schindel W, Dove R (2016) Introduction to the Agile Systems Engineering life cycle MBSE pattern. Paper presented at the international symposium, international council on systems engineering, Edinburgh, Scotland
19. Dori D (2016) Model-based systems engineering with OPM and SysML. Springer, New York
20. Broy M, Feilkas M, Herrmannsdoerfer M, Merenda S, Ratiu D (2010) Seamless model-based development: from isolated tools to integrated model engineering environments. Proc IEEE 98(4):526–545
21. Denno P, Thurman T, Mettenburg J, Hardy D (2008) On enabling a model-based systems engineering discipline. Paper presented at the 18th INCOSE international symposium
22. Wymore AW (1967) Model-based systems engineering: an introduction to the mathematical theory of discrete systems and to the tricotyledon theory of system design. Wiley, New York
23. Cole B, Delp C, Donahue K (2010) Piloting model based engineering techniques for spacecraft concepts in early formulation. Jet Propulsion Laboratory (INCOSE), California Institute of Technology, Pasadena, CA

Chapter 4
Analyzing Requirements in an Operational Viewpoint

4.1 From Requirements to Architecture

4.1.1 Requirements Sources and Categories

The first step of MBSAP as outlined in Chap. 3 is to transform customer requirements into an architectural context, represented by the artifacts of an Operational Viewpoint (OV). This creates the indispensable foundation for system design, and insufficient attention to this initial stage makes it almost certain that errors, omissions, and inconsistencies will crop up later in the development. In an environment of rapid, relentless change in technologies and operational needs, a customer's statements about desired system capabilities are likely to be incomplete and ambiguous, sometimes even internally contradictory. In the course of building an OV, they are subjected to rigorous analysis to resolve these issues, to provide a high-quality basis for an effective and affordable solution, and to establish a common reference point for the various engineering specialties involved in developing a complex system. The OV is therefore a key tool for dialog with stakeholders to shape the program and communicate the features and operational behaviors of the evolving solution, especially when customer personnel are not familiar with formal architecture methods.

Customer requirements, especially for Government programs, can be expressed in a variety of forms. The following paragraphs describe some of the more common types of requirements statements:

- *Operational concept*—the customer will sometimes develop a concept of operations (CONOPS) or a similar description in advance of or in addition to a specification. A CONOPS expresses the needs and assumptions of the operational end user; describes processes, missions, and constraints; and provides context for trade studies and design decisions. It includes all the functions, behaviors, and characteristics a system must exhibit to be satisfactory in its intended usage. Often, it will also describe the organizational context, important elements of the

J. M. Borky, T. H. Bradley, *Effective Model-Based Systems Engineering*,
https://doi.org/10.1007/978-3-319-95669-5_4

operational and support environments in which the system must function, constraints such as policies and operating rules, and other factors that are important to the system architecture. If a CONOPS is available, it is important in modeling system behavior to ensure that Use Cases and other model elements capture all its content. In Chap. 11, we return to this idea in discussing the validation of the architecture against the needs of system stakeholders.

- *Specification*—provides verifiable definition of system characteristics and functions. Ideally, these should be performance specifications that leave the design details to the developer, but customers issue specifications with varying degrees of design guidance. A specification may define both Threshold (minimum acceptable) and Objective (desired) levels of performance, effectively establishing a trade space within which performance, cost, schedule, risk, reliability, and other variables can be analyzed to optimize the solution. When the correct system parameters and their values are unknown or incompletely known at the outset of a program, an initial concept development phase may be needed to develop a good specification.
- *Statement of work (SOW)*—defines the tasks required in a development program. These may include de facto requirements because tasks may impact the system and its development process in assorted ways. It is increasingly common in Government procurements for the SOW to direct the development of an architecture that complies with policy and meets program-specific needs, along with delivery of architecture models and other artifacts. Sometimes the customer issues a more general statement of objectives and requires a potential system developer to propose a SOW to which the developer agrees to be contractually obligated.
- *Policies and mandates*—earlier chapters have alluded to a variety of policies that are articulated by both public and private sector organizations and often documented in terms of architecture frameworks, reference architectures, and mandated standards. These are imposed as requirements on development programs falling within their scope. Chapter 12 gives more details.
- *Miscellaneous requirements statements*—useful insight into requirements and customer priorities is often available through white papers, industry day briefings, lessons learned documents, analysis of customer business processes, demonstrations and experiments, reports from study programs, and various other sources. During the early planning and shaping phases of a program, such materials may be the best description of what the customer wants, and they are often a vehicle for customer-supplier dialog in support of program definition, budgeting, risk analysis, and other planning activities.

Chapter 3 talked about the evolutionary spiral approach, which was originally created to replace the traditional "waterfall" in software development. Spiral development is motivated by situations in which requirements can be expected to evolve based on experience gained in the course of system development, integration, and testing. In a spiral strategy, requirements and design evolve together, with provisions to adjust requirements at the end of each spiral. Spiral or incremental development has been criticized as leading to requirements "creep," that is, undisciplined growth in requirements resulting in cost increases and schedule slips. An alternative is a block strategy in which successive, relatively large groups of requirements go all the way through development to deliver a new or significantly upgraded system product. In software efforts especially, the spiral has largely given way to a variety of "agile" methods that stress flexibility, small increments of development, adaptation to change, and stakeholder involvement. In any such program, the requirements baseline in the OV will be frequently updated and must be carefully controlled.

We can illustrate the complexity that can arise in creating and managing requirements using a typical space system development program of the National Aeronautics and Space Administration (NASA). Such a program normally goes through phases that include Concept Studies; Concept and Technology Development; Preliminary Design and Technology Completion; Final Design and Fabrication; System Integration, Assembly, Test, and Launch; Operations and Sustainment; and Closeout

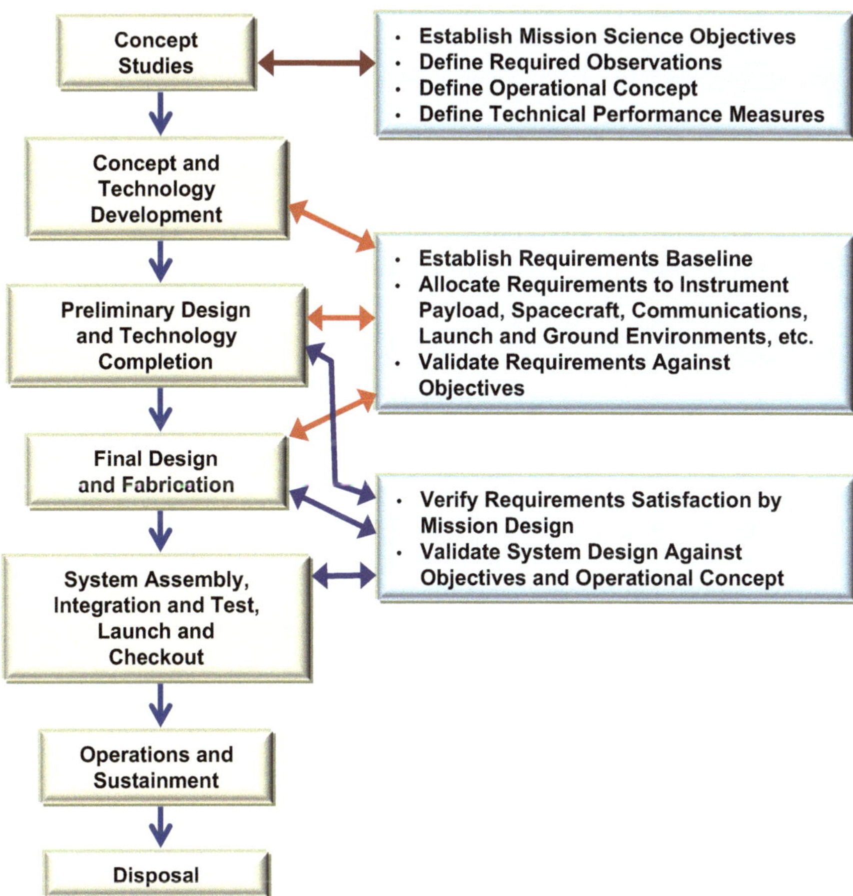

Fig. 4.1 Requirements development in the early phases of a space science mission

[1]. Figure 4.1 shows how requirements development aligns with this project life cycle for a typical space science mission.

Using another approach to defining requirements, the Department of Defense (DoD) links system requirements to needed operational capabilities through the Joint Capabilities Integration and Development System (JCIDS) process [2]. In this system, a development program moves through Concept Refinement, Technology Development, System Development and Demonstration, Production and Deployment, and, ultimately, Operations and Support phases. Requirements are documented in an Initial Capabilities Document (ICD), Capability Development Document (CDD), and Capability Production Document (CPD). In addition, a Capstone Requirements Document (CRD) is used to describe standards and other aspects applicable to a class of systems. Other agencies, as well as major private sector organizations, have similar requirements development processes.

An important strength of SysML is the means it provides to explicitly allocate requirements to the elements of a system. In MBSAP, two distinct requirements allocation procedures are essential. Requirements must be associated with system *structure* and with system *behavior*. The first of these is meant to ensure that every requirement is supported by resources included in the system design baseline, and these allocations are flowed down the structural hierarchy of the system, ultimately to one or more components that satisfy the requirement. The second ensures that every FR is accounted for in one or more behaviors that implement the required function and are also flowed down as the behavioral hierarchy is defined, typically through decomposition of Use Cases as described in Chap. 2. In an architecture model, these allocations are documented in the specifications for the relevant structural and behavioral elements.

4.1.2 Requirements Database

Step one in architecture development is to compile the available information on customer requirements from sources like those just described. A requirements baseline is then created. Many organizations use a specialized tool to define and manage requirements in a database. The alternative is to use SysML Requirements Diagrams. The effort involved in crafting a high-quality requirements baseline may be substantial, especially for a system with thousands of discrete but interdependent requirements, but the benefits will richly repay the investment. A well-designed requirements database or model is one of the prerequisites of traceability from contractual and other customer requirements through to the elements of the delivered solution, including linkages to:

- Special features like those involved in safety and information security
- Test, verification, and validation methods to be used for each requirement
- Documents such as design specifications and drawings

- Technical performance measures (TPMs), measures of performance and effectiveness (MOPs/MOEs) resulting from operational analysis, and other metrics
- Program management and control systems

Every author of an SE text has a favorite scheme for categorizing requirements. In MBSAP, we use two broad categories: functional requirements (FR) that prescribe system functionality and performance (sometimes described as what a system *does*) and nonfunctional requirements (NFR) that prescribe system characteristics not specifically associated with performance (sometimes described as what a system *has*). Some FRs have quantitative measures (speed, throughput, capacity, etc.) and can be called performance requirements. Others call for the system to be able to do things, e.g., detect an alarm condition, accept a set of configuration parameters, or provide a data display, that don't have associated numerical values. Performance requirements are verified by measuring the appropriate parameters and comparing them to required values. Other FRs are verified by showing that the system provides the required functions.

NFRs often represent the key concerns of various stakeholders. Taking an environmental sensor system as an example, typical FRs might involve the operational modes, range, distinguishable environmental conditions, and coverage rate. By contrast, typical NFRs deal with overall aspects of the system such as reliability, maintainability, and availability (RMA); environmental tolerance; safety; security; and policy compliance, especially with respect to open architecture. NFRs may be verified through inspection of the system and its documentation, analysis of compiled data on things like failure and repair rates, feedback from operational users, design audits against applicable standards, and many other sources. Both FRs and NFRs require rigorous analysis and management in the requirements database or model, and both must be verified before a system can be accepted by its users. Chapter 11 goes into requirements verification in more detail.

FRs usually have Threshold (and possibly Objective) values that can be measured by properly constructed tests, demonstrations, and analyses, and the system either does or does not satisfy each requirement. NFRs, on the other hand, may be satisfied to varying degrees and may have acceptance criteria that ultimately come down to "good enough" when assessed by the involved stakeholders. In modern architecture and system engineering practice, the preferred approach to dealing with NFRs is the use of Quality Attributes (QAts) that can be decomposed into things like design features and best practices and can be objectively assessed using qualitative measures and sometimes quantitative metrics. QAts are closely tied to architecture governance; Chap. 15 discusses this key subject in more detail. As one example, Suppakul and Chung [3] have proposed a UML Profile for representing NFRs in a UML model.

Lattanze [4] identifies a set of issues and factors affecting a system architecture that become sources of requirements, including:

- *Stakeholders*—may include executive managers and policy setters, program managers, and operational users, as well as staffs concerned with engineering

and technology, integration and test, maintenance and logistic support, training, and supply chain, and contracting
- *Business or mission requirements*—typically establish many FRs
- *Business or mission processes*—may generate both FRs and NFRs through the need to preserve and enhance them
- *Environment*—another source of a variety of NFRs
- *Enterprise data structures*—can lead to both FRs and NFRs depending on the specific requirements for implementing and interoperating with them
- *Infrastructure*—especially existing hardware and software that must be reused, interoperated with, supplemented, or replaced

These can be distilled into four broad categories of architecture drivers:

- High-level requirements, which will be a prime source of Use Cases in the OV
- Business constraints
- Technical constraints, especially those resulting from a priori design decisions
- Quality Attributes describing, prioritizing, and measuring the NFRs

Kossiakoff and Sweet [5] give another useful scorecard for good requirements. They describe a well-stated requirement as:

- Independent of specific technology or design
- Clear and unambiguous
- Phrased so that there is one requirement per statement (typically, one requirement for each "shall" in the requirements document)
- Verifiable, normally by test, demonstration, analysis, or inspection
- Vetted by stakeholders
- Validated through data and analysis (to establish that the requirement is needed to achieve a satisfactory solution to customer/stakeholder needs)
- Prioritized, e.g., to distinguish "essential" from "desirable" requirements
- Consistent with the system context, so that each requirement can be feasibly met

Chapter 3 discusses issues associated with the need for an integrated tool environment to support efficient MBSE processes. A good example arises here when both a requirements management tool and SysML Requirements Diagrams are used. It's extremely helpful for the requirements and architecture tools to have automatic linkages so that changes in one are automatically reflected in the other. Some SysML tools incorporate this function. Such linkage has a side benefit in forcing careful attention to the first stages of traceability because each requirement must be explicitly associated with one or more architecture elements in the OV. Once this is done, when the customer updates the requirements, these additions, deletions, or changes are captured in the requirements database. Then changes, e.g., in performance parameters, are passed to the architecture model using a script, export/import transaction, or other mechanism to create the linkage. Added or deleted requirements require analysis of OV artifacts to determine if corresponding additions or deletions are required. In any case, a disciplined process provides the basis for traceability to keep requirements and architecture consistent.

4.2 First Example Problem: The E-X "Overseer"

Before continuing, it's time to introduce the two principal examples used in this book. Our goal is to present realistic MBSAP applications without drowning the reader in details. We have chosen these specific systems based largely on our personal experiences in the aerospace and energy communities. However, MBSAP is equally effective in any other domain where complex, multi-technology, and information-intensive systems are employed.

Our first example is a system architecture for an airborne sensor platform that derives its characteristics from a category called intelligence, surveillance, and reconnaissance (ISR) systems. We refer to our example as the E-X "Overseer." This represents a class of systems used in both military and civil operations, gathering real-time information on objects ranging from downed aircraft to wild fires. Figure 4.2 gives a sketch. Another example is a "hurricane hunter" aircraft, where skilled pilots guide a plane through the violence of a storm while specialists gather data using radar and other instruments. Appendix C is an example of an architecture document whose primary content is exported from the model that gives much more detail than is possible in the main body of the text.

The E-X has two radio frequency (RF) sensors and one electro-optical (EO) sensor plus onboard processing, sensor management, and communications. The RF sensors

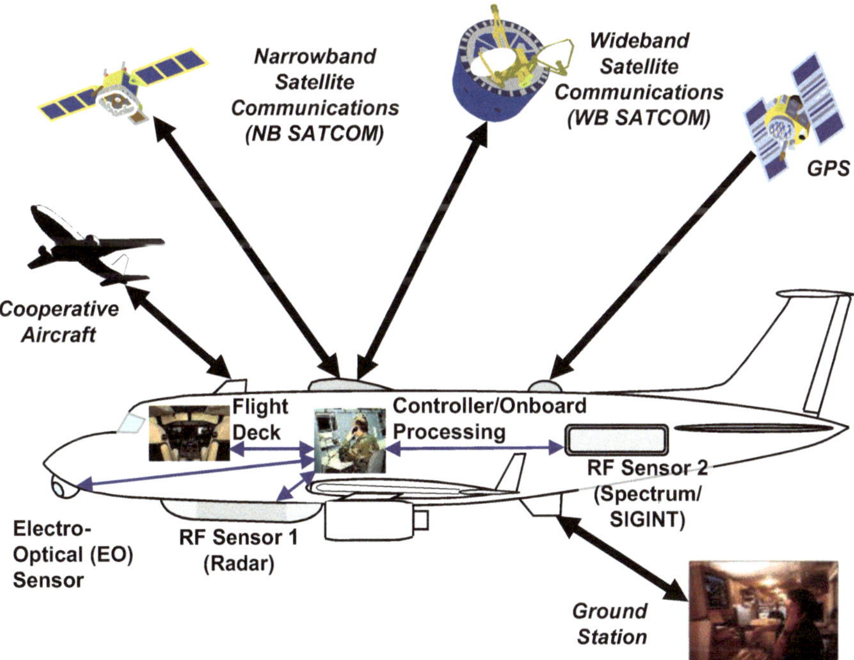

Fig. 4.2 The E-X "Overseer" Airborne Multi-Sensor System

are a multimode radar and a passive (receive-only) spectrum surveillance system (also called Signals Intelligence, or SIGINT), while the EO sensor collects imagery in both visible (usually called EO) and infrared (IR) spectral bands. The E-X interacts with a ground station and with other nodes of a larger sensing and information processing enterprise using data links, wideband and narrowband satellite communications (SATCOM), and other radios. It relies on the Global Positioning System (GPS) for position and timing data. The Flight Crew, in addition to ensuring safe operation of the aircraft, work with the other members of the Mission Crew in the back of the plane to ensure it is properly positioned for employment of its sensors to execute the tasks assigned to it and thereby satisfy the system's "customers." We will develop additional characteristics of the E-X example as the discussion proceeds through Operational, Logical/Functional, and Physical Viewpoints (OV, LV, and PV), following the MBSAP methodology.

The E-X example can be used to illustrate the application of the architecture taxonomy defined in Chap. 1. The E-X is a node (or "platform"), lying at the second level of the Axis of Organization. In terms of the Axis of Categories, the E-X includes real-time systems or subsystems like sensors and their controllers, but its primary characteristics are those of a networked decision aiding system that creates, communicates, stores, processes, and displays information to operators performing a variety of planning and control activities on and off the plane. Initially, real-time systems can be treated as "black boxes" with interfaces through which their capabilities are accessed by the main system; they will, in practice, have their own architectures that apply real-time principles. The architecture development must address all levels of the Axis of Abstraction.

These considerations place the E-X in architecture space as shown in Fig. 4.3. Once the system is decomposed into subsystems, there will be architectures at the next level down that fall into both the Hard Real-Time/Embedded and Process/Decision Real-Time performance categories (e.g., Fig. 1.2). Considering the fourth dimension of the taxonomy, time, the first instantiation of the system will be near-term, using products and technologies that are off the shelf or in an advanced state of development. The likely evolution of the architecture in mid- and long-term epochs is one of many important factors to be taken into account in defining it, especially in placing emphasis on an open architecture.

4.3 Second Example Problem: A Smart Microgrid

The second example we will use to illustrate the architecture principles and techniques discussed in this book is drawn from the world of electric power utilities. We will develop a basic architecture for a Smart Microgrid (SMG) which combines two trends that are increasingly prominent: the use of higher levels of data processing and automation to create a Smart Grid [6] and the use of relatively small-scale groupings of energy generators and loads that are interconnected and controlled to create a microgrid. We will actually look at three versions of an SMG to show the

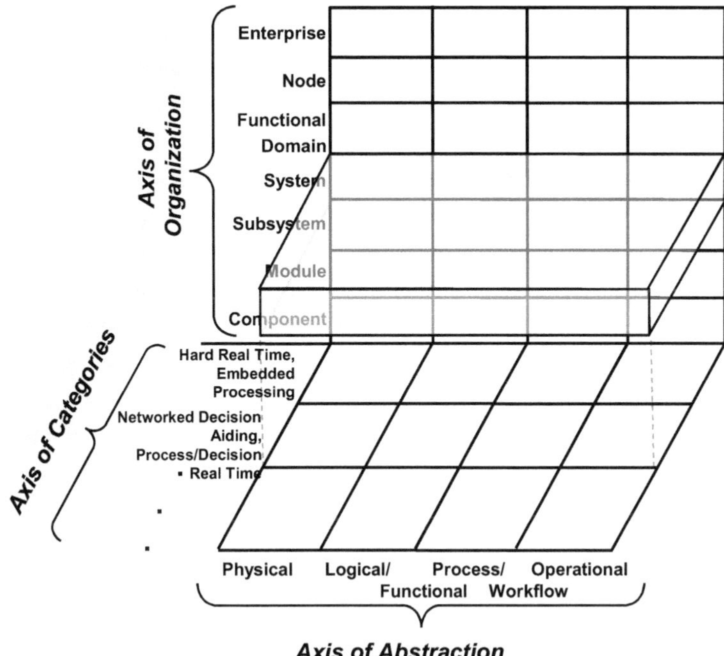

Fig. 4.3 The E-X example system in the MBSAP architecture taxonomy

power of a model-based approach in scaling and tailoring a basic architectural construct to a wide range of practical applications. Our example includes:

- *Microgrid A*—a community microgrid incorporating a variety of energy sources and consumers that is grid-connected so that it is capable of supplying power to an overall Primary Grid (PG) as well as drawing power from that PG
- *Microgrid B*—a grid-connected microgrid co-sited with a high-value load such as a hospital, which may include local or "distributed" power generation and storage and which actively manages the distribution and use of electricity within the facility
- *Microgrid C*—a non-grid-connected microgrid powered by a centralized installation of photovoltaic (PV) panels or other sources that is designed to provide a basic level of electric service to a group of homes in a rural or remote environment

Figure 4.4 is a simplified sketch of the SMG example. When a primary power grid exists, it transmits power to local substations for distribution to customers. Those customers could include the first two variants of our architecture example, Microgrids A and B. The graphic also shows a non-grid-connected Microgrid C.

The primary characteristic of a smart grid is the use of information technologies to improve the efficiency and reliability of the grid while reducing costs. Smart grid technologies enable improved integration of traditional, renewable, and distributed

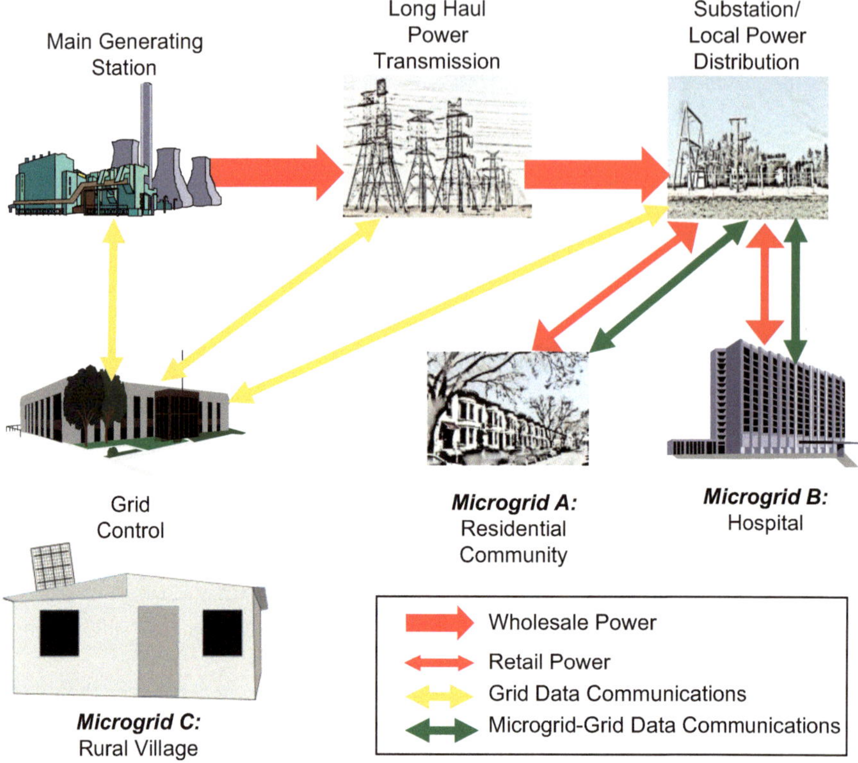

Fig. 4.4 Simplified diagram of a primary electric power grid and various microgrids

power sources. For example, such technology can allow instantaneous local adjustment of power production and distribution to match local demand. Hebner [7] has provided a summary of the impact these technologies are expected to have on electric power grids in the coming decades.

Five key smart grid technologies identified by the US Department of Energy (DoE) are:

- Integrated, bidirectional, real-time data communications
- Sensing and measurement
- Advanced components
- Advanced control methods
- Improved decision support

An SMG can be thought of as a localized version of a primary smart grid that uses the same concepts of intelligent energy control to achieve local optimization. This level of sophistication makes a smart grid or an SMG an excellent opportunity to apply modern MBSE. A grid-connected SMG must be able to quickly and seamlessly connect to or disconnect from the PG and must support the PG's overall balancing and control strategy. Another important consideration is that electric power is at the top of any list of critical infrastructure resources, and the growing concern for protection against natural disasters and hostile attacks creates serious security challenges. We will use this example to show the application of MBSE, and specifically of the MBSAP methodology, in areas such as the integration of a PG and SMG, cybersecurity for critical infrastructure protection, and behavioral simulation.

Microgrids A, B, and C obviously have somewhat different requirements, and each can be captured in a capabilities database or table and modeled in SysML Requirements Diagrams. In each case, requirements are dominated by considerations of reliability and safety, electricity quality of service parameters, efficiency, and cost. Much of the requirements baseline for a given microgrid depends on the number and type of loads and generating resources, applicable standards, details of a PG interface (for Microgrids A and B), business functions such as customer billing, and compliance with laws and regulations, e.g., dealing with emissions and other environmental concerns.

In terms of the architecture taxonomy, Microgrid A is best characterized as an enterprise on the Axis of Organization composed of systems for power generation, distribution, and control and with significant human operator involvement. Microgrids B and C would probably be classified as systems, depending on specific details. All of these are sufficiently unique in their characteristics that a distinct architecture category should be defined, although they will share aspects of each of the categories that were discussed in Chap. 1.

4.4 Modeling Requirements

A real system specification for the E-X would be too large and complex for this book, but Table 4.1 gives some high-level FRs, allowing us to show how MBSAP transforms them into an architectural context. The sensor modes, onboard processing, and communications are typical of a wide range of sensor platforms. We will add detail selectively in later stages of the development of the architecture. NFRs such as open architecture, operational availability, compliance with an overarching enterprise architecture, and ease of use by mission personnel are also part of the requirements baseline and will be considered in developing the OV. NFRs will be assessed using appropriate Quality Attributes.

As the first step in building a SysML model of the E-X, the requirements in Table 4.1 are transformed into a Requirements Diagram in Fig. 4.5. An overall Requirements Block at the top of the diagram is decomposed into the six requirements areas of the table, and one of these, Sensing Requirements, is further broken down into requirements for the E-X sensors and associated sensor data processing. The following descriptions of terms used in Table 4.1 and elsewhere may be useful to readers not familiar with systems of this kind:

- *Multimode radar*—an active (transmitting and receiving) RF sensor that can operate in multiple ways ("modes") to collect various kinds of information.
- *Synthetic aperture radar*—a mode that creates images of an area of interest; SAR uses the motion of the platform to create an effective antenna size much larger

Table 4.1 High-level requirements for the E-X system

Requirements area	High-level requirements
R1.1 Sensing	R1.1.1 The system shall configure and control sensors, perform quality control, create collection reports, and perform sensor cross-cueing R1.1.2 The SIGINT/ELINT sensor shall collect and analyze RF emissions in specified bands R1.1.3 The EO sensor shall perform imaging of surface targets in specified bands R1.1.4 The Multimode Radar shall perform surface surveillance using synthetic aperture radar (SAR) and ground moving target indicator (GMTI) modes
R1.2 Sensor Exploitation	R1.2.1 The system shall detect objects and events of interest in sensor data R1.2.2 The system shall perform tracking, geolocation, and identification of objects of interest R1.2.3 The system shall manage and update tracks R1.2.4 The system shall fuse sensor reports from onboard and off-board sensors to create estimates of track state R1.2.5 The system shall declare and report targets and target updates based on track parameters
R1.3 Mission Management	R1.3.1 The system shall perform preflight mission planning and in-flight dynamic replanning R1.3.2 The system shall monitor execution of mission plans and report status R1.3.3 The system shall generate alerts, notifications, and situation reports R1.3.4 The system shall coordinate flight and mission operations
R1.4 Communications and Networking	R1.4.1 The system shall manage integrated onboard and off-board networking R1.4.2 The system shall manage prioritized communications and implement Quality of Service (QoS) policy R1.4.3 The system shall plan and manage employment of communication and networking resources
R1.5 Information Management	R1.5.1 The system shall provide storage and access control for mission data, including plans, battlespace objects, terrain, etc. R1.5.2 The system shall implement information assurance at multiple levels of classification R1.5.3 The system shall provide an efficient and ergonomically sound human-machine interface (HMI) with visualization services and operator decision aids
R1.6 Flight Operations	The system shall perform aircraft operations and navigation for safe flight and to enable execution of mission plans
R1.7 Support Systems	R1.7.1 The system shall provide mission planning R1.7.2 The system shall support maintenance activities R1.7.3 The system shall support training activities

than the actual antenna, and since resolution is inversely proportional to antenna dimensions, this allows high-quality pictures to be created.

- *Ground moving target indicator*—a mode that uses the motion of a target over the ground to discriminate it from other things within the radar beam that reflect

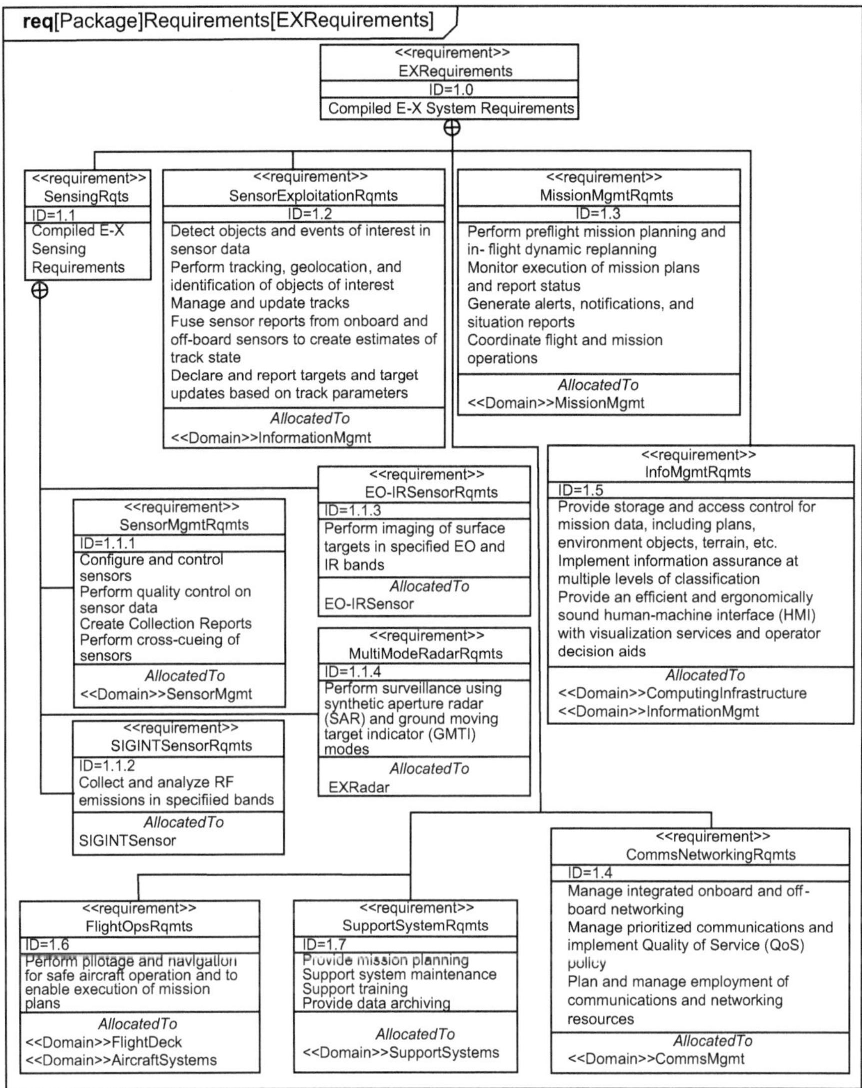

Fig. 4.5 Top-level requirements diagram for the E-X

back energy (referred to as "clutter"). This enhances the radar's ability to detect and track moving ground targets in complex environments.

- *Tracking*—the process of determining and maintaining the location and motion of a target.
- *Geolocation*—the process of correlating a target location to a point in a coordinate system, analogous to sticking a pin in a map where the target is.
- *Quality of service*—a measure of how well a guarantee of certain performance parameters is honored in delivering results to a client or customer.

In a real system, requirements decomposition like that shown for Sensing Requirements in Fig. 4.5 would be carried to whatever level of detail is needed to link requirements into other elements of the architecture. The figure also includes Allocated To compartments to show the linkages of requirements to Domains, which are defined and described later in this chapter. The structure in the diagram should mirror that in a requirements database maintained in a dedicated requirements tool. Among other things, this applies to the numbering scheme used for Requirements Block identification numbers, which should correspond to requirements IDs in the database.

4.5 Architecture Styles

The next step is to translate these requirements into an OV. In creating an OV, a natural extension of the thought process involved in sorting an individual architecture into its place in the architectural universe, defined by the taxonomy of Chap. 1, is to think about the common characteristics it will share with other architectures in the same category. This leads to the idea of architecture styles. Chapter 5 talks about design patterns, which are proven solutions to commonly occurring problems. These design patterns enable design reuse and reduce time and risk in solving equivalent challenges in new architectures. Architecture styles are high-level patterns for an enterprise or system and can be useful in the early stages of developing an OV. This can help the architect identify commonality of the new architecture with other, similar systems and enterprises and to think about broad categories of solutions that may apply. Because a style is accompanied by an overall design and implementation approach, it can help focus the architecture effort, especially in the early stages, and can help identify required specialized skills that will be needed, suggest candidate product families for the implementation, and bring to light experiences and lessons learned from previous examples that employed the style.

Rozanski and Wood [8] suggest the following as a representative list of architecture styles for high technology, information-intensive systems that are the focus of this book. They are not mutually exclusive, and a given architecture may have characteristics of several styles:

- *Pipes and filters*—applies to an architecture that processes one or more streams of data using a sequence of "filters" connected by "pipes." Individual filters can be changed and recombined to create specific transformations of the data. This style implies that the processing is "stateless," meaning that the same steps are applied to the incoming data regardless of prior history. A strength of this style is support for parallelization to improve throughput.
- *Client/server*—an extremely widespread style with many variations, all sharing the basic characteristic that functions such as computing and data management are implemented on shared "servers" and accessed when needed by multiple "clients." In a real-time system, a variant called an Integration Hub is often used where the shared resources are some kind of integrated core processing suite that

is used by multiple system elements such as sensors, communications, and human-machine interfaces.

- *Tiered/layered*—another widely used style organizes functionality into "tiers," which are levels or "layers" of related services. Each layer creates an abstraction of the implementation details contained in lower layers, to which it gives access. An enterprise architecture and a complex information system are two examples that are likely to use both of these concepts. Tiering and layering are discussed in later chapters.
- *Peer-peer*—consists of nodes connected by networks (any-to-any). Nodes can discover and exchange content with each other.
- *Publish/subscribe*—a specific messaging style in which information providers "publish" using registered topics to which consumers can "subscribe" in order to get the specific data in which they are interested when it becomes available.
- *Asynchronous data replication*—occurs in systems that broadcast information or must synchronize content across multiple data stores.

Many other architecture styles can be identified. A system like the E-X presents architecture challenges in multiple categories, including information processing, sensing, platform control, and communications, and therefore a variety of architecture styles might be applied. One, which has become widely used in modern airborne systems, is commonly called Integrated Modular Avionics (IMA). In this style, a core suite of processing, networking, communications, human-machine interface, and other resources is implemented with a small inventory of hardware and software modules, creating an Integration Hub. A common core is assembled by combining the appropriate types and numbers of these building blocks and then shared among the various functional areas of the system, reducing cost, weight, and power compared to an architecture in which every subsystem has its own dedicated resources. Chapter 8 discusses this further as a particular aspect of the larger subject of real-time performance, where successful designs exhibit common characteristics that can be viewed as a distinctive style. Another widely encountered architecture style, Service-Oriented Architecture, is the subject of Chap. 7. More examples could be cited. The point is that an understanding of proven and successful styles can be a significant advantage to a system architect in expeditiously arriving at a good system solution to a customer's requirements, and we will illustrate this concept at a number of points throughout this book.

4.6 Structural Perspective

We now have the background to start actual architecture modeling. Chapter 3 noted that the OV and LV are conveniently organized into Structural, Behavioral, Data, Services, and Contextual Perspectives, and this is a reasonable sequence in which to develop them. In practice, the Perspectives interact, and the architect should expect to go back and forth in an iterative process of discovery and refinement affecting all these Perspectives, until a representation of the OV is obtained that accounts for all requirements and satisfies the basic architectural principles.

4.6.1 Defining Domains

We consider the Structural Perspective first by examining the nature of the system[1] and looking for a natural partitioning into groups of related functions and resources, labeled Domains. These are often treated as subsystems, and the `<<Domain>>` and `<<Subsystem>>` stereotypes can be used to identify them. To begin, the Structural Perspective is modeled using a Block Definition Diagram (BDD). For this initial, top-level partitioning of the system, some architects prefer to use Packages because these represent collections of related model elements, which is very much the nature of a Domain. However, it's generally preferable to model Domains as Blocks so that the power of SysML in representing structure starts right at the top. We'll show this using the E-X and SMG examples. The emphasis in constructing a Structural Perspective is on system resources, functions, information, and external entities with which the system interacts.

Domains are defined such that each includes related or closely interacting functions. The structural organization also considers where information is generated, transformed, stored, and used and how it is exchanged among functional areas and with the external environment. Indeed, many authors use the term "domain" to mean specifically and only a category of information, much like the top-level structure of the Conceptual Data Model that is presented later in this chapter. However, MBSAP broadens the term to mean a grouping of information, resources, User Roles (actions and responsibilities of system users), and functions that collectively represent a major portion of a system or enterprise. If done correctly, this initial partitioning of the system or enterprise into groups of related capabilities, responsibilities, user interactions, information content, hardware, and software establishes the foundation for an open, robust, and effective architecture.

There are no hard-and-fast rules about the number and content of Domains, and this is an area where judgment and experience are required. As we noted in Chap. 1, there are many instances in practical architecting where heuristics are valuable, and this is a place where some rules of thumb will help the architect in getting the model off to a good start.

- The number of top-level or system-level Domains will generally be in the range of 4–12, depending on system complexity and diversity of functions and resources. Fewer than this probably means they are too highly aggregated and

[1] From this point onward, we will generally speak of architecting a "system" and use the E-X and SMG as system examples, even though taxonomically these fall higher on the Axis of Organization. The reader should remember that the principles apply equally to an enterprise, system, subsystem, or any other level on the Axis of Organization, but talking about "systems" makes the discussion less cumbersome.

lack coherence; more tends to obscure the fundamental structure of the system unless it is very large and complex.

- Well-chosen Domains generally result in a clear association of each with one or two types of external users or systems, because the logical grouping of functions generally corresponds to individual User Roles or specific external interactions. If User Roles interact with multiple Domains, the partitioning should be examined to ensure it is an optimum grouping of system content.
- Good functional partitioning results in a situation where a large percentage of interactions among system elements happen inside the boundaries of single Domains, while a much smaller fraction crosses those boundaries. This localization of activity is an aspect of the general principle of separation of concerns and is the foundation for loose coupling in an architecture.
- Whenever possible, a Domain should contain resources and functions from the same category of the architecture taxonomy in Chap. 1. In the E-X example, it is convenient to group real-time system elements into separate Domains from those representing processes involving human interactions and decisions. Another common partitioning rationale, applicable to systems that handle information at multiple levels of sensitivity, is to define Domains in which data and processes are at or below a given level of trust and required protection.

4.6.2 E-X Functional Partitioning

We'll apply this logic first to the E-X example. Figure 4.6 shows an initial decomposition in a BDD, and Fig. 4.7 shows Domain interactions in an IBD. A basic hierarchy of User Roles, described in Sect. 4.6.4, was created at the same time so that these Actors could be used in the Block Diagrams. The Domains correspond to the primary functions of the sensor platform. Some mission crew members perform planning, analysis, and data management, and others directly control sensor operations. The platform also needs communications and the information infrastructure to support these functions. An overall E-X Block, stereotyped as `<<block>>` and `<<System>>`, anchors the structure, and Composite Associations declare the other Domains to be Part Properties (Parts) of this system Block. Domains are stereotyped as `<<block>>` and `<<Domain>>` and have Parts compartments that list the results of the next level of decomposition. Blocks modeling Hard Real-Time Domains have a gray background. User Roles have multiplicities to indicate how many instances of each Role may be in place, and off-board roles have gray shading.

The IBD in Fig. 4.7 uses the model elements defined in the Domain BDD to show the top-level interactions of the system. Onboard and, in some cases, corresponding off-board personnel interact with the various Domains. If it's important to show explicitly the elements that are part of the system and those that are external, a graphical system boundary could be added to the figure. Figure 4.7 is laid out to emphasize the central role of the Mission Management, ISR Management, and

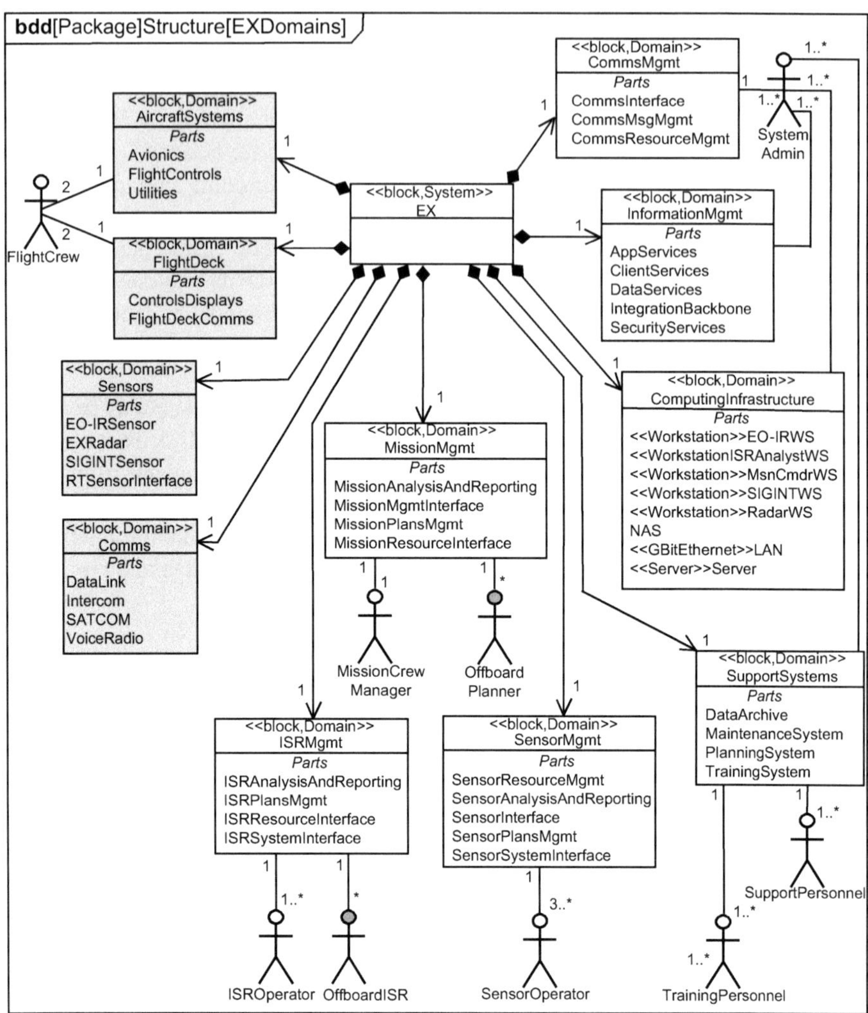

Fig. 4.6 BDD showing E-X system Domains with associated User Roles

Sensor Management Domains. The Mission Management Domain supports both the onboard command function and interaction with off-board command and control personnel. The ISR Management Domain supports onboard ISR Operators and also provides for interaction with ISR Managers at other nodes of the enterprise within which the E-X platform operates. The Sensor Management Domain provides the functions through which onboard Sensor Operators directly control the operation of the RF and EO sensors.

Many systems possess an intrinsic hierarchy of functions and User Roles that is similar to this. In general, tasking flows down the hierarchy, while data products and status reports flow upward. For example, the Mission Management Domain, used

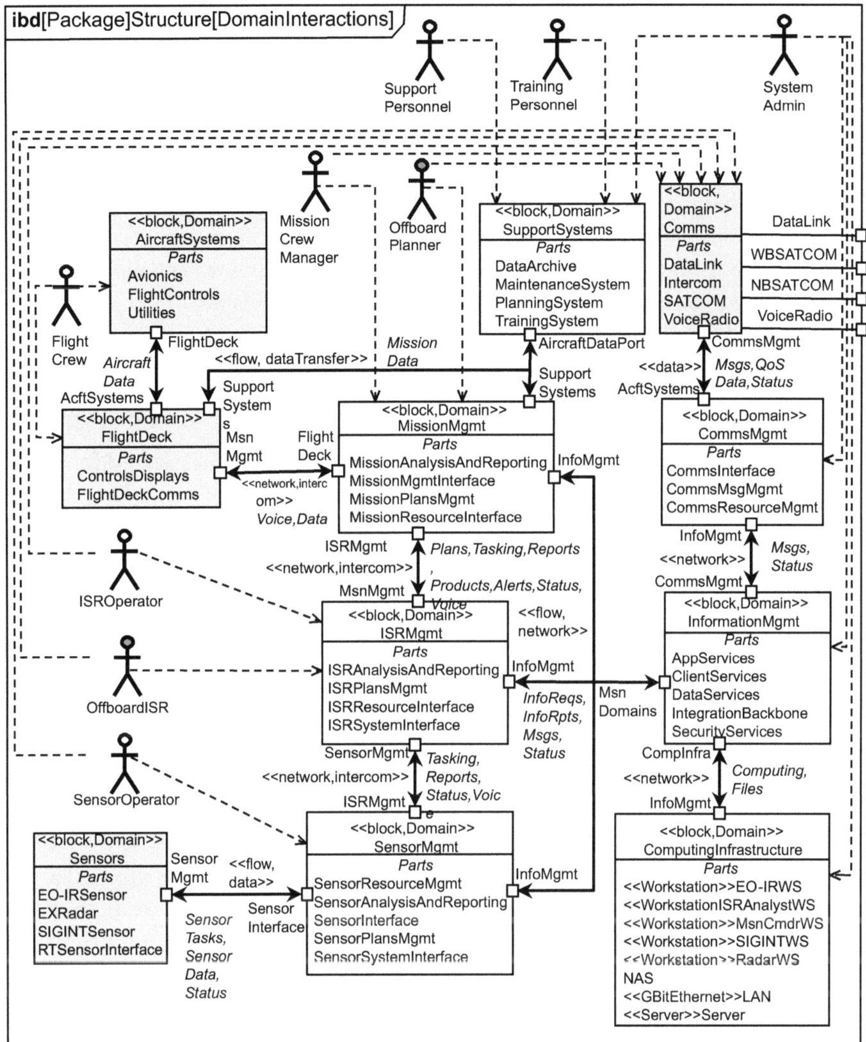

Fig. 4.7 IBD showing interactions among E-X Domains

by the Mission Commander or Supervisor, tracks the execution of a Mission Plan and creates new or modified mission tasks in response to events as they occur. These tasks flow down to the ISR Management Domain, and ISR Operators turn them into specific sensor collection and data processing tasks. Collection tasks then flow to the Sensor Management Domain where Sensor Operators execute them using the E-X platform's sensors. Then raw and processed sensor data flows up the chain, along with status information about system resources and task executions, to be used by the Supervisor and shared with other platforms and with the system's ultimate customers. Because these three Domains are responsible for the main E-X operational activities, we refer to them as "mission" Domains. Regardless of whether the system being architected is military, civil, or commercial in nature, a basic understanding of fundamental processes and relationships, like those just described for the E-X, is an essential part of a good OV, and therefore of the foundation for a successful system development.

Domains in Fig. 4.7 have Ports that model interfaces through which the Domains interact. Initially, in the OV and LV, Ports are defined functionally. Ultimately, in the PV, these will be structural interfaces that are realized as connections to the network that ties the information resources of the system together. Flows among these Ports are labeled with high-level Flow Items to show the general nature of the information being exchanged. These will be documented in detail in subsequent artifacts of the model as well as in Interface Control Documents and possibly other forms. These Flows are the beginning of identifying and defining the major interfaces within the system.

The E-X is an information-intensive system, and it has an Information Management (IM) Domain that aggregates the integrated processing, data management, and human-machine interface (HMI) functionality for the entire system. The IM Domain uses the resources of a Computing Infrastructure (CI) Domain, including servers and associated software, local area networks, and bulk data storage. Together, IM and CI constitute an infrastructure or execution platform for the application software of the mission Domains. This separation of concerns (mission applications vs. execution environment) is an important application of the tenets of good architecture, including openness. As the high level of connectivity makes clear, the IM Domain is the information crossroads of the system. It creates a service interface through which other Domains and system users access shared resources. IM also creates a generalized HMI interface that is personalized to meet the specific information display and control input needs of each User Role through User Workstations.

The IM Domain manages the flow of information on and off the platform through a Communications Management (CM) Domain. The IM Domain sends both outgoing traffic and control information such as current Quality of Service (QoS) policy to tell the Communications Domain how to use available communications resources to support prioritized messages. The CM Domain creates an abstraction of the external communications environment and isolates the other Domains from the impact of continuing evolution in external networks and channels, which would otherwise send a steady stream of changes rippling throughout the system. It allows communications functions to be accessed as services. One further User Role, labeled System Administrator, interacts with the IM and CM Domains to perform tasks such as radio frequency management and supervision of the computing and network infrastructure.

Coordination between the Flight Crew and other Mission Crew members occurs between the Flight Deck and Mission Management Domains. The pilots are responsible for aircraft controls and systems. The Flight Deck Domain can access the IM Domain through the Mission Management Domain to send data about flight and aircraft status that the mission Domains need and to access mission data when required to manage flight operations. This creates a single, clean interface between the Flight Deck and the mission Domains.

One final Domain, Support Systems (SS), is not installed on the aircraft. It provides training, preflight planning, postflight data analysis, and other ground-based services. Support and Training personnel interact with this Domain, e.g., maintenance technicians and crew training instructors. The SS Domain exchanges data with the airplane through the IM Domain and creates an HMI to be used by crew members for ground training. There are other important interactions between the aircraft and the installed mission equipment, e.g., providing power and environmental conditioning, but we omit them for the present because this discussion is focused on the information processes of the E-X. Of course, in a real system, such utilities would be of great importance.

The E-X example could be partitioned in a number of other ways, but the general concept is clear. The key idea is to group system functions into Domains that localize related activities and present coherent interfaces to each other and to operators performing well-defined roles. Shared functions such as communications and information processing are strong candidates to be treated as Domains. The reader may find it interesting to experiment with alternative structures. For example, it might make sense to simplify the architecture by merging the Mission Management and ISR Management Domains since both are primarily concerned with managing plan execution and interpreting the results of mission operations. Another possibility, especially for a highly secure system, could be to declare a distinct Cybersecurity Domain that interacts with Information Management, Communications Management, and other Domains; this would focus attention on the resources and functions involved in the critical role of protecting sensitive data and processes. Many other variations on the structure in Figs. 4.6 and 4.7 are possible.

4.6.3 Domain Specifications

Drawing diagrams is only the beginning of developing an architecture model. All of the major model elements must be documented in specifications, and popular architecture tools provide utilities to make this easy to accomplish. This is another example of an architecting activity that can be time-consuming initially but pays off in the long run by promoting Domain structures that are completely understood, consistent with each other, and easy to communicate to customers and system developers. Appendix C has specifications for a number of E-X Domains.

A typical Domain specification includes:

- *Owner*—identification of the organization or individual(s) who have approval authority for the definition of the Domain and its specification.
- *Description*—a summary statement of the purpose and functions of the Domain.
- *Definitions*—any terms required to understand the Domain definition (these can be conveniently collected in an Integrated Dictionary for the entire architecture).

- *Operations*—a listing of the behaviors the Domain produces, which are best divided into Public or externally observable operations and Private or internal operations; in some cases, it may be appropriate to suppress the publication of Private operations under the principle of encapsulation.
- *Data*—a listing of the primary data entities that the Domain produces or uses; these can be collected for convenience in the Conceptual Data Model, discussed later in this chapter.
- *Interfaces*—a listing of the external and internal (inter-Domain) interfaces created or used by the Domain.
- *Allocated Requirements*—once requirements are mapped to the Operational Viewpoint, each Domain specification tabulates the FRs that are to be satisfied in whole or in part by that Domain and the NFRs that impact the Domain and must be addressed in its implementation.

4.6.4 User Roles

As noted earlier, it's a good idea in conjunction with the Domain Diagram to define the structure of User Roles early in the OV development and to model them as Actors in a dedicated diagram. Figure 4.8 is a simple E-X example that groups the roles from Figs. 4.6 and 4.7 into Mission Crew, Support Crew, and Offboard Crew. These may or may not correspond one-to-one to human operators; a given Operator may perform various roles at various times, and multiple Operators may be needed for a heavily tasked role. Often, specific Roles will be specializations of more general or abstract roles. For example, a generic Role of System Operator might specialize through Inheritance to roles like Supervisor, Data Analyst, Planner, Equipment Operator, and so forth. The specifications for both generic and specific roles present a good opportunity to capture training requirements, skills inventories, job classifications, and other information for easy access, and this information is then carried into every diagram where the Actor representing a role is instantiated. As an example of the utility of inheritance, the Mission Crew role specification could define general aircrew qualification requirements, which would then be part of the attributes of each specific role. Similar common characteristics can be defined for the Support Crew and Offboard Crew Roles. As before, off-board Actors are shaded.

4.7 Behavioral Perspective

In parallel with working out the Structural Perspective, it's important to start defining the behaviors of Domains and their interactions with each other, with Users, and with the external environment in general. The Behavioral Perspective formally represents the observable behaviors exhibited by the system and its Domains. Behavior

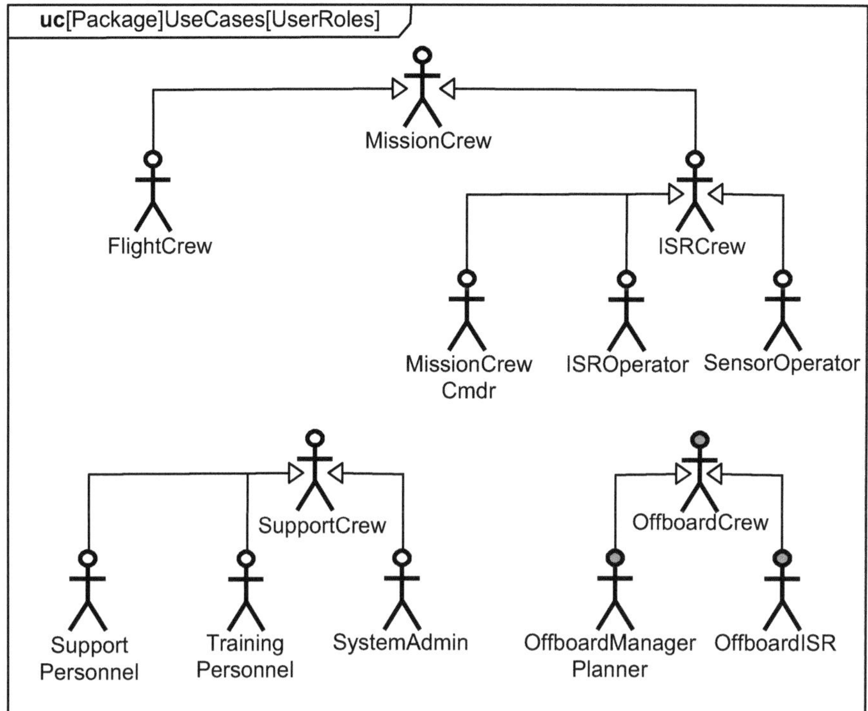

Fig. 4.8 Example of a User Role Diagram for the E-X

and structure are complementary, and the evolving content of each frequently helps in refining the other. The OV uses primarily Use Case, Activity, and State Machine Diagrams, and there are some situations in which Sequence Diagrams are also appropriate. The choice of diagram for a specific behavioral modeling task depends on the specifics of the behavior. In this section, we explore the creation of these diagrams.

4.7.1 Defining Use Cases

Organizing Behavior: The starting point in modeling behavior is to develop a set of Use Cases (UCs), each of which is an abstraction of a behavior or group of related behaviors that the system or other entity being modeled can exhibit. An architect might prefer to focus on behavior first and then address structure, for example, when behavior strongly influences things like optimum Domain partitioning. A well-chosen UC models a coherent increment of system functionality and activities, for example, the functions or services that the system provides to a particular Operator or external system [9]. If the system has been effectively partitioned, there will be a clear mapping of behavior to structure and a logical flow of system-level behaviors through interacting Domains.

Collectively, the UCs must account for all of the functional content of the requirements database or model, that is, all the things the system will actually do. Accordingly, mapping FRs to UCs is a natural way to establish the essential linkage between the requirements baseline and the architecture. NFRs, on the other hand, are typically associated with the system as a whole and are captured in QAts and in supplemental artifacts such as reports exported from the requirements database. In the process of defining UCs, it's common to discover better ways of partitioning the architecture, and thus to refine or replace an initial set of Domains, on the basis of clear and logical associations of behaviors with structural groupings.

The behavior of a UC is expressed in the form of one or more Scenarios, each of which is a specific sequence of activities that yield the behavior. There will generally be a primary Scenario that describes what the system does under normal circumstances, plus alternative Scenarios dealing with behavior under other conditions, which might include things like an equipment failure, an unusual event in the external environment, or a change to a secondary mode of system operation. As with the definition of Domains, certain amount of judgment is involved in deciding how many Scenarios are needed to explore the interesting range of system behaviors within a UC.

In this book, we use terms like "thread," "mission thread," "task," and "business process" refer to sequences of system activity, usually to satisfy FRs. A thread can be thought of in much the same way as a Scenario, with two important nuances:

- Threads are commonly assumed to represent primary mission- or process-related behaviors, while Scenarios must also be defined for a wide range of supporting activities.
- Threads often involve large-scale system behaviors that span multiple UCs and are therefore often composed by stringing together successive UC Scenarios.

Like Domains, UCs are hierarchical. The primary UCs are defined at system level,[2] but it's often useful to decompose these. Lower-level UCs may be helpful in identifying important specific behaviors. These may be Domain-level UCs that capture the primary behaviors of individual Domains. When this decomposition and mapping to system structure can be done easily and logically, the architect has another indicator that Domains and UCs have been well chosen and are consistent.

Figure 2.4 illustrated the format and content of a UC Diagram. Such diagrams only list the UCs; full definition and analysis is the next step in the process. Even so, a UC Diagram is a handy index and reference, especially for a complex system with many functions and user relationships. It helps the architect ensure that all the required behaviors and their relationships have been identified and to confirm them with the system's stakeholders. Like Domains, UCs can and should be mapped to the requirements database as further assurance that all the things the system must do

[2]Again, these principles apply at any level of the Axis of Organization, but we use "system" for simplicity.

are explicitly accounted for. External Actors such as system users have associations with the UCs that support their roles and functions. As described in Chap. 2, dependencies stereotyped as `<<include>>` and `<<extend>>` allow decomposition of UCs and definition of optional behaviors.

Use Case Specifications: Like Domains, UCs require detailed specifications as part of the architecture model. A reasonable outline for such a specification includes the following sections:

- *Description*—a summary that allows a quick check by a user of the architecture model as to whether the UC deals with a particular behavior of interest:
- *Preconditions*—things that must be present or true for the UC behavior to execute.
- *Trigger*—events or circumstances that cause the UC behavior to execute.
- *Postconditions*—things that are present or true when the UC behavior completes.
- *Data Objects*—contents of the Data Model that are involved in the UC behavior.
- *User Roles*—the User Roles associated with the UC.
- *Scenarios*—sequences of activities and events that make up the UC behavior and are normally written from the perspective of the User Roles in terms of the behavior as it is observed from outside the system; there will always be a Primary Scenario, which is the default, and almost always one or more Secondary Scenarios that define excursions in response to other than normal circumstances.
- *Allocated Requirements*—just as Domain specifications map requirements to the system elements responsible for meeting them, UC specifications map FRs to the system behaviors that exhibit the required functionality and must meet quantitative performance parameters. UCs can be written for NFRs, but the use of Quality Attributes has a number of advantages in defining and assessing these characteristics.

4.7.2 E-X Use Cases

Returning to the E-X example, and referring to Table 4.1, some initial system-level behaviors or threads suggest themselves. The following are examples, with references to requirements numbers in the table:

- *Conduct Aircraft Operations*—prepare the plane for flight, execute the flight plan, navigate in the mission area, and recover the aircraft after flight (R1.6).
- *Conduct Sensing (ISR) Operations*—prepare the mission systems, perform sensor data collection and processing, create and disseminate information products from collected and stored data, and download and archive the information post-flight (R1.1–R1.5).

- *Conduct Support Operations*—perform ground-based functions such as mission planning, data archiving and distribution, training, and aircraft maintenance (R1.7).

In the E-X example, the association between system-level behavior and structure is fairly clear. The Flight Deck and Aircraft Systems Domains are involved in air operations, the Support Systems Domain provides the capabilities for support operations, and the other Domains are concerned with various aspects of airborne sensing operations. One way to improve the readability of these diagrams for a large, complex system is to break the UCs into those associated with mission or business operations and those associated with support activities. Figures 4.9 and 4.10 show this approach and include decomposition of major UCs into smaller, more easily analyzed increments of behavior. Sorting the UCs into Tiers is another way to enhance diagram readability.

The Conduct Aircraft Operations UC combines pre-, post-, and in-flight activities. Both Flight Crew and Support Personnel are associated with these UCs. It might be useful to split the Support Personnel Actor into Maintenance Personnel and other Support Personnel; for now, they are kept together assuming that a common pool of skilled individuals will share the duties of overall system support and of preparing and recovering aircraft before and after missions. Dozens of subtleties like this will arise in a real-world architecting exercise and should be discussed with the customer.

The Conduct ISR Operations Use Case will naturally have the richest set of behaviors, and Fig. 4.9 reflects this. A Coordinate Mission Operations UC accounts for the interaction between the Flight and Mission Crews. The various Actors from Fig. 4.8 are shown, some associated with high-level Use Cases because their roles involve the entire area of behavior and others who have more specialized associations with subordinate UCs. The Actor representing External Communications Networks is not included in Fig. 4.8 because it doesn't model a User Role, but it is an important external entity with which the E-X interacts.

The System Administrator User Role is involved in both Mission and Support Use Cases. In flight, these personnel work with the Domains of the computing infrastructure as well as with communications resources. They are thus associated with both data and communications management UCs. On the ground, System Administrators are associated with the UCs shown in Fig. 4.10. In practice, some E-X behaviors are involved in both flight and ground operations. For example, system administration functions such as restarting a crashed server can be needed both in the air and on the ground, and such actions are considered part of the Manage Data UC in Fig. 4.10. The importance and unique nature of training is reflected in a Perform Training UC with a dedicated Training Personnel User Role. All other system operators have associations to this UC because each receives training support in specific areas relevant to their roles.

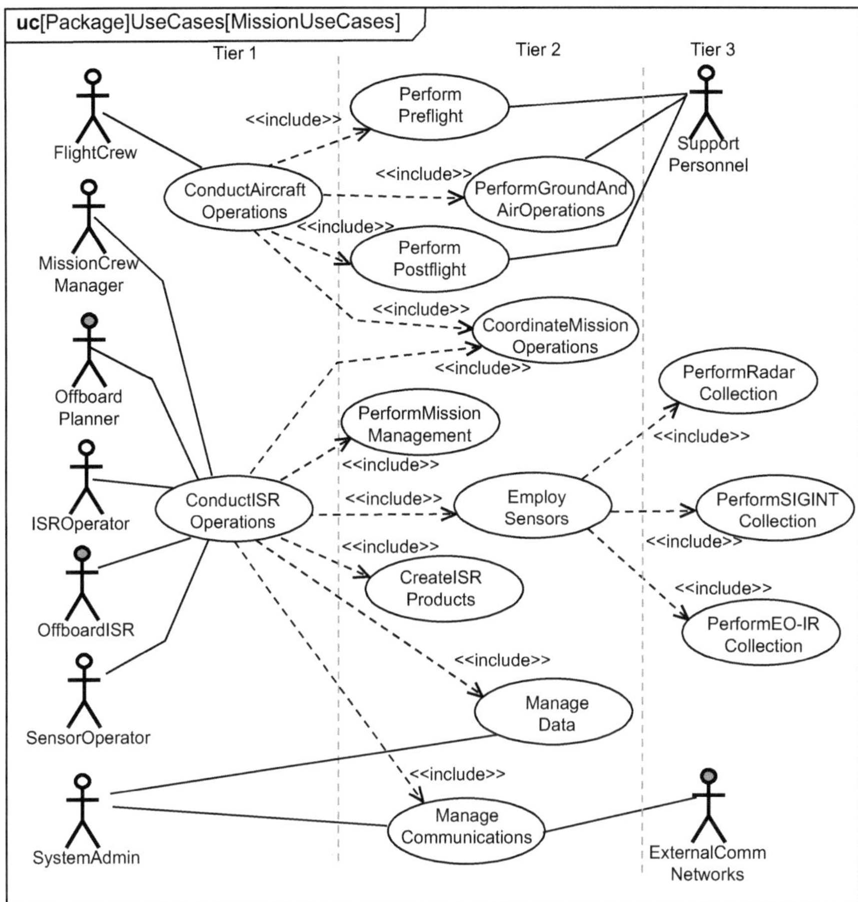

Fig. 4.9 E-X Mission Use Cases with decomposition

4.7.3 *Scenarios and Activity Diagrams*

It is now possible to unite the Structural and Behavioral Perspectives as we flesh out the content of Use Cases. In MBSAP, UC Scenarios are developed using Activity Diagrams (ADs), and these are organized with swim lanes that correspond to system Domains, User Roles, and external systems. We should note that it is also possible to use Sequence Diagrams to model these scenarios, but they are generally better suited to the more fine-grained behaviors of groups of Blocks in the LV. The principal exception to this rule of thumb occurs when Scenarios have important content involving interactions with external entities, especially using messaging. In this case Sequence Diagrams showing message exchanges between system Domains and external entities can be very effective. In almost all cases, we find it more convenient to model the OV Behavioral Perspective using ADs.

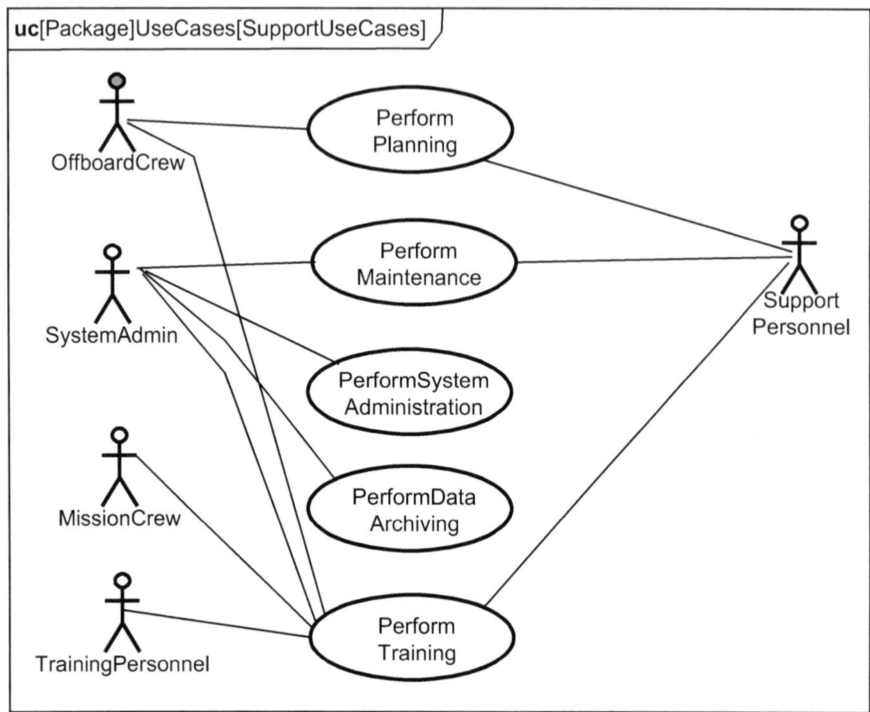

Fig. 4.10 E-X Support Use Cases

Figure 2.5 presented the basics of Activity Diagrams (ADs). Control and Object Flows show how a given Scenario involves the structural entities that are allocated to the swim lanes. An Action is created in the swim lane of the Domain or other model element that is responsible for performing it, and an Object is created in the swim lane of the entity that creates or otherwise owns it. Objects include the primary data items or messages that are created and used as the Scenario executes. In addition to adding useful detail to the Scenario definition, this establishes a link to the Data Perspective. Our experience has shown that the process of thinking through and diagramming Scenarios is a great opportunity for data discovery, that is, to identify and define the information that is inherently imported, created, transformed, and exported as part of the behavior. This provides a start for defining the Conceptual Data Model, which is discussed in Sect. 4.8. Forks and Joins allow clear portrayal of parallel activities, and Branches show alternative paths based on conditions such as the state of the system, events in the external environment, operator decisions, and the like. In this way, a single AD can be used to show both primary and alternative Scenarios.

4.7.4 E-X Scenario Examples

Figure 4.11 shows a simple example of this approach to portraying a Scenario for an E-X UC, Create ISR Products, in an AD. It begins when the Mission Management Domain sends a task to the ISR Management Domain to develop a product such as an image of an object or area on the ground or to perform and report analysis of a specified set of sensor data. An ISR Operator assesses the task and provides the appropriate inputs to create a sensor task. This is passed to the Sensor Domain, which returns the requested data in the form of a Sensor Report. The operator evaluates the data and uses the functions of the ISR Management Domain to create the requested product.

A more complex AD in Fig. 4.12 illustrates an OV artifact that is often very valuable in the early stages of developing an architecture. It can be quite helpful at the start of behavioral modeling and analysis to construct a high-level thread or process flow that shows primary system activities, typically in the context of an end-to-end business operation such as a manufacturing cycle for an industrial system or a mis-

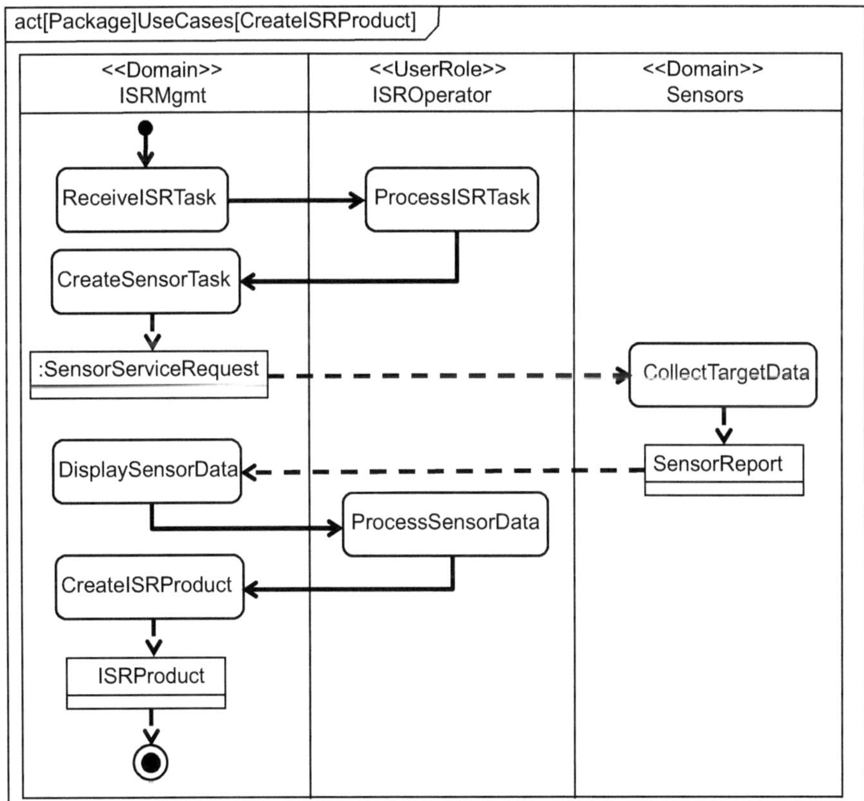

Fig. 4.11 Scenario Activity Diagram for the Create ISR Report Use Case

sion for a system like the E-X. This representation of typical overall system functioning is commonly called a "Day-in-the-Life" diagram because it represents a "normal" period of system functioning. It's a bit like an artist blocking out the overall structure of a painting before filling in the details.

Figure 4.12 is labeled a "Basic Mission," and it begins with tasking or guidance from the E-X Organization that owns and operates the system, modeled by an initial swim lane in the diagram. The resources of the Support Systems Domain convert the Mission Tasking into a Mission Plan that includes both a Flight Plan and a Collection Plan that spells out the use of sensors and onboard processing to produce required ISR Products. The Flight Crew loads the Flight Plan via the Flight Deck Domain, while the other Mission Crew personnel load Collection Plan Data into the various Mission Domains and configure them for operation. Once the aircraft is on station in the designated area for sensor operations, the pilots fly the Flight Plan, and the other crew members carry out planning, tasking, sensor data collection, data processing, and ISR Product generation and distribution.

Both the E-X Organization and the Mission Crew Commander monitor the progress of the mission and assess the system's products. They can then modify the overall Mission Plan and adjust specific ISR tasks as required. Once the Mission Plan has been completed, the Flight Crew returns the aircraft to base. Then ground personnel using the resources of the Support Systems Domain perform postflight activities, conduct a debriefing (post-mission interview) of the Mission Crew to identify problems or failures, and schedule any required maintenance. Other exceptional conditions such as an aircraft problem or an urgent request for information from a customer of the system's products could be included in the diagram if they are important to understanding the overall scheme of system operations.

This Day-in-the-Life establishes a pattern for thread and Scenario modeling that helps ensure consistency in portraying more detailed behaviors, as well as in communicating with customers, checking on the completeness of the requirements baseline, and completing other early architecture tasks. We maintain traceability to requirements through the specifications of behavioral model elements. Many of the Actions in a primary behavior like that in Fig. 4.12 will, in fact, be UCs, and we have already seen that UC specifications have an Allocated Requirements topic. Other behavioral elements such as Actions in an AD that are not themselves UCs also require specifications, and these can identify allocated requirements in the same way. Wherever more detail is needed, Actions are routinely broken out in subordinate behavior diagrams. The diagram includes a number of decision points such as a recurring check by the E-X Organization to determine if a revision to the original Mission Plan is required.

The effort of developing ADs like this one pays benefits by allowing the architect to improve and validate the consistency of the Structural and Behavioral Perspectives, to ferret out subtleties in required system functions, to confirm that the Domains do a good job of grouping functions, and, very often, to discover derived requirements that are not in the customer's requirements specification or the capabilities database. If the flow through the Scenario is hard to follow, it may be a sign that the Domains aren't well partitioned. If the Scenario implies the need for a function that isn't in

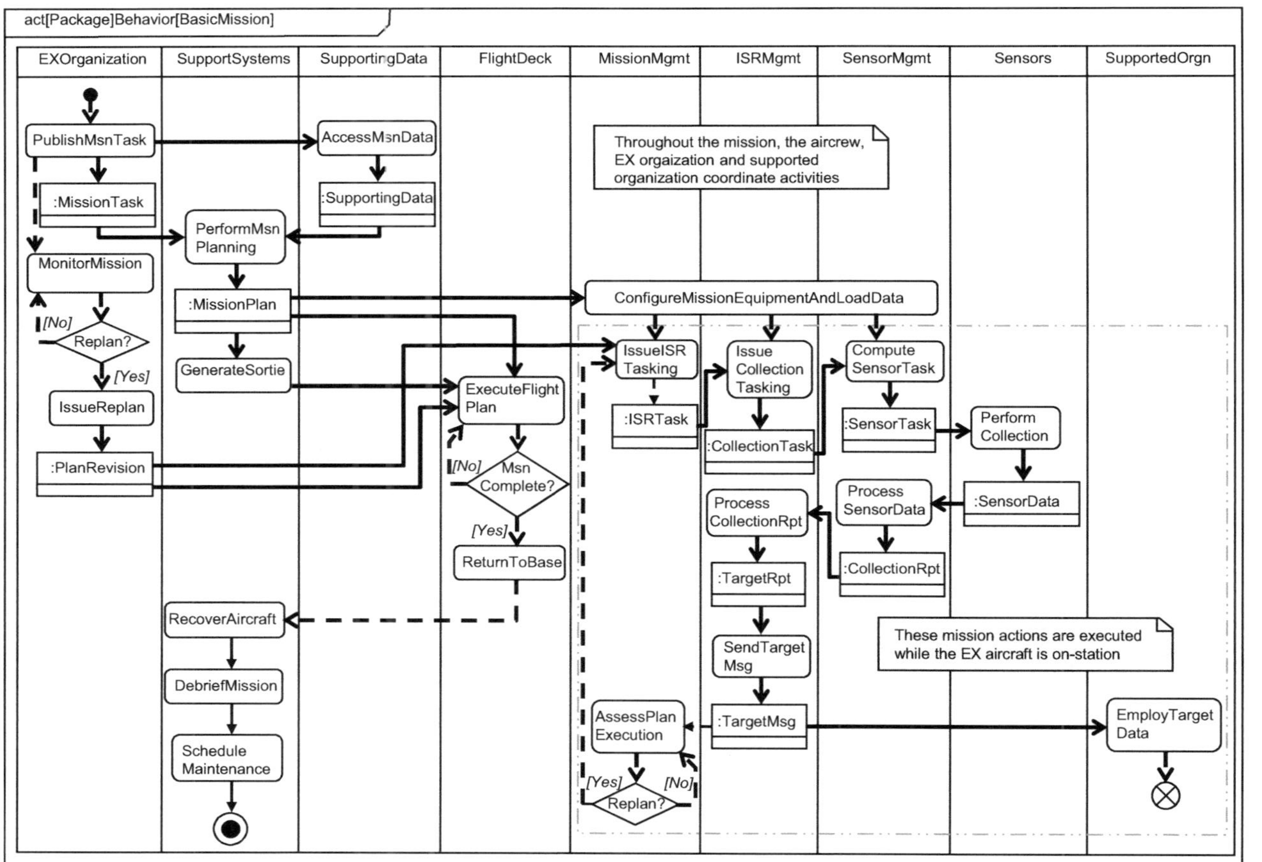

Fig. 4.12 "Day-in-the-Life" basic mission for the E-X

the customer requirements, it may well signal a derived requirement that should be fed back to make the capabilities database more accurate and complete. Traditional requirements analysis deals with flowdown, allocation, and consistency checking. MBSAP improves the analysis by making explicit the ways system elements interact in delivering the required behaviors and providing a more rigorous method to uncover derived requirements.

Every data Object declared in the Scenarios shown in these ADs is a required data entity in the Conceptual Data Model, which we will introduce next. These ADs can also be used as inputs to Process and Workflow Modeling to begin constructing an executable representation of the OV. Constructing, analyzing, and refining these diagrams are hard work, but the results are invaluable in ensuring that customers and developers have a common understanding of the required system behaviors and in avoiding problems when the time comes to actually design and build the system.

A further aspect of the Behavioral Perspective comes into play when it is important to model the states that the system can be in and the conditions that trigger transitions among these states in a State Machine Diagram (SMD). Figure 4.13 shows this for the E-X example. The states have shown progress through the phases of a typical mission, consistent with Fig. 4.12, and an additional state is provided for times when the aircraft is taken out of service for maintenance. In any given state, the system is in a kind of "behavioral configuration," meaning that various system elements will be active or inactive, User Roles will involve actions specific to the state, various external interactions will be needed, and so forth. When dealing with complex behaviors like those of the E-X, ADs and SMDs complement each other and present a more complete and understandable view of system functioning.

4.8 Data Perspective

4.8.1 Conceptual Data Model

Next in the development of an OV is an activity that is extremely important but all too often neglected at the start of an architecture development: data modeling. Data is obviously the essential content and product of an information (or information-enabled) system, and it must be considered at the same time as the system's structure and behavior. As with other aspects of the architecture, the data model begins in the OV with an abstract representation, a Conceptual Data Model (CDM), which will become progressively more concrete and specific as a Logical Data Model (LDM) in the LV and finally a Physical Data Model (PDM) in the PV. This hierarchy of data models was originally defined in the Joint Command, Control, and Collaboration Information Exchange Data Model (C3IEDM) [10] and is summarized in Fig. 4.14. The CDM establishes the overall scope and structure of a system's information content and its representation in a system data model.

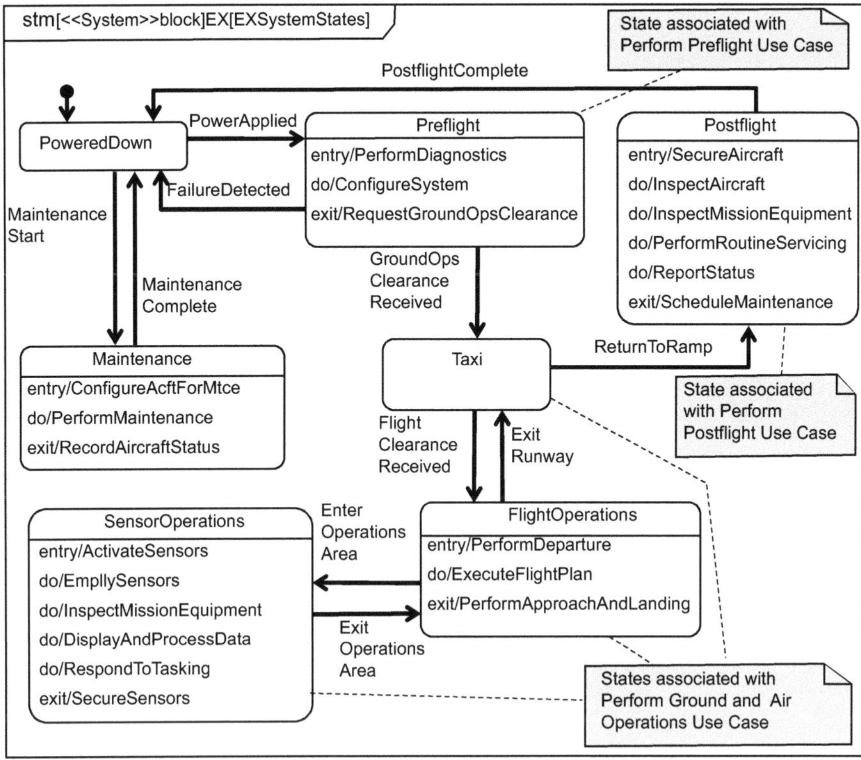

Fig. 4.13 State Machine Diagram showing how the E-X progresses through a sequence of states during a mission and can be taken out of service for maintenance

In his classic text on database design, Date [11] gives an excellent summation of the importance of conceptual data modeling, noting that a conceptual schema is an important and lasting basis for the development, operation, and evolution of a system or enterprise. This schema is an abstract description of data and other system content and should only be changed as necessary to keep the data model aligned with the real world. As discussed in Chap. 3, we begin by building a CDM using BDDs.[3]

The Data Perspective thus becomes an integral part of the overall architecture model, and, among other benefits, this makes it easy to associate data entities with Block Values and Operations and to show their involvement in system behaviors by including them in Activity Diagrams. One of the great strengths of OO is that it unites in a single construct what in earlier architecture methods were separate data and functional models. A good data model, starting in the OV and LV, also eases the

[3] Historical note: the original UML Class Diagram is a refinement and generalization of the older Entity-Relationship Diagram (ERD); SysML Block Diagrams offer more powerful and convenient data modeling techniques than older constructs such as ERDs and the Integration Definition for Functional Modeling (IDEF) method.

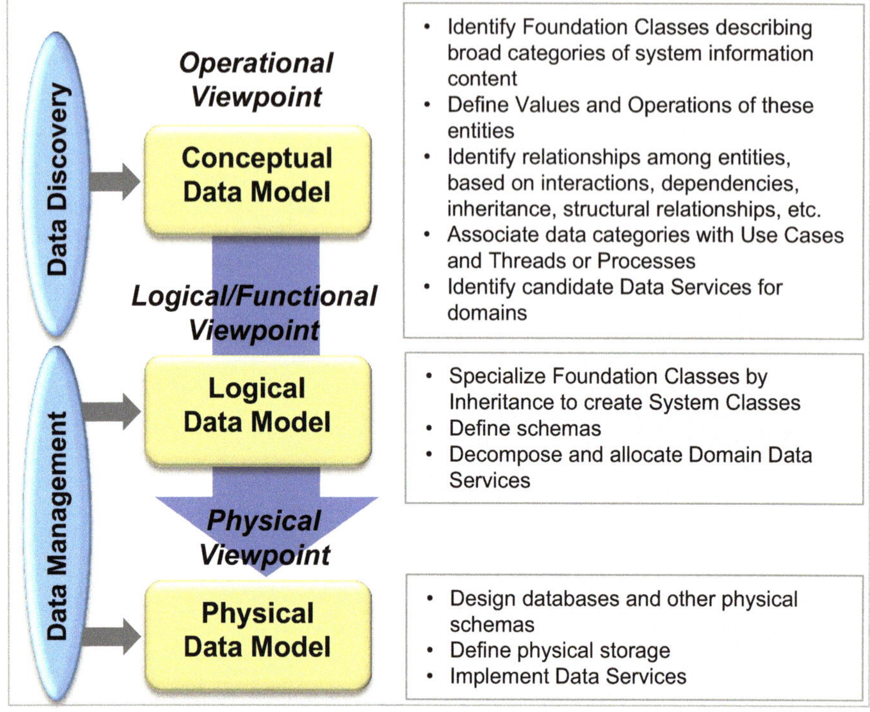

Fig. 4.14 Data model hierarchy Starting with a Conceptual Data Model and progressing through Logical and Physical Data Models

task of developing physical data schemas and the perennially difficult tasks of normalizing and renormalizing databases.

In addition to establishing rules and conventions for data modeling, the CDM represents the product of what can be called "data discovery." This happens in conjunction with requirements analysis and development of the Structural and Behavioral Perspectives of the OV, all of which yield insight into both general and specific data entities that the system must import, create, use, store, and export. The CDM may begin with a simple spreadsheet in which such data entities can be listed with initial descriptions, but it should be migrated to BDDs during OV development. Fundamentally, there are three important aspects to consider in building the CDM:

- Identification of information entities, which may be concrete or abstract
- Definitions of the characteristics of these entities, including identity, location, time parameters, content and format, ownership and pedigree, data precision, and similar properties
- Identification of relationships among entities, based on interactions, dependencies, inheritance, structural relationships, and so forth

Once data entities are identified and defined as part of OV development, the architect must determine where they are used in the Activities of various Domains. Other important considerations include where information is stored, temporarily and permanently; when and by what processes information entities are created, transformed, and destroyed; and how they are exchanged, for example, by a publish/subscribe mechanism on a data server. The CDM can deal with this by associating data entities with Domains, defining appropriate Values and Operations in Blocks that model information entities, placing Data Objects in Activity Diagrams like Fig. 4.13, and coordinating the CDM with data services in the Services Perspective. In addition to these diagrams, the CDM can include supporting documentation such as an initial Data Dictionary (sometimes called a Data Element Dictionary), predefined message catalogs like those associated with data link protocols, and an Information Exchange Requirements (IER) matrix.

The core concept in building a CDM in MBSAP is the use of Foundation Classes. The actual data entities encountered in a system are very often specialized instances of a fairly small number of general types or categories, which can have Value Properties, Part Properties, Reference Properties, Operations, Associations to other elements, defining behaviors, and many other characteristics. Foundation Classes model these and thereby establish the overall structure of the data model. The term "Foundation Class" is potentially confusing, because in SysML, these "classes" are actually modeled as Blocks. We have retained the name that is widely understood in the architecture community, and the reader should simply make the translation from Foundation Class to Block.[4] As we will show in our examples, it's helpful to give these Blocks a stereotype such as `<<InfoElement>>` to make it absolutely clear what their purpose in a model is.

A Foundation Class is `<<abstract>>` because it is never directly implemented as a Block Instance. It serves as the root of an inheritance tree through which the shared characteristics of the Foundation Class, especially Values and Operations, are flowed down to the actual, individual data Blocks within the information category. Data Blocks are further defined with additional characteristics as needed to model various information entities. This refinement of Foundation Classes is the main activity required to transform the CDM to a Logical Data Model (LDM) in the LV, as discussed in detail in Chap. 5. We refer to the Blocks in the LDM as System Data Blocks to reflect the fact that unlike Foundation Classes, they are actually instantiated in the system and represent identifiable information entities such as parameters, files, messages, and records.

High-level BDDs capture Foundation Classes and the relationships among them, sometimes including a degree of decomposition by Generalization/Inheritance to show their nature clearly. Because the shared characteristics of a category of data or other entities are defined once, in the Foundation Class, there is an assured level of consistency and accuracy in the System Data Blocks that inherit from it. Moreover, this method reduces the total data modeling effort by eliminating repeated entry of the same information. This helps ensure important characteristics are not over-

[4] Also, the term "Foundation Block" seems to conjure up an image of the masonry under a house.

looked or inconsistently used, as well as facilitating an effective and normalized design for data storage and access.

When modeling data, it's common to define Blocks that have only basic Operations, such as `get()` and `set()` or `create()`, `retrieve()`, `update()`, and `delete()` (abbreviated CRUD), because the emphasis is on the Values that define the information content and on the relationships among the entities being modeled. In fact, a Foundation Class typically contains most or all of the Operations needed by the Data Blocks in its category, so that detailed data modeling can concentrate on the particular Values of individual Data Blocks. Among the Values commonly encountered in conceptual data modeling are:

- *Type*—numerical value (integer, double, floating point, etc.), text, structured ("struct") and unstructured data, enumerated lists ("enum"), telemetry, imagery, alerts, coordinates, data and time values, etc.
- *Timing*—time stamp, period of validity, minimum and maximum process durations and latencies, etc., which are especially important in real-time applications
- *Security*—classification or sensitivity, authentication values, and integrity or error checking
- *Quality and accuracy*—e.g., a precision parameter or error statistics
- *Source*—identity and characteristics of the data producer
- *Format*—e.g., text and imagery standards
- *Storage and posting*—locations where data is stored temporarily or permanently ("persisted") and where it is posted for discovery and access

It turns out to be very useful in some situations to define another type of Foundation Class that generalizes system resources other than data. In the E-X example, these would be things such as sensors, communications devices, and computers. These are stereotyped as `<<abstract>>` and `<<Resource>>`.[5] An additional stereotype can be used to identify the category of resources modeled; e.g., a given Foundation Class that accounts for all the computing devices in the system could be stereotyped `<<Processor>>`. It may appear unnatural to include these with traditional data constructs in a CDM, but this approach yields a coherent and intuitive way to model the content of a system. The important distinctions between data, which ultimately becomes values stored and used in the system, and resources, which translate into the components from which the system is constructed, become clear as the Foundation Classes are specialized into Blocks and then implemented as code, databases, hardware, and other physical elements. Once a Foundation Class is defined, in the OV, the architect can use it wherever needed in the architecture model by simply copying it into a new diagram.

Figure 4.15 shows the general idea, using only Data Foundation Classes, and illustrates the power of SysML to portray the details of a data structure in a very flexible and intuitive way. At the top of the data model hierarchy is a Generic

[5] In general, stereotypes that are defined in SysML are left lower case, while those defined in MBSAP start with a capital letter.

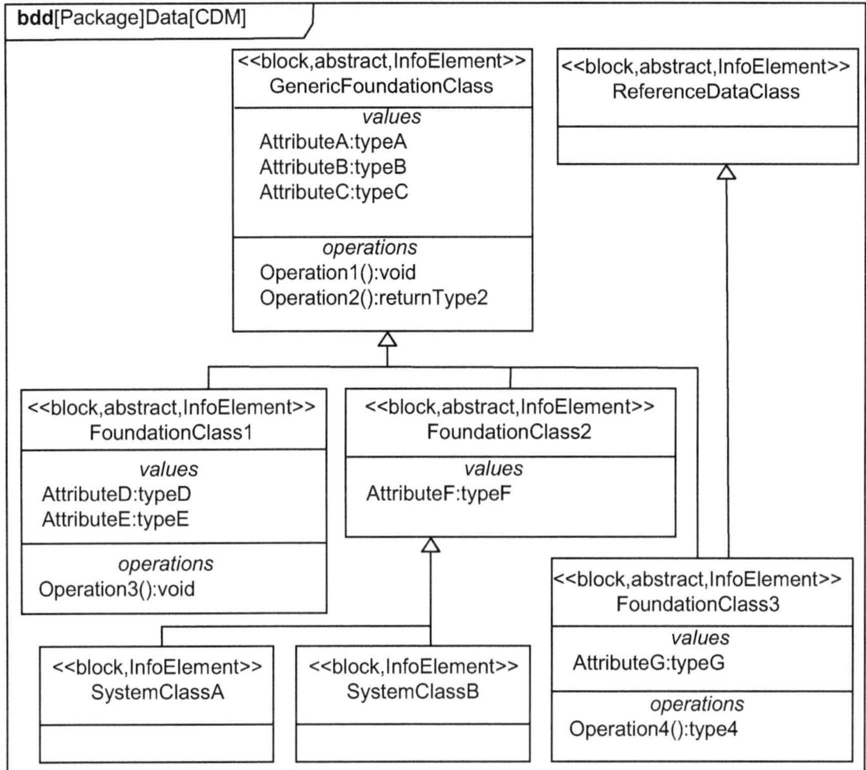

Fig. 4.15 Format of a generic Conceptual Data Model including a Reference Foundation Class

Foundation Class that has the Values and Operations needed by every information entity in the model. Three Foundation Classes defining categories of information inherit from the generic class, and two System Data Blocks are shown inheriting from Foundation Class 2, although this would often be deferred to the LDM. A final refinement is that Foundation Class 3 has dual inheritance, including the characteristics of a Reference Data Block that could be part of a library of reusable architecture artifacts or of a Reference Architecture as discussed in Chap. 12. Again, in a real architecture, all these Blocks would display Values and Operations.

4.8.2 E-X Conceptual Data Model

Returning to the E-X example, Fig. 4.16 shows a possible set of Data Foundation Classes. As in the previous figure, the root of the data model structure is a completely generic Data Foundation Class that includes the Values and Operations that any Information Element in the system is expected to possess. These shared

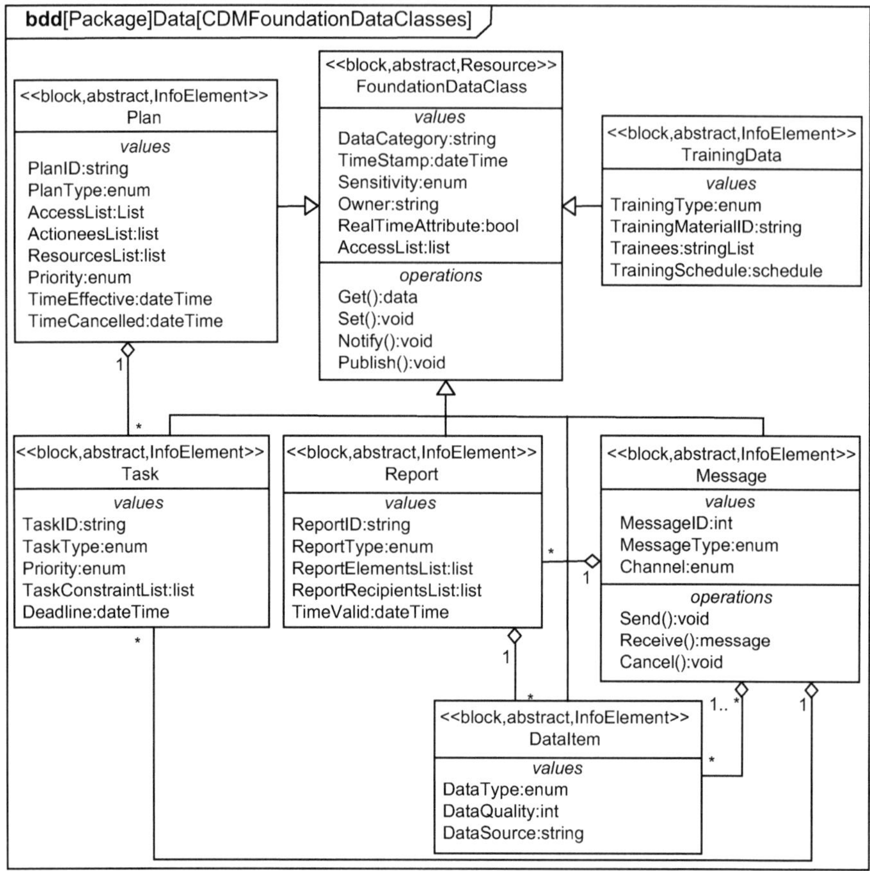

Fig. 4.16 E-X Data Foundation Classes

characteristics are inherited by `<<abstract>>` Classes for Plan, Task, Report, Message, and Training, as well as one simply named Data. These Foundation Classes model the primary categories of information used by the E-X. The Data Item Class at the bottom of the diagram abstracts various information categories, including sensor data, analysis data, display data, and so forth, that are encountered in the primary mission activities of the system.

Reference Associations show that these data items may be incorporated in reports and messages. The Plan Foundation Class aggregates tasks and adds information such as times, resources, priority, and a list of the "actionees" who are required to execute the plan. The training function for a system like the E-X is sufficiently complicated and diverse that a Training Data Foundation Class is included. Obviously, the types and relative size and importance of information vary tremendously with the nature of a specific system, but this approach to a CDM is generally applicable.

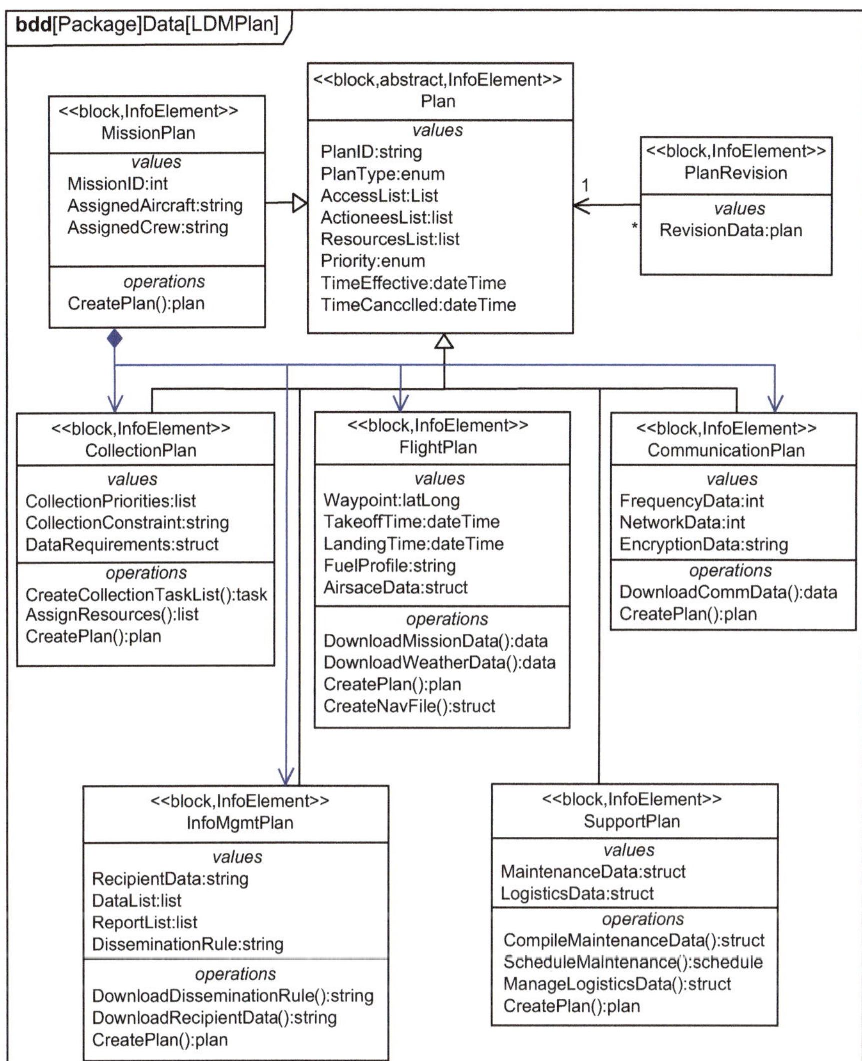

Fig. 4.17 Plan Foundation Class specialized by Inheritance to define particular plans in the Logical Data Model

The next chapter discusses decomposing Foundation Classes to create an LDM, but as an early example, Fig. 4.17 illustrates the concept and shows how Foundation Classes work in a data model. For example, a plan is a common data type in many systems. Before an E-X mission, the crew will build plans for things like the flight path, the intended use of the sensors, and the setup and operation of the aircraft communication systems. Any such plan has Values such as the originating authority, the times the plan becomes and ceases to be effective, the persons or organizations to whom the plan is directed, the resources it requires for execution, and security parameters. Similarly any plan object has basic Operations to obtain content, modify content, and notify recipients of the arrival or modification of the plan. This is modeled as an abstract Foundation Class called Plan. Then Flight, Sensor, and Communications Plan Classes inherit these universal characteristics and add specific Attributes/Values and additional Operations pertinent to their nature. Composite Associations show that Collection, Flight, Information Management, and Communications Plans are Parts of a Mission Plan, and a Plan Revision can be applied to any of the Plans. Note that the data Types shown for Attributes and Operations are notional; for example, `struct` is a placeholder for a complex data type that hasn't been defined yet. It should be clear that development of the LDM typically starts in the course of defining the CDM as candidate Foundation Classes are explored to see if they are appropriate for the system.

4.9 Services Perspective

In the era of SOA, interactions among humans, computers, and networks are moving away from reliance on rigid, intricately detailed protocols and point-to-point interfaces toward a model based on finding, requesting, and receiving services that are defined in much more functional and intuitive terms. Instead of asking for a piece of data at a predefined physical address, a client might broadcast a request over the network for any available information that matches a set of plain text attributes. The client does not need to know in advance what node on the network has a given item of information or can perform a particular computation; an easily formulated query will turn up whatever is available from any source, and a simple protocol will request and deliver the service. Chapter 7 goes into much more detail about services and SOA, but a preliminary discussion is included in this chapter and the next two in terms of the Services Perspective of the OV, LV, and PV.

This same service-oriented approach is frequently valuable inside the boundary of a system to establish how the Domains or subsystems interact. Our discussion of the E-X example identified services created by the IM and CM Domains. Once the fundamental principles of modularity and encapsulation are adopted, such that the internal workings of one Domain are not visible to other Domains, these interactions must occur in terms of functions and information exposed at Domain boundaries. For example, one Domain might send a request to another over the system's internal network asking for a computation to be performed, a piece of data to be provided, or a control action to be taken for an item of machinery.

It's very natural to view such interactions as the invocation and delivery of services, and the requestor, or client, usually has no need to know how the provider, or server, goes about creating the service. A little thought will show that this is a powerful step in implementing the critical open system architecture tenet of loose coupling because it greatly reduces dependencies, such as the need for a priori knowledge, among the Domains. The Services Perspective of the OV identifies the external and internal services that the system uses and provides, including the very important step of assigning responsibilities for service creation and consumption to Domains and associating services with the interfaces through which they are requested and delivered. Chapter 7 goes into service modeling in detail, including examples of diagrams and service interface definitions.

In these architectures, Data Services provide the means by which information is managed, stored, accessed, and shared. Accordingly, the data discovery and high-level data modeling that produce the CDM should be accompanied by early attention to these services. Said another way, a networked enterprise using a SOA foundation requires a networked data strategy. A representative set of overall data characteristics that are important in any SOA includes:

- Visible, available, and usable when and where needed for decision-making
- Tagged with metadata[6] to enable discovery by authorized users
- Posted to shared spaces with appropriate controls
- Compatible with the achievement of interoperability through network-centric, many-to-many data exchanges

Both discovery metadata and descriptive metadata will be important elements of continuing data modeling as the architecture progresses into the LV and PV because they are key enablers of data management and sharing. Chapter 7 continues this topic.

It's often the case that Data Services, and the interfaces through which they are accessed, are among the most important elements of a SOA, and therefore of the Services Perspective of the OV. Indeed, the Data and Services Perspectives are frequently related, just as the Structural and Behavioral Perspectives interact closely with each other. This creates a data modeling challenge that would be very difficult to deal with using earlier data modeling methods. In addition, Government agencies and other customers frequently mandate standards for data modeling and Data Services; an example is a discovery metadata standard as a key aspect of achieving interoperability across systems and consistency in using information.

Most current data strategies stress data tagging with the eXtensible Markup Language (XML) to promote common data definitions and compatibility with open standards for Web Services, especially for data discovery and dissemination using standardized mechanisms such as publish and subscribe. The development and use of XML schemas is part of the LDM, but a thorough job of collecting the key information in the Conceptual Data Model makes that task far easier.

The basic architecture strategy of treating the behaviors exposed by system Domains as services that can be invoked and delivered through Service Interfaces at Domain boundaries or at the system boundary will become much more concrete in the LV. As the evolution of the architecture moves into functional design, Domains are expanded using Blocks that realize their behaviors, including services, as well as additional details of Service Interface definitions. We addressed the mapping of

---

[6] Metadata, in this context, is information expressed in a defined format or standard concerning a data entity; it can describe content, source, location, purpose, and many other aspects.

customer requirements to Domains earlier in this chapter. This can now proceed a step further by attaching requirements to services in each Domain. It's important to note that the mechanisms through which services are implemented range from very simple to quite complex. Although many system architects think of "services" as being synonymous with "Web Services" over the Internet, it's generally the case in real-time systems that services must use lightweight, low overhead protocols rather than the processing-intensive and time-consuming protocols associated with the World Wide Web.

If the customer's requirements include, either explicitly or implicitly, a tabulation of required services, then the Services Perspective should obviously follow and build on that input. Architecture frameworks (Chap. 12) frequently prescribe specific artifacts for defining services. Services might be spelled out in a set of service specifications or definitions in an Integrated Dictionary. Alternatively, service behavior can be explicitly described in Activity or Sequence Diagrams, and other forms of service representation are possible. Even when the customer does not address services explicitly, the architect may need to analyze system requirements, along with information on the customer's business processes, to develop a services taxonomy. Developing a Services Perspective is yet another exercise that can be very helpful in defining a good Domain structure on the basis of logical grouping of related or interacting services.

A simple way to start is to construct a spreadsheet listing required services and associating them with the model elements of the OV. The matrix will acquire columns and more detailed entries in the LV and PV as the architecture spells out the implementation of system services from the initial high-level tabulation down to the physical components that ultimately do the work. In effect, the Services Perspective becomes a high-quality compliance matrix showing traceability from requirements

Table 4.2 Sample of a Services Taxonomy and mapping, beginning in the OV

System service	Use cases	Domains	Domain services
Sensing:: Perform radar surface surveillance in SAR and MTI modes	Employ Sensors:: Perform Radar Collection	Sensor Management	Maintain sensor schedule Control sensors Process sensor reports Generate track reports
		RF1	Collect in SAR mode Collect in MTI mode
Generate, update, and manage tracks	Create ISR Products	ISR Management	Process track reports Correlate and fuse track reports Maintain track database

to implementation while maintaining consistency with the evolving architecture and system design.

Table 4.2 shows a simple example for the E-X. Referring to Table 4.1 and Fig. 4.12, for surface surveillance we use the radar and target track management as examples of typical services to be provided by the system. The first of these involves the RF1 Sensor (the radar) and the Sensor Management Domain, while the latter is contained entirely within the ISR Management Domain. The Actions in the Activity Diagrams for the Use Cases provide important content for defining Services. This structure would be extended to all Domains and services as part of a complete OV.

4.10 Contextual Perspective

Thus far, the emphasis has been on applying formal modeling, focused on a SysML model, to build the OV. There are some additional important artifacts that don't fit as cleanly into this mold, and they are lumped together in, for want of a better term, a Contextual Perspective. In general, this is important material that isn't well represented by the kind of artifacts used in the other Perspectives. Many of these artifacts deal with NFRs, and they are often either text documents or graphics.

- *Architecture policies and mandates*—In many cases, an architecture must conform to constraints associated with a larger enterprise and associated policies and directives. A company or Government agency may have an enterprise architecture or reference model that establishes rules, standards, and conventions applicable to specific systems within that enterprise. Examples include enterprise services that must be supported. They do not in themselves create architecture artifacts, but it will be useful to document in the architecture description and in the specifications for various model elements the way these policies are addressed in developing the architecture.
- *Additional nonfunctional requirements*—A variety of NFRs that do not fit naturally into other Perspectives may be accounted for under Context. These might include available resources and processes for system support; operational doctrine and procedures; material controls and accountability; reliability, maintainability, and availability (RMA); and many others. All such requirements should be captured in the requirements database, and many should be mapped to the Structural and Behavioral Perspectives, e.g., in Use Cases dealing with maintenance, repair, and system preparation generation times. Design requirements such as intrinsic failure-free component lifetimes and redundant implementation of critical functions need to be propagated to the PV, where they will impact both physical design and product selection. In the OV, NFRs can often be captured as text documents linked to the requirements database.
- *Customer-defined products*—If the architecture is required to conform to customer architecture frameworks, there are a typically data items, diagrams, and other artifacts that fall into the Contextual Perspective because they aren't inher-

Table 4.3 Representative functional requirements for smart Microgrid A

Requirements area	High-level requirements
R1.1 Interact with Primary Grid	R1.1.1 The system shall be able to provide power to the primary grid R1.1.2 The system shall be able to consume power from the primary grid R1.1.3 The system shall manage financial exchanges with the primary grid based on power provided and consumed
R1.2 Manage Microgrid	R1.2.1 The system shall monitor connected devices and loads R1.2.2 The system shall perform business management for microgrid operations R1.2.3 The system shall control microgrid energy storage R1.2.4 The system shall control microgrid power generation R1.2.5 The system shall support interactions with microgrid customers dealing with electrical service, billing, account management, et al.
R1.3 Support Microgrid	R1.3 The system shall support installation, maintenance, and repair of microgrid equipment and resources
R1.4 Perform Customer Activities	R1.4.1 The system shall process customer bill payments R1.4.2 The system shall support addition and deletion of customer devices R1.4.3 The system shall be able to change customer loads

ently part of the MBSAP architecture methodology and aren't generated directly from the architecture model. These typically deal with program planning and financial information, the system's operational context, rules and other constraints affecting system operations, architecture governance, and potentially other subjects. Examples are sprinkled through this chapter, and Chap. 12 discusses this further in terms of architecture frameworks.

- *System baselines*—Experienced system engineers will ask how this approach corresponds to the familiar Requirements Baseline, Allocated Baseline, and Product or Configuration Baseline. Basically, the OV, LV, and PV are the architectural equivalents and serve essentially the same purposes. However, each Viewpoint is developed and documented using artifacts from a single model, which helps ensure consistency and traceability among the baselines. If required as documentation, the various baselines can be exported from the model and would typically be archived as part of the Conceptual Perspective.

4.11 Smart Microgrid OV

We will use our second system example both to show MBSAP in application to a very different system category and to illustrate some additional modeling techniques. We will primarily examine Smart Microgrid A (SMG A) since it has the scope, complexity, and characteristics we need for the discussion. As a basis for creating an OV for SMG A, we need to establish some requirements. Table 4.3 contains examples of major functional requirements for this community microgrid. We will then go through the Functional, Behavioral, and Data Perspectives, in less depth than for the E-X example, but with emphasis on additional modeling concepts and techniques. We will not create a Services Perspective because, while there are some interactions that could be implemented as services, that is not the conventional approach for these systems. As usual, the Contextual Perspective is a container for relevant information that doesn't fit elsewhere. The architecture modeling process and the sequence of creating model content are substantially the same as for the E-X.

4.11.1 Smart Microgrid Functional Partitioning

Compared to the E-X, a somewhat simpler Domain Diagram results from the partitioning of Smart Microgrid A because there are fewer inherent resource types and functions. Figure 4.18 shows one possible top-level structure. The Domains names are used to identify "levels" to suggest a hierarchy, which extends from the centralized supervisory level at the top, to the Domain holding the various devices making up the microgrid, and finally to the physical level that models the actual generation, flow, and consumption of electric power. Again, we have defined a set of basic User Roles. An Association models the critically important relationship of this microgrid

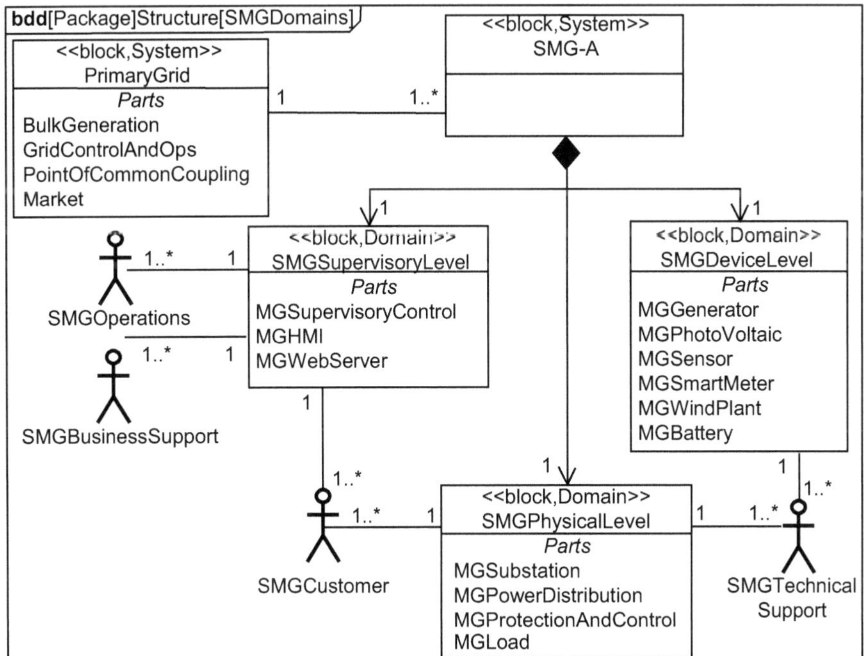

Fig. 4.18 BDD showing the Domains of Smart Microgrid A

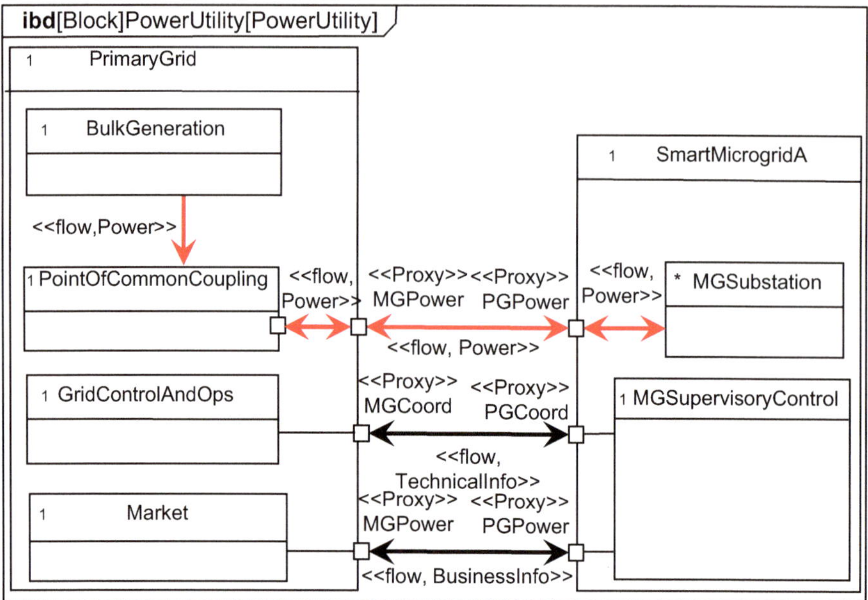

Fig. 4.19 IBD showing the primary interactions between a primary grid and a Smart Microgrid

to the Primary Grid (PG), which will later be refined into Flows of information and electricity. The relevant Parts of the PG are listed. Many of the details shown in this diagram are more fully explained in Chap. 5. An IBD showing interactions within the SMG is part of the Logical/Functional Viewpoint, also in Chap. 5.

Another diagram that is often valuable as part of an OV is one that establishes the context within which the system being architected will operate. One way to do this is shown by the IBD in Fig. 4.19. Here, the Primary Grid and some number of Smart Microgrids are collectively labeled a Power Utility. Nested Part Properties model the elements of the PG and SMG that are responsible for interactions between them. Power flows between the PG Point of Common Coupling and the SMG Substation and can flow in either direction. As a way to emphasize that there are two basically different kinds of exchanges, one of electrical power and the other of information, the Flow arrows for power are colored red. The SMG Supervisory Control is responsible for processing both technical and business information. Technical data such as the status of SMG components is exchanged with the PG Control and Operations Block, while business data such as electricity consumption by customers is exchanged with the Market, ultimately with one or more Service Providers who create the Market.

4.11.2 Smart Microgrid Use Cases and Scenario

The UCs associated with the SMG, shown in Fig. 4.20, mirror the requirements in Table 4.3. They allow us to illustrate some other interesting aspects of behavior modeling in the OV. This diagram includes some Actors that are associated with User Roles and others that model additional external entities. One of the latter is the PG, with which the SMG exchanges power. Another is an Actor labeled Environment, which accounts for the fact that both solar and wind power generation depend heavily on current weather conditions. There is also a Service Provider Actor that models the interaction with the business entities that create the electric power market. A Microgrid Customer Actor accounts for SMG interactions with the ultimate power consumers, and an Interact with Customers UC deals with things like electric service requests, billing for energy usage, and problem reports.

Fig. 4.20 Use Cases for Smart Microgrid A, including interactions with the Primary Grid, customers, companies in the electric power market, and the environment

There are four Tier 1 UCs, and these are decomposed as shown in the diagram. For example, the Interact with Primary Grid behavior has two subordinate UCs for power transfers to and from the PG, plus a Manage Financial Exchange UC that models billing, payments, financial planning, and other business activities. In several cases, a User Role interacts with some but not all of the Tier 2 UCs associated with a given Tier 1 UC. For example, the Microgrid Business Support User Role is associated purely with the business-related UCs.

Another subtlety is shown by the chain of `<<extend>>` dependencies starting with a customer Pay Bill behavior and ultimately leading to the Manage Financial Exchange UC. This reflects the fact that paying for electricity use starts with individual customer payments and works its way up to an aggregated financial transaction between the SMG and the PG. The Manage Financial Exchange UC is also extended by the technical activity of Monitor Devices and Loads. In the full SMG architecture model, each of these UCs has a specification and a scenario Activity Diagram. The primary SMG behavior is modeled as a Sequence Diagram and included in the LV, as discussed in Chap. 5.

The equivalent to the "Day-in-the-Life" scenario shown in Fig. 4.12 for the E-X example is a main control loop that represents the normal behavior of Smart Microgrid A. Figure 4.21 is an AD showing how the Supervisory Controller interacts with the various equipment and devices of the microgrid. Once an electric service connection is established with the PG, the Supervisory Controller initializes its application software and takes any other actions needed to configure the SMG for operations. The main loop starts with polling of SMG devices, and the values received update a current status picture. The Controller checks for alarm conditions; if such are detected, the Process Alarm Action represents a separate behavior that deals with anomalies.

Next, the Compute Power Balance Action determines if SMG power production is such that a surplus can be stored or sold to the PG. There is a fundamental invariant in a system like this that available power generation must always equal the instantaneous load. The fact that this SMG has energy storage, assumed to be a battery, affects this invariant and allows time for the system to react to a change in available power. However, energy discharged from the battery cannot exceed its instantaneous capacity. The Controller determines any needed adjustments to the operations of SMG Generators and the Battery and sends appropriate commands. Power from PV panels is current-controlled. Power can also be sent to the SMG Battery for energy storage, depending on its current state of charge. To smooth customer service in the face of momentary outages, the battery is connected with current control to the Feeder Loop to supply power to loads. The Controller coordinates with the PG Grid Control and Operations center to manage the bilateral exchange of power.

The Controller uses Smart Meter data to measure power usage of retail customers and may implement other functions such as time-of-use billing and emergency load shedding to deal with an outage. This main loop behavior also assumes that the SMG is required to periodically generate reports on its operations and send them to the PG, including the appropriate Service Provider, to regulatory agencies, and per-

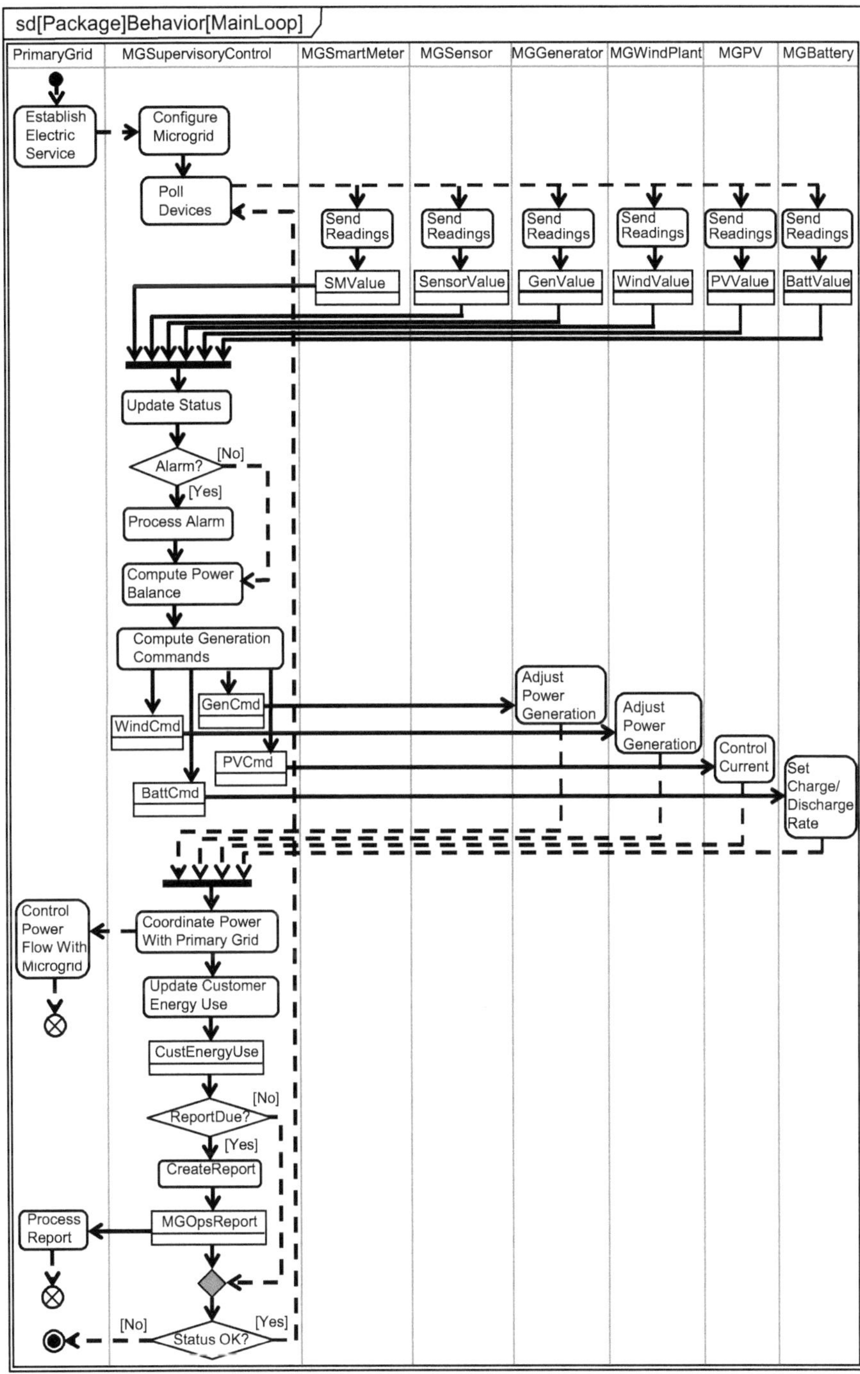

Fig. 4.21 Main control loop for Smart Microgrid A

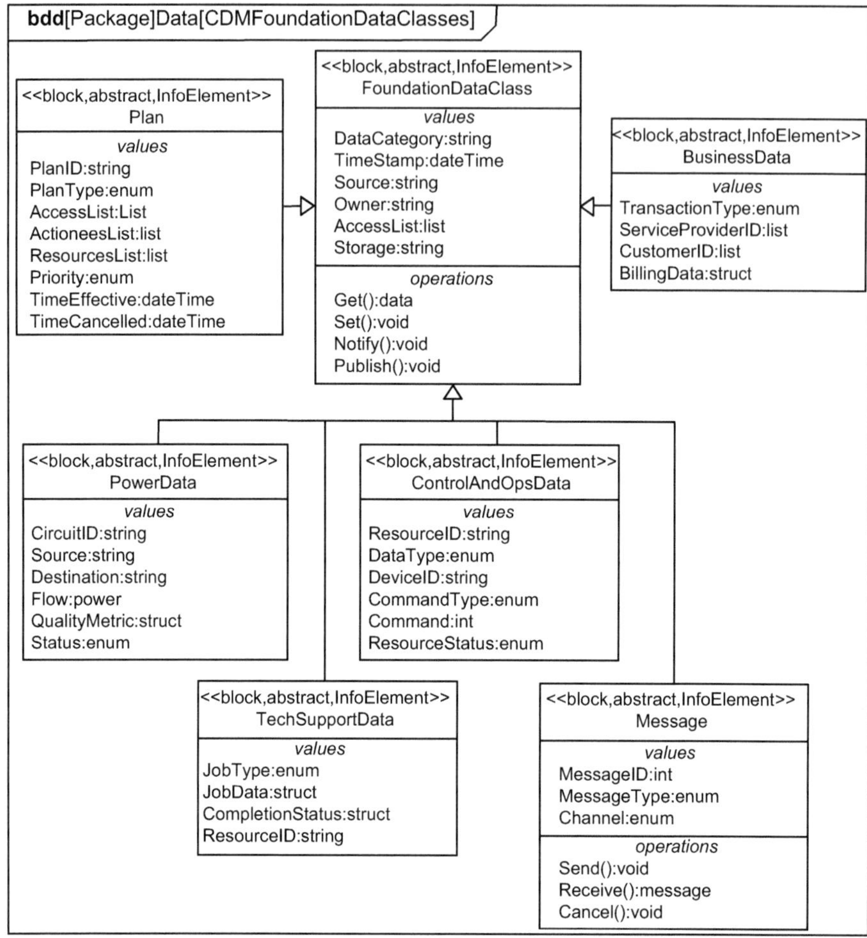

Fig. 4.22 Data Foundation Classes for Smart Microgrid A

haps to others. Assuming all conditions are normal, the main loop closes back to the Poll Devices Action. The loop can execute at whatever rate is needed for optimization of SMG functions. A typical rate of device polling is 1 Hz.

The microgrid Supervisory Controller is a classic example of basic real-time behavior and is typical of many closed loop digital equipment control designs. The controller accepts status and command data, computes and sends equipment control values, performs logging or reporting in accordance with a stored schedule, and repeats its cycle at a constant frequency consistent with equipment operating timelines and overall system deadlines. Like many such controllers, it regularly tracks and reports overall system status and coordinates with conditions in the external environment. In short, the pattern embodied in Fig. 4.21 can be used in a wide variety of real-world equipment control situations.

4.11.3 Smart Microgrid Conceptual Data Model

A second CDM example, this time for Smart Microgrid A, is shown in Fig. 4.22. It highlights some additional features of this early data modeling strategy. Figure 4.22 is remarkably similar to the E-X CDM in Fig. 4.16, showing that even for two such diverse systems, good data structures have many common features. Foundation Classes like Plan and Message are almost identical. The same very basic set of Operations, defined in the root Foundation Data Class, applies in both models. Another key point is that a comparison of Figs. 4.20 and 4.22 shows significant parallelism between the organization of behaviors in a UC taxonomy and of information in a CDM. A Foundation Class for Power Data models the information involved in generation, distribution, and consumption of electricity. Another for Business Data generalizes a set of Data Blocks for billing and payments, both directly for customers (who have their own Foundation Class) and between the SMG and PG. The SMG Technical Support Staff has a dedicated Foundation Class to efficiently handle information about their tasks. This kind of consistency across Perspectives within a model is a good indicator of model quality.

4.12 Executable Architecture

A unified discussion of virtual, as well as physical, prototyping is deferred to Chap. 11, but it's important to recognize that the development and use of simulations starts as part of the OV. The OV interacts with the top two levels of the M&S hierarchy: Operational Modeling and Process Modeling. An early start on these simulations will provide tools that enhance the quality and efficiency of the process that yields the OV. No amount of staring at Domain and Activity Diagrams can replace the insight yielded immediately by a simulation of an executing process model and the statistics it can generate. This is doubly true in the context of discussing the architecture, and the requirements refinement it supports, with the customer.

At the Operational level, a simulation typically models the system as a single entity with the primary characteristics and behaviors involved in its interactions with the external environment. For the E-X, this could include things like the speed and altitude of the aircraft, the field of regard and detection capabilities of various sensors, and communications with other platforms. There may be things like tool-supported scenario generation, digital terrain mapping, static and dynamic targets and other "players," weather effects, and so forth. The goal is to visualize how a system performs and reacts when placed in a realistic operational situation. This can support refinement and validation of requirements and operational concepts. It can also be valuable in presenting the system and exploring alternatives in a way that is familiar and comfortable to operational stakeholders.

One level down the Axis of Abstraction, a Process/Workflow simulation is based on the sequences of actions carried out by the system, interactions with operators,

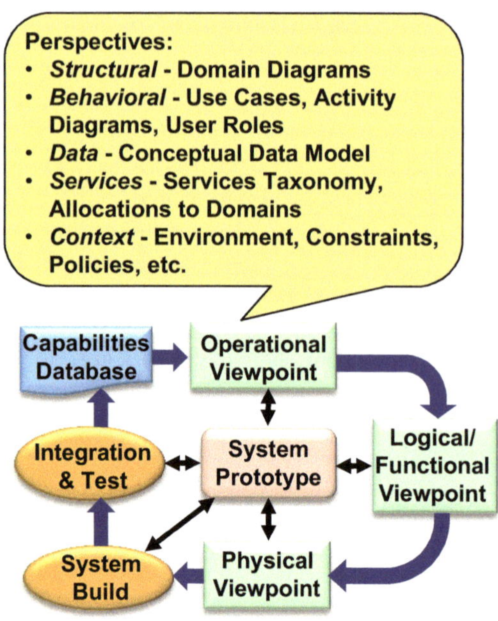

Fig. 4.23 Summary of the Operational Viewpoint

information storage and retrieval, rates at which task queues build up and are serviced, and other aspects of system functioning. This can be helpful in determining the optimum allocation of tasks to human operators, spotting potential bottlenecks in the flow of activities through the system, and surfacing dependencies among system resources and functions that might not be apparent from architecture diagrams alone. Not every architecture development can justify the time and cost of executables at these levels, but when such analysis is appropriate, the payoffs in establishing a firm requirements baseline, getting initial system sizing correct, and validating the structural and behavioral content of the OV can be substantial.

4.13 Summary

This chapter has taken the leap from customer requirements and desires, likely to be expressed in a wide range of forms, to a formal, analyzable architecture context. Figure 4.23 summarizes the content of the OV. Chapter 1 noted that architecture is fundamentally about managing complexity, and the first step is to frame that complexity in a modeling environment that accounts for all its characteristics and allows designers to attack the right problems in the right order, supported by analysis and traceability. The OV provides a top-level functional structure of the enterprise or system, the behaviors each functional area and the enterprise or system as a whole exhibit, and a start on defining information content and services. This stage of MBSAP also brings along an updated requirements database with flowdown and allocation of requirements to whatever level of decomposition the OV reaches as well as clarification of customer requirements and identification of derived requirements. Matching the architecture modeling, the executable dimension of the architecture starts with Operational and Process/Workflow M&S. OV artifacts support customer dialog and provide the point of departure for the design of a mission solution that meets the customer's needs, both initially and for the long term, at the lowest feasible cost and in the shortest feasible time. The next chapter addresses the transition from this initial definition of the system into a functional design in the LV.

Exercises

1. What is the difference between a CONOPS and a performance specification in defining system requirements?
2. Describe the verification approach for performance requirements, general functional requirements, and nonfunctional requirements.
3. Considering a Domain Diagram as the top-level partitioning of a system, describe some of the considerations involved in creating Domains that facilitate an open, efficient, resilient (change-tolerant), and evolvable architecture.
4. Why is a Domain structure modeled in both a BDD and an IBD?
5. Write down a Domain specification for SMG supervisory Level Domain in Fig. 4.17.
6. Write down a Use Case specification for the Control Microgrid Power Generation Use Case in Fig. 4.19.
7. Sketch an Activity Diagram for the scenario of the Control Microgrid Power Generation Use Case in Fig. 4.19.
8. In data modeling, what is the difference between a Foundation Class and a System Class?
9. List some typical content of the Contextual Perspective of the OV for smart Microgrid A.
10. Describe possible Operational and Process/Workflow simulations for the E-X Operational Viewpoint, including the content of the simulation and the primary desired outputs.

Student Projects Students should start their system models by developing the content of the five Perspectives of the OV. Depending on the scope and complexity of each project, it may be in order to "drill down" in only one or two areas to demonstrate the various MBSAP activities. If an architecture modeling tool is used, an essential early activity is to create the model file and to structure it (generally in the tool Browser) in terms of Requirements, Use Cases, Structure, Behavior, Data, Services, and Context, aligning with the MBSAP framework.

References

1. NASA (2007) Systems engineering handbook, NASA/SP-2007-6105
2. Department of Defense (2009) Chairman of the Joint Chiefs of Staff Instruction (CJCSI) 3170.01G. https://acc.dau.mil/CommunityBrowser.aspx?id=28947. Accessed 19 May 2017
3. Suppakul S, Chung L (2005) A UML profile for goal-oriented and use case-driven representation of NFRs and FRs. Paper presented at SERA 05
4. Lattanze AJ (2009) Architecting software intensive systems: a practitioner's guide. CRC Press, New York
5. Kossiakoff A, Sweet W (2003) Systems engineering, principles and practice. Wiley, Hoboken, NJ
6. Department of Energy (2016) The smart grid: an introduction. https://energy.gov/oe/downloads/smart-grid-introduction-0. Accessed 14 Apr 2017
7. Hebner R (2017) The power grid in 2030. IEEE Spectr 54(4):50–55
8. Rozanski N, Woods E (2005) Software systems architecture: working with stakeholders using viewpoints and perspectives. Addison-Wesley, New York
9. Jacobson I, Christenson M, Jonsson P, Overgaard G (1992) Object-oriented software engineering: a use case driven approach. Addison-Wesley, Reading, MA
10. NATO (2009) Standardization Agreement (STANAG) 5255
11. Date CJ (1981) An introduction to database systems, 3rd edn. Addison-Wesley, Reading, MA

Chapter 5
Designing in a Logical/Functional Viewpoint

5.1 Foundation for Design

In the Operational Viewpoint (OV), we carefully kept the architecture abstract, independent of any particular design, and focused on overall structure, behaviors, and information content, including the introduction of services as a model for interactions. The Logical/Functional Viewpoint (LV)[1] starts the process of designing a solution to meet customer requirements. This involves decomposing the structure, behaviors, and high-level information entities (Foundation Classes) established in the OV to arrive at individual Blocks or groups of Blocks that will ultimately be instantiated as the actual components of the delivered system.[2] The LV does not address specific technologies and products, although it takes into consideration practical concerns such as the state of the art in available products that will affect implementation in the Physical Viewpoint. As a result, LV artifacts represent the system design in a functional, technology-agnostic form.

Over time, as the relentless pace of technology improvements and inevitable changes in requirements impact the system over its lifetime, the LV provides the starting point for making required updates. Taking advantage of modularity, loose coupling, and open standards, the LV facilitates a process of technology refreshment by allowing selected system components to be re-implemented with additional or improved products. This minimizes a common dilemma with previous SE methods in which the only way to deal with obsolete and no-longer-available products, as well as major changes in requirements, has been to reverse engineer the original design and start over, which, even if it can be done, is expensive. Architecture is thus the key not only to an efficient and affordable initial solution to customer needs but also to long-term supportability and evolvability under the stresses of the real world,

[1] For present purposes, "Logical" and "Functional" are equivalent, and both are in common use; we use the abbreviation LV for conciseness.

[2] Continuing the convention of the previous chapter, we use "systems" generically to also refer to enterprises, nodes, subsystems, or any other levels of the Axis of Organization.

J. M. Borky, T. H. Bradley, *Effective Model-Based Systems Engineering*,
https://doi.org/10.1007/978-3-319-95669-5_5

a central goal of agile SE. MBSAP seeks to deliver systems that will remain interoperable, maintainable, and operationally capable over many generations of technology advancement and requirements evolution.

The five Perspectives that are defined first in the OV (Sect. 3.4) carry forward into the LV, where they are further analyzed to develop the artifacts that will serve as the specifications used to design or procure the actual system components in the Physical Viewpoint (PV). In the course of this, it's common to discover a need for revisions to the OV to implement refinements and fix problems. The functional architecture that originates in the OV is documented in the LV using SysML diagrams and associated model element specifications. There may also be supplemental artifacts for layered infrastructures, customer-specified frameworks and models, and other design aspects. MBSAP maps functions, behaviors, and services from the OV to operations of individual Blocks and to cooperative behavior among system elements that work together to realize required functionality. In Sect. 5.5, we will explore how the Conceptual Data Model (CDM) forms the basis for creating a Logical Data Model (LDM). The LDM defines actual data entities in the system with their and Operations. We also capture the messages exchanged internally and with external Actors.

As with every stage of architecture development, the architect must constantly apply judgment, dealing here with matters like the appropriate level of functional decomposition, the effort put into defining the characteristics of Blocks, and the types of executable models that should be developed. A particularly important outcome of LV analysis is to identify areas within the architecture where questions about throughput, precision, simultaneity, resource loading, timing, and similar performance issues call for physical modeling to ensure requirements are satisfied with adequate margins. Once the LV is modeled and analyzed, the resulting specifications and other artifacts provide a solid foundation on which to build the system.

5.2 Design Patterns

In developing the LV, we place special emphasis on design patterns. These are proven solutions to commonly encountered design problems, and many books have been written on design pattern theory and applications [1, 2]. We do not attempt a full treatment of this complex subject, and some readers will find it useful to explore the theory and use of patterns in greater depth. This chapter illustrates existing, widely used design patterns and extends them with others that have proved useful.

Design patterns are powerful tools for a number of reasons.

- They establish a common vocabulary based on commonly understood terminology in a field notorious for the propensity of various communities to use the same name for different things and different names for the same thing; patterns make designs easier to understand and communicate.

- They often allow the architect to save time in arriving at a good design and to profit from lessons learned in earlier work that have been used to develop new patterns and refine existing ones.
- They assist in achieving system qualification and showing compliance with requirements.
- They promote a consistent approach to basic architectural tasks like defining interfaces, and they use OOD principles.
- They are important in deriving and documenting Reference Architectures (RAs) as discussed in Chap. 12, including providing a convenient way to capture new lessons learned from ongoing projects by adding content to existing RAs.

MBSAP uses the following three types of patterns.

- *Core or Foundational Patterns*—these are drawn from the literature and widely used in information systems. The discussion begins with the Foundational Patterns defined in the classic work on software design patterns by Gamma et al. [1]. These cover a surprisingly high fraction of the Block combinations, both hardware and software, encountered in complex systems, as this chapter illustrates. These patterns can be divided into three categories:
  - *Creational Patterns*—these provide the basis for the actual Blocks used in a system by abstracting commonly required information content and specific or small-scale behaviors.
  - *Structural Patterns*—these describe the ways Blocks are composed into larger structures.
  - *Behavioral Patterns*—these focus on algorithms, assignment of responsibilities or roles among a group of Blocks, and the way they interact to deliver large-scale behaviors.
- *Applied Patterns*—these are derived from experience with real-world systems, and they document solutions to problems commonly encountered in system categories like those addressed in this book. Applied Patterns are translated into SysML models as Block collaborations,[3] each consisting of a group of Blocks connected in a certain way. In general, they have emerged as good designs for performing resource management, dynamic planning, information dissemination, decision support, and other challenging functions.
- *Foundation Classes*—as introduced in the previous chapter, these are data entities created in the Conceptual Data Model that define the fundamental information content and resources of the system and its environment. They establish the characteristics required in every instance of a given category of system information content and resources, promoting consistency and enforcing system-wide policies for things like information security.

[3] This use of the term "collaboration" should not be confused with the UML Collaboration Diagram, which is a behavior diagram. It was retitled a Communication Diagram in UML Version 2 and is not used in MBSAP.

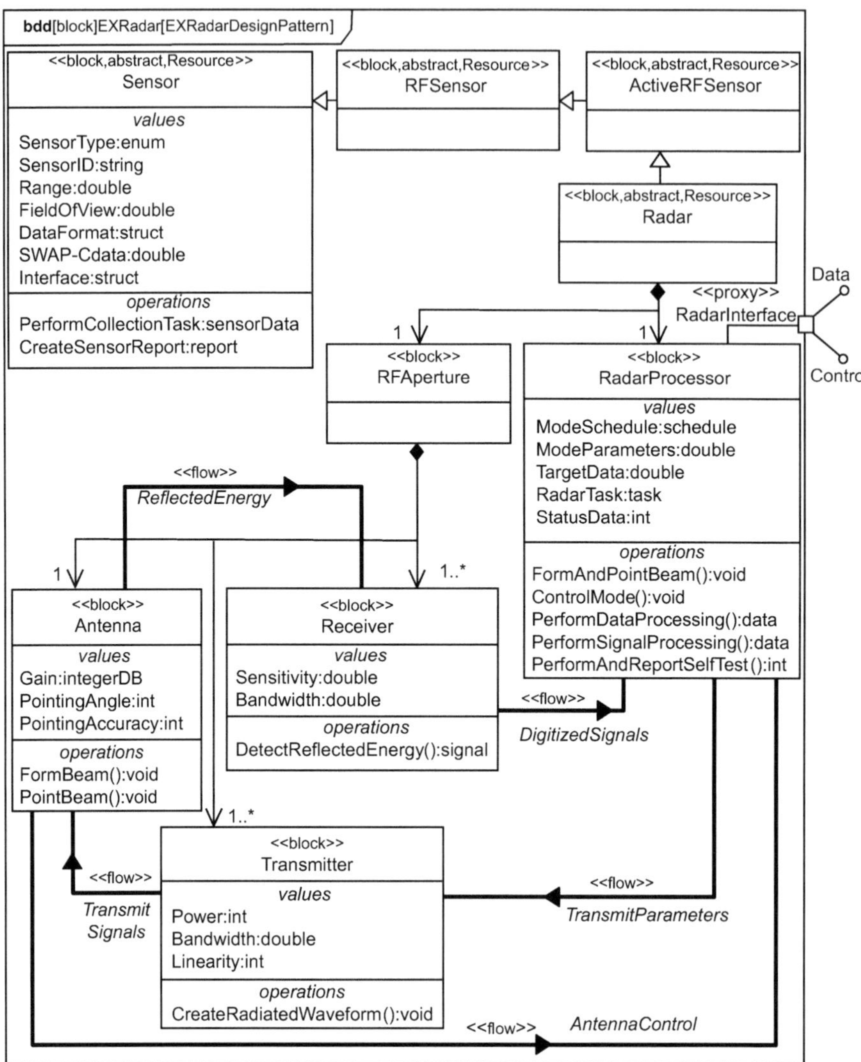

Fig. 5.1 Applied design pattern for a radar with inheritance from abstract sensor types

To illustrate these concepts, the multifunction radar for the E-X gives us an opportunity to examine an Applied Design Pattern. Figure 5.1 is a BDD that starts with a completely abstract Foundation Resource Class called Sensor and shows successive inheritance down to a generic radar. Next, the design pattern, drawn from hundreds of actual designs [3, 4], breaks the radar down into two broad types of resources: a radio frequency (RF) Aperture with the elements that handle RF signals and energy and a Radar Processor that is a specialized computer performing digital signal and data processing and supporting the primary interface through which the

Fig. 5.2 Examples of gimbaled (mechanically pointed) and phased array (electronically pointed) radar antennas

radar interacts with users and perhaps other systems. In current technology, a radar RF Aperture will generally fall into one of the following two categories (Fig. 5.2).

- Originally, radars used a mechanical antenna structure designed so that when stimulated by energy from a transmitter, it reflects that energy into space as a narrow beam. Working inversely, the antenna then collects energy reflected by a target and directs it into the "plumbing" that leads to a receiver. Such an antenna is often pointed using a gimbal or, alternatively, fixed in a particular direction of interest. Variations on this basic approach include a bistatic radar that uses separate antennas to radiate and receive.
- More recently, many radars employ phased array antennas consisting of multiple antenna elements and using adjustment of the phase of the signals transmitted or detected by each element to synthesize a beam with the desired shape and direction. When transmit/receive (T/R) elements are used, such an aperture is called an Active Electronically Steered Antenna (AESA). Passive (receive only) arrays are also sometimes used, and an array can be mounted on a gimbal to increase the range of possible beam pointing directions.

This is shown in the design pattern of Fig. 5.1. The RF Aperture consists of an antenna structure, a receiver, and a transmitter. The latter two can be implemented by discrete receiver and transmitter units or by some number of transmit, receive, or T/R modules, indicated by the 1..* multiplicities in the diagram. In a phased array, the antenna structure provides the mounting for the RF modules along with connections for RF, control, and other signals and for power and cooling. A passive RF sensor would have a very similar pattern, omitting the transmitter function. We can continue this pattern to any desired level of detail, e.g., by including provisions to generate and distribute prime power and to cool the aperture and other electronics.

The value of such a design pattern lies in helping engineers concerned with specifying, designing, or selecting a radar to account for the important design alternatives, to structure trade studies, and to compare the current radar requirements against successful (or unsuccessful) solutions to similar needs in earlier systems.

For example, it might be determined that an existing radar processor, perhaps with some software modifications, can meet the needs of the new system, while a new or modified RF aperture is needed to satisfy performance or reliability requirements. In the case of the E-X, the multifunction radar could be designed using a gimbaled mechanical antenna, a passive-phased array with a single transmitter aperture, or an AESA. In practice, many such systems use an active array mounted on a pivot so that it can point at the ground on either side of the aircraft. This approach to Applied Design Patterns can be applied to virtually any commonly used system elements, including sensors, processors, communications devices, equipment control and monitoring, and operator workstations.

5.3 Structural Perspective: E-X Structural Decomposition

LV analysis begins with the Domain Diagram from the OV and considers how the functionality in each Domain can best be realized with system elements. Initially, it's appropriate to defer most decisions about allocating functions to hardware or software, about choices of specific products, and about which standards to apply. The goal is to identify the parts of the system, how they interact, and what they must do individually and collectively. In general, functions are allocated to system elements (Blocks, Parts, etc.), and inputs and outputs are allocated to interfaces. This approach is well matched to Object-Oriented Design, because the first principle in defining Blocks is to discover the entities comprising the system being architected, define their associations and characteristics, and focus on their interactions via well-defined interfaces.

Although most physical design decisions come later, a good LV is influenced by the practical realities of implementing the design in hardware and software. In decomposing Domains, the objective is to arrive at Blocks whose complexity and functionality are appropriate for eventual realization using feasible and available components. A sense of the types and capabilities of products and other system elements that can be achieved in the current state-of-the-art is important to arriving at a system structure that supports an effective and affordable solution. Often, two levels of Domain decomposition are involved, first into Subdomains and then into Blocks, but this is only a guideline. As always, judgment and experience, including the application of design patterns, are called for.

To illustrate the approach, we break out the Sensor Management Domain of the E-X System from Figs. 4.6 and 4.7 into a set of five Subdomains: Sensor System Interface, Sensor Plans Management, Sensor Resource Management, Sensor Analysis and Reporting, and Sensor Resource Interface. This is, in fact, an Applied Design Pattern because it deals with a set of related behaviors that are found in many systems that rely on a hierarchy of functions and User Roles. In the E-X, these system functions come down to being given something to do and resources to do it with, plus some rules or constraints, all centered on gathering, processing, and using information. Figures 5.3 and 5.4 show the decomposition and internal interactions

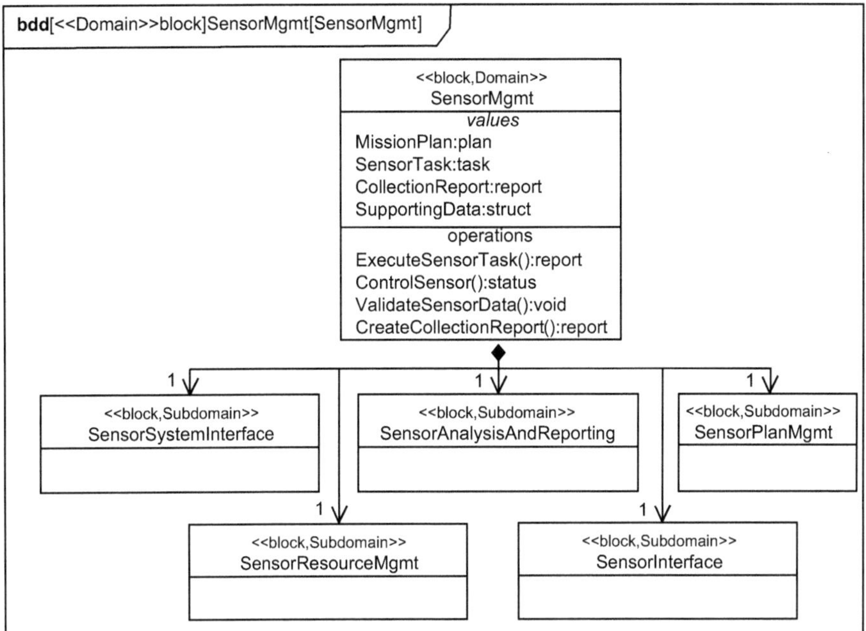

Fig. 5.3 Decomposition of the Sensor Management Domain

of the Domain using a SysML BDD and IBD. As usual in this decomposition process, a BDD is created first to define these lower-level Blocks. The Subdomains become Part Properties of the Sensor Management Domain Block. Ports on the boundary of the Domain support interactions with other Domains.

The following paragraphs further describe the Subdomains to show how the overall design pattern is employed in accomplishing the specific tasks of the Sensor Management Domain.

- *Sensor System Interface*—responsible for interacting with system information resources (the Information Management Domain) and the ISR Management Domain. This provides a stable interface to the rest of the system that is independent of the implementation of other Subdomain functions and manages the flow of information in and out of the Sensor Management Domain. It also provides the logic to drive the Sensor Operator interface through the Information Management Domain. The HMI processing associated with controlling sensors is placed here because the majority of the information needed by the Sensor Operator is held here and any other necessary information can be readily accessed. The actual operator workstation on which the HMI is implemented is part of the Computing Infrastructure Domain.
- *Sensor Plans Management*—accepts and stores plans from the ISR Management Domain, formulates and provides status of plan execution, performs feasibility testing and other processing of plan contents, and translates collection tasks from plans into sensor tasks to be passed to Sensor Resource Management.

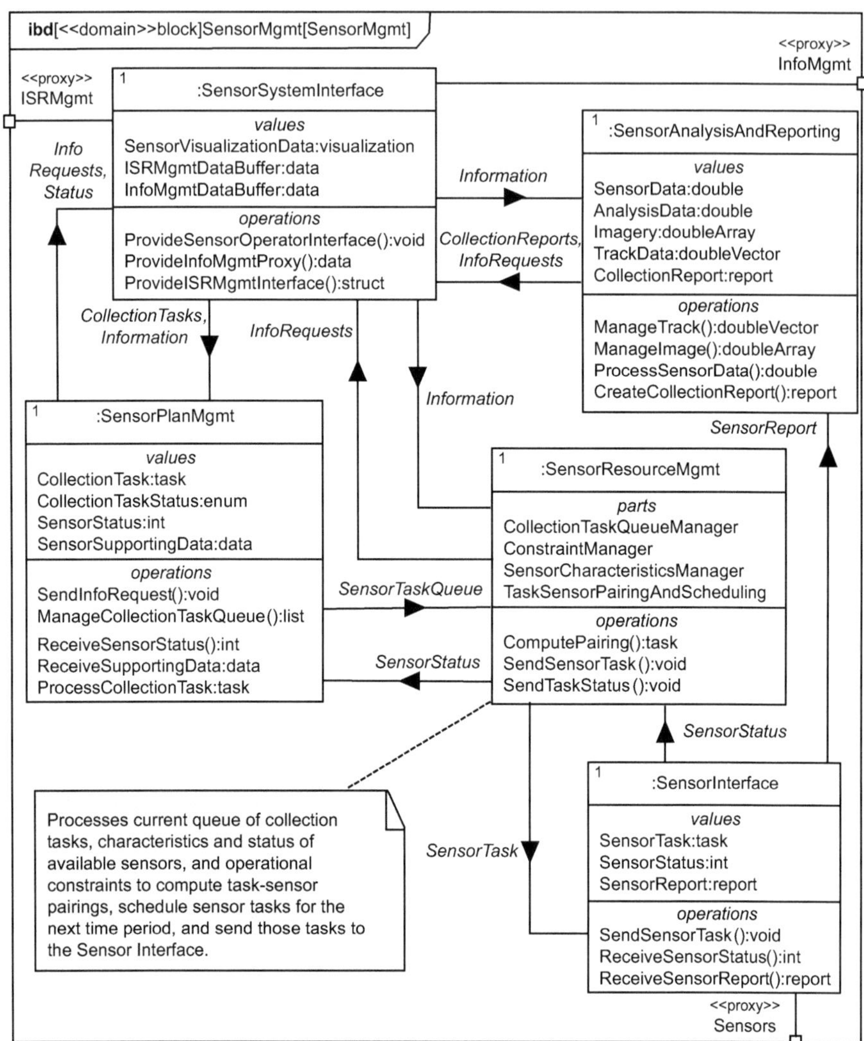

Fig. 5.4 Internal information flows within the Sensor Management Domain

- *Sensor Resource Management*—pairs sensing tasks from the plans manager with available sensor resources, including scheduling time slots when collections can be performed. The resource manager issues Sensor Tasks (also called Sensor Service Requests (SSRs)) and monitors the health of the sensors and the status of task execution. Later paragraphs continue the analysis of this Subdomain to show the next level of decomposition into Blocks. By abstracting sensor control in this way, control logic can be employed that is tailored to the specific sensors involved each time the pattern is instantiated in a system. This helps to keep the design pattern completely general.

- *Sensor Analysis and Reporting*—processes sensor data into the reports and other information products needed by Sensor Operators and the rest of the system. Like the Resource Manager, this Subdomain abstracts logic appropriate to handling data from the particular sensors being used so that the design pattern is valid across all equivalent design situations. The E-X radar might output already-formed images, and this Subdomain would do things like formatting and time-tagging individual images to create reports. Alternatively, the radar might simply produce digitized signals, in which case the Sensor Analysis and Reporting Subdomain would do the computations to create the images.
- *Sensor Resource Interface*—abstracts the specific interface associated with control of the E-X sensors. This functionality could be combined into the Sensor System Interface Subdomain, but partitioning the Subdomain this way allows the designer to focus on the details of interacting with a specific set of sensors and to deal with the fact that these control and data exchanges are typically more time-sensitive than general system information flows.

These Subdomain summaries are examples of responsibility statements. It's good OOD practice in general to explicitly associate with the elements of a system design concise statements about what they are expected to do. In later stages of system development and integration, it can be all too easy to forget this aspect of the thinking that went into the early discovery phase of design. Responsibility statements are important for Blocks as well. Statements of this kind are often used as an overall description at the start of a model element specification.

The partitioning rationale for the E-X OV included the realization that there is a group of closely related activities and data associated with receiving and managing information collection tasks, sending specific commands to sensors, turning the resulting sensor data into reports that other Domains can use, and monitoring the operation, health, and status of sensors and associated processing resources to ensure satisfactory performance. This kind of specialized workload in the Sensor Management Domain merits designation of a crew User Role, titled Sensor Operator. Typically, a set of tasks is formulated as a plan. The current state of system resources and the status of plan execution are reported using status messages. Information entities received from and provided to other Domains and Actors are modeled as data items. Plan, Message, and Data Item are examples of the generalized information constructs that are defined as Foundation Classes in the CDM in Fig. 4.15, along with Report, Task, and Training Data. In Sect. 5.5, we will look at additional examples of deriving system data items from Foundation Classes.

The next level of decomposition involves realizing these Subdomains with Blocks and other structural elements. At this point, system elements can often be identified that will ultimately be implemented as hardware or software components. In transforming the LV to the PV, these may be either procured as available products or developed as new products, depending on the results of "make-vs.-buy" decisions in design analysis. In fact, they will commonly be identified as configuration items in the course of developing the PV. A rule of thumb is to decompose to a level of complexity at which a designer or small team can be given the job of implementing

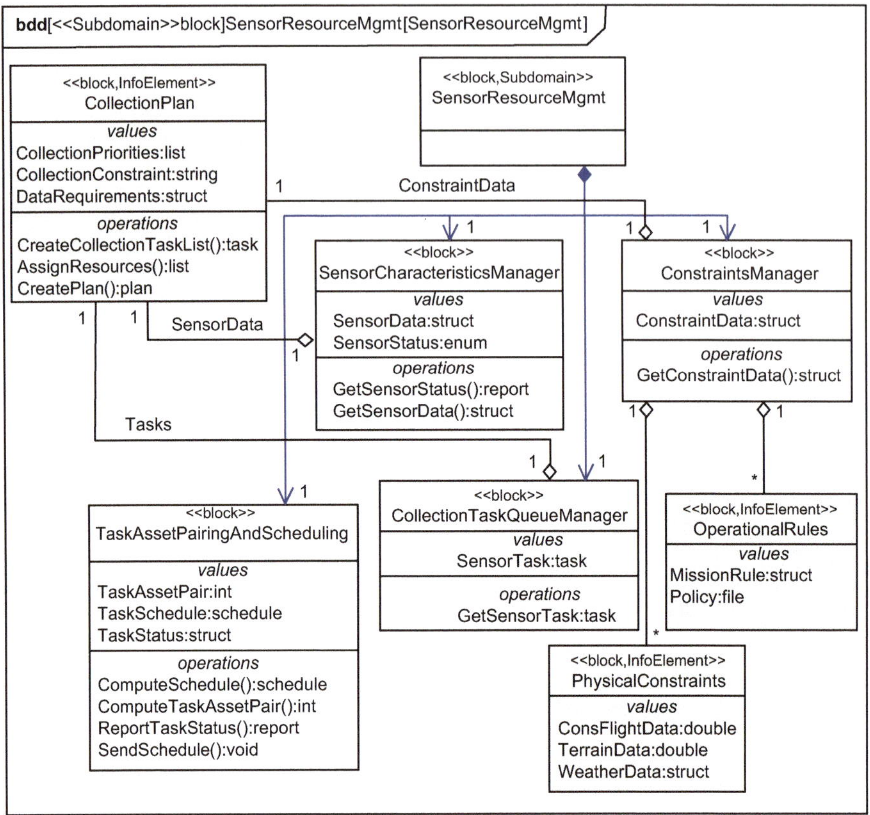

Fig. 5.5 Internal structure of the Sensor Resource Manager Domain with Reference Properties showing external information inputs

a Block or group of Blocks in hardware and software. This means that in some cases further decomposition will be needed. As the architects discover and define Blocks, they look for additional Foundation Classes, including opportunities to apply those that have been created in previous programs or are available in Reference Architectures (RAs), which are discussed in Chap. 12.

Figures 5.5 and 5.6 show the Resource Manager design pattern in action as part of the Sensor Management Domain. This pattern is suitable for systems that must meet "ilities" requirements such as Adaptability and Reconfigurability. There are three Blocks whose function is to hold the principal data inputs to task/resource pairing. One manages the current queue of sensor collection tasks, another tracks the currently available sensor assets and their status, and the third stores applicable constraints such as terrain blockage that could affect the feasibility of executing a given task with a given sensor. Each of these has the Collection Plan from the Sensor Plans Management Subdomain as a Reference Property as shown by the lines with open diamond heads. The nature of the reference is indicated for each on the diagram.

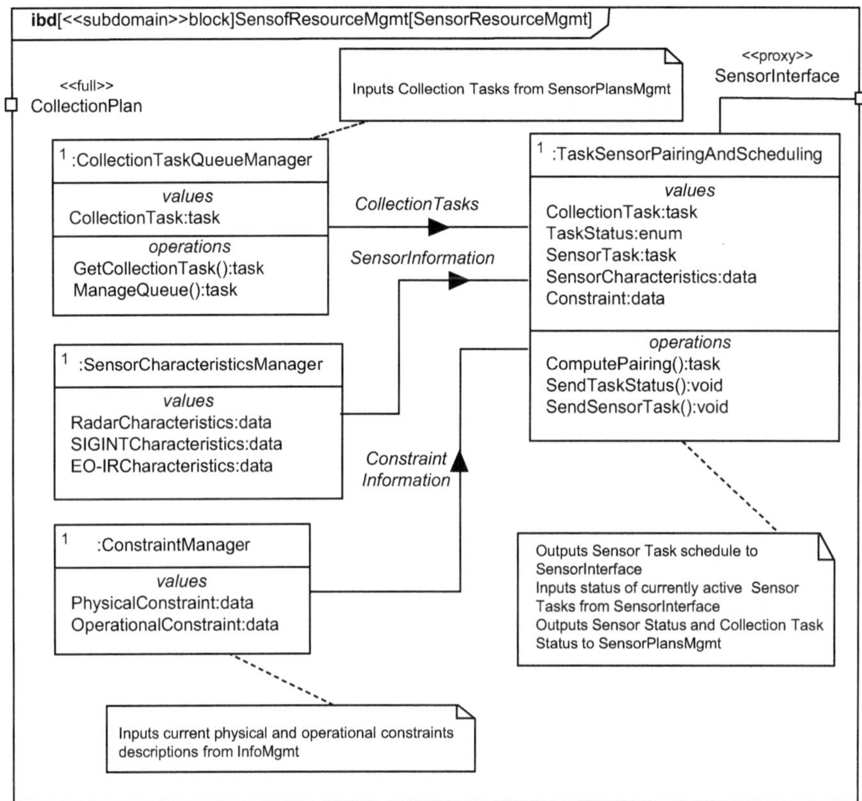

Fig. 5.6 Internal information flows of the Sensor Resource Management Subdomain

These three plus the Task Asset Pairing and Scheduling Block are the Parts of the Sensor Resource Management Subdomain, and these Blocks form a compositional hierarchy for the Subdomain. The Collection Constraints Block also has Reference Properties in the form of two additional Information Elements. Operational Rules include overall policy restrictions as well as specific rules applicable to an E-X mission. Physical Constraints represent information about conditions in the external environment that affect the system such as terrain and weather, along with information about the aircraft position and flight path. Information about the earth itself is often referred to as Geospatial Information System (GIS) Data. The Blocks modeling the various external information sources (References) are shaded.

The Task/Asset Pairing and Scheduling Part is primarily concerned with using the inputs from the data Blocks to develop an optimum set of task/sensor pairings. It then formulates and sends the resulting Sensor Tasks and keeps track of task execution via status reports fed back by the sensor. This process is cyclic and continuously adjusts the schedule as new tasks are received, existing tasks are executed or reassigned, and various constraints become active or inactive. Each planning cycle results in a set of Sensor Tasks to be sent to the Sensors Domain in the next time interval of the sensor schedule.

The IBD in Fig. 5.6 shows the data input flows into the task asset pairing logic. Communications with the Sensors Domain go through the Sensor Interface Subdomain through a<<proxy>> Port that is delegated to the Task/Asset Pairing and Scheduling Part. Communications with the Sensor Plans Management Subdomain use a <<full>> Port that talks to the Collection Plan. As with any Full Port, this will require that the Sensor Resource Management Subdomain has the means to implement the Port, in this case presumably a software component that manages the information exchanges and distributes the appropriate content to the three data Blocks. The reason for using a Full Port here is that none of the Parts of the Sensor Resource Management Subdomain has sole responsibility for supporting this Port and its interface. Instead, the Port supports the entire Subdomain, and it models additional resources that the Subdomain must provide to create it. As discussed in Chap. 2, a Full Port is a kind of Part Property of the Block on which it is declared, and it is distinct from the other Parts of the Block. These details should be spelled out in the specification for the Port.

This example illustrates several important points. One is the fact that Blocks often fall into three general types: (1) those that primarily model information entities and have relatively few Operations (referred to as "data" Blocks), (2) those that primarily model processing logic or other behaviors with only the Attributes/Values necessary to define that behavior (referred to as "computational" or "algorithmic" Blocks), and (3) those that model noninformation components such as electrical, mechanical, or hydraulic equipment (referred to as "mechanical" Blocks). The Sensor Task Queue, Sensor Assets, and Collection Constraints Classes are examples of the first type, while Task Asset Pairing and Scheduling is clearly a computational Block. A Block that has both extensive Values and many Operations may be too complex for direct implementation and should be examined for further decomposition.

This resource management example can be elaborated in a number of additional ways. For example, the Collection Plan information element is a specialization of the Plan Foundation Class and inherits a set of generic Values and Operations in addition to the specific characteristics shown in Fig. 5.5. The Operation of the Task Asset Pairing and Scheduling Block that is named Compute Task Asset Pairing could be implemented with a variety of algorithms such as a rule base, a neural network, a fuzzy logic engine, or some other mathematical form selected as best suited to the decisions involved. Regardless of the details, the Resource Manager in Fig. 5.6 is an Applied Design Pattern that offers a solution to a very general class of problems. In the E-X architecture another likely application of the Resource Manager pattern is in the Communications Management Domain where the task is to dynamically pair message traffic with available communications channels.[4]

A complete Structural Perspective of the LV for the E-X requires decomposition of all Domains and Subdomains across the architecture, using the techniques presented in this section. The overall partitioning of the E-X architecture that was

[4] This and other Applied Design Patterns in this chapter were developed by the authors and their colleagues and are being published openly here for the first time.

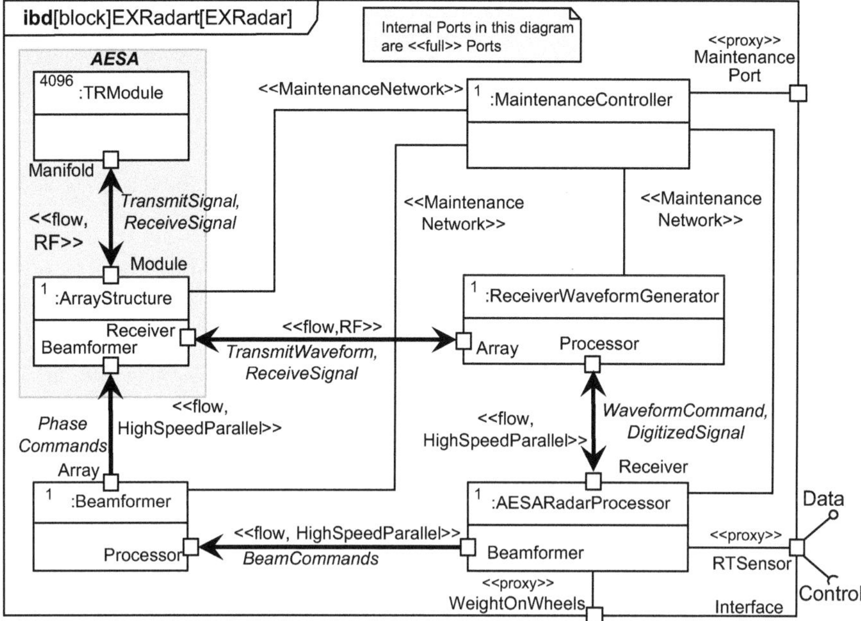

Fig. 5.7 Basic design of an Active Electronically Scanned Antenna (AESA) radar for the E-X

described in Chap. 4 is such that Mission Management, ISR Management, and Sensor Management Domains will probably be realized as software. The Information Management and Computing Infrastructure Domains provide the hardware and middleware that create the computing platform on which these software components will execute. The resulting layered architecture is further discussed in a later section of this chapter. Once all Blocks are fully defined in functional (or logical) terms, the architecture will produce the specifications that can be used to implement them using hardware and software components.

A final structural example takes the discussion into the real-time Sensors Domain and uses the Radar Applied Design Pattern from Fig. 5.1. The goal here is to show that the same basic architecting techniques apply whether the area being modeled involves information processing, sensing of the environment, machinery controls, or many other kinds of systems. Figure 5.7 shows a somewhat simplified version of the E-X radar in an IBD. It's assumed that the E-X uses an AESA with 4096 transmit/receive modules, as indicated by the multiplicity of the Transmit/Receive (TR) Module Part in the figure. The following are the major elements.

- *T/R Modules*—RF components that radiate and receive energy with controlled phase.
- *Array Structure*—mechanical assembly that mounts the modules and distributes signals via an RF "manifold," as well as power and cooling, completing the AESA.

- *Beam Former*—specialized electronics unit that converts a commanded beam pointing command into the individual phase commands for the T/R modules.
- *AESA Radar Processor*—the radar's embedded computer that performs a variety of functions associated with carrying out scheduled sensor tasks. These include:
  - Computing beam commands
  - Selecting or computing the parameters of the waveform to be modulated onto the transmitted signal
  - Digital signal processing for detected target reflections
  - Digital data processing to determine target locations and movements (MTI mode)
  - Creating images of areas on the ground (SAR mode)
  - Managing interactions with the Sensor Management Domain via an RF Sensor Interface
- *Receiver-Waveform Generator*—specialized electronics unit that generates the actual waveform to be applied to the transmitted energy during a specific time interval and performs detection and initial signal processing of reflected target signals.
- *Maintenance Controller*—specialized electronics unit that monitors the health of the radar and performs fault detection and other diagnostics, which are reported through a Maintenance Port.

To ensure safety, a signal from the Weight on Wheels switch inhibits the radar from transmitting when the aircraft is on the ground. Many of these elements are obviously "mechanical" Blocks. Proxy Ports are defined for the main Radar Interface, the Maintenance Controller, and the Weight on Wheels discrete signal. In summary, the E-X example lets us look in some detail at the way MBSAP tackles the definition and structural decomposition of a complex system. Along the way, design patterns allow us to profit from successful prior experience. Although every system architecture is unique, the overall MBSAP approach illustrated by the E-X applies to most situations.

5.4 Behavioral Perspective: E-X Behaviors

Whereas the Structural Perspective sought to define the allocation of functions to systems elements, and of I/O to interfaces, the Behavioral Perspective defines the sequences of activities that are carried out by individual Blocks and groups of Blocks, including alternatives that may be triggered by events, conditions arising in the system and its environment, or information exchanges. If Subdomains correspond to actual subsystems and are treated as LV elements, their behaviors must be modeled also. There may be interesting behavior within a Domain that is easiest to see and understand when portrayed as a series of actions and exchanges among Subdomains like those illustrated under the Structural Perspective. Such behavior might arise in conjunction with decomposition of system-level Use Cases to discover one or more subordinate Use Cases that happen entirely within particular Domains. Looking back to Fig. 4.9, the Perform Mission Management Use Case is such an example; the behavior is implemented entirely within the Mission Management Domain. In a situation like this, the same kind of Activity Diagram used in Chap. 4 is often a good modeling approach, with Subdomains at the heads of the swim lanes.

More commonly, the interesting behaviors in the LV involve individual Blocks or groups ("collaborations") of Blocks. These behaviors are generally either stateful or non-stateful. A state is a distinct condition of a system or system element in which its parameters have certain values and in which it satisfies constraints, performs particular actions, and responds in a certain way to events. As discussed in Chap. 2, stateful behavior is driven by both the data and inputs that are currently being acted on and the history of past activities as reflected in the current state of a Block, collaboration, Domain, or System. Depending on the state, quite different behaviors may be triggered in response to an event such as the arrival of a message or notification of an occurrence elsewhere in the system or its environment. Non-stateful behavior, by contrast, depends only slightly or not at all on history and tends to follow a given flow regardless of the circumstances, perhaps with some intermediate decision points. SysML shows stateful behavior using State Machine Diagrams (SMDs) and normally uses Sequence Diagrams (SDs) for non-stateful behavior.

We will illustrate this Perspective with representative behaviors in the E-X example. First, we assume that the Communications Management Domain from Fig. 4.6 has been decomposed and a Foundation Resource Class has been discovered that generalizes (abstracts) all of the system's communications device controllers. This Foundation Class will be specialized by inheritance for the various communications components on the E-X. One of these is a Wideband Satellite Communications Controller, modeled as a WB SATCOM Controller Block, which manages the Wideband SATCOM Terminal. The controller creates the interface required to talk to the actual terminal and passes the required control data and messages. Figure 5.8 is a BDD showing the important relationships for this example. The Foundation Class is stereotyped as `<<abstract, Resource, Controller>>`, and the WB SATCOM Controller specializes it with specific Values and Operations. A Comms Resource Management Subdomain dynamically allocates traffic to this and other channels. A Comms Message Management Subdomain passes the allocated messages to and from each communication device. The controller is shown in the diagram with `<<Usage>>` Dependencies on these Blocks.

This Wideband SATCOM Controller provides an illustration of stateful behavior based on the need to establish and maintain a link to the wideband communications satellite. The Block is in one State when a valid link is in operation and in another State when the link is not available. We will model how the Controller reacts to events like the arrival of new messages depending on which State it currently occupies.

The wideband SATCOM terminal must establish a link and synchronize timing with the satellite before it can pass message traffic. If the link is lost, the first priority

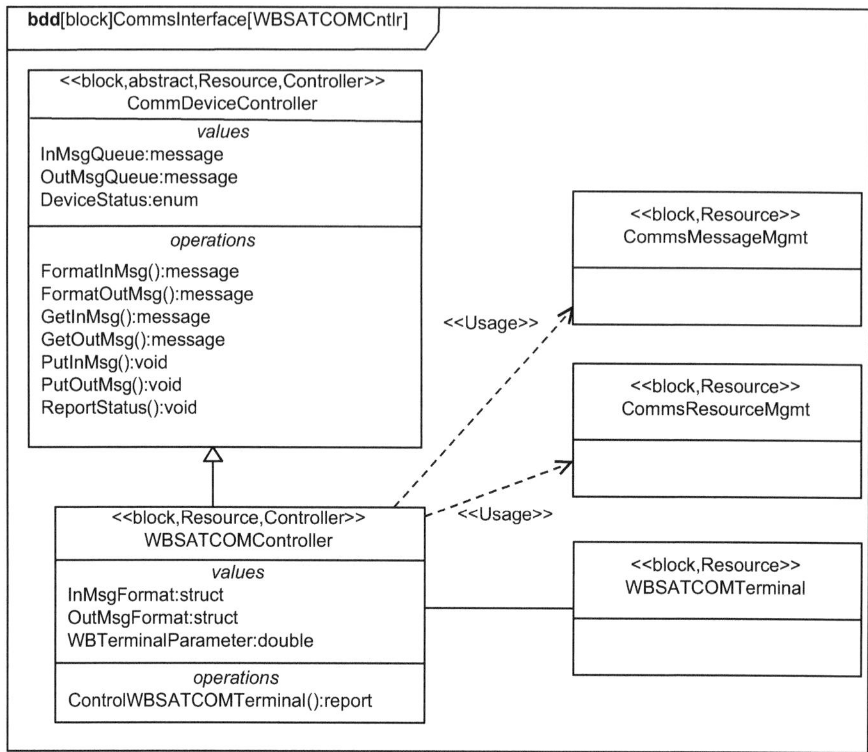

Fig. 5.8 BDD for the Wideband SATCOM Controller

is to re-establish it. The WB SATCOM Controller can therefore actually be in one of three States as shown in Fig. 5.9. At system startup, and whenever there are no messages to be handled, the Class will be in an Idle State. Every time the Class enters this State, the terminal is commanded to perform a self-test, and the resulting status is reported to the Comms Resource Manager. A Link Flag variable is set to true when the terminal is synchronized and the link is active, based on a discrete signal from the terminal, and false otherwise. A New Message event is issued when an incoming or outgoing message is sent to the controller and Comms Message Manager, respectively. In order to track the status of message traffic, Message Queue Flag variables are true if any messages are waiting to be forwarded and false otherwise, in which case the controller returns to the Idle State.

With these assumptions and definitions, the logic in the SMD in Fig. 5.9 is straightforward. Whenever the link is lost, the Block enters the Establish Link state and cycles there until the link is restored, at which point a transition to the Active state or back to Idle is triggered, depending on whether a message is pending. If the Block is in the Idle state and the link is up, a New Message event triggers the transition to the Active state, and the message is sent to the corresponding interface. If the message queue becomes empty, the flag is set to false and the class transitions back to the Idle state to self-test the terminal and await the next message.

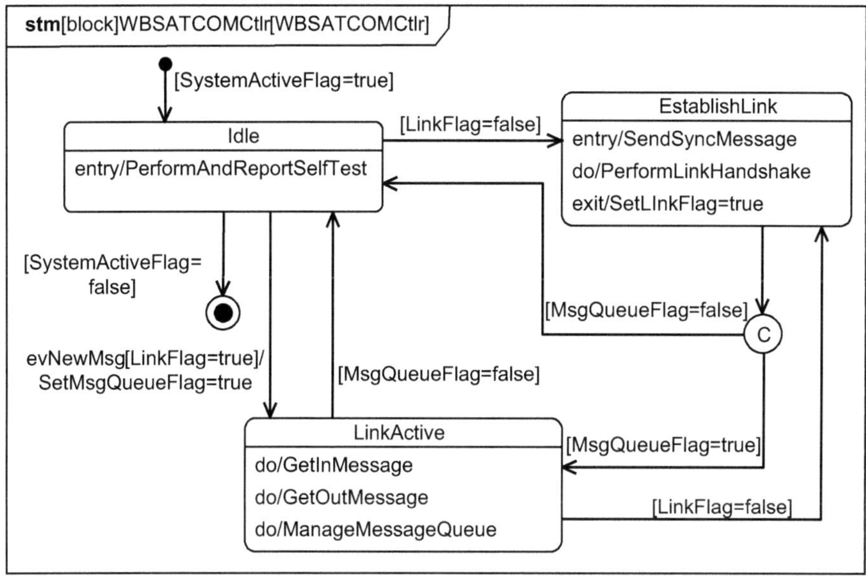

Fig. 5.9 SMD for the Wideband SATCOM Controller Block

Chapter 2 showed some of the ways SMDs offer a very powerful and flexible means of modeling behavior. They can also be executed (run as simulations) in many architecture tools, as part of the Logical Architecture level of modeling and simulation to visualize and evaluate the behavior. To recapitulate, some of the key aspects of SMDs include:

- Modeling concurrent behaviors such as threads that execute simultaneously
- Using nested states (substates within superstates) to decompose complex behaviors, either on a single diagram or on separate diagrams
- Accounting for Actions that can be very precisely associated with specific points in a behavior
- Synchronizing behaviors using transitions among states that are triggered by events and controlled by guard conditions
- Representing complex sequences of behavior with conditional transitions, history, and other modeling subtleties

SMDs are particularly important Data-Driven and Event-Driven architectures, which are common in complex systems and are discussed later in this chapter. Additionally, it is often important that the sequence of actions and the conditions under which they occur are precisely specified and controlled in order to satisfy performance, safety, security, and other requirements. Real-time systems often require behavior to be linked to a system or universal time source (clock). In all of these situations, SMDs are the usual way to model the behavior, especially using events to synchronize behaviors across subsystems within a system and across systems within an enterprise.

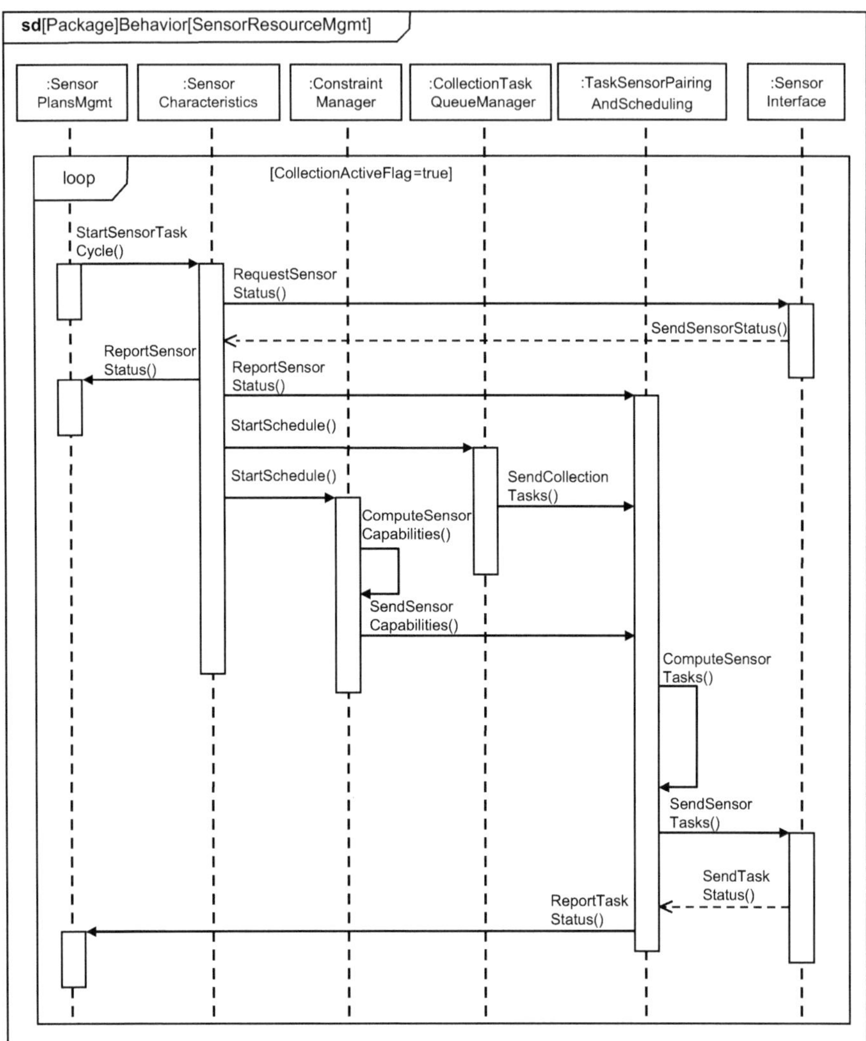

Fig. 5.10 SD for the Sensor Resource Management Subdomain

The Sensor Resource Management Subdomain (Figs. 5.5 and 5.6) furnishes another example of E-X behavior, this time involving non-stateful behavior diagrammed in an SD. Figure 5.10 shows how the Parts of the resource manager, now represented as Block Instances, interact to accomplish the sensor control task. This was summarized earlier in the Structural Perspective section. The Interaction Fragment with a Loop operator shows that the resource manager will cycle as long as a Collection Active Flag is set to true. At the beginning of each cycle, the Collection Plan updates the inputs. The Sensor Assets Block queries the sensors via the Sensor Interface Subdomain for their current status. Then the three data Blocks

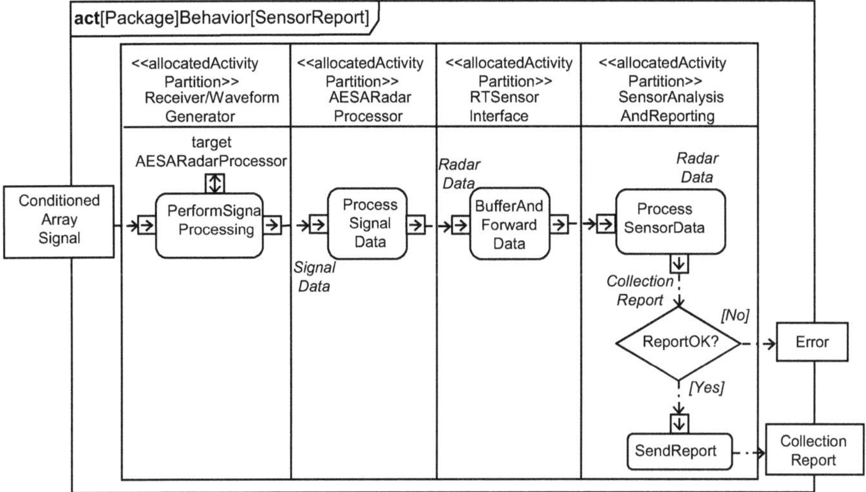

Fig. 5.11 Sensor Report Development modeled in a SysML AD

feed the latest information to the Task/Asset Pairing and Scheduling Block, which computes a new set of optimum assignments of collection tasks to sensors and creates the corresponding SSRs. These are sent to the Sensors Domain via the Sensors Interface, the status of sensor task execution comes back for use in the next planning and scheduling cycle, and the same status is reported to the Collection Plan Block so that it can track progress. The cycle repeats until collection activity ceases.

A third example, diagrammed in Fig. 5.11, illustrates SysML's enhancements to UML Activity Diagrams (Fig. 2.14), including Activity Parameter Nodes on the boundary and Call Operation Actions. An AD does not explicitly distinguish stateful vs. non-stateful behavior. The behavior modeled by an individual Action can be detailed out in an SMD or SD depending on its nature. The Actions in a detailed AD like this map to Operations of structural elements like Blocks and Interfaces, while Objects represent units of system content and are defined in the Data Perspective, which we consider in Sect. 5.5. For example, the Receiver/Waveform Generator (RWG) in Fig. 5.7 would have an Operation to process data from the array.

Figure 5.11 shows more explicitly what various Parts within the Sensor and Sensor Management Domains do to develop a Collection Report. This is a particular type of report that inherits from the Report Foundation Class and contains the formatted results of an individual sensor interaction with the external environment. We assume that the modules of the radar AESA do initial signal conditioning, such as filtering, and perform analog-to-digital conversion. This "conditioned" array data is fed to the RWG which performs signal processing to create a basic radar measurement data set with content such as target range, elevation, and azimuth (direction angles to the target). The AESA Radar Processor supports this signal processing by computing values based on the original collection task that set various parameters in the signal processing algorithms. The RWG invokes this as a Call Operation of the radar processor, and this is modeled by the "target" Pin on the top of the Action icon. The signal data produced by the RWG goes to the AESA Radar Processor to be formatted as radar data. That, in turn, is sent via the Real-Time Sensor Interface to the Sensor Analysis and Reporting Subdomain within the Sensor Management Domain. There, the data is processed to create a Collection Report and checked against criteria such as a maximum estimated target location error. If the report passes, it is provided as an output parameter of the Activity to be passed to the ISR Management Domain. If it fails, an error is generated.

Our final example introduces the use of an SD for initial timing analysis, as discussed in an earlier section. Figures 5.12 and 5.13 illustrate this for the E-X radar. An individual collection event within a radar schedule is commonly called a "dwell." The two SDs model a set of notional actions during a dwell and allocate times to three primary segments. First, the radar must configure the AESA by applying the correct phase setting to each T/R module to form the desired beam. Next, the AESA radiates energy using the waveform sent by the Receiver-Waveform Generator and detects energy reflect by any targets in the spatial volume of interest. Finally, these detected signals are subjected to amplification, filtering, and possibly other techniques. These may be accomplished by the AESA modules or in the receiver-Waveform Generator. The resulting conditioned signals go through analog-to-digital conversion to create a digital data stream that is sent to the Radar Processor. Figure 5.15 models the Radiate and Receive activity that is pointed to by a Ref frame in the primary diagram.

Figures 5.12 and 5.13 also contain timing marks, which are explained in Sect. 5.8; these define aspects of the allowable times for various parts of the behavior. The majority of behaviors in the systems of interest in this book can be modeled and analyzed at a level of detail appropriate to an LV using the techniques presented in this section.

5.5 Data Perspective: E-X Logical Data Model

The next task in building the LV is to transform the CDM into an actual data design in an LDM that is compatible with MBSAP. For many systems, the LDM looks forward to data dissemination and sharing in a service-oriented architecture (SOA) environment as discussed below and in Chap. 7. The first step, and the majority of the effort in constructing an LDM, is to model the data in BDDs showing inheritance of data entities from Foundation Classes, decomposition of complex entities into their parts, and associations among them. This provides the essential starting point for the precise description of the information content (and, in some cases, the resources) of the system being modeled. In many architecture efforts, an important objective is to provide the metadata that describes this information in a formal structure so that it can be processed automatically by XML parsers and other software to enable storage, retrieval, search, posting, exchange, and other Data Services.

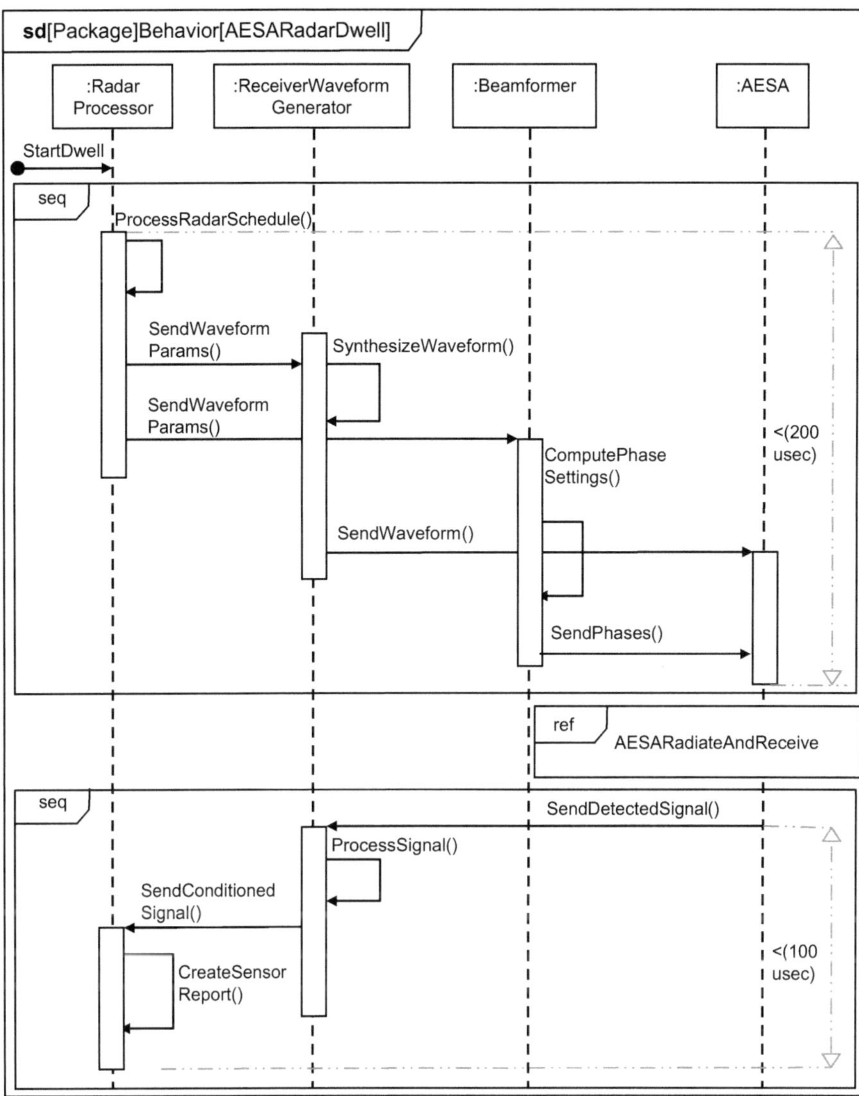

Fig. 5.12 Radar operation during a dwell interval

The data structure is commonly defined in eXtensible Modeling Language (XML) schemas, which are typically documented in XML Schema Definitions (XSDs). XML is a language, similar on some ways to software programming languages, but with the purpose of encoding documents in a format that is readable by both people and computers. The language is widely used to facilitate information exchanges over the Internet, especially of data structures, text documents, and other files. An entire family of other languages has been developed using XML as their

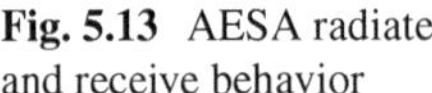

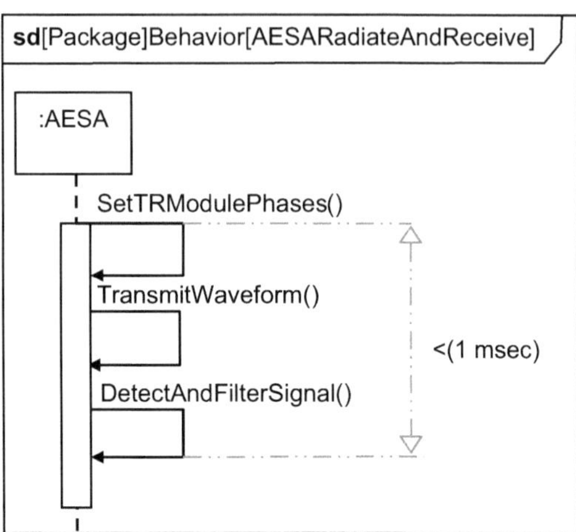

Fig. 5.13 AESA radiate and receive behavior

foundation. Many excellent references on XML are available [5]. Here we limit the discussion to some factors that impact data modeling in MBSAP.

Figure 4.20 showed an initial LDM diagram for the E-X example, starting with Plan Foundation Class and specializing it by Inheritance. Completing the LDM requires that all these specific System Data Blocks be fleshed out with specifications, including Values and Operations, to complete their descriptions. It's quite common for initial Block descriptions to be refined as the architecture development proceeds, e.g., when content must be added to a data Block in the form of additional Values or specialized Operations. Figures 5.14 and 5.15 give two more examples for the Report and Sensor Data Foundation Classes.

In Fig. 5.14, Inheritance is used to define the specific reports that inherit from the Report Foundation Class. In addition, Composition Associations show that three different kinds of reports can be Parts of a Sensor Report. All of the Operations needed by these Blocks were defined by the Foundation Data Class in Fig. 4.18 and thus are not repeated here. Since all these Blocks are part of the data model, they have the `<<InfoElement>>` stereotype.

Figure 5.15 does the same sort of thing for the Sensor Data Foundation Class. Onboard sensor data comes in the form of reports from the radar, SIGINT, and EO sensors, and collections performed by sensors on off-board platforms come in as reports through various communications channels. Blocks are defined for RF and EO Sensor Data that are further specialized in System Data Blocks for different kinds of sensor reports. The Off-board Sensor Report inherits from the Report Foundation Class. Note again that the E-X is notional and these diagrams are for illustration; the details in a real system would be more complete and would vary with the specific design. Appendix C gives additional E-X LDM examples.

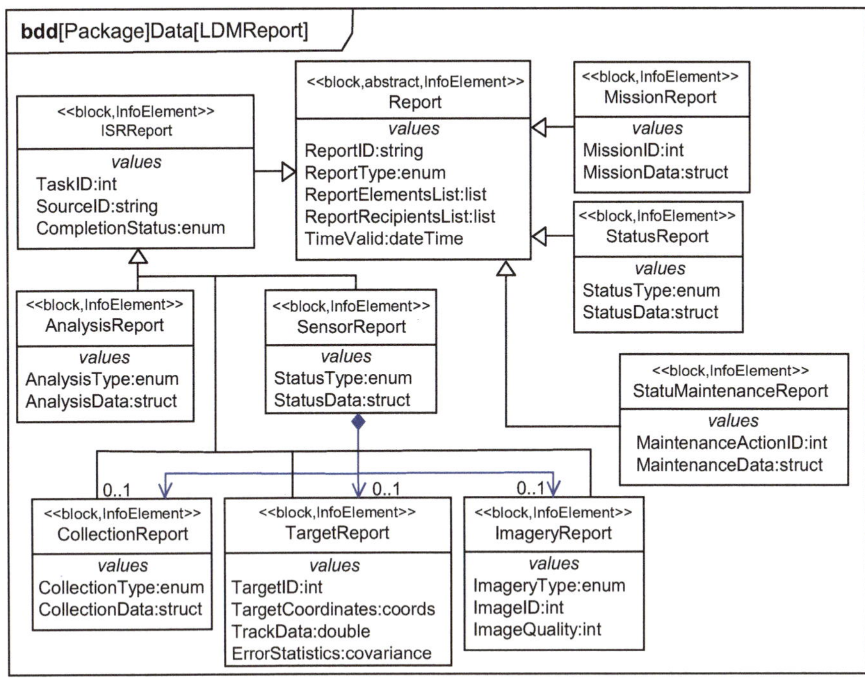

Fig. 5.14 System Data Blocks for the E-X Report Foundation Class

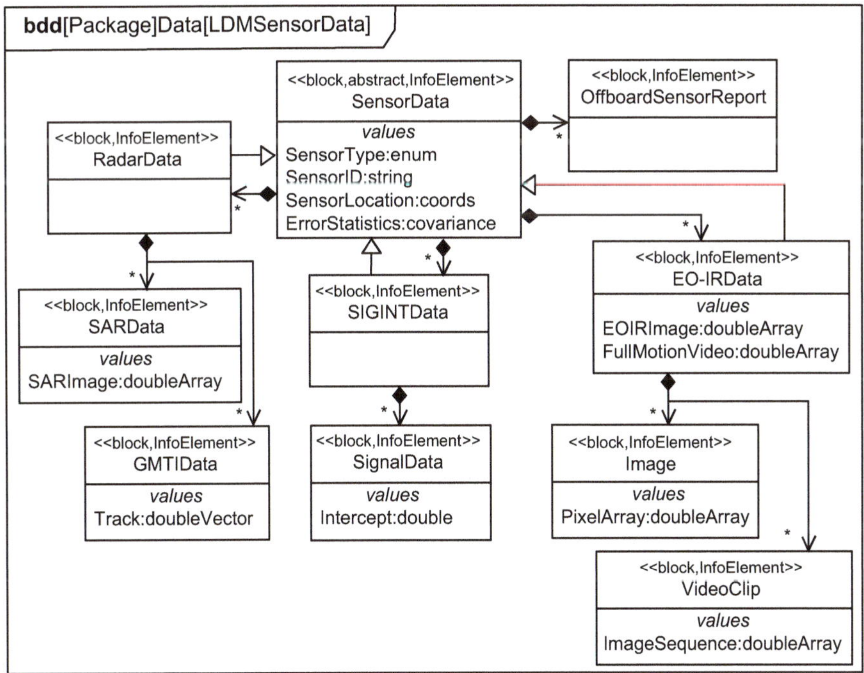

Fig. 5.15 System Data Blocks for the E-X Sensor Data Foundation Class

5.6 Services Perspective

The LV addresses Services in two distinct senses. The first continues the taxonomy begun in the OV by further decomposing System and Domain Services and allocating them to Blocks and their Operations. The second involves systems that employ a SOA. These have become very common; a system architect must have the skills and insights to effectively deal with them. In this chapter, we define the typical features of a layered structure that organizes various kinds of services in a way that promotes efficient system operation. Such structures can be designed at the enterprise, node, or system level, depending on the architecture being developed, along with the services provided by the various layers that enable the system to function as required. These two aspects of the Services Perspective are related, and an important part of the continued development of the services taxonomy is to identify the layers where various services are mapped. The Axis of Categories from Fig. 1.1 comes into play here as well, because different categories of systems, especially real-time and non-real-time, are best matched to different layering schemes. Some of the content of this section could have been deferred to the SOA discussion in Chap. 7, but it is so intimately connected to the LV that we have elected to introduce it here.

We'll illustrate this topic using the E-X example, which has many behaviors that can be implemented as services. In starting the development of the Services Perspective of the LV, it's helpful to recall the basic organization of the system. The E-X Domain structure, Figs. 4.6 and 4.7, has three primary Mission Domains: Mission Management, ISR Management, and Sensor Management. Two others, Information Management (supported by a Computing Infrastructure) and Communications Management, are associated with the system infrastructure. Application software in the primary Domains accounts for the bulk of E-X information processing and decision support functions. The hardware and software resources of the infrastructure Domains create the environment in which the software of the primary Domains can execute and communicate. Stated simply, the infrastructure hosts the applications. The Support Systems Domain accounts for all the non-platform system elements and may include a variety of hardware and software components, some of which may be common with the Mission Domains.

5.6.1 E-X Services Taxonomy

Building on the foundation established in the OV, Table 5.1 takes the top line of Table 4.2 and maps the Domain Services into the LV structure established in this chapter. This simple example illustrates a number of important points. To begin with, the effort spent building a services matrix or table provides an invaluable independent check on the definition of Blocks, both in terms of how well the architecture collects related Values and Operations and in terms of whether it accounts for

Table 5.1 Continuation of the E-X services taxonomy and mapping from the OV into the LV

System service	Use cases	Domains	Domain services	Subdomains	Blocks	Operations
Sensing:: Perform radar surface surveillance in SAR and MTI modes	Perform Sensing	Sensor Management	Maintain Sensor Schedule	Sensor Resource Management	Collection TaskQueue	GetTask() UpdateTaskQueue() CreateTask()
			Control Sensors	Sensor Resource Management	TaskQueue	GetTask() UpdateTaskQueue() CreateTask()
					Sensor Characteristics	GetAllocatedSensor() GetResourceStatus() ReportResourceStatus()
					Constraint Manager	GetExternalConstraints() GetResourceCapabilities() ReportResourceCapabilities()
					TaskSensor Pairing And Scheduling	GetTask() GetResourceStatus() GetResourceCapabilities() GetSSRStatus() ComputeUtilityFunctions() CreateResource/Task Pairing() SendResource/TaskPairing()
					Sensor Service Request	GetResource/TaskPairing() CreateSensorServiceRequest() SendSensorServiceRequest()
					SSRStatus Report	GetSSRStatus() ReportSSRStatus()

(continued)

Table 5.1 (continued)

System service	Use cases	Domains	Domain services	Subdomains	Blocks	Operations
			Process Sensor Reports	Sensor Product Development	SensorData Processing	GetSensorData() ProcessSensorData() SendProcessedData()
					Sensor Report Processing	GetProcessedData() CreateSensorReport() SendSensorReport()
				Sensor Info Management	Sensor Information Management	GetSensorReport() PublishSensor Report() UpdateSensorDataBase()
			Generate Track Reports	Sensor Product Development	Sensor Report Processing	CreateTrackReport() SendTrackReport()
				Sensor Info Management	Sensor Information Manager	GetTrackReport() PublishTrackReport()

all required behaviors. This may feel like doing the design job twice, but the Block diagrams look at the analysis from the standpoint of a logical and coherent decomposition of the system, while the services matrix stresses requirements traceability and compliance.

Discoveries made in each category help refine the other. For example, the way the matrix displays Block Operations helps the designer spot poor partitioning of functionality, overlaps and duplications, and other problems. Poor partitioning might be seen when a System Service can't be cleanly decomposed into lower-level services that are logically associated with Domains. Overlapping or duplicative responsibilities among the Blocks in a Domain will show up when common Operations that are needed to implement a service must be assigned to multiple Blocks. The general principle is that when service decomposition is logical and sensible, and when lower-level services can be associated cleanly with structural elements such as Blocks and Interfaces, both the structural partitioning and the service definitions are likely to be well done and consistent. When that's not the case, the architect should take that as a strong indicator that the affected areas of the architecture should be revisited and refined.

Comparing the Behavioral and Services Perspectives, it's common to find that system elements such as Blocks have been declared that are not directly traceable to explicit customer requirements. In the E-X example, the requirements set included "Maintain Sensor Schedule," but not "Manage Collection Plans." Yet experience prompted the use of a design pattern with a Sensor Plans Management Subdomain. The functions associated with this Subdomain are *derived* requirements that are, in fact, essential for the system to perform properly. This kind of information about gaps in the stated requirements is fed back to the requirements baseline and, if it is important to fully describe system behaviors, to the affected Domain specifications and scenarios in the OV. Derived requirements are often very important in ensuring that the customer and developer share a common understanding of the problem and of the emerging solution. Once approved and incorporated in the baseline, they have the same authority as original customer requirements.

A complete services taxonomy for a major system like the E-X could be large and requires effort to maintain as the system design evolves. Systematically examining system functions and deciding which of them should be implemented as services, which is discussed more completely in Chap. 7, is generally the most efficient approach to building an inventory of system services. Some SysML tools can export a tabular artifact like Table 5.1, minimizing the manual effort to create it. In addition, the scope of the task can often be limited in the following three ways.

- An important consideration in defining Services is granularity, which is also further discussed in Chap. 7. A fine-grained services taxonomy has many, relatively simple Services, while a coarse-grained one has fewer, more complex Services. By making the service granularity as coarse as is consistent with effective service implementation, which usually means a set of software code entities that implement the functions and interface of the services, the architect can minimize the magnitude of the task.

- When the customer provides detailed requirements for the services to be implemented, it may be sufficient to limit the analysis to the top two or three levels of the system structural hierarchy. This is especially true if specific services are associated with Block Operations or methods and can be accounted for in functional specifications for those behaviors.
- It may also be sufficient to limit the analysis to the particular Domains and Subdomains where allocation of service requirements is not obvious and where derived requirements are important to discover, especially system areas with stressing performance, high levels of parallel task execution, complex control flows, or other design challenges.

5.6.2 Layered Architecture

Chapter 7 discusses Service-Oriented Architecture (SOA), including layered services structures, in more detail, but a brief introduction is important for a complete description of the Services Perspective of the LV. There are many ways to organize Services, but one useful approach places them in the following four broad categories.

- *Mission/Business Services*—associated with the primary purpose for which a system is developed. These services provide the functionality that individual users and the using organization as a whole employ to implement its business processes and accomplish its mission. Typical examples are planning, tasking, analysis, reporting, resource control, and data visualization.
- *Data Services*—provide access to and management of data from internal and external sources, e.g., by managing access to databases and communications.
- *System/Infrastructure Services*—associated with the infrastructure that creates the environment for Mission and Data Services. They may be considered part of the system itself, or they may be independently implemented by the node or platform on which the system and frequently other systems are hosted. They enable execution of applications and other software, support human-machine interaction, provide monitoring and fault management of system resources, implement security protections, and do many other things.
- *Enterprise Services*—associated with the infrastructures of nodes and systems forming a networked enterprise. They create the framework for network information exchanges, collaborative activities and decisions, and management of the enterprise as a whole, and they extend the other Service categories so that they can be shared across the network.

We need a few more basics for this discussion of services. For each of the service categories just listed, services are exposed, discovered, and delivered through service interfaces, which should be defined by fully detailed specifications that may include Service-Level Agreements (SLAs). An SLA can be thought of as a contract between the provider and the consumer or client of a service that contains certain guarantees about services to be delivered. In dealing with the integration of application software onto an execution platform, a service interface can be thought of as an evolved form of a traditional Application Programming Interface (API). Separating the service interface from the service implementation reduces dependence on low-level detail, making the function (service) easier to invoke and use and less dependent on details of how it's realized. A key SOA concept is that a client does not need to know where a service is physically provided nor how it is designed and built. The client only needs the address where the service can be found and the protocol for invoking it. The service interface establishes the protocol and guarantees the content and quality, e.g., timeliness, of the service.

When a layered architecture is to be employed, some important decisions need to be addressed in the LV. Maier [6] has noted the overall trend toward layered abstractions in composing a system vs. the more traditional hierarchical structures based on functional and structural decomposition. As was shown in Chap. 2, in SysML, structural decomposition is shown using Composite Associations that declare some Blocks to be Part Properties of other Blocks. These relationships build up a structural or compositional hierarchy of the system as we saw in Sect. 5.3. In today's advanced systems, a functional hierarchy of services and their interfaces increasingly complements this structural tree. This is due in large part to the fact that complex functions are realized by composing lower-level services to create more powerful, higher-level ones. We will see an early version of this sort of layered model in Chap. 9 when we examine networking. A good LV, then, defines the layering and places System Services into the appropriate layers on the basis of grouping related services and enabling these service groups or layers to interact in a simple and logical way. In doing this, a cardinal principle is that each layer presents the capabilities of all the layers below it to the layer above it via one or more service interfaces while abstracting the details of their implementation.

We can make these ideas clearer with a basic example. We'll start with some conceptual graphics to show the main concepts and subsequently look at capturing these structures in SysML. Figure 5.16 shows the basic context of a services layering for a node or platform. This organization follows the four broad service categories defined earlier. At a very general level, the nature of the layers, each of which may include both hardware and software, can be summarized as follows.

- *User Services Layer*—provides controls and displays plus clients, proxies, agents, and other functions needed to support the HMI. Mission Services may create their own User Services such as visualization or may integrate with User Services provided by the infrastructure.
- *Mission/Business Services Layer*—provides the primary functionality that enables the node and its systems to perform mission tasks, including business processes in a commercial or civil system; in many cases, the software components that create these services are hosted and run on a "computing platform" like that modeled by the Computing Infrastructure and Information Management Domains of the E-X architecture (Figs. 4.6 and 4.7).

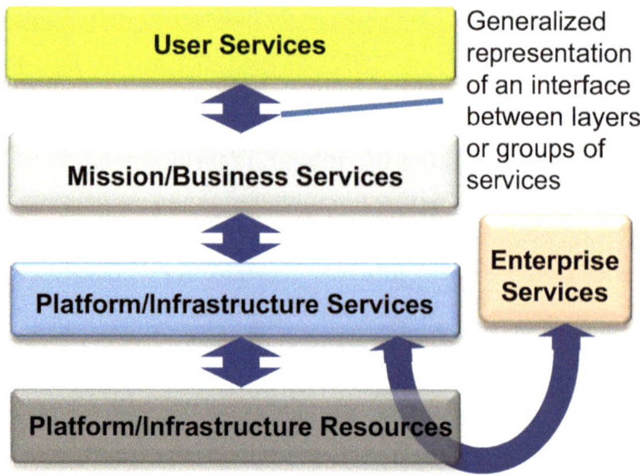

Fig. 5.16 Basic model for a layered architecture with service interfaces between layers or groups of services

- *Platform and Infrastructure Services Layer*—exposes the capabilities of the underlying computing resources of the system, along with associated utilities and supporting software components, as services, creating the environment in which the Mission Services can operate.
- *Platform and Infrastructure Resources Layer*—provides the computers, networks, storage, communications, equipment controls, and other hardware and software components needed by the system. This and the preceding layer create a computing platform.
- *Enterprise Services*—external functions that provide the enterprise-level capabilities to enable an ensemble of systems to work together to accomplish overall goals. Commonly, each participating system or platform implements these Services in its infrastructure. As the figure suggests, the Platform and Infrastructure Services layer works through Platform and Infrastructure Resources, especially networking, to interact in the enterprise in order to make these capabilities available to Mission/Business and User Services.

Among the many benefits of this kind of architecture layering is the fact that a software application that provides a Mission Service has no need to know the details of the underlying computing platform and thus is not affected when technology refreshment[5] or other circumstances cause the infrastructure to change, provided the integrity of each service interface is maintained. The arrows between layers in Fig. 5.16 represent these service interfaces. It's important to distinguish these from the physical connections that we model in SysML Block Diagrams. A common rule is that only downward dependencies are allowed: a higher layer invokes services from the layers below, but a lower layer cannot depend on services from the layers above. Among the most important open system standards are those defining these services and their interfaces so that many product developers can pursue innovative

[5] Technology refreshment is the common term for a process of replacing components that have become obsolete, are no longer available, or threaten to become so, with newer or more capable products to maintain system performance, reliability, and supportability.

and competing solutions to specific problems while still allowing their products to be integrated in a system.

5.6.3 Process/Decision Real-Time Nodes and Systems

We noted in Chap. 1 that the category of a system affects many aspects of architecture, and this includes services and service layering. The basic model in Fig. 5.16 can be expanded as shown in Fig. 5.17 to create a layered SOA that applies most directly to the system categories labeled "Process/Decision Real-Time Systems" and "Management Information Systems" in Chap. 1. These are precisely the systems that are generally candidates for use of a "conventional" SOA that does not have to deal with hard real-time performance requirements. Figure 5.17 adds detail to the basic layered model and identifies a representative set of services found in the various layers. Again, the blue arrows represent generalized interactions among groups of services, not physical connections among Blocks. Modern layered architectures in this category are the product of evolution from early client-server networks to today's SOAs and are sometimes referred to as n-tier architectures because they refine the original client-server structure by inserting layers or tiers that group and expose various services. The Platform and Infrastructure Services layer is often labeled middleware, because it mediates between system applications software (Mission Services) and the underlying data repositories, computing infrastructure, networking, and other resources.

There is no single "right" way to portray this, but Fig. 5.17 shows a typical structure. This architecture draws heavily on successful SOAs that have been developed and used in a variety of system types. Chapter 7 goes into more detail about services, but for now, the following paragraphs summarize some important features.

- A common HMI uses the human performance design for both local and remote (via the network) users. This layer makes maximum use of presentation standards such as a style guide to achieve a common look and feel to promote ease of use and shared understanding of information displays across the enterprise in which the system or node participates.
- Combined Client and Presentation Services can incorporate a variety of client models, including Web browsers or portals and more complex clients that often accompany complex mission applications, especially when legacy products are reused. These and the preceding layer create the User Services Layer of Fig. 5.16.
- Mission Services, created by software applications, are incorporated as components on an Integration Backbone (IB), which is described in Chap. 6. The IB provides whatever services the various components require and is based on an established SOA design pattern within the Mission Services Layer. A Groupware Framework takes care of important functions such as visualization of information from various components, collaboration tools for system users, and user identity services.

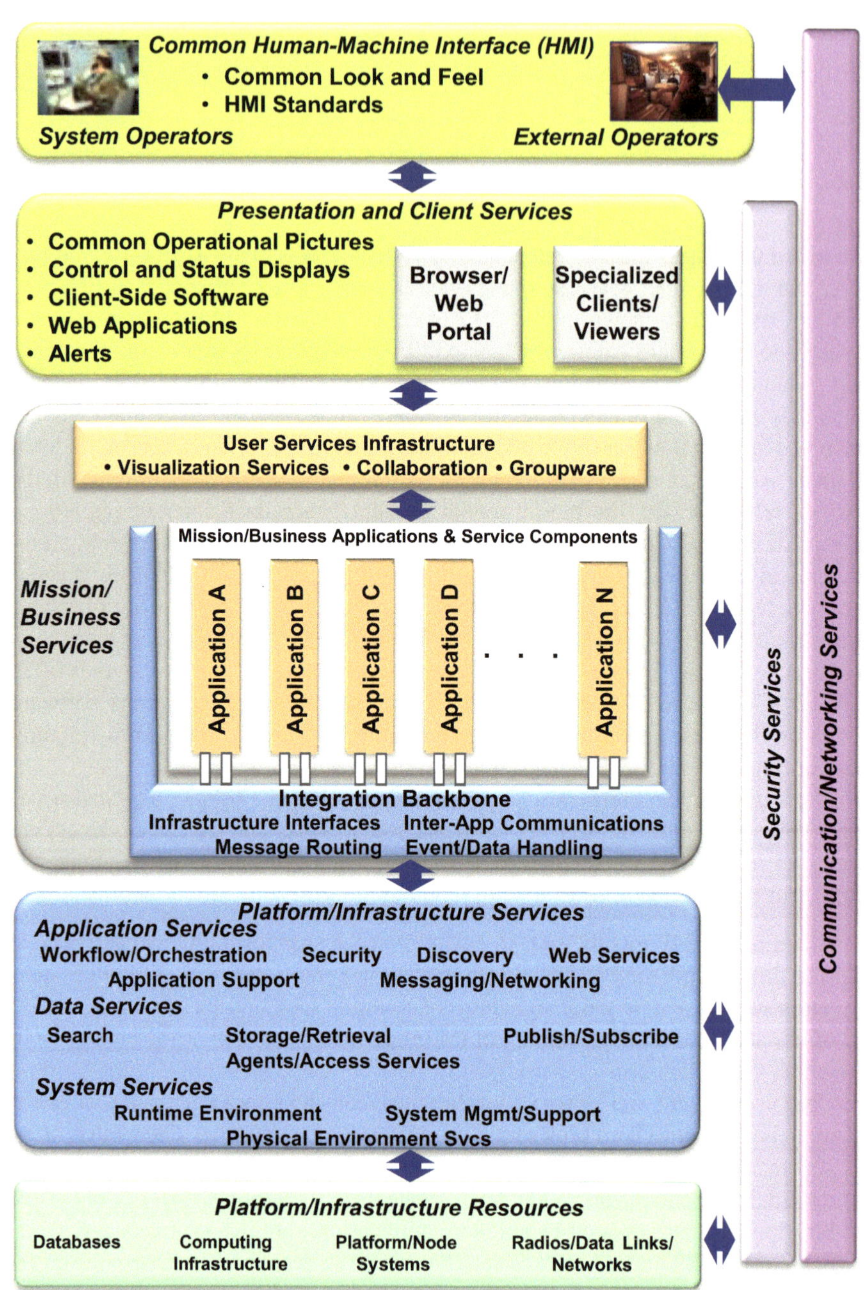

Fig. 5.17 Representative layered SOA for a Process/Decision Real-Time System or Node

- A Platform/Infrastructure Services Layer is further broken down into a hierarchy of Application Services, Data Services, and System/Support Services, with typical examples of each shown in the figure. Services such as Orchestration and Workflow are explained in Chap. 7.
- A Platform/Infrastructure Resources Layer includes the data repository, which will often be assembled from a mixture of enterprise-wide or legacy databases and node-specific databases. If an MDC is part of the design, it will be included here, and other resources include the servers and local networks comprising the Computing Infrastructure, communications equipment, and interfaces to or proxies for any systems associated with the platform hosting the system that need to be exposed as services.
- Communications/Networking Services and associated Security Services are shown on the right side of the figure and span the other layers. The first of these provides information paths among the layers and to external users. These paths are protected by Security Services that include everything from virus checking to sophisticated protection measures as discussed in Chap. 10.

This class of architecture is optimized for information sharing among a group of users, whether at a given system or node or across an enterprise. There are often deadlines associated with key mission threads, making them, strictly speaking, real-time systems, but the timelines tend to be driven by human decision-making rather than by physical phenomena and thus on the order of seconds or minutes rather than milli- or microseconds. In secure systems of this type, computing and communications security is often a challenge especially when multiple levels of data sensitivity and of user information access must coexist on a shared infrastructure. Yet another difficulty arises because the nodes within an enterprise often vary widely in development timeframe, technology baseline, use of commercial products, and concept of operations, making the implementation of enterprise services and infrastructures that meet the needs of all participants a very difficult proposition. Chapter 13 goes into this in more detail.

Modeling a layered construct like Fig. 5.17 in SysML uses the same basic approach that has been applied up to this point. It's important to remember that this is not really a structural diagram; rather, it is concerned with grouping system elements, especially services, into layers so that dependencies among them and common interfaces can be dealt with. This alternative viewpoint can be helpful in conveying the role each component and service plays in the context of the overall system. A simple way to capture this is illustrated in Fig. 5.18. Here, Packages represent layers and contain the services in each layer. The Associations among Packages are `<<usage>>` dependencies, and for the primary layers, they are required to be downward-oriented. Security supports the primary layers as well as Communication/Networking Services. The diagram is annotated to show correspondence to the E-X Domains in Figs. 4.5 and 4.6. An alternative presentation that would convey more information would use a BDD to define the layers using Blocks and then an IBD to show their exchanges of information and services. Services in each layer could be modeled as Part Properties of the corresponding Blocks.

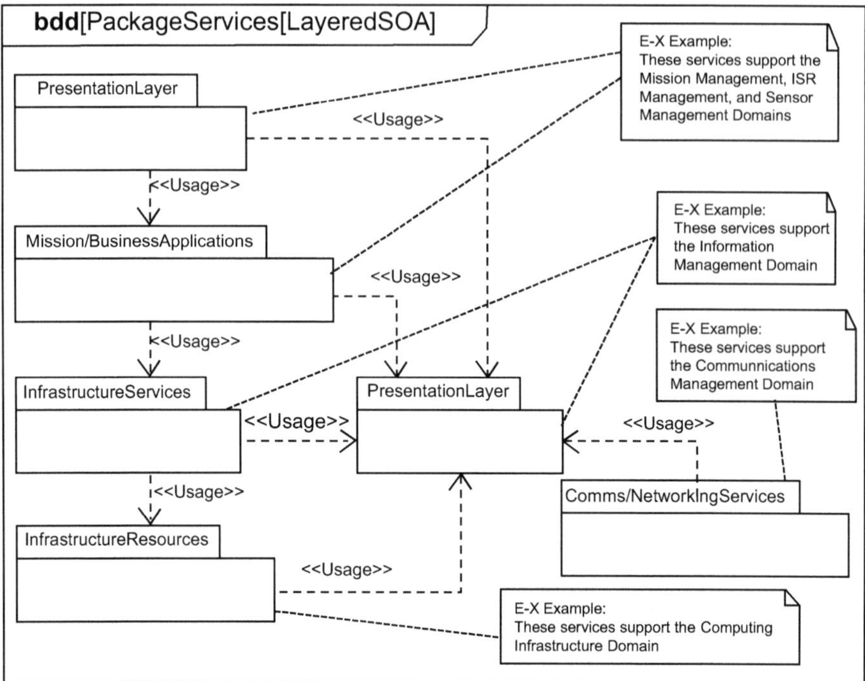

Fig. 5.18 Modeling a layered SOA structure with Dependencies

Figure 5.19 shows what such a BDD would look like; the services shown are not a complete set, but are representative. Block specifications contain descriptions of services and their interfaces, which are further discussed in Chap. 7.

The functional design of the infrastructure uses Block Diagrams, just as was done for the Mission Domains. As a simple example, Fig. 5.20 shows a possible design for the Integration Backbone (IB) from Fig. 5.17. The IB has four primary functions: Adaptation, to integrate a Mission Application with the Infrastructure, further discussed in Chap. 6; Messaging; Routing; and Mediation, which deals with things like format and protocol translations and resolution of conflicting values. The diagram shows the Ports that support the IB's principal interfaces. The Messaging Part handles information exchanges between the Information Management and Communications Management Domains and realizes Ports for both of these. It works with a Routing Part that determines which messages should go through which channels. The Application Adapter Part primarily creates the interfaces used by one or more Mission Applications. Multiple adapters might be needed in a system using applications from a variety of sources.

A SOA typically includes a variety of Application Services such as Orchestration and Workflow (see Chap. 7) that control system functions and interact with applications and users through the IB. The Application Adapter also supports a Client Services Port for interaction with system users. In the Physical View, it will be essen-

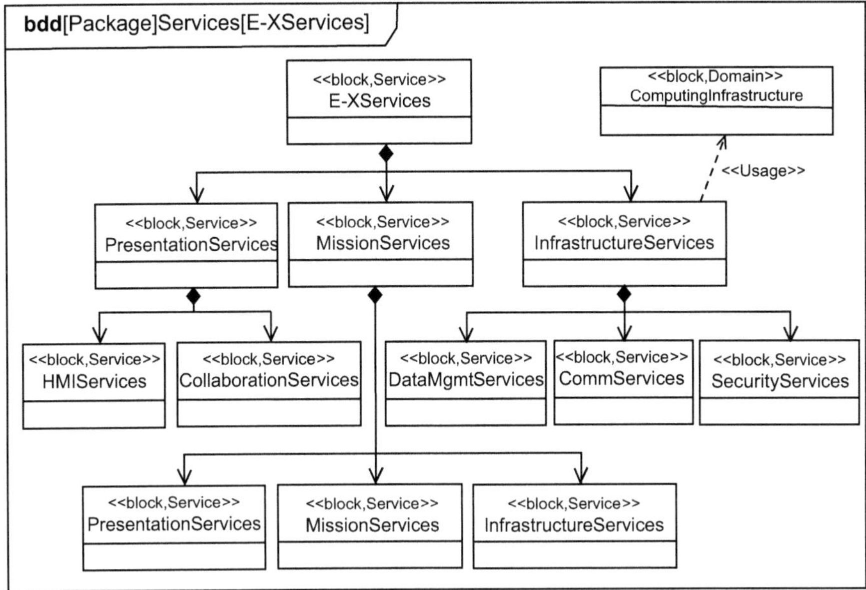

Fig. 5.19 Representative E-X services modeled as Blocks, with individual services as Part Properties of service groups or layers

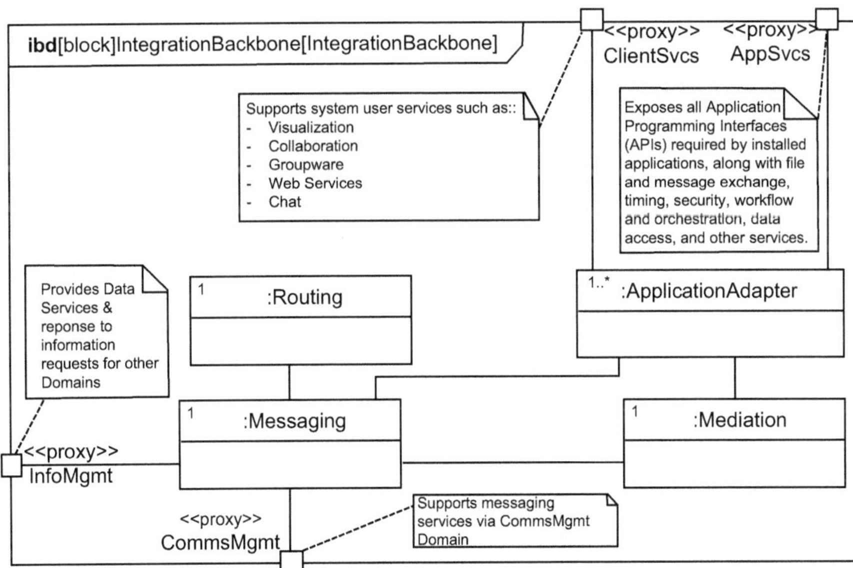

Fig. 5.20 Basic Integration Backbone design modeled in an IBD

tial to carefully implement interfaces like that between an application and the IB. A similar structure of Blocks and their Parts can be designed for the resources within each layer of the SOA. Chapter 7 shows approaches to extending these diagrams in various ways to account for the full range of required services, the components that support and use them, and the interfaces through which they are exposed and accessed.

It is frequently the case that the architecture for a networked node must deal with various software execution environments in order to accommodate the full range of Mission Services required. For example, although the basic node is non-real time, it may need to include some real-time components and to create a suitable execution environment for them to run in. This is definitely true of the E-X example, where hard real-time Domains for sensors and communications interact with soft- or non-real-time Domains that do core information processing. Other common situations involve reuse of legacy components that have native execution environments different from that of the core architecture and a requirement to connect components that process information at different levels of sensitivity or classification. In such cases, a common solution is to extend the basic architecture using an Enterprise Service Bus (ESB) as shown in Fig. 5.21 to span a range of environments.

The ESB concept was originally developed mainly as a way to link distributed sites or nodes in a commercial network having a variety of computing resources, databases, networks, and so forth. At its most basic, an ESB provides a messaging service to allow otherwise non-interoperable nodes to communicate and collaborate. It can be expanded into a framework within which messaging, data transformations, Web services, and intelligent routing, as well as specialized interfaces for particular components, especially Mission Services, can be configured and managed. In the layered structure we are discussing, the Integration Backbone is merged with an ESB, which is natural because both are dominated by exchanges of messages and data. This structure provides the interfaces needed by the heterogeneous environments shown in Fig. 5.21. The interface to a privileged or restricted environment, especially one processing highly sensitive information, would include a guard to protect against unauthorized information exchanges between different levels of trust.

This chapter emphasizes the importance of patterns in the LV, and Fig. 5.22 shows another aspect with regard to this layered SOA by illustrating the application of design patterns to various elements of the architecture. Reference [1] spells out the patterns in detail, and Table 5.2 gives a brief summary. As one example, the Façade and Observer patterns that are annotated for the Integration Backbone highlight the essential nature of any such interface, including the functionality provided by an ESB, and the use of these patterns facilitates the merger of elements of the integration interface into a single, coherent framework that provides the Infrastructure Services required by all Mission Service components. The Façade pattern generalizes the notion of creating a simple and uniform interface through which clients can access one or more objects, while the Observer pattern provides a uniform way to notify specified "observers" when specified changes of state, such as events, occur in the system. A typical example would be delivery of a message from one application to another. Across the board, use of patterns helps ensure a consistent, high-quality design that leverages the experience of the community in solving similar problems and facilitates reuse of existing components, Service definitions, and infrastructure.

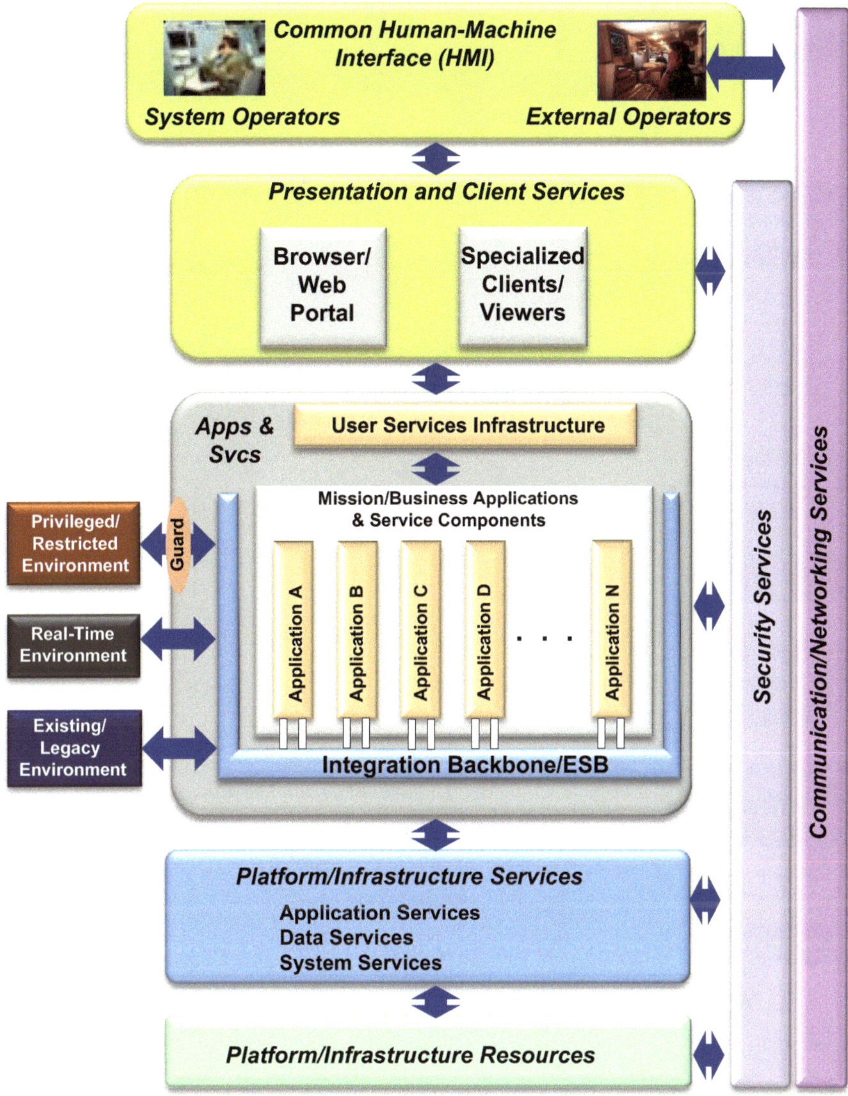

Fig. 5.21 Layered SOA with an Integration Backbone built around an Enterprise Service Bus to accommodate heterogeneous execution environments

There are many variations on this general layered structure, depending on the nature and requirements of a particular system. The simplest architecture is a single computer or group of computers that does everything and outputs its results for other systems or users to access. One step-up in complexity would be the addition of an HMI for a single user. Here, the infrastructure is little more than the computer and its operating system. Such low-end systems can be mission critical and require high-quality designs, but they scarcely merit the effort of fully applying MBSAP.

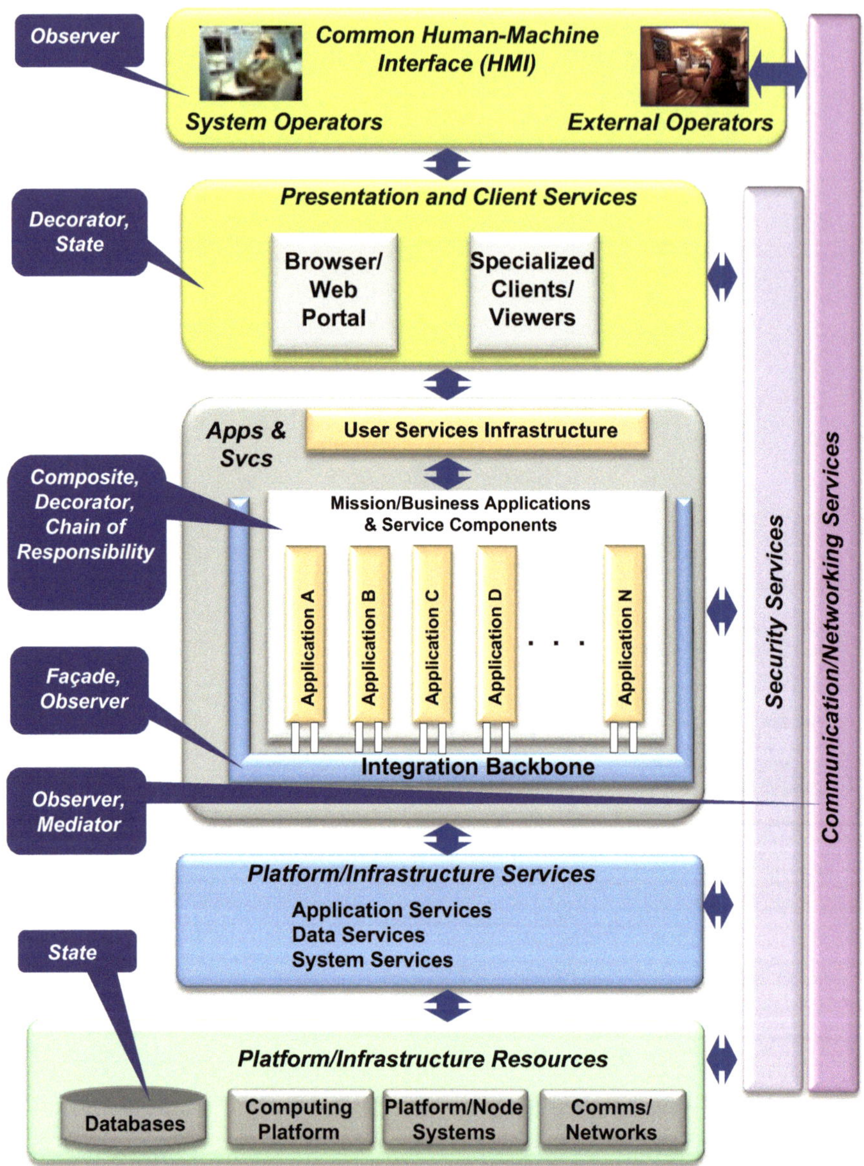

Fig. 5.22 Design patterns for the Layered SOA

It is when multiple system operators perform diverse tasks and access shared resources that the kind of architecture discussed here applies. Even then, there are ranges of complexity, and the following patterns summarize a typical hierarchy of such architectures. The individual computers employed in systems like these may use various versions of the layered architectures that we have just discussed.

Table 5.2 Patterns used in the layered SOA architecture

Design pattern	Description
Observer	Establish a one-to-many dependency such that when one entity changes, dependent entities update automatically
Decorator	Add responsibilities dynamically
State	Allow an entity to change behavior when the internal state changes (equivalent to changing the entity's classifier)
Composite	Create tree structures that represent whole-part relationships
Chain of responsibility	Allow multiple entities to handle a request by passing the request along a chain until it is handled
Façade	Provide a unified, higher-level interface to a set of interfaces in a complex entity such as a subsystem
Mediator	Define how a set of entities can interact and promote loose coupling by avoiding direct entity-to-entity references

- *Basic Client/Server (C/S)*—applies to any architecture in which users employing workstations or other clients use a network to access information and other resources on a shared computer that is called a server. The oldest C/S model is a File Server that allows individual user workstation computers to access central file storage. This corresponds to a "federated" architecture in which each workstation hosts a set of applications and users operate largely independently except for exchanging data and messages. One step-up from this replaces the File Server with a true Database Server providing more sophisticated services such as queries and reports in addition to data management.
- *Application Server*—a slightly more complex pattern in which one or more computers host software applications as well as shared files that are accessed by multiple users. This architecture variant transfers complexity from the client to one or more, typically more powerful, servers and allows any user to employ any application without having a private copy. Software vendors have licensing provisions to deal with this sort of shared use of their products.
- *Transaction Server*—with this advancement, a client can invoke remote procedures (as services) using a transaction mechanism, and the server provides online transaction processing (OLTP). Early transaction servers soon evolved to the next level of complexity which uses an Object Application Server that provides an Object Request Broker (ORB). This allows a client to locate an instance of a server object such as an application, which may be local or remote, invoke methods on that object, and get back the result.
- *Groupware Server*—in addition to any of the above, the system can include a server that supports e-mail, bulletin boards, chat rooms, and other collaboration tools.
- *Web Server*—in the currently common generation, a Web-based architecture delivers content over the Internet using a very simple client such as a Web browser.[6]

[6]The original term "World Wide Web" is now universally contracted to just "Web."

As the architecture moves up this scale of complexity, there is a tendency to move from "thick" to "thin" clients and from "federated" to "integrated" system functions. A thick client is essentially a stand-alone workstation that uses network connectivity to access and share basic information such as files and applications, while a completely thin client would host nothing but the HMI and get all other computing support from servers. This migration promotes a more interactive environment for operators because everyone is using the same resources, looking at the same operational pictures, potentially sharing every transaction, and collaborating via groupware. In practice, most mission and business applications take a middle position in which one or more server-side components carry the primary computation and data access load, while a client-side component supports more sophisticated operator interactions than a browser can handle and manages access to server resources. Many workstations include a portal that provides focused access to Web Services, including Web-enabled applications, i.e., those that provide content using Web pages.

5.6.4 Real-Time Systems

The "hard" real-time (RT) system category introduced in Chap. 1 has a different set of driving requirements, and these systems tend to have different infrastructures and layering. They have stringent deadlines for mission or business threads and also require deterministic execution (schedulability), preemption of lower priority threads by higher-priority ones, complex responses to events and messages, and many other features. All this is the topic of Chap. 8. For now, it's enough to point out some of the key aspects of layered SOAs for RT systems. A typical networked RT system is built around an RT object request broker (RT ORB), and a number of commercial products are available for this. It may employ a Real-Time Operating System (RTOS), depending on requirements. The layered SOA is likely to look like Fig. 5.23, with RT ORB middleware providing the critical services of the environment for the mission applications. Allowing for differences in applications and other details, this structure can be found in RT systems in aerospace, industrial controls, networked multimedia, and other areas. The following paragraphs summarize some important features.

- Mission or business applications create the primary system behaviors, working with resources like sensors, communications, and equipment or vehicle controls. Applications also include the logic that creates content for the HMI. For maximum performance, these applications may be grouped into Domains, each of which provides specialized services.
- The Mission Applications Layer may also provide Framework Services like those shown in the figure when these are common to multiple applications or needed for time-sensitive inter-Domain interactions.

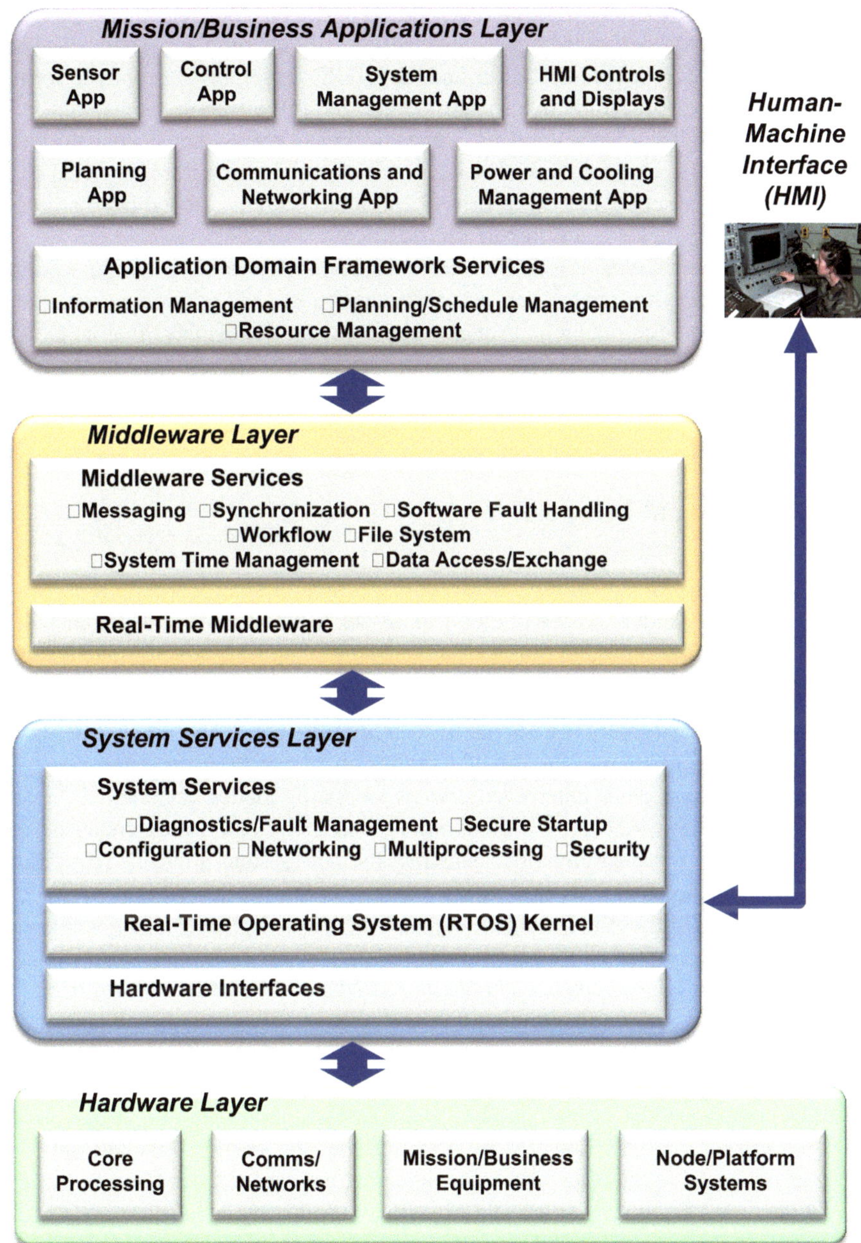

Fig. 5.23 Layered architecture for a real-time system using real-time middleware services

- The Middleware Framework includes the RT ORB or other RT middleware and the associated services. In some designs, especially for simpler systems, the separate middleware layer is not used, and the RTOS creates the environment for the mission applications directly or using specialized service components.
- An RTOS with its kernel and associated utilities creates the System Services Layer. Because interaction with resources is especially important in a real-time application, a set of hardware drivers, protocol engines, and other hardware interfaces is a major factor in performance.
- The Hardware Layer has all the resources controlled or used by the system, including computing, interconnects, communications, and mission systems such as sensors and equipment controllers. The controls and displays that create the HMI (shown here as a cockpit) are treated as system resources rather than as a client and integrated via this hardware interface layer.

5.7 Contextual Perspective

As with the OV, the remaining architecture artifacts of the LV are collected in a Contextual Perspective. Also as before, these are generally text and graphics that capture aspects of the design not readily represented by the kinds of modeling discussed so far. This Perspective might include a wide variety of artifacts.

- *Enterprise Design Drivers*—When a system participates in a larger enterprise, the LV needs to capture important design information from the enterprise context. This may include enterprise services, operating modes, data models, enterprise standards, policies of various kinds, and many others that create design requirements or constraints. For example, the new system might be required to implement specified activities within a larger, enterprise-level business process and to do so without requiring changes to the other, legacy systems within the enterprise.
- *Nonfunctional Requirements and Quality Attributes*—NFRs flow down from the OV to the LV and are explicitly associated with model elements and their characteristics using notes, model element specifications, Block Values and Operations, and other contents. The Contextual Perspective should capture this information to help ensure complete and consistent representation in the model. For example, specifications for Subdomains and Blocks should include application of Quality Attributes and associated metrics, derived from sources in the Contextual Perspective, to assist designers in developing or selecting fully compliant products. Taking reliability as an example of an NFR, when a behavior that has a requirement such as a minimum acceptable failure-free lifetime is realized in a Block or group of Blocks, the architect may want to allocate failure rates (or, equivalently, lifetimes) to individual Blocks as part of the design specification for the ensuing Physical Viewpoint. Then the implementation using hardware and software components will be required to meet the specified reliability.

Standards such as bus protocols that are part of the customer requirement should be associated with the model elements and interfaces to which they apply. Any other model annotations or supplemental text and graphics required to capture aspects of the design that are not represented in the basic model simply become part of the LV library.

- *Customer-Specified Artifacts*—Customers commonly have internal processes and standards that require the creation of architecture artifacts that are not part of the basic model. In general, these documents, tables, graphics, or whatever is called for are generated in an appropriate tool and either imported into the Contextual Perspective of the model or indexed there for easy access and retrieval.
- *Architecture Policies and Mandates*—Compliance of the Logical Data Model with a variety of Government standards has already been raised as a common constraint. Another place where policies and mandates are important is in the layered SOA. For provable compliance, the layering structure will either have to reflect the particular layers dictated by the customer or have a clear mapping showing that the design is fully equivalent to the required structure. When a customer network-enabled policy is imposed, it may be useful to include artifacts such as a networking checklist filled out to show how the design achieves compliance and where in the design Enterprise Services are instantiated.

5.8 Performance and Timing

As the design takes form in the LV, an initial analysis of process or thread timelines, parallel task execution, process synchronization, resource utilization, and other performance and timing aspects frequently becomes important. This analysis is essential for any system that has performance requirements, whether it is considered real-time or not, to ensure that the design is feasible and balanced and to generate quantitative performance requirements for the Physical Viewpoint. Both static analysis using diagrams and dynamic analysis using simulation may be useful, depending on the issues to be addressed. This initial performance analysis will become progressively more detailed and accurate as architecture development proceeds into the Physical Viewpoint, supported by selective physical modeling, and as architecture models are calibrated with measurements on a physical system prototype.

A typical question concerns allocating the total time allowed for a given thread to the activities within the thread in order to establish parameters like allowable execution times of individual Operations. It may also be important to account for physical phenomena beyond the architect's control such as the data rates and latencies of external communications paths. Sequence and State Machine Diagrams furnish the main tools for an initial examination of timing and performance. Figures 5.12 and 5.13 showed a simple example of initial static timing analysis, and this section gives a slightly more complex illustration. We assume that the E-X has a requirement to display and publish certain time-sensitive target tracks produced by the

MTI mode of the radar to a remote ISR Operator within a maximum time, starting with the creation of a track report. The thread to which this deadline applies starts when the Sensor Management Domain sends an MTI target report to the ISR Management Domain. It includes the processing of that report to create a new track or to update an existing one if the target has been reported before, as well as the formation and transmission of a track update message. The timeline ends when the update is displayed to the remote ISR Operator.

The next four diagrams provide the details for this example. Figures 5.24 and 5.25 show the decomposition of the ISR Management Domain into Subdomains, using the same design pattern as was employed in Figs. 5.3 and 5.4, and Subdomain information flows. The Resource Management Part is omitted because this Domain doesn't use it. Figures 5.26 and 5.27 further decompose the ISR Analysis and Reporting and ISR System Interface Subdomains using IBDs; these structures were first defined in BDDs similar to Fig. 5.24. Block Values and Operations can be shown in either type of diagram. These Subdomain Blocks provide most of the functions involved in the behavior being examined.

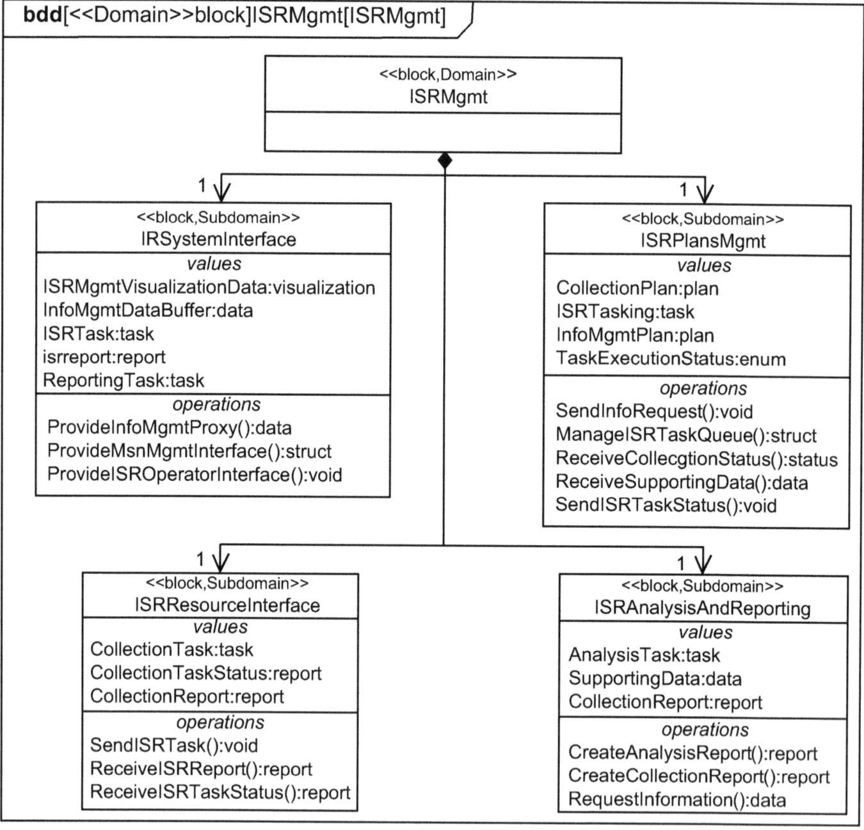

Fig. 5.24 BDD showing decomposition of the ISR Management Domain into Subdomains

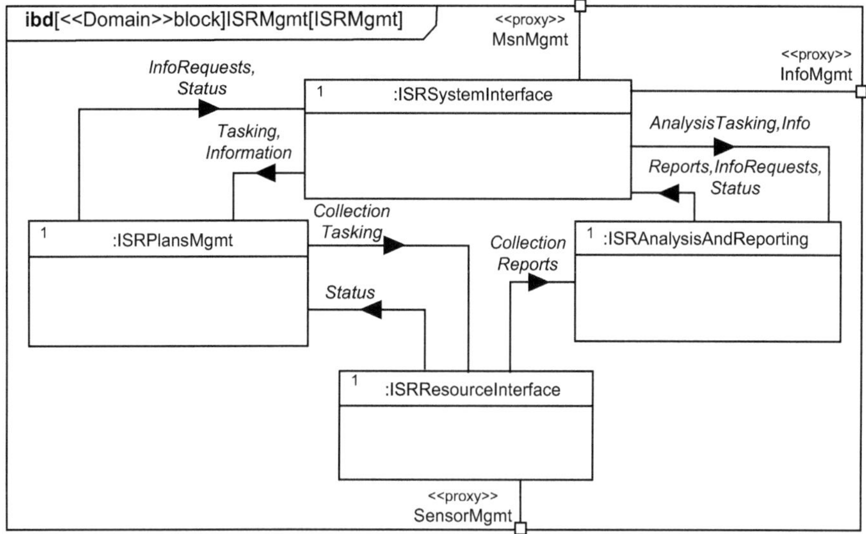

Fig. 5.25 IBD for the ISR Management Domain

The ISR System Interface Subdomain includes an ISR Information Manager that contains the primary logic associated with routing information into, out of, and within the Domain. An Information Management Proxy provides a local (inside the Domain) interface to the Information Management Domain, while an ISR Visualization Block takes care of the HMI for the ISR Operators, both onboard and off-board, with the same functionality on the aircraft and at the ground station.

The ISR Analysis and Reporting Subdomain has MTI Processing and SAR Processing Blocks that transform radar sensor reports into new or updated tracks and images, respectively. Similar Blocks process EO/IR and SIGINT data. A Track Manager that maintains the current inventory of moving object tracks supports the MTI Processing Block. These tracks can be numerous for a system like this and require constant updating. An ISR Product Block manages incoming product requests, packages the data from sensor processing in the correct formats, and sends out ISR Products via the ISR System Interface Subdomain.

With this structure, the thread of behavior to be analyzed can be shown in a Sequence Diagram like Fig. 5.28. The flow of messages is essentially linear, and the usual Return messages acknowledging the completion of an Operation are omitted in order to keep the diagram from becoming too cluttered. Thread timing is modeled with the times at which Messages are sent; Messages in the figure are annotated with Constraints such as {<500 μS} and Time Intervals showing the allowable range of time durations for various parts of the sequence. Downward sloping messages have latencies associated with network speed and message handling overhead. Timing annotations can be as detailed as desired. Horizontal messages happen within a Subdomain on a single computer and are assumed to be effectively instantaneous.

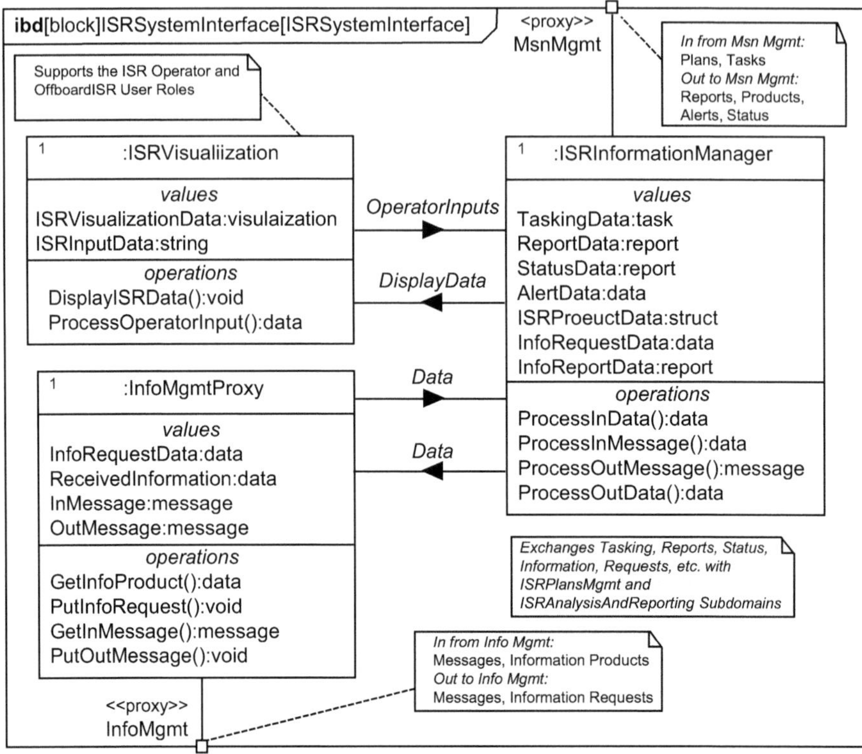

Fig. 5.26 IBD decomposing the ISR System Interface Subdomain

Depending on the stage of architecture development at which such timing analysis is conducted, the results might be used to establish performance parameters such as worst-case execution times as part of the specifications of system elements. Individual time intervals within an overall allocated deadline might be compared to equivalent times measured on existing systems as a reasonableness check. If there is reason to suspect that one or more time constraints can't be satisfied, it may be necessary to revisit the design.

In the present example, the total end-to-end time is dominated by the 5–10 s allotted for a human operator to review a nominated track, decide if it is a time-sensitive track, and take action to forward it. If the process were fully automated, for example, by having the Track Manager apply filters on track type and location, this decision could be much faster, and the other times would become more significant. Suppose this initial timing analysis discovers that overall system requirements such as tracks processed per unit time will not be met, due to an earlier design decision to have operators make these decisions. This could prompt a change to use a more powerful Track Manager that can automate this task, perhaps with an option for operator review. In that case, the diagram would change to show an additional processing step by the Track Manager to apply the rules for time-sensitive tracks

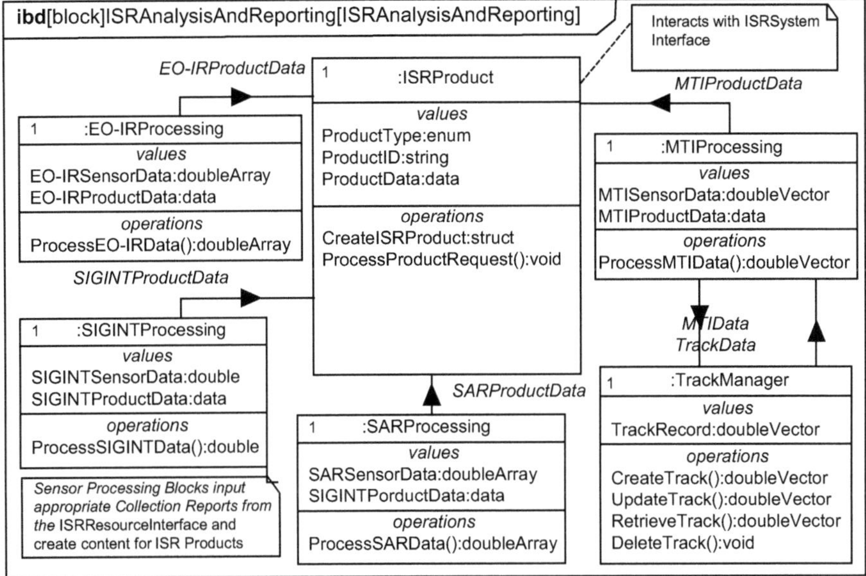

Fig. 5.27 IBD Decomposing the ISR Analysis and Reporting Subdomain

followed by a message directly to the Info Management Proxy Object to trigger distribution of tracks that qualify. An intermediate solution could be an enhanced HMI that presents track information in a way that allows a faster operator decision. Then there would be additional processing time, but a reduction in the allocation of time to operator action. In short, this timing analysis supports a range of design choices.

This kind of analysis requires some reasonable basis for associating timing with various steps in the sequence. Experience with previous systems or early prototyping work may shed light on typical times for at least some of the steps. For others, simple calculations such as the time to push a given track report message through a channel with a given data rate may be good enough, along with reasonable assumptions based on the complexity of various information processes. Among other things, this can yield a sense of the relative importance of external communications delays in satisfying the deadline.

Even this early and somewhat crude analysis may give insight into things like the need for broadband communications. Assuming that in some circumstances multiple targets will be detected and must be processed like this, the analysis may also produce an estimate of the simultaneity the system must handle and thus for the speed at which a single process instance must execute in order to meet the deadline for all targets under worst-case task loading. Finally, this initial static analysis often helps identify potential bottlenecks and other performance issues where higher fidelity is needed, and thus to focus M&S efforts in the next step, which is the executable representation of the LV as discussed in Sect. 5.11.

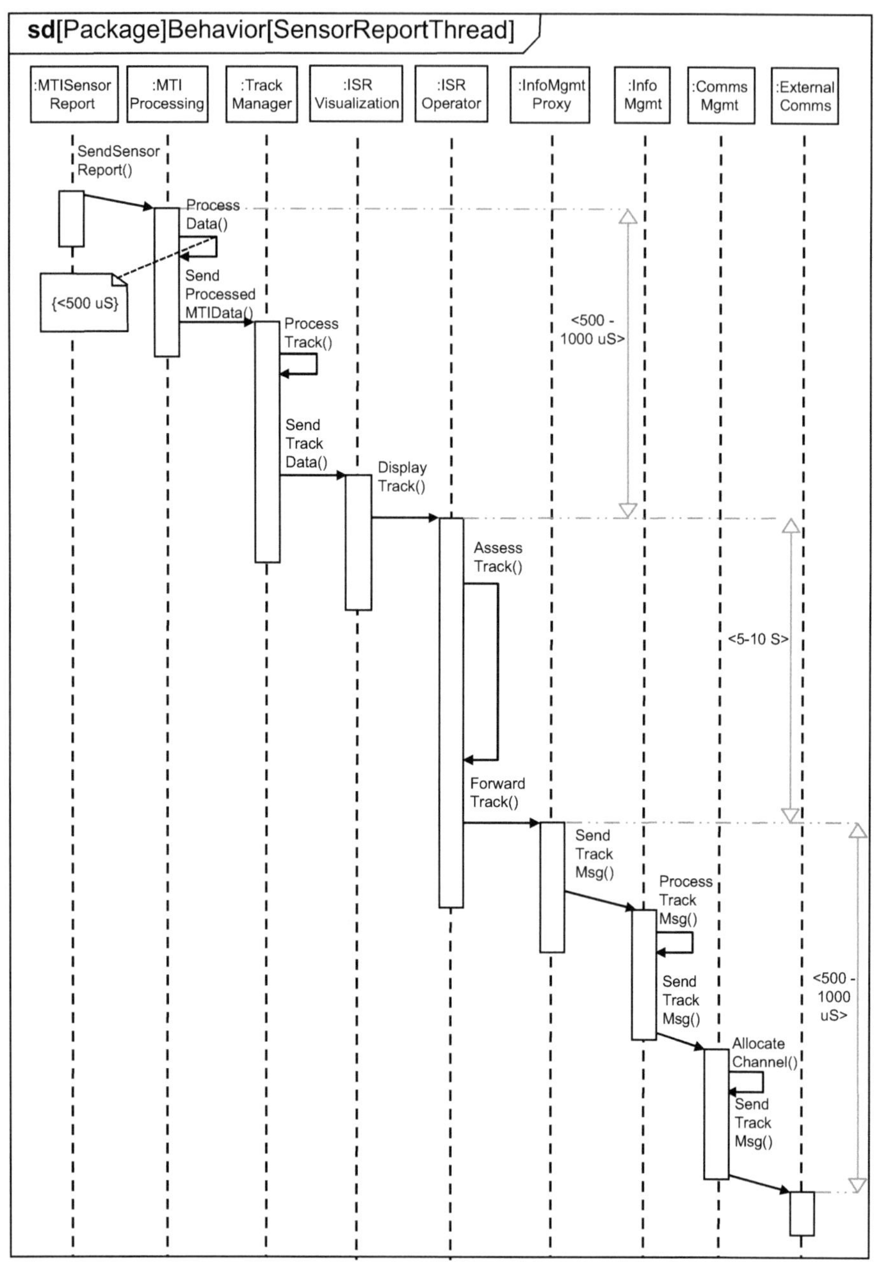

Fig. 5.28 Initial timing analysis of a Sensor Report Handling thread

5.9 Smart Microgrid LV

Having used the E-X system example to illustrate the flow and techniques of LV development, we will now present the equivalent content for our second example, the Smart Microgrid.

5.9.1 Smart Microgrid Structural Decomposition

Smart Microgrid A: We will explore the LVs of architecture models for the three variants of the Smart Microgrid (SMG) that were introduced in Chap. 4 to show how common model elements can be applied across a wide range of variations within a system category. Starting with the most elaborate, Smart Microgrid A for a community, Fig. 5.29 is a BDD that decomposes the Domain structure shown in Fig. 4.17. Graphical annotations using boxes with dashed outlines denote the Domains. Both the SMG and the Primary Grid (PG) are decomposed into their principal Parts. These were shown in Parts compartments in Fig. 4.17 and used in Fig. 4.18 to model the primary interactions between the SMG and PG. We will continue to use the idea of levels that was introduced in Fig. 4.17 as a way to organize the SMG structure.

A traditional "dumb" microgrid would have only the Physical Level with a substation to transform grid voltage to the voltage used by customers, a set of power distribution circuits, some protective devices, and the customer loads. The Supervisory and Device Levels add the functionality to optimize power generation and consumption. In modeling this SMG, we generally follow IEEE Std 2030-2011, *IEEE Guide for Smart Grid Interoperability of Energy Technology and Information Technology Operation with the Electric Power System (EPS), End-Use Applications, and Loads* [7]. The "intelligence" of the SMG is focused in the Microgrid Supervisory Control Block, which deals with the primary information and functions. A distinct Microgrid Human-Machine Interface (MGHMI) allows SMG personnel to monitor and direct operations, and a Microgrid Web Server provides an interface to the PG and to the world at large.

This SMG is implemented with a variety of devices for generating, storing, measuring, and controlling electric power. The microgrid has internal resources for power generation, including fossil fuel-powered generators, photovoltaic arrays, and wind plants. Energy storage in an array of batteries supplies power to smooth variations in generating demand and maintain service during outages. Smart meters apply sophisticated measurement and control of electricity consumption by loads, and various sensors are available to monitor conditions throughout the SMG. This is a very general SMG model that can be tailored to a wide range of high-end microgrids.

In Fig. 5.29, all of the Blocks comprising the SMG and PG are declared as Part Properties using Directed Composition associations. A Block named Device and

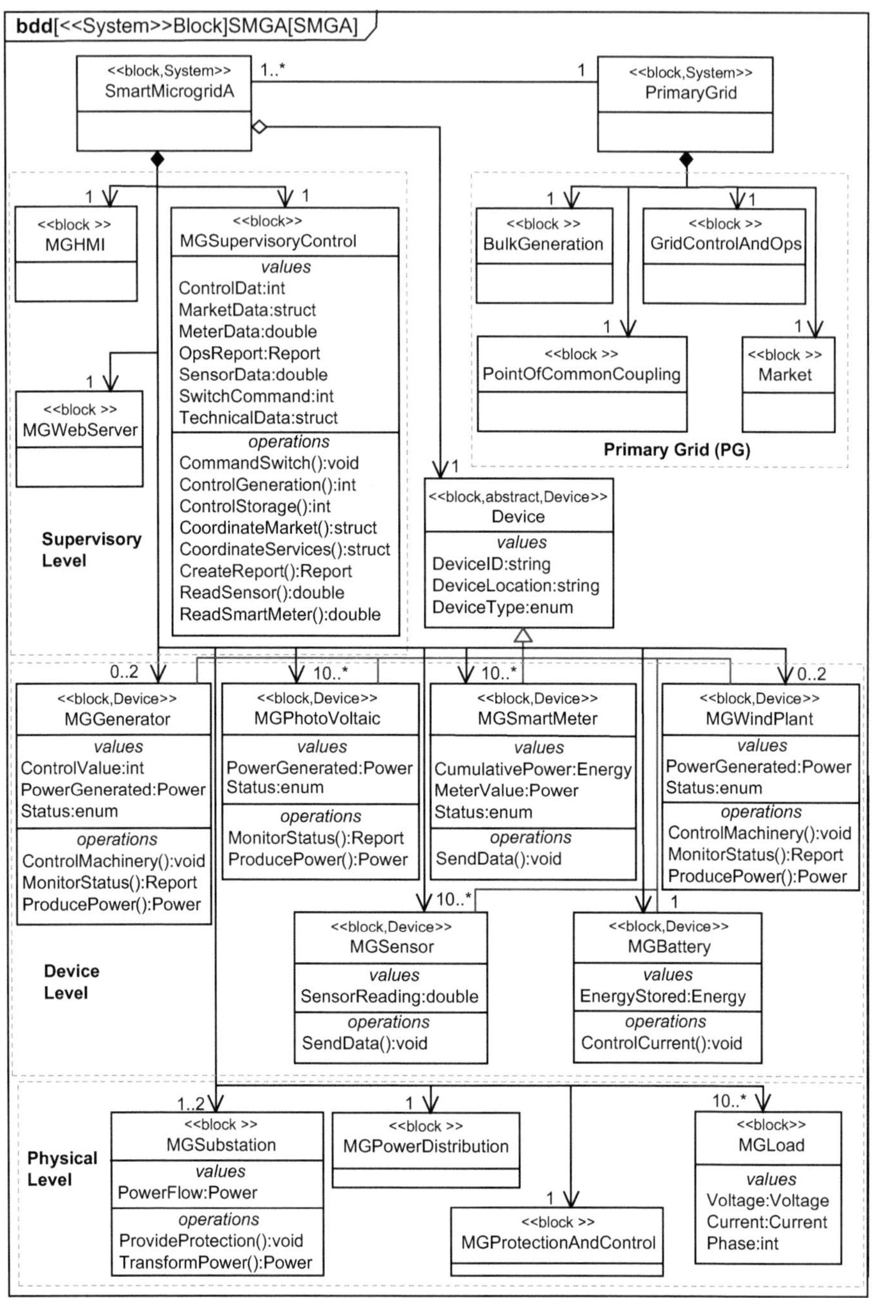

Fig. 5.29 Decomposition of Smart Microgrid A for a community, including interaction with a Primary Grid

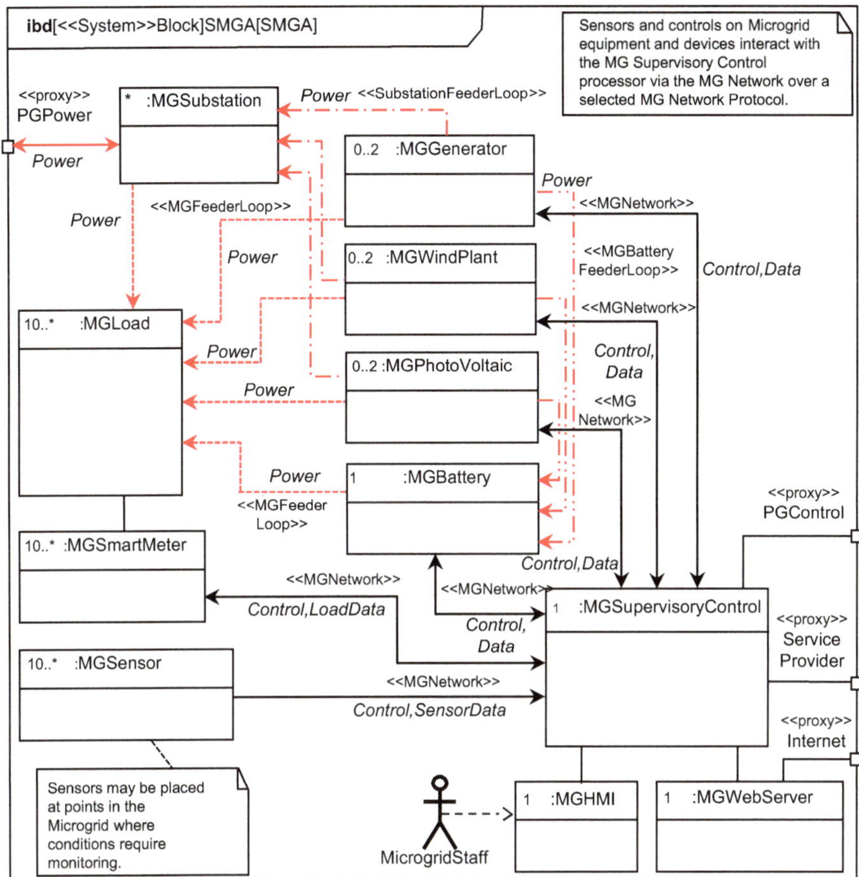

Fig. 5.30 Interactions among the Parts of Smart Microgrid A

stereotyped as `<<abstract>>` provides a convenient way to capture the common characteristics of SMG devices, which otherwise span a wide range of Values and Operations. For additional clarity, these Blocks are stereotyped as `<<Device>>`. Multiplicities in the diagram convey a sense of the widely varying numbers of devices, loads, and other microgrid elements that are possible, showing that this model can be applied to a broad assortment of actual situations.

Figure 5.30 is an IBD that complements the BDD by showing how the Parts of the SMG interact. Once again, red Flow arrows denote electrical power, and black arrows show exchanges of data. In addition, various line styles make it easier to distinguish the several feeder loops within the SMG. The Supervisory Control processor constantly monitors SMG devices and issues appropriate commands to power production resources. These power sources can connect to the SMG's loads via the MG Feeder Loop or to the Substation via a separate Substation Feeder Loop when surplus power is being delivered to the PG. Each load has an attached Smart Meter, and a variety of sensors can be deployed to monitor assorted conditions of interest such as weather parameters, battery charge level, and equipment health and status. The controller supports the Ports initially shown in Fig. 4.18, and the Web Server connects the SMG to the Internet.

The HMI used by the Microgrid Staff could be a laptop, a Web terminal, or some other workstation. Technical support personnel can manually invoke commands to supervised devices. For example, the staff might launch a different SMG management algorithm in response to a weather forecast indicating that PV or wind power availability is about to change. They can also act in response to unexpected events such as equipment failures or PG outages. A note on the IBD reminds the architecture team that a suitable network protocol must be selected for the MG Network as part of creating the PV. Use of widely supported open standards is essential to effective integration of the SMG as well as to smooth interaction with the PG. The Microgrid Supervisory Control computer and the Microgrid Web Server have the resources to realize the three Proxy Ports shown on the left of the diagram, while the Microgrid Substation implements the PG Power Port through which power is exchanged with the PG.

Smart Microgrid B: The next variant in this example, Smart Microgrid B for a large facility like a hospital, gives us the opportunity to examine a more quantitative approach to defining the LV. Without prematurely making design decisions that belong to the PV, we can nevertheless look at the way system requirements influence the functional architecture. The main motivations for the hospital SMG are to ensure uninterrupted electrical service to critical functions and to reduce costs by limiting power purchases from the PG. The key characteristics of Microgrid B are therefore:

- A PV installation able to meet all or most of the hospital load during daylight hours when facility power demand peaks, minimizing PG power consumption during the most expensive time periods
- Backup fossil fuel generators that can be quickly brought on line to deal with PG outages and that have their own battery-powered starters
- Battery storage to further smooth demand and help deal with conditions such as cloudy weather that reduce PV output
- A switched connection to the PG, but not a separate Substation as included in Microgrid A
- A facility power distribution design that groups internal hospital loads into zones that are prioritized by the criticality of the functions each supports

Representative values for this SMG, subject to refinement to satisfy performance requirements, are as follows:

- Facility size: 40,000 sq ft
- Electric load density: 8–10 W/sq ft (daytime peak), 6–8 W/sq ft (average)
- Total load: 0.3–0.4 MW (peak), 0.25–0.3 MW (average)
- Renewable energy from PV panels:
  - 1 MW for 8 h, bright sunshine
  - 0.5 MW for 4 h at low sun angles in bright sunshine

– Lower power levels under reduced solar flux conditions

- Battery capacity: nominally 0.5 MWh

The hospital electric load per square foot is assumed to be approximately twice that of a typical office or residential building due to a combination of extensive power-consuming equipment and energy-inefficient environmental systems that are needed to provide controlled temperature and humidity on a per-room basis. We assume the facility is sited in a location with prevailing sunny weather, making PV the renewable energy source of choice. Electricity demand is high during business hours when most patient care procedures such as surgery, diagnostic testing, and therapy take place. The normal operating mode has the PV panels providing all or most facility power during this peak period with battery and PG power brought in as needed to meet instantaneous demand. At other times, and when solar flux is reduced by clouds for extended periods, the hospital purchases PG power. Most such transactions will be at off-peak reduced rates. In case of an outage, the storage battery allows electrical service to continue for at least an hour, allowing for orderly termination of noncritical activities and startup of the backup generators and, if necessary, shedding low-priority loads to preserve critical activities.

Figure 5.31 shows a structure based on these considerations, and Fig. 5.32 shows the primary flows of information and power. Comparison to Fig. 5.29 shows that this is effectively a subset of the Smart Microgrid A architecture, and the Blocks have similar Values and Operations to their counterparts in Smart Microgrid A. We have deleted the Wind Plant since this is not feasible in an urban setting. Key con-

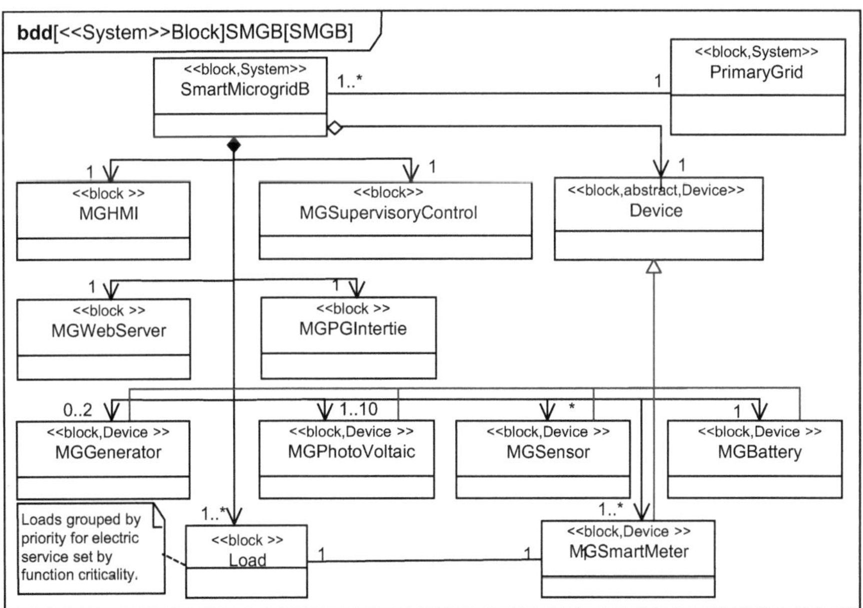

Fig. 5.31 Decomposition of Smart Microgrid B for a major facility

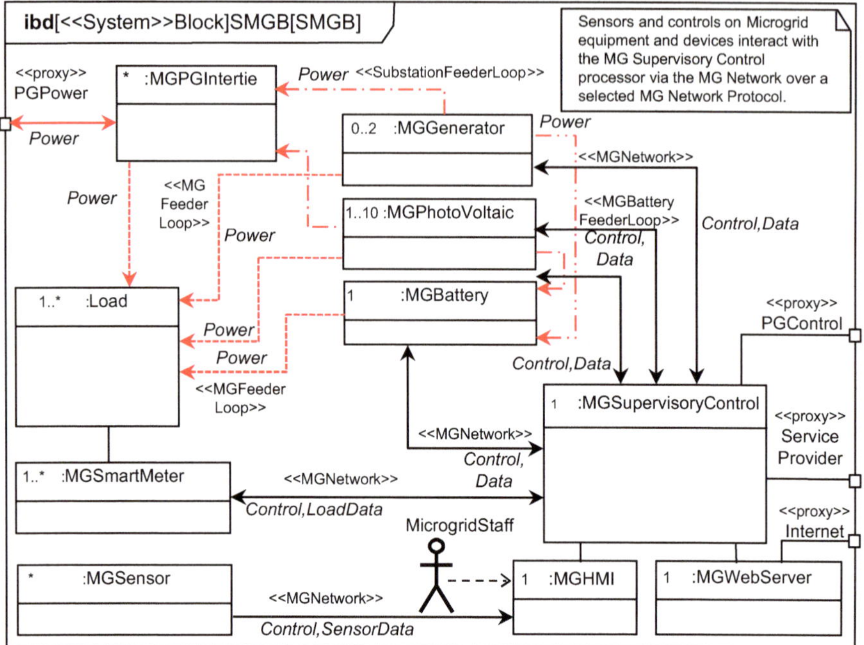

Fig. 5.32 Information flows and external interfaces of Smart Microgrid B

siderations such as the area of solar power arrays and backup power generation capacity required are reserved to the point design in the PV, but feasibility checks are very much in order at this point. For example, 1 MW of solar power would typically require substantial space for photovoltaic arrays. If the area is too great for a rooftop installation, the architecture team will want to confirm that adjacent space is available before proceeding with this approach. A typical compromise would be to relax the requirement for distributed generation equal to the full facility load in favor of supplementing PG power in a fashion that minimizes its cost, especially in peak periods. In any case, much of the LV content is common to both grid-connected SMGs, including lessons learned in early designs and the basic functions of the Supervisory Control processor.

Smart Microgrid C: The final SMG variant in this example is designed to provide a level of electrical service to a group of homes and small shops that do not have access to a PG. Such microgrids are especially valuable in developing countries where they can have a significant impact on quality of life [8]. Designs vary over a wide range of SMG sizes, types of service, costs, and other parameters. An important goal is to achieve high reliability of electrical service at the lowest feasible cost to customers. Figure 5.33 shows the structure of this microgrid. Although it shares important characteristics with the larger microgrids, a system this small and relatively simple needs its own tailored design approach to achieve both customer service and economic goals. Figure 5.34 shows the flows within SMG C.

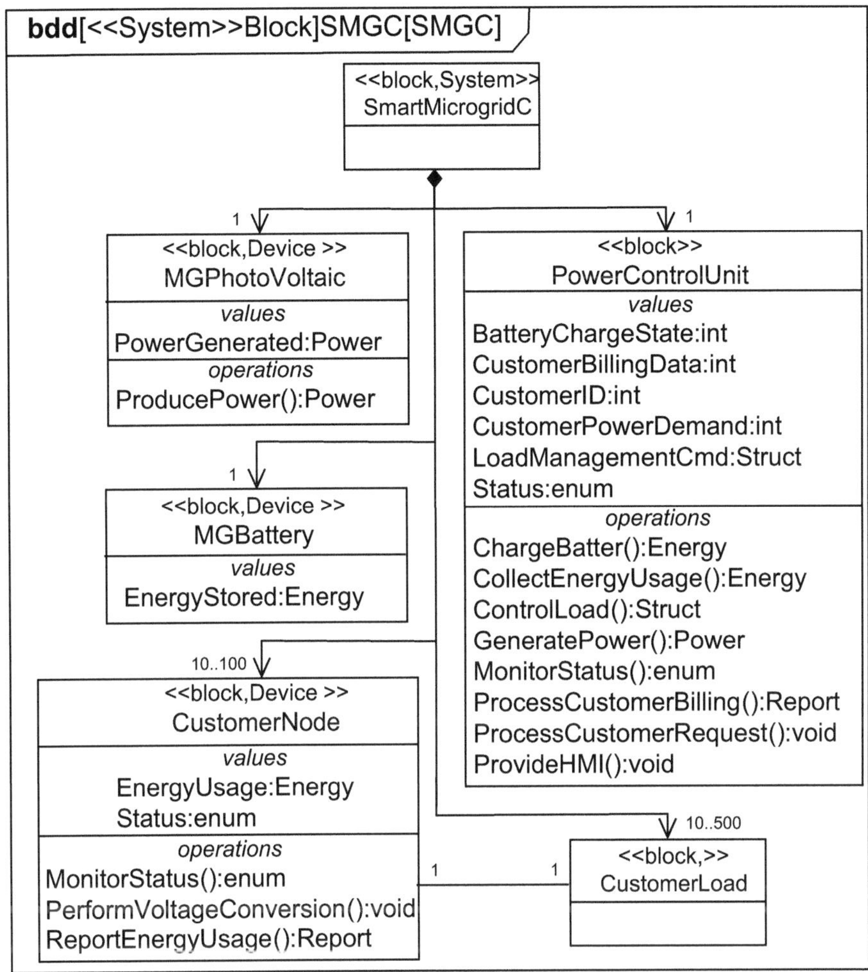

Fig. 5.33 Decomposition of Smart Microgrid C for a remote or rural community with no access to a Primary Grid

Smart Microgrid C consists of a centralized generation, storage, and control system, distributing power to 10–100 homes. The central facility is referred to as a Base Station. Photovoltaic panels produce 300–500 W of power each and can be combined as needed to supply the customer community. Power can be distributed as either AC or DC. Each home has a customer node that incorporates a smart meter that monitors power usage and sends usage and status data to the Base Station. The customer node may also perform voltage conversion, e.g., to output 5 VDC to USB jacks. In a typical SMG of this type, the battery is sized to provide 2.5 kWh of energy storage for each 500 W of PV generation, and power is distributed as 48 VDC. As with PV capacity, battery size can be scaled up depending on the number of customers. Each home is limited to 50 W of consumption, which provides enough

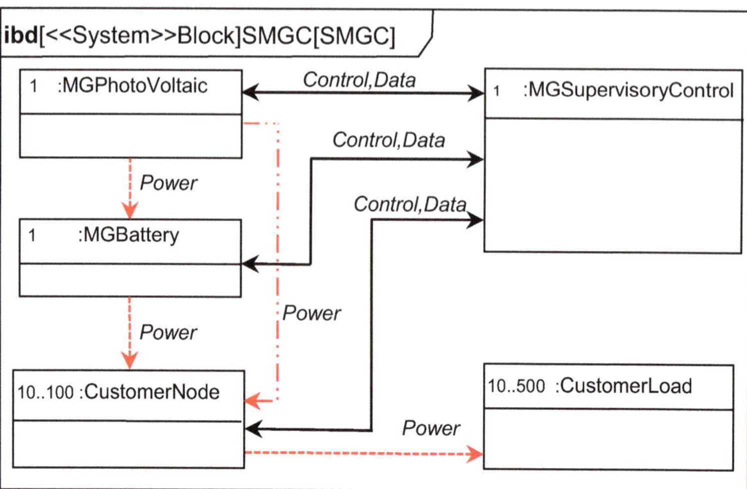

Fig. 5.34 Power and Data Flows in Smart Microgrid C

power for lighting and cell phone or laptop computer charging. The SMG uses a wireless network implemented with mobile cellular devices to connect customer nodes and the Base Station. Within the Base Station, a Power Controller Unit manages power distribution; performs intelligent load management to maximize service reliability, including on cloudy days; and handles billing and other business aspects of the SMG. It provides a basic HMI with indicators and controls to enable a local manager to monitor health and status and to manually intervene in SMG operations when necessary.

The basic operational concept of this SMG is to charge the battery during the day and supply a period of power for lighting (assumed to be efficient LED lights) to customers after dark. Several control strategies, selectable by the SMG manager, can be implemented, one example being to disconnect certain loads in order to maintain highest priority service during periods of curtailed PV power output. An attractive addition would be to extend the wireless network to enable Internet access in areas where the appropriate service is available.

This microgrid is obviously much simpler than the two preceding varieties. However, the fundamental architecture pattern is similar, and there are opportunities to provide the most, and the most flexible, service to the remote consumer by intelligently managing the limited available power. By including the components that make up this isolated microgrid in a model library or reference architecture, a supplier of such systems spanning a range of applications can capture product data in a common repository, share and exploit lessons learned, and reduce the time and cost to develop new systems tailored to various circumstances. As one example, the lower-power microgrid could be scaled up to provide more power generation and energy storage and enhanced further to provide power for economic development purposes, such as running a grain mill or providing refrigeration for agricultural products. Then if PG power becomes available at a future time, the installed microgrid base could be upgraded at less effort and cost to take advantage of this higher level of service.

5.9.2 Smart Microgrid Behaviors

As an example of behavioral modeling for the SMG example, we will examine how the Microgrid Generator of SMG A responds to commands from the Supervisory Control processor as shown in Fig. 4.20. Figure 5.35 is an SMD for the generator Block. We assume the generator has a controller that receives and carries out commands from the Supervisory Control Block. When the generator is commanded to start up, it enters a Startup State with an entry Action to execute the sequence of steps involved in safely spinning up the machine to prepare to generate power. The specifics of the startup could be modeled in a yet more detailed behavior diagram. Once the generator is running, a set of self-test routines verifies that operation of the machinery is within acceptable limits. If one or more of these tests fail, the generator control automatically transitions to a Shutdown State and sends a failure report to the Supervisory Control.

Given a successful self-test, the generator can transition either to a Generate Power State for normal operations or to a Maintenance State for scheduled maintenance or a repair task, with either transition initiated by a command from the Supervisory Control. In the Generate Power State, the generator first obtains the

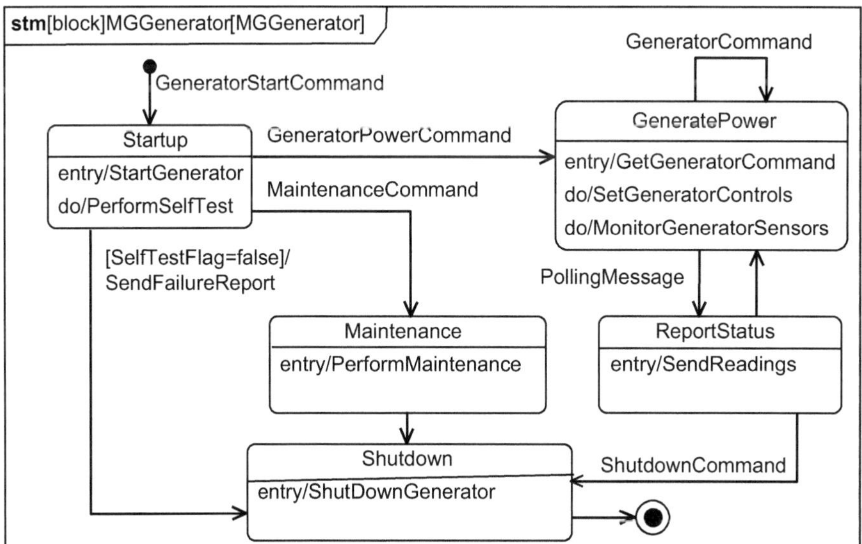

Fig. 5.35 State Machine Diagram showing the behavior of the SMG A generator

current Generator Command, which causes the generator controller to set machinery controls to provide a specified level of power to the Microgrid Substation or to the Battery. If a new Generator Command arrives, there is a self-transition back to the entry Action to obtain and implement the new command. In the State, the generator controller adjusts the machinery to maintain the desire power output and continuously monitors sensors, e.g., to detect an over-temperature condition. Referring again to Fig. 4.20, when the Supervisory Control executes its Poll Devices Action, a command goes to the generator to report its current sensor readings, triggering a transition to a Report Status State. When a maintenance task is completed, the generator transitions to the Shutdown State. The Supervisory Control can also send a command to cause shutdown. Once the generator executes a safe shutdown sequence, the behavior ends.

5.9.3 Smart Microgrid Logical Data Model

Figure 5.36 shows the Business Data Foundation Class from the CDM of Smart Microgrid A (Fig. 4.21) specialized into a set of System Data Blocks. Business data is used throughout the PG and SMG, making a consistent and effective model of that data very important. The diagram shows three System Data Blocks that account for data involving customers, the SMG as a whole, and a Business Plan that guides

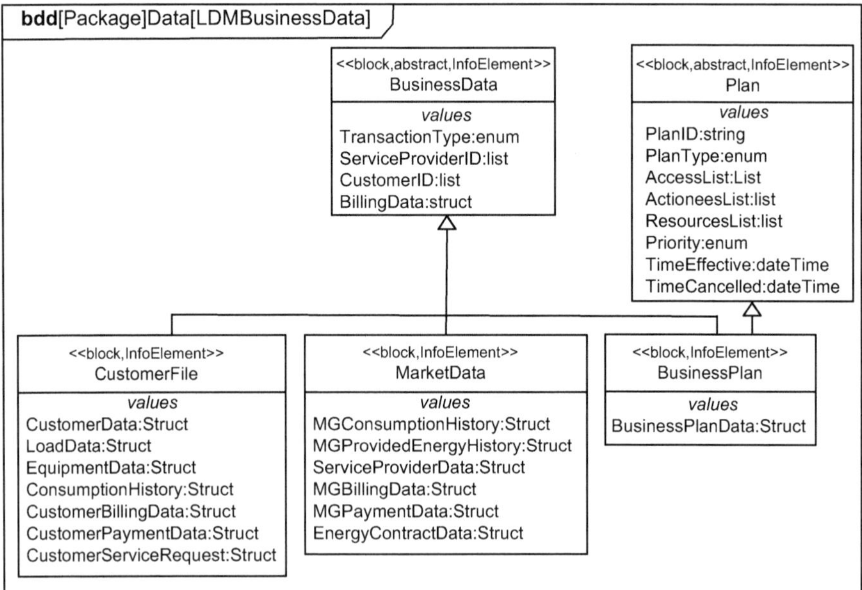

Fig. 5.36 Business Data Foundation Class from the CDM of Smart Microgrid A specialized in System Data Blocks

long-term growth and evolution. The Business Plan Block inherits from both Business Data and Plan Foundation Classes, which must be correctly and consistently defined to avoid a conflict. This sort of multiple inheritance is quite common in a complex data model. A complete LDM would contain similar diagrams specializing the other Foundation Classes in the CDM to define the full information content of the system. For a microgrid that interacts with a PG, care must be taken to accurately reflect data definitions used by the PG, which is one example of the general concern with developing and maintaining a system data model that supports correct and unambiguous interactions with its environment.

5.10 Interface Modeling and Control

Interface management is a core SE responsibility, and MBSE supports it in a number of ways. Proper interface control is an essential element of successful system integration as well as long-term system evolution. As discussed in Chap. 1, interfaces and the standards they implement are central to achieving open architecture. Among the concerns associated with interfaces are:

- Identifying all internal and external interfaces and the interactions that each supports (e.g., exchanges of messages or files)
- Ensuring that all aspects of an interface are defined and documented
- Ensuring that an interface between system components that are controlled by different organizational elements is understood and agreed to by the parties involved
- Maintaining rigorous configuration management in the face of inevitable changes in interfaces over the course of system development and operation
- Ensuring that appropriate open interface standards are identified and implemented

Interface definition in MBSAP follows the flow of architecture development. In the OV, attention is focused on interfaces among system Domains and between the system and its environment. The LV identifies the actual points of interaction associated with Blocks and defines their identities, Values, Operations, and other essentials. The E-X and Smart Microgrid examples have furnished multiple examples. Finally, the PV completes interface definition with details ranging from components such as connectors and fittings to voltages, waveforms, and flow items.

In SysML, interfaces are most commonly associated with Ports as described in Chap. 2, and a Port is typed by the associated interface or interfaces. This can be modeled in several ways. The specification of a Port is a convenient place to document the interface(s) and to provide details, either directly in the element specification or by a link to an interface definition document. In addition, SysML provides a specialized `<<interface>>` Block to type a Proxy Port in which the details of an interface, including Values and Operations, can be captured. These Blocks should be defined in a designated area of a model so that they can be used whenever a Port needs to be typed. In Chap. 6, we will return to this topic and discuss physical modeling of interfaces.

5.11 Data- and Event-Driven Architectures

In the early stages of design in the LV, the architect may find it appropriate to apply either or both of two general approaches or styles that are usually called Data-Driven and Event-Driven Architecture (DDA, EDA). The first of these, DDA, can be a useful way of thinking about systems where the availability and management of data are major drivers and of system design and functioning. DDA complements SOA and typically depends on a well-defined set of Data Services. There are two primary aspects:

- The LV starts with definition of data entities, modeled as System Data Blocks, and the Structural and Behavioral Perspectives are then developed primarily on the basis of data import, export, storage, and processing.
- System activities are largely triggered by data availability; a processing step is suspended until all required data inputs are present and then executes to create an output, which may, in turn, trigger the next step. This sort of behavior is effectively modeled using Activity Diagrams with synchronous Join nodes that combine the various data inputs to an Action.

In contrast to DDA, EDA is appropriate when system behavior is heavily driven by asynchronous events that may be internal or external to the system. The following are some key features of an EDA design.

- System activities are triggered by receipt of signals associated with event occurrences, which can include completion of predecessor activities, circumstances in the external environment, operator inputs, and many other things. Activities can be composed into workflows designed so that specific events cause successive steps to execute, including conditional or alternative paths.
- EDA supports a high level of loose coupling because dependencies among system elements are largely reduced to generation of and response to events.
- EDA behavior is often well modeled using State Machines in which events trigger changes of state (transitions) that can have associated Actions (including issuing new events) and can be guarded when conditions must be enforced in enabling a transition.
- An architecture function called Complex Event Processing is based on analyzing patterns of simple event occurrences to infer complex events. Such a process typically maintains a buffer of messages or signals and compares these to a set of templates that identify a complex event like the loss of a network node or termination of an external process.

5.12 Executable Architecture

As discussed in more detail in Chap. 11, tools are available to make a SysML model executable in a variety of forms, essentially by converting static behavioral diagrams into simulations. Many of these are based on animating Activity, Sequence, and State Machine Diagrams, and there is a trend in the IT community toward extending these executables to include automatic code generation (autocoding). This requires that the model be complete and consistent, with action semantics to fully describe the intended system behavior. More details can be found in [9]. Additionally, the execution environment or platform on which the software will run must be completely described to the tool. This is the basis for a common strategy for model validation because an attempt to simulate a diagram will fail if there are errors, and the tool can highlight the points where the simulation halted. As a less ambitious first step, executables can be viewed as a way to refine the early performance and timing analysis and to visualize performance issues for a variety of stakeholders.

One approach is to extend a Process/Workflow model from the OV. By selectively decomposing model elements to insert greater detail where performance and timing questions have been identified, the analyst can often get useful results while keeping the total modeling effort within bounds. For example, if the static timing analysis indicates a possible problem with getting large data files into and out of disk storage or through the network, the executable can add elements to represent the buffers, network interfaces, internal data paths, file processing, and other steps in the data transfer chain. Such a model can then be driven with statistical representations of things like message sizes and file transfer rates to explore the behavior of the various parts. A typical result would be to establish that the internal bus of a file server needs a higher than anticipated throughput, which then becomes part of the requirements that drive the actual computer configuration in the Physical Viewpoint.

Another kind of executable that may be useful at this stage is built using State Machines to model the behavior of a Block or collaboration of Blocks. Modern tools for state machines allow the modeler to insert actual code or algorithms and to calibrate the time associated with a process or event sequence using, for example, temporal operators that count "ticks" of a master clock. Some tools allow mixed modeling of continuous and discrete processes so that, for example, a state machine representing information processing can be combined with a continuous model of physical processes. The architect can model the execution of a complex process by associating it with a state and programming an appropriately scaled time delay representing the estimated run time. Some architecture tools support linkage of a Parametric Diagram to an external mathematical "solver" that evaluates the equations expressed as Constraints in the SysML model. Collectively, techniques like these can allow a model with adequate fidelity to answer important design questions to be built and run with a relatively modest level of effort and before physical system components are available for measurements and experiments. Chapter 11 has additional discussion about these topics.

5.13 Summary

The development of an architecture using MBSAP has now reached the point of constructing, modeling, and analyzing a functional design. Structure is defined down to Blocks that represent actual system elements, and we have examined the behaviors of individual Blocks and groups of Blocks working together. In the process, the rich literature on design patterns has been mined in search of proven solutions to equivalent challenges. LV development has also considered the kinds of layered SOAs that are appropriate to various system performance categories, and the E-X example system illustrates tailoring of the best-matched design patterns to meet the needs of a particular system. The Conceptual Data Model from the OV has been elaborated into more detailed diagrams defining System Data Blocks, and from these XML schemas can be created. This level of architectural detail has allowed early performance and timing analysis with both static and executable representations of interesting areas of the architecture.

The artifacts of the LV, summarized in Fig. 5.37, become design requirements and specifications for the actual system implementation in the Physical Viewpoint. In fact, the ability to create design and procurement specifications that allow physical system components to be developed or procured is the best evidence that the LV

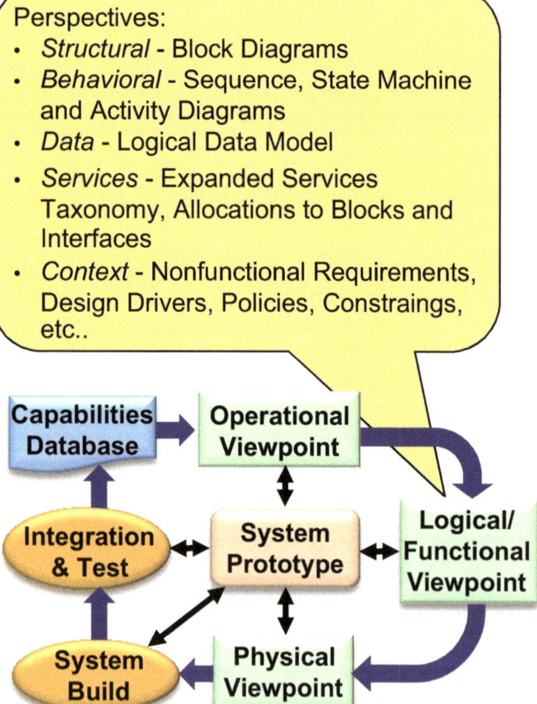

Fig. 5.37 Summary of the Perspectives of the Logical/Functional Viewpoint

is complete. If a decision is made to implement the functionality associated with a Block as a software component, the Block specification, including Values, Operations, inputs and outputs, descriptive notes, and other details, can serve as a Software Requirements Specification (SRS). Similarly for a hardware component, a Block specification becomes the basis for a new design or for the procurement of an existing product, often based on the outcome of product trade studies. If a commercial-off-the-shelf/Government-off-the-shelf (COTS/GOTS) infrastructure is to be used in the layered SOA, the Service definitions in the LV provide the basis for selecting and configuring the products to be integrated. Results of performance and timing analysis are important factors in specifying and selecting hardware components. Chapter 6 explores this next set of steps in translating design into implementation.

Exercises

1. List some reasons for using design patterns in developing a system design.
2. What would be a reason to decompose a Domain from the OV in two stages to Subdomains and then to Blocks?
3. Take the applied design pattern embodied in Figs. 5.3 and 5.4, and show how it could be used to decompose the Control Room of an electric power plant that schedules, monitors, and controls the operations of generators and other equipment.
4. What are the primary uses of Sequence and State Machine Diagrams in the LV?
5. Consider the autonomous navigation function of a self-driving car. What type of behavior diagram would be best used to model its operation in responding to external events, and why is this the best choice?
6. For the State Machine Diagram in Fig. 5.7, what would happen if a new outbound message is received while the Class is in the Establish Link State?
7. What does it mean in a BDD when a subordinate Class or Block has both Generalization/Inheritance and Composite Associations with a higher-level Block?
8. List categories of Services and identify Service layers in a typical layered SOA.
9. Why might it be important to include a layered SOA in the LV?
10. Consider a SOA for the information system used in a public library to maintain the catalog, support check-in and check-out of materials, exchange requests with other libraries, and so forth. What service layers would be appropriate, and what would be representative services in each layer?
11. What is the difference between an Application Server and a Transaction Server?
12. In a real-time SOA, why is the HMI likely to be integrated directly through a hardware interface rather than via a User Services or Presentation Layer?
13. Consider the information system supporting a major hospital. Sketch a CDM with Foundation Classes for the primary types of information involved in the system.
14. How can a Sequence Diagram be used for initial timing analysis of the behavior of a Block or Group of Blocks?

15. Redraw Fig. 5.18 modeling the layered SOA with Blocks instead of Packages and showing services as Part Properties of each Block.
16. Suppose the Microgrid Supervisory Control unit in Smart Microgrid A (Fig. 5.29) were dual-redundant for safety and reliability reasons. Further suppose that a Primary Control unit normally operates the microgrid with a Backup Control unit monitoring the Primary and taking control in the event of a failure. Sketch a behavior diagram showing this.

Student Projects
Students should model the content of the five Perspectives of the LV by decomposing and supplementing the OV. Good model structure should be maintained, e.g., by creating more detailed diagrams beneath the associated higher-level diagrams.

References

1. Gamma E et al (1995) Design patterns: elements of reusable object-oriented software. Addison-Wesley, Boston, MA
2. Shalloway A, Trott J (2001) Design patterns explained: a new perspective on object-oriented design. Addison-Wesley, Boston, MA
3. Constant JN (1972) Introduction to defense radar systems engineering. Spartan Books, New York
4. Stimson GW (1998) Introduction to airborne radar, 2nd edn. Scitech, Mendham, NJ
5. World-Wide Web Consortium (W3C) (2015) eXtensible Markup Language (XML). http://www.w3.org/XML/. Accessed 20 May 2017
6. Maier MW (2006) System and software architecture reconciliation. Syst Eng 9(2):146–159
7. IEEE (2011) Std 2030-2011. http://standards.ieee.org/findstds/standard/2030-2011.html. Accessed 21 May 2017
8. MeshPower (2016) Connecting communities with clean energy. https://www.meshpower.co.uk/how-it-works.html. Accessed 20 May 2017
9. Mellor SJ (2013) Executable UML. http://www.stephenmellor.com/uploads/XTU,%20doc,%20 Executable%20UML.pdf. Accessed 6 Feb 2013

Chapter 6
Implementing in a Physical Viewpoint

6.1 General Considerations

6.1.1 From Functional to Physical Design

In this chapter, we consider the process that transforms the functional design in the LV into a physical implementation composed of hardware, software, other equipment, data, user interfaces, communications, networks, and all the other elements of the actual system or enterprise, including applicable standards. We use the generic term "product" to refer to the pieces of a system. In some cases, suitable products already exist and can be procured "off-the-shelf" from vendors. In others, something new or different may be needed, and new products may be developed using internal resources or obtained by contracting with external organizations. Quite often, a series of "make/buy/modify" decisions must be addressed to settle on the optimum approach to meeting specific system needs. In general, the functional specifications that are a key outcome of the LV are the basis for all these activities. The PV models and documents both the characteristics of the selected products and the way in which they are integrated and used to satisfy system requirements.

Figure 6.1 summarizes some of the primary elements, beyond the obvious hardware and software components, of completing a successful system or enterprise solution to a customer need. We use the term "mission" to emphasize the importance of maintaining focus on delivering the capabilities needed by the customer to meet organizational goals. Because the decisions at this stage depend heavily on the specifics of the system being architected, a procedure as detailed and uniform as those we've laid out for an OV or LV is not practical. Nevertheless, some important general principles can be applied that will improve the chances of a satisfactory outcome and that efficiently exploit the foundation established in the earlier Viewpoints.

We pointed out in Chap. 3 that MBSAP combines the Structural and Behavioral Perspectives of the OV and LV into a Design Perspective because both structure and

J. M. Borky, T. H. Bradley, *Effective Model-Based Systems Engineering*,
https://doi.org/10.1007/978-3-319-95669-5_6

Fig. 6.1 Elements of a mission solution

behavior are embodied in individual products and in their integrated configurations. At the same time, we add a Standards Perspective to handle this vital aspect of a physical design. Integration of components into subsystems and subsystems into a system is usually an essential part of constructing the PV, and we use the layered architectures from Chap. 5 as a framework for discussing integration. This is a natural way to consider the creation of the system infrastructure, the integration of mission services or business logic onto the resulting platform, and the crafting of human-machine interfaces (HMIs) and other user services. The Data, Services, and Contextual Perspectives complete the basic structure of the PV.

An essential task at this stage of MBSAP is to model, analyze, and document the physical components that implement the functional Blocks from the LV. In SysML, a Block becomes a specific tangible entity in the form of a Block Instance. Integration of these products is typically directed first to one or more system prototypes and, ultimately, to the delivered solution. In an incremental, spiral, or agile development, each prototype adds functionality until the full set of system requirements is satisfied, and the results obtained with each interim prototype are used to refine the overall requirements and to choose the specific capabilities (new or refined) that will be allocated to the next development period.

We need to be careful with the terminology we use to describe the transformation of a functional design in the LV to a physical implementation in the PV. Key terms are "instantiation," "realization," and "implementation." It can be tempting to use these words interchangeably, but they mean different things. In MBSAP, we use the following definitions, consistent with the SysML language standard:

- *Instantiation*—A Block Instance is a particular use or example of a Block. As noted above, it is defined by giving numerical or other quantification of Block Values, which may be done using a Block Instance Specification, and it may extend the characteristics of the parent Block. A Composition Association makes a Block Instance a Part Property of another Block Instance. Block Instances also appear at the heads of lifelines in a Sequence Diagram. Creating Block Instance Specifications is commonly an aspect of transforming the LV to the PV.
- *Realization*—A Realization Dependency shows that one model element, in the role of a client, realizes a specification that is defined for another model element in the role of supplier. This is commonly used to model an Interface being realized by a Block or a design package being realized by an implementation package. Such dependencies can be useful in relating PV content to LV content. It's correct, but not very useful, to say that the PV realizes the LV.
- *Implementation*—Implementation is the generic term we use for the process of creating a physical system, modeled in a PV, from its functional definition in an LV. Accordingly, physical products implement functional model elements and must satisfy their requirements.

Hardware Blocks are implemented by the various items, ranging from sensors, controllers, vehicles, and machinery to processors, networks, data storage, and workstations, that comprise the hardware resources of the system, and the PV incorporates the physical product specifications that define these items. Software Blocks are instantiated as executable code that can be stored in the system and loaded into computer memory to be run. The code that was specified in functional requirements in the LV must be written, tested, and integrated, with the resulting documentation and software files captured in the PV. A model element that describes a complete functional entity including both hardware and software should be decomposed to clearly define these in subordinate Blocks, Parts, and Block Instances. Although SysML eliminates the UML Deployment Diagram, it provides equivalent means to show the deployment of software entities onto hardware resources, primarily using Allocations.

Critical system functions may require the use of products or systems that have undergone various kinds of testing, certification, or accreditation. Two common situations are systems trusted to process sensitive (including classified) information and those with safety, accuracy, or reliability requirements. Such systems must typically incorporate trusted components ranging from secure computer operating systems and mission applications to hardware with guaranteed reliability or precision in order to meet their criteria. Subsequent chapters consider these issues in more detail.

In terms of the model itself, much of the content of the PV results from adding this physical detail to the specifications of the model elements that were defined in the LV, thus creating Block Instances, rather than from creating a new "physical" model. This physical detail can be captured in the appropriate specifications within the model or in separate documents that are linked to those specifications. If product selection or design decisions were based on trade studies, it may be useful to incorporate the documentation, typically by reference to a design archive, to ensure this rationale remains readily available so that future engineers and system users can understand the design decisions that were made.

Sometimes, functions must be shared across physical components, and there may not be a perfect one-to-one mapping of model elements from the LV to the fully detailed PV, but this will be the exception if the LV is well partitioned. Another basic consideration is appropriate use of physical modeling to diagnose problems and find solutions. The complete development cycle includes testing and the opportunity to fix the things that prove to be wrong or inadequate.

6.1.2 Trade Studies

A major benefit of MBSE comes from its ability to support a wide range of trade studies. Trades can involve setting the values of performance requirements, comparing alternative system designs and operating concepts, choosing among candidate physical components, and many other things. For purposes of the PV, the primary trade studies deal with selecting the best products with which to implement the system. These may be preceded by trades aimed at make/buy/modify decisions. Figure 6.2 shows the flow of a typical product trade study, and architects and designers should be familiar with the basics of such an analysis.

With MBSE, a system model supports a trade study in a variety of ways, including:

- Supplying context to frame the trade, including identifying alternatives, relevant factors, system-level constraints, etc.
- Identifying influences, e.g., dependencies on internal and external resources
- Modeling the structure and behavior of the alternatives
- Providing technical data for the trade analysis

Existing product data, supplemented when necessary by research and supplier contacts, can be archived in or accessed by the model and helps establish the set of available alternatives. Mandatory characteristics like those discussed earlier will often narrow the field of possible choices. Then the implementation team, coordinating with program management and the affected stakeholders, defines the criteria that are most important. These typically include cost, technical maturity (or its inverse, risk), reliability, performance against the most stressing requirements, support that the product requires from the host system and its environment, long-term product evolution plans, and compatibility with the overall architecture. The selected factors are weighted, and candidate products are scored to arrive at the best choice or to narrow the field to a small set of candidates for further evaluation.

An important factor that can be hard to assess simply from product data and specifications is a product's ease of use, as seen both by system developers and by

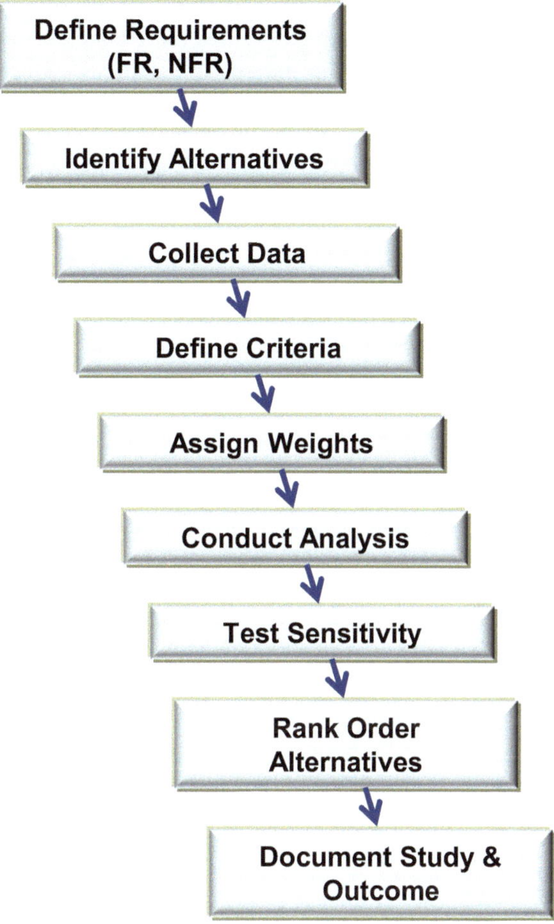

Fig. 6.2 Elements of a trade study

end users of the system. Vendors are unlikely to highlight the fact that their offerings require excessive effort or special skills and training in order to integrate and use them. For these and other reasons, the best product among several similar candidates may not be initially obvious, and an evaluation in the laboratory or as part of the system prototype may be called for. One of the great benefits of an open, modular, loosely coupled architecture is that products that don't work out despite this early evaluation, or that are superseded by better ones, can be replaced without wrecking the overall system design.

In short, the primary decisions associated with the PV have to do with product selections and related design decisions such as the choice of technical standards. These decisions are usually addressed through trade studies. Selections may come down to make/buy/modify decisions, but they are often complicated and involve much more than purely technical factors. Economics, commonality and interoperability among systems, and customer policies and requirements are among the considerations that can constrain the designer's choices. For example, the customer or the developing organization's policies may require conformity to reference architectures and enterprise standards or use of standard products. Even without such constraints, reuse of existing products, often in the form of commercial- or Government-off-the-shelf (COTS/GOTS), may prove to be the best answer. When a project involves modernizing or otherwise modifying an existing system, there are often difficult choices involving retention or replacement of legacy products. As an example of economic influences on product trades, established vendor and partner relationships may lead to selection of a particular product or product family because of advantageous pricing, access to proprietary product data, or other aspects of an established vendor relationship.

6.2 Design Perspective

SysML provides the tools to effectively model a physical architecture, especially when there are significant design issues associated with hardware, including integrated system performance, requirements allocation, physical interface management, and so forth. Earlier chapters have discussed the use of Blocks, Ports, Parts, and other structural model elements. The emphasis in creating a PV is on Block Instances.

6.2.1 Physical Detail in Block Instances

In a diagram, a Block Instance has a Name (or Namespace) compartment with the general form: `<name>:<type>` where `<type>` is the name of the Block that types the Instance, and `<name>` is a unique identifier for the Instance. It's often not necessary to give Instances their own names, and so the full syntax is commonly elided to simply: `<type>` to show that what's being represented is an Instance of the named Block. Various architecture modeling tools have various provisions for declaring Block Instances and labeling them in diagrams. A common example occurs when a Block is placed in a Sequence Diagram to create a lifeline.

An Instance Specification showing the particular values for the Block's Value Properties is shown in a BDD as a rectangle with the same name as the Instance, except underlined, as in: <instance name>:<block name>, with a single compartment giving values for individual properties; this may be important if these Instance values override default values specified for the parent Block or if additional properties must be defined that aren't in the parent Block. In addition to a Block Instance Specification, which is defined in a BDD and deals with a Block's Values, the model normally includes a more comprehensive Model Element Specification that captures additional information. Typical content of a Model Element Specification for a Block includes:

- *Nomenclature*—product model number and name, revision number, and date.
- *Description*—summary of features and functions, product family, intended usage, implemented standards and practices, etc.
- *Links to drawings and specifications*—traditional design data may be useful.
- *Physical characteristics*—dimensions, weight, power consumption, installation, cables and connectors, etc., sometimes including a Size, Weight, and Power-Cooling (SWAP-C) table (this could be placed in an Instance Specification).
- *Software documentation*—depending on the software methodology employed.
- *Environment*—temperature range, shock and vibration, power quality, etc., in which the component is guaranteed to perform correctly and achieve specified reliability.
- *Maintenance, servicing, and repair data*—essential for system support.
- *Other*—additional information required to install, integrate, operate, upgrade, support and, potentially, replace the component at a future time.

SysML Allocations have a significant role in the PV. An Activity or other behavioral element can be allocated to a structural element such as a Block, Part, Interface, Block Instance, etc. that is responsible for the behavior. Ultimately, the Allocation is to a physical system element in the PV. This is an example of declaring the mapping of design to implementation. Another important use is in modeling the deployment of software and data to hardware. Instead of the UML Deployment Diagram, SysML uses `allocatedFrom` compartments within Block Diagrams to show the software and data deployed to a given Block or Part, and the corresponding Blocks modeling software resources can have `allocatedTo` compartments.

SysML is particularly useful when an architecture model needs to deal with complex design and implementation details of both hardware and software. BDDs can be carried to whatever level of detailed decomposition is needed, and the use of Allocation, Verification, Satisfaction, and other Dependencies keeps structural elements tied to requirements, functions, test cases, and other parts of the model. Well-constructed physical diagrams are both rich in content and easy to understand, making the model a powerful communication tool.

6.2.2 Physical Interfaces

Chapter 5 described functional interface modeling in the LV. Among the most important content of the PV is the fully detailed description of interfaces. Both internal interfaces among system components and external interfaces to other systems or networks must be documented. An architect can add these details to the model using the same mechanisms as for other system element specifications. A network interface might need an expanded specification containing the protocols and other standards that it implements, parameters such as data rates and latencies, functional information about things like error handling, and the physical media and connectors. Again, the information can be entered in the model using the description window for a given model element or in a separate document with a reference in that window.

If the modeling tool being used explicitly supports formal contracts associated with Ports, those are the preferred places to document a Port's interfaces since they are the essence of the contract. Interface definitions can take a variety of forms, depending on the nature of a given interface. The Interface Description Language (IDL) that was developed as part of the Common Object Request Broker (CORBA) standards family is widely used to define interfaces even when the underlying CORBA solution is not used.

One of the most common forms of interface documentation is the Interface Control Document (ICD). ICD content typically includes:

- *Nomenclature and description*—unique identifier and overall summary.
- *Physical Data*—specific design of connectors, mounting hardware, vibration isolation, safety provisions, cooling transfers, etc.
- *Electrical Data*—specific design of electrical power (voltage, current, power quality, etc.), interlocks, overload protection, and other details of power transfer across the interface.
- *Functional Data*—interface standards, data/message formats, protocols, timing and data rates, and other specifics of information transfer across the interface; this can also address electromechanical device characteristics, fluid pressures and flow rates, controls, flow definitions, etc.

Another type of interface documentation applicable to a network is a Ports and Protocols Matrix, which is described in Chap. 9.

As part of the Services Perspective, physical specifications for service interfaces must be defined and modeled. In addition to basic ICD content, these Service Level Agreements (SLAs) or other interface specifications include all the protocols and formats required to discover, bind to, and invoke the service and to receive the resulting product or service. For example, the SLA for a service that requires a set of parameters from the client that is invoking it would spell out the order and data formats for those parameters. A very important element of an SLA is an appropriate set of Quality of Service (QoS) parameters that define things like guaranteed minimum and maximum response times, provisions for error handling, data accuracy, and other "contract terms" that the service provider promises to honor. See Chaps. 7 and 9 for more details.

6.2.3 Node and System Infrastructure

As discussed at length in the previous chapter, information-intensive systems can usually be approached in terms of layers, each of which provides services to the layer(s) above it, ultimately to system users. As later discussions will show, this overall construct applies to all of the system categories introduced in Chap. 1. So far, the discussion has been general, applying architecture principles to any level from an Enterprise down to Systems or Subsystems or even lower. However, for the rest of this chapter, we will focus on Nodes and Systems, because this is where physical architecture is primarily realized. The Enterprise is an operational and organizational construct that actually exists in the form of the Nodes that comprise it together with the Systems that create the functionality of each node, and it is for these that physical architecture must be created. An organization responsible for implementing a Subsystem may employ similar PV techniques, but this would be simply a scaled-down version.

Our focus continues to be on high-technology, information-intensive systems, recognizing that these often also include elements such as controllers, sensors, mechanical devices, communications devices, electrical power components, vehicles, and so forth. The foundation for a system's information processing capabilities is an infrastructure, made up of both hardware and software resources that create a computing platform, together with a suite of software components that are hosted on the infrastructure and with interfaces to other system elements. The other parts of the system must be integrated with these computing resources and so must also be fully represented in the architecture.

For the present, we focus on the computing and networking infrastructure. These components must work together to achieve the following capabilities:

- Create the environment that will host the hardware and software components that deliver Mission Services, i.e., do the primary work of the system. If the architecture is not built on a service model, these can be thought of simply as functions. A critical requirement is that the infrastructure be able to support diverse, flexible, and easily changeable combinations of Mission Services or functions.
- For a participant in a larger enterprise, implement the interfaces and at least a portion (sometimes all) of the functionality through which a Node or System provides and uses Enterprise Services. Selection of both components (products) and standards is likely to be heavily influenced by the need to efficiently participate in Enterprise Services such as Discovery, Information Dissemination, Security, and Collaboration.
- Perform housekeeping, accounting, resource management, cybersecurity, and other general support activities required for the satisfactory operation of the Node or System. These services are essential to reliability, maintenance and failure management, information protection, performance auditing, system administration (e.g., managing user accounts and enforcing authentication and access privileges), and a host of other mundane but critical functions.

The specifics of an infrastructure depend heavily on the requirements to be satisfied by a particular node or system. Accordingly, this discussion is confined to a set of general principles and guidelines.

Architectural Layering: Careful consideration should be given to the layering within the infrastructure (Figs. 5.16–5.23). Through its interfaces, each layer should isolate the layers above from its own implementation, as well as those of lower layers and the external environment. To recapitulate the discussion in Chap. 5, it's conventional to depict hardware such as computers, networks, storage, communications, sensors and effectors, equipment and controls, and other resources as the lowest tier within the infrastructure layer. Next come the hardware and software that create the interfaces through which these resources are accessed and controlled (e.g., device drivers and network interfaces). The operating system and its utilities, referred to as System Support Services, are often considered the next tier. Then come tiers with the remaining infrastructure services, arranged in accordance with their mutual dependencies. At the top of the infrastructure is the complex interface, called in this book an Integration Backbone as shown in Fig. 5.20, that exposes all services required by the Mission Service Components as well as supporting the User Services (HMI and Client/Presentation) layers.

Implementing Products: The infrastructure will frequently be assembled from off-the-shelf products, which may be commercial, customer-supplied, or free/open source, with the appropriate setup and tailoring to the needs of an individual system design. Adopting a proven product family, especially if supported by a vendor partnership agreement, can reduce the time and expense of integration. That product line needs to enforce open architecture principles so that specialized components and services can be easily added. Product selection criteria should include the pedigree and field service history of each candidate, the availability and cost of vendor support during system development and integration, compatibility with architecture tools and libraries, and, ideally, information about vendor roadmaps and future product plans to help ensure long-term evolvability of the system.

Operating System: For real-time systems, the key product is the Real-Time Operating System (RTOS) with its associated framework products, and the system typically needs customized infrastructure components to meet performance, safety, and other requirements. For Process/Decision Real-Time architectures, the infrastructure is typically centered on a Component Framework and a suite of middleware, while the OS and System Support Services may be bundled with the servers or procured separately. Important factors in product selection are typically support for workflows, performance under worst-case task loading, support for heterogeneous data bases, implementation of Enterprise Services, and cost (e.g., license fees). A Management Information System will almost certainly use an all-COTS infrastructure with some product tailoring for local requirements like system resource management and data replication.

Resource Margins: Infrastructure implementation must provide adequate margins for processor throughput, network throughput, storage allocation, and utilities like power and cooling. There is an obvious cost impact to provide more than the minimum required resources to meet requirements, but the long-term consequences of a fragile design with inadequate margins will inevitably be even more costly to correct and will often make the customer dissatisfied with the system. Initially, good margins greatly facilitate the integration task. Over time, requirements and utilization will inevitably grow, and generous initial margins reduce the risk that a significant modification or technology refreshment will be needed early in the life of the system. Together with modularity and loose coupling, solid margins reduce the risk of timing faults, make the system robust in the face of varying workloads, facilitate scaling to add capacity, and generally make the system easier and less costly to operate and maintain. There are no hard and fast rules, but a typical guideline for a system expected to be in service a long time and to experience requirements growth is 50% spare capacity in the as-delivered system plus provisions for modular hardware expansion, e.g., through empty rack slots with power and cooling that can accommodate additional processor cards, blades, or other modules without increasing the system's installed footprint.

Implementation Constraints: Product choices may be further constrained by special requirements such as cybersecurity. The design may have to incorporate a trusted operating system (TOS), at least on processors that enforce security functions, which may, in turn, limit the choices of infrastructure product families to those compatible with a mature and accredited TOS. Other factors may be maximum weight and size for an airborne installation, intrinsic reliability for mission-critical systems, environmental tolerance, and commonality with existing infrastructures to minimize operations and support costs.

6.2.4 E-X Computing Infrastructure

Figure 6.3 from the E-X example shows a typical PV design artifact. The diagram shows the Computing Infrastructure (CI), and similar diagrams arise in a wide range of systems that incorporate servers, networks, storage, workstations, and similar resources. This diagram would be created initially in the LV, and the specifications for the various Blocks and Interfaces would become key content in performance specifications to be used in creating or selecting products for the PV. Once these procurement and development activities are complete, the resulting physical details can be captured in the model and exported in reports to create PV documentation.

For example, suppose an Item Flow between two Blocks or Parts consists of a stream of messages and can be traced to an Object Flow in an OV Activity Diagram. In the PV, the Object Node describing the messages being exchanged needs to be given a specification describing the messages along with their format, timing requirements, message or data rate, and other parameters of the information. This detail is part of transforming the LDM to the PDM. Then the item flow path, which might be the local area network (LAN), must be described in terms of the physical medium, the protocol, electrical and signaling parameters, and other characteristics. Chapter 9 has more specifics on network modeling. The ports through which the flow passes in and out of the Parts might be described by the same specification as the flow path or might require additional data, e.g., to describe message buffering, message acknowledgment, error correction, and other functions they provide. Depending on the tool being used to build the model, various means may be used to record this information, most often using notes attached to a model element or a Model Element Specification. Many tools allow the specification to include links to other files, making it easy to bring in a product data sheet, standards documentation, or other information.

In Fig. 6.3, a variety of workstations, each configured for a particular Crew Role, a system server and network-attached storage (NAS), and an interface from the CI to the sensors, are integrated on an Ethernet LAN. Each of these is a Part Property

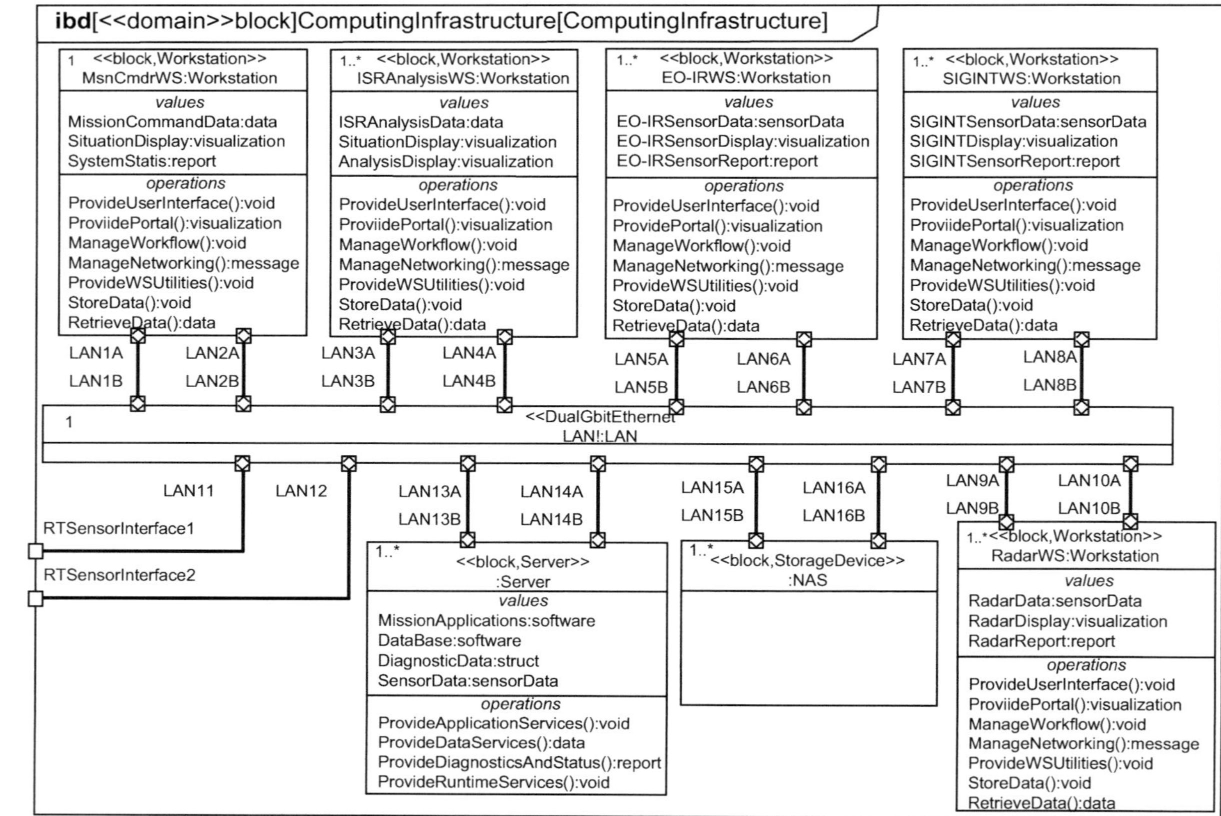

Fig. 6.3 IBD of the E-X Computing Infrastructure

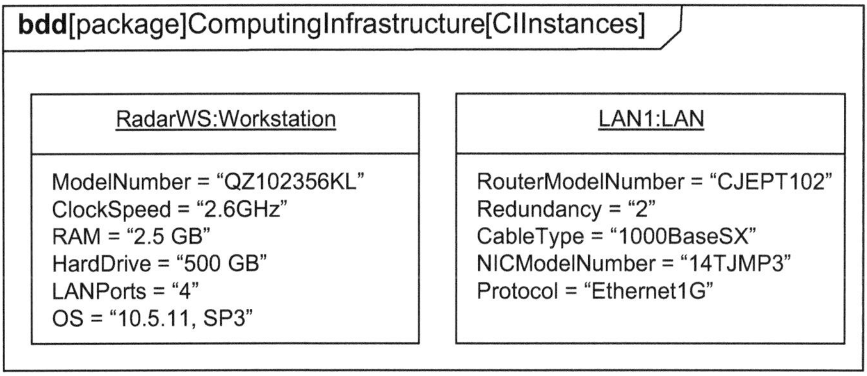

Fig. 6.4 Example of Instance Specifications for the E-X Computing Infrastructure

of the CI Domain Block. The various operator workstations are named Instances of a generic Workstation Block, while the other Parts are unnamed instances. Each Part has two independent LAN connections, corresponding to a dual redundant network for high reliability. Full Ports are used for network traffic, and each has a placeholder Port name (e.g., LAN1A) that will ultimately be mapped to an actual Port address. Additional Parts such as communication devices that attach to the LAN can also be added as the design matures. As one element of fully defining the Block Instances in Fig. 6.3, Fig. 6.4 is a BDD containing Block Instance Specifications for two Parts, the Radar Workstation and the Local Area Network. Other Parts have similar specifications. Sometimes, an effective diagram can simply document technical details using Notes. Full technical descriptions are often best entered into Model Element Specifications, possibly using pointers to embedded or library documents.

The diagram in Fig. 6.3 can be the basis for a focused Network Viewpoint as discussed in a later section and might be used to produce such a graphic by replacing the block icons with clip art images of the actual devices used to instantiate the various Blocks. If it later becomes necessary to add, replace, or upgrade elements of the CI, this diagram and the accompanying LV and PV specifications provide a solid basis for a trade study, integration and test planning, and other Systems Engineering actions to accomplish the upgrade.

6.3 Integrating Mission Components

As we saw in Chap. 5, many systems have a basic structure that involves integrating mission components onto an infrastructure, and in such cases the details of that integration are a central element of the Design Perspective. The interests of a system's operators are naturally focused on the functions and services that support the system's operational goals. Those services are generally created by application software, which in a complex system is likely to come from multiple sources and to employ a range of software technologies. Further complications arise when both hard real-time and soft or non-real-time components must be included. MBSAP seeks to provide a flexible integration strategy for coping with these realities. Looking back to the system categories presented in Chap. 1 for the architecture taxonomy, the following paragraphs discuss the challenges of integrating components with varying degrees of real-time behavior. We will use the layered architectures from Chap. 5 as the framework for this topic.

The architect of a Process/Decision Real-Time system will often have limited opportunities to develop new Mission Services Components. Much more commonly, the principal challenge will be to integrate a mixture of legacy, modified, and newly developed components, including COTS/GOTS, driven by considerations of cost, required product accreditation, user familiarity, and standards compliance. In an information system, the primary hardware components are in the infrastructure, including connected devices. Mission Services are implemented mainly in applications software and are exposed to system users through the Presentation/Client layer and to the external enterprise through interfaces created by the infrastructure. Accordingly, the Mission Services layer will generally be the result of a combination of software product trades, some software development and modification, and, most importantly, integration of the resulting components onto the infrastructure.

The situation in dealing with Management Information Systems is a variant of this. Such systems involve extensive use of COTS/GOTS products for both Infrastructure and Mission Services, sometimes with a certain amount of new application software for unique requirements. Many such systems also require extensive effort in customizing and configuring COTS/GOTS Mission Service products and in migrating legacy data to the new environment, but they generally have fewer performance and timing issues to deal with. For both of these latter two system categories, the discussion is focused on the integration task. Since modifying COTS hardware and software is generally expensive and risky,[1] the architects and system integrators must meet system requirements with the capabilities of the selected products. The required facilities and staff skills are different in character from those of the real-time world, stressing software integration and the associated instrumentation and coding capabilities, but they are no less indispensable to a successful program.

The hard real-time category introduces special considerations. The architect of a system such as a sensor, machine controller, or real-time network typically has considerable leeway in defining Mission Services Components and deciding how to obtain them. Reuse of existing products is often attractive from a cost and schedule viewpoint, and this figures into product selection trade studies. Nevertheless, a significant level of new development is often inescapable to meet a challenging set of performance requirements. Real-time behavior also complicates the integration task, especially when a new real-time system must be integrated into an existing

[1] It's been said that modifying a COTS product actually has the effect of creating a custom product with a very limited production run—the worst of all worlds.

node and made to interact with older systems. If a good job has been done developing the LV, performing the required performance analyses and simulations, doing early prototyping of critical components, and developing high-quality design and procurement specifications, the architect will be able to give the system development and integration team a solid basis for implementing the solution.

6.3.1 System Integration Challenges

In many situations, the main factors affecting success or failure in system integration are likely to be availability of adequate facilities and staff to complete the technically difficult task of procuring or developing components and of integrating and testing the system. A large system may require multiple integration and test laboratories, corresponding to the organizational hierarchy of the development team, to support progressive integration from components and assemblies through major functional areas to subsystems and finally the complete system. Failure to properly resource the lower levels of the integration "tree," especially at subcontractor and vendor facilities, has in the past led to major problems. Similarly, adequate design, integration, and test engineer staffing with all the specialized expertise involved by a particular system are a prerequisite to successful completion of system development on time and within budget.

Integrating a variety of products that have an assortment of technology baselines, required execution environments, visualization approaches, databases, and other characteristics requires a robust, open architecture supported by a full toolkit of integration methods. The following paragraphs give examples of the associated architecture challenges:

Heterogeneous Infrastructure: When existing business process or mission applications and other components are selected for a system, it's likely that the architect will find they have been designed to run on different operating systems, use different messaging protocols, employ different database management systems, output their data in unique formats, or otherwise require different and incompatible infrastructure capabilities. The infrastructure will then need the flexibility to provide the required spectrum of runtime environments, messaging mechanisms, databases and other features. One widely used approach is an Enterprise Service Bus (ESB) as introduced in Chap. 5 that implements a variety of interfaces and does things like protocol and format translation to allow heterogeneous environments to interact. Other functions of a SOA can also be used, such as Data Services that abstract different databases to a common service interface. SOA and ESBs are further discussed in Chap. 7.

Legacy Product Integration: A number of techniques can be used to host heterogeneous mission applications on a shared infrastructure. Perhaps the best known is the use of wrappers and adapters that transform the native interface of the application to the common interface of the infrastructure. A wrapper is a software component that is linked to the application at build time to deal with things like parameter passing and infrastructure service calls. Its purpose is to make the application appear native to the infrastructure. An adapter, by contrast, is a runtime component that performs the necessary transformations on each interaction between the application and the infrastructure, e.g., to translate incompatible data structures. An adapter is often either a built-in function of an ESB or a plug-in used with an ESB.

A more radical approach is refactoring of an application, which involves modifying the code to run in native mode on the infrastructure. This eliminates the processing overhead of wrappers or adapters and the associated performance loss but is typically far more expensive to do. For other than very simple applications, refactoring is likely to be feasible only when done by the original software developer and only when a small percentage of the code needs to be changed.

Prototyping: As essential as a good laboratory/prototyping environment is to a new development project, it is arguably even more essential to the integration of existing products. This is because the architect and integration team will almost never have full knowledge or documentation of the products they deal with, will have fewer ways available to solve integration challenges, and will sometimes be reduced to a "cut-and-try" strategy of changing things until a given application and its interface are working. Legacy products may be "black boxes" whose internal details are completely hidden or "white boxes" where some details of structure and internal functions are known, but complete data (e.g., source code) is unlikely to be available unless the integration organization was also the original product developer. To the extent that the product is well designed and has a fully defined, standards-compliant interface, these problems will be minimized. But in the real world, an architect facing the integration of products from external providers should insist that the program plan include schedule and resources for a robust prototyping activity to find and fix such problems as early and inexpensively as possible.

Virtualization: Advances in IT make more and more techniques and products available to deal with integration challenges. One of the most important is virtualization (see Chap. 14). In the most general sense, this simply refers to abstracting information resources so that a user need not be concerned about the details of their implementation. It can be applied to servers, workstations, storage, networks, and other computing resources. A virtualized server farm presents the aggregate capability of a group of computers as, effectively, a single, allocable resource pool that can support fault tolerance, load balancing, and other efficiency and reliability features. A somewhat simpler approach is server clustering in which a group of servers are networked together; this can also include failover and load balancing. Workstations can be virtualized to allow remote tailoring and manipulation, e.g., by a server or a security device. Virtualized storage presents a uniform interface using standard protocols and may be helpful in implementing data services on heterogeneous databases. Chapter 14 has more on this topic and related subjects such as cloud computing.

6.3.2 Integration Framework

One way to attack the legacy application integration challenges listed above is an integration framework that offers a range of techniques, from very simple to quite complex, from which the architect can select the best approach to integrate a given software application, database, computer, or other system element. This approach derives from a number of earlier systems such as the Distributed Common Ground System (DCGS), which employs an integration backbone (IB) like that shown in Fig. 5.20 [1, 2]. The first systems using this approach were built on the Java Platform, Enterprise Edition (JEE, formerly J2EE) component platform, but other environments, including an open-source component framework, can also be used. The approach embodies principles on which this book and MBSAP are based, including:

- Layered, service-based, loosely coupled, and component-based architecture with an infrastructure that creates an IB for flexible integration of new, modified, and existing Mission Service components; the strategy is built on best commercial practices, open standards, and validated design patterns.
- A Metadata Framework (MDF), employing a Metadata Catalog (MDC) for efficient Data Services involving large, heterogeneous repositories and spanning local, regional, and global networked information environments and implements; the MDF provides data management and interchange services and supports "smart-push/intelligent pull" data dissemination for fast, tailored information access across an enterprise; an evolved version called a Distributed Data Framework (DDF) [1] adds significant new data management capabilities.
- System functions exposed as integrated Web services through a portal; the Presentation/Client layers can be extended to include other models such as rich clients when required to support legacy Viewer/Presentation functions.
- Comprehensive system services for user interfaces, graphics, node/platform communications, system management, and distributed computing.
- Supporting applications for multimedia, data exchange, and business processes.
- Security functions that implement "defense-in-depth," including boundary security, communications/network security, authentication and access control, data security and labeling, and security activity management, logging, and analysis (see Chap. 10).
- Implementation of enterprise services so that a node or system is network-enabled and complies with required enterprise architecture standards and customer policies. In general, the components of the infrastructure provide the required functionality and standards compliance; Table 6.1 lists some representative services and the associated standards (current as of the date of this writing), and Table 6.2 describes some additional relevant standards.
- Compatibility with emerging standards that are expected to be the basis for future mainstream information technology.

Table 6.1 Enterprise services and implementing standards

Enterprise service	Standards
Security	– Public Key Infrastructure (PKI) – Secure Sockets Layer (SSL) Protocol – Security Assertion Markup Language (SAML)
Discovery	Universal Description, Discovery, and Integration (UDDI)
Service Description	Web Services Description Language (WSDL)
Transport	– Simple Object Access Protocol (SOAP) – Direct Internet Message Encapsulation (DIME) – Hypertext Transmission Protocol (HTTP) – File Transfer Protocol (FTP)
Portlet	JSR-168
Data Description	– eXtensible Markup Language (XML Registry) – Lightweight Data Access Protocol (LDAP) – Open DataBase Connectivity (ODBC) – Java Database Connectivity (JDBC) – Structured Query Language (SQL)
Resource Management	Simple Network Management Protocol (SNMP)

Table 6.2 Additional standards

Standard	Description
Web Services for Remote Portals (WSRP)	Defines the interface and semantics and provides a Web service standard for content sources with portals and other Web apps
XML Key Management Specification (XKMS)	Interface to PKI enabling Web services to register and manage keys for digital signatures and encryption
Extensible Access Control Markup Language (XACML)	XML schema designed for creating and automating policies to control applications access on a network
UDDI Version 3	Expands UDDI 2.0 to include registry-to-registry interaction between both public and private registries
Internet Protocol Version 6 (IPv6)	Next-generation Internet protocol designed by the IETF to replace the current IPv4

Table 6.3 and Fig. 6.5 summarize the strategy, showing more complete levels of integration, from leaving a system element to operate stand-alone at Level 0 to making it fully native to the shared infrastructure at Level 4. New components/services should be developed at Level 4. Existing components that were implemented with modern software engineering methods and have long projected lifetimes may justify the refactoring effort to achieve Level 3 and possibly long-term migration through redesign to achieve Level 4. When a trade study concludes that a legacy application does not warrant these higher levels, adapters allow effective integration with workflow, data services, and presentation services. When an older component must be incorporated and it is impractical to adapt or refactor the design, "Level 0" integration allows the application to be hosted on a node and used in a stand-alone

Table 6.3 Description of integration levels

Level	Description
Level 0: Peer-to-Peer Interface	– Most basic integration approach – Message only interface – Specific data exchanges tightly coupled to the applications and documented in a Peer-to-Peer Interface Control Document (ICD) – Proprietary format allowed (not preferred) – No use of Integration Framework functions
Level 1: Data Store Integration	– Use of Resource Adapters to discover and access data – Use of the Metadata Catalog to publish data – Use of Web Portal and Web Services to access metadata and data – Non-intrusive to legacy component/application
Level 2: Application Integration	– Builds on Level 1 – Use of workflow/event management for application integration – Legacy application commands and status exposed via resource adapters – Use of publish/subscribe to alert users and applications of application status – Non-intrusive to legacy component/application
Level 3: Refactored Legacy Application	– Builds on Level 2 – Parts or all of Legacy Applications refactored as applications in a JEE Enterprise Application Integration (EAI) environment or equivalent – Refactored application information accessed via Web Portal and Services – Some legacy functionality may remain within the legacy application
Level 4: Native Application	– Service-based application – No service adapters required – Applications run in homogeneous JEE environment (or equivalent) – Full application access via the Web Portal and the Web Services

mode. This is not a real-time environment, due partly to the processing overhead associated with the powerful services of the MDF/DDF.

In Fig. 6.5, the IB has been generalized to an integration framework that provides additional techniques and resources for dealing with heterogeneous system components. This integration strategy is an example of the architectural tenets defined in Table 1.2. These general principles must be enabled by an architecture-centric methodology and tool set, a full complement of implementing products, a comprehensive standards profile, and the required facilities and technical skills. The result is a flexible capability to craft new and evolved mission solutions using a mix of legacy/existing, modified, and newly developed components. This approach also facilitates compliance with customer policies such as open architecture and networked interoperability, and delivery of functionality tailored to the needs of a particular problem domain.

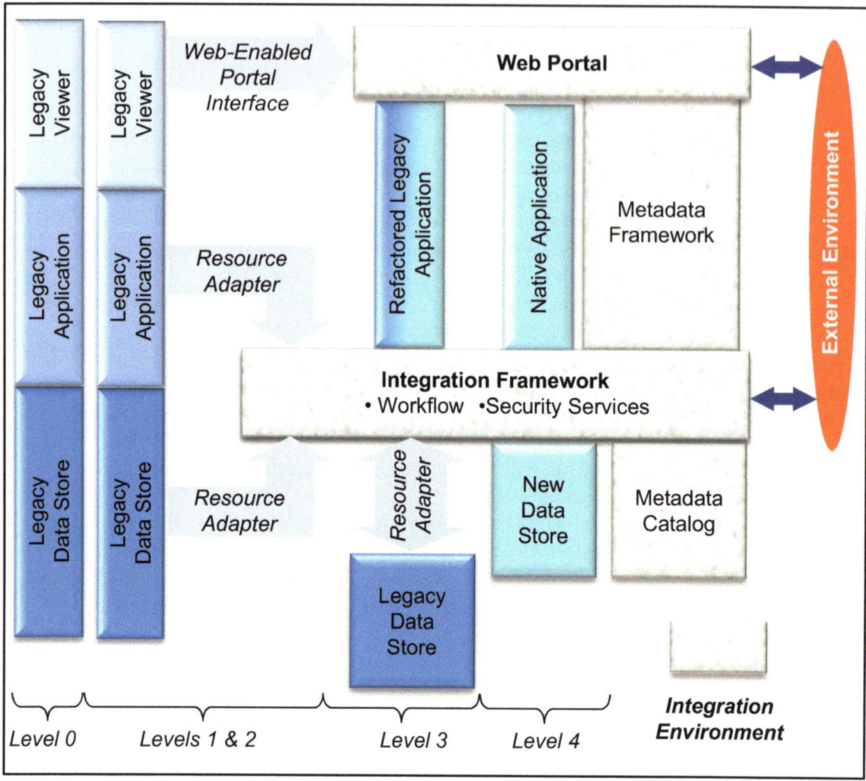

Fig. 6.5 Levels of integration in a comprehensive integration framework

6.3.3 Interfacing Real-Time and Non- or Soft Real-Time Domains

A system like the E-X, in which real-time mission equipment such as sensors and data communications interacts with a human-in-the-loop computing environment, presents another integration challenge. The hard real-time Domains incorporate embedded processing that can keep pace with the data rates, resource scheduling deadlines, data and signal processing, and other time-sensitive requirements of the equipment. The Process/Decision Real-Time Domain is built primarily on the kind of general purpose information technology discussed in the preceding section, with a variety of applications software integrated on a computing platform. There are situations where the progress in general purpose computers has been such that they are fast enough to do real-time processing for some types of mission equipment and dedicated real-time processors are not needed, but the methodology needs a way to deal with the situation shown in Fig. 6.6.

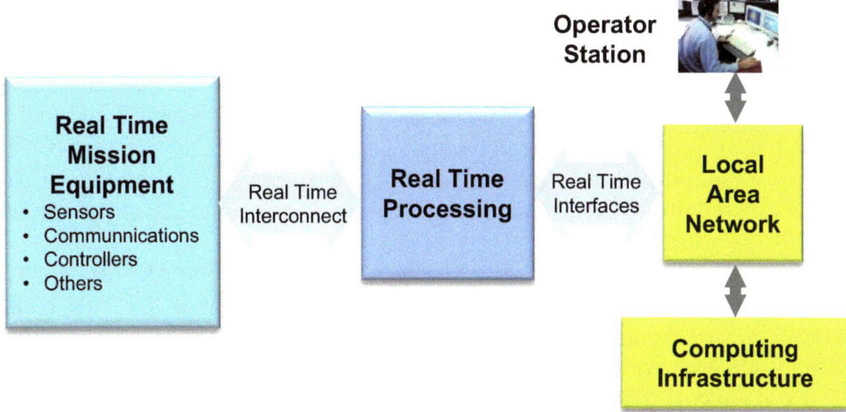

Fig. 6.6 Integration of Hard Real-Time and Non- or Soft Real-Time domains using one or more real-time interfaces to mediate interactions between them

The figure gives a simple representation of the integration problem, highlighting a Real-Time Interconnect that ties that Domain together and one or more Real-Time Interfaces between the Domains. Data rates, maximum allowable process times and latencies, data formats, timing, and many other parameters are likely to be very different on the two sides of the interface, requiring functions that allow the Domains to interact correctly. Furthermore, when off-the-shelf mission equipment like sensors and radios is reused and integrated in a new system, there is seldom the option of modifying designs, so that responsibility for the interface tends to fall on the non-real-time side. The Real-Time Interface will usually consist of both a hardware adapter to handle any electrical and mechanical aspects and one or more software adapters running as mission applications to handle the data. Typical interface functions include:

- Generating and synchronizing timing signals on both sides of the interface
- Buffering data and commands to accommodate different rates and schedules
- Performing any required data format and message protocol translations

6.3.4 *Visualization and the Human-Machine Interface*

The top layer of the architecture construct for Process/Decision Real-Time and Management Information System nodes and systems delivers User Services, both those created locally and those accessed across a larger enterprise. It consists of one or more HMIs and the Client/Presentation layer that powers them. Systems in many application areas increasingly have complex HMIs with a wide range of display and command functions. The quality and ease of use of the HMI is likely to be one of the most important factors in user satisfaction or dissatisfaction with the design. The following paragraphs discuss some key implementation issues:

- Architects and designers must select one or more client models based on system requirements and the nature of the Mission Services components. Typical alternatives are Web pages accessed through a browser, thin clients that may consist only of a browser or may include additional client-side visualization functions, and rich or "thick" clients that do significant processing and can create specialized HMI capabilities.
- The integration strategy for the Client/Presentation layer must be defined. The previous section described the basic options, which range from designing Mission Service components to expose their functions as Web services to providing a Client/Presentation layer adapter to hosting a stand-alone client on the appropriate user workstations.
- In addition to the basic client model, designers must define the HMI content, especially the screens, menus, buttons, alerts, and so forth. This is driven by program requirements, user preferences, operational doctrine and procedures, training, and other factors. The Presentation Layer will typically need logic to accomplish any required data and format transformations from Mission Services to the HMI.

Many HMIs present information in both spatial and status formats, recognizing that human understanding of information is assisted by use of the proper visual paradigm. Spatial screens use maps, images, multidimensional data plots, and other visual modes where situational context is essential to understanding. They routinely build up a composite screen using data overlay layers on a background map or image and often provide pan, zoom, nesting, annotation, and other features of a graphical user interface (GUI). When the central user task is monitoring and controlling processes, a tabular display showing status information, such as the currently active task in a workflow, activity rates and completion states, color-coded symbols showing current status vs. a plan, and associated cues and alerts, is preferred. Such HMIs are often called "dashboards." A user workstation may need to switch between spatial and status formats or even to combine them in a composite display. Framework products are available that allow the composition of this class of powerful HMI by integrating multiple data sources and processes.

Thick clients come into play when lower integration levels are used with older legacy Mission Service components, or when user requirements dictate that a unique HMI be carried forward into a new system. Unless dedicated user positions are allocated to these applications, the workstation must have both the resources to run the client-side applications and the ability to switch among the various client models. This is routine with current workstation products.

Turning briefly to the Hard Real-Time architecture category, the Client/Presentation layer construct can be applied in at least some cases. However, as shown in Fig. 5.23, performance issues and the highly specialized nature of the HMI often make it preferable to eliminate the overhead associated with this layer. The HMI is then driven directly from the infrastructure using data from the Mission

Services, which may include an HMI Domain to fuse data from other Domains and format it for the operator environment. For example, a modern aircraft cockpit uses one or more multifunction displays to present a rich assortment of operational and system information in one place. This greatly facilitates pilot performance of diverse simultaneous tasks such as maintaining situational awareness, flying the plane, communicating with controllers and other aircraft, and managing airplane systems, all of which are time sensitive and may have safety implications as well.

6.4 Standards Perspective

We have consistently emphasized compliance with mainstream open standards as an important element of an effective open architecture strategy, although it is far from the only consideration. Unfortunately, there are those who believe that mandating standards is enough to ensure an open system, regardless of the structural and behavioral properties of the design. It's essential to recognize that choosing and applying standards constitute a set of implementation decisions that should only be made once the basic system or enterprise design is established in the Logical View with attention to modularity, loose coupling, interface definition, and other key architecture tenets. Once this is done, the standardization task becomes one of:

- Identifying system elements, interfaces, activity and service definitions, data models, and other constituents of the PV to which standards potentially apply
- Identifying and assessing candidate standards for each
- Establishing a Technical Standards Profile that documents the resulting standardization decisions

Suitable standards are defined and maintained by recognized bodies through an open process that includes publication of drafts, working papers, and approved standards definitions, with balloting to resolve issues. The resulting "consensus" standards ideally represent the best judgment of the community as a whole. Even under these conditions, there are often competing standards, especially in areas that are emerging and where the choice of a particular standard can have significant financial implications for commercial products and markets. Moreover, even successful and established standards can be volatile and short-lived. The rapid pace of technology evolution drives a matching rate of change in standards as entirely new kinds of products emerge. A current example is the large amount of effort being devoted to enriching and updating the primary standards for Web services. The result of all this is that the architect must consider the pedigree, stability, and likely evolution of standards in choosing which ones to apply and when. It may be important not only to define a current standards profile but also to carefully consider a forecast of technology evolution and future standards.

Table 6.4 summarizes the primary standards categories with examples that are in wide use at the time of this writing. The list is illustrative but not exhaustive. Many more standards, often representing competing approaches to the same processing

Table 6.4 Primary categories of open standards

Standards categories	Typical standards
Information Processing: – User Interface – Data Management – Data Interchange – Graphics, Imagery and Video – Operating Systems – Internationalization – Distributed Computing – Environment Management – Biometrics – Component Frameworks	*Goal—Commonality, Interoperability and Reuse* – Operating System User Interface Standards – DataBase Management System Standards/Languages – Data Formats and Markup Languages – Graphics, Imagery and Video File Formats – Portable Operating System Interface (POSIX) – Character Sets, Language Service Standards, etc. – Naming, Events, and Transactions Standards – Workflow and Records Management Standards – Common Biometric Exchange File Format (CBEFF) – JEE, .NET, and CORBA
Information Transfer: – Host Systems – Mail – Directory Services – Graphics and Imagery – File Transfer – Remote Terminals – Time Synchronization – Web Services – Transport Services and Protocols – Networking (Local and Enterprise) – Quality of Service (QoS) – Voice over IP (VoIP) – Video Teleconferencing – Facsimile – Global Positioning System (GPS) – Identification Friend or Foe (IFF) – Transmission Media – Network & Telecommunications Management	*Goal—Interoperability Among Nodes and Systems* – IETF Internet Host Standards – Electronic Mail Protocol and Format Standards – X.500 and Other Directory Service Standards – Graphics, Imagery and Video Transmission Protocols – File Transfer Protocol (FTP) – Telecommunications Network (TELNET) Protocol – Network Time Protocol (NTP) – URL, SOAP, LDAP, UDDI, WSDL, and other Protocols – TCP, UDP, and other Transport Protocols – IP versions 4 and 6, Ethernet, ATM, FibreChannel, etc. – Integrated and Differentiated Services QoS Standards – VoIP Standards – Video, Audio and Control Protocols – Digital Facsimile Type I and II – GPS Standards – FAA and International IFF Standards – Wired, Wireless and Optical Media Standards – Simple Network Management Protocol (SNMP, Telecommunications Management Network (TMN)
Information Modeling: – Data, Activities and Objects – Data Architecture – Data Definitions – Information Exchange	*Goal—Interoperability and Consistency for Data* – Unified Modeling Language (UML) – DoD Data Architecture (DDA) – Multiple Data Models, Schema and Dictionaries – Message Catalogs and Formats
Human-Machine Interface (HMI): – User Interface Design – Style Guides – Symbology	*Goal—Uniformity and Interoperability for Users* – Multiple Graphical User Interface (GUI) Standards – X-Windows, Mission-Oriented Style Guides – Symbology Standards for Various System Categories

(continued)

Table 6.4 (continued)

Standards categories	Typical standards
Cybersecurity: – Applications – Operating System Security – Cryptography – Enclave Boundary Security – Public Key Infrastructure – Intrusion Detection – Network Security – Physical Security – Evaluation Criteria	*Goal—Confidentiality, Integrity, and Availability* – Standards for Secure Comms, Data Services, etc. – Standards for OS Security Services – Encryption/Decryption, Key Management, Hashing, etc. – Firewalls and Associated Security Services Standards – PKI Family of Standards and Protocols – Intrusion Detection Exchange Protocol (IDXP) – Internet Protocol Security (IPsec) Suite – Security Standards for Media, TEMPEST, etc. – Evaluation Criteria for IT Security (Common Criteria)
Mission-Specific Standards	Multiple Standards Aimed at Consistency, Interoperability, and Openness in Various Domains

and information exchange services, are associated with things like computing component frameworks, data exchange formats, programming languages, networks and data links, and system resource management. Information processing standards deal with computing environments and user interfaces. Information transfer standards govern networking, voice and data communications channels, collaboration tools like Email, and other interactions. Information modeling standards cover data representation and exchanges. HMI standards seek to create effectiveness and consistency in the interaction of humans with computing environments. Information assurance standards capture proven approaches to secure information usage and exchanges. A wide variety of standards have been defined for various system categories. Finally, it's important to remember that although, given the focus of this book, the emphasis here is on information standards, many architectures need equivalent emphasis on mechanical, electrical, materials, and other standards categories. Appendix E gives a representative list of core IT standards in current use.

System customers, especially in the public sector, are often both adamant about standardization and naïve about the suitability and stability of their favorites. Careful coordination of this critical aspect of the evolving architecture with the customer is therefore likely to be important. Together with volatility in technologies and products, this can result in the need to refresh the Technical Standards Profile just as obsolescent system components need replacement or updating over time. Once again, a technology- and standards-independent design captured in the Logical View, together with a sound basic architecture, makes this far less troublesome than in systems where there is critical dependence on specific standards so that changing them forces an expensive redesign.

A major customer such as a Government agency will typically have a list of preferred or approved standards for use in the systems it procures. As an example, technical standards mandated for use in defense systems are available in the DoD Global Information Grid (GIG) Technical Guidance Federation (formerly the DoD Information Technology Standards Repository). This lists something on the order of a thousand individual standards and is updated regularly. Customer policy commonly requires the use of applicable standards from these sources unless an exception is warranted based on performance or other factors. The preference goes to established commercial standards, supplemented by customer standards in areas where no commercial equivalent exists. The Federal Enterprise Architecture Framework (FEAF) [3] addresses standards as a key element of the Reference Models that make up the framework. Open standards bodies include:

- Organization for the Advancement of Structured Information Standards (OASIS)
- International Organization for Standardization/International Electrotechnical Commission (ISO/IEC)
- Internet Engineering Task Force (IETF)
- International Telecommunications Union (ITU)
- World Wide Web Consortium (W3C)
- Institute of Electrical and Electronics Engineers (IEEE)
- Object Management Group (OMG)
- The Open Group (TOG)
- Distributed Management Task Force, Inc. (DMTF)
- National Institute of Standards and Technology (NIST) and others

6.5 Data Perspective—: Physical Data Model

6.5.1 Physical Data Model Content

The LV includes a Logical Data Model that defines the information content of a node or system in the form of Block Diagrams, often accompanied by one or more XML schemas. In the PV, this is extended with one or more physical schemas that add physical metadata and storage specifications. The specifics vary widely with the nature of a given system, but the overall idea is that the PV documents all the physical details required to implement data storage, access, retrieval, duplication, dissemination, safeguarding, and, ultimately, deletion.

Hard Real-Time systems may require custom-designed data stores to meet performance and data integrity requirements. In other cases, the system data repository is implemented using a set of databases, usually built on one or more Relational DataBase Management System (RDBMS) products. Many data architects equate a Physical Data Model simply with the definitions of tables and rows in the RDBMS, but this should often be extended to a much richer construct involving metadata-enabled Data Services.

Continuing with the earlier example of a multilevel integration strategy, the Metadata Framework (MDF) abstracts the details of the repository from the Mission and Presentation Services layers, creating a semantic interface that provides richer access methods while minimizing the need for detailed knowledge of database structures. It's also important to capture the details of data storage, access, and management as the basis for defining a suitable suite of hardware and software. The following paragraphs address some of the principal issues associated with a PDM:

- A key early decision involves selecting the RDBMS standards and products that will be used and defining the required table and column designs, operations, queries, and other details. Especially in large systems with significant reuse of legacy components, the data layer is likely to be heterogeneous both in the products it incorporates and the basic database models it must support. A major function of an MDF is to manage such situations and present an efficient and stable data interface to the higher layers of the architecture.
- Another important topic is designing the physical repository using products like Storage Area Networks (SANs) or Network-Attached Storage (NAS) and ensuring the required level of fault tolerance through techniques like Reliable Array of Independent Disks (RAID) products.
- The design team must ensure definition of any Metadata Catalog (MDC) content that has not been completed as part of the Logical View, including physical data attributes in the XML schema.
- Other important details include defining and implementing a storage map and mechanisms for data replication, data discovery and dissemination through Enterprise Services, data archiving and analysis, and any other functions needed to meet system requirements. For example, the system security policy may require that data at rest (stored in nonvolatile form) be encrypted before being stored.
- A special challenge confronted by systems that must handle information at multiple levels of sensitivity concerns the mechanisms to be used to exchange data across boundaries between areas that are trusted to different levels. This commonly uses products called guards that can handle data security labels, upgrading and downgrading rules, and provisions for any required human intervention in these transactions.

Physical schemas for legacy databases are often represented using the IDEF1X formalism [4] or the SQL Data Description Language (DDL) [5]. These are readily translated into XML. Another consideration is that legacy data often requires normalization to ensure consistent representation and usage in a new repository and schema. The Joint Collaboration, Command, and Control Information Exchange Data Model (JC3IEDM) [6], which is the basis for the CDM, LDM, and PDM in the MBSAP methodology, defines a Data Exchange Schema (DES) as part of the physical schema to be used in conjunction with replication methods to ensure consistent and interoperable data exchanges. The DES includes physical properties of the data to be exchanged such as database table and column names and physical data types.

6.5.2 E-X Physical Data Model

The E-X deals with a large quantity and variety of information, and its processes depend on fast, efficient data services. The Computing Infrastructure Domain includes the resources to handle large databases and to meet requirements for high-frequency onboard and off-board data transactions, error detection and correction, data security, and other functions. A well-designed PDM mirrors the structure of the system, starting with the Mission Domains and continuing down to individual software applications, user interfaces, messaging protocols, and other data-related features. This design of the system data model, following this principle, begins in the CDM and LDM, and the PDM transforms System Data Blocks into a physical storage and retrieval implementation. There would, for example, be databases dedicated to Mission Management, ISR Management, and Sensor Management Domains where the heaviest use of data services occurs. Assuming an RDBMS is adopted as the foundation of the PDM, the rows and tables designed for the various data categories are crucial PDM content.

Some of the important considerations that go into the PDM include:

- Use of an MDC with the XML schemas required to support sophisticated data services, as described in Sect. 6.6
- High-speed mass data storage devices with automatic error detection and correction
- Database designs that incorporate all the Values defined for Data Blocks in the LDM as well as all required database operations
- Security features to guard against erroneous database access and other error conditions such as simultaneous attempts to write to a common data item
- Application of appropriate standards such as Structured Query Language (SQL) and Open Database Connectivity (ODBC)

6.6 Services Perspective

Earlier sections of this chapter talked about incorporating services in interface definitions and about physically implementing a layered SOA infrastructure. One further aspect of the Services Perspective in the PV concerns the actual functionality of services that must be provided or consumed by system elements. The PV completes the traceability of compliance from customer requirements to delivered solution by mapping the services in the Logical/Functional Viewpoint to the components that implement them, including the interfaces through which services are accessed and their accompanying interface specifications or SLAs. For example, a Data Service would typically be allocated to one or more databases and their management systems, and the PV should document the design details, including references to the PDM and to a MDC, if one is used.

A service defined in the LV may use several components, and a given component may support several services, but the table that was developed as part of the OV and LV (Table 5.1 is the example) can be extended with additional columns to document the physical associations. It may also be useful, at least for the primary system functions, to develop a matrix that associates services with the Block Operations that implement them when instantiated physically. Chapter 7 has additional discussion of ways to model services.

A system like E-X needs a comprehensive suite of data services to meet performance requirements in areas such as application execution timing, simultaneous task loading, and interactions with off-board entities. These would begin with basic database functions like Create, Retrieve, Update, and Delete and would typically add more sophisticated services such as Publish and Subscribe, Perform Local and Global Search, Send and Receive Updates, and Mediate Data Conflicts. The mission services defined in the LV need fully detailed service interfaces and any additional details to define how they are executed and invoked. For example, a Services Directory specifying the details of each service and its interface is a necessity before the application can be written and tested and then during actual system operations. The full PV Services Perspective also includes any other services implemented or used by the system. These would include infrastructure services such as user account management, time generation, system health and status monitoring, security functions, and many others. Communications and networking capabilities can also be presented as services. Whatever the ultimate set of services turns out to be, the PV must capture the physical details allowing those services to be created and used.

6.7 Contextual Perspective

As with the Operational and Logical/Functional Viewpoints, architecture artifacts that do not fit cleanly into other Perspectives are captured here. This may include:

- Supporting documentation about the design and the products used to build it, including the focused views described earlier and specific design data such as interface control documents (ICDs), if not elsewhere accounted for
- Operational, maintenance, training, and other user information
- Customer-mandated artifacts that have significant physical content

The amount and variety of information in the Contextual Perspective will necessarily vary tremendously with the nature of the system, the users and other stakeholders, required documentation, and other factors. It's important to strike a balance between dumping everything connected to the system into the architecture model library and failing to include important content. A useful way to think about this is to consider the model as a kind of "yellow pages" of the system, meaning that a well-structured model can both incorporate all important system information and facilitate the finding of specific items when needed. When viewed in this way, the model can be identified as a Single Source of Technical Truth (SSTT) for the system. Browsing the primary structure of Viewpoints and Perspectives is straightforward. More careful thought is needed to structure the Contextual Perspective and to select its contents with the aim of making the model a useful tool with which many kinds of stakeholders can efficiently find information.

6.8 Focused Viewpoints

Many of the Focused Viewpoints introduced in Chap. 3 pertain to the PV and capture specific aspects of the architecture that are of interest to various stakeholders and specialists. These may use combinations of graphics, tables, and text, depending on the subject, to complement the SysML diagrams. Much of the content of these focused views originates in the LV, for example, in the functional descriptions of infrastructure components and services. The following paragraphs expand on the summary descriptions given in Chap. 3 for several Focused Viewpoints:

- *Network Viewpoint*—concentrates on details such as network physical media, data links, switches and routers, data terminals, network Quality of Service (QoS) mechanisms, and tabulated ports and protocols for connected hardware. This Viewpoint may document both local and wide area networks (LANs/WANs) as appropriate. Chapter 9 has much more detail on these subjects. A Network Viewpoint normally includes both graphics and tables and can also include supplemental information such as product documentation. Graphics can be simple block diagrams but often use pictures of actual components for clearer communication. For a real-time system, this Viewpoint would show the high-throughput, low-latency interconnections among system components with details of protocols, data rates, interfaces, and traffic types. The Network Viewpoint should also spell out any QoS functionality of the various networks, especially external, e.g., the ability to handle traffic in accordance with message priorities to maximize the probability that the most important traffic will get through when a network is overloaded.
- *Hardware/Installation Viewpoint*—shows the layout and mounting of hardware devices, cable runs, power and cooling distribution, and other important hardware details. For example, this View might contain graphics like Fig. 6.7 showing the arrangement of hardware components in racks. These will usually be drawings and other documents that are familiar to hardware designers, installers, and maintainers.
- *Runtime Environment Viewpoint*—collects in one artifact the details of the software infrastructure, including operating systems, utilities, database managers, middleware components, and other elements of the software environment. Important details include product versions, installed service packs, and other data required by application developers and system integrators. This Viewpoint can be developed using BDDs to show the software installed on each box, but a supplemental tabulation giving products, versions, installed service packs, configuration scripts, and other details can be useful in activities ranging from procurement to system installation.
- *Communications Viewpoint*—extends the data connectivity that may be covered in a Network Viewpoint for systems that need documentation of voice radios, telecommunications, data links, and other external communications. This is similar to the Network Viewpoint and may be combined with it, except that it deals

Fig. 6.7 An installation graphic is typical content of a Hardware Focused Viewpoint

with other types of channels and equipment. This view will generally focus on connectivity diagrams and specifications for external communications devices.

- *Messaging Viewpoint*—describes the various messaging modes and associated protocols and content used by a system. This is important in systems and enterprises where messaging is an important aspect of meeting requirements. It covers the messaging types and formats, which might include Publish and Subscribe, broadcast, large file transfers, content streaming, Web services, and others, as well as topics like mediation, format translation, store and forward mechanisms, synchronous and asynchronous modes, and reliable messaging. Table 6.4 lists some of the candidate messaging standards, and Chap. 9 has more details.

- *System Management Viewpoint*—describes the mechanisms and products used to manage system resources and maintain correct and reliable system functioning. The primary customers of this Viewpoint are System, Network, and Security Administrators and organizations responsible for approving operations with sensitive data. Complex systems need both automated and manual methods of monitoring hardware and software operations, e.g., to detect and deal with network or computer overload conditions; detecting, diagnosing, and correcting faults and failures; event logging and analysis; and maintaining configuration control. The functions described in this Viewpoint work closely with those in the Security Viewpoint (Chap. 10) in dealing with things like user accounts and privilege management.

6.9 Executable Physical Architecture

Many system engineers, when thinking about the role of simulation, focus on physical models. This is an incomplete view, but it is certainly true that effective physical modeling and simulation can be a powerful technique for design optimization, problem solving, stakeholder dialog, and many other SE activities. However, careful choice and design of physical simulation experiments are vital because a complete, high-fidelity executable model is likely to be prohibitively expensive and time-consuming.[2] The executable representations of the OV and LV can feasibly include the entire architecture at their respective levels of abstraction. With the PV, it's generally necessary to become more selective in applying simulation. The objective is to use these tools where they provide the most value, e.g., when they are the best means to ensure that the architecture meets requirements with adequate margins under all conditions. This requires judgment about the places where the architecture has critical performance requirements or timing relationships, heavily stressed resources, or other conditions that may create bottlenecks, intolerance for variable task loading, insufficient performance, or other problems.

A wide variety of tools, both commercial and private, may be useful. Individual tools are optimized for network analysis, performance analysis of processor/software configurations, reliability analysis, and other aspects of physical architecture. It is highly important that physical modeling be closely coordinated with system integration and test. In introducing virtual and physical prototyping as part of MBSAP in Chap. 3, we made the point that data from an integration lab can calibrate simulation models, while the models can extend the range of configurations and test scenarios to be explored beyond what is feasible with the actual hardware and software. Models will typically include simulators for external systems that are not available in the laboratory environment. It's also important that the modeling

[2] It has often been noted that the ultimate high-fidelity model of a system is the system itself.

process be flexible to support problem solving in the laboratory, analysis of requirements changes, and other day-to-day needs of a program. Because physical simulation is an integral part of an overall prototyping strategy, we defer a detailed discussion until Chap. 11, where we address closely related subjects of system verification, validation, prototyping, and test with examples of computed results for the E-X and Smart Microgrid.

6.10 Planning for System Evolution

It's especially important in the PV for the architect to think about the ways in which the system is likely to be employed and to change over its operational lifetime. A robust architecture, built on the principles that were discussed in Chap. 1 and using Quality Attributes to continuously assess their application, facilitates the task of keeping a system effective, supportable, and affordable. A best practice for overall program planning is the development and use of a Master Evolution Plan (MEP) that orchestrates all the activities associated with maintaining operational effectiveness and supportability as user needs, operational environments, technologies, and other factors inevitably change.

Two particular aspects of planning for the long term deserve mention. The first is a technology refreshment and obsolescence mitigation strategy and process. The goal is to keep the system current with successive technology generations and to minimize the economic, operational availability and performance impacts of technical obsolescence. Such a process can use the architecture as a framework for tracking things like vendor support for products and the predicted times when individual products go out of manufacturing, off warranty, and out of vendor support. One strategy is to preplan and budget for technology refreshment events in which specified hardware and software, especially in the infrastructure, will be updated. Some organizations maintain a watch list of products approaching obsolescence or non-availability and plan for their replacement or for alternative means of long-term sustainment. This is where the LV can be invaluable by providing a functional definition of a system component as the basis for upgrading or replacing it with minimal impact to the rest of the system. The PV complements this information with physical details that may be essential to planning and executing such projects.

The second long-range planning consideration concerns standards. Predicting the lifetime of a standard is at least as hard as forecasting the rate of technology change, but a good architecture process includes activity to track emerging standards and tries to identify when they are likely to require implementation in the system. Standards forecasting is important both in terms of the internal architecture of the system and in terms of the external environment with which it must interact.

Once again, the E-X example can illustrate these ideas. Some typical evolutionary planning examples include:

- Plan for long-term upgrading of sensors, communications, and other mission equipment; this might include things like upgrading sensor components, adding software for new or modified sensor modes, and replacing data links to remain interoperable with external nodes and systems.
- Plan for technology refreshment of the computing infrastructure as computers, networks, storage, and other hardware items become obsolescent and higher-performance replacements become available.
- Plan for evolution of enterprise services that the E-X supports and uses.

6.11 Summary

The PV completes the basic cycle of architecture development and establishes a design baseline for implementation, integration, and test. Figure 6.8 summarizes the primary artifacts of this Viewpoint. The focus is on point design decisions about

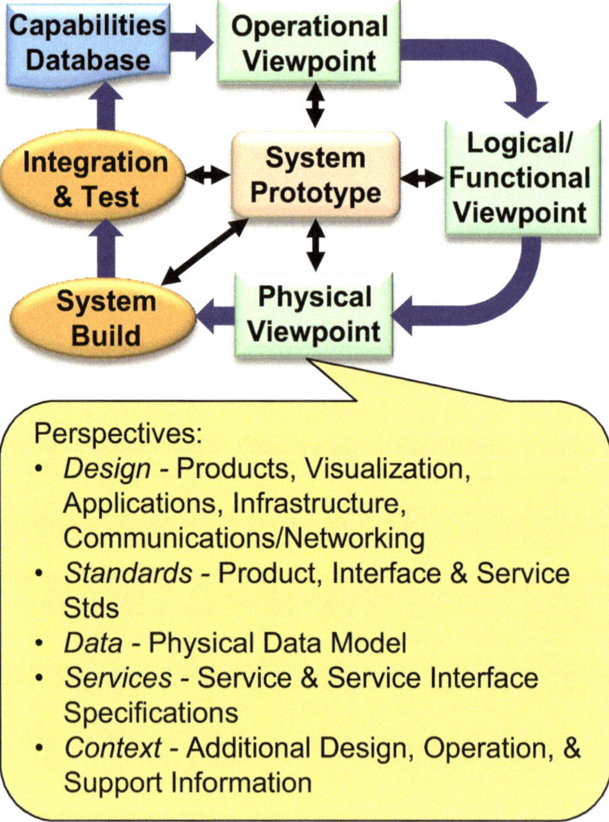

Fig. 6.8 Summary of the Perspectives of the Physical Viewpoint

products, standards, integration methods, HMI details, and other aspects of the system. The layered SOA construct enables a powerful integration strategy that accommodates a wide variety of legacy, modified and new system components. The View is documented in both the core SysML architecture model and an assortment of supplementary documentation. Additional issues involving Cybersecurity, Networking, Enterprise Integration, and other important aspects are dealt with in later chapters.

Exercises

1. Considering the Smart Microgrid A example in Fig. 5.30, list examples of physical detail that would be added to Block Specifications for the Supervisory Control, Battery, and Smart Meter Blocks in transforming the LV into the PV.
2. For the E-X Computing Infrastructure Domain, list some standards that could be applied to promote an open architecture.
3. What are some advantages of a layered structure?
4. Create an Instance Specification for the Server Block of the E-X Computing Infrastructure in Fig. 6.3.
5. For the three types of Resource Adapters shown in Fig. 6.5, list the functions each must perform to achieve the desired system integration.
6. For the E-X, list the applicable Focused Views, and identify Stakeholders who are supported by each of them.
7. Considering the challenges of integrating Mission System Components from multiple sources onto a shared infrastructure, list some ways in which a prototyping environment could be used to help meet them.
8. For an operator workstation on the E-X (Sensor Operator, ISR Operator, etc.), list some considerations that apply to designing the human-machine interface.
9. List some typical functions of the interface between Hard Real-Time and Soft or Non-Real-Time Domains.
10. Summarize typical content of a Physical Data Model (PDM) for the E-X.

Student Projects

Students should complete their basic architecture models by adding physical detail to the elements of the LV. For basic projects, it may be sufficient to capture the kinds of physical data to be included rather than to do extensive product research to produce realistic content.

References

1. DCGS Met Office (2012) DCGS integration backbone (DIB) v4.0 overview. http://www.afei.org/PE/4A07/Documents/DIB%20Description.pdf. Accessed 24 May 2017
2. Raytheon Co (2017) Distributed common ground system. http://www.raytheon.com/capabilities/products/dcgs/. Accessed 24 May 2017
3. US Government (2016) The Federal Enterprise Architecture Framework (FEAF), ver 2. https://obamawhitehouse.archives.gov/sites/default/files/omb/assets/egov.../fea_v2.pdf. Accessed 24 May 2017

4. Noran OS (2005) UML vs IDEF: an ontology-oriented comparative study in view of business modelling. Paper presented at the 6th international conference on enterprise information systems (ICEIS 2004), Porto, Portugal, 2004
5. Groff J, Weinberg P (2002) SQL: the complete reference, 2nd edn. McGraw-Hill, New York
6. MIP-NATO Management Board (MNMB) (2009) Standardization Agreement (STANAG) 5255

Chapter 7
Implementing Service-Oriented Architectures for Enterprise Integration

7.1 The Importance of Service-Oriented Architecture (SOA)

In recent years, the growing use of SOA has been one of the most prominent trends in the world of information technology (IT). Like many emerging technologies, SOA passed through an early period of hyperbolic enthusiasm, and it was all too often described as the cure for all the ills of information-based enterprises. It has since settled into an appropriate role as a mainstream approach to creating and maintaining enterprise business processes on a foundation of geographically and technologically diverse information resources. The commercial world has adopted this approach on a large scale as a way of integrating distributed facilities and systems, especially when heterogeneous assets that were originally developed in isolation must be harmonized to work together. An early example was the insurance industry which explored SOA constructs as a way to create integrated business processes from large, existing data systems that were incompatible as written but too expensive to replace or rewrite.

The primary motivation for SOA comes down to a few basic advantages over older approaches to applying information technology in modern systems and enterprises. These include:

- *Agility*—the ability to rapidly combine and configure capabilities in new and adaptive ways to create flexible and efficient business processes or mission threads to deal with changes in requirements and operational environments
- *Interoperability*—the ability of resources in various physical locations to employ various technologies and products to work together smoothly and to cooperate in accomplishing overall organizational and mission objectives
- *Robustness*—the ability of a system or enterprise to react quickly and affordably in the face of inevitable and often rapid changes in implementing technologies and products

J. M. Borky, T. H. Bradley, *Effective Model-Based Systems Engineering*,
https://doi.org/10.1007/978-3-319-95669-5_7

Network-enabled enterprises and SOA are inextricably linked. The operational concept of enterprise networking calls for systems and personnel to interact via network connectivity to share information, to collaborate in decisions, and to coordinate activities. This is seen as the key to enhance effectiveness through common understanding of operational situations and tasks, mutually supportive actions by various organizations and individuals, and the ability to rapidly adjust to changing circumstances. The examples are many. We have already mentioned the insurance industry, and similar considerations motivate SOA approaches in manufacturing, healthcare, financial services, education, and many other communities. Air traffic control is moving rapidly to networked operations based on SOA. The Department of Defense (DoD) has adopted a SOA model for the DoD Information Network (DoDIN), the connecting fabric of the Global Information Grid (GIG). Other agencies of the Federal Government have taken similar initiatives. Virtually any organization that relies on information technology and processes has adopted SOA or, at the very least, uses services over the Internet or over private networks.

Under the service paradigm, enterprise networking and integration are achieved primarily by requiring platforms and systems to expose data and functions as services and to participate in local and geographically distributed SOAs. For example, a number of space-based Earth observation systems make information that they collect available as standards-compliant services and implement capabilities like communications, sensor data management, and satellite control as services. Standardization bodies like the World Wide Web Consortium (W3C) develop protocols and specifications that enable interoperability, security, management, and other key service attributes. These standards support interoperability of services, including SOA implementations that combine products from various sources.

The SOA concept itself has in recent years undergone yet another profound transformation, this time to an information processing approach that is generally called cloud computing. Now, some or all of the hardware, software, networking, and other resources needed by an organization to execute business processes are provided as services in a data center operated by the organization itself or by a Cloud Services Provider (CSP). Chapter 14 goes further into this important topic. For the moment, it is enough to note that the SOA concepts and practices described in this chapter are rapidly moving to a model in which the services used by a networked enterprise are procured from a vendor rather than being provided by the participants in the enterprise.

In this chapter, we provide an introduction to this large and complex subject. We offer a set of defined terms to try to bring some order out of the welter of inconsistent terminology that plagues the field, and we discuss the primary ways services and SOA are represented in the MBSAP methodology, including the SOA Modeling Language (SoaML) [1]. We also highlight some of the important standards that apply to various aspects of SOA. An enormous library of SOA books and articles has been generated that look at the subject from every conceivable viewpoint and can be consulted for further detail [2–5]. Erl [2] provides a concise history of the evolution of SOA.

7.2 Foundational SOA Definitions

As with any engineering subject, SOA needs a clear, consistent, and accepted terminology, but, as may be inevitable with a rapidly evolving technology, there is currently a lack of standard terms and definitions. More than 30 definitions of SOA have appeared in various publications. The Organization for the Advancement of Structured Information Standards (OASIS) has published a SOA Reference Model that seeks to normalize concepts and terminology applicable to any SOA [6]. What follows is an attempt to capture consensus on the more important terms, drawing on many sources and applying some common sense and practical SOA experience. These definitions are sometimes drawn from, and are always compatible with, the SoaML specification. They are used consistently in addressing the process of designing, modeling, and building SOA solutions to enterprise architecture challenges.

7.2.1 Basic Concepts

We begin with the basic terms and ideas that enable a service strategy and make it both feasible and powerful.

Service Orientation: The starting point for SOA is to adopt a design philosophy or point of view for a system or enterprise that calls for:

- Implementing functions, or refactoring existing functions, that are needed in multiple places and processes across an enterprise in the form of shared, reusable services
- Instantiating a service once in such a fashion that it can then be used repeatedly
- Making the availability and use of a service independent of the location where it is physically instantiated

Service Orientation can be implemented to varying degrees, from including selected capabilities as services in a traditional architecture along with more traditional functions and data access all the way up to fully service enabling a system or enterprise.

Service: Fundamentally, a service is an offer of value by one participant in a community to another participant. It is enabled by one or more capabilities of the provider, exposed through a well-defined (generally standards-compliant) interface, and available to the community as a whole (which may be the general public). When invoked, a service causes the provider to do work, which may be as simple as retrieving a piece of information or as complex as an elaborate computation, and the results to be delivered to the receiving party. In terms of implementation, a service is an entity that possesses the following characteristics:

- It encapsulates components and other resources that create capabilities, it is instantiated at a known location, and it is exposed via a service interface.

- It provides functionality that can be discovered, invoked, and delivered, usually over a network, to provide value to a consumer or client.
- It has the resources and performs the activities required to provide the requested functionality or product.
- It may be subject to constraints or policies governing the conditions under which it can be used, e.g., by being available only to approved recipients.
- It is composable, meaning that it can be combined with other services to create larger processes or more complex services.

Figure 7.1 is a simple graphic showing the essence of how a service model simplifies a basic interaction between a system user (client) and a server that provides files, applications, or other content. In the conventional transaction model, the client must have detailed knowledge of the network location and functions of the server and typically sends a series of server-specific commands to obtain the desired content. When this server capability is enabled as a Web service, the client only needs a functional identifier such as a Uniform Resource Locator (URL) or other Web address and can send a request message and receive the service product. This is the essence of the discover, invoke, and deliver service model.

Service Categories: Services can be organized in a few broad categories such as the following, which are similar to those introduced in Chap. 5 in the context of architectural layering:

- *Mission or Business Services*—services that carry out functions and deliver products associated with executing mission or business processes.
- *Data Services*—services that expose or ingest data and perform related functions such as create/retrieve/update/delete (CRUD), broadcast, security, synchronization, and backup and restore; these services do not include business logic, i.e., computations that create or transform content.

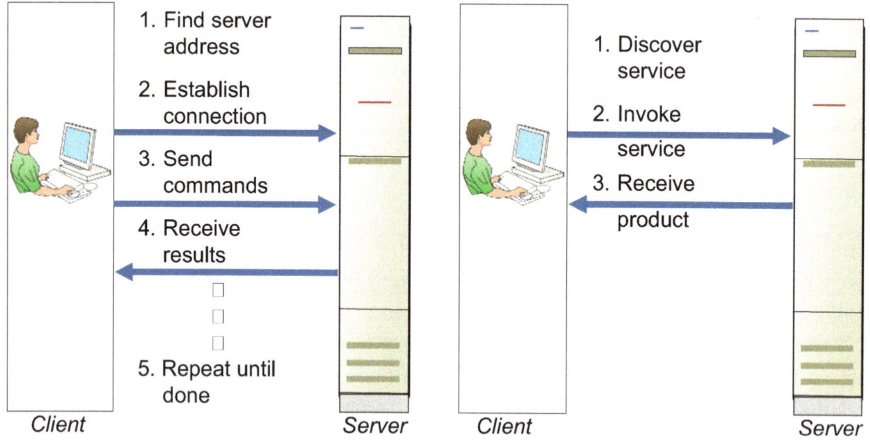

Fig. 7.1 Conventional vs. service-enabled transaction

- *System Services*—services that provide supporting functions such as messaging, networking, discovery, search, storage, and system and resource management; in general, System Services create the environment for Mission and Data Services and are sometimes also called Infrastructure Services.
- *Enterprise or Community of Interest Services*—services that create the framework for information exchanges, collaborative activities and decisions, and management of activities and resources across an enterprise or community as a whole.

Service Interface: A service interface creates a representation of the service that is accessible at a specified location, which is most commonly identified by a network address. A service interface should conform to the following principles:

- It defines service capabilities and the specifics of how they are invoked.
- It mediates service invocation and product delivery between the service implementation (providing party) and the consumer or client (receiving party).
- It is separate from the service implementation and abstracts the encapsulated functions and data of the service from other entities on the network so that a change to the internals of the implementation does not affect its use by clients/consumers, provided the interface is unchanged.
- It may allow clients to access variations on the service depending on specifics of the invocation (called polymorphism in Object-Oriented terminology).
- It should conform to established open standards such as the Web Service Description Language (WSDL).
- It is governed by a contract such as a Service-Level Agreement (SLA) between the service provider and consumer or client or by some other suitable interface definition.
- It defines the service endpoint (point of contact) with details for forming invocation messages and the physical location (network address) of the service itself or of a broker or other agency that provides access to the service.

Service-Oriented Architecture: SOA is not a predefined product but rather an architectural *style* that embraces service orientation; it is built on services and service interfaces, in which:

- A community of participants is defined who agree to honor service contracts governing how they will interact to accomplish a purpose.
- Individual functions of participants (e.g., application logic/computation) and their data are exposed as services that can be accessed at various physical locations, which are usually identified using network addresses, i.e., via service interfaces.
- Higher-level mission or business functions (business processes or mission threads) of the community or enterprise are implemented by composing services that are distributed across a networked enterprise, loosely coupled, and interoperable.
- Flexibility and agility in creating processes are achieved by dynamically assembling (composing) them from local and distributed services.

- There is a separation of concerns between what is to be done and how, where, and by whom or by what it is accomplished, as well as between architecture and its implementation and technology; separation of concerns also generally implies the desirability of restricting a service to a single capability (delivered value) or at least to a set of closely related capabilities.

7.2.2 Common SOA Features

Next, we need definitions for some things that are commonly encountered in a SOA implementation.

Web Service: A service that is:

- Implemented using protocols associated with the World Wide Web, especially WSDL and Universal Description, Discovery, and Integration (UDDI) [7]
- Registered at a known network location for discovery by a client
- Invoked using SOAP[1] messaging

Web services are the most common, although not the only, service model for SOA. In a later paragraph, we introduce Representational State Transfer (REST), which is an alternative approach to enabling interactions over a network such as the Internet. Still other service models are used in real-time systems. However, Web services are the basis for most SOAs, especially those involving Management Information Systems.

Orchestration: In a SOA context, orchestration is the combination, coordination, and use of services to create business processes. Orchestration is a widely used approach to service composition. A SOA in which orchestrations are easily created and modified supports process agility, which is a primary motivation for this architectural paradigm.

Workflow: Typically, a workflow is a service orchestration that involves interactions with system users. Orchestrations and workflows are generally implemented using infrastructure components such as a "workflow engine," which executes process models that define the sequence of service invocations, user interactions, etc.

Choreography: Choreography is a mechanism that is similar in some respects to orchestration but focused on patterns of collaboration among service providers based on public messages exchanged by the participants rather than on sequences of service invocations.

Enterprise Service Bus (ESB): As introduced in Chap. 5, an ESB is an infrastructure component that is often, but not necessarily, an element of a SOA. The term ESB is applied both to an architecture pattern for exposing and accessing services and to a software product or group of products. An ESB itself typically performs System/Infrastructure Services that support interactions among systems and system

[1] This was originally the Simple Object Access Protocol but is now so ubiquitous that it is simply identified as SOAP.

components, including messaging, routing, transformation, mediation, invocation, event processing, and orchestration/choreography.

SOA Modeling Language (SoaML): SoaML is a UML profile specifically designed to facilitate modeling services. SOA has become so prominent that the Object Management Group has published a specification that provides a metamodel and profile for the specification and design of services within a SOA. It extends UML 2 to support identification and definition of services, definition of service providers and consumers, service policies, and other key aspects of SOA. It can be used equally well in a SysML model with the usual translation of Classes to Blocks.

Rich Internet Application (RIA): An RIA is a Web application that resembles a traditional desktop application. RIA environments can aggregate (fuse) data from multiple sources in a unified display that appears to the user to be driven by a unitary application. User interface logic resides in a Web client, while data is maintained on servers.

Business Process Management (BPM): BPM is a systematic approach to defining sequences of activities that make up a business process. It provides orchestration tools for designing and implementing applications that optimize and coordinate business processes with the goal of making them more efficient, resilient, and effective. Although they originated in separate lines of development, with BPM being application-centric rather than service-centric, BPM and SOA have begun to converge as it has become clear that they have very common objectives in implementing robust, agile business processes.

Really Simple Syndication (RSS): RSS is a family of Web feed formats used to publish content updates to Web sites that can be read using an RSS reader or aggregator. Assembling complex objects typically requires interaction with multiple content providers, each having a distinct network address or Uniform Resource Locator (URL). RSS may be useful as an efficient way to generate alerts across multiple sites, which can then be followed by the use of SOAP to access content.

Representational State Transfer (REST): REST is the primary alternative to SOAP for implementing loosely coupled, scalable, distributed processes [8]. It is a simpler paradigm, generally based on exchanging Extensible Markup Language (XML) documents over the most common Internet protocol, Hypertext Transmission Protocol (HTTP). It seeks to emphasize basic architectural principles such as scalability, generalized interfaces that encapsulate resources, independence in deploying components on a network, minimum latency in interactions, and support for transaction security. REST emphasizes interactions with stateless resources rather than exchanges of messages. It therefore defines an alternative Web Service model to SOAP services, and these are commonly referred to as RESTful services. REST has had a devoted following, especially in the academic community, and is coming into wider use. An important development is the fact that Version 2 of the WSDL standard implements binding to all the basic operations of HTTP (GET, POST, PUT, DELETE, etc.) and thus facilitates implementation of RESTful services. As commercial tools provide more robust support for REST in general and for WSDL 2 in particular, RESTful services will become more common, especially in SOA situations where they offer important advantages over traditional SOAP-based services in areas such as service interoperability.

7.2.3 SOA Characteristics and Principles

One further set of definitions deals with principles that have proved to be important to successful SOA implementation. Good SOA practice is based on a set of tenets that are closely related to the general architecture measures of effectiveness that were discussed in Table 1.2. To be fully effective, the SOA paradigm calls for some elaboration of these best practices as discussed in the following paragraphs. Some of these ideas were introduced earlier as part of foundational SOA concepts.

Loose Coupling: Services are stateless whenever possible (RESTful services are required to be stateless) and have minimum dependencies on each other and on the client. There are several implications of this approach to loose coupling.

- Only service descriptions are required to enable interaction, and a small set of simple interfaces is available to all participants. Descriptive messages are constrained by an extensible schema (so that new ones can be created) and do not define system behavior, which is part of separating the service interface from the service implementation.
- Service implementations are compatible with heterogeneous and asynchronous communication channels. Components interact primarily through open, self-sufficient messages (or basic HTTP operations in REST) rather than remote procedure call (RPC) protocols that require detailed knowledge of the service implementation.
- Workflow coordination can be used to manage interactions among independent entities (see Autonomy).
- Error management is provided to deal with problems that occur in the context of interactions among autonomous, loosely coupled entities.

Autonomy: A service is self-contained and can control and govern the encapsulated logic and data.

Composability: Services can be combined to create higher-level mission or business functions.

Reusability: The service and its interface can be deployed in multiple locations and contexts where the capability is required and can be accessed by many clients.

Governance: Mechanisms are provided to enforce service interfaces and overall rules for:

- Policies that define rules, standards, guidelines, and best practices to promote consistency, interoperability, reliability, security, and performance across a service-oriented enterprise
- Processes to verify compliance, control configurations, verify and validate component designs, audit and assess operations, and control service component deployment and support

- Metrics to quantify and track status and trends in service implementation and delivery
- An organization with roles and responsibilities to provide governance

Quality of Service: Services advertise and honor, as part of the service contract, certain guarantees in areas like timeliness, accuracy, and policy enforcement, including:

- Services preserve security, integrity, and reliability in executing transactions.
- Services honor applicable client and enterprise priorities and policies governing their functions.

Extensibility: Service functionality can be expanded, e.g., by adding new features or polymorphic service variants, without breaking the interface.

Coarse Granularity: Services generally provide relatively complete, high-level functions that:

- Minimize the messages that must be exchanged to complete a transaction
- Rely on rich, document-oriented messaging with high content

Security: Trusted mechanisms are in place to ensure confidentiality, integrity, availability, non-repudiation, and access control, including authentication and authorization of participants in a service-oriented enterprise and protections against internal and external threats. Chapter 10 addresses security in more detail.

Federation: Mechanisms are in place to allow interaction among heterogeneous systems via nonproprietary, standardized communication mechanisms to allow cooperative functioning; the interacting systems may be geographically dispersed.

7.3 SOA Strategy

With the basic principles established and the jargon defined, we can now address the practical aspects of planning, implementing, and employing a SOA to meet the goals of an organization. There are as many approaches as there are authors, and they range from general guidance to detailed, step-by-step recipes. We necessarily lean toward the former, partly due to space and partly because there is really no single strategy that fits all situations. Indeed, one of the most important and difficult decisions in any large-scale SOA project is selection of the method to be followed and the tools, products, consultants or contractors, and other resources to be employed. The objective of this discussion is to raise awareness of the issues and pitfalls and to suggest the general outline of a SOA strategy.

The first consideration is a solid foundation in policy, organizational strategy, planning, and advance preparation for a SOA project. An organization that plunges into a SOA implementation without doing the essential homework incurs unnecessary risks that can be avoided by a more thorough and deliberate approach. This early work need not consume a great deal of time and will usually actually save time by making the rest of the project go more smoothly.

SOA projects broadly are divided into two categories: new system development and transformation of a legacy system to a SOA. There are many common features between these two, but the latter imposes additional challenges that make careful planning even more essential. In this section, we summarize the matters that should be accounted for in preparing for a SOA implementation, recognizing that the details will vary with individual projects.

7.3.1 Overarching Challenges

Publicly available information on SOA efforts that failed to meet requirements or achieve desired business outcomes indicates that the failures often result from a small set of root causes. Generally, these come down to a lack of commitment by an organization, its leadership, and its staff to a different way of doing business. This is especially problematic when a legacy system or enterprise is to be service-enabled, because this can entail far-reaching changes to the roles, responsibilities, and work procedures that are familiar and comfortable to the staff of the organization. Cultural issues can be harder to deal with than technical ones, and dealing with them depends critically on top-down executive support for the program. A common indicator of insufficient commitment is the failure to budget adequate time and resources to do a good SOA implementation. The following paragraphs describe some specific areas where shortfalls can lead to problems or failure.

Vision and Leadership: Credible approval of a SOA project and empowerment of the team that is to execute it from the highest levels of the organization are essential. This includes senior management participation in planning and execution, manifested in support for education, process transformation, and organizational changes. Securing this commitment may require a complete and approved business case for implementing a SOA.

Policy: A project is likely to either fail or be longer and more expensive than necessary if the organization does not establish policies that articulate and enforce key areas of the SOA strategy. Policy areas typically include:

- Technical standards to be implemented
- Governance, including management of the services portfolio, enforcement of the architecture, versioning as the architecture evolves, and other critical activities (see Chap. 15)
- Service architecture and taxonomy
- Interoperability within a service-enabled enterprise
- Reuse of applicable legacy processes and resources
- Security, including SOA-unique aspects, as discussed in Sect. 7.4
- Provisions for recovery from incidents such as failures or natural disasters

The ultimate goal is improved operational effectiveness through the payoffs SOA offers to an enterprise, and sound policy is an indispensable element of overcoming obstacles to realize those gains. SOA architects working on programs for the

Department of Defense must cope with a plethora of policies, and Appendix K provides a summary of these.

Strategy and Roadmap: Another way to add unnecessary risk to a project is to fail to develop and promulgate a SOA strategy that is well-matched to project goals and content. For example, it is often important to plan for incremental transformation, program alignment, and creation of a cross-enterprise architecture. Related needs include creating a common vocabulary and developing plans to enforce policies and security. Since one key payoff from SOA is process agility, the strategy must include controlled ways to change processes and services over time. A SOA implementation that involves transforming a legacy enterprise to a SOA model is typically best carried out through an incremental deployment to minimize day-to-day impacts to ongoing processes and functions. The project roadmap normally includes demonstrations, pilot projects, standards-based process modeling, development and maintenance of a services catalog, development of SLAs, and provisions to manage and monitor services and their usage.

Up-Front Analysis: Part of the essential preparation involves early attention to analyzing existing or planned business processes or mission threads and the improvements required to meet organizational goals. It's especially important to identify and define a good set of services. The service categories in Sect. 7.2.1 are a good structure for the analysis of system content and functions that are candidates to be service-enabled. In a SOA transformation of a legacy system, it's equally important to define the process changes and the accompanying organizational adjustments that will be required, e.g., as manual processes are automated. Another important early analysis involves measures of effectiveness to define a way to gauge progress and assess whether each increment of SOA deployment is yielding the expected benefits. Properly done, this early analysis creates a basis for setting goals and expectations, developing a good project plan and schedule, and preparing the organization and its staff for the changes the transition to SOA will bring.

7.3.2 Basic SOA Strategies

One possible starting point is to think in terms of top-down, bottom-up, and hybrid approaches, as sketched in Fig. 7.2. The following paragraphs summarize these alternatives.

Top-Down: A top-down strategy starts with goals and requirements, first analyzing the business processes or mission threads that are to be service-enabled and then focusing on the capabilities needed to meet system and enterprise objectives. This can be effective when the project has a reasonably clean slate and is free to define and implement services with few constraints arising from things like reuse of existing assets or conformity to a larger enterprise.

Bottom-Up: A bottom-up approach is based on analyzing the (often pre-existing) resources used to implement the system or enterprise and seeking to transform capabilities into services. This happens when it is too hard or too expensive to

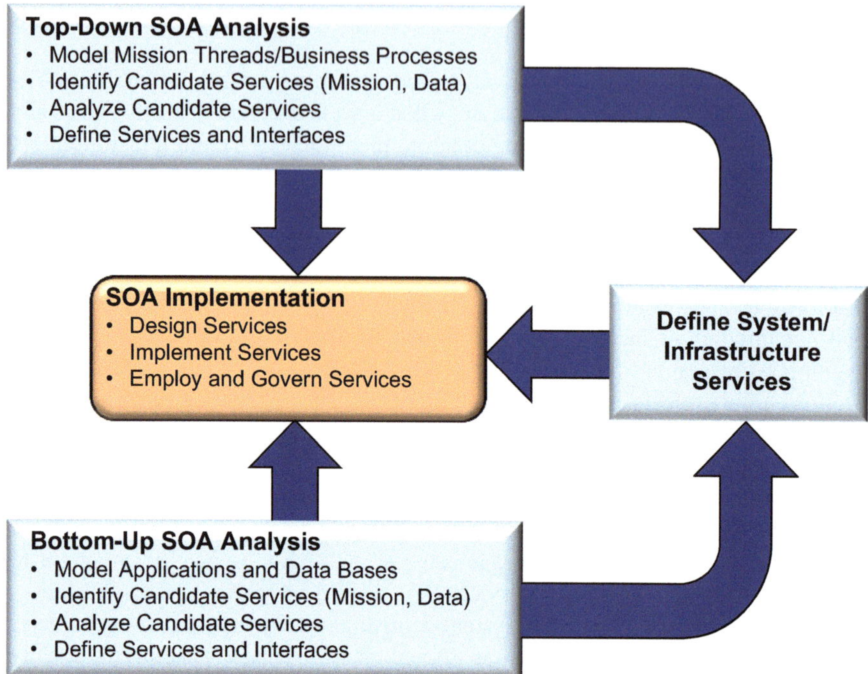

Fig. 7.2 Basic SOA strategies

replace an existing body of assets. The goals here often include replacing manual workflows with more automated (sometimes called "machine-to-machine or M2M") information exchanges and improving the sharing of resources such as data and high-end computing hardware.

Hybrid: A hybrid approach has elements of both top-down and bottom-up and can yield an agile methodology that can be optimized for the specifics of an SOA project. Realistically, this will usually be the approach of choice because most projects involve a mixture of new capabilities to be developed as services and old resources to be reused and service-enabled. A rule of thumb might be that if a project does both top-down and bottom-up, and the results meet in the middle to derive largely the same set of services, the resulting answer is probably fairly good.

Regardless of the approach, the problems cited earlier must be dealt with. Any SOA project must begin with up-front analysis to derive an optimum set of services and make provisions for service refinement as experience is gained. Buy-in and, ideally, participation at all levels of the organization will significantly improve the odds of success, especially when the new approach will force people to accept new ways of doing familiar tasks and require changes in organizational structure to realize the benefits of SOA. The project needs to be adequately resourced in terms of funding, scheduling, and staffing. It must also have the flexibility to experiment with alternative solutions, make and recover from mistakes, accomplish training and documentation, and phase in the new architecture in an orderly sequence to minimize cost and disruption. Governance needs attention from the outset and must be a fundamental element of the solution.

7.3.3 SOA Payoffs and Measures of Success

We mentioned previously that early attention to a scorecard for assessing SOA progress and results is important. The following are some candidate attributes and outcomes that might be considered in this regard; most of these trace to the SOA characteristics that were defined earlier.

Agility: This involves demonstrating the ability of organizational business processes to respond to changes such as different tasking and long-term system or enterprise evolution. An agile system has dynamic operational capabilities, and this might be assessed through experiments and early pilot projects to see how the system responds to induced changes in processes, technologies, rules, and other variables.

Interface Stability: Another indicator could be to determine whether responses to changes break the associated service interfaces or force requestors to change methods for discovering and invoking a service.

Composability: This can be assessed using experiments that show how services can be orchestrated or otherwise composed to efficiently implement complex processes, including new ones. Composability often involves achieving a higher level of process automation, and the desired outcomes may include reduced end-to-end time to complete a task or produce a product as well as reduced staffing.

Reuse: Reuse can be determined by showing that services can be employed across systems or enterprise nodes. This is important because reuse increases the return on their original development investments and is an important element of long-term enterprise cost reduction.

Other: Additional measures could include verifying that service interfaces are platform neutral and independent of the execution environment. It might be important to establish that heterogeneous IT environments can be integrated via standards-compliant services, including new, modified, and legacy components that use various technologies and designs.

7.3.4 Service Partitioning Guidelines

The burgeoning SOA trend has provoked a number of debates, none more intense than the argument over how to partition capabilities into well-formed, highly reusable services. Based on the literature and on practical experience, the following guidelines may help in dealing with this issue.

Define Service Granularity: The goal here is to partition resources and functions at a level of granularity, meaning the size and complexity of individual services, which maximizes composability, reuse, and ease of governance. Big, multifunction services minimize the total service count and the number of interfaces to be managed. However, those interfaces may become complex and hard to use, and it can be very difficult to prove that even a small change to the service implementation has done nothing to invalidate the interface. At the other extreme, services that are too fine-grained mean that anything more than trivial interactions will require multiple service invocations. A well-partitioned service is usually one that performs a commonly needed task or produces a commonly needed product like a plan, a sequence of machine instructions, an operator interface, a report, a response to a query, or the actions that respond to a task. Most SOA architects lean toward relatively coarse-grained partitioning to promote effective service management and minimize the overhead associated with excessive XML processing, which can consume a surprisingly large fraction of server throughput.

Ensure Services Are Well-Formed: The characteristics of a well-formed service include:

- Encapsulation of an appropriate set of logic and data, meaning that the service performs one or more logically complete activities that are likely to be needed in multiple processes; this is the heart of the service granularity argument.
- Conformity to service-oriented principles as described earlier in this chapter.
- Satisfaction of the needs of the processes in which it is employed.
- Feasibility of implementation with available technology.
- Inclusion of features and functions that satisfy cybersecurity requirements.
- Ability to deliver required QoS, including response time.
- Ability to be administered or governed with available tools and methods and without excessive effort.
- Use of appropriate naming conventions for the service endpoint, service operations, and message values.
- Use of operations that are designed to be inherently extensible for long-term service evolution.
- A fully defined service description documented in an interface specification or SLA.

Link Services to Organizations and Processes: Another way to approach service partitioning is to ensure that there is a clear and logical mapping to organizational entities, external service users and providers, sources and products involved in business processes, supporting technologies and infrastructure, information classification levels, and similar factors. If a function that is to be service-enabled requires participation by multiple organizations or uses a wide range of IT resources, it's probably best to compose it from lower-level services that have cleaner mappings to resources and organizational elements.

7.3.5 Design for Change

A robust SOA strategy deals with inevitable changes in services, service interfaces, business processes, implementing technologies, and other factors. The following are some of the ways to do this.

Design for Ambiguity: The program should start with a prototype SOA infrastructure, implement it rapidly, and make best estimates of the initial services. The infrastructure should ensure that the design can quickly accommodate changes as the initial service catalog changes in response to problem discovery, changing requirements, and other facts of life.

Design for Flexibility: To realize the full benefit of a SOA implementation, the system or enterprise must be able to exploit the service model to effectively respond to changes in system processes, products, environment, priorities, customers, and many other circumstances. This can involve revising the services themselves, but that is likely to require expensive software reengineering. The fastest and least expensive way to deal with a change often involves different ways of composing basic services into higher ones. A related technique uses new or modified orchestrations and workflows that don't require major revisions to the underlying services. The SOA strategy should stress abstraction and loose coupling and other features that facilitate changes.

Plan for Iterations: The complexity of most service-enabled enterprises is such that it's unlikely that an initial SOA will prove satisfactory in all respects. The SOA strategy should then explicitly provide for the necessary experiments and refinements to optimize the design and implementation. Accordingly, the SOA strategy should call for planning, budgeting, and adjusting to organizational structures that accommodate continuous change in a spiral model.

Measure Frequently: Given the likelihood that final success in a SOA project will result from trying alternatives and refining the architecture and services, it's important to ensure the effort stays on track and converges to a fully satisfactory solution. The program should assess each development increment or spiral against success criteria that are relevant to business or mission objectives. This is an early stage of the overall governance process that is discussed in Chap. 15 and should evolve into a continuing assessment process over the system's lifetime.

7.4 SOA Implementation

Once a SOA strategy founded on organizational policies and goals and accounting for agility, security, and other concerns has been developed, attention turns to the design, development, testing, documentation, and deployment of a service-enabled system or enterprise. Once again, we focus on general guidance and issues because the specifics depend heavily on the nature and objectives of a given SOA project. The Open Group publishes an extensive body of SOA source material [9].

7.4.1 SOA Structures

In Chap. 5, we introduced the idea of architectural layering in which user, mission, and infrastructure services are grouped in layers, each depending on the ones below. A SOA is also commonly described in terms of a layered structure, which can be mapped to that in Chap. 5, but is different in the way it aggregates resources and functions. Figure 7.3 shows a typical three-layer portrayal of a SOA.

Mission Threads/Business Processes Layer: The top layer represents the processes that do the work of the system or enterprise to achieve its business or mission objectives. Typically, these are created by service orchestrations or workflows and may be called "business applications." The graphic shows the general character of a workflow with processing rectangles, decision diamonds, database accesses, and user interactions.

Service Interfaces Layer: The middle layer contains the interfaces (represented as colored circles) that expose functionality and data provided by resources at various network locations. Interface locations are called service endpoints. Some interfaces expose individual services associated with specific resources. Others expose more complex services that compose basic services into higher-order services and may include additional business logic. Various methods may be used to expose

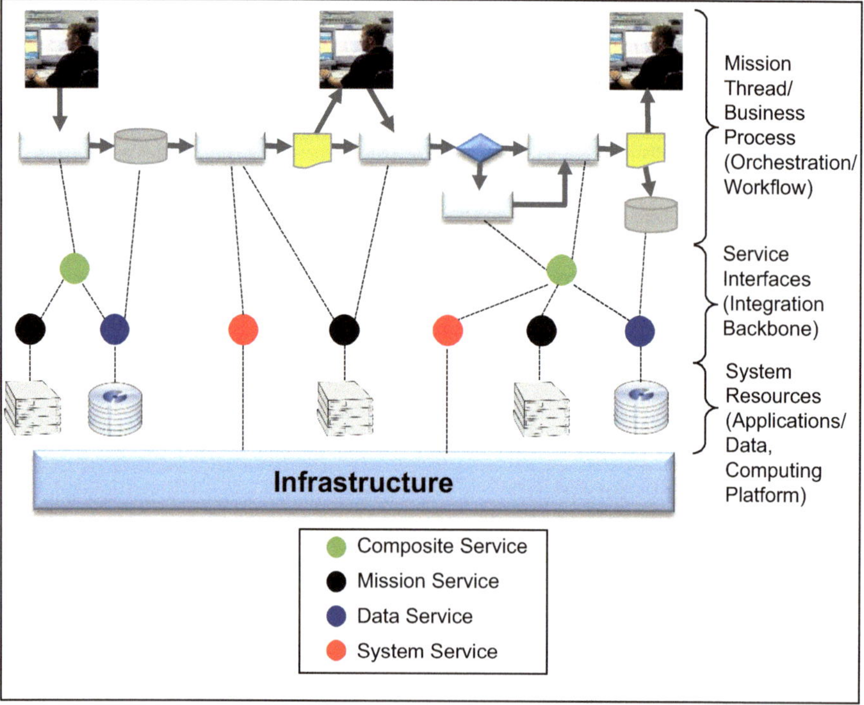

Fig. 7.3 Basic SOA layered structure

capabilities as services including applying specialized software interfaces called wrappers to components and other resources to service-enable them. As discussed in Chaps. 5 and 6, these service endpoints are available in the architecture layer called the Integration Backbone (IB).

System Resources Layer: In this construct, the bottom layer in Fig. 7.3 collects most of the functionality of the layers shown in Fig. 5.17. It contains the resources, primarily applications and data, which are exposed as services and used to implement business processes. An application and its data are often composed to create a complete function or service but may also be accessed individually. This layer also accounts for the infrastructure resources that are exposed as services and that must sometimes be directly accessible to business processes.

In principle, services can be implemented at any level of the Axis of Organization introduced in Chap. 1. Figure 7.3 describes services in terms of a functional organization. The resources in the bottom layer are physical assets, the interfaces in the middle are abstractions of resource capabilities, including composition to produce higher-level services, and the workflows in the top layer are a purely functional construct describing sequences of service invocations. It's also important to think about SOA in terms of how this organization is actually translated into an implementation architecture with mission applications hosted on an infrastructure. Frequently, mission applications are developed by or procured from sources other than the infrastructure provider. The architecture must support the integration of all these components into an operationally suitable solution as discussed in Chap. 6. Then the capabilities that are to be service-enabled must be provided with appropriate service interfaces located at known service endpoints.

Figure 7.4 is a simplified version of Fig. 5.17 that shows one way of mapping the SOA concept onto a practical architecture built with available IT products. This example focuses on the Decision/Process Real-Time category of the taxonomy in Chap. 1. It's common with this class of systems to think in terms of hosting mission applications, which typically come from multiple sources and are a mix of existing (legacy), modified, and new software components, on a shared infrastructure. In a typical system or enterprise, that infrastructure is established on one or more servers and connects through a local area network (LAN) to workstation computers that give users access to mission capabilities. The color coding in Fig. 7.4 shows where Mission and System/Infrastructure Services reside, while Data Services are provided both by mission applications (e.g., mission data products) and by the infrastructure (e.g., database access and search).

The client layer is hosted on user workstations or other devices (such as smartphones or tablet computers). It commonly includes components such as a browser for connecting to Web sites and services along with various viewers. It may also host other client-side software that is associated with mission applications. These components interact with the presentation logic of applications to create displays and accept user inputs via a graphical user interface (GUI).[2] A portal is often used to provide a single, user-customized means of accessing Web content and services.

[2] To oversimplify slightly, browsers and viewers correspond to what are typically called Thin and Thick Clients.

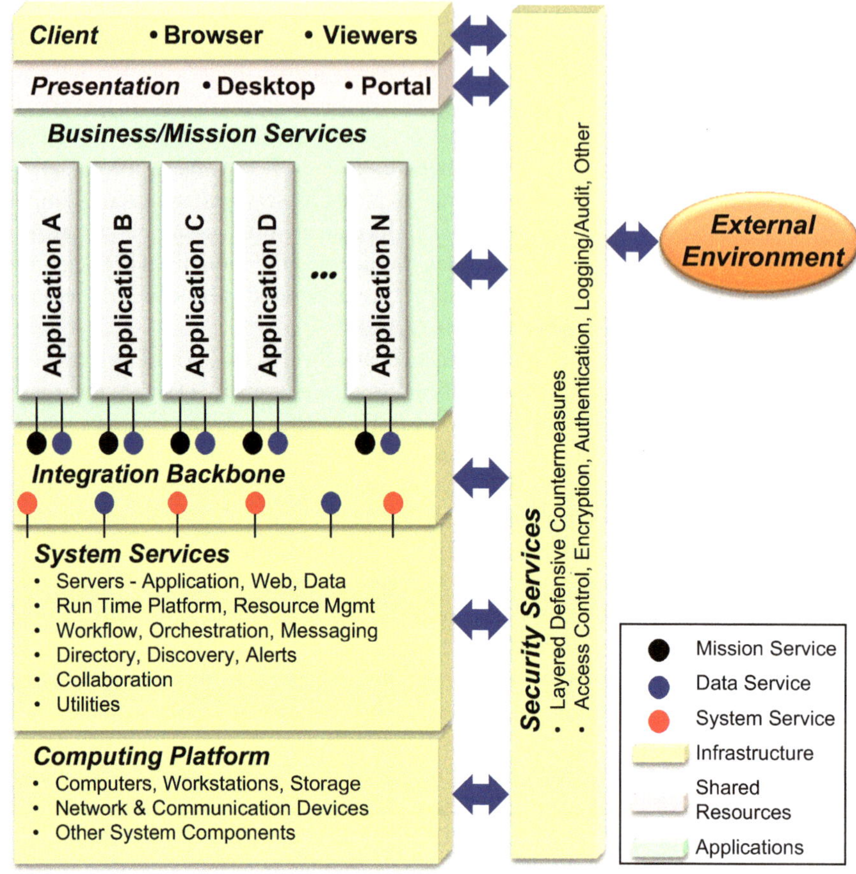

Fig. 7.4 Service Categories mapped to a Layered Architecture

Mission applications are grouped in a Business/Mission Services layer. They typically expose both Mission and Data Services as defined earlier in this chapter, and in the architecture design pattern of Fig. 7.4, an IB creates the shared space in which service interfaces are accessible. The IB holds service endpoints and is fundamentally the architectural domain where infrastructure, mission applications, and external interactions come together. The IB provides services and resources that allow mission applications to interact with each other and with the external environment, and in the process it enforces key principles like loose coupling.

As Fig. 7.4 shows, Infrastructure Services (red circles) are exposed in the IB where they can be discovered and invoked by mission applications; in fact, a catalog or registry of both Infrastructure and Mission Services is one of the most important items of information needed by application developers in order to efficiently account for them in their code. Viewing the IB in this way, it's natural to think of it as the place where service orchestration or workflow is implemented, through sequences of Mission Service invocations and user interactions via service interfaces, and it makes available all the other capabilities of the infrastructure.

The software components of the infrastructure, which provide Mission Services, are in the layer below the IB. In current technology, this layer will typically use COTS products in an n-tier middleware configuration. Figure 7.4 lists some common components and services, including things like local and enterprise search engines for discovery, utilities that support Continuity of Operations (COOP) by implementing data backup and handoff of mission functions to a backup site in the event of failure, and management tools for computers, networks, services, and other resources. As in Chap. 5, the bottom layer contains the hardware resources on which the system is built.

7.4.2 SOA Elements

Although it's important to stress that SOA is an architecting style or template, rather than a specific design, there are some things that are so commonly used that they are, in effect, default SOA elements and will be included unless there is some compelling reason not to. The following are some of the most important elements of a typical SOA. The descriptions are highly condensed and intended mainly as a terminological reference. A SOA designer will need a much more in-depth familiarity with these elements than we can present here.

XML Foundation: Service and data definitions, SOAP messaging (or the REST alternative), and other key aspects of the architecture are built on XML core standards. These include XML itself, Extensible Stylesheet Language Transformations (XSLT), XML Schema Definition (XSD), WSDL, and others. These are referred to as the XML Foundation. XML defines perhaps the most widely used form of "metadata," which can potentially describe both the form (syntax) and content (semantics) of an information entity using a highly stylized and standardized "markup language." These markup "tags" are generally not displayed to the user of the document or other information but enable automated processing and handling.

Protocols: The service model is built on a set of protocols. For Web services, these include WSDL, SOAP, UDDI, WS-I Basic Profile, etc., while RESTful services rely on HTTP. For other service models, for example, real-time services that must honor strict deadlines, equivalent standards, and protocols must be defined; these typically involve simple messaging between service providers and clients whose network locations are known a priori along with high levels of integrity and fault handling.

Security: The service model, especially when the Internet is involved, creates significant security risks if not properly protected. This critical SOA element is discussed in Chap. 10 as part of the overall cybersecurity dimension of system architecture.

Messaging: Participants in a networked SOA interact using messages and file exchanges. The following are some typical interaction mechanisms within a service-enabled enterprise. Chapter 9 treats the overall topic of networking, in which a messaging model plays a key role.

- Common messaging paradigms include SOAP for Web services and other appropriate protocols such as Java Messaging Service (JMS) or remote procedure calls (RPC) for other service categories.
- Headers and metadata are used to control information exchange and security functions.
- A SOAP message is packaged in an "envelope" and handled by sender, receiver, and intermediary "nodes." The protocol allows calls between components to be formatted ("serialized") into XML documents suitable for transmission by HTTP. A message body is frequently an XML-formatted payload. The basic message pattern is unidirectional, but others can be implemented via header blocks.
- Header blocks contain meta-information to make the message as independent as possible of messaging technologies and environments; these blocks provide information on context, processing, routing/workflow, security, reliability rules, correlation with other messages, and other aspects.
- Attachments can be used to bundle non-XML data into a message in its native format.
- Fault handling involves providing a way to specify exception handling rules, e.g., "send error message."

Figure 7.5 shows some of the elements of a service messaging transaction, once the requestor has discovered and bound to (established a connection or session with)

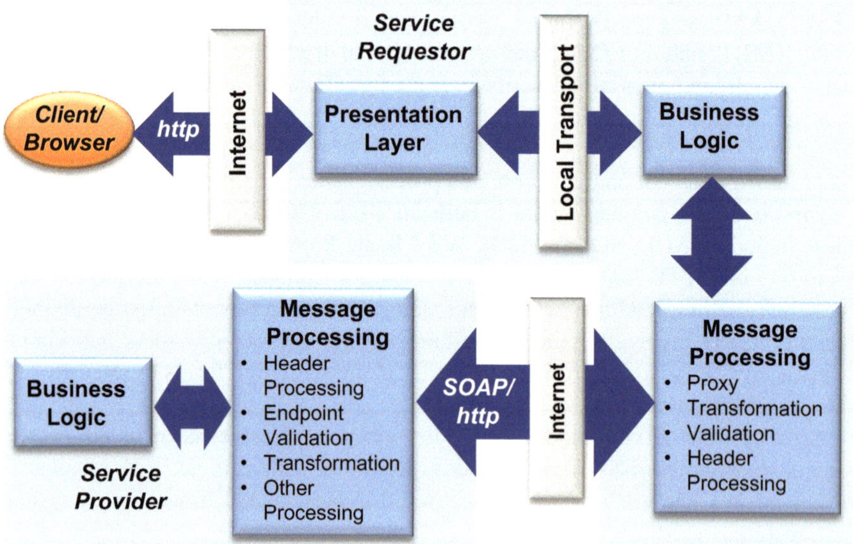

Fig. 7.5 Typical message-based service transaction

the provider's service. The transaction employs various networking protocols, flows through the layers of the architecture, involves business logic (Mission Services) of both the requestor and the provider, and requires a number of successive steps in the Message Processing service at both ends of the Internet connection. The Messaging Focused Viewpoint in Chap. 6 is a convenient way to document the messaging solution employed in a particular system or enterprise.

Roles: Participants and services in a service-oriented enterprise may assume roles that include Provider, Requester, Intermediary, Initial Sender, Ultimate Receiver, and others; these may be temporary classifications valid for a specific transaction and changeable for subsequent transactions.

Processes: As noted earlier, a major purpose of SOA is to enable higher-level business or mission activities to be created by composing or assembling lower-level services to create complete business or mission processes.

Orchestrations: This was defined earlier and is frequently standardized by the Web Services Business Process Engineering Language (WS-BPEL) specification.

Choreographies: This was also defined earlier and is frequently standardized by the Web Services Choreography Description Language (WS-CDL).

Collaboration: This is a general term for a wide variety of services found in most SOAs that allow users to interact at a variety of levels; they range from simple services such as Email and shared calendars up to chat rooms, electronic blackboards, video teleconferencing, and collaborative decision tools.

Service Models: A variety of models have been defined. Prominent classifications of services include Business/Mission Service, Data Service, Utility/System Service, Controller Service, and others.

Service Contract: For a consumer to be willing to use a service, it will generally be necessary that the service provider makes binding commitments on the content, format, quality, and other key aspects of the service, exposed at its interface. A SLA is one common form of documentation of such a contract and specifies both functional and nonfunctional requirements for requestor-provider interactions.

WSDL Definition: Since Web services are the predominant service model, the WSDL standard is essential in many SOAs; the standard defines a markup language that includes both abstract and concrete elements.

- *Abstract definition*—includes portType/interface (high-level description of the service interface and of the messages a service can process into operations), operations (specific action(s) performed by the service, including input and output messages), and message exchange patterns.
- *Typical message exchange patterns*:
  - *Unidirectional*—messages that are sent with no requirement for acknowledgement or response, typically in the form of "fire and forget," broadcast, or multicast transmission.
  - *Request-Response/Solicit-Response*—pattern that models a basic dialog between a provider and a consumer.

  - *Publish-Subscribe (Pub/Sub)*—the provider establishes a list of topics to which consumers can subscribe; consumers are then notified of new or changed content in a subscription.
  - *WSDL 2.0 Message Patterns*—include in-out (request response), out-in (solicit response), in-only (one way), out-only (event notification), robust in-only, robust out-only, in-optional-out, and out-optional-in.

- *Concrete definition*. This includes:

  - *Binding*—defines the physical connection to a service or operation
  - *Port/endpoint*—defines the physical address of the service interface
  - *Service*—specifies a group of related endpoints/addresses

XML Schema: The schema or structure of an XML document is normally expressed in an XML Schema Definition (XSD), which is a recommendation of the World Wide Web Consortium (W3C). An XSD formally specifies the elements of an XML document, allowing automated processing of content. It thereby enables more efficient and error-free interactions among services.

UDDI Registry Entry: A UDDI server is the most common, though not the only, means used to advertise services so that clients can discover and bind to them. Figure 7.6 shows the usual graphical representation of a UDDI Registry Entry. The elements are:

- *Business entity*—information on the service provider
- *Business service*—information about the service
- *Binding template*—information for clients to bind to a service (establish a connection to use the service)
- *tModel*—for a Web service, a set of pointers to service descriptions

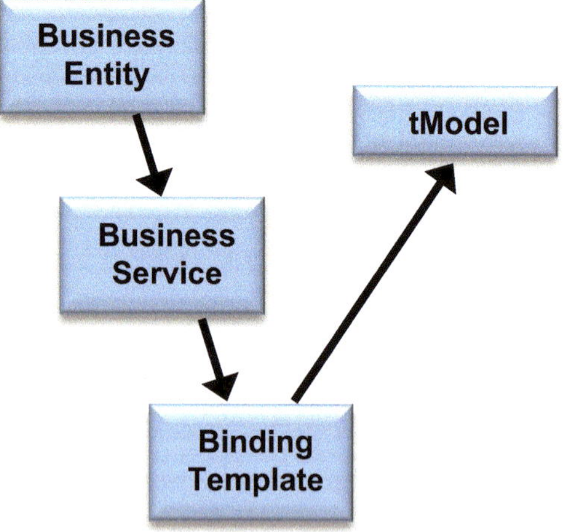

Fig. 7.6 Graphical Portrayal of a UDDI Registry Entry

Related Documents: This is a generic term for any additional information related to a service that may help developers implement instantiations of it and help consumers make decisions about whether and how to use it. Such documentation is very important for a SOA that employs locally developed services that are not widely familiar. It can take many forms and can be used in system integration and testing, user training, creation of orchestrations and workflows, and many other contexts.

7.5 Evolution of the World Wide Web

The World Wide Web (W3), or more simply the Web, is the primary means of interaction and information exchange over the Internet. Users access resources in the form of Web pages using a great variety of Web browsers. Resources available over the Web are tagged with URLs. This ubiquity makes the Web a fundamental part of most SOAs whose participants are geographically dispersed and thus a subject of importance to most system architects. Since its invention by the English scientist Tim Berners-Lee in 1989, the Web has continuously evolved and has been an important factor in the exponential growth of traffic on the Internet. Many of the standards and practices of contemporary SOA trace back to what is often referred to as the Second Generation of the Web ("Web 2.0"). We summarize the concepts involved in transforming the Web to the new paradigm before talking about using these technologies to build systems and enterprises. Before leaving this topic, we will briefly summarize some current trends that are leading to Web 3.0 and even beyond.

The original Web 1.0 model was based on consumers finding and accessing information from centrally built and managed nodes. Fundamentally, Web 2.0 replaces this with a far more participatory model in which content is distributed, constantly changing, and primarily created and shared by the community. A typical list of Web 2.0's basic principles would include:

- The client *is* the server.
- The user experience *is* the service.
- The service *is* the added value.
- The fragments *are* the file.
- The participants *are* the architecture.
- The peer contributions *are* the application.

Similarly, a list of Web 2.0 characteristics would usually include:

- Platforms (Windows, SOA, Open Source, etc.) are more important than applications.
- Web 1.0 hyperlinks (pointers to Web sites) evolve to become participant contributions and collaboration (Open Source, Wikis, Blogs).
- Value moves up the layers of an architecture (often called "the stack"); services are more important than software.

- Algorithmic data management is the key to an information-based community; a search provider is the exemplar.
- Participants/clients become virtual servers; an electronic marketplace is the exemplar.
- A "long tail" of many small users offers more opportunity than a few big users.
- Files consist of fragments on many servers that are assembled dynamically.
- Pages and links become dynamic, based on ever-changing content.
- Traditional periodic software releases become continuous updates.

There are obvious challenges in making the Web 2.0 paradigm work when considerations of security, safety, and guaranteed deadlines are important, which is commonly the case in industrial control environments, financial networks, aerospace systems, and many others. However, the enhanced collaboration, inherent redundancy, dynamic processes, and other aspects of Web 2.0 are attractive in many situations. In any case, the Web 2.0 model increasingly influences the products and standards from which such systems will be assembled.

It's widely believed that a transition to Web 3.0 is underway. Although there is no accepted definition, Web 3.0 will almost certainly feature a high level of machine intelligence, enabled in large part by very extensive research efforts in artificial intelligence (AI). Computers, especially when connected in networks, will be increasingly able to understand and reason about the content those networks exchange. Just as Web 2.0 replaced the static data access model of Web 1.0 with a far more dynamic and participatory model, Web 3.0 is likely to enhance human collaboration with far richer information services, and the computer is likely to become less a tool and more a teammate. As this next generational transition gradually takes place, system architects and their tools will need to keep pace.

7.6 The Web Service Framework (WSF) and Second-Generation Web Services (WS-*)

Any SOA built on Web services will use the WS-* family of specifications, which have become so pervasive that some of them are mandated by customer architecture policies. A comprehensive discussion is beyond the scope of this introduction to SOA, but some of the more prominent WS-* specifications are summarized below. This is a highly dynamic field, and a SOA architect will want to stay currently with new and updated service definitions and protocols. As with the discussion of SOA elements in Sect. 7.4.2, we provide here a condensed description, this time of the primary WS-* protocols and specifications, intended primarily as a terminological reference. Detailed information is available from the W3C [10].

Messaging and Notification: The following protocols support the messaging dimension of Web services as summarized above under SOA elements.

- *WS-Addressing*—provides transport standardization independent of network transport protocols (discussed in Chap. 9); it includes:
  - *Endpoint Reference*—provides a service address, usually a Unified Resource Locator (URL), as well as reference properties, reference parameters, port type, and policy
  - *Message Information (MI) Headers*—provide destination, source endpoint, reply endpoint, fault endpoint, message ID, relationship, and action (overall message purpose)
- *WS-ReliableMessaging*—protocol that provides QoS features for guaranteed delivery or failure notification of SOAP messages and serves as a fundamental WS-* extension for many other specifications
- *WS-Notification Framework*—family of protocols that collectively provide key elements of messaging, including:
  - *WS-BaseNotification*—establishes standard interfaces used by services on either end of a notification exchange
  - *WS-BrokeredNotification*—standardizes the broker intermediary used to send and receive messages on behalf of publishers and subscribers
  - *WS-Topics*—standardizes delivery of notification messages to services that have subscribed to topics
- *WS-Eventing*—standardizes an event-oriented messaging model for Pub/Sub
- *WS-MetadataExchange*—standardizes request messages (Get Metadata) asking for some or all meta-information concerning a specific endpoint address, including WSDL, schema, policy, and other information

Policy/Security: This family of protocols defines publicly accessible service metadata for key aspects of service use; these are essential elements of achieving adequate security when using Web services, as further described in Chap. 10.

- *WS-PolicyFramework*—part of WS-Security Framework which includes WS-Policy, WS-PolicyAttachments, and WS-PolicyAssertions
- *WS-Security Framework*—the heart of Web security, dealing with identification, authentication, authorization, confidentiality, and integrity. This specification includes WS-Security, WS-SecurityPolicy, WS-Trust, WS-SecureConversation, WS-Federation, eXtensibleAccessControlMarkup Language (XACML), eXtensibleRightsMarkupLanguage (XRML), XMLKeyManagementSpecification (XKMS), XML-Signature, XML-Encryption, SecurityAssertionMarkup-Language (SAML), NETPassport, TransportLayerSecurity (TLS, formerly SecureSocketsLayer (SSL)), and WSI-BasicSecurityProfile. The basic security profile further breaks down into:
  - *WS-Security*—parent specification that defines language elements for security metadata, including header blocks, usernames, passwords, and tokens
  - *XML-Encryption*—defines constructs for encrypting XML documents

  - *XML-Signature*—defines the assembly and processing of parts of a document to create a digital signature, which is central element of message authentication and the principle of non-repudiation (see Chap. 10)

Service Management and Coordination: Service management is a critical aspect of governance and is the object of a number of WS-* specifications, including the following:

- *WS-Coordination*—framework based on a coordinator service model, including Registration, Activation, and Protocol services. This specification supports complex Service Activities involving many participants and services and allows propagation of identifiers and context information for activity *correlation* among services.
- *WS-AtomicTransaction*—extension to WS-Coordination for ACID (atomic, consistent, isolated, durable) transactions, including Completion, Durable Two-Phase Commit (2PC), and Volatile 2PC protocols. A transaction is an activity performed within a database management system (or similar system) against a database and treated in a coherent and reliable way independent of other transactions, including steps to prevent data corruption due to simultaneous database accesses. The key terms are:
  - *Atomic*—describes an indivisible transaction; all changes succeed or none do; a failed transaction requires rollback to the original state for a retry.
  - *Consistent*—requires that all database changes honor associated data models; a violation requires rollback to an earlier valid data state.
  - *Isolated*—requires that concurrent transactions cannot interfere with each other.
  - *Durable*—requires that changes from successful transactions survive later failures.
- *WS-BusinessActivity*—another extension to WS-Coordination for long-running, complex (non-atomic) tasks involving both a primary process and a compensation process to deal with exceptions. It includes BusinessAgreementWith ParticipantCompletion and BusinessAgreementWithCoordinatorCompletion protocols.
- *WS-BPEL or BPEL4WS, Business Process Engineering Language*—basis for building protocols that define business processes using *orchestrations* of basic activities, structured activities, sequences, flows, links, etc. Such an orchestration is itself a service, and organization of activities can include sequences, flows, and links or dependencies as well as *correlation sets* for service coordination.
- *WS-CDL, Choreography Description Language*—similar standard for defining orchestrations with public messaging patterns that create relationships and channels among distinct service-providing entities and the rules and constraints that apply to interactions and work units accomplished by each participant.
- *WS-I Basic Profile*—set of mature, well-supported standards that create a typical Web services platform, including current versions of WSDL, SOAP, UDDI, XML, and XSD.

7.7 Service Modeling

This book is about MBSE using architecture modeling as the central process, and so it is essential that the modeling approach include adequate means of representing services and SOA. Until a few years ago, this was somewhat problematic, because none of the main modeling languages and tools dealt directly with services. Fortunately, the emergence of the SOA Modeling Language (SoaML) [1] goes a long way to solve the problem. SoaML will follow the path of SysML in being progressively implemented in mainstream architecture modeling tools. The SoaML profile is mostly concerned with some modeling conventions, service-related design patterns, and a set of stereotypes that facilitate descriptions of services and their implementations. These features can be manually declared in any SysML model, so SoaML constructs can be used to deal effectively with service modeling even before it is fully supported in tools. The remainder of this chapter is devoted to describing and illustrating such an approach.

7.7.1 Basic Principles

As always, it's essential to start with some fundamentals. Effective service modeling needs to enforce the principles and best practices we have presented in earlier section. These include:

- Decompose, represent, analyze, and optimize services using a formal, tool-based method.
- Ensure each individual resource such as an application program or database is accounted for in a single basic service, which may then be composed into multiple higher-level services.
- Group services by similarity of function and content to develop an effective taxonomy and help ensure consistency and completeness; this is the basis for a highly usable service catalog and registry.
- Account for reuse of services across business processes or mission threads and across nodes and systems of an enterprise, following the principle of "instantiate once, use everywhere."
- Account for service dependencies among themselves, with the supporting infrastructure and with the external environment.
- Map system and enterprise requirements to services to ensure comprehensive satisfaction of objectives.
- Validate and refine models based on integration, test, and deployment outcomes; plan to update services based on results of operations and on changing requirements.

7.7.2 Describing a Service for Modeling

Ultimately, any service should be modeled and documented in the architecture using the techniques in the following paragraphs. However, a useful technique in the early stages of SOA development, generally in conjunction with filling out a services taxonomy table as described in Chaps. 5 and 6, is to sketch out an initial service specification. This helps ensure that each service planned for development or adoption is fully understood and that all appropriate services have been identified. Among other uses, such a taxonomy is convenient in dialog with system stakeholders to ensure consensus is achieved on the services to be provided. It's also useful in creating and refining a layered architecture on the basis of grouping services into layers.

The essentials of a service definition can be captured using a template like the following:

- Unique identifier/name
- Summary description of the purpose and products for which the service is to be used
- Location of the service endpoint, if known, which may be a network address or a server port, along with any other details required to bind to the service such as the mechanism to validate the requestor's access and privileges
- Format of an invoking message, including protocol and any required variables
- Format of the product(s) produced by the service
- Operations or methods of the service that the requestor can invoke, including polymorphic variations
- Quality of Service (QoS) parameters such as guaranteed response time, product accuracy, and assured handling of priorities

7.7.3 Overview of the SOA Modeling Language

SoaML gives the architect a very useful toolkit for incorporating services in a model. As described in the SoaML Profile specification of the Object Management Group, the main goals are:

- Intuitive and complete support for modeling services in UML (equally applicable in SysML)
- Support for bidirectional asynchronous services between multiple parties
- Support for Services Architectures where parties provide and use multiple services
- Support for services defined to contain other services
- Ability to be easily mapped to and made part of a business process specification
- Compatibility with UML/SysML, the Business Process Definition Meta-model (BPDM), and the Business Process Model and Notation (BPMN) for business processes

- Direct mapping to Web services
- Top-down, bottom-up, or meet-in-the-middle modeling
- Design by contract or dynamic adaptation of services to specify and relate the service capability and its contract
- No changes to UML/SysML

SoaML extensions to UML/SysML consist primarily of stereotypes that define model elements that are of greatest importance in dealing with services. The following are among the most commonly used of these.

Participant: A <<Participant>> is a service provider or consumer that:

- Is modeled as a Block or Component.
- Can have specialized Ports called Service Points and Request Points.
- Can have a Participant Architecture that defines its characteristics.
- Plays one or more Roles under the terms of applicable Service contracts; the concept of Roles is central to the SoaML representation of services.

Agent: An <<Agent>> is a specialized Participant that can interact with and adapt to its environment, usually to perform automatically tasks that would otherwise be done manually by a system user; instances of an Agent share features, constraints, and semantics.

Message Type and Attachment: The <<Messagetype>> and <<Attachment>> stereotypes identify specialized elements that model service interactions.

Capability: This <<Capability>> stereotype models either the general capabilities of a Participant or a specific ability to provide a service.

Request Point: <<RequestPoint>> is a key service-related stereotype that models the use of a service by a Participant and defines the connection point through which a Participant makes requests and uses or consumes services; it's equivalent to a Required Interface in a basic SysML model.

Service Point: Another key stereotype is <<ServicePoint>>, which models an offer of a service by one Participant to others using well-defined terms, conditions, and interfaces. This model element defines the connection point through which a Participant offers its capabilities and provides a service to clients and is equivalent to a SysML Provided Interface.

Service Channel: A model element with the <<ServiceChannel>> stereotype represents a communication path between Service and Request Points.

Service Contract: The <<ServiceContract>> stereotype formalizes a binding exchange of information, goods, or obligations between parties; in essence, it defines a service.

Service Interface: This is the SoaML construct for the service interfaces that we have stressed throughout our discussion of SOA. It defines the interface to a Service Point or Request Point. The interface is normally typed by a Participant Role in a Service Contract, and it can be bidirectional and can incorporate a Protocol.

As use of SoaML matures, it will be increasingly harmonized or integrated with other business modeling standards. These include BPMN, PBDM, and the Business

Motivation Model (BMM). These trends, and the evolution of modeling tools that incorporate them, are matters of continuing interest to system architects dealing with SOA.

7.7.4 Simple Service Modeling

With this quick SoaML introduction, we can now introduce some basic techniques that can be very useful in representing services. The first method is just annotation of primary architecture model artifacts. The following are some of the ways this can be done.

Service Documentation: Perhaps the most intuitive way to represent services is to document their implementation and use in the model element specifications for individual system components. For example, services provided or used by a component can be documented as behaviors or interactions in Activity, Sequence, or State Machine Diagrams to show service delivery and use. One approach would be to develop BDDs that represent the service catalog and then instantiate individual services in behavior diagrams where they are employed.

Data Modeling: A similar idea is to document Data Services in data models as Operations of the Blocks representing individual data elements. For example, a Data Service may provide full CRUD (create, retrieve, update, and delete) functions to data consumers, participate in Pub/Sub, allow read-only access, or provide other Operations.

Service Interfaces: The architecture model can declare interfaces on Blocks where services are exposed, label them with the service(s) name(s), and attach notes or interface specifications giving the details, which might include links to WSDL files or registry entries. A useful clarification is to stereotype these interfaces as `<<Service>>` or even a specific category of service.

Supplementary Model Content: Supplemental views and artifacts can be created that may be helpful to fully describe the service dimension of the architecture. The following are three ways this could be approached.

- A service catalog can be written that defines provided services with WSDLs and possibly other definitions such as links to authoritative data references. The catalog can be organized into Business/Mission, Data, and System Services and should reflect the services taxonomy. It may be used at build-time (static bindings to services) or at runtime (discovery and dynamic binding).
- Specific architectural artifacts can be used. These may include:
  - *Services taxonomy*—much like data, services can be modeled using BDDs.
  - *Capability-to-service mapping*—it may be useful to create BDDs associating capabilities with services.
  - *Other techniques*—in many cases, Sequence diagrams are well suited to capturing the message exchanges involved in service invocation and delivery. Another technique is to model constraints in a compartment of a

<<Service>> Block. Stateful services, which are generally to be avoided in order to preserve loose coupling but do occasionally arise, may be modeled using State Machines.

- A Services Focused Viewpoint can be created that collects services information in a format designed for ease of use by, e.g., software engineers who will implement the services and their interfaces.

Annotations: The architect can simply include service annotation on network, runtime, backup/restore, and other Focused Views. For example, Ports and Protocols annotations can show physical details of service interfaces such as a server port where a service is accessed. Other annotations can be placed in primary architecture diagrams to identify such things as where services are supported and exposed, where they occur in various system behaviors, and how they are associated with User Roles and other Actors.

7.7.5 Complex Service Modeling

The remainder of this discussion introduces a richer approach to service modeling based primarily on SoaML. The first step is to identify the participants in a SOA, primarily people, organizations, technology components, and systems. Participants offer services through Ports stereotyped as <<ServicePoint>> and request services through Ports stereotyped as <<RequestPoint>>. A Service Point may be typed as a basic Interface for simple services or a Service Interface for complex services, especially when these are bidirectional.

Figure 7.7 illustrates the use of basic Interfaces on a Block that models a SOA Participant, using provided and required interface icons for services the Block provides and consumes. Operations associated with these services are defined in an Interface Definition Block. For example, the Participant might be a data manager that provides a data object retrieval service and consumes a data update service.

Figure 7.8 is a SysML BDD showing a more complex service description. Participant B has a port exposing both provided and required interfaces. The Service Interface Block contains interacting Parts stereotyped as <<BlockProperty>> that define the interfaces. For Interface 1, the Service Interface Block is responsible for creating the Interface and its Operations; and this is shown as a realization (arrow with dashed stem and closed head). For Interface 2, the Service Interface Block does not implement the Interface and its Operations; a <<usage>> Dependency links it to a separate Interface 2 Block. This would be typically implemented as a Call Operation to that component. Finally, the Service Interface Block incorporates a protocol defined by the Protocol 1 Block.

SoaML advocates modeling SOA at (at least) the following two levels.

Community or Enterprise Architecture: This is the high-level contextual view that represents the overall community of Participants, with the services they provide and consume and their roles with respect to these services. A service is modeled as

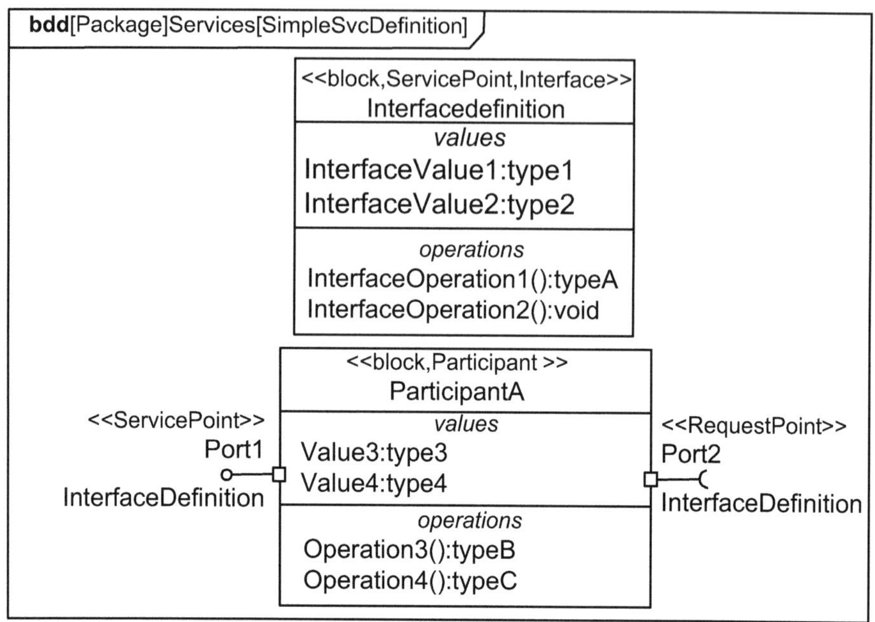

Fig. 7.7 Simple Service Points of a Participant typed by Interfaces as defined by an Interface Definition Block

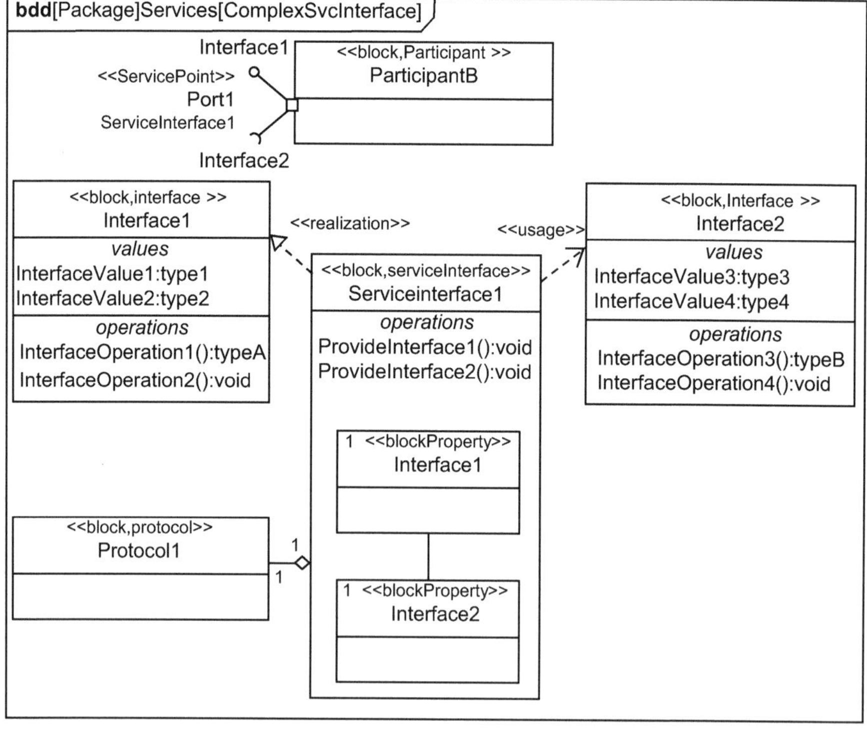

Fig. 7.8 Complex service interface that realizes Interface 1, uses Interface 2, and incorporates a protocol governing use of the service

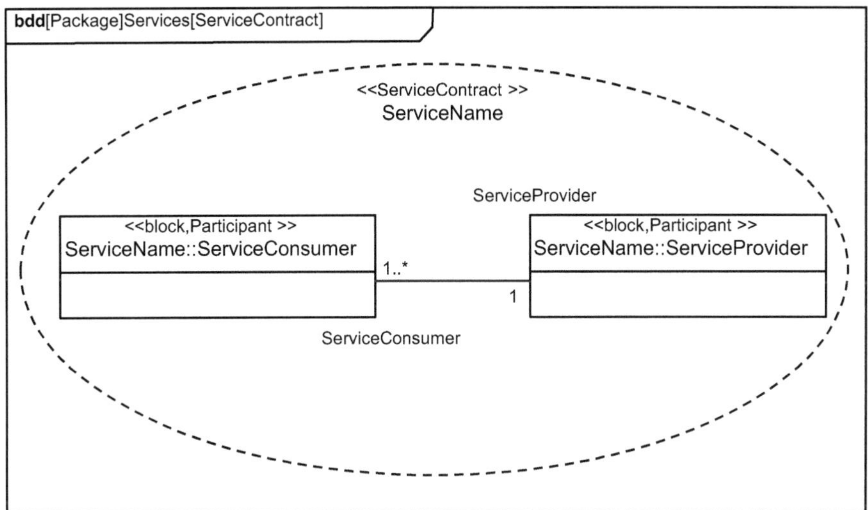

Fig. 7.9 Simple Service Contract modeled as a Collaboration among Participants who exchange a service

a Block Collaboration (an oval with a dashed outline) that defines a Service Contract among interacting entities along with one or more Sequence or Activity Diagrams showing the exchanges between service providers and the consumers. The specification for the Collaboration can incorporate, or provide a link to, service documentation. Details can be modeled by drilling down into the participants and services. For example, a high-level service can be modeled as a Collaboration, stereotyped as `<<ServiceContract>>`, and then decomposed into smaller services with more specific roles assigned to Participants, as illustrated in Fig. 7.9. Fine-grained services are modeled using orchestrations that show the specific exchanges among participants as shown in Fig. 7.10. The behavior associated with a service can be modeled using any behavior diagram. Mission Services are typically modeled using Activity Diagrams, while System/Infrastructure Services are often best represented using Sequence Diagrams as illustrated in Fig. 7.11. Participant roles are typed by the Service Interfaces associated with each role.

Participant Architecture: This is modeled using Block Diagrams. It includes sub-participants, services, and Service Points through which Service Contracts that define Service Interfaces are implemented. Figure 7.12 shows an example in an IBD. This Participant provides Capability 1 as a service at Port 1 and uses content from Port 2 to enable Capability 2.

As the discussion in this section shows, modeling services and a SOA in which they can deliver their promised benefits to an organization is a complex and nuanced task. SoaML, often in conjunction with a business process modeling language such

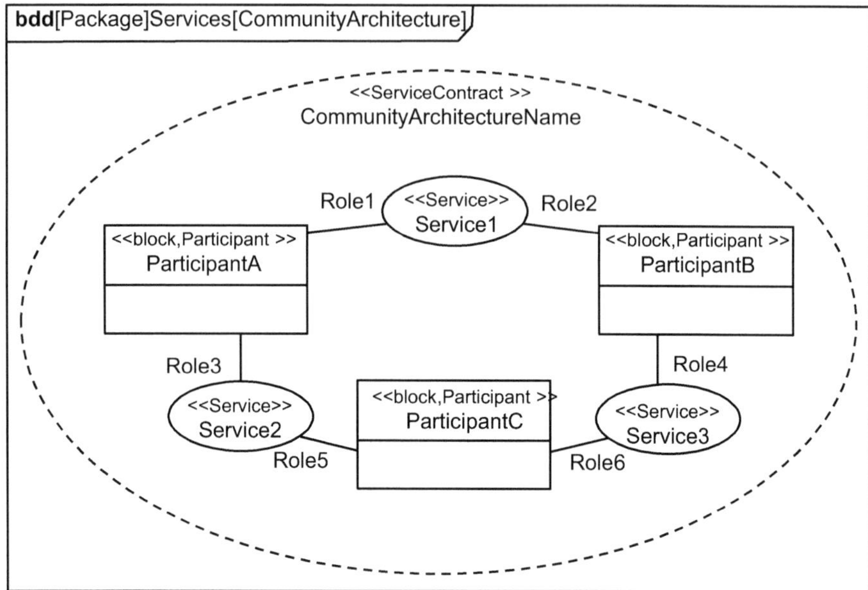

Fig. 7.10 Community Architecture modeled as a Collaboration of Participants and services with roles associated with service production or consumption

as BPMN, extends the basic UML/SysML framework to provide the necessary constructs. While every situation is different, a general flow of service development progresses through:

- Identification of participants in the service collaboration
- Identification of functions and processes to be service-enabled
- Definition of services, their interfaces, and their exchanges among participants
- Implementation of the SOA and the individual services, following the principles laid out in this chapter

This should be accomplished within the overall MBSAP methodology to ensure the resulting architecture has a solid foundation, clear traceability to requirements, and provisions for long-term management and governance. Both structure and behavior must be modeled, and the service model may have significant data content as well. In a general architecture model, all this is placed in a Services Perspective. In cases where a dedicated service architecture is what is needed, the model can simply be set up with Structure, Behavior, Data, and Context Perspectives. The underlying principles and techniques remain unchanged.

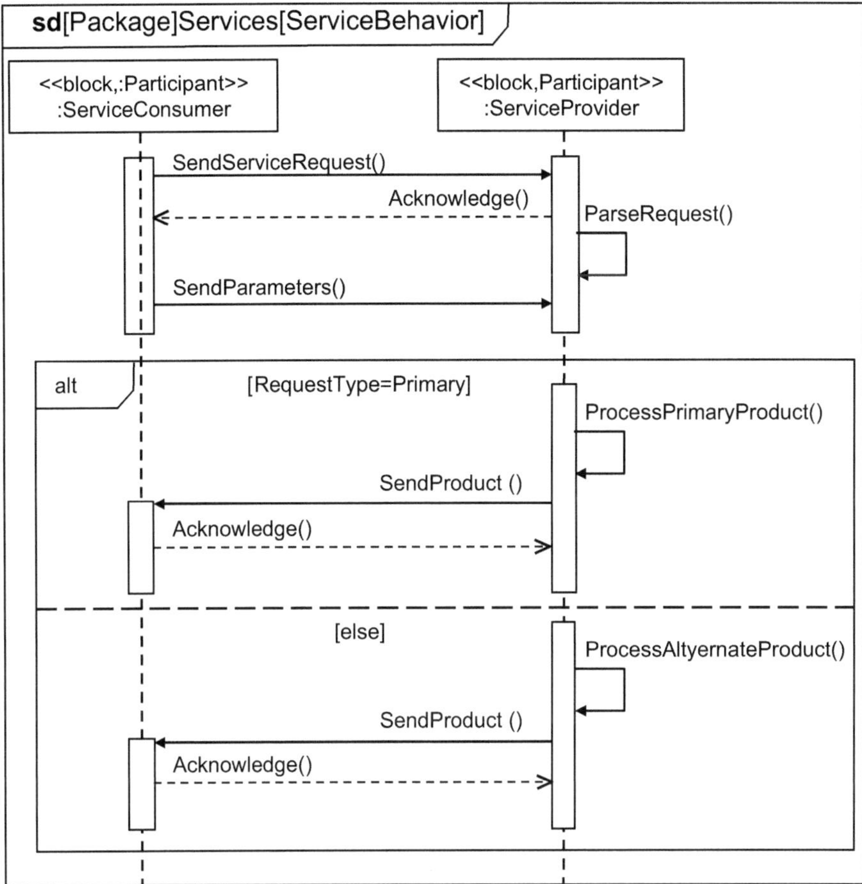

Fig. 7.11 Service Behavior modeled as a Sequence Diagram with Primary Interactions

7.8 Smart Microgrid Services

We can illustrate these SoaML techniques using our Smart Microgrid (SMG) example, specifically the interactions between the microgrid and its Primary Grid (PG). The exchanges of technical and business data that were shown as data flows in Fig. 4.19 are excellent candidates to be service-enabled. In Fig. 7.13, we show a community architecture with the SMG exchanging services with the PG Grid Control and Operations center and the PG Market that is created by one or more electricity service providers. The Control and Ops center and the Market use a Market Planning Data service by which service providers publish data such as current and projected

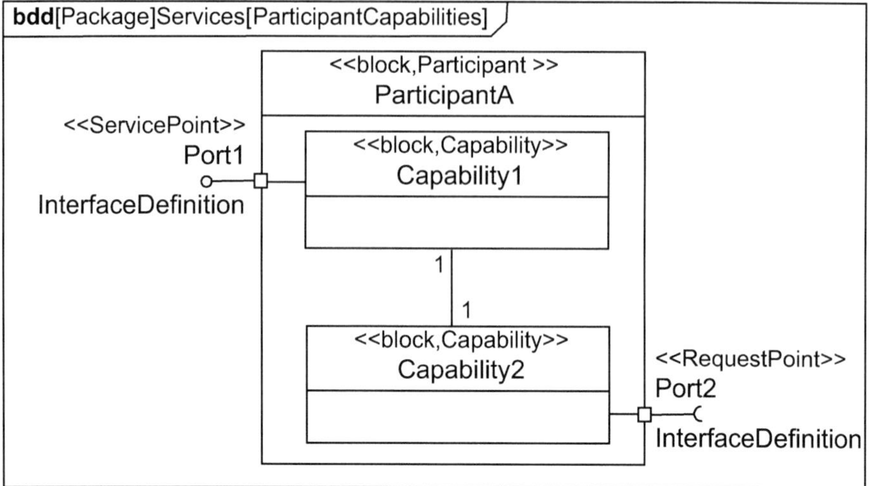

Fig. 7.12 Capabilities associated with a Participant modeled using an Internal Block Diagram and showing Exposure via Service Points

power consumption by their customers. Data exchanges between the SMG and PG are bidirectional, and Fig. 7.13 models this with pair of basic services. SoaML calls such a combination of a Compound Service. For each of the services shown in the figure, one participant serves as a provider and fulfills a "publish" role, while another serves as recipient and fulfills a "consume" role.

Figure 7.14 shows one of the many ways detail can be added to this basic picture, specifically for the Microgrid Technical Data (MGTechData) service. Here, the Service Contract is between the two participant roles that were declared in Fig. 7.13. Each of these roles is typed by a Service Interface. Each of these realizes a data interface that defines the specific data to be exchanged. The Service Interfaces have `<<Uses>>` dependencies on each other's data interfaces. In Fig. 4.19, we defined Ports on the various Parts of an Electric Utility that are involved in SMG-PG interactions. We can now give these Ports the appropriate `<<ServicePoint>>` and `<<RequestPoint>>` stereotypes, and we can attach Service Interface specifications to them. The model will then give a development team the information needed to correctly implement the services.

7.9 Summary

Service orientation now stands alongside object orientation as the established and preferred way to approach many information-intensive systems and enterprises. This chapter has introduced the large and complex subject of services and SOA with three main objectives: (1) to convey a sense of the nature and importance of the

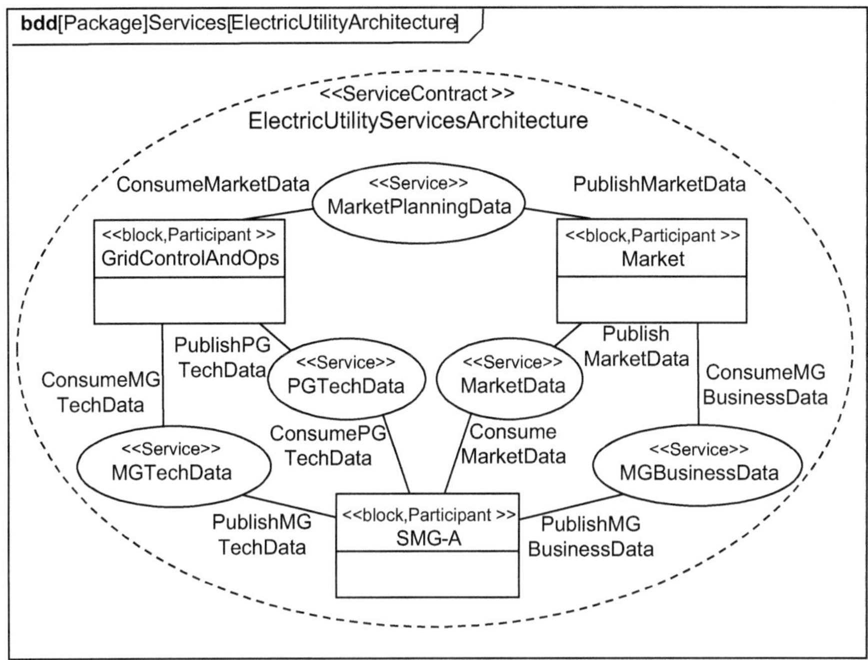

Fig. 7.13 Community Architecture showing Service-Enabled Data Exchanges between Smart Microgrid A and its Primary Grid

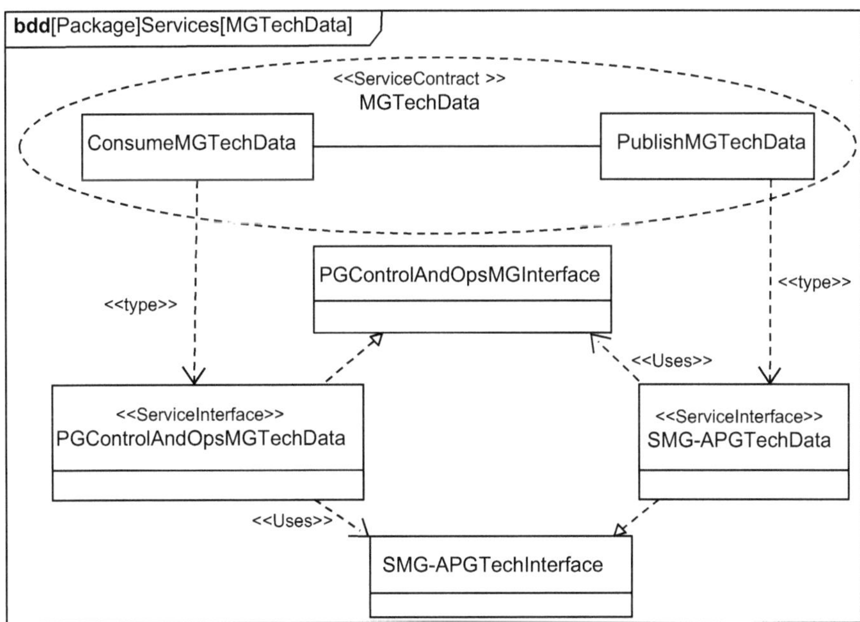

Fig. 7.14 Details of the Service Contract between the Participant Roles involved in the Microgrid Technical Data Service

trend, (2) to show how services are accounted for in MBSAP, and (3) to suggest some of the ways services can be modeled. There is far more to the subject than this chapter could treat, and the serious architect will want to pursue greater depth, especially as tools that fully support SoaML become the standard. A number of organizations maintain the standards and specifications that underpin SOA practice and publish a variety of tutorial materials. In subsequent chapters, we will continue to expand on the use of services in discussing real-time behavior, information assurance, and other topics.

Exercises

1. Consider a large financial services corporation with geographically distributed locations that currently uses many manual, paper-based processes to accomplish its business. Assume that various locations have computer systems and they have been independently developed using a variety of hardware and software technologies. (a) List some benefits that could be obtained by service-enabling the functions and data at these locations, and (b) list the primary elements of a strategy to migrate the corporation to a service-oriented architecture.
2. What are the differences among the concepts of Service Orientation, Service Interface, and Service-Oriented Architecture?
3. Assume that the corporation in Question 1 has products that are created through a sequence of steps carried out in various locations. Describe how Orchestration and Workflow could be used to create these products more quickly and efficiently.
4. Assume that a specific location of the corporation in Question 1 has an information system that has been built up over a period of time using software products from multiple vendors and that it has extensive interactions with other companies and customers over the Internet. Referring to Fig. 7.4, list some functions of the Integration Backbone that can make the information system as a whole work more effectively.
5. What is the common meaning of Collaboration in a SOA?
6. Referring to Sect. 7.6, list some WS-* standards and specifications that could be used in a SOA for the corporation in Question 1 and the contribution each would make to improving the processes of the corporate enterprise.
7. List some commonly cited reasons for the failure of SOA projects to achieve their objectives and the steps that should be taken to minimize the risk of failure.
8. List some characteristics of a SOA for the corporation in Question 1 that can help ensure its ability to deal with inevitable changes over time.
9. For the corporation in Question 1, identify a Business/Mission Service, Infrastructure Service, and a Data Service, and create a description of each using the template in Sect. 7.7.2.
10. Describe the concept of a Participant in SoaML.

Student Projects

Students should build on the Services Perspective created in earlier architecture modeling by adding detail as described in this chapter, including more complete descriptions of services and service interfaces, application of service-related specifications and standards, and diagrams describing the structure and behavior of selected services.

References

1. Object Management Group (2012) Service Oriented Architecture Modeling LanguageTM (SoaML®). http://www.omg.org/spec/SoaML/. Accessed 27 May 2017
2. Erl T (2005) Service-oriented architecture: concepts, technologies, and design. Prentice Hall, New York
3. Marks E, Bell T (2006) Service-oriented architecture: a planning and implementation guide for business and technology. Wiley, Hoboken, NJ
4. Manes AT (2006) Service-oriented architecture: developing the enterprise roadmap, ver 2.0. Burton Group, Bloomingdale, Midvale, UT
5. Keifer MA, et al. (2008) Acquisition of information services and SOA systems. http://www.afei.org/Pages/default.aspx. Accessed 27 May 2017
6. OASIS (2005) SOA reference model TC. http://www.oasis-open.org/committees/tc_home.php?wg_abbrev=soa-rm. Accessed 27 May 2017
7. Barry DK, Dick D (2013) Web services, service-oriented architectures, and cloud computing. Morgan Kaufman, Waltham, MA
8. World Wide Web Consortium (2004) Web Services Architecture. https://www.w3.org/TR/2004/NOTE-ws-arch-20040211/#relwwwrest. Accessed 11 May 2018
9. The Open Group (2017) The SoA source book. http://www.opengroup.org:80/soa/source-book/intro/index.htm. Accessed 11 May 2018
10. World Wide Web Consortium (2004) Web services glossary. https://www.w3.org/TR/2004/NOTE-ws-gloss-20040211/. Accessed 11 May 2018

Chapter 8
Extending Architecture to Real-Time Domains

8.1 The Nature of Real-Time Systems

In Chap. 1, we suggested that the "real-timeness" dimension of performance is a very fundamental way to distinguish basic architectural categories, and this chapter specifically addresses what makes real-time (RT) systems different from others. Douglass [1] has written extensively on architecting RT systems in the Object-Oriented paradigm and describes an RT system as one in which correctness depends chiefly on timely execution, the ability to schedule tasks for assured on-time completion, and the overall performance needed to satisfy such a schedule under worst-case operational conditions. In other words, in an RT system, a result that arrives late or an action that takes longer than a scheduled time constitutes a failure that is as bad as, if not worse than, one that is erroneous. RT systems must execute processes, produce outputs, make or support decisions, monitor and control assigned resources, and do other time-sensitive things within maximum allowable times (deadlines) that derive from the operational characteristics of the controlled equipment or external environment. An action that completes within an allowable time is *timely*, while one that does not is *late*. Frequently, timeliness is accompanied by requirements for safe and provably correct behavior because of the consequences of errors. For example, an aircraft flight control system that is too slow to react to atmospheric turbulence or pilot control inputs could cause a crash.

Chapter 1 suggested "Hard," "Decision/Process" and "Non-" RT system categories. This Chapter deals with the first two, but most of the emphasis is on Hard RT. The term "embedded processing" is often to these systems because computing is a built-in element of the system. There is also a system category that is usually labeled "Soft" RT in which minor or occasional deadline misses may be acceptable, especially when automated recovery mechanisms are in place such as retransmitting a lost message or when the system meets its deadlines *on average* or by some other criterion. For example, an equipment controller designed with digital inner and

J. M. Borky, T. H. Bradley, *Effective Model-Based Systems Engineering*,
https://doi.org/10.1007/978-3-319-95669-5_8

outer loops might have an inner loop designed to strict hard RT criteria but allow less stringent rules for the outer loop.

Figure 8.1 shows a fairly elaborated RT taxonomy based on a number of schemes that have been used over the years in RT Systems Engineering, with four levels of RT behavior and examples of system categories in each. This scheme adds a "Tempered" RT category where lateness is less disastrous than with hard RT but more so than with soft RT. Some of the system types straddle two RT levels, embodying some of the characteristics of each. These initial general notions about real-timeness can be made a bit more definitive by listing some of the key things that set RT apart from the much larger and more familiar universe of non-RT systems. The following are some of these important differences:

- RT processes have *deadlines*, and exceeding these timing constraints causes adverse consequences up to and including mission failure and catastrophic loss.

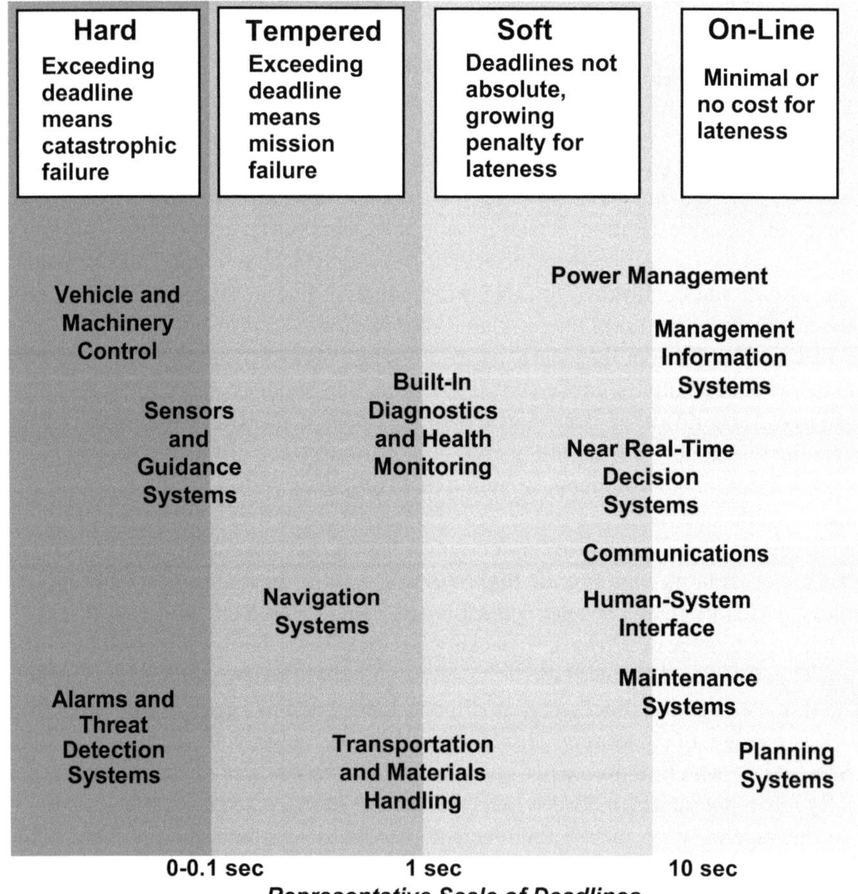

Fig. 8.1 Spectrum of real-time categories with system examples

- RT system operation must be *schedulable* and *deterministic*, such that processes will execute in accordance with a policy or set of rules, and the timing of that execution can be assured and known in advance.
- RT systems are usually *event-* or *time-driven* because they operate interactively with controlled resources or external environments, rather than data-driven as is more common in non-RT systems. Similarly, these systems are often required to assure synchronization of concurrent processes, both internally and externally. Thus, events, signals, responses, actions, and timing are essential elements of an RT system architecture model.
- RT systems are likely to be required to implement one or more *quality of service (QoS)* policies dealing with assured timeliness, honoring of priorities, accuracy of results, and other measures of correct operation specified by designers and users.
- Systems with stressing RT performance requirements will generally be built on a *Real Time Operating System* (*RTOS*) that is different in fundamental ways from the kind of general-purpose OS found in a non-RT system or enterprise. Indeed, RT software development in general is more difficult because of the constraints that apply to it and the consequences of errors in the code.
- Although not strictly an aspect of RT operation, such systems routinely have stringent *safety and security* requirements because they are involved in critical operations.

Examples of RT systems include machinery and industrial process controls; guidance systems for aerial, surface, and subsurface vehicles; controllers for sensors, communications, and other RT subsystems; and embedded controllers used in instrumentation. Clearly, this kind of application is less frequently encountered than non-RT systems in the Information Technology community as a whole. A single brief chapter can only introduce the important characteristics and issues that RT system engineers and architects must deal with, and we refer the reader to the large body of published work in this area for deeper insight. In particular, application and extension of UML to RT domains have been underway for many years, typified by the early work of Selic and Rumbaugh [2], much of which has matured in UML 2 and SysML. A later section of this chapter, along with Appendix D, deals with the Modeling and Analysis of Real-Time and Embedded (MARTE) System Profile of UML, which provides the current basis for formally representing time, resources, scheduling, performance, and other aspects of RT in a model.

8.2 Characteristics of Time

The timeliness of actions in an RT system can be described by a variety of parameters such as the following.

- *Deadline* – the maximum allowable time (or final instant of time) by which an action must complete; a deadline can be:
  - *Absolute* – based on an independent measure of time such as a clock
  - *Relative* – measured against another event or process or used to describe a soft deadline
  - *Hard* – resulting in a failure if exceeded
  - *Soft* – representing a desired completion time that may be satisfied statistically (e.g., on average or in accordance with a specified probability density function) or may have some number of permitted violations
- *Action Priority* – not a time value per se but applied when multiple actions compete for shared resources and used to schedule their execution so as to make most effective use of those resources and minimize impacts of exceeding deadlines.
- *Explicit Time Parameters* – these may include:
  - *Start, Stop, and Duration* – clock values associated with process execution
  - *Ready Time* – the instant that a process is ready or eligible for execution
  - *Blocking Time* – the time when an action is prevented from execution while awaiting the availability of required resources

Process timing is often associated with events to which an RT system responds and thus with the arrival of messages or signals communicating such events. Depending on the nature of the system and the mode in which it is operating, events may be periodic, such as a scheduled frequency of execution, or they may be aperiodic or random when they arise from unpredictable occurrences inside or outside the system. It may be important to characterize the arrival patterns of random event messages in order to ensure the system's performance is adequate to meet its deadlines. Sources such as Reference [1] go into these aspects in more detail.

In many cases, system time is defined by a timing device. This is usually a clock that samples a time base and produces the time value associated with a particular instant. A clock may also output a succession of "ticks," each representing the elapse of some basic time increment. An RT system will generally have an internal time reference and may also use an external time source to obtain time values directly or to synchronize an internal clock. A common example is the use of time derived from the Global Positioning System (GPS) to correct accumulated errors in a system clock. Some timing device characteristics that may be important to account for in a model include:

- *Current Value* – how far the time reported by the device has progressed from some origin (also called metric time).
- *Reference Clock* – the identity of a time source to which the device is somehow related.
- *Origin* – some clearly identified timed event that represents the initial instant from which the device proceeds to measure time.
- *Maximal Time Value* – the greatest indicated time which the current reported value cannot exceed; a clock is commonly designed to "roll over" and resume

measuring from the origin if the maximal value is reached and may generate an event or error when this happens.

- *Resolution* – the minimal time interval that can be recognized by the mechanism; this is the *precision* of the device.
- *Stability* – the ability of the mechanism to measure the progress of time at a consistent rate.
- *Skew* – a measure of how well the mechanism tracks a reference clock or other recognized time source or standard; this is essentially the *accuracy* of the device.
- *Drift* – the rate of change of the skew.

8.3 Events, Threads, and Concurrency

Ideally, a given time-sensitive process would be assigned a dedicated set of resources, which could then be sized to ensure that the required performance and timing are satisfied. However, this is often unaffordable in a practical system, so that multiple processes must execute on shared resources, which are generally called "execution platforms." This, in turn, implies a need for some means of optimizing their scheduling, which comes down to allocating individual processes to particular resources and determining the sequence and timing of their execution. One of the key concerns in RT system development is to assure "schedulability," meaning some guarantee that all processes can be executed within their deadlines under all operational conditions using available resources.

Processes, whether they use the same or different resources, must often be synchronized, and this is frequently accomplished using events that are exchanged via messages. An Event generated in one Block could trigger an operation or a transition between states in another Block. For example, an event might be generated in response to the occurrence of a condition such as detection of a fault or receipt of a new urgent tasking. This condition might require interrupting a process that is currently running in order to allow a different one that has now become higher in priority to be executed immediately. Sending and receiving of events may be synchronous or asynchronous, and the resulting event patterns must be accounted for in scheduling. For example, a synchronous event that requires acknowledgment might suspend all system functions until that response is received. Another event pattern could result from periodic event occurrences. The schedule analysis must account for the finite time required to handle events and for other overheads like the time to switch processes within a computer. Event generation and handling are a key element of overall timing in many RT systems, and an RT execution platform must provide the mechanisms for generating, propagating, receiving, and responding to events.

A unit of system behavior has thus far been called a process, but other commonly used terms, especially in the context of an RTOS, are *thread* and *task*. The MARTE Profile makes these concepts more formal and exact. Any complex system is likely to be multi-threaded, i.e., to support concurrent processes running on one or more execution platforms. In this situation, the basic unit of system execution is the thread or task, and the central challenge is to ensure these can be scheduled in accordance with a "scheduling policy" such that they all meet their deadlines. In UML, especially when the focus is on software, a thread is owned by an Active Object and composed of Objects that are involved in the thread behavior. A SysML model does the equivalent using Blocks and Block Instances. In RT systems, careful definition of threads, establishing valid priorities, and optimized scheduling are essential to the ability of the system to meet all deadlines without incurring excessive costs through inefficient resource utilization.

The basic approach to identifying and modeling threads is the same as we discussed in Chaps. 4 and 5. Concurrency is readily shown on Activity and Sequence Diagrams. However, RT design demands particular attention to a number of specific aspects of system behavior, including the following:

- Resource contention among tasks raises the possibility that high-priority tasks may deny access to shared resources by lower-priority tasks and, conversely, that lower-priority tasks may block resources and thus prevent higher-priority tasks from executing, a condition known as "priority inversion."
- Task synchronization may be needed whenever tasks have mutual dependencies. Interacting threads require a means to coordinate control and exchange data, which means that one task may be balked (prevented from continuing) while waiting on an input from another. This communication, which may use events or other signals, is often called a "rendezvous," and can be modeled by a Block that is attached to the Association between interacting Blocks. Such an Association Block deals with preconditions, the response when a precondition is not met, data sharing, and other aspects of the rendezvous. Douglass [1] gives a Rendezvous design pattern in his chapter on Mechanistic Design.
- Overall schedulability must be assured in the presence of concurrency, including the possibility of transient surges of system activity and tasking. Any system that is subject to widely varying task loading must have adequate margins on resource capacity and timing. This is discussed further in a later section, with emphasis on annotating a model to allow manual or automated analysis of schedulability or performance.

8.4 Real-Time in the E-X Example

Before continuing the RT modeling discussion, it may be helpful to illustrate the concepts presented thus far using the E-X example. The Sensors and External Communications and Networks Domains in Figs. 4.6 and 4.7 contain RT resources. Figure 8.2 is an IBD showing the three E-X sensors communicating through an RT Sensor interface Block that takes care of data buffering, timing and control, message translations, and any other aspects of allowing the sensors to interact with the Sensor Management Domain via the local area network (see Sect. 6.3.3). The network

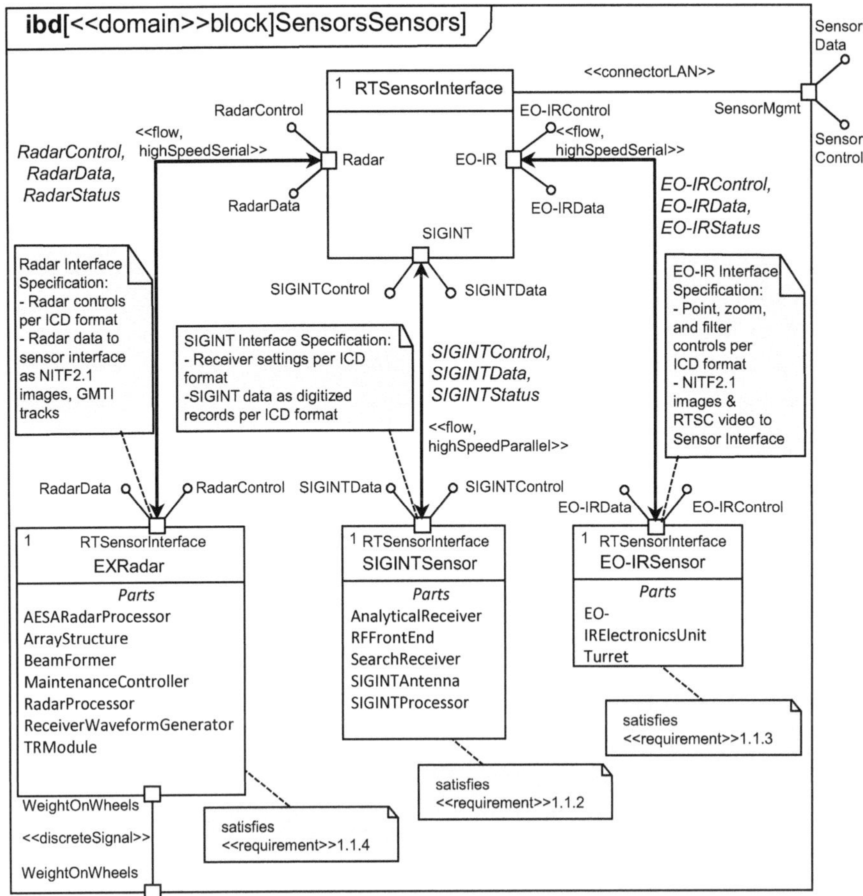

Fig. 8.2 E-X Sensors interaction with the rest of the E-X system via a sensor interface

interface is modeled as a Port with a required interface for sensor control messages from Sensor Management and a provided interface for sensor data. The interfaces to the three sensors are annotated with brief placeholders for detailed interface specifications; for example, radar imagery will be provided in compliance with standards like the National Imagery Transmission Format, version 2.1 (NITF 2.1).

The RT behaviors of the sensors and their associated electronics are driven by a number of factors, including the physics of various sensing mechanisms, dynamics of aircraft motion, available time to process collected energy, and required response times to sensor tasks. For example, the EO/IR sensor turret typically includes stabilization to ensure blur-free images and video in the presence of aircraft maneuvers, atmospheric turbulence, or other disturbances. The turret must be able to detect and compensate for these unpredictable movements in a time that is effectively instantaneous compared to the motion of the aircraft. Additional deadlines apply to processes such as radar beam forming, SIGINT receiver tuning, and many others. The sensors incorporate embedded processors employing an RTOS, and the Sensor Interface Block is responsible for allowing interactions between this RT embedded processing and non-RT software in the Sensor Management Domain that executes on the resources of the Computing Infrastructure Domain.

Figure 8.3, repeated from Chap. 5, takes the level of structural detail a step further by decomposing the radar into its constituent line-replaceable units (LRUs or "black boxes") and showing key interfaces. In this example, the radar employs an Active Electronically Steered Antenna (AESA) with 4096 transmit/receive modules. The radar has embedded processing for beamforming computations (to determine the phase setting for each module in each transmit/receive period in order to create and point the desired beam), as well as waveform synthesis, signal and data processing, and various radar control and maintenance functions. A less expensive (and less capable) E-X design might use a gimbaled reflector antenna with separate transmitter and receiver boxes. In either case, the radar computes and synthesizes an appropriate waveform, generates control signals for the antenna and RF electronics, detects and filters reflected target energy, converts these signals to digital data, performs signal and data processing, and creates reports containing the desired information such as a target track or image.

A multimode surface surveillance radar can intersperse modes such as moving target indication (MTI), synthetic aperture radar (SAR) imaging, inverse SAR (ISAR) motion detection, and others. This requires fast, sophisticated RT scheduling

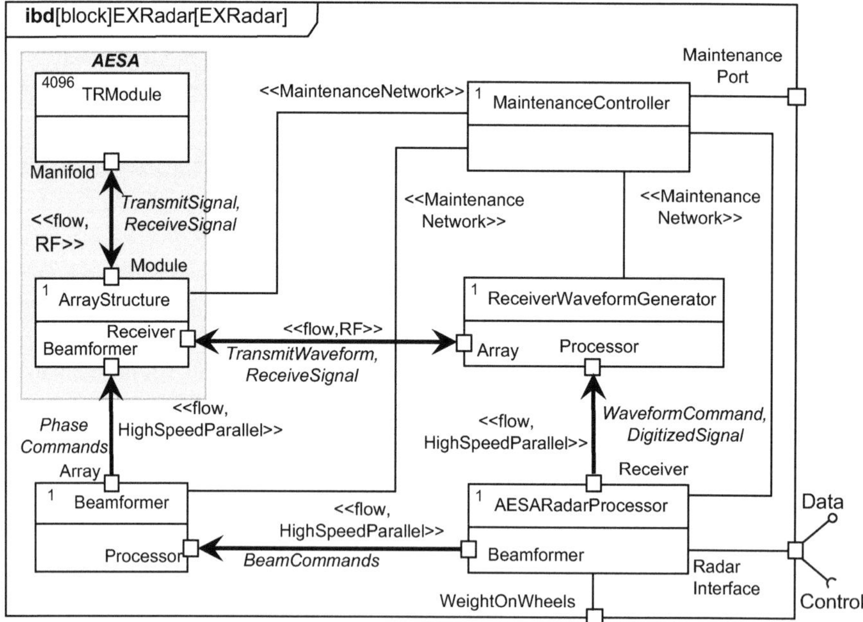

Fig. 8.3 E-X radar modeled using blocks for the primary boxes connected via labeled ports with flows and a maintenance network

to go with the high performance processing tasks already listed. The other sensors have different RT processes but present similar challenges in scheduling, sensor control, and signal and data processing.

Figure 8.3 illustrates some additional features that are important to capture in a model and some key points for the remainder of this chapter. Via a separate proxy port, the Radar Processor accepts discrete signals such as a weight-on-wheels signal that causes the transmitter to be inhibited to prevent a high power RF safety hazard when the aircraft is on the ground. There is also a Maintenance Controller that monitors built-in diagnostics in the various elements of the radar and supports maintenance and health data logging via its own interface. Since these tasks are not essential to radar behavior, the Maintenance Controller might well be designed to Soft RT criteria.

8.5 Real-Time Requirements

Good architecture begins with good requirements, and this is especially true in the world of RT systems. The basic flow of developing an Operational View is the same as described in Chap. 4, but RT introduces additional considerations. The following are some things the RT architect should pay attention to in developing Requirements, Use Cases, Domains, Domain Services, a Conceptual Data Model, and Mission Threads.

- *Performance* – perhaps most obviously, any time-sensitive aspect of system performance needs to be fully specified. Whereas non-RT system requirements are often written as "The system shall be able to perform a function," an RT system is likely to have quantitative constraints on the function execution time, latency in delivering a product, number of task executions per time interval, and so forth. This may be complicated by incomplete customer performance requirements, and these may even be internally inconsistent, requiring careful analysis and good stakeholder communications to deal with required revisions.
- *Requirements Analysis* – there will usually be greater emphasis on early performance modeling and timing analysis to ensure that all aspects are accounted for and that the initial system partitioning and mission threads are consistent with requirements. In practice, many, perhaps most, timing requirements will be derived rather than explicitly stated by the customer, and as the LV design effort proceeds, frequent updating of the requirements baseline can be expected as subtle aspects of system design and performance are discovered. If the customer does not provide complete and adequate operational scenarios to support requirements analysis and refinement, the developing organization may have to fill this gap and coordinate the results with the customer. This is a situation where timing and performance analysis, including simulation, are often very important in refining and validating detailed timing and performance requirements.

- *Architecture Modeling* – behavioral diagrams with timing are much more important in RT architecture modeling than with non-RT systems because of the need to document timing details. We've already noted their importance in conjunction with developing, analyzing, refining, and coordinating RT requirements. Activity diagrams are useful in showing concurrent thread execution, for example, by using swim lanes to model individual execution platforms with parallel flows of behavior, which can be synchronized using Events. Earlier chapters noted that timing can be shown on Sequence Diagrams. Sequence Diagrams may be more useful than Activity Diagrams in elaborating RT Use Cases because they allow timing allocation and analysis and can include State Invariants as a linkage to State Machine Diagrams. SMDs are also valuable for time-sensitive behavior analysis because they can include time operators such as elapsed time and clock ticks and can allow detailed modeling of reaction to events and messages. Earlier chapters, along with Appendices A and C, have more examples.
- *Functional and Requirements Allocation* – careful allocation of time-sensitive functions to system components is the heart of the RT scheduling challenge and may have a significant impact on the difficulty and cost of satisfying RT requirements. For example, the RT processing associated with a sensor is often divided between signal and data processors, with the former designed for high-throughput computation of complex algorithms and the latter for more general-purpose scheduling and control functions. Typical allocation decisions involve choosing the processors (execution platforms) to which various steps in the processing chain should be assigned for best overall performance. The allocation process works hand in hand with continuing requirements analysis to optimize the system design.
- *External Interfaces* – an RT system is likely to participate in time-sensitive processes involving external systems and resources. There may also be significant issues with interface standards, security, and other aspects of interacting with the external environment. Interfaces such as Web Services that are built on non-RT principles will often fail to meet the needs of an RT system, but other, more efficient and deterministic service models are available.
- *Safety* – RT systems are often safety-critical, and requirements must capture both overall safety requirements such as standards to be applied and specific requirements associated with safe operation of the system in its intended environment. Section 8.9.2 has more on this subject, including the basics of the primary safety standards.
- *Quality of Service (QoS)* – it's particularly important that any QoS parameters that the system must meet in order to perform correctly in a larger enterprise be captured in the requirements. As discussed in Chap. 7, between a resource that offers some product or service and a client or consumer, there exists the equivalent of a contract, and the QoS parameters of the resource must equal or exceed the needs of the client. In an RT system, three of the most important QoS characteristics are time, schedulability, and performance. Others can include accuracy, correctness, policy compliance, and many more. This is a central concept in RT modeling and analysis.

8.6 Modeling Real-Time Systems: The UML Profile for Modeling and Analysis of Real-Time and Embedded (MARTE) Systems

The original UML formalism for RT modeling was the *UML Profile for Schedulability, Performance, and Time (SPT)* developed and maintained by the Object Management Group (OMG) [3] and popularly called the Real-Time Profile.

More recently, the OMG has published the MARTE Profile [4] that replaces the RT Profile and adds both rigor and enhanced semantics to the modeling process. Appendix D summarizes the essentials of the profile, which runs to over 750 pages with its Annexes. This chapter and Appendix D can only highlight key features of MARTE, and a reader who is concerned with RT design, analysis, and modeling should refer to the full document. It should also be noted that both MARTE and its predecessor are concerned with RT software execution, primarily with scheduling application software tasks to run on execution platforms. More generally, RT behavior also includes communication delays, actuator response times, equipment acceleration and deceleration, and a host of other physical phenomena that may be important in end-to-end timelines that are constrained by deadlines. These factors are generally straightforward to include in a model as timed steps in a process flow or scenario.

Any approach to RT modeling, including MARTE, must deal with the following aspects of RT systems and their performance.

- *Model Element* – a unit of model content that can be a Classifier, such as a Block, or an Instance (see Chap. 2). The modeling method must support representation of all varieties of content, especially a wide range of Resources, which are important in RT analysis and design. MARTE has extensive provisions for modeling hardware and software entities and their characteristics, as well as the other kinds of content listed below.
- *Resource* – a Model Element representing a physical or logically persistent entity in the system. It offers one or more services such as computation, storage, communication, or control and is often characterized by the capabilities provided and by other properties such as reliability, safety, and security.
- *Service* – a capability or value offered by a Resource (in the role of server) and invoked by a client or consumer; this is consistent with the Chap. 7 definition.
- *Time* – a core element of RT analysis; the strategy must support modeling key aspects of time as introduced in Sect. 8.2, including the allowed forms for expressing time values, various predefined event arrival patterns (including common probability distribution functions), and timing marks; modeling must also account for the properties of devices such as clocks that measure and report time values as discussed earlier in this chapter.
- *Behavior* – another core element; the strategy must support modeling the ways in which a system acts under RT constraints, including simple and complex Actions, Scenarios, and Events, as well as concurrency, causality, and many other aspects.

In MARTE, with its focus on RT software, behavior modeling begins with Causality that specifies how things happen at runtime and establishes the foundation for runtime semantics.
- *Allocation* – the way in which processes, especially software applications, are associated with resources, especially execution platforms, which the strategy must account for in modeling. The MARTE profile specification states that allocation of functional application elements onto the available resources (the execution platform) is the main concern of real-time embedded system design.
- *Schedulability* – the degree to which a system is known to be able to execute all tasks within their deadlines. The strategy must support commonly used schedulability analysis methods, including task priorities, resource utilization, blocking behavior, worst-case task completions, and other factors that determine whether a set of tasks can be successfully scheduled.
- *Performance* – similarly, the degree to which a system is known to be able to meet performance requirements. The strategy must support analysis of system performance parameters to ensure the system can meet its requirements.
- *Quality of Service* – the degree to which a system is known to be able to honor QoS requirements. The strategy must support analysis of QoS parameters as described earlier to ensure satisfactory service responses to clients. QoS associated with a Resource and a service provided by that resource may include allocation of thread timing, allocation of policies to the Resource Management function, scheduling of all resources involved in providing the service, prediction and measurement of performance, resource loading, and margins, and potentially other aspects such as communications reliability and the cost of delivering a given level of QoS. The OMG has published a UML Profile for Modeling QoS and Fault Tolerance Characteristics and Mechanisms [5].

The most common use of MARTE is to incorporate stereotypes and constraints in models in order to define and document the RT aspects of a system. At the time of this writing, MARTE is only beginning to be supported by modeling tools. As a result, it may be necessary to create stereotypes, manually insert constraints, and otherwise add MARTE content to a model in a tool that does not make these features available automatically. Individual tools will provide various ways to include MARTE content, so that diagrams will vary somewhat depending on the tool being used. The following are a few illustrative examples, and Fig. 8.4 shows a typical diagram annotated with constraints. These examples barely scratch the surface of MARTE, and serious use of the profile requires careful study of the full specification. For basic RT modeling, Appendix D is intended to provide enough information at least to get started.

- System elements can be stereotyped as `<<ComputingResource>>`, `<<CommunicationsMedia>>`, `<<StorageResource>>`, `<<DeviceResource>>`, `<<SchedulableResource>>`, `<<MutualExclusionResource>>`, and many others. A resource is typically further defined using descriptive Constraints of the form `{speedFactor = <value>}`, `{capacity = <value>}`, or other parameter/

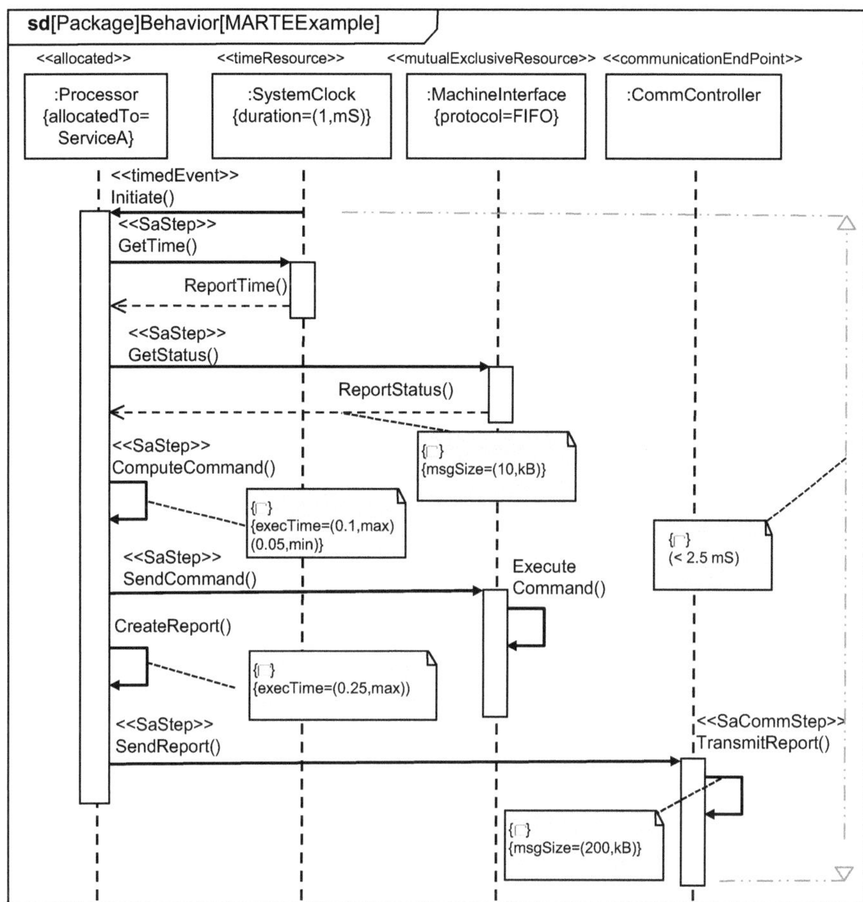

Fig. 8.4 Typical MARTE profile content applied to a basic sequence diagram

value pairs. These resource descriptions can be effective in both structural and behavioral diagrams.

- States in a State Machine Diagram can be stereotyped as <<mode>> and transitions as <<modeTransition>> to explicitly show modal behavior.
- An Activity Diagram used to support analysis of a workload involving concurrent flows can be identified with a stereotype such as <<GaWorkloadBehavior>> to show that it is covered by the Generic Quantitative Analysis Modeling (GQAM) subprofile of MARTE. GQAM establishes the foundation for behavior modeling. With the diagram identified as representing a workload, swim lanes representing parallel behaviors (flows) would be labeled as <<SaEndToEndFlow>> <FlowName> {<Constraints>} where <FlowName> is the unique identifier for a given Flow and a typical Constraint would be an end-to-end duration such as {endToEnd = (<value>,<units>), for example, (5, ms). In the typical situation

where a flow is initiated by an Event, the Event would be labeled `<<workloadEvent>> <EventName> {<Constraints>}`, and the Constraints could specify things like an Event arrival pattern or other timing.

- A Sequence Diagram could be stereotyped `<<GaScenario>>` to indicate that it models a GQAM Behavior Scenario. Then the Objects for the diagram's lifelines could be described with various MARTE stereotypes and associated Constraints, just as for any Resources, while the Messages and durations within the diagram can have stereotypes such as `<<saStep>>`, `<<saCommStep>>`, `<<saRelStep>>`, etc. along with the usual Actions and associated Constraints such as `{execTime = (<value>,max) (<value>,min)}` to show the allowable limits on the time to execute the Action.

Figure 8.4 is a simple Sequence Diagram showing some common MARTE content. The `<<Step>>` stereotype from GQAM identifies an increment of behavior, typically the Action associated with a Message. It can be further specialized as `<<SaStep>>` which adds Attributes from the Schedulability Analysis Modeling (SAM) Subprofile associated with schedulability and as `<<SaCommStep>>` when the behavior involves a communications action. The flow starts with a `<<Timed Event>>` message from the System Clock. The Block Instances heading the diagram's lifelines are stereotyped with various Resource categories, and several have Constraints that declare important properties. Constraints in the diagram spell out things like minimum and maximum allowable execution times and message sizes. A conventional time interval for the entire diagram shows that the end-to-end time cannot exceed 2.5 ms.

8.7 Schedulability

As noted earlier, once decisions about allocation of time-sensitive functions to resources have been made, ensuring schedulability is typically the primary challenge in RT system design. The MARTE Profile deals at length with various aspects of allocation and scheduling. Essentially, RT processes and applications are broken up into jobs or tasks, usually packaged as threads for execution, that are assigned to execute on resources such as an execution platform in specified time intervals. In the simplest cases, such as an embedded microcontroller dealing with a single process on a single resource, no schedule is required – the software simply runs as fast as the hardware will support.

More commonly, in complex systems, multiple processes or software applications compete for the use of shared resources. When multiple concurrent threads must be allocated to execution resources with assurance that deadlines will be met even under worst-case system operating conditions, the design includes a scheduler. This is a kind of resource broker that implements a scheduling policy. A scheduling policy is a method that establishes the order in which task executions are scheduled.

It is based on a combination of one or more optimality criteria and a scheduling algorithm that seeks to optimize the schedule against the criteria.

Early RT systems used simple scheduling algorithms such as a cyclic executive in which jobs run on a pre-established periodic timeline. Modern scheduling algorithms achieve much better efficiency in using resources as well as improved stability and responsiveness to dynamic conditions. Popular examples include Rate Monotonic Scheduling (RMS), Deadline Monotonic Scheduling (DMS), Least Laxity (LL), Maximum Urgency First (MUF), and Earliest Deadline Scheduling (EDS). A sophisticated scheduler may implement more than one algorithm. Douglass [1] gives a detailed description of scheduling policies in terms of characteristics like fairness, complexity, responsiveness, stability, robustness, and predictability. Sha, Rajkumar, and Sathaye [6] published a summary of Generalized RMS in 1994 that remains a valuable reference.

We can illustrate scheduling computations with a brief description of RMS, which is one of the most widely used scheduling algorithms and one that is supported by a variety of tools. RMS applies to tasks that are time-driven (periodic). As the name implies, it assigns scheduling priorities as a function of the frequency with a given task or job needs to execute, weighted by task criticality. Accordingly, the algorithm assigns a task priority as:

$$P_i = C_i / T_i \tag{8.1}$$

where P_i is the priority of the ith task, C_i is an assigned importance or criticality parameter for the task, and T_i is the period of task execution or the deadline, whichever is shorter. If tasks are independent (non-blocking), then an RMS-based system is guaranteed to be schedulable if:

$$\Sigma\left(E_i / T_i\right) \leq n\left(2^{1/n} - 1\right) \tag{8.2}$$

where E_i is the execution time of the ith task. E_i/T_i is the Utilization Parameter for the ith task, i.e., the required fraction of available time when the task must execute to meet its deadline. If tasks can block each other, the left side of the inequality must include the worst-case blocking time, but the result is similar. These are strong conditions, and in fact a system may be schedulable even if the condition is not completely satisfied, which might be important to investigate using modeling if the required resources to fully comply are expensive, unreliable, or otherwise undesirable.

Given task priorities from Rate Monotonic, Deadline Monotonic, or some other algorithm, the resource manager of the execution engine must load and run tasks, including preemption of a running task when a higher-priority task is ready. The design must ensure tasks do not contaminate each other's data, must ensure that low priority tasks do not become "stranded" by never being allowed to run, and must have adequate margins on resource utilization without incurring excessive costs by "gold plating."

A schedule can be static, meaning that it is determined during system design and applied in all scenarios, or dynamic, with adjustments in response to changing operational priorities or circumstances. Familiar algorithms like RMS are static. Dynamic scheduling is also called value- or utility-based scheduling and periodically recalculates scheduling priorities based on some function that evaluates the situation. For example, task priorities might be very different when a system is in different states or in different phases of a business process or mission. Schedulability is a far more complex topic than this chapter can adequately treat and remains an area of active research to come up with better ways to optimally employ resources under RT constraints.

As a simple example, Table 8.1 describes five tasks or threads running on a shared processor. Each has a known or estimated execution time, E_i, and needs to execute cyclically with a period, T_i. Utilization and Priority values are computed using Eqs. 8.1 and 8.2. Deadlines are typically somewhat shorter than the task period, and a Utilization value is simply the ratio E_i/T_i. The Importance Parameter, C_i, is assigned based on factors like the criticality of the task to system functionality or the consequences of missing a task deadline. Task 1 is clearly the most important, based on both its assigned importance and its execution frequency ($1/T_1$). The priorities of the other tasks are calculated in the same way. A dynamic scheduler could revise the C_i values for a scheduling period prior to running the RMS algorithm.

From Inequality 8.2, the maximum total Utilization that represents guaranteed schedulability can be calculated with results as shown in Table 8.2. As the number of tasks goes up, the cumulative Utilization that is schedulable goes down. From this simple analysis, it's guaranteed that Tasks 1 and 2 are schedulable, because their cumulative Utilization is 0.8, less than the limit of 0.828. Beginning with Task 3, the criterion is violated, so the low priority tasks are very unlikely to make their deadlines. The analysis suggests that the system needs additional or upgraded processing

Table 8.1 Example of rate monotonic scheduling

Task	Execution time, *E*	Period, *T*	Deadline	Utilization	Importance, *C*	Priority, *P*
1	0.1 µS	0.2 µS	180 nS	0.5	10	50
2	0.15 µS	0.5 µS	0.45 µS	0.3	8	16
3	0.2 µS	1.0 µS	1.0 µS	0.2	5	5
4	0.3 µS	2.0 µS	1.8 µS	0.15	2	1
5	0.5 µS	3.0 µS	2.5 µS	0.17	1	0.3

Table 8.2 Maximum resource utilization

# Tasks	Maximum utilization
1	1
2	0.828
3	0.780
4	0.757
5	0.743

resources to ensure all tasks will be schedulable with adequate margins, especially under transient conditions of heavy task loading. A faster processor will shorten execution times, which makes everything better, although the cost may increase. Alternatively, the tasks might be divided between two processors.

Figure 8.5 shows a basic timing analysis of the situation created by Table 8.1. This example assumes a preemptive scheduling technique where a higher-priority task that becomes ready to run, basically by arriving at the time of its next periodic execution, can suspend the execution of a lower-priority task. The figure confirms the prediction of Inequality 8.2 that Tasks 1 and 2 will always meet their deadlines. Moreover, Task 3 also meets its deadline, albeit with no slack and with the assumption of perfectly efficient task suspension and resumption by the scheduler and the operating system. This illustrates the fact that Inequality 8.2 computes a very hard limit on schedulability, and tasks that modestly exceed the prediction may still meet their deadlines. However, Tasks 4 and 5 never even launch because there is never a time window when a higher-priority task is not executing. If tasks are allowed to block the execution engine until they run to completion, the situation is even worse, and even Task 1 will occasionally miss its deadline.

As a somewhat brute force approach to fixing this problem, suppose the throughput of the shared processor hosting these five tasks is quadrupled, shortening their execution times by a factor of four while leaving the other parameters unchanged. The result is shown in Fig. 8.6. Now, all five tasks meet their deadlines with robust margins and only minor preemption of the lowest priority task. However, the figure also shows significant idle time for the processor, so this is almost certainly expensive overkill. The optimum design probably involves either a less dramatic investment in additional computing resources or the allocation of additional tasks to this execution platform, possibly including background processing tasks such as built-in test and system monitoring.

A classic problem in RT scheduling, mentioned earlier, is priority inversion, a condition in which lower-priority tasks, by locking up a resource, prevent higher-priority tasks from executing. There are known techniques for preventing or minimizing this, such as the Priority Ceiling Protocol. In modern RT architecture practice, the normal approach is to develop and annotate an architecture model for

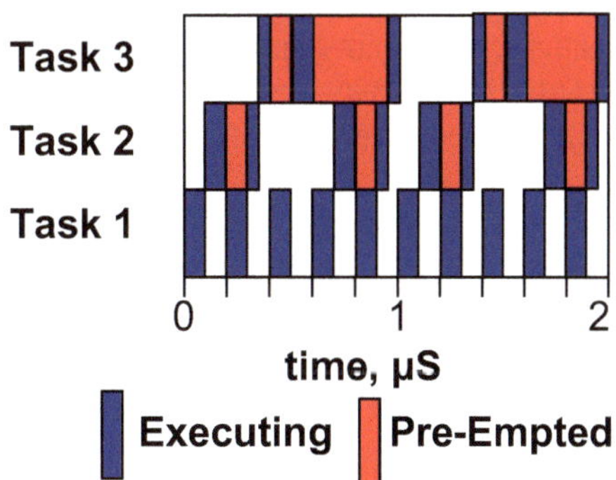

Fig. 8.5 Timing analysis of the tasks in Table 8.1

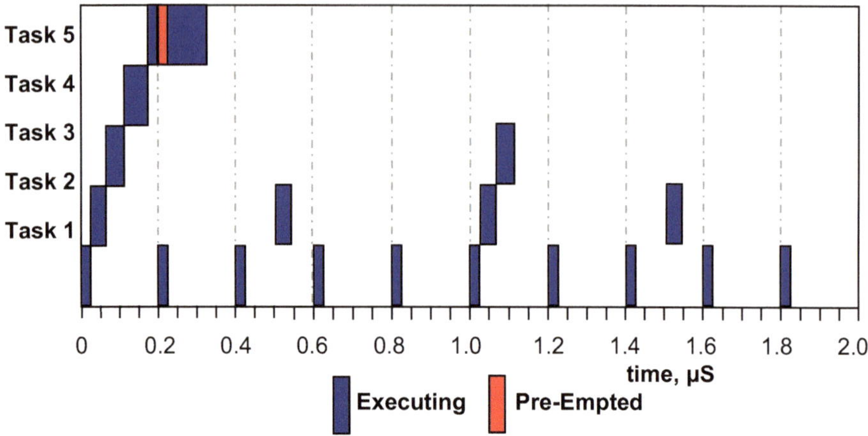

Fig. 8.6 Timing analysis reflecting a fourfold increase in processor throughput

analysis by a tool that determines the schedulability and highlights areas where refinements are necessary to meet system requirements.

A related but distinct issue is analysis of system performance, and the Performance Analysis Modeling Subprofile of MARTE provides stereotypes and tags for annotating a model for this purpose. Performance analysis is primarily applied to "best-effort" or soft RT systems and is basically concerned with statistical measures of system outcomes. Performance is generally characterized in terms of parameters like response times to service invocations, resource utilization and margins, task throughput, or task queue servicing rate. The analysis seeks to identify bottlenecks that limit the aggregate performance of a set of resources executing a scenario. All of these are likely to be characterized statistically, at least in terms of average and worst-case values, and to be compared to measured data once a physical system implementation is available. The two primary performance analysis methods are Queuing Models and simulations, which may use tools that animate an architecture model or physical models that represent the actual behavior of resources such as computers and networks.

As the MARTE Profile becomes more widely used, it can be expected that tools will be available that can read, or "parse," an annotated model and perform automatic Schedulability and Performance analysis. Such tools already exist for the RT Profile. Appropriate implementation of the MARTE Profile in tools is an essential element of applying its elements (stereotypes, models, etc.) in building and analyzing an architecture model. In fact, the practicing architect will probably encounter the Profile mainly in terms of modeling and analysis tools that conform to it. A number of tool vendors have products for uses like Rate Monotonic Analysis that integrate with the mainstream architecture modeling tools [7]. Refinements of RT modeling in UML remain an active area of research. A representative example is in a paper by Graf, Ober, and Ober [8] who have proposed a profile called OMEGA-RT based on treating an event as representing an instant of state change with expressions for duration constraints between event occurrences.

8.8 Basic Timing Analysis for the E-X Radar

We now return to the E-X example. The processing tasks of the E-X Radar Processor and Receiver-Waveform Generator during the execution of a radar task give a basic illustration of timing analysis. We assume the radar is controlled through a schedule of collection tasks. These are commonly described using terms such as modes, dwells, or beams. Each such task involves synthesizing transmit and receive beams (in the case of a phased array), radiating a specific waveform that is optimized for the target or imaging area, detecting and processing the reflected RF energy, forming an image (SAR mode) or track (GMTI mode), packaging this as a radar report, and returning it to the Sensor Management Domain. Chapter 5 introduced the idea of annotating Sequence Diagrams for initial timing analysis, and Fig. 5.11 showed time intervals for the major portions of a radar dwell. Figure 8.7 continues this by adding MARTE stereotypes and parameters to the first part of the figure, identifying each action as an `<<SaStep>>` and decomposing the overall 200 μs interval into times for the main processing steps. These processing tasks deal with analyzing the tasking associated with the next radar dwell and computing the parameters of the waveform and the phase settings for the modules of the array.

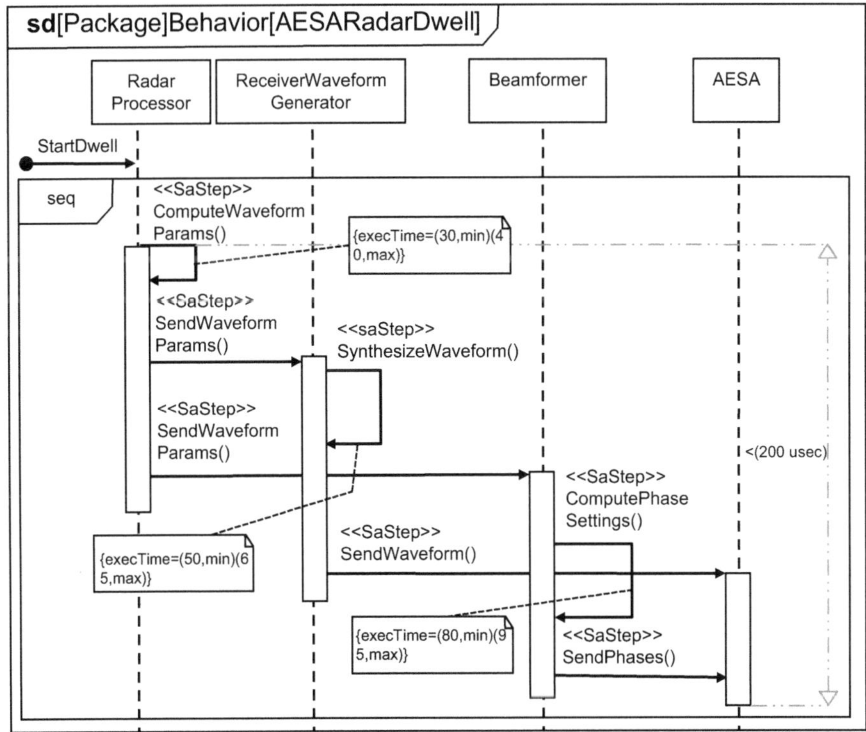

Fig. 8.7 AESA radar dwell with MARTE annotations for timing analysis

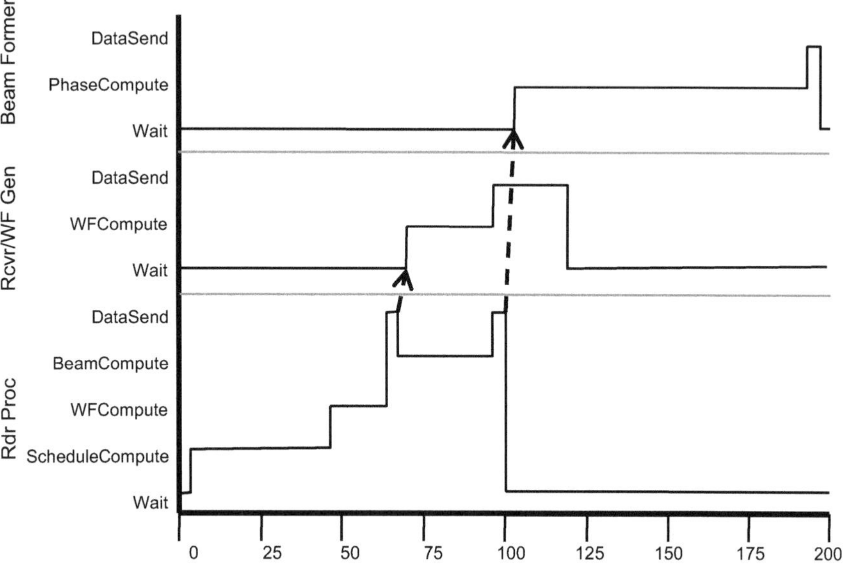

Fig. 8.8 Timing diagram for radar dwell computations

The `execTime` minimum and maximum values might be derived from top-down analysis of overall radar performance parameters such as pulse repetition frequency (which sets a value for the overall duration of a dwell), or they might result from bottom-up analysis of measured or estimated times for the algorithms involved in these computations. In the latter case, it would be typical to examine the resulting end-to-end time estimates vs. the overall radar performance requirements to get a sense of whether the design is likely to be satisfactory.

SysML formally deprecated the Timing Diagram from UML. However, some architecture analysts prefer Timing Diagrams like that in Fig. 8.8 in dealing with RT behavior. The figure shows timelines for the Radar Processor, Receiver/Waveform Generator, and Beam Former, each having defined states for the various processes they execute. This diagram contains the same timing information as Fig. 8.7, but the time scale is now linear, making relative timings, transitions, idle times, and other key aspects easier to see. Timing Diagrams can be elaborated with variable processing times, network message timing, and other details as appropriate to achieve the desired level of detail.

8.9 Implementing Real-Time Systems

8.9.1 Real-Time Hardware and Software Considerations

This is a book about system architecture and MBSE, not hardware or software engineering. However, the architect needs to understand the constraints imposed by technology, tools, and methods on the developers who will implement an architecture, and in this section, we briefly highlight a few of these issues. The early applications of digital computers in RT systems like flight and equipment control were hampered both by limitations on hardware performance and lack of software tools adequate to deal with RT challenges like rigorous schedulability. As a result, much early RT software was written in Assembly Language (ASM) or even directly in machine code. To cite one example based on the authors' personal experience from the 1960s, the main computer for the F-111 fighter aircraft had 65,000 words of magnetic core (not semiconductor) memory and was programmed in hexadecimal number strings. To meet RT timing and other requirements while fitting into available memory and throughput, programmers played elaborate games with memory management and bit-level manipulation.

Several decades of progress in computer design, underpinned by technology advances and Moore's Law, have largely eliminated the need for this kind of manpower-intensive and error-prone software development. Today, many RT processes can run on general-purpose, reduced instruction set computer (RISC) architectures. Specialized hardware to accelerate the execution of computationally intensive tasks such as pipeline algorithms for image processing can be built using application-specific integrated circuits (ASICs). Even allowing for the inevitable tendency of system complexity to increase to fully occupy whatever level of hardware performance is available, the challenge of RT system implementation is now primarily on the software side. Important aspects of RT software include real-time operating systems (RTOS), addressed in the next section, programming languages, and design patterns.

Over time, higher order language (HOL) programming for RT systems has become feasible, and a good deal of existing RT code is written in C or C++ as well as earlier languages like JOVIAL. Even when an HOL is used, it is not uncommon to code especially time critical (and usually small) software modules in ASM.

An RT programming language must support a number of basic principles:

- *Task or Thread Scheduling* – efficient support for the scheduling mechanisms used in various RT systems
- *Task Synchronization* – management of multiple tasks/threads, including high-integrity priority handling, protection against priority inversion, and support for signals, semaphores, and other communications between tasks or threads
- *Asynchronous Messaging and Control* – efficient mechanisms for handling and responding to unpredictable events in the external environment
- *Memory Management* – a deterministic and fully predictable way to allocate and reclaim memory ("garbage collection") while minimizing the risk of error conditions such as memory leaks

Efforts continue to develop better RT languages that balance high software productivity and quality with assured satisfaction of RT requirements. One recent initiative is the Real-Time Specification for Java™ (RTSJ) [9]. Java, a "write once, run anywhere" language, has been popular for years in general-purpose computing but was unsuitable for RT systems because of high execution overhead due to runtime interpretation of intermediate bit code as well as nondeterministic behaviors such as random garbage collection. The RTSJ seeks to provide a basis for Java tools to fix these problems. Java relies on a large standard library of classes that were not designed to be time-deterministic. Dautelle [10] has published an alternative Javolution Library that provides time-deterministic and RTSJ-Safe alternative implementation of the standard library interfaces. To date RT Java has seen limited application in less stressing situations such as soft RT systems, but as the technology matures, it is likely to penetrate additional RT domains.

RT systems offer fertile ground for design patterns, because equipment controllers, communications devices, sensors, and other typical RT applications share a great deal of common functionality. The Resource Manager pattern introduced in Chap. 5 is a typical example. Douglass [1] describes a number of important RT design patterns, including Observer, Proxy, Reliable Transaction, Smart Pointer, Guarded Call, Container, and Rendezvous.

Safety critical software typically requires a more rigorous development and testing process than noncritical software, with controls and metrics aimed at driving latent errors of design and coding to the achievable minimum. At the same time, software productivity must be kept high enough to control development costs. Most companies that do RT development have evolved processes based on the specifics of their products and their experience, which can include some very painful and expensive lessons learned. Schneidewind [11] has recently described a production model incorporating the idea of process reliability, both current and projected, and including feedback to correct anomalies in both process and product.

8.9.2 Real-Time Operating Systems (RTOS)

An operating system (OS), generally incorporating a set of utilities that create System Services, turns a "bare metal" suite of computing hardware into an execution platform for applications. It begins with a kernel that takes care of the most critical functions like booting the system, scheduling tasks, managing memory, and controlling access to hardware resources. The OS abstracts the underlying hardware plant from the applications software and provides a library of standard functions and services that can be used by applications and infrastructure software components. An RTOS must do this in a way that satisfies RT criteria and provides the services needed by RT software. The hardware, RTOS, and applications work together to meet system performance, safety, and other requirements. The following are some key RTOS characteristics, many of which evolved in the development of the computer and software for the Project Apollo Lunar Lander.

- *Task or Thread Scheduling* – the RTOS implements the scheduling policy of the system by dispatching, preempting, suspending, terminating, or otherwise controlling tasks in accordance with the established priorities and rules.
- *Hardware and Data Management* – the RTOS manages shared access across tasks to hardware and data resources to prevent conflicts, errors, priority inversions,

and unsafe conditions; this may involve mechanisms like semaphores to block a resource until an active task completes or locking a data store during a transaction to prevent storage of conflicting or inconsistent values.

- *Message and Event Handling* – the RTOS manages internal and external messages to and among tasks and handles arriving events to provide appropriate task notifications.
- *Memory Allocation and Protection* – the RTOS allocates fixed or variable blocks of memory to active tasks and has mechanisms to deal with memory fragmentation and to prevent memory leaks and other errors caused by dynamic use of shared address spaces.
- *Exception Handling* – the RTOS requires trusted mechanisms to deal with error conditions, especially those that may create unsafe operation.
- *Load Shedding* – the RTOS may have provisions to drop certain functions under a priority rule when they cannot be executed within time constraints.
- *Security and Safety* – the RTOS also requires trusted mechanisms to implement applicable security and safety policies.
- *Overall Efficiency* – the RTOS must have minimum overhead and latency in performing tasks like memory management and interrupt handling such that it has only minor impacts on meeting execution deadlines.

An early RTOS generally used a clock-based scheduling scheme in which tasks were assigned to run in specific time frames; an example is the Cyclic Executive. Variations included variable time frames, frame stealing, slack time management, and so forth. These are simple but inflexible schedulers, hard to maintain, and best suited to small numbers of tasks. Later, more elaborate RTOS schedulers use various kinds of preemption based on task priorities to try to keep the shared resources busy and to optimize their utilization in accordance with system priorities.

The primary safety standard affecting RTOS, as well as applications software, is DO-178, *Software Considerations in Airborne Systems and Equipment Certification* [12]. DO-178 establishes Design Assurance Levels from Catastrophic to No Effect along with processes for developing and certifying safety-critical systems (including more than just aircraft). The levels are:

- *Catastrophic* – failure may cause a crash.
- *Hazardous* – failure has a large negative impact on safety or performance, reduces the ability of the crew to operate the plane due to physical distress or a higher workload, or causes serious or fatal injuries among the passengers.
- *Major* – failure is significant but has a lesser impact than a Hazardous failure (e.g., leads to passenger discomfort rather than injuries).
- *Minor* – failure is noticeable but has a lesser impact than a Major failure (e.g., causing passenger inconvenience or a routine flight plan change).
- *No Effect* – failure has no impact on safety, aircraft operation, or crew workload.

The long-standing B revision of the standard, which was issued in 1992, has recently been replaced by DO-178C which revises the content to resolve ambiguities and cover modern software engineering methods that have emerged since the previous version [13]. DO-278A is a companion standard governing development of ground-based software. They are supplemented by a number of other documents, including:

- DO-248C, "Supporting Information for DO-1789C/DO-278A"
- DO-330, "Software Tool Qualification Considerations"
- DO-331, "Model-Based Development and Verification Supplement to DO-178C/DO-278A"
- DO-332, "Object-Oriented Technology and Related Techniques Supplement to DO-178C/DO-278A"
- DO-333, "Formal Methods Supplement to DO-178C/DO-278A"

These standards emphasize evidence-based software certification using data on software planning, requirements, development and life-cycle processes for verification, configuration management, quality assurance, and the like. They are applied to both executable code objects and to Parameter Data Items that define aspects of execution. The approach emphasizes bidirectional traceability among requirements, source code, test cases, test results, software development and management procedures, and other key aspects of software specification, development, verification, and management.

Most modern RTOS products are now developed to be compliant with DO-178, and some also implement cybersecurity features and functions, as discussed in Chap. 10. Tribble and Miller [14] describe an approach to software safety analysis based on executable models and formal methods that may contribute evidence to support certification of both an RTOS and safety-critical applications.

Another key standard, applicable to complex, high-end RT systems, is ARINC 653 [15, 16]. RTOS and RT applications products are commonly certified to both ARINC 653 and DO-178 criteria. Originally developed for integrated, modular avionics (IMA) systems (see Integrated, Modular Electronics below), 653 is a specification for an Application Executive (APEX) that abstracts the execution platform from the RT applications running on it. It allows the creation of space (memory) and time partitions within which memory management, scheduling, multitasking, warm and cold startup (initialization), error handling, and other software management tasks can be performed. This is an RT version of virtualization, which is further discussed in Chap. 14, and the APEX can be thought of as a kind of RT hypervisor with a partition functioning as a kind of virtual machine. Partitions use preemptive scheduling internally.

When a system is integrated from applications provided by multiple developers or vendors, it may be advantageous to create partitions for each to maintain loose coupling and minimize conflicts and impacts resulting from inconsistent design methods. Each partition can hold a different application programming interface (API) to create a native execution environment for the applications hosted there. The APEX API provides a set of services in the categories of partition management, process management, time management, sampling and queuing port management, inter-partition communication, intra-partition communication, buffer management, blackboard management, event management, semaphore management, and error handling. Individual partitions can be certified to various DO-178 levels. Figure 8.9, adapted from a paper by Wilson and Preyssler [17], shows the general idea of creating a group of partitions at various DO-178C levels of criticality and each with an RTOS that supports its level, all running on a 653-based execution platform hosted on a single computing module or board. These authors advocate the use of XML configuration files to define the partitions, with the XML segregated in accordance with DO-297 to facilitate system safety certification [18].

These standards arose and are mainly used in the aviation community, where the required levels of assured system safety are very high. Other safety-critical system categories have their own supporting standards. For example, the International Atomic Energy Agency publishes standards for nuclear power facilities [19]. However, DO-178 is widely used, and any such safety governance must include elements such as levels of criticality and procedures for software development, testing, and maintenance.

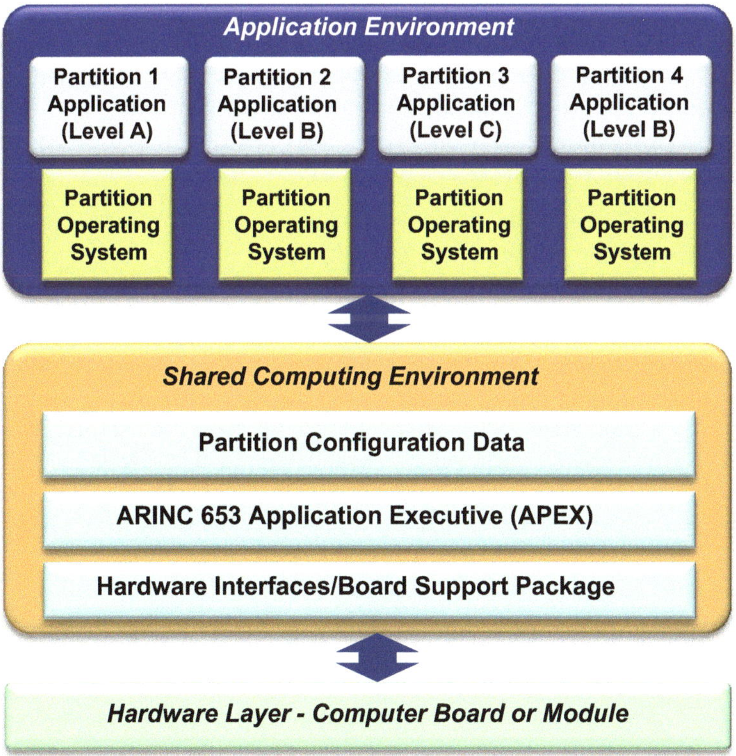

Fig. 8.9 Basic organization of an RT computing solution controlled by ARINC 653 and partitioned by level of DO-178C criticality

8.9.3 Real-Time Middleware

Chapter 5 briefly discussed the application of Service-Oriented Architecture (SOA) to RT systems; Fig. 8.10 repeats Fig. 5.19, showing the essential elements. RT middleware builds on the functions of an RTOS to create a richer execution environment and, especially, to support hosting and execution of heterogeneous application software components or objects. The core of such an architecture is middleware that, like the RTOS, is compatible with RT requirements. One approach to hosting dissimilar applications uses ARINC 653 as described in the preceding section. RT middleware offers a more general approach to implement a multi-application platform based on another category of products.

The traditional approach to RT systems involves a single development organization using a single process and technology baseline to create the entire system, emphasizing performance and usually resulting in little potential for reuse as well as significant challenges in system modification and upgrading. Just as in non-RT systems, middleware offers a path to evolve toward integration of heterogeneous and even distributed components with greatly enhanced scalability, evolvability, and component-level reuse across multiple systems.

For a long time, the RT community was skeptical that this is feasible, but a growing number of products and successful system applications show that in many cases RT middleware is an effective architectural strategy. Krishna, et al. [20], describe approaches to optimize RT middleware for distributed real-time and embedded (DRE) systems with emphasis on product lines that seek to reduce cost through reuse. They describe the role of middleware as:

- Functionally bridging the gap between applications and platforms
- Controlling many aspects of end-to-end QoS
- Simplifying integration of components from multiple suppliers

They describe the key challenge in terms of balancing the desire for general, standard-based middleware platforms and the needs of specific systems to tailor the middleware in order to meet performance requirements. The overall way to achieve this is partial evaluation (PE) in which specific areas of the middleware are specialized for a specific need without invalidating the underlying general model. This is consistent with familiar software practices in which standard library classes or existing components are specialized by adding extensions for a given system while retaining the basic characteristics of the original.

The architecture patterns in Figs. 8.9 and 8.10 can be combined; a partition can include RT middleware running on the Partition Operating System and configured to support the partition's level of criticality. The best-known RT middleware standards are RT CORBA [21] and the Data Distribution Service for Real-time Systems (DDS) [22, 23]. CORBA supports composition of reusable software component services and provides resources for managing processing, network, and memory resources. DDS standardizes the application programming interface (API) by which a distributed application can use Data-Centric Publish-Subscribe (DCPS) as

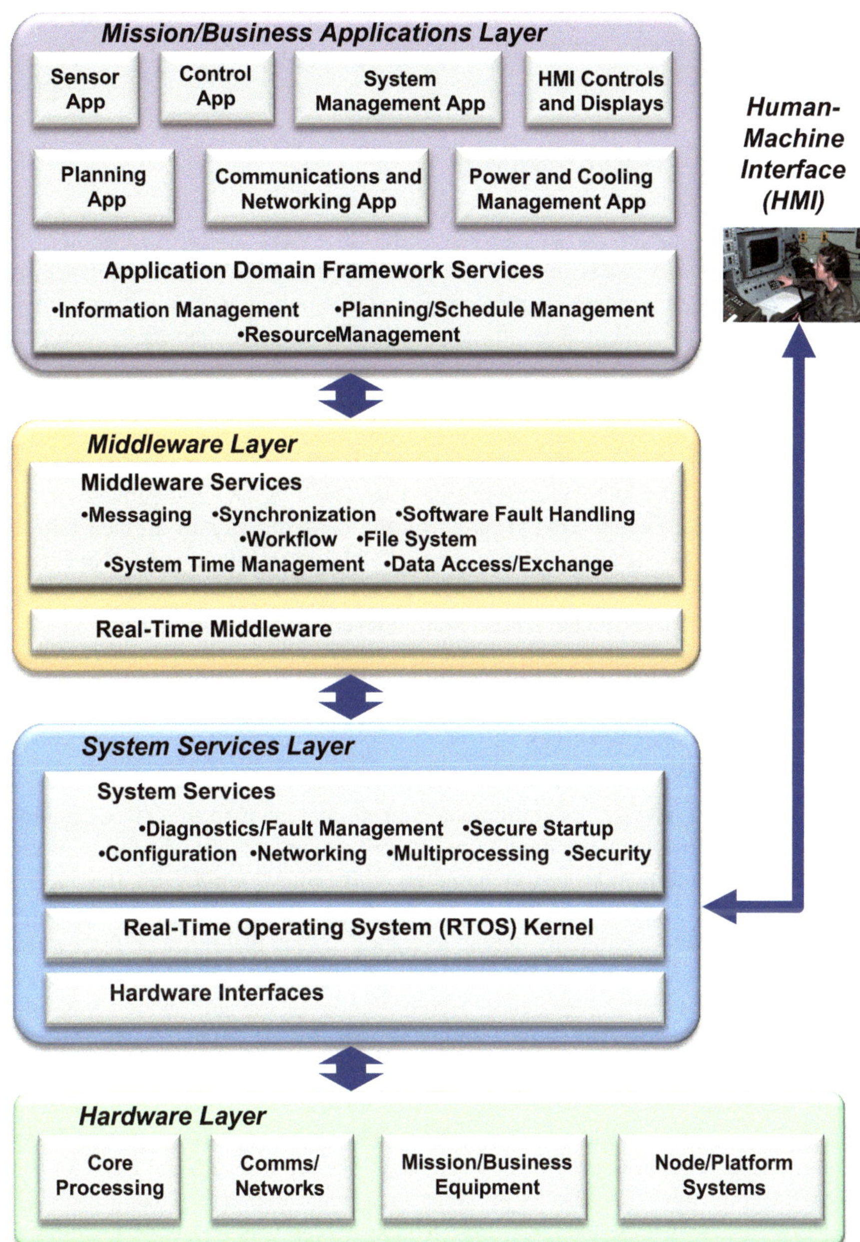

Fig. 8.10 Basic structure of a real-time layered architecture showing typical functions of the various layers

a communication mechanism. Middleware that complies with RT CORBA and DDS is available from a number of companies and universities.

8.9.4 Safety-Critical Hardware

The hardware complement to DO-178 is RTC/DO-254, *Design Assurance for Airborne Electronic Hardware* [24]. Its stated purpose is to provide guidance for design assurance of airborne electronic hardware, beginning with initial conception and continuing through initial certification and subsequent post certification product improvements, which may be needed to ensure continued airworthiness of aircraft using these products. Like DO-178, the standard was originally developed for military aircraft but is increasingly used more generally in connection with safety-critical hardware. Forsberg [25] has described the challenges in updating this kind of system under DO-254 governance. Like DO-178, DO-254 defines a set of Design Assurance Levels (DALs) based on the consequences of a given failure: A. Catastrophic, B. Hazardous/severe-major, C. Major, D. Minor, and E. No Effect. It then prescribes the activities and controls to be implemented in a hardware design life cycle to achieve certification at a required level.

8.9.5 Real-Time Services

Like middleware, service orientation is a concept that many in the RT community believe is inapplicable due to limitations on performance and schedulability. If services are restricted to Web services based on XML, SOAP, and UDDI, this view is probably correct. However, the services construct can be generalized to any design approach that puts functionality behind an interface such that it can be discovered, invoked, and delivered. As enterprises become increasingly networked and interactive, RT resources can be expected to become "nodes on the net" and to be used by remote clients. Even within a local node, the services paradigm may be a useful integration tool, especially when heterogeneous resources are involved [26].

An important aspect of extending SOA to RT domains is the inclusion of all required QoS parameters in a service interface. A number of interface protocols can be used, especially those provided by standards like RT CORBA and DDS for distributed architectures. The service provider must create an interface that is discoverable over a network, honors priorities, handles exceptions and error conditions, implements any applicable security policies, and has governance to ensure long-term integrity and compliance. By adopting the principle of separating the interface from the implementation, a RT SOA promotes long-term operational effectiveness of RT resources by allowing them to evolve while still presenting a stable service end point for interaction with the larger enterprise.

8.9.6 Integrated, Modular Electronics

In the 1970s, the Air Force Avionics Laboratory[1] conducted a technology program called Pave Pillar that built on early work with "glass cockpits" to develop an integrated avionics architecture. The basic principles of Pave Pillar have been refined and applied to a number of military, and more recently civilian, aircraft, under the general title of Integrated Modular Avionics (IMA). The main features include:

- Hardware packaging in relatively small modules installed in racks with high performance connectivity fabrics such as non-blocking crossbar switched networks
- Software implemented in components, preferably employing an object-oriented paradigm
- Shared core processing resources for economy and to facilitate interactions among subsystems such as sensors and communications; the original Pave Pillar explicitly distinguished between signal and data processing, which were assumed to use different hardware technologies
- Integrated human-system interfaces to reduce aircrew workload by presenting synthesized and tailorable information displays that reduce cockpit clutter and minimize the number of places the pilot or system operator must look to find information
- Support for integrated diagnostics, fault tolerance, and system health monitoring and recording
- Standardization of key system elements and interfaces

As integrated system technology matured and began to be applied, an important distinction arose that is commonly identified by the words "integrated" and "federated." This comes up most frequently in the context of RT systems, but the terms can be used in many system categories. Basically, a federated system is composed of a set of essentially independent subsystems, which may exchange information and use a common user interface but perform and control their individual functions with few or no mutual dependencies. An integrated system, by contrast, uses features like a shared core processor and synthesized human-machine interfaces in an effort to achieve overall capabilities greater than the sum of the capabilities of the individual subsystems. The downside of such integration is that it can violate the principle of loose coupling unless there is very strict control of the interfaces to the subsystems and careful definition of mission threads to minimize or eliminate propagation of errors and design changes. A middle ground that might be called "functional integration" is based on preserving the physical integrity and encapsulation of subsystems for loose coupling while using their products in centralized processes such as data fusion and optimized task scheduling to realize the benefits of integration.

These architecture principles are now widely used in system categories other than avionics, so that it's appropriate to speak of Integrated, Modular Electronics. IME represents the practical application of the tenets and methods that are the core

[1] Now the Sensors Directorate of the Air Force Research Laboratory

of this book, and that content, in turn, has evolved in large measure through experience over several decades in the evolution of IME. A system architect working in aerospace systems is likely to deal with development and upgrading of systems designed in accordance with the IME model. In particular, a group of DoD open architecture initiatives that were mentioned in Chap. 1 are especially relevant to IME. These include the Navy Future Airborne Capability Environment (FACE), the Air Force Open Mission Systems (OMS) project, and the Army development of "Victory Standards" for integrating capabilities from multiple systems onto a common platform. For example, a recent news report cited the use of an OMS "wrapper" on software modules that greatly accelerated the migration of new capabilities from one aircraft to another, which is one of the most important intended payoffs from open architecture in RT systems [27].

8.9.7 Security in Real-Time Systems

Chapter 10 focuses on the security aspects of system architecture, including the special challenges posed by RT systems. For now, it's enough to note the unfortunate fact that accepted approaches to highly secure system designs are often in conflict with the demands of time-sensitive performance. Conventional methods for user authentication, secure data access and communications, encryption and decryption, and other processes can involve overhead and latency that are incompatible with RT system deadlines. There are ways to deal with this and to implement trusted RT systems that can be accredited to process highly sensitive information, but the system architect needs to be sensitive to this need and must ensure security considerations are part of the foundation of the architecture.

Overall requirements for Confidentiality, Integrity, and Availability must still be satisfied, but the mechanisms used are typically different in RT vs. non-RT systems. For example, the fact that an RT system typically has many fewer authorized users than non-RT systems means that very strict access controls can be imposed. Similarly, relatively expensive high-speed encryption/decryption using specialized hardware products that would be unaffordable in most large non-RT systems may be feasible and necessary to meet the timing demands of an RT system.

One very important element of RT system security involves secure RT kernels. In general, an RTOS has features such as memory protection that are very useful in reducing cybersecurity risk. Highly secure RTOS kernels are generally based on a Multiple Independent Levels of Security (MILS) architecture, which is a concept that has evolved under the oversight of the National Security Agency (NSA). The key concept is called a Separation or Partitioning Kernel, a small, mathematically verified piece of software trusted to guarantee separation of time and space partitioning by level of trust [28]. This is obviously related to the ARINC 653 approach to space and time separation.

8.9.8 Real-Time Prototypes

Given the focus on timing, performance, and deadlines in an RT system, it's not surprising that the primary issue in simulating such a system concerns the Physical Level. Both high-fidelity virtual prototyping and the construction of an integration laboratory are typically essential to ensuring an RT design is satisfactory in meeting requirements and in maintaining adequate margins on timing and resource utilization. In addition to the higher level timing analysis techniques we have introduced in earlier chapters, specialized simulation tools that account for factors like network data rates and latencies, protocol processing overhead, context switching times in a multi-threaded processor, and other subtleties are often valuable. Accurate measurement of software loading and execution times, response times associated with equipment controls, and any other critical timing is essential. A typical problem that can be addressed in a prototype involves an interface between subsystems or components that were developed independently and prove to have incompatible timing, inconsistent implementation of an interface specification, or other issues. It's not uncommon to compute statistical measures of varying network traffic ("burstiness"), event occurrences in the external environment, and other random processes as part of ensuring that a prototype is tested under realistic conditions. Prototyping can be a significant expense item in the budget of a system development effort, but the risk mitigation and problem solving benefits it delivers justify the cost.

8.10 Summary

RT systems occupy a fairly small niche when compared to the overall world of information systems. Nevertheless, satisfying RT performance requirements is not an excuse to dispense with architectural discipline or to take expedient shortcuts. RT architecture demands skills and methods over and above those typically needed for non-RT systems. The primary challenge is to ensure RT threads/tasks can be scheduled for guaranteed execution within established deadlines. Technology progress over the past half century has made it increasingly feasible to apply proven architecture approaches like object orientation, layering, modularity, loose coupling, and services to all but the most stringent RT situations. Indeed, when system developments fall behind schedule, see cost overruns, and fail to meet user expectations, a common root cause is that the critical architectural foundation was either not laid at all or compromised in the name of expediency. Expertise in selecting or developing system components with the necessary RT attributes is another essential ingredient. A frequent cause of problems is the failure to provide adequate physical and virtual prototyping. RT architecting is undoubtedly harder in many respects, but proven tools and methods exist, and the critical nature of most such systems makes good architecture, if anything, more important than in more typical and less demanding applications.

Exercises

1. What are the three fundamental considerations that set Real-Time systems apart from Non-Real-Time Systems?
2. Name some parameters that can be used to describe the timeliness of an Action in a RT system.
3. What is the key challenge in achieving schedulability in a multi-threaded RT system?
4. Consider a RT system that is to be developed to control safety-critical machinery in an electric power generating station. Write down a set of verifiable requirements for such a system.
5. Modify Fig. 5.28 by annotating it with appropriate content from the MARTE Profile.
6. List some factors that are important in selecting a Scheduling Policy to be implemented by a Scheduler?
7. Suggest an approach that could be implemented in an RT scheduler to prevent priority inversion.
8. List some aspects of RT systems that a good RT programming language must account for.
9. Considering the required characteristics of an RT system, list some specific features and functions of RT middleware needed to meet RT requirements.
10. Consider an RT system for an autonomous (self-driving) vehicle. Sketch an Integrated Modular Electronics design that incorporates sensors, actuators, communications, navigation, and power management for such a vehicle.
11. For the IME design in Exercise 10, write down a set of RT requirements applicable to various components of the system.

Student Projects

Students should extend their system architecture models to include time-sensitive behaviors. This should include real-time requirements, basic timing analysis, timing annotation of behavior diagrams, selection of a scheduling policy, and other RT architecture aspects as discussed in this chapter.

References

1. Douglass BP (2004) Real-time UML: advances in the UML for real-time systems, 3rd edn. Addison Wesley, New York
2. Selic B, Rumbaugh J (1998) Using UML for modeling complex real-time systems. Rational Software Corporation, Cupertino, CA
3. Object Management Group (2005) UML™ Profile for schedulability, performance, and time specification, ver 1.1. http://www.omg.org/spec/SPTP/. Accessed 28 May 2017
4. Object Management Group (2012) Modeling and analysis of real time and embedded systems. http://www.omgmarte.org. Accessed 28 May 2017
5. Object Management Group (2008) UML profile for modeling quality of service and fault tolerance characteristics and mechanisms™ (QFTP™). http://www.omg.org/spec/QFTP/. Accessed 28 May 2017

6. Sha L, Rajkumar R, Sathaye SS (1994) Generalized rate-monotonic scheduling theory: a framework for developing real-time systems. Proc IEEE 82(1):68–82
7. Martins P (2013) Integrating real-time UML models with schedulability analysis. https://scholar.google.com/scholar?q=Integrating+Real-Time+UML+Models+with+Schedulability+Analysis&hl=en&as_sdt=0&as_vis=1&oi=scholart&sa=X&ved=0ahUKEwicq4uaypPUAhUU7mMKHc8VCawQgQMIJDAA. Accessed 28 May 2017
8. Susanne Graf S, Ober I, Iulian Ober J (2006) A real-time profile for UML. Int J Softw Tools Technol Transf 8(2):113–127
9. Dibble P (2006) The real-time specification for Java™. http://www.rtsj.org/docs/rtsj_1.0.2_spec.pdf. Accessed 28 May 2017
10. Dautelle J-M (2008) Fully time deterministic java. Paper presented at JavaOne Conference, Moscone Center, San Francisco, 23–24 June 2008
11. Norman Schneidewind N (2008) Software production process for safety critical software. AIAA J Aerosp Comput Inf Commun 5:72–83
12. RTCA (2011) DO-178c. http://www.rtca.org/. Accessed 28 May 2017
13. Youn WK, Hong B, Oh KR, Ahn AS (2005) Software certification of safety-critical avionic systems: DOP-178C and its impacts. IEEE A&E Syst Mag 20(4):4–13
14. Tribble AC, Miller SP (2004) Software intensive systems safety analysis. IEEE A&E Syst Mag 19(10):21–26
15. Airlines Electronic Engineering Committee (AEEC) (2008) Avionics Application Software Standard Interface, ARINC Specification 653-2. ARINC, Inc., Annapolis
16. Rufino J, Craveiro J Robust Partitioning and Composability in ARINC 653 Conformant Real-Time Operating Systems. Paper presented at the 1st INTERAC Research Network Plenary Workshop, Braga, Portugal, July 2008
17. Wilson W, Preyssler T (2009) Incremental certification and integrated modular avionics. IEEE A&E Syst Mag 24(11):15–22
18. RTCA (2005) RTCA DO-297: Integrated Modular Avionics (IMA) development guidance and certification considerations. http://standards.globalspec.com/std/2018378/rtca-do-297. Accessed 28 May 2017
19. IAEA Standards Home Page. http://www-ns.iaea.org/standards/documents/default.asp?s=11&l=90&sub=10. Accessed 3 May 2018
20. Krishna AS et al (2006) Towards highly optimized real-time middleware for software product-line architectures. ACM SIGBED 3(1):13–16
21. Giddings B, Beckwith B (2003) Real-time CORBA tutorial. http://www.omg.org/news/meetings/workshops/RT_2003_Manual/Tutorials/T1_RTCORBA_Giddings.pdf. Accessed 30 May 2017
22. Pardo-Castellote G, Farabaugh B, Warren R (2005) An introduction to DDS and data-centric communications. http://www.omg.org/news/whitepapers/Intro_To_DDS.pdf. Accessed 28 May 2017
23. Object Management Group (2015) Data Distribution Service™ (DDS™). http://www.omg.org/spec/DDS/. Accessed 28 May 2017
24. DO254 User Group (2005) RTCA/DO254 Design assurance guidance for airborne electronic hardware. http://ajo21.hol.es/rtca-do-254.pdf. Accessed 28 May 2017
25. Forsberg H (2010) Challenges in updating military safety-critical hardware. IEEE A&E Syst Mag 25(9):4–10
26. IEEE (2009) 2nd International workshop real-time service-oriented architecture and applications. Seattle, 20–24 Jul 2009
27. Warwick G (2016) Senior sensors. Aviation week and space technology, June 6–19, 2016. p 42–44
28. VanderLeest SH (2018) Is formal proof of seL4 sufficient for avionics security? IEEE A&E Syst Mag 33(2):16 21

Chapter 9
Developing the Network Dimension

9.1 Roles and Varieties of Networks

Any information-intensive system or enterprise requires networking both for internal connection of subsystems and other components and for interaction with the external environment. This is a complex engineering specialty in its own right, and a SE team will need the appropriate expertise to implement this dimension of the system. In this chapter, we briefly survey the issues with which a system architect should be familiar in order to deal effectively with networking in terms of the many options for implementing internal and external connectivity and their impact on system performance, reliability, and cost. For readers to whom this is an unfamiliar topic, Dordal [1] provides an excellent general tutorial on networking basics.

One common way to characterize networks is based on their scope. A Local Area Network (LAN) connects systems and their components at a given location such as a building, a ship, or some other node in an enterprise. A Metropolitan Area Network (MAN) connects nodes and systems located in separate facilities within a city, campus, operational site, or other locality. A Wide Area Network (WAN) supports long-distance communication, which is increasingly global and occasionally reaches beyond planet Earth. In addition, specialized network categories are used for demanding connectivity needs; an example is a Storage Attached Network (SAN) that is optimized for providing access to shared data storage, usually via one or more ports on a LAN.

Yet another way of classifying networks involves topology, the geometric arrangement of connections that comprise the network. Figure 9.1 shows some basic possibilities. The simplest is a linear bus in which a central communications medium such as a terminated cable is tapped into by the connected systems and components. This is limited to relatively small networks and subject to disruption by a failure at any point. It is the basis for widely used multiplex bus networks in vehicles, aircraft, and space systems; a familiar example in aerospace systems is the MIL-STD-1553B standard. A variation is the ring topology in which point-to-point

J. M. Borky, T. H. Bradley, *Effective Model-Based Systems Engineering*,
https://doi.org/10.1007/978-3-319-95669-5_9

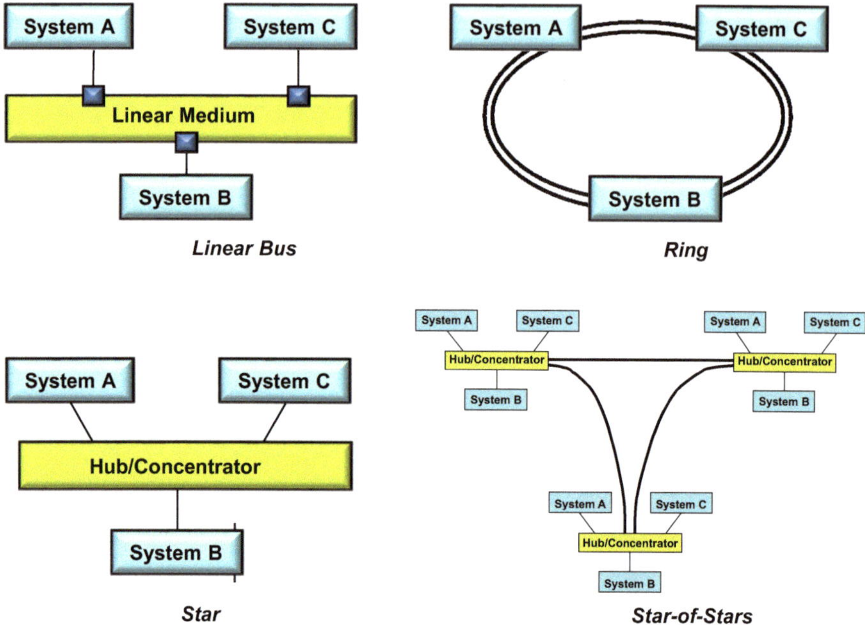

Fig. 9.1 Basic network topologies

links connect participating systems and messages flow around the ring from source to destination. This has largely disappeared in favor of the star topology, in which connected systems have dedicated paths to a central hub or concentrator that either rebroadcasts traffic or forwards it based on addressing by creating a path between input and output ports. A tree topology, not shown in the figure, is a hybrid that uses a core linear bus to connect a set of stars that in turn connect clusters of systems. The most common topology today, including the Internet, can be called a star-of-stars in which concentrators such as Internet Service Providers (ISPs) connect individual systems and exchange traffic with other concentrators. This provides a flexible, highly extensible networking foundation but has the important consequence that any given message will pass through a number of intermediate network nodes which can create issues with both performance and security.

Finally, networks can be classified on the basis of the protocols that define their operation. Most networks use packet switching in which the data in a message is encapsulated in discrete packets and then further packaged into frames, each with a header, that are sent individually (often by different routes) and reassembled at the receiving end to reconstruct the message. The overwhelming choice today for general purpose networking is the Internet Protocol stack, frequently running on Ethernet media for a LAN, as discussed in the next section, and on various telecommunications media for a MAN or WAN. Older protocols such as Token Ring, Frame Relay, Asynchronous Transfer Mode, Internet Packet Exchange, and many others survive in legacy systems but are seldom used in new designs. Many specialized protocols have been developed for things like Virtual LANs (VLANs), business messaging, boundary gateway routers, and wireless networks. Reference [2] is an example of the huge number of available references on this subject [2]. Security aspects of networking, including Virtual Private Networks, are addressed in Chap. 10.

9.2 Network Reference Models and Associated Network Devices

After more than 30 years, the seven-layer model of a network developed by the Open Systems Interconnection (OSI) Initiative still provides the basic vocabulary for networking. Table 9.1 lists the OSI model layers. Today, a simpler structure that is associated with the Internet is commonly used, and Table 9.1 shows how the OSI layers are consolidated into the four Internet layers. The third column lists some associated networking protocols; this tabulation is very incomplete but identifies the most prominent examples. Appendix J expands on the OSI layers and many commonly encountered networking acronyms, including those in the table. The Web Services protocol stack (Chap. 7) has a different structure, with standards in categories such as resource management, business processes, security, messaging, portals, and metadata. However, Web Services use the Internet Protocols both for connectivity and for functions such as addressing, routing, security and error detection, and correction.

Figures 9.2 and 9.3 expand on the nature of the OSI and Internet layering structures. Figure 9.2 shows the specific exchanges between two network hosts that implement the seven OSI layers, from the data or messages to be exchanged at the Application layer down to the physical signals and binary data that are actually sent over the network's physical medium. Figure 9.3 emphasizes the way data and messages are progressively encapsulated with header information defined by the Internet protocol stack to create segments, packets, and finally frames. The following sections present some practical details of interest in applying MBSAP.

Table 9.1 Network layers and protocols

OSI model layers	Internet layers	Internet protocol stack
7. Application	Application	HTTP, HTTPS, TSL/SSL, FTP, DNS, VoIP, E-mail (e.g., SMTP), chat (e.g., IRC)
6. Presentation		
5. Session		
4. Transport	Transport	TCP, UDP
3. Network	Network	IPv4, IPv6, IPSec
2. Data link	Link	Ethernet, FDDI, SONET, ISDN, wireless (e.g., 802.11)
1. Physical		

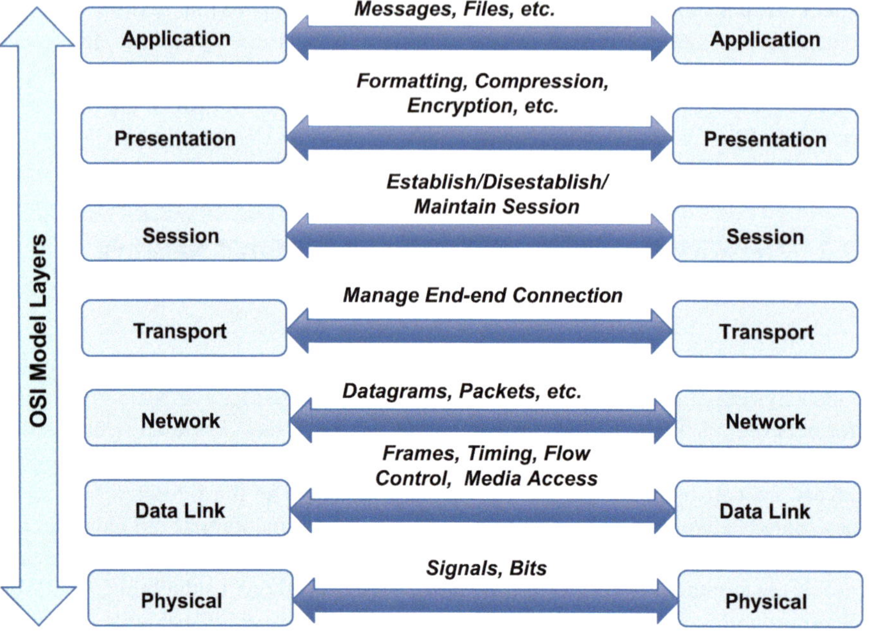

Fig. 9.2 Content exchanges by OSI layer

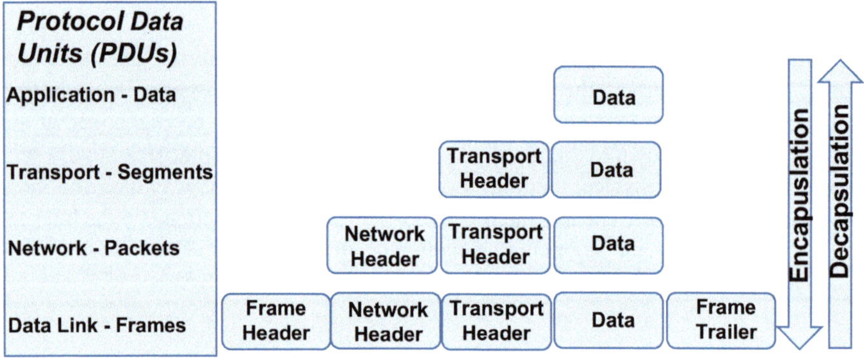

Fig. 9.3 Data structures by Internet layer

Physical Layer Media A key factor in the increase in network speeds from a few hundred bits per second (bps) to tens or hundreds of Gbps has been the availability of affordable media that support ever-greater signal frequencies and thus higher data rates. Copper, optical fiber, and coaxial cabling have all been used. Most LANs today use unshielded twisted pair (UTP) wiring, and fiber can be used for connections requiring higher data rates. There are seven categories of UTP cable with progressively higher data rates. Category 5 supports 100BaseT and is widely used in LANs; Category 6 allows 10 Gbps and is the standard for Gigabit Ethernet.

Table 9.2 Ethernet categories

Ethernet category	Cabling type	Speed range
Basic Ethernet	10BaseT – UTP 10Base2 – Thin coax 10Base5 – Thick coax	10 Mbps
Fast Ethernet	100BaseT – UTP 100Base FX – Fiber 100BaseBX – Single-mode Fiber 100BaseSX – Multimode Fiber	100 Mbps
Gigabit Ethernet	1000BaseT – UTP 1000Base FX – Fiber 1000BaseBX – Single-mode Fiber 1000BaseSX – Multimode Fiber	≥1 Gbps

A great advantage of the Ethernet standards package is that it supports any of the topologies in Fig. 9.1 as well as a wide range of speeds, as summarized in Table 9.2. The table identifies the various cabling types available in each category. Wireless technologies such as the ubiquitous WiFi™, based on the IEEE 802.11 standards, offer great flexibility and convenience, but at slower speeds, over short distances, and with less reliability and significant security vulnerabilities. WANs are supported by satellite communications, high-speed microwave links, terrestrial cabling (including undersea), and, selectively, fiber optics. Physical layer protocols define the electrical characteristics of media and the signals they carry.

Physical Layer Devices Network media are connected by a variety of hardware devices such as switches, routers, bridges, repeaters, gateways, and other appliances that create the network topology, connect the physical links, and transmit and receive data. Computers and other devices connect to the network using Network Interface Cards (NICs). A hub works at the physical layer to connect a group of twisted pair wire or fiber cables into a network segment; an active hub also amplifies the signals passing through. A hub simply retransmits every arriving packet on any port into all the other ports, and an active hub can be thought of as a multi-port repeater. A repeater is a device that boosts signals so that they can be sent over longer distances. For example, a repeater might be placed between hubs that connect widely separated clusters of network-attached devices where the distance between would otherwise exceed the maximum allowable cable length.

A high-end system is likely to use a dual redundant switched Gigabit Ethernet LAN with some combination of UTP and fiber cabling that delivers high performance, high margins for traffic surges, and high reliability, all at an affordable cost. Most mainstream computer operating systems today provide the features of a Network Operating System (NOS), usually combined with the function of a network file server, to control network traffic among connected systems. NOS features are described more fully later in this chapter.

Data Link Layer OSI Layer 2 deals with encoding and decoding data packets for transmission and reception plus functions like timing, synchronization, error detection and correction for the physical layer, and avoiding contention for the network due to simultaneous attempts to transmit by different attached devices ("collisions"). The Ethernet standard covers both Layers 1 and 2 and is the primary approach for LANs. The basic unit of Ethernet data transmission is a frame, as shown in Fig. 9.3, which includes a preamble, source and destination addresses, a length field, the data payload, and a frame check sequence (FCS). Ethernet uses the Carrier-Sense Multiple Access/Collision Detection (CSMA/CD) method to deal with attempts by multiple devices to send packets simultaneously.

Each network device has a unique Media Access Control (MAC) address that is programmed by the manufacturer and usually shown as an alphanumeric string in the form aa-bb-cc-dd-ee-ff. A packet header contains MAC addresses for both source and destination. A device called a switch functions at Layer 2 much like a hub but with the intelligence to associate MAC addresses with ports and send incoming packets only to the destination device. A switch can also transfer packets across network domain boundaries, which are set up to segment the network and reduce the probability of collisions between simultaneous packets, as well as between different network types such as Basic and Fast Ethernet. Switches can be daisy-chained to achieve the required number of ports.

Switches can also have dedicated ports for maximum data rates to and from the busiest or most critical computers or other connected devices. Some switches support full-duplex operation in which separate NICs are provided on a critical device for inbound and outbound traffic to essentially double the data rate. A rule of thumb is that when network loading reaches 30–35% of capacity, performance will be improved by segmenting it with switches to reduce collisions. High-performance networks may justify the cost of non-blocking switches that maximize data flow through all ports by setting up separate paths for simultaneous traffic between pairs of ports.

Most modern switches support the Simple Network Management Protocol (SNMP) [3] that allows a central network management tool to control a network and monitor health and status of connected devices. The term bridge is also used for a device that connects networks based on Layer 2 information and physical addresses and can be used to partition a large network into smaller networks to improve performance. Unfortunately the terminology in this area is imprecise, and some writers assert that "switch" has no clear definition, despite being the most commonly used name for these devices. In any case, a bridge generally connects two LANs.

Network Layer Layer 3 handles the routing and switching of data among nodes on the network, including addressing, exchanging traffic across boundaries between networks ("internetworking"), and forwarding and sequencing data packets. This is the domain of the Internet Protocol (IP), arguably the most successful standard in the history of technology. A less frequently used alternative is Internetwork Packet Exchange (IPX). IP prescribes the encapsulation of data in packets or datagrams (both are self-contained packages of information with the addressing and other data to allow them to be routed to a destination). The primary IP protocol for some time has been version 4 (IPv4), but an updated version 6 (IPv6) is now being deployed. IPv6 has many improvements, including:

- An enormous expansion of the number of IP addresses available by increasing address size from 32 to 128 bits

- Support for multicasting and increased host control of traffic paths
- A simpler header with some fields made optional
- Quality of Service labeling of packets
- Security extensions for authentication, data integrity, and confidentiality
- Improved options for routing, fragmentation/reassembly, and other functions

Some organizations, especially in Government, have mandated a transition to IPv6, and it's not uncommon for a system to find it necessary to run both versions to maintain network connectivity while the transition is in progress.

Devices communicating over an IP network are generically called hosts, and each is uniquely identified by a logical IP address composed of a network ID plus a host ID. IP addresses can be assigned dynamically when a host joins the network, generally using the Dynamic Host Configuration Protocol (DHCP), or statically if there is a reason to ensure a given device always has the same address, perhaps due to security concerns. With IPv4, an IP address is a 32 bit binary number divided into four octets and usually represented as four digital numbers, AAA.BBB.CCC. DDD. An IPv4 address can be embedded in the lowest 32 bits of an IPv6 address. The Layer 3 protocol, whether it's IP or IPX, translates logical addresses to physical MAC addresses to allow packet delivery. When a network is partitioned into subnets, a subnet mask identifies which part of an IP address represents a subnet network ID; the remaining bits are then the host ID.

At Layer 3, switching gives way to routing (which is sometimes referred to as Layer 3 switching) in which both wired and wireless connections can be used to transfer data from one device to another across the boundaries of two or more networks and their segments. Routers use protocol-dependent Layer 3 information in conjunction with dynamic routing tables and are used with network/transport layer protocols like Transport Control Protocol over Internet Protocol (TCP/IP) and Sequenced Packet Exchange over Internet Packet Exchange (SPX/IPX). A router sends each packet to the destination device instead of simply transferring it to the appropriate network segment and has the computing power to determine the best available path through a set of network nodes and links that may offer many possible routes. A routing function can be combined with a switch in a single device.

IP assumes that the network is unreliable, i.e., has no guarantees of traffic delivery, and that no centralized control exists. IP packet routing is based on forwarding to the next available network node that has a path to the destination. This constitutes a "best effort" approach to packet delivery, and errors may arise due to outages, traffic conflicts, and other practical circumstances, especially with heavily loaded long-haul communications. IP routing within a LAN is typically fast and reliable, but across a WAN, there can be significant latencies, packets can arrive out of order, and data can be lost. IPv4 computes packet checksums that can be used to detect data drops and other errors, while IPv6 relies on higher-level protocols for this in order to reduce routing overhead.

Routers located at system boundaries or other major network connection points use the Boundary Gateway Protocol (BGP), or its external (eBGP) or internal (iBGP) variants, to exchange routing information. Large, high-performance networks may need a hierarchy of routers, with simpler routers connecting local groups of devices, intermediate routers aggregating traffic from these, and one or more very fast central routers creating a network backbone for an enterprise. A common application of a router is to connect a LAN, which is typically interconnected by a hub, to the Internet. A router connecting a subnet to a larger network has an address called the Default Gateway Address. A router or a routing function in a Network Operating System is likely to include firewall functions to block unauthorized packets from the local network.

Multiprotocol routers support mixed network layer protocols, which can be useful when connecting legacy networks that use protocols other than TCP/IP. High-end models can implement functions like traffic shaping for network Quality of Service. Hybrid Layer 2/Layer 3 switches, also known as routing or IP switches, combine packet switching with more sophisticated physical layer traffic flow control. There also exist devices called brouters that combine the functions of bridge and router. In fact, devices are now available at affordable prices that combine the functions of router, bridge, hub, and switch in one compact box. Figure 9.4 suggests how various devices might be used to create a network of distributed clusters and individual hosts.

Recently, a new protocol, called Advanced Message Queuing Protocol, has emerged from the banking industry to deal with some of the drawbacks of IP [4]. AMQP is an open protocol that stresses security, reliability, and interoperability and is aimed at messaging between applications and organizations in the business community.

Transport Layer Layer 4 deals with the flow of network traffic from source to destination nodes or hosts. For example, a Transport Layer protocol ensures that packets that arrive out of order due to varying paths through the network or other circumstances are reassembled correctly to reconstitute the message. The two primary protocols are TCP and the Uniform Datagram Protocol (UDP), both running over IP. TCP implements messaging by using acknowledgments to confirm packet delivery or to trigger retransmission of dropped or corrupted packets. UDP is unreliable but more efficient. TCP/IP is the dominant network standard and likely to remain so. SPX/IPX may also be encountered.

Higher Layers Layers 5, 6, and 7 of the OSI Model are largely concerned with the functions of connected devices, the software running on them, and the user interfaces they support. These layers are typically lumped together in an Application Layer that is responsible for accessing and using the network, establishing logical connections, formatting data to be fed into the Transport Layer, and supporting applications. These layers are responsible for abstracting the network and presenting an interface that can be easily employed by systems, resources, and users to exchange information.

The following are some of the more common protocols spanning the higher layers:

- *DNS (Domain Name System or Server)* – translates network addresses such as IP addresses to and from readable forms such as Uniform Resource Locators

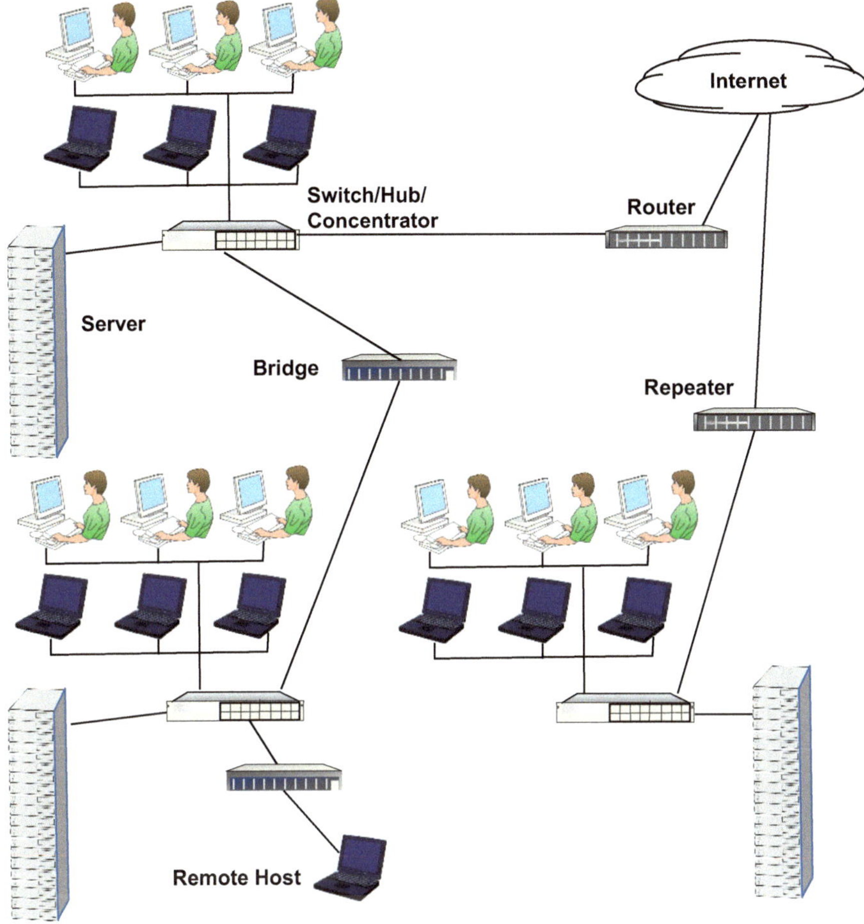

Fig. 9.4 Various devices used to create a network for a specific host configuration

(URLs), which are descriptive labels that are entered into browsers and search engines.
- *DHCP (Dynamic Host Configuration Protocol)* – automatically assigns Internet addresses to computers and users; key to the use of wireless routers where connected users change frequently.
- *FTP (File Transfer Protocol)* – a streamlined protocol for transferring and manipulating (typically large) files.
- *HTTP (HyperText Transfer Protocol)* – the basic Internet protocol for defining and exchanging information in the form of Web pages; HTTPS is a more secure variant using Transport Layer Security (TLS), and even more secure protocols have been developed (see Chap. 10).
- *IMAP (Internet Message Access Protocol)* – a basic Internet e-mail messaging protocol.

- *IRC (Internet Relay Chat)* – used for Internet chat and other communications.
- *POP3 (Post Office Protocol, Version 3)* – used by e-mail clients to retrieve messages from remote servers.
- *SMTP (Simple Mail Transfer Protocol)* – the standard Internet e-mail protocol.

9.3 Ports and Protocols

In addition to the physical network ports and IP addresses discussed above, logical ports called TCP/IP or UDP/IP ports are identified by assigned numbers and allocated to user sessions and server applications to control network information exchanges. A port number and IP address together constitute a socket. Port numbers are included in packet headers and used in routing. There are standard port numbers, generally in the range of 0 to 1023, which are used the same way in all systems; examples are Port 25 for SMTP traffic, 53 for DNS, and 80 for http[1] traffic. Vendors and other organizations can register other port numbers for proprietary applications.

A common technique for maintaining network control and, especially, enhancing security is to restrict the port numbers that are active and to ensure they are used consistently. As a result, a frequently required element of a Networking Focused Viewpoint is a Ports and Protocols (or, sometimes, Ports, Protocols, and Services) Matrix that documents the key characteristics of each network port. Government agencies routinely publish standard versions of this matrix that developers are required to use. A typical entry in the matrix would include:

- Port name
- Transport protocol
- Application protocol
- Allocation (static or dynamic)
- Port number or range of numbers
- Source – server name, client name, or other using process
- Destination – same
- Direction – in, out, or bidirectional
- Rule – e.g., "Use default http proxy"
- Usage – time period when active, indefinite, etc.
- Description – brief statement of functions supported

9.4 Wireless Networks

The convenience of wireless data communications in reducing the need for expensive network cabling and in supporting mobile users means that it will inevitably be called for in more and more applications. However, current wireless technology has

[1]As an acronym, HTTP should normally be capitalized, but in a URL it's commonly entered in lower case.

two significant drawbacks compared to conventional wired networks: lower data rates and significant security and interference vulnerabilities. Wireless connectivity depends on the individual user being within range of an Access Point, which is usually limited to 50 meters or less for basic wireless devices, especially if there are intervening structures. However, the ability to add roving network access to a large facility and to allow users to connect while off-site through commercial wireless points is increasingly important. Range can obviously be extended using repeaters and higher-performance wireless hubs.

Wireless networking is governed by the IEEE 802.XX® family of standards [5]. Data speeds have improved significantly in recent years, and better security measures are also available. The following are some particulars of interest to a system architect.

- Wireless hubs and routers generally conform to the standards endorsed by the WiFi Alliance [6], which certifies product compliance. They serve as gateways to wired networks and come in multiple varieties, including:
  - Single and dual band (2.4 GHz, 5 GHz, or both).
  - Various data rates and signal ranges; the most popular as this is being written is the IEEE 802.11n (N) high speed router. Portable routers are usually slower.
- WiMAX (Worldwide Interoperability for Microwave Access) is a long-range, broadband telecommunications technology for extending wired networks, based on IEEE 802.16d and e.
- Various security mechanisms are available, including:
  - *MAC filtering* – allows only specific machines to connect, considered easy to crack today because packet headers include MAC identifiers.
  - *WiFi Protected Access (WPA2)* – a WiFi Alliance certification program; WPA uses advanced encryption that implements IEEE 802.11i; this replaces the now-obsolete Wired Equivalent Privacy (WEP).
  - *WiFi Protected Setup (WPS)* – creates a security key for access, less secure than a full WPA implementation.
  - *Virtual Private Network (VPN)* – secure access to an organizational network using Public Key Encryption (see Chap. 10).

9.5 Virtual Local Area Networks (VLANs)

Regardless of the physical implementation of a network, and which computers and other devices are connected to which network equipment, it is often desirable to create connected groups of resources that have heavy interaction, share policy rules such as restricted information access, provide mutual backup, or are otherwise closely related. A common way to deal with this is the creation of a VLAN of network hosts that act as if they were connected to a common domain even when not wired to a common switch. Both Ethernet and WiFi networks can support VLANs.

A VLAN provides through software the kind of network segmentation that would be done physically with a router. This is really a Layer 2 construct, equivalent to an IP subnet. The most common protocol is IEEE 802.1Q [5]. A VLAN operates by tagging Ethernet frames to indicate that they represent VLAN traffic and provides the appropriate identifiers.

VLANs can serve multiple purposes. They give a System or Network Administrator (SYSAD/NETAD) a tool to optimize traffic flow in a heavily loaded network, including allowing traffic that would otherwise have to go through routers to be handled more efficiently by switches. They are often set up to interconnect participants in a particular functional area of an organization such as Human Resources or Finance. When used this way, they can make a contribution to cybersecurity as discussed in Chap. 10 by reducing the risk that information that should be restricted to a defined set of system users or devices could be sent outside that boundary.

9.6 Network Management

Because the performance and reliability of the network is so vital to the operation of most information-intensive systems, the architect must pay special attention to the means provided for network management. Network care and feeding is normally the responsibility of SYSADs or NETADs who have special training and use a variety of tools to perform functions that include:

Network configuration and provisioning – providing and configuring network resources to ensure reliable performance under the loads imposed by all connected systems, devices, and users.

Network monitoring – tracking network traffic to compile data on network use, to spot bottlenecks or other performance issues, and to anticipate or diagnose problems for quick resolution; this often includes logging and analysis to assess network performance and establish a baseline against which trends and anomalous events can be compared.

Network maintenance – a wide range of tasks that includes replacing failed units, restarting network devices after an outage, installing software and hardware upgrades and problem fixes, and adjusting network device parameters to "tune" the network for best performance.

Network security – configuring and applying security mechanisms to protect the network and its connected equipment; this function also routinely includes logging and auditing to detect unauthorized activities such as penetration attempts (see Chap. 10).

A variety of tools are available for network management. Most use SNMP, which uses agents running on network-connected equipment to collect health and status data and detect problems along with device information stored in a Management Information Base (MIB). Starting with SNMP Version 3, the standard includes security mechanisms that significantly reduce the vulnerabilities introduced by earlier versions [3]. Network managers also use methods such as sampling and analysis of packet flows and retransmissions, synthetic monitoring (simulated network transactions), and user monitoring. A modern tool provides the SYSAD/NETAD with a graphical display of network status and performance that greatly assists in spotting problems and making resource management decisions such as allocating bandwidth to individual users and virtual network segments (VLANs).

Most networks also employ a NOS. A separate computer running an NOS can be used to control and manage a large network, but it's more common today for this functionality to be built into a server OS. Even a single-user computer OS today will usually include at least basic NOS capabilities. Essentially, an NOS enables one computer to provide services to another over the network and manages access to shared devices such as printers and storage. Typical functions include:

- Manage user access to the network and data, including security functions such as password management, data encryption, and digital certificates used to authenticate users and files.
- Provide naming and directory services.
- Provide file sharing services with appropriate controls, possibly including mirroring or other automated backup for data protection.
- Support multiple simultaneous users, sometimes called multitasking
- Provide network resource management, fault management, administration, and auditing; a NOS can include capabilities such as server clustering and automated backup.
- Provide interfaces to multiple vendor protocols by stacking them between a transport-independent interface to applications and a logical interface to network device drivers.
- Provide logical network driver interfaces to all network adapters used by the equipment selected to implement the network.
- Provide transport-independent Application Programming Interfaces (APIs) for use by mission applications, again depending on the products comprising the network.
- Provide protocol translation to allow applications written for one transport to use others.

Looking ahead to the next chapter, it's important to note that the NOS may be an important element of a cybersecurity solution. As such, the NOS must be trusted to the appropriate level and tested as part of the overall security assessment supporting a system's authorization to operate. It must also be monitored to detect abnormal network behavior possibly associated with an attack and regularly scanned for vulnerabilities.

9.7 Quality of Service (QoS)

The concept of QoS was originally applied to networking and has since been generalized to any category of services. In the context of networking, QoS is concerned with making the best possible use of available connection capacity to accomplish information exchanges in accordance with a set of rules and priorities. Most operationally critical systems have both process priorities and deadlines, and in these situations networking QoS seeks to:

- Ensure deadlines are satisfied using policy- and rule-based decisions based on established process priorities and ensuring that resources are reserved or can be preempted to service time-urgent communication needs.
- Optimize the use of constrained resources by eliminating congestion points and supporting dynamic resource management to match operational needs.
- Adapt to the dynamic operational environment, including survivability and graceful degradation of networks.

Effective QoS results from careful analysis and design for a specific networking situation. First comes operational and process analysis to establish process requirements, priorities, and timing or deadlines. The results are embodied in a networking design that supports policy-based dynamic management to allocate resources where they are needed at a given point in time for successful system and enterprise operation. This is easier said than done. Policy-Based Network Management (PBNM) has been pursued for some time and has proven harder than expected to implement because the rules governing a sophisticated network supporting a diverse and competing set of operational processes are difficult to define and optimize. In many respects, PBNM resembles policy-based security, described in Chap. 10, and is implemented using a Policy Enforcement Point, Policy Decision Point, Policy Repository, and other elements of a policy structure. In most cases, network QoS requires integrated end-to-end management of internal (LAN) and external (WAN) network resources and must use the network management resources provided by each.

The Internet Engineering Task Force (IETF) [7] defines two basic QoS models. Integrated Services (IntServ), also called the Reservation Model, uses the Resource Reservation Protocol (RSVP) to reserve resources such as bandwidth for designated traffic flows. This seeks to guarantee the reliability and timeliness of message delivery, but it imposes significant penalties on lower priority traffic, has limited scalability, and involves high routing overhead for information dissemination mechanisms such as publish and subscribe (Pub/Sub). The more popular scheme is Differentiated Services (DiffServ) which applies sophisticated packet-level processes for marking, policing, and shaping message traffic to optimize message queue servicing and channel use in accordance with a prioritization scheme. DiffServ offers weaker QoS guarantees but has high scalability and low overhead. A hybrid approach also exists in which an IntServ "tunnel" can be created through a DiffServ "cloud" to create a "fat pipe" for assured delivery of the highest priority messages. Modern network routers have the embedded logic to support all of these approaches. As one example of the many analyses of QoS mechanisms that have been published, Kim and Sebüktekin [8] analyzed an integrated QoS architecture using a bandwidth broker network management scheme and admission control applied to applications entering a given class of service.

9.8 Mobile Ad Hoc Network (MANET)

An Enterprise that includes participation by mobile or intermittent nodes and users has membership and a network configuration that change constantly as participants enter and exit the network and move around within it. Examples include a collection of air or surface vehicles that interact in performing their functions and a space system with satellites that pass into and out of the field of view of various ground stations. This kind of situation has motivated the creation of self-configuring networks that allow participants to enroll and disenroll and that maintain a dynamic picture of the network geometry to support message routing. The generic name for this class of networks is Mobile Ad Hoc Network (MANET) [9, 10].

A MANET is a network composed of mobile network components (routers/hosts), connected via a wireless medium, which move freely and thereby introduce frequent connectivity changes. Two key challenges are to maintain a current network participation table and to dynamically adjust message routing based on the constantly shifting network geometry. Since wireless links tend to have bandwidth limitations, a MANET is typically limited in scalability (number of participants) and in the ability to optimize routing, which requires a good deal of background traffic to locate network nodes and analyze routing paths. MANET research has focused on specialized routing algorithms based on automating the process and achieving the most efficient routing possible within the constraints of the network. In general, distributed routing approaches enhance both scalability and performance compared to centralized or global routing algorithms. With these algorithms, each node discovers its neighbors; performs some kind of optimization based on QoS, throughput, or latency; and adapts to circumstances such as loss of links. The goal is to maximize the delivery of information across the network in the presence of these variables and disturbances.

A MANET or other network involving mobile and transitory participants may use specialized communication channels called data links. For example, these are increasingly important in Air Traffic Control and have been used for years to communicate with satellites and airborne sensor platforms like the E-X. Data links use specialized radios or free-space optical systems to exchange data (and sometimes voice) traffic. The equipment set for a data link includes a terminal, which processes the data and controls the link, and a receiver/transmitter for the signals. A data link standard includes definition of waveform parameters such as frequency and modulation type, a protocol for data link use, and a catalog of standard messages. Data links have been developed with throughputs ranging from a few hundred bits per second to hundreds of megabits per second.

Another mobile networking problem arises when wireless systems are unable to communicate because they use different frequencies, waveforms, message formats, and other parameters. An approach that is growing in popularity is the so-called Software-Defined Radio (SDR). In an SDR, a suite of general purpose radio and signal processing components is configured and controlled by a processor in such a way that they can emulate a variety of systems. Then, for example, the SDR can receive messages or other transmissions using one setup and retransmit the content using a different configuration to bridge the gap between sender and receiver. Perhaps the most widely known standard for defining an SDR is the Software Communications Architecture (SCA) [11]. Such a system can support improved reliability and interoperability among diverse wireless networks without requiring wholesale replacement of legacy equipment by:

- Performing data translation and reformatting
- Implementing QoS rules
- Providing store-and-forward functions between otherwise incompatible data links
- Doing sophisticated routing via available channels to maximize delivery of high-priority traffic

The following general principles are useful to a system architect in dealing with heterogeneous or incompatible data links and other channels:

- Determine the data links to be used, and obtain the most complete information possible on performance, interfaces, physical characteristics, and other features; it's worth remembering that older data link equipment that was developed for limited uses is likely to be both expensive and unreliable.
- Incorporate data links in architecture modeling and performance analysis, including sensitivity analysis to determine the impact of their performance on end-to-end process execution.
- Account for reliability in terms of both the equipment itself and the reliability of the protocol, based on error detection and correction, detection and retransmission of dropped messages, and other features.
- Provide for integration of non-IP data link traffic with LAN and WAN architectures, which usually requires a specialized processor or software application; the objective is to make a heterogeneous network of IP and non-IP channels transparent to the end users and software applications of the system or enterprise.

9.9 The Future of Networking

The proliferation of networking has given rise to several significant trends in information technology. One is the rise of distributed computing, which refers to a variety of schemes in which computers at various locations interact over a network to function as a single computational engine. A related trend is the explosive growth in cloud computing, in which a set of shared resources is provided by a cloud services provider (CSP) and used by multiple clients. These topics are explored in Chap. 14.

A potentially more profound future trend involves the prospect that the familiar Internet, based primarily on the TCP/IP protocols, which has been in use for decades, may give way in whole or in part to a new model. Research into fundamentally new networking schemes is driven by the known problems of the current

Internet in handling massive growth in information exchange volumes plus serious security deficiencies (Chap. 10). As one example, Edens and Scott [12] describe a Content-Centric Network (CCN) aimed at dramatically improving the performance and reliability of the global network using a new traffic routing concept that is based on the way information is organized within the network rather than on IP addresses of network hosts. The future global network is likely to use diverse formatting and routing to keep pace with growing demand for digital communications, improved security, and collaborative computing.

9.10 Summary

This chapter has focused on some aspects of networking that impact the development, analysis, and optimization of architectures. An architecture team will normally need networking expertise, especially when a combination of local and wide area networks are involved, and must be integrated into a seamless connectivity solution. In critical and time-sensitive systems and enterprises, the QoS attributes of the network are frequently prominent. Similarly, the ability of a MANET to provide reliable connectivity in the face of dynamic network topology and participation is especially important as networking among mobile users proliferates. As the world becomes increasingly connected and network-enabled, and as the operational payoffs of networked operations and distributed computing become more compelling, the importance of networking as a key element of system and enterprise architectures will only increase.

Exercises

1. Sketch a LAN for a neighborhood branch of a bank. Include bank teller workstations, a central file and applications server, connectivity to a WAN, and connected equipment such as an Automatic Teller Machine (ATM) and devices for counting coins and dispensing paper money. Include the primary protocols and standards that are applied to the network.
2. Assuming a decision is made to use a wireless LAN for the network in Question 1, list some factors involved in arriving at a network with adequate performance, security, and long-term growth and evolution.
3. For the network in Question 1, write down ten or more requirements that are clear and verifiable and can be incorporated in a requirements baseline for the bank information system.
4. How do current WiFi security techniques improve on older approaches?
5. List some typical areas of responsibility for a Network Administrator.
6. For the network in Question 1, list the functions of a Network Operating System (NOS).
7. For the network in Question 1, identify the primary Quality of Service concerns that the network design must satisfy.

Student Projects

Students should add networking detail to their architecture models reflecting the nature of their specific systems. This can be done by adding detail to structural and behavioral diagrams in the LV and PV, e.g., to outline product specifications for network devices and media. A start should be made here on defining the networking solution needed by the enterprise in which each student's system participates, which is further pursued in conjunction with Chap. 13.

References

1. Dordal PL (2017) An introduction to computer networks, rel 1.9.0. http://intronetworks.cs.luc.edu/current/ComputerNetworks.pdf. Accessed 30 May 2017
2. JAVVIN (2005) Network protocols handbook, 2nd edn. http://bkarak.wizhut.com/www/lectures/networks-07/NetworkProtocolsHandbook.pdf. Accessed 30 May 2017
3. Digital Ocean (2014) An introduction to SNMP (Simple Network Management Protocol). https://www.digitalocean.com/community/tutorials/an-introduction-to-snmp-simple-network-management-protocol. Accessed 30 May 2017
4. OASIS (2017) Advanced Message Queuing Protocol (AMQP). https://www.amqp.org/about/what. Accessed 30 May 2017
5. IEEE Standards Assoc (2017) IEEE 802 Standards. http://standards.ieee.org/getieee802/portfolio.html. Accessed 30 May 2017
6. WiFi Alliance (2017) WiFi specifications. http://www.wi-fi.org/. Accessed 30 May 2017
7. IETF (2009) Intserv and Diffserve. https://datatracker.ietf.org/wg/intserv/documents/ Accessed 30 May 2017
8. Kim B, Sebüktekin I (2002) An Integrated IP QOS Architecture – Performance. Paper presented at MILCOM 2002, Anaheim, 7–10 Oct 2002
9. Cheng T, et al. (2002) Ad Hoc mobility protocol suite (AMPS) for JTRS radios. Paper presented at the SDR 02 technical conference and product exposition, San Diego, 7–10 Oct 2002
10. Young K, et al. (2003) Ad Hoc Mobility Protocol Suite for the MOSAIC ATD. Paper presented at MILCOM 03, Boston, 8–11 Oct 2003
11. Joint Tactical Networking Center (2017) Software Communications Architecture, ver 4.1. http://www.public.navy.mil/jtnc/sca/Pages/sca1.aspx. Accessed 30 May 2017
12. Edens G, Scott G (2017) The packet protector. IEEE Spectr 54(4):42–48

Chapter 10
Protecting Information with Cybersecurity

10.1 The Cybersecurity Challenge

This chapter provides a brief, high-level summary of a very large and complicated topic: protecting information-intensive systems against attack, compromise, corruption, theft, unauthorized use, and other malicious acts. The overall term for this is cybersecurity, which was referred to in the past as Information Assurance.[1] Scarcely a day passes without a major news report on computer crime or other incident, from things as basic as defacing Web sites to identity theft, crippling virus attacks, diverted bank accounts, ransomware, and compromise of sensitive operational data and intellectual property. Systems with large numbers of users, geographically distributed locations, and networked access are especially vulnerable. Security professionals, many of whom have specialized training and hold credentials like Certified Information Systems Security Professional (CISSP), fight a never-ending battle to defeat criminals, hackers, terrorists, foreign intelligence services, and deranged people who derive perverse satisfaction from releasing viruses, Trojans, worms, and other "malware." Even a system with robust safeguards against external attack may be vulnerable to the "insider threat" from personnel with authorized access who have become disaffected, have taken money from a criminal or hostile agency, or are simply poorly trained and careless. Any organization or enterprise that relies on information processing is a potential target, and the reality is that most systems of any size or significance have already been penetrated. Figure 10.1 suggests the range of threats confronting secure systems.

Appendix G tabulates some common attack methods and possible mitigations. A few examples from recent reports on cyberattacks and data breaches will serve to highlight the challenge [1].

---

[1] We will not use the obvious CS acronym since this is so widely taken to mean computer science.

J. M. Borky, T. H. Bradley, *Effective Model-Based Systems Engineering*,
https://doi.org/10.1007/978-3-319-95669-5_10

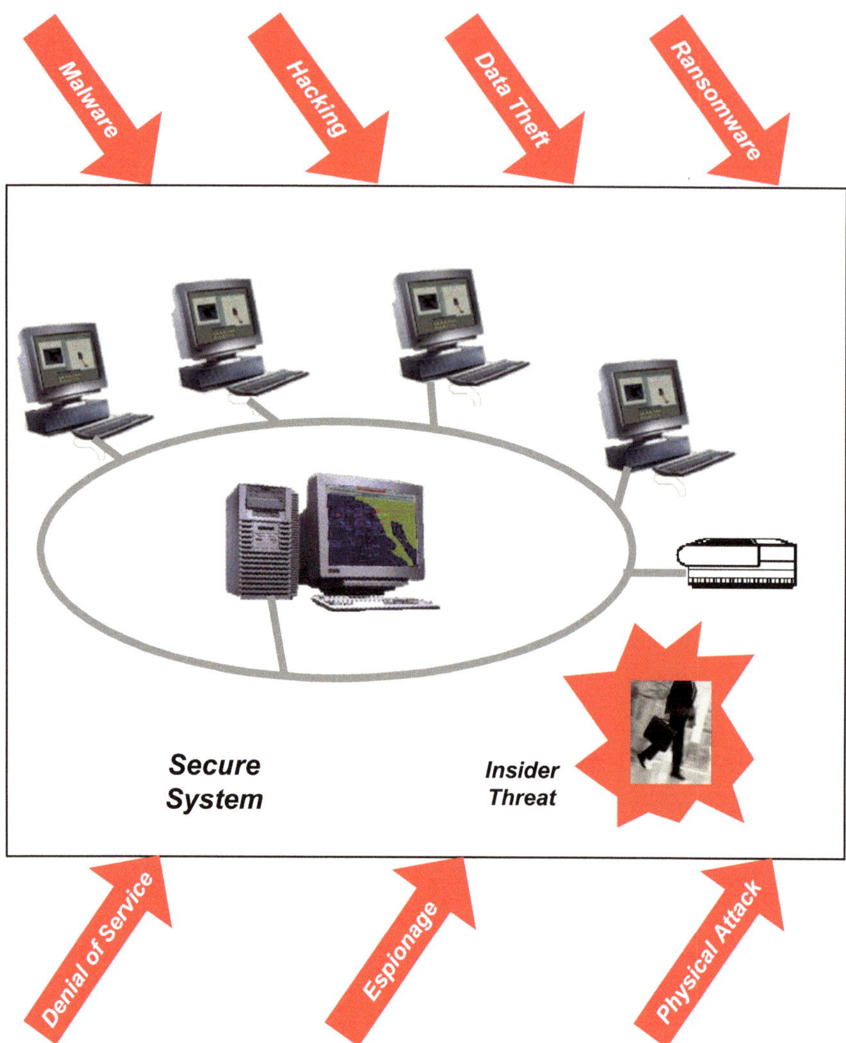

Fig. 10.1 Multiple threats to secure systems

- Common estimates are that the number of new threats is doubling every year, and the global cost of data breaches is measured in trillions of dollars per year.
- Many, perhaps most, common attacks could be defeated with available and affordable safeguards that victims simply fail to deploy; as an example, the average time from publication of a software patch or other mitigation to close a newly discovered vulnerability to the actual installation of the protective measure by both private and public sector organizations is months or even years, when it should be hours or days.

- The top ten system vulnerabilities account for approximately 85% of data breaches; some have been around for as long as several years but continue to be exploited.
- Roughly 80% of successful attacks originate with external threat agents, but the majority also involve either deliberate or accidental actions by "insiders," i.e., members or employees of the victimized organization; a common example is an attack that starts with "phishing" to trick an insider into revealing information or downloading malware, giving the attacker access to the system.
- A cybersecurity provider recently reported that ten out of ten mobile devices and applications used for payments at stores, theaters, hotels, etc. had major security weaknesses such as failing to encrypt Social Security Numbers.
- Relatively new and more sophisticated threats are proliferating; one such is a "ransomware" attack in which a criminal attacker encrypts a victim's databases and demands payment to restore them.
- Online security across the board is so bad that most successful hackers don't actually need to employ hacking methods; they simply enter the victim's network through unprotected public interfaces.

Just as a perfectly safe airplane is one with so much strength and redundancy that it's too heavy to fly, a perfectly secure information system would simply deny access to anyone. Cybersecurity is therefore a balancing act involving an adequate level of protection against known or postulated threats while still allowing systems and their users to carry out their legitimate functions and accomplish their business objectives or operational missions. For example, a classic dilemma facing a security architect today is the unanticipated user, an external party who is not known to a system a priori, and may very well not even be physically located at the system, but who nevertheless has permissions granted by a higher authority to access functions and data. The security design must then ensure that such legitimate users are granted access, while imposters are blocked with acceptably low error rates.

Since this is a book about system architecture and architecture-centric systems engineering, the emphasis of this chapter is on ways to implement adequate safeguards against cyberattacks within the context of an overall balanced solution to customer needs. The discussion is primarily about architectural approaches, especially layered defense design patterns for security, as well as selection of security controls (safeguards or countermeasures) and the products that implement them. We necessarily pay less attention to procedural aspects of security such as enforcing strong passwords and providing security awareness training, although these are absolutely essential elements of an effective cybersecurity posture.

Given that absolute security is impossible, the approach taken today by security policy-makers and architects is based on risk management. A typical security architecture begins with the most thorough analysis possible of the sources and methods of system attack. The security architect then follows an orderly process of selecting and applying protective measures while ensuring that a system can still accomplish its purpose with acceptable impacts on performance, cost, reliability, and long-term evolution. Security design often uses products with defined assurance levels and employs design patterns, security protocols, and standards to achieve an acceptable level of security risk. This chapter describes a typical security architecture process and introduces some terminology and design approaches. The objective is to enable a general system architect to appreciate the dimensions of the security challenge, to interact more effectively with security specialists, and to ensure compliance with security policies. There are frequent references to Department of Defense (DoD) security policies and practices both because of the importance of protecting classified information and because much of the security technology and many standards and implementation methods, which are used throughout information technology (IT) systems, derive from the experience and investments of the Defense community.

10.2 Basic Concepts

10.2.1 Elements of an Attack

Increasingly sophisticated attacks on information systems are often characterized as cybercrime, cyber-espionage, and even cyberwarfare. Figure 10.2 is a graphical portrayal of the main elements of a cyberattack and introduces some standard terms. An *Asset* is any information, process, or other system content that requires protection. Assets are potentially at risk if the system has *Vulnerabilities* that an attacker can exploit. Collectively, these vulnerabilities constitute the system's *Attack Surface*, and a primary goal of secure architecture is to minimize this. An asset such as a database typically has *Attributes* that are the specific items sought by an attacker. A *Threat* is created by a *Threat Agent* who seeks to exploit a vulnerability to steal, corrupt, delete, or otherwise harm an asset. Then if the system has safeguards, properly called *Security Controls*, that adequately mitigate the vulnerability, the asset is protected; otherwise, there can be an *Exposure* or data breach. The structure in Fig. 10.2 is the basis for the Vocabulary for Event Reporting and Incident Sharing (VERIS) [2] which asks "What Threat Actor took what Action on what Asset to compromise what Attribute," referred to as the 4 As, and is widely used in reporting and compiling information on cybersecurity events.

Published work on information security, of which a paper by Whitmore [3] is a typical example, generally highlights the following as the principal factors in the success or failure of a system that is trusted to handle sensitive information:

- Clarity and completeness of requirements, including validation by stakeholders, especially the ultimate system users
- Trustworthiness of the components and policies used in system implementation
- Satisfaction of requirements by the system design and implementation
- Correct operation and maintenance of the system to preserve security characteristics

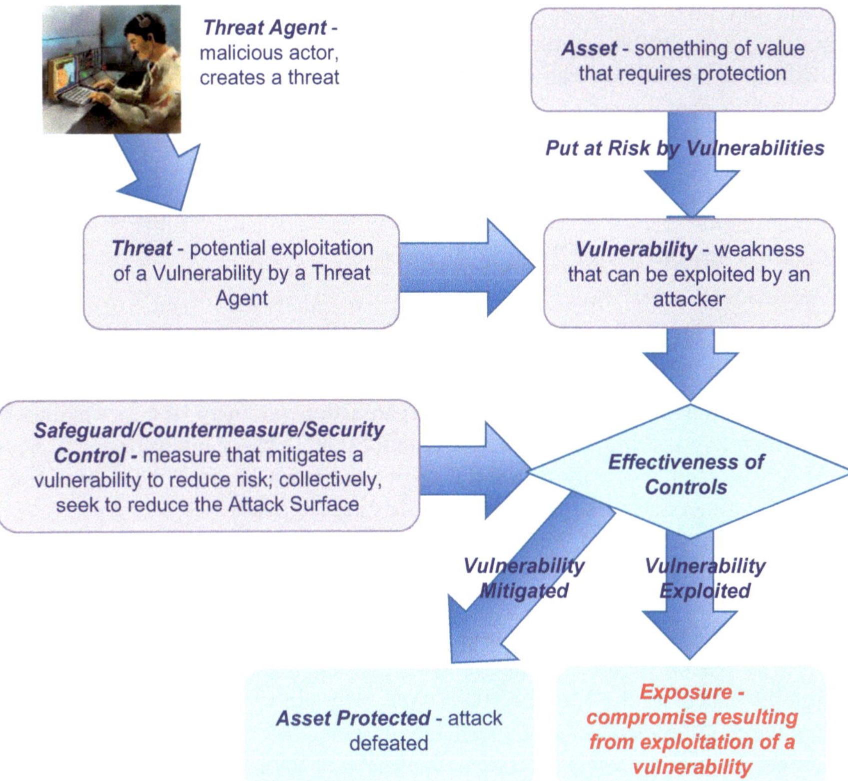

Fig. 10.2 Elements of a cyberattack

- Understanding of the environment in which the system operates, including the threats that security mechanisms must counter

The remainder of this chapter explores these aspects of system security and discusses some of the current approaches in use by security architects and engineers.

Although classified information is most often associated with military systems, any organization that has sensitive data needing protection should apply a consistent scheme for sorting that information into levels of importance or sensitivity and clearly labeling it so that appropriate safeguards can be applied. The US DoD defines classification levels, such as For Official Use Only (FOUO, also called Controlled Unclassified Information), Confidential, Secret, and Top Secret, primarily on the basis of the damage to national security that would result from information compromise. Intelligence agencies typically require more stringent control and protection for their information, which is often achieved through "compartments" that have very strict access requirements and higher levels of physical, administrative, and technical security measures. Civil agencies and commercial enterprises typically define sensitivity levels, e.g., Restricted, Confidential, Private, etc., for information such as intellectual property, personal and financial data, strategic planning, and compliance with laws governing information protection. A good data classification system concentrates the most stringent security controls on the most sensitive information, especially when it is impractical to give everything the highest level of protection.

10.2.2 Categories of Threat Agents

The threat agents in Fig. 10.2 come in a wide variety of forms. The following is a representative tabulation.

- *Insiders* – people within the targeted organization who may be either malicious (deliberately seeking to do damage, commit theft, etc.) or inadvertent (careless, poorly trained, etc.); these are the most dangerous because they are already inside system defenses and have access to targeted assets.
- *Hackers, Thrill Seekers, and Individual Criminals* – individuals or small groups whose motivation may range from ideology to financial gain to an adrenaline rush from cracking a system.
- *Organized Crime* – organizations looking to compromise systems for purposes of theft, blackmail, data ransom, or other criminal objectives; stolen data is commonly traded on the Dark or Black Web, a group of clandestine peer-to-peer networks ("darknets") using the public Internet but with measures to control access and prevent users from being identified or traced.
- *Terrorists* – a variety of criminal organization that, in addition to cybercrime, may seek to compromise target systems as part of a political, ideological, or simply psychopathological campaign.
- *Advanced Persistent Threat (APT)* – the most sophisticated category of attackers, often state-sponsored for purposes of military or commercial espionage; APTs commonly have extensive financial and technical resources to execute elaborate campaigns extending over long periods, employing mixtures of tactics, and seeking to thoroughly infiltrate, and even take control of, target systems.

10.2.3 Fundamental Security Concepts

System, network, and enterprise security is customarily discussed in terms of a set of fundamental concepts. These include:

- *Confidentiality* – the ability to preserve secrecy by preventing unauthorized access to sensitive data, whether stored ("at rest"), being communicated ("in transit"), being processed ("in use"), or being created (e.g., by being intercepted by an attacker's tool such as a keyboard sniffer)

- *Integrity* – the ability to preserve content by preventing an unauthorized party from modifying, corrupting, inserting, deleting, or duplicating data
- *Availability* – the ability to ensure timely access to protected data and functions by authorized recipients
- *Authentication/Identity* – the ability to prove with adequate certainty that a party is who the party claims to be and has an identity that is known to the system or enterprise for purposes such as Access Control
- *Authorization/Access Control* – the ability to restrict access to sensitive data to authenticated recipients who possess the necessary access permissions and need-to-know
- *Non-Repudiation* – the ability to prove that a given party took part in an information transaction despite that party's attempts to deny involvement
- *Audit* – the ability to detect, inspect, record, analyze, and report events associated with security mechanisms, which may be important in determining that a compromise or other unauthorized action has occurred, verifying that security procedures have been followed, and detecting events or trends that may reveal hostile activity

Confidentiality, Integrity, and Availability are the primary cybersecurity concerns and are, with a certain amount of irony, referred to as CIA. An additional term and concept that is increasingly stressed is *Resilience*, which is the overarching characteristic of a secure system that implements all of the listed concepts, is able to defeat both existing and emerging threats, can maintain essential functions during and after an attack, and can maintain an acceptable level of risk as the threat environment evolves.

10.2.4 Elements of Resilient Cybersecurity

To achieve a cyber-resilient system architecture and implementation while maintaining acceptable system performance, cost, and reliability, the security architect employs proven designs, mechanisms, products, and procedures. All too often, organizations and managers assume that cybersecurity begins and ends with technical safeguards such as firewalls. In reality, an effective security solution requires all three of the elements sketched in Fig. 10.3. They include:

- Personnel who are trained, motivated, and empowered to perform their duties in a secure and reliable fashion, including all levels of management, system users, and system support staff
- Processes and procedures that implement a security policy and maintain the effectiveness of safeguards
- Technologies that implement security controls and that evolve to keep pace with an ever-changing threat environment

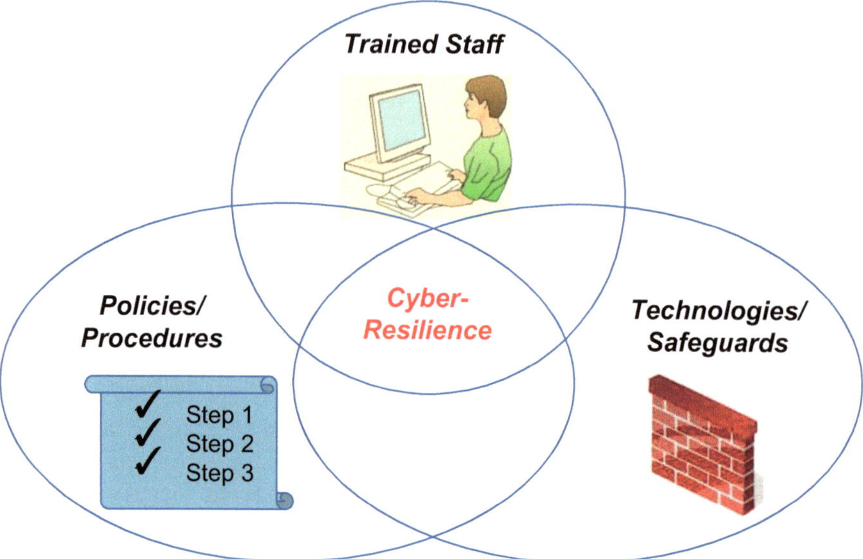

Fig. 10.3 Essential elements of an effective cybersecurity solution

A second threefold taxonomy is useful in describing the security controls or safeguards employed to protect a system. Securing a system that has human involvement normally requires a mix of technical, physical, and procedural security measures.

- *Technical measures* – these can be quite diverse and may include:
  - Firewalls and other protective devices at the system boundary to protect against unauthorized information flows in or out of the system
  - Intrusion detection devices to detect, log, and analyze unauthorized attempts to access a system, often resulting in alarms so that timely defensive measures can be taken
  - Public Key Infrastructure (PKI)
  - Auditing of activity logs, system configuration changes, and other events to detect and respond to suspicious or prohibited actions
  - Security testing, including simulated hostile attacks, to verify the robustness of information protection and identify deficiencies that must be corrected
  - Policy servers that automate the application of rules governing the operation of security mechanisms
  - Access control mechanisms to enforce user privileges and restrict unauthorized use of resources and data
  - Lockdown ("hardening") of servers and other computers by disabling any operating system functions, ports, utilities, or other capabilities that are not absolutely necessary and that may create security vulnerabilities

- *Physical measures* – these seek to place sensitive information or other protected resources behind barriers ranging from facility access controls such as badges, door locks, and guards to locked enclosures and backup power and air conditioning. These can range from property fences and controlled access points to alarms and locked equipment enclosures.
- *Procedural measures* – these deal with secure operations and practices aimed at maintaining the effectiveness of security controls and eliminating vulnerabilities caused by human error. A typical example is a requirement to erase sensitive software and data from a computer at the completion of a work period or task. Others include automatic locking of computers after a period of inactivity and regular, timely system scans to detect suspicious behavior or altered system configurations. Such procedures may be documented in an Automated Information System (AIS) directive. A very important category of procedural security involves training system users to practice good security "hygiene" such as frequently changing passwords and detecting and defeating "phishing" attacks that try to trick a user into disclosing sensitive information or loading malicious software.

Any protective measure is likely to eventually be compromised or circumvented. As a result, traditional defensive configurations that tend to be static and to protect only the system boundary are evolving toward more dynamic and sophisticated approaches. Layered defense approaches such as Defense-in-Depth (DiD) and Zero-Trust Architecture, described below, are an important advance over previous security implementations. The basic idea is that an attacker, to reach sensitive system content, must penetrate a series of barriers, which can be individually managed and updated to deal with newly detected threats and methods of attack. Another important cybersecurity trend involves anti-malware tools that go beyond detecting a threat based on comparing a message or file to a library of threat signatures and seek to analyze a message and any attached files to determine its functions, spot the presence of malicious code or Web sites, and block an attempted attack.

Yet another evolving security approach seeks to be proactive and is based on continuous, even real-time, monitoring of user behavior and cybersecurity events coupled with dynamic responses to minimize or contain attempted intrusions [3, 4]. This could involve deploying agents (see Chap. 14) at various locations within an information enterprise to measure events such as password change attempts (which may indicate a password breaking attack), failed log-on attempts, blocked data packets, or data objects failing integrity checks. Agent reports can alert operational personnel to an abnormal situation and can trigger automated responses such as blocking suspected network addresses. Logging and analysis of messages and other traffic, particularly amounts, combinations, or timing of data downloads that have not been previously seen, can reveal patterns that may indicate the early stages of a cyberattack. A number of commercial products based on inspection and sophisticated statistical analysis of network activity are now available and have proved effective. Continuous testing of security controls and application software is essential to detect and mitigate vulnerabilities, especially new ones from the ever-evolving threat environment.

10.2.5 Cybersecurity Domains

Security professionals typically organize the cybersecurity discipline using the domains of the CISSP Common Body of Knowledge (CBK). The following paragraphs briefly summarize these[2] and give the equivalent of an executive summary of the subject.

- *Access Control* – measures to ensure only authorized persons or other entities ("subjects") can get possession of protected content ("objects"). They are based on file permissions, program permissions, and data rights that are tailored to each individual system user's need for such content and embedded in a user account. Access control includes the process by which individuals receive authorizations to access particular system content and authentication of the identity of a subject requesting access to an object.
- *Telecommunications and Network Security* – measures to ensure cybersecurity when protected objects are transmitted. This includes network architecture and design, trusted network components, secure channels, and protection against network attacks.
- *Information Security Governance and Risk Management* – security policy, guidance, and direction that a system must follow to attain an Authority to Operate (ATO) with sensitive data. This commonly dictates security constraints and procedures that must be followed, security functions vs. operational or business goals and missions, policy compliance and enforcement, phases of the information life cycle, and governance of third parties such as component vendors, personnel security, education and training, metrics, and resources, especially budgets and skilled personnel.
- *Secure Software Development Life Cycle (SSDLC)* – building cybersecurity into system and application software from the outset. This includes determining information protection needs, establishing security requirements, developing secure software architecture and design, software security testing, and assessing protection effectiveness. Software security controls must be applied and tested in a development environment that is isolated from production systems.
- *Cryptography* – converting plaintext into unreadable ciphertext using complex algorithms. This domain includes tailored applications of cryptography, the life cycle of cryptographic materials such as encryption keys, basic and advanced cryptographic concepts, encryption algorithms ("ciphers"), public and private keys, digital signatures, non-repudiation (generally using digital signatures), attack methods, cryptography for network security, cryptography for secure application software, PKI, issues with digital certificates, and information hiding (what and when to hide). The protection of the crypto keys is more important than the strength of the cryptography itself; hence a good Key Management Plan means more than the strength of a cipher.

[2]Actually, this discussion uses the CBK Domain structure that precedes the most recent revision because it is much more logical and useful for tutorial purposes.

- *Security Architecture and Design* – embedded features and functions that implement security controls and eliminate or mitigate vulnerabilities. This includes models and concepts, model-based evaluation, security capabilities, vulnerabilities, countermeasures, selection of trusted components, and mechanisms to establish and maintain the required level of system security over time.
- *Security Operations* – ongoing activities to enforce policies and procedures, detect and prevent or mitigate attacks, and maintain a system security posture. This includes operations security, resource protection, incident response, attack prevention and response, management of software vulnerabilities and patches to mitigate them, change and configuration management, and procedures to preserve system security resilience and fault tolerance.
- *Business Continuity and Disaster Recovery Planning* – measures to minimize the organizational and mission impacts of cyberattacks and natural disasters. This includes requirements, impact analysis, backup and recovery strategies, disaster recovery, and testing of plans for continuity and recovery.
- *Legal, Regulations, Investigations, and Compliance* – legal issues, ethics, investigations and evidence, forensics, compliance procedures, contracts, and procurements.
- *Physical Security* – measures to establish and maintain a secure environment for sensitive resources and processes. This includes site and facility design, perimeter security, internal security, facility security, equipment security, privacy, and safety.

10.2.6 Cybersecurity Foundations

An effective, affordable, and resilient cybersecurity solution begins with two fundamental principles.

Security Policy and Requirements The first essential is a policy that concisely describes the goals assigned to the security function of an organization and how those goals will be met with available or planned resources, together with specific, verifiable security requirements. A good starting point is a Security System Concept of Operations (SECOPS), which complements an overall Concept of Operations (CONOPS) by providing a focused definition of security requirements levied on the components or functions of a system with rationale for each. For example, the SECOPS would describe the required level of access control and what the components responsible for user account management, privilege control, and user authentication must do. A SECOPS often includes "Abuse Cases," which are analogous to normal Use Cases except that they describe behaviors associated with a cyberattack.

The SECOPS may incorporate or be the basis for a Security Policy Document (SPD). A security policy should be stated in a form that is useful to security stakeholders and includes an overall security policy and governance strategy. The policy is typically elaborated in a Security Plan (SP) that spells out specific policy for all system elements, defines an acceptable risk threshold, and gives an overview of system security requirements and the existing or planned controls to satisfy those requirements. Special Publication SP 800-18, *Guide for Developing Security Plans for Federal Information Systems*, Rev. 1, from the National Institute of Standards and Technology (NIST) [5] has general guidance applicable to security planning in both public and private sectors.

Ultimately, a secure system needs clear, unambiguous, effective, and verifiable requirements for security features and functions as part of an overall requirements baseline to support development, acquisition, integration, test, operational support, and other activities. Federal Information Processing System Publication 200 (FIPS 200) [6] lays out minimum acceptable requirements in 17 areas and can be used as a template for developing security requirements. Figure 10.4 suggests the flowdown from overall organizational policy and goals to focused security policy, a SECOPS, and ultimately security requirements and implementation guidance.

Secure System Life Cycle The best and least expensive cybersecurity solution results when security is "baked in" to a system design from the outset. A secure life cycle starts with making security requirements a core part of the overall system requirements baseline and continues through all phases of development, integration, test, production, fielding, and operational support. All too often, an attempt is made to graft cybersecurity safeguards onto an existing system, but this is virtually guaranteed to be much harder to do and to yield a less resilient and more costly result. A secure life cycle is particularly important for application software and is further discussed under that topic later in this chapter.

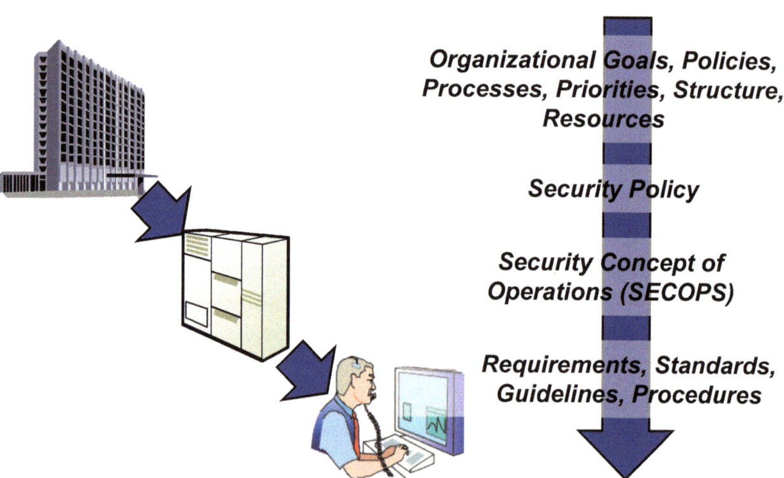

Fig. 10.4 Effective cybersecurity originates in overall organizational policy and flows down to security requirements and implementation

These principles may seem obvious but they are, in fact, widely ignored. A recent analysis of the cybersecurity of US Air Force systems uncovered two pervasive trends in risk management [7]:

- Failing to adequately capture the impact to operational missions, a violation of the first basic principle
- Implementing cybersecurity as an add-on to existing systems, versus designing it in, a violation of the second

The report goes on to recommend better definition of cybersecurity goals; clear roles and responsibilities with adequate authority to act; more comprehensive and better prioritized security controls across systems and enterprises; improved access to cybersecurity expertise; continuous assessment and reporting of security status; better threat data; and, perhaps most significantly, holding individuals responsible for violation of cybersecurity policies. None of this is new or surprising. Similar findings come up repeatedly in reviews of nominally secure systems in Government, industry, and elsewhere, especially after cyberattacks. Clearly, the security community has a long way to go in getting system owners to take the cyber threat seriously and to deal with it effectively.

It should be clear from this preliminary survey of cybersecurity that this is a multidimensional problem and a process that continues for the entire operational lifetime of a system or enterprise. Just a few of the matters requiring diligent attention and adequate resources are promptly installing software patches, managing user accounts, acting quickly to detect and mitigate new vulnerabilities, maintaining the security awareness and defensive training of personnel, validating the trust level of new system components before installation, and monitoring system resources and activities to detect and defeat hostile actions. A good basic agenda for establishing and maintaining cyber resilience is spelled out in the Critical Security Control (CSC) list maintained by the Council on CyberSecurity [8]. The remainder of this chapter explores the aspects of cybersecurity that are most prominent in system architecture and in MBSE.

10.3 Cybersecurity Risk Management

Risk management is a core activity of Systems Engineering, based on assessing the probability of occurrence of potential adverse events or conditions and the consequences of such occurrences. A risk management process usually consists of:

- Risk identification
- Risk assessment in terms of probability of occurrence and consequence of occurrence
- Identification of risks requiring mitigation
- Planning and budgeting for risk reduction activities

- Tracking the progress of risk mitigation until a risk is retired or a decision is made to accept it

In dealing with cybersecurity, risk management is focused on protected assets, threats, vulnerabilities, and risk mitigation using security controls. It begins with risk identification and analysis, which then supports selection of security controls to achieve an acceptable risk level. This, in turn, is the basis for securing approval to operate the system. Accordingly, risk analysis is an essential component of defining security policy and requirements. The goal is to apply security controls intelligently to mitigate the most important risks while preserving an operationally effective, affordable, and supportable system. Risk management continues over the life of a system to deal with changes to the system itself and with the emergence of new vulnerabilities and threats.

Cybersecurity risk management deals with four basic elements.

- *Risk* – a measure of the likelihood of occurrence of an adverse event such as loss of sensitive information and of the potential resulting consequences to a system or organization that is the target of an attack. This is conceptually equivalent to the probability and consequence of occurrence used in general risk management.
- *Threat* – the potential danger of a compromise that is associated with a vulnerability. The term is commonly associated with an entity (Threat Agent), event, or circumstance that has the potential to adversely affect an organization/target by exploiting a vulnerability.
- *Vulnerability* – a flaw in a target that can potentially be exploited by a threat. This is often due to the absence or weakness of a security control or countermeasure.
- *Exploitation* – an event in which a threat agent leverages a vulnerability to carry out an attack or otherwise induce adverse consequences for a target. An exploit typically results in an Exposure/Compromise/Data Breach, i.e., an instance of the target being exposed to loss.

The basic risk relationship can be expressed by considering an individual risk as a function of a specific vulnerability and a threat that arises from the possibility that the vulnerability may be discovered and exploited. We can state this simply as:

$$\text{Risk}_n = f\left(\text{Threat}_n, \text{Vulnerability}_n\right) \tag{10.1}$$

where the nth risk, Risk_n, results from the nth threat associated with the nth vulnerability.

Cybersecurity risks are associated with information assets (data, systems, processes, etc.) whose value is such that they require protection. The sum of vulnerabilities (or, equivalently, of potential attack vectors) constitutes an Attack Surface. Managing cyber risks then comes down to minimizing the Attack Surface. This is made difficult by the reality of a diverse, growing threat environment; a full listing of known vulnerabilities and attack methods would fill a small book, and a complete description would fill a large one.

The first step in cybersecurity risk management is risk identification (RI). This begins with asset valuation, meaning a consistent approach to determine both quantitative (cost) and qualitative (relative importance) values. Factors include costs to acquire or develop an asset, initial and recurring data maintenance costs, importance of assets to the organization and to others (including criminals), and public value in terms of things like intellectual property. Next comes threat analysis, involving identification and definition of known and potential threats, consequences of an exploitation, the estimated frequency of threat events, and the probability that a potential threat will materialize. The third element of RI is vulnerability assessment. This can draw on a variety of sources, including published lists of vulnerabilities, a history of security events, and results of vulnerability testing.

Now comes risk analysis (RA), which can be quantitative or qualitative. Quantitative RA has advantages in devising an optimum cybersecurity solution but requires more information and more effort. It attempts to assign a numerical value or cost to assets and threats to support cost/benefit analysis and more clear and concise characterization of particular risks. NIST SP 800-30, *Risk Management Guide for IT Systems* [9], defines a nine-step risk assessment process. The analysis begins with the asset valuations and assessments of threats and vulnerabilities from RI. For each combination of an asset, a vulnerability, and a threat, an Annualized Loss Expectancy (ALE) is calculated as:

$$\text{ALE} = \text{SLE} \times \text{ARO} \tag{10.2}$$

where SLE is the Single Loss Expectancy and ARO is the Annualized Rate of Occurrence, stated as the expected number of occurrences of a specific threat in a 12-month period. SLE is determined as:

$$\text{SLE} = \text{Asset Value} \times \text{EF} \tag{10.3}$$

where EF is the Exposure Factor, consisting of the fraction of total Asset Value expected to be lost in a single security event. Obviously, many of the values used in these calculations are estimates, although careful tracking of published security data and analysis of an organization's security history and of recent event logs can make the estimates more accurate.

The next step is to evaluate the feasibility and cost of possible security controls. It can be difficult to arrive at the true *total* cost of a safeguard. In addition to the cost of developing or purchasing a product that implements a control, there are likely to be costs associated with planning, implementation, and maintenance; operating costs (including skilled personnel); and costs associated with economic impacts such as effects on productivity. Once a reasonable total cost is obtained, the value of a given control against a given threat can be computed as:

$$\begin{aligned}\text{Control Valuc} = \text{ALE}(\text{without control}) - \text{ALE}(\text{with control}) \\ -\text{Total Cost of Control}\end{aligned} \tag{10.4}$$

Qualitative RA is a heuristic method (Delphi technique) based on scenarios. It does not attach absolute numerical costs, but typically uses scoring by knowledgeable people on a 1–5 or 1–10 scale of threat severity, potential loss, effectiveness of protective measures, etc. These subjective assessments can be compiled to get some insight into the relative importance of risks and the relative cost and effectiveness of countermeasures. When quantitative RA is not feasible, qualitative analysis is better than nothing.

With the results of RI and RA in hand, risk management moves to decisions about how each risk should be treated. There are four basic options, representing the available safeguards:

- *Risk Reduction* – application of security controls or countermeasures to mitigate vulnerabilities and thereby reduce risk
- *Risk Assignment or Transference* – movement of a risk to a third party, e.g., by using insurance
- *Risk Avoidance* – action to eliminate the situation that gives rise to a risk, e.g., by disposing of assets or modifying vulnerable processes
- *Risk Acceptance* – deciding to live with a risk when the cost of mitigating it exceeds the potential consequences of a compromise

Several factors may bear on risk treatment decisions. In addition to the cost/benefit associated with implementing a security control, which may indicate accepting the risk or pursuing risk avoidance, some of these considerations are:

- *Legal Liability* – becomes important when there are statutory requirements or civil liabilities associated with failure to implement security controls.
- *Standards* – publications from NIST and other standards bodies offer guidance on optimum risk treatments.
- *Operational Impacts* – may be a factor if failure to mitigate a risk has potential operational impacts such as added cost for security surveillance, reduced productivity, or the need to implement work-arounds in affected processes.
- *Technical Factors* – a situation in which a proposed security control to deal with one risk actually introduces new vulnerabilities and thus leads to other collateral risks.

From all these considerations, the last step is to decide which risk treatments to apply based on an integrated assessment of the most effective protection strategy within the constraints of available resources and system impacts. Another way to say this is that we seek to reduce a system's Attack Surface to the feasible minimum. The remaining *residual risk*, which might be quantified by summing the ALEs of the untreated risks or simply described in qualitative terms, must correspond to acceptable risk level as defined in policy and approved by senior management of the organization. Security policy and requirements are ultimately based on achieving this level, and it is essential that responsible authorities in the organization approve it as being acceptable. This is then the basis for granting or withholding permission to operate the system with sensitive data and processes. RI, RA, security controls, and assessment of residual risk should be repeated as necessary in the face of the constantly changing threat environment, specifically whenever new threats, new vulnerabilities, and changes to system content and processes make the previous analysis no longer valid.

A number of systematic threat modeling approaches are available. One is the Microsoft Threat Modeling Process, which is supported by a free tool and is complemented by threat analysis tools [10]. Once a set of security objectives for a system has been established, this process goes through a sequence of characterizing and decomposing applications to discover trust boundaries, data flows, entry and exit points, and external dependencies that potentially introduce vulnerabilities. These are then analyzed in terms of known threats and attack vectors, using methods such as attack tree analysis, resulting in a prioritized threat list. Finally, actual and potential vulnerabilities are identified, and the analysis results are fed back to the application characterization stage to support design changes, addition of countermeasures, or other steps to mitigate the threats. Other threat modeling methodologies include:

- *STRIDE* – this model uses six basic threat categories: Spoofing Identity, Tampering with Data, Repudiation, Information Disclosure, Denial of Service, and Elevation of Privilege; it then identifies mitigation strategies for each.
- *DREAD* – this approach scores each identified threat in terms of Damage Potential, Reproducibility, Exploitability, Affected Users, and Discoverability from 1 to 10 and then computes overall risk by dividing the sum of these scores by 5.

10.4 Cybersecurity Guidance and Resources

Organizations that develop, own, and authorize secure information systems commonly establish policies and other directives governing these systems and their operations. The flowdown of security policy from overall organizational policies and objectives was described earlier. These policies often draw upon a growing body of standards published by Government, industry, and other bodies. System architects and engineers need to be familiar with this aspect of cybersecurity.

Table 10.1 lists some of the more prominent sources of cybersecurity data and guidance. The list contains Government agencies, industry consortia, professional societies, and others. Collectively, they offer current cybersecurity intelligence, tools and methods, standards, training, and many other resources of great value to the secure system architect, as well as later to the security staff that must support and maintain adequate security over a system's operational lifetime.

As Table 10.1 suggests, there is a large and diverse community striving to cope with the ever-growing and constantly changing cyber threat environment. As part of these activities, a number of standards and taxonomies have been developed.

- *Vocabulary for Event Recording and Incident Sharing (VERIS)* – a set of metrics designed to provide a common language for describing security incidents in a

Table 10.1 Sources of cybersecurity information and guidance

Source	Information provided
Web Application Security Consortium (WASC)	Best practice security standards, tools, resources, and information, e.g., Web application scanner evaluation criteria
Open Web Application Security Project (OWASP)	Tools, articles, other resources; development/testing/code review procedures, list of top risks
Department of Homeland Security (DHS)	Best practice tools, guidelines, rules, principles, other resources; Build Security In (BSI) initiative; provides CWE (below) as a service; SW assurance Common Body of Knowledge; comprehensive guidance on secure SW development at buildsecurityin.us-cert.gov
Institute of Electrical and Electronics Engineers (IEEE), Computer Society (CS), Center for Secure Design (CSD)	Recent initiative aimed at providing a variety of artifacts for secure system development, e.g., "How to Avoid the Top Ten Software Security Flaws"
ISO/IEC 27034	Standard guidance on integrating security in processes for managing applications
System Administration, Networking, and Security Institute (SANS)	Education, certification, reference materials, conferences dealing with information security
Council on CyberSecurity	Established in 2013; publishes a list of prioritized Critical Security Controls (CSCs) that are widely used as a measure of cyber resilience (www.counciloncybersecurity.org)
National Institute of Standards and Technology (NIST)	Federal Government lead for cybersecurity, e.g., Risk Management Framework (RMF); multiple standards and best practice publications, including SP 800 series and Federal Information Processing Standards (FIPS)
National Security Agency (NSA)	Published the original "Rainbow Books" on computer and network security; operates the Center for Assured SW
SW Assurance Metrics and Tool Evaluation (SAMATE)	Provided by DHS and NIST
Common Attack Pattern Enumeration and Classification (CAPEC)	Sponsored by DHS and MITRE (http://capec.mitre.org/index.html); lists common attack patterns, comprehensive schema, and classification taxonomy; ~500 entries (so far)
National Vulnerability Database (NVD); NIST SP 800-53	Federal Government repository of standards-based vulnerability management data; uses the Security Content Automation Protocol (SCAP) which includes the Extensible Configuration Checklist Description Format (XCCDF); supports automation of vulnerability management, security measurement, and compliance; includes databases of security checklists, security-related software flaws, misconfigurations, product names, and impact metrics; supports the Information Security Automation Program (ISAP); provides a list of Common Vulnerabilities and Exposures (CVEs) and scores CVEs to quantify the risk of vulnerabilities, calculated from a set of equations based on metrics such as access complexity and availability of a remedy; includes a Computer Platform Enumeration (CPE) dictionary

structured and repeatable manner [2]. VERIS helps organizations collect and share useful incident-related information.

- *Common Weakness Enumeration (CWE)* – a common language for describing software security weaknesses in architecture, design, or code; this is the standard measure for SW security tools targeting these weaknesses and a common baseline standard for identification, mitigation, and prevention of software weaknesses [11].
- National Vulnerability *Database (NVD)* – a Government repository of standards-based vulnerability management data, described in NIST Special Publication 800-53.
- *Vulnerability Description Standards* – some of these are listed in Table 10.1; they include:
  - Common Vulnerabilities and Exposures (CVE)
  - Common Configuration Enumeration (CCE)
  - Open Vulnerability Assessment Language (OVAL)
  - Computer Platform Enumeration (CPE)
  - Common Vulnerability Scoring System (CVSS)
  - Extensible Configuration Checklist Description Format (XCCDF)
- *Common Attack Pattern Enumeration and Classification (CAPEC)* – a dictionary and classification taxonomy of known attacks that can be used by analysts, developers, testers, and educators to advance community understanding and enhance defenses.
- *Other Public Materials* – multiple publically available security taxonomies, research reports, checklists, newsletters, blogs, and other data sources.

10.5 Secure Architecture and Design

Given an effective cybersecurity policy and associated set of requirements, a sub-discipline of Systems Engineering that is generally referred to as System Security Engineering (SSE) comes into play as part of the overall MBSAP flow from requirements to an OV, LV, and PV. At every step, security requirements are allocated to a progressively more detailed system structure, and security functions and processes are modeled as part of the Behavioral Perspective. The overall strategy recommended in this book can be called "layered defense," and the layers of this structure are defined in parallel with other architecture development activities. Layered defense has two complementary elements:

- *Defense in Depth (DiD)* – deployment of diverse and mutually supportive security controls at various points in the architecture to create, in effect, concentric shells of protection that an attacker must defeat to reach protected assets. Later paragraphs and Appendix F go into more detail on this approach. DiD can be thought of as security in depth.

- *Zero-Trust Architecture* – a strategy that extends and refines DiD architectures with fine-grained segmentation, strict access control, strict privilege management, and advanced protective devices, all based on a rule that no person and no resource is trusted without adequate verification. Zero-Trust represents a broad approach to secure systems.

10.5.1 Secure Architecture and Design Process

Figure 10.5 shows the primary steps of a typical end-to-end security architecture process. As with architecture strategy in general, this flow is highly iterative, and results at any step may cause a return to an earlier step to correct a problem, refine a requirement, perform additional analysis on a discovered issue, and so forth. By making this an integral part of an MBSAP effort, security architecture becomes an integral part of the overall architecture of a secure system or enterprise, not something that is ad hoc or independent. Cybersecurity requirements, functions, and standards will impact many aspects of the system and may be the dominant factor in decisions such as product selections. Security testing involves a distinct set of events, but it must be integrated with the overall test plan to ensure all required data is collected at minimum cost and effort. Similar considerations apply throughout the MBSAP process.

The following paragraphs summarize the primary actions at each step in Fig. 10.5.

- *Planning and Initiation* – as part of basic program planning, establish the methodology, resources, schedule, and other basics of the security architecture development; this includes securing commitment from program management, customers, and other stakeholders.
- *Threat Analysis and Risk Assessment* – perform the analysis activities described earlier under Risk Management.
- *Cybersecurity Policies and Requirements* – once the RI, RA, and risk treatment analysis is complete, create a security policy, SP, SECOPS, and specific security requirements.
- *Cybersecurity Architecture* – define the security architecture under MBSAP, including:
  - Selecting and prescribing security boundaries and mapping them to system structure and partitioning.
  - Defining and modeling the implementation of selected security controls at each level of a layered defense structure.
  - Defining and modeling specific security functions and services; for example, security services like user authentication and access control are usually made available to both infrastructure and application components of a system.
  - Defining metrics to be used to assess the quality and policy compliance of the security architecture; another NIST Special Publication, 800-55 [12], offers guidance on the selection and application of security metrics.

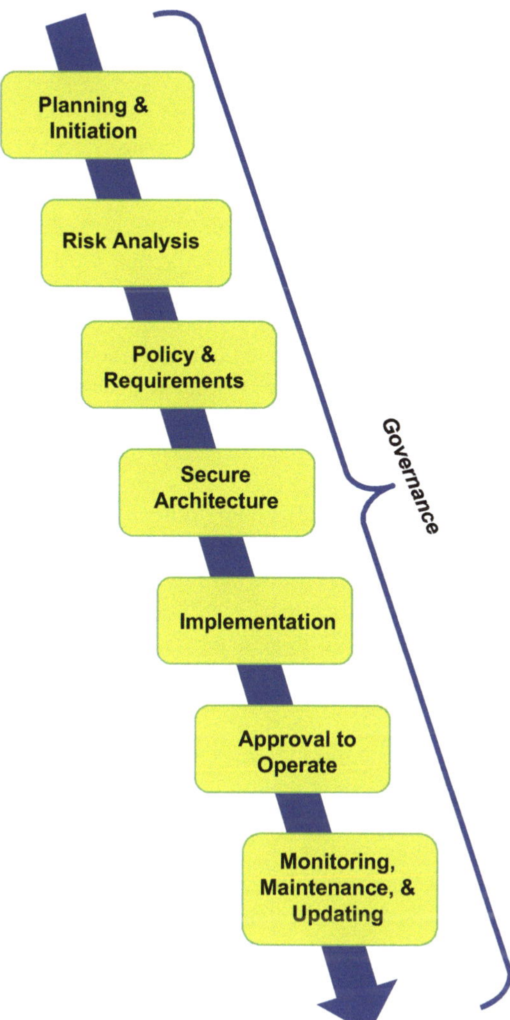

Fig. 10.5 Basic flow of a security architecture process

 - Defining roles and responsibilities of system users, security managers, and others.
 - Defining other aspects of the security architecture such as required trust levels for various system components; this must obviously be done in close collaboration with the system architecture development as a whole, and there will be security content in the OV, LV, and PV.

- *Implementation* – as described in more detail in the next section, implement the security architecture within the PV, including:
  - Selecting and configuring components and processes that implement security controls

  - Verifying that system components have appropriate required trust levels based on the nature of the threat and the sensitivity of protected information and that selected products comply
  - Developing security-related architecture artifacts and documentation
  - All other aspects of building the system in conformity to requirements and the security architecture

- *Approval to Operate* – once the implementation is complete, accomplish the analysis and testing to verify that the required risk level has been achieved. Regardless of whether a system has a Government or a private customer, adequate testing must be planned and conducted as an integral part of the total test program. Ultimately, a body of evidence is presented to an appropriate authority to support an approval to operate decision. This is further discussed under System and Component Trust Evaluation below.
- *Monitoring, Maintenance, and Upgrading* – during system operation, provide the resources and procedures for monitoring functions such as event logging, configuration audits, and vulnerability testing. Since technology, requirements, and threats continuously evolve, even the best security architecture and implementation will require periodic updating and retesting to maintain assurance that the required risk level is still being met. Plans for long-term system support and upgrading need to account for the inevitable updates and problem fixes associated with system security components and functions. Specific activities in this area include:
  - Configuration management to protect against unknown or unapproved content
  - Prompt installation of software patches as soon as released to deal with new vulnerabilities
  - Technology refreshment as cybersecurity components, like any other products, evolve, become obsolescent, and require replacement or upgrading
  - Addition, upgrading, or reconfiguration of protective measures to deal with new or changing threats and policies
  - Education, training, and security awareness of personnel to ensure the human element of the cybersecurity posture does not degrade as personnel and system functions change

- *Governance* – as shown in Fig. 10.5, implement governance at every step and throughout a system's lifetime. Valuable guidance is provided in NIST Special Publication 800-100 [13], which is targeted to Government agencies and policies but provides useful insight for private security architects and managers as well; Chap. 3 is especially relevant to secure system development. This document stresses the importance of clear individual and organizational roles and responsibilities, regular monitoring of security features and policy compliance, periodic reassessment of the effectiveness of security controls and procedures using measurements and metrics, rigorous configuration management and change control, and continuous analysis of incidents, supported by auditing of system logs.

A number of Government and private sector organizations issue guidance on design of a resilient cybersecurity solution. Rozanski and Woods [14], writing about software architecture from a largely commercial point of view, suggest the following elements of a security approach:

- *Principle of Least Privilege* – grant system users only those accesses and capabilities required to perform their roles.
- *Secure Weakest Link* – focus protection on system elements that are most vulnerable and most likely to be attacked.
- *Defend in Depth* – this is essentially layered defense.
- *Compartmentalize Content* – structure and manage information such that access to a given data area does not accidentally expose information that should be separately controlled.
- *Keep Security Simple* – avoid implementations that are hard for system users and administrators to work with and therefore invite cheating, shortcuts, or work-arounds.
- *Establish Secure Defaults* – ensure that if failures or anomalies force a system to revert to default configurations and settings, these still provide adequate protection.
- *Fail Secure* – similarly, ensure that if a system fails or shuts down, it does so in a state that does not expose sensitive content; a typical example is to encrypt all data at rest (i.e., in nonvolatile storage); this is the security equivalent of "fail-safe."
- *Start Up Secure* – the inverse principle is that upon startup, a system must activate all protective measures before accessing or exposing sensitive information.
- *Ensure Trusted Externals* – implement protections that provide adequate assurance of the identity, trust level, and access privileges of any external entity trying to use the system.
- *Log, Analyze, and Audit* – provide the capability to detect and log all security-related events, and implement a rigorous process of analysis and auditing to make full use of the recorded event data.

The Department of Defense focuses on the following essential aspects of cybersecurity policy for the DoD Information Network (DoDIN), supporting the Global Information Grid (GIG); these apply to most private sector cybersecurity situations as well:

- *Identification and Authentication* – the policy requires both non-forgeable credentials and the use of a Strength of Mechanism assessment to ensure acceptable risk that unauthorized individuals will be able to gain access to classified systems and data.
- *Access Control* – the policy directs that traditional Role-Based Access Control (RBAC) give way to more powerful and flexible approaches such as Risk-Adaptive Access Control (RAdAC) as described later in this chapter.

- *Privilege Management* – the policy calls for adaptive and flexible privilege management to ensure all authorized users have timely access to the resources needed for mission accomplishment in accordance with the local operational situation.
- *Trusted Computing Platforms* – the GIG requires that participants implement TCPs, also described below.
- *Secure End-to-End Communications* – the policy prescribes migration to a "black core" in which sensitive information is always encrypted when it is outside a secure environment; ideally, data is encrypted at the computer that produces it and decrypted at the computer that consumes it.

Such policy requirements could drive anything from the selection of biometric means for user authentication to the development of new algorithms for real-time adjustment of individual privileges. For example, most security guidance documents now call for two-factor or multifactor authentication, involving two or more independent means of verifying identity, in place of sole reliance on passwords that are all too frequently compromised. In a private sector context, a security architect might well be called upon to implement controls that give citizens of multiple countries in a global company or enterprise access to appropriate data and processes while ensuring no export control laws or company policies on information release are violated. These examples illustrate the importance of very thorough analysis of policies and security requirements early in the architecture cycle, followed by allocation to the elements of an emerging system architecture and design, to achieve the objective of embedding security features and functions in the system.

10.5.2 Layered Defense Architecture

To provide a basis for discussing layered defense, Fig. 10.6 shows a notional networked enterprise to which a variety of risk mitigations can be applied. This is at least representative of the contents of a wide assortment of real-world information systems. As used in MBSAP, layered defense has two components: a DiD structure of successive layers of protective mechanisms and Zero-Trust Architecture, which enforces segregation and strict access controls for all sensitive content.

Layered defense begins as a dimension of the secure system architecture and continues in the selection and deployment of products and processes for security controls at various points. This is a much larger topic than can be fully treated in a single chapter. Accordingly, the emphasis here is on key aspects of secure system implementation that are likely to be important to a general systems engineer charged with satisfying the full scope of system requirements, including security. The following paragraphs describe some important considerations in devising a DiD structure for a system like that in Fig. 10.6. Table 10.2 lists a typical set of layers in a DiD structure for such a system with representative safeguards and functions. Appendix F gives an extensive set of security control definitions for these layers, many of which are referenced in Fig. 10.6.

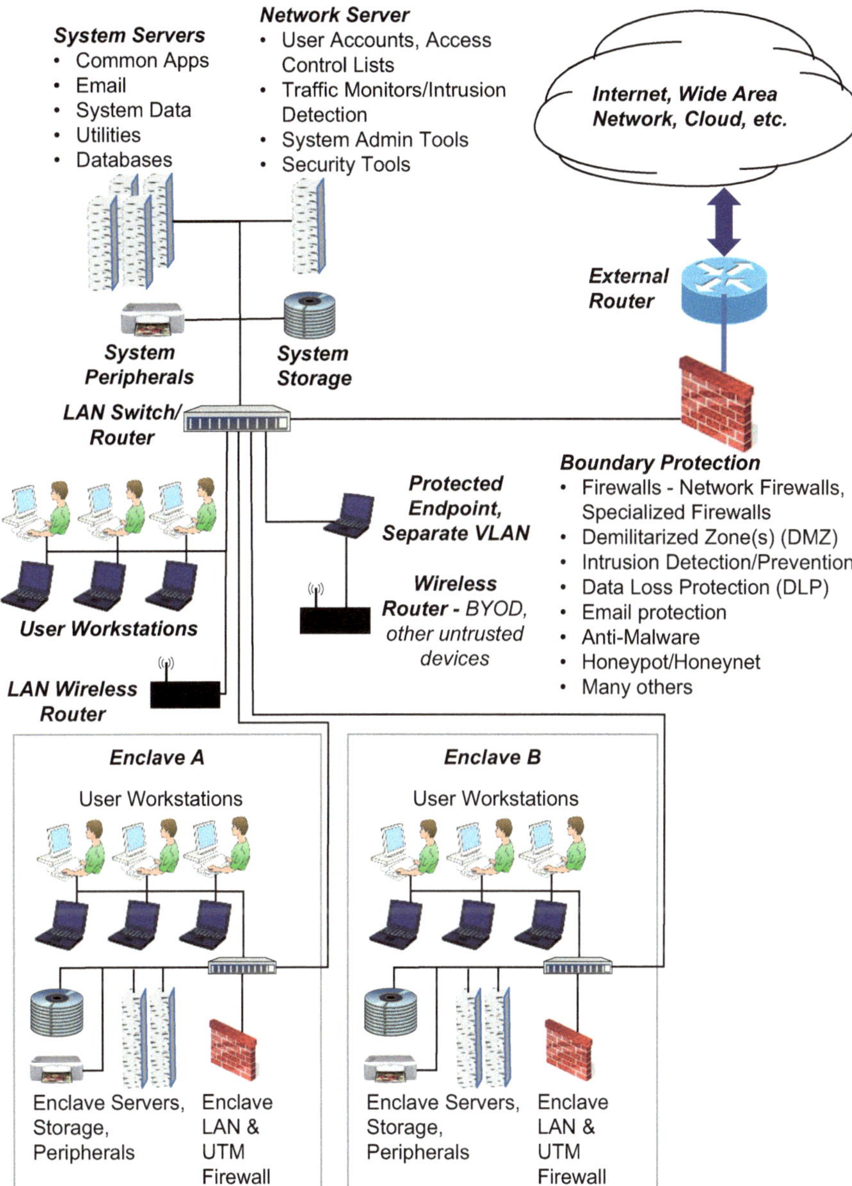

Fig. 10.6 Notional networked enterprise requiring defense-in-depth protection

Boundary Security A Boundary Security System (BSS) protects the point of entry from the external network environment and implements functions like those listed in Fig. 10.6. Some specific examples include:

- *External (Boundary) Router* – may provide some protective functions such as address filtering on messages.

Table 10.2 Examples of defense-in-depth protective measures

Protective layers	Protective measures	Protective functions
System Boundary or Perimeter	• Firewall(s) • External Networking • Demilitarized Zone (DMZ) • Intrusion Detection System/ Intrusion Prevention System (IDS/IPS) • Encryption/Decryption in boundary Protection Devices • Data Loss Prevention • Honeypot/Honeynet • Anti-Virus/Anti-Malware	• Block unauthorized information transfers and scan for malicious or prohibited content • Allow connections only to trusted addresses • Isolate system resources from external access; provide address space for protective devices and tools • Monitor external activity and log unapproved traffic; IPS blocks untrusted messages • Ensure sensitive ("red") content is converted to unreadable ("black") before release from a protected environment • Detect and block unauthorized efforts to extract ("exfiltrate") data • Trap would-be attackers in a "sacrificial" server; gather data for attack analysis • Detect/defeat malware at system boundary
Network	• Network-Based Intrusion Protection/Access Control • Enclave Segregation with Firewalls • Anti-Virus • Labeled/Protected Messaging • Message Integrity Checking • Encryption	• Monitor network traffic, detect suspicious or prohibited use, block unauthorized use • Provide customized protection for separate network enclaves • Detect/defeat malware in network traffic • Restrict traffic to authorized participants, prevent/detect message corruption or tampering • Detect tampering with network traffic • Eliminate unprotected data on the network
Endpoint	• Operating System (OS) Lockdown/Hardening • Host System Security (HSS) • User Accounts • OS Control File Protection • Administrative Separation of Privileges • Firewalls/Intrusion Detection	• Disable functions not required for system operations that may create vulnerabilities • Detect/prevent unauthorized or suspicious access attempts, malware, other hostile actions • Enforce access controls, user authentication, principle of least privilege, etc. • Prevent unauthorized changes to controls • Restrict access to security controls and settings • Customize protection for endpoint devices

(continued)

Table 10.2 (continued)

Protective layers	Protective measures	Protective functions
Application Software	• Logging and Auditing • Software Testing and Trusted Software • Specialized Firewall(s) e.g., Web Application and XML • Content Monitoring/Filtering	• Detect suspicious/prohibited access to apps • Rigorously analyze and test software to preclude malicious code and other vulnerabilities; detect/prevent unauthorized modifications • Apply customized protection to applications • Detect suspicious or unauthorized processing
Data	• Logging and Auditing • Encryption of Data at Rest • Data Loss Prevention (DLP)	• Detect suspicious or prohibited access to data • Ensure even theft or other physical compromise cannot obtain sensitive data • Block unauthorized data access/ exfiltration

- *Firewalls* – a variety of types can be deployed to inspect various aspects of in- and outbound traffic and block suspicious or unauthorized messages.
- *Demilitarized Zone (DMZ)* – internal and external firewalls can be used to create a “buffer zone”[3] between the untrusted external environment and sensitive system resources; this can host a variety of protective measures.
- *Intrusion Detection System/Prevention System (IDS/IPS)* – specialized device or software that detects and (if an IPS) blocks and records suspicious activity.
- *Data Loss Protection (DLP)* – identifies, monitors, and protects data in use, data in motion, and data at rest through deep content inspection and other sophisticated techniques.
- *Proxy Servers* – servers that allow external users to request and receive services such as Email and applications without being granted direct access to the actual servers and data.
- *Malware Protection* – software that tries to detect malicious software and prevent it from executing.
- *Honeypot* – a “sacrificial” server that mimics real system resources to attract an attacker and gather information; multiple honeypots create a “honeynet.”

System Resources There is a set of overall system resources, including servers, workstations, storage, peripherals, and a local area network (LAN) router. These provide functions needed by all system users such as account management and common software applications. All these components are candidates for protection using tailored security controls.

[3] The implied parallel to the situation between North and South Korea is quite deliberate.

Protected Enclaves The specialized functions of the enterprise are isolated in enclaves, each with its own LAN and a Unified Threat Management (UTM) firewall. Only two enclaves are shown, but there would typically be an enclave with restricted access for each individual department or other organizational element of the enterprise. This segmentation is a core principle of Zero-Trust Architecture. As with all cybersecurity devices, it's important that candidate UTM products be carefully evaluated for their protective capabilities, especially against the assessed threats to a system.

Protected Wireless Connection The system LAN has a wireless router for connection of approved (trusted) devices. There is also a wireless router for other access, such as by personal (Bring Your Own Device, or BYOD) devices. This router is protected by a dedicated processor that implements a variety of security controls and establishes a virtual LAN (VLAN) as described in Chap. 9 that is separate from the system network.

The other key element of layered defense is a Zero-Trust (ZT) Architecture [15], and Table 10.3 gives some of the main features for a system like that in Fig. 10.6. The name "Zero-Trust" derives from the basic principle that no one and nothing is trusted beyond a single authenticated instance. ZT complements and strengthens DiD by adding stringent controls, especially for access to sensitive content. For example, a system that makes an authentication and authorization decision for every request for access to sensitive content provides a higher level of assurance than one that requires only one-time access approval for an entire user session.

All of the DiD and ZT features in Tables 10.2 and 10.3 could be applied to a system like the one in Fig. 10.6, depending on the assets to be protected and the risks to be mitigated. One of the major objectives of adding a Zero-Trust network to a DiD architecture is to minimize the damage an attacker can do after successfully

Table 10.3 Features of a Zero-Trust Architecture

Protective features	Protective functions
Zero-Trust	• No system users or resources (applications, databases, etc.) are trusted a priori • Every resource access, installation action, etc. requires explicit authentication and authorization
Address control	• All resources are protected by strong access controls, regardless of location (e.g., multifactor authentication of users) • Enforces highly granular access to resources (specific permissions)
Network segmentation	• Network is divided into multiple segments enclosing related resources and functions • Enclaves have distinct protected perimeters and enforce access controls • Enclaves are protected by multifunction devices such as a Unified Threat Management (UTM) firewall[a]
Network security	• Implements a secure packet forwarding engine • Monitors all network traffic for suspicious activity

[a]Forrester Research has coined the term "Segmentation Gateway" for this device

penetrating a system. Every attempt at "lateral movement" from one resource area or enclave to another must defeat a layered defensive structure within the security perimeter of that enclave. Additionally, every attempt to access a protected asset requires strong authentication of the party requesting access and fine-grained authorization (access permission) for every specific asset. DiD with Zero-Trust, using advanced safeguards such as those which apply statistical behavior analysis and artificial intelligence, represents the current state-of-the-art in secure architecture.

As an example of secure architecture strategy based on deployment of multiple, complementary safeguards, Jones and Horowitz [16] have published an approach to protecting critical utility infrastructure, especially nuclear power plants. Basic elements of their proposal are:

- Use of diverse and redundant subsystems, increasing the effort an attacker must make to defeat a given layer of protection; this can also draw out a "signature" by which an attacker can be identified and recognized.
- Configuration "hopping" to restore a compromised system by switching to an alternative configuration.
- Data consistency checking using techniques such as voting among parallel diverse components.
- Tactical forensics to distinguish true attacks from false alarms or natural failures.

10.6 Secure System Implementation

Once an approved security architecture is in hand, the next stage of the process in Fig. 10.5 is implementation. We assume here that a set of security controls and other risk treatments has been selected and embodied in an architecture with the appropriate features and functions. These must now be realized with hardware and software components, procedures, staff training, and other elements of a cybersecurity solution. The system must then be tested to confirm that security controls function as designed, that no known vulnerabilities remain open, and that the required residual risk level has been achieved. The following paragraphs discuss implementation of these controls in the context of the enterprise in Fig. 10.6.

10.6.1 Boundary Security

A DiD design begins with protective measures at the system or enclave perimeter as described in Sect. 10.5.2. In an older security architecture, boundary security may well be the primary defense. This approach has been called "crunchy outside, chewy inside" and is now thoroughly obsolete. There is no such thing as a standard BSS design, but Fig. 10.7 shows some typical components in an Internal Block Diagram.

The Blocks associated with the BSS have a corresponding stereotype. The system connects to the outside world, which might be a public network like the Internet, a private wide area network (WAN), a cloud environment provided by a third party, or some other external environment, through a Boundary Router. There may also be other channels such as encrypted point-to-point links, and Fig. 10.7 includes a Virtual Private Network (VPN) server used by authorized parties outside the enterprise. This and other networking approaches are discussed in Chap. 9. There may be an external Network Switch if the Boundary Router needs this functionality. The Boundary Router normally uses the Border Gateway Protocol to establish a session with another autonomous system on the network. It may participate in the security scheme by filtering incoming traffic, e.g., by applying Time of Day limits to exclude messages that fall outside established time parameters.

Perhaps the most ubiquitous security component is a firewall, which is a specialized computer, running trusted software, that applies a range of techniques to

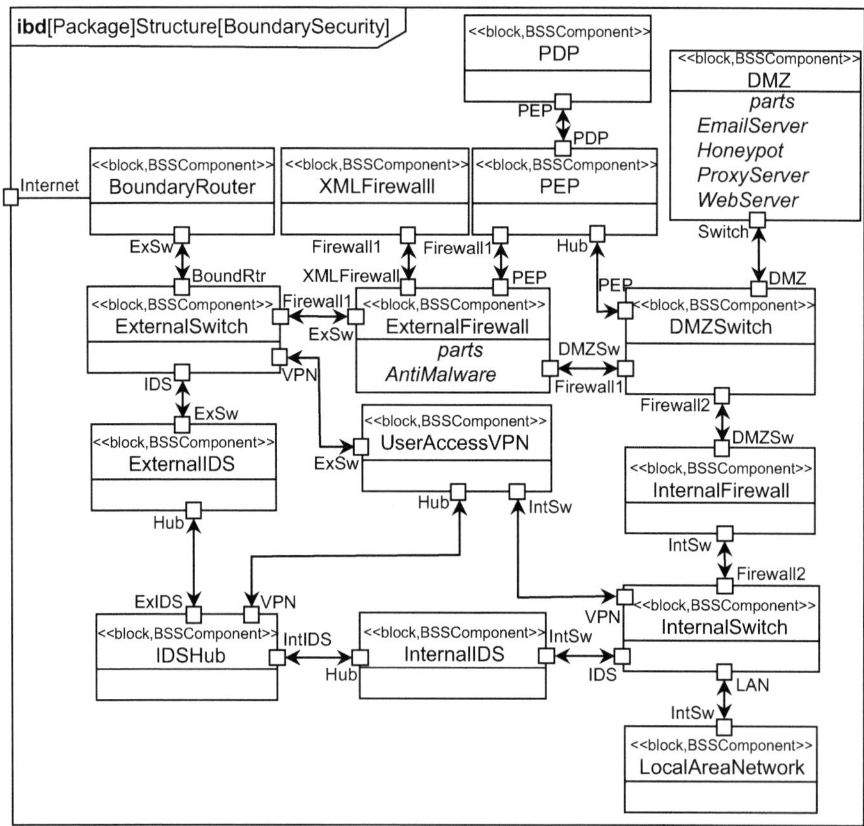

Fig. 10.7 Typical elements of a boundary security system, forming the perimeter security layer of a DiD implementation

incoming and outgoing traffic to identify and block unauthorized or threatening messages. The firewall has rule sets that can be programmed by security managers to establish the basis for allowing or preventing messages from passing, as well as an Access Control List (ACL) that identifies authorized external sites and users.

Firewalls apply a variety of filtering techniques, usually at the packet and protocol or application levels, based on user-supplied rules and parameters. A packet filtering firewall, which can be a software component running in a sophisticated boundary router, permits, rejects, and drops packets based on parameters such as IP address and port number. A proxy firewall runs proxy gateways for mail and other kinds of traffic using rules for context, authorization, and authentication. These, and other firewall types, can be combined to meet specific security requirements. A sophisticated boundary security system may have a DMZ with both external and internal firewalls, which is further described in a later paragraph. We also note that any firewall will have a certain number of false negatives (legitimate access requests that are blocked) and false positives (the opposite mistake). System and Network Administrators may be required to "tune" the firewall rules to achieve acceptable levels of these as defined by organizational security policy. For example, it might be acceptable to have 1% of legitimate traffic denied if that's the consequence of blocking 99.9% of malicious traffic.

As Fig. 10.7 indicates, a firewall may be supported by a variety of specialized security controls. The External Firewall is shown as including an Anti-Malware Part and interacting with an XML Firewall that specifically scans XML traffic and protects XML-based interfaces. The Internal Firewall is shown as incorporating DLP, which is often deployed at multiple layers of a DiD structure. The dual firewalls in Fig. 10.7 are complemented by Internal and External Intrusion Detection Systems, connected by a Hub. An IDS detects unauthorized traffic by determining patterns of activity and matching them to known signatures that are identified with known vulnerabilities or attack vectors. It can apply a variety of algorithms at the packet, message, protocol, and application layers and may be set up to respond to network attacks, data-driven attacks, attempts to subvert access controls and privilege management, unauthorized log-ons, and unauthorized access requests. The IDS may also supplement malware protection software in dealing with Trojans, worms, and other attacks. A more powerful component is an Intrusion Protection System (IPS) that actively blocks "bad" traffic. To remain effective, an IPS must receive regular updates as new threats are identified; this is analogous to keeping malware detection libraries current.

Both public and private sector systems have begun to implement the DMZ concept as a way to make information available to authorized external users while protecting core resources. A DMZ typically provides a data buffer and task queue. It resides in a separate enclave with its own address space and may have an external Domain Name System (DNS) server to process external user identities. A DMZ can be implemented with one or two firewalls, the latter being more secure. Figure 10.7 shows the DMZ, accessed through a DMZ Switch, lying between the External and Internal Firewalls and hosting functions like an Email Server, a Web Server, and one or more Proxy Servers that expose data and processes from system servers without allowing direct access to those resources. An Email Gateway is a server that accepts, forwards, delivers, and stores messages on behalf of system users. It typically also scans messages and attachments for viruses and other malicious content and can perform filtering to eliminate unwanted traffic (the notorious "spam"). A DMZ typically creates a "holding pen" for traffic to which the firewall can send suspicious incoming messages for additional processing before delivery or deletion. A further feature shown in the figure is a Honeypot that mimics actual system resources to trap malicious intruders and allow defenders to record information and to observe the details of an attempted penetration.

For many years, cybersecurity was based on applying rigid rules governing which users could access what information and processes. This makes it hard to accommodate the dynamic situations common in modern enterprises subject to sophisticated cyberattacks and can create a large administrative burden to fine-tune the privileges of a large and constantly changing user base. The current trend is toward policy-based security management in which rule sets take into account a variety of circumstances and make near real-time adjustments. For example, a specific individual might need access to specific data for a specific period of time, after which this privilege should be revoked. Or the access rules might need to become more stringent in response to a detected hostile penetration attempt. One implementation uses a Policy Enforcement Point (PEP), shown interacting with the External Firewall in Fig. 10.7, which is usually a dedicated computer that mediates user access requests, message traffic, and other activities on the basis of a set of policy attributes. In a typical situation, a user would request access to an application or data set, and the PEP would obtain the user's certificate and other attributes (see Access Control below); would apply current policy, provided by a Policy Definition Point (PDP), to determine if the requested access is allowed; and, if so, would authorize the requested activity. The PDP server gives System and Security Administrators important tools to manage access control.

On the inner edge of the BSS, an Internal Switch connects traffic that passes the Internal Firewall, along with VPN traffic and data from the IDS, to the system LAN. This is the next DiD level, with security controls like those listed in Table 10.2. The other layers have obvious mappings to endpoint devices connected to the network, to application software running on servers in the enclaves, and to data, wherever it exists and whether it is in use, in motion, or at rest. Space prevents a description of these layers as detailed as that just given for the Boundary or Perimeter Security Layer. However, Appendix F gives more detail about the candidate controls at each level that are summarized in the table. Because many cyberattacks seek to exploit weaknesses in Web-facing software, an important example is the use of a Web Application Firewall (WAF) using Deep Packet Inspection (DPI) either as part of the overall system BSS or in conjunction with the access point to an enclave.

10.6.2 Trusted Computing Platform (TCP)

The term TCP identifies a class of resources that satisfy stringent cybersecurity criteria and thus can contribute to highly trusted system designs. It was originally defined for Defense systems but can be applied to any secure system implementation. A TCP implements the following characteristics:

- The system boots to a secure state before accessing any applications or data.
- The system is trusted to load only signed software that is encoded to ensure that no modifications have been made to the code from the original developer.
- The system authenticates each client or user to the infrastructure.
- The system implements host-based (endpoint) intrusion protection.
- The system enforces separation of content by level of sensitivity/classification.
- The system encrypts data at rest and in motion.
- The system enforces security policies.
- The system sanitizes memory and storage upon shutdown or other defined conditions.
- The system performs client/user attestation.
- The system labels data with sensitivity and access control parameters.

The Trusted Computing Group (TCG) [17], composed of a number of prominent IT providers,[4] requires a TCP to enforce security features in hardware and software, including a secret encryption key that protects a computer even from its owner. The TCG has defined a Trusted Platform Module (TPM), which is a specialized chip using encryption to provide a hardware-based approach that is "a secure cryptographic integrated circuit (IC) [that] provides a hardware-based approach to such sensitive functions as user authentication, network access, and data protection." For example, the TPM allows the computer to furnish a certificate to a third party proving that trusted software and hardware have not been altered. To fully comply with the TCG specification, a computer must implement the following five specific concepts:

- *Endorsement Key* – a public/private key pair (see Encryption, below) that is embedded in the TPM
- *Secure Input/Output* – techniques that minimize security vulnerabilities associated with system I/O
- *Memory Curtaining* – extensions to normal memory protection to fully isolate sensitive memory areas
- *Sealed Storage* – binds private information to a specific hardware/software platform
- *Remote Attestation* – allows authorized third parties to detect changes to the hardware/software platform

[4] It is important to note that the TCG also seeks to improve data rights management, especially to prevent unauthorized copying and dissemination of copyrighted material.

10.6.3 Quality of Protection (QoP)

By analogy to Quality of Service (QoS, Chap. 9), security architects have defined a QoP construct that defines attributes of a data object describing how it is to be protected at rest and in transit. Interestingly, QoS and QoP are often in conflict, because the steps taken to secure information generally add overhead processing that can increase the latency in communicating the data. A QoP profile will typically define requirements for encryption, allowable routing, access controls, and destruction of a sensitive data object.

10.6.4 Encryption and Certificates

Schemes for converting information into a form that is only intelligible to an intended recipient using some form of encryption are as old as civilization. The original message or information is generally referred to as clear text or plaintext, and the encrypted version is called ciphertext. The encryption algorithm is commonly known as a cipher. "Code breaking" has played a decisive role in warfare for millennia, and commercial espionage that involves breaking into secure communications among companies and nations is widespread.

Cryptography has three primary uses:

- *Confidentiality* – make plaintext into unreadable ciphertext.
- *Integrity* – use hashing and message digests to ensure integrity.
- *Authentication and Non-Repudiation* – use digital signatures, certificates, and PKI with public key cryptography to prove the identity of the source of a message or transaction.

Cryptography is an entire discipline in its own right, and only a brief overview is possible here. Details of the types and classes of ciphers, the mathematics of encryption algorithms, tools and methods used to create and break codes, and other aspects are beyond the scope of this book. The following paragraphs discuss some aspects of particular interest to system architects and engineers.

Fundamentally, encryption methods are divided into secret key and public key (also called asymmetric key) approaches as summarized in Fig. 10.8. With a secret key, the recipient of secure communications provides an encryption key to any potential sender using secure means such as using a courier or sending parts of the key through multiple independent channels. Keys might be valid only for specific time windows or geographical areas and should normally be changed frequently and randomly to complicate the task of a code breaker. This can be very secure if a high-quality key is used, but it has significant overhead. By contrast, public key schemes eliminate the need for secure key distribution. The recipient publishes a public key while retaining and safeguarding a private key. The encryption mathematics is such that only the private key can decrypt ciphertext produced with the public key.

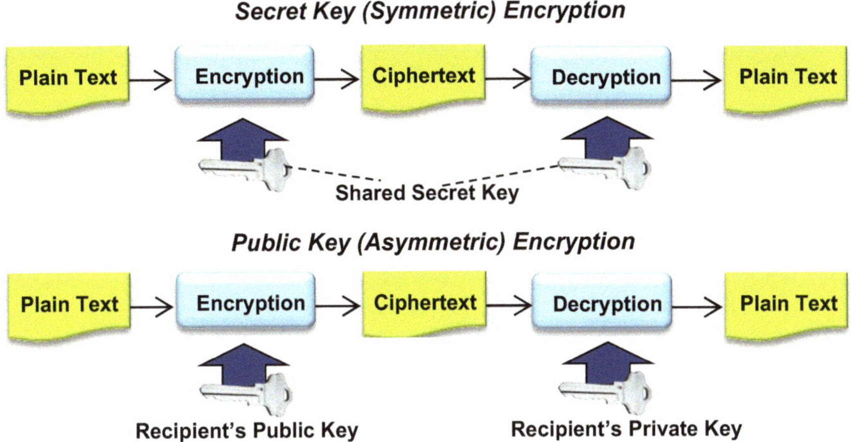

Fig. 10.8 Basic encryption schemes

A digital signature uses public key encryption to allow a recipient of a message to verify the identity of the sender and confirm that the content has not been altered. The sequence is:

- The sender computes a hash value[5] for the message and encrypts it with a private key; this is called "digital signing."
- The recipient also computes a hash value and decrypts the hash value that was sent, using the sender's public key.
- The recipient compares the hashes to confirm the sender's identity and message integrity.

Many encryption schemes involve trusted third parties, especially in the form of a Certificate Authority as sketched in Fig. 10.9. The Kerberos protocol, defined in RFC 4120 from the Internet Engineering Task Force (IETF), is the most common standard for mutual authentication between computers [18]. A Digital Certificate (or Public Key Certificate) is a document that binds a public key to one or more identification attributes of the key issuer, who may be the recipient of secure messaging or the client of a service provider. An enterprise can create a PKI that provides for issuing authentication credentials, such as an identity card with an embedded electronic certificate, to participants and can provide services such as certificate revocation if a credential is withdrawn or compromised, using a Certificate Revocation List (CRL). An alternative to PKI is called a Web of Trust scheme in which individual participants in an enterprise digitally sign their own certificates which are then attested by a third party. A popular implementation of this approach is Pretty Good Privacy (PGP), which allows use of email digital signatures and lets participants publish their public keys. A Web of Trust can interoperate with a PKI if

[5] Hashing is the application of an algorithm to a digital data object of arbitrary size to create a value of fixed size that can serve as a kind of fingerprint of the original; many hash algorithms are used.

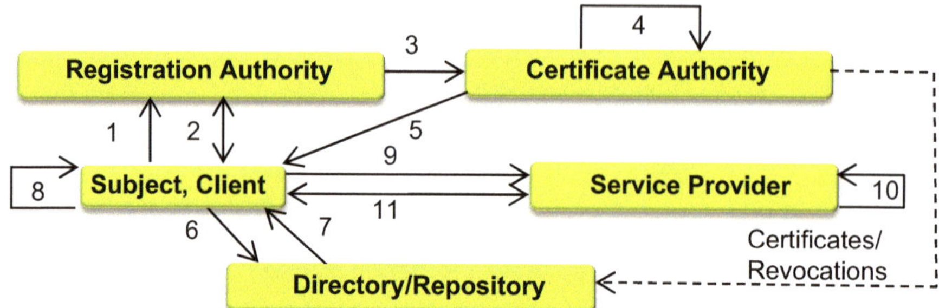

1. Request certificate, provide subject public key
2. Request and receive ID credentials, verify subject ID
3. Forward request
4. Create and sign certificate with subject public key and ID
5. Send subject certificate
6. Request provider's public key
7. Receive provider's digital certificate, extract public key
8. Create session key, encrypt with provider's public key
9. Send session key and subject's digital certificate
10. Verify subject's certificate, decrypt session key
11. Communicate using session key

Fig. 10.9 Digital certificate process

all parties trust the Certificate Authority. Multiple CAs can be used to create a distributed ledger of certificates. Certificates can be used as part of a Non-Repudiation implementation, e.g., by using proof-of-origin and proof-of-receipt, as evidence that an individual participated in a transaction. There is precedent that an RSA signature (see next paragraph) with non-repudiation is legally binding.

Many encryption algorithms exist, and NSA develops and applies suites of algorithms for protecting US Government information. Two of the most widely used in private information systems are the Advanced Encryption Standard (AES) [19] and the RSA encryption algorithm, named for its inventors Rivest, Shamir, and Alderman [20]. An algorithm is often described in terms of the length of time a code breaker with high-end computing resources would need to penetrate it, which may range from minutes to years.

AES and RSA allow keys of varying lengths, the longer keys being harder to break, to allow system designers to trade off encryption overhead vs. level of communications security. A cryptographic checksum can be used to detect attempts to modify ciphertext in transit. Exotic new encryption schemes are emerging such as Quantum Key Distribution (QKD) which uses quantum mechanical effects to create secure keys and detect any attempts at tampering. It seems likely that the means to securely encrypt even the most sensitive information will be adequate for the foreseeable future, at the cost of implementing complex and expensive encryption devices. FIPS 140-2, *Security Requirements for Cryptographic Modules* [21], is the Government standard for encryption devices.

NSA classifies encryption devices on the basis of the algorithms they incorporate and the level of sensitive information they are trusted to protect. The basic categories are:

- *Type 1* – trusted to cryptographically secure classified information
- *Type 2* – endorsed for telecommunications and information technology systems to protect unclassified national security information using classified NSA algorithms
- *Type 3* – endorsed to protect sensitive but unclassified information using unclassified algorithms
- *Type 4* – exportable devices using less robust encryption algorithms and keys

Encryption can be applied to multiple layers of a network protocol stack described in Chap. 9. At the Network or Internet Layer, the IETF defines the IP Security (IPSec) protocol suite that encrypts and authenticates each packet in an IP data stream. IPSec protocols can also perform mutual authentication and key negotiation between network participants, including individual users, servers, firewalls, routers, and others. At the Transport Layer, IETF defines the Transport Layer Security (TLS) protocol, successor to the earlier Secure Sockets Layer (SSL). In addition to securing the Transport Control Protocol (TCP), TLS has variants for web browsing, mail, instant messaging, and voice over IP (VoIP) digital voice communications. At the Application Layer, the Hypertext Transmission Protocol – Secure (HTTPS) combines HTTP with TLS using RSA public key encryption to create secure channels over nonsecure networks like the Internet. A number of schemes have been devised for securing Email, including Privacy-Enhanced Email (PEM) and PGP.

The black core concept mentioned earlier involves packet-level encryption and can be applied at the system boundary, at enclave boundaries within the system, or at individual computers, workstations, and storage devices. One approach uses individual encryption/decryption devices that conform to the NSA High Assurance IP Encryptor (HAIPE) specification and can, at some cost, be applied at any of the boundaries just listed.

10.6.5 Access Control

Like other aspects of cybersecurity, the mechanisms used to control the data and processes a user can access have become more sophisticated and flexible as threats have increased. Access control has major impacts on Confidentiality, Integrity, and Availability. A secure architecture can incorporate administrative or procedural, physical, and technical security controls to achieve acceptable risk that only authorized subjects can gain access to protected objects. Access control can be centralized for a system or enterprise as a whole, or decentralized, e.g., at the level of individual enclaves. A closely related security control, noted in Fig. 10.7, is the use of IDS/IPS. A cybersecurity best practice, mentioned earlier, is to enforce the principle of least privilege, limiting each subject's access to categories of objects required by the subject's role or duties, and the even more stringent concept of need-to-know, which literally imposes access rules for individual objects. This is an important protection against the most damaging access control errors. Important elements of an access control strategy include:

- *Identity Management* – using mechanisms such as directories, password and other credential management, and user account management to verify that a subject has an identity known to the system
- *Authentication and Authorization* – verifying that a subject is who the party claims to be and determining the subject's permissions to a given object
- *Accountability and Auditing* – recording access control activities and analyzing these logs to support non-repudiation, threat warning, enforcement of policies and procedures, and other security goals

For many years, the standard was Role-Based Access Control (RBAC) under which a user who successfully authenticated his or her identity was given access according to the role contained in a user profile. More recently, RBAC has begun to give way to Attribute-Based Access Control (ABAC), especially for large systems with many unanticipated users [22]. Under ABAC, both users and resources have defined attributes, and policies, which can be adjusted dynamically, support access decisions. It was mentioned earlier that two-factor authentication, or more generally multi-attribute access control, which might combine a password with a biometric factor such as a fingerprint or with possession of a physical credential such as a token, is highly recommended and growing in use to strengthen access control. NIST has published a model for Risk Adaptive Access Control (RAdAC) in which an estimate of security risk and an estimate of operational need are computed in each situation and then used to make an access control decision [23]. Still another variant uses ABAC first to authenticate a user to a system, followed by RBAC to control the specific privileges the user will have.

The Organization for the Advancement of Structured Information Standards (OASIS) supports the eXtensible Access Control Markup Language (XACML) for defining schemas for authorization policies and access requests and responses. OASIS also supports the Security Assertions Markup Language (SAML), which is a profile of XACML that provides needed assertion and protocol mechanisms [24]. A secure system can use SAML tokens to authenticate the source of a message or file.

A party (a person or system) requesting access to protected data and resources under ABAC can have a variety of attributes, including:

- *Identity* – established by authentication, "who you are"

- *Certificate or Token* – used for authentication but can also contain access information, "what you have"
- *Password or Passphrase* – also used for authentication but can access a user profile with access parameters, "what you know"
- *Role* – "what you do"
- *Location or Source* – "where you are"

Data attributes can include classification or sensitivity, releasability rules, timeliness, a digital signature, and many others.

The standard model for an ABAC implementation has the following elements, some of which are illustrated in Fig. 10.7:

- *Policy Enforcement Point (PEP)* – receives access requests, forwards them for decision, and enforces decisions
- *Policy Decision Point (PDP)* – maintains current policy, compares requests to policy, and returns access decisions to the PEP
- *Policy Administration Point (PAP)* – stores and manages policies
- *Policy Information Point (PIP)/Policy Repository* – delivers policy data to the PDP
- *Attribute Authority (AA)/Attribute Repository* – provides attribute data

As enterprises become more networked and dynamic in their functions and participants, continuous evolution of access control mechanisms is likely. In particular, means to dynamically adjust access policies, for example, to accommodate unanticipated users or to clamp down system access in response to an attack, will be increasingly important in maintaining required levels of security while allowing systems and enterprises to function effectively. The ramifications of a networked enterprise and its environment include the need for effective access controls for a SOA and strong authentication and authorization for remote system users.

10.6.6 Virtual Private Networks (VPNs)

A VPN is popular way to create secure communications channels over nonsecure networks, especially the Internet. It is essentially a virtual network composed by using IPSec to tunnel through the unsecured transport layer of the public network. Users have a VPN client on their computers and have a token (often in the form of a key fob) that provides a frequently changing encryption key for a session. A basic VPN may simply consist of a set of point-to-point channels, but a modern VPN supports more complex and changing virtual network topologies. A VPN can be used within a system or node as a way to segregate communications among a certain set of users from the overall network. There are many variations on these basic VPN concepts. Increasingly, VPN functionality is available as a feature of browsers and operating systems.

10.6.7 Separation Kernels

One approach to using off-the-shelf hardware and software components in accreditable designs of secure systems is called a separation (sometimes partition) kernel, which was first mentioned in Chap. 8 [25]. Combined with the concept of virtualization (see Chap. 14) to create a Separation Kernel Hypervisor (SKH), this approach isolates hardware resources, host software, and guest operating systems in independent partitions with controlled information flows. The goal is to enable secure processing on commodity computing platforms whose design details are generally not available because they are proprietary to their vendors.

10.6.8 Multiple Levels of Security

In discussing Risk Identification, a key point was the need to characterize information by its sensitivity and required level of protection. Many systems are required to handle information at multiple levels and to ensure that more sensitive data is not exposed by being commingled with less sensitive data that has a lower level of protection. Such a system must be accreditable, and that means that there must be high confidence that information at a higher level of classification cannot be exposed at a lower, less secure level. Because an individual system user is likely to need to see data at multiple levels, the ideal solution would be a full Multi-Level Security (MLS) architecture with trusted mechanisms to ensure that every internal and external system user can only see data for which they have access permission and a need-to-know. This has presented an intractable challenge for many years, and the problem becomes even more difficult in situations where a system has international users and where legal considerations such as export control laws apply.

Full MLS is notoriously difficult, and the most common solution with current technology is a Multiple Independent Levels of Security (MILS) or Multiple Security Levels (MSL) architecture. In this approach, the system or enterprise is partitioned into enclaves or domains, each of which is trusted to store, process, and display information at a given level of classification and releasability. Such an enclave is said to run "system high." Each enclave or domain has a network, and it's common to use the network to label the domain. Thus, a company network might have enclaves for Company Private, Personally Identifiable Information, Public Information, and other kinds of restricted or confidential content.

When information must be exchanged across domain boundaries between different levels of sensitivity or trust, each having specific rules for user access, the normal approach uses specialized computers called guards. A guard applies a set of rules to a message, file, document, or other entity to be exchanged to validate that the exchange is allowed. Passing information from a lower to a higher level is much easier to deal with than the reverse. However, sensitive information can often be "downgraded" by removing especially sensitive content such that it can be released in a less trusted domain. Guard rule bases, sometimes supplemented by human examination of the information, are the basis for this, and it is often highly desirable that the overhead processing and latency associated with a high-to-low transaction be minimized to maintain system performance. Increasingly, the trend is to base guards on XML documents, taking advantage of the fact that:

- XML documents are well structured, which facilitates automated rule processing.
- Technologies are available to describe XML messages and documents and to validate content against descriptions.
- Relatively mature security mechanisms exist for XML, e.g., for SOAP messaging, which may apply to cross-domain information exchanges.

In a MILS architecture, a user who needs to see data at multiple levels may need a separate computer or terminal for each domain network. In an MLS system, the user would have transparent access to any authorized information at any security level on one screen.[6] Accrediting such a system has historically been too challenging and expensive to be feasible. However, technology advances such as increasing use of trusted XML mechanisms may eventually solve the problem. This is an area where message and system access logging are also helpful.

10.6.9 TEMPEST

Because digital electronics can emit radio-frequency signals associated with the rapid switching of their logic, there exists a possibility that an adversary might detect and decode such signals and thereby obtain sensitive information. TEMPEST is the name usually given to design and test activities aimed at preventing this. Commercial computing equipment may require special enclosures with improved shielding and grounding. Buildings, aircraft, and vehicles that house classified computing environments may require testing to ensure detectable emissions are not present. When TEMPEST requirements are levied on a system, the architect has the responsibility to ensure that they are properly flowed down to the appropriate components and that TEMPEST test requirements are accounted for in overall test planning.

10.6.10 System and Component Trust Evaluation

The approach to specifying and verifying a level of trust for an information system has evolved over time in response to both the growing complexity of these systems and the increasing sophistication of cyber threats. Many of today's principles and

[6] It's been suggested that the litmus test for MLS would be the ability to drag and drop from a window at one security level to one at another level.

practices, in both the public and private sectors, originated in early Government work to protect sensitive information [26]. An early, widely used attempt to achieve defined levels of trust in information systems was documented in the "Rainbow Books" from the National Security Agency (NSA), so called because the books, more than 20 in all, each dealing with a specific aspect of security, had distinctive cover colors. The central volume, the Orange Book, or more formally *A Guide to Understanding Audits in Trusted Systems* [27], defined a set of security classes, with A1 representing the highest level of trust and D2 the lowest. A set of Trusted Computer System Evaluation Criteria (TCSEC) was published as DoD 5200.28-STD, December 1985, to provide a basis for evaluating commercial computer systems in terms of their security features and development processes. TCSEC has been largely supplanted by the Common Criteria approach, described below, and the entire field of secure system evaluation has evolved considerably.

Before being authorized to store, process, and exchange sensitive data, i.e., to "go live," a secure system should undergo a formal, documented process to verify that an acceptable level of trust, the inverse of security risk, has been established. Traditionally, such processes tend to be variants of the Certification and Accreditation (C&A) process that has been used for many years within the Federal Government. C&A has three basic elements:

- *Evaluation* – technical assessment of the protection features of the products used in a secure system design against stated criteria, independent of any mission or operational environment, based on analysis of design documentation and on testing, which ideally includes expertise from independent external test organizations
- *Certification* – a finding by system evaluators that security policies and requirements have been satisfied
- *Accreditation* – approval by a designated official or organization for the system to operate with sensitive content

Originally, C&A was a static process under which a system received an Authority to Operate (ATO) or something equivalent for a period of time, usually with a requirement for periodic security assessments to ensure adequate security was preserved. Accordingly, a C&A process typically proceeds through phases of system development, certification testing, accreditation, and reaccreditation.

Today, the familiar C&A approach is giving way to a more rigorous strategy based on *continuous* security evaluation, sometimes called Assessment and Authorization (A&A). Government policy now requires the use of what is known as a Risk Management Framework (RMF) [28]. All Government agencies are moving to implement RMF, which is defined in a series of publications from the NIST Computer Security Division [29, 30]. This improved strategy is also appropriate for a wide range of private sector systems and enterprises. RMF is similar in many ways to earlier C&A processes, including an emphasis on managing the risks to an organization and to the individuals and systems that implement information processes. It is different, however, in that periodic discrete C&A events are replaced by a continuous process of monitoring and assessment to ensure the integrity and policy compliance of the information protection solution. Under NIST guidance, RMF also calls for deploying more, and more effective, security controls. The RMF approach has six steps:

- Categorize the system and its information content.
- Select the baseline security controls, with tailoring and supplementation as needed.
- Implement the controls.
- Assess the controls as implemented for effectiveness and appropriateness.
- Authorize system operations based on a determination of acceptable security risk.
- Conduct ongoing monitoring and reassessment, including documentation of changes to the system, its environment, and the security impact of changes; report the security state of the system to appropriate authorities.

An effective process of continuous monitoring and assessment can entail significant effort and cost, both in the form of automated tools and in the need for human expertise, to carry out functions such as analyzing event logs, conducting frequent security testing, and conducting regular configuration audits. These must be planned for and provided if the benefits of RMF are to be realized. An approach that is growing in popularity is to create a virtual computing environment that matches the actual system but is completely separate from it, commonly labeled a "sandbox." This allows software to be hosted and thoroughly tested before it is deployed in the production environment and to be retested as necessary in response to newly discovered vulnerabilities without disrupting operations. Continuous monitoring is also helped by the availability of increasingly sophisticated tools for analyzing network traffic patterns, event logs, and other indicators. A sandbox can be used in conjunction with a honeypot or honeynet to perform forensics such as identifying an attacker.

NIST and the National Security Agency (NSA) play key roles in enhancing cybersecurity for both the public and private sectors, including guidance that covers many aspects of implementing layered defense. NIST releases a series of FIPS publications [31]; two widely used examples are FIPS Publication 191, *Guidelines for the Analysis of Local Area Network Security*, and FIPS Publication 197, *Advanced Encryption Standard (AES)*. Another excellent security reference is NIST Special Publication 800-27, *Engineering Principles for Information Technology Security (A Baseline for Achieving Security)* [32]. NSA supports primarily Government system security in a wide range of areas from cryptographic technologies and devices to recommended system configurations. A complete compendium of policies, directives, guidelines, standards, and other documents dealing with cybersecurity would be very large. The documents identified in this paragraph contain additional references that can help the reader find specialized information about particular topics.

Since an important element of implementing security is the use of products that are trusted to exhibit certain secure attributes and to be free from vulnerabilities, a method of evaluating and certifying such products has been key to cybersecurity for a long time. The primary approach today is the Common Criteria for Information

Technology Security Evaluation, usually shortened to Common Criteria or simply CC, which is based on the international ISO/IEC 15408 standard [33]. The CC structure allows security requirements for an item to be specified in a way that allows a product to be tested against them. Security requirements for a class of devices are captured in a Protection Profile (PP), and one or more PPs are used in compiling a Security Target (ST) that defines the security properties of a product and a set of Security Functional Requirements (SFRs). PPs are oriented toward allowing users to define their needs, while STs allow providers to spell out the security their products provide. The CC standard provides a catalog of SFRs as well as Security Assurance Requirements (SARs) that deal with product development and testing.

The most common use of CC is to assign an Evaluation Assurance Level (EAL) from 1 to 7 to a product. Essentially, an EAL corresponds to a set of SFRs, and the higher levels represent more complete (and more expensive) testing and validation. The highest levels can consume so much time and money that products are either not economically feasible to evaluate or approaching obsolescence when the process is complete, which is one of the major criticisms of CC. Government agencies typically mandate an EAL for a component of a secure system based on the sensitivity of the information being protected and the nature of the threat [34]. For example, information whose compromise would cause some damage (roughly equivalent to a classification of Confidential or Company Private) and that faces a threat characterized as unsophisticated but willing to take risks (perhaps a basic hacker) warrants an EAL of 2. Information whose compromise would cause exceptionally severe damage (e.g., a classification of Top Secret or key intellectual property) with a sophisticated adversary who has reasonable resources and will take significant risks (perhaps a major terrorist organization) calls for an EAL of 6, which many commercial products cannot meet. Security architects working with highly sensitive information are likely to have to deal with this aspect of system design.

No product evaluation is perfect. As an example of emerging threats, Yang described a "hardware Trojan" that can be inserted in a processor chip that is undetectable by currently known methods and that can give an attacker complete access to a computer's operating system [35]. Protection against such a threat literally demands a completely secure and trusted chip fabrication process that prevents an attacker from gaining physical access to parts.

MBSE can support system security evaluation and accreditation in a number of ways, as summarized in the following paragraphs.

- A formal architecture model allows capture, maintenance, and visualization of the configuration baseline and processes or behaviors of a system or system of systems. Artifacts exported from such a model can be focused on specific aspects such as security features and functions and can effectively provide the information needed by security testers, analysts, and accreditation authorities. For example, architecture data can support planning of effective vulnerability assessments and penetration testing.

- An architecture model allows rigorous and auditable flowdown of requirements, including security requirements, to components, interfaces, and processes, to ensure all requirements have been accounted for in system design and to provide a basis for requirements verification and validation.
- As service-oriented architecture (SOA) becomes the preferred approach to implementing large information systems and enterprises, the power of architecture modeling in defining services and, especially, service interfaces is invaluable in system accreditation involving security services, both for internal services and for those a system provides or consumes as a participant in a larger enterprise.
- An architecture model provides a basis for configuration management, data management, and change management. For example, a model captures dependencies among system elements to support defining required regression testing following a design change. The model also serves as a searchable repository of system and component information, including functions such as archiving data from previous security testing to keep subsequent testing to the feasible minimum.
- Modern system architecture modeling tools allow simulation of behavioral diagrams to visualize the processes that they represent. This "executable architecture" technique is valuable in assessing the operation of security controls.

10.7 SOA, Web, and Cloud Security

Any situation in which sensitive information is transmitted outside a secure environment raises cybersecurity concerns. The two most prominent cases involve the Web, which essentially means the Internet, and the use of Cloud Computing. As discussed in Chap. 7, Service-Oriented Architecture is predicated on allowing multiple participants in an enterprise, who may come and go unpredictably and who use multiple information technologies, to interact in implementing business processes. However, proliferating access to data and resources inevitably creates security vulnerabilities. In addition to all the other cybersecurity measures discussed in this chapter, a SOA needs robust mechanisms for assuring confidentiality, integrity, and access for the services that are exposed and consumed by enterprise participants. Because Web services are the overwhelming choice today for SOA interactions, they are the focus of efforts to achieve security.

For a long time, the primary mechanism for protecting SOAP messages has been HTTPS, the secure version of the Hypertext Transmission Protocol. This encrypts an entire message and is adequately secure for a point-to-point transaction. However, Web traffic increasingly includes intermediate nodes that must read parts of a message in order to perform their functions and therefore, with HTTPS, must decode the entire message. With the sensitive content now exposed in multiple places, the level of security risk goes up significantly. The inflexibility of HTTPS and the overall need to move to higher QoP have motivated the creation of the WS-Security Framework, anchored in the WS-Security standard. Repeating the list from Chap. 7, the WS-Security Framework deals with identification, authentication, authorization, confidentiality, and integrity and includes WS-Security, WS-Security Policy, WS-Trust, WS-Secure Conversation, WS-Federation, eXtensible Access Control Markup Language (XACML), eXtensible Rights Markup Language (XRML), XML Key Management Specification (XKMS), XML-Signature, XML-Encryption, Security Assertions Markup Language (SAML), .NET Passport, Secure Sockets Layer (SSL), and WSI-Basic Security Profile.

WS-Security defines language elements for security metadata, including header blocks, usernames, passwords, and tokens. It provides an overall means of associating security tokens with messages, incorporates security features in the header of a SOAP message, and provides integrity (anti-tamper), confidentiality (encryption), and various authentication and authorization mechanisms, including proof-of-origin certification. WS-Security allows selective encryption/decryption of message elements. In conjunction with WS-Security message encryption, a Kerberos or PKI service is commonly used to distribute session keys so that participants do not have to create, store, and protect large amounts of key material.

Working together, the standards and protocols of the WS-Security Framework provide the tools to implement cybersecurity concepts and requirements for the exchange of services in a SOA. Cybersecurity protective measures implemented at the nodes of the enterprise must be complemented by those implemented for the enterprise as a whole. When service models other than Web services are used, e.g., to meet performance requirements, equivalent mechanisms for authentication, data integrity and confidentiality, and access control must be provided. Table 10.1 lists organizations that are concerned with security for Web applications.

Additional methods used to improve Internet security include:

- *Transport Layer Security (TLS) (formerly called Secure Sockets Layer (SSL))* – a session-based protocol operating at the Transport Layer (see Chap. 9), especially for secure access to Web applications; TLS is also used for VPNs.
- *Secure Hypertext Protocol (S-HTTP, not to be confused with https)* – connectionless-oriented protocol that encapsulates data after session security has been negotiated; this approach is not widely used.
- *IPSEC* – operates at the Network Layer (see Chap. 9) to provide encryption and authentication; this is the primary VPN protocol.
- *Multiprotocol Label Switching (MPLS)* – fast packet forwarding using labels inserted between Layer 2 and 3 headers in a packet; it is protocol-independent and scalable and provides Quality of Service (QoS) with multiple Classes of Service (CoS) and Layer 3 VPN tunneling.
- *Secure Shell (SSH2)* – secure remote access using an encrypted tunnel between an SSH Client and an SSH Server; it can authenticate a Client to the Server (note: SSH1 is used but has exploitable vulnerabilities).
- *Wireless Transport Layer Security (WTLS)* – security services for the Wireless Application Protocol (WAP); classes of security include:
  - Class 1 – anonymous authentication
  - Class 2 – server authentication only

 – Class 3 – client-server authentication using Service Set Identifiers (SSID) and Wired Equivalent Privacy (WEP) keys or a more secure encryption such as WAP2

Cloud security has all the challenges of other secure architectures plus the facts that data and other resources are not under the physical control of the owning organization and that those resources are shared with other cloud clients. Attackers seek to exploit vulnerabilities in the virtualization technologies used in a cloud, in access control mechanisms, and elsewhere. One example is called a Side-Channel Attack, which has several variations, all aimed at hijacking a cloud client's account or files. Another attack method is called Hyperjacking and is based on compromising the hypervisor that creates and manages virtual environments. Still another is a Virtual Machine Escape in which an attacker breaks out of a virtual machine to attack the underlying hosting environment.

The Cloud Security Alliance [36] is a nonprofit organization that seeks to promote best practices and education for improved cloud security. The following are some of the considerations involved in achieving an acceptable risk level when employing cloud services.

- It is incumbent on the Cloud Service Provider (CSP) to provide adequate assurance that personnel with access to resources and data are trustworthy, and clients should obtain explicit proof of this. Typical measures include security screening and intensive education of staff and access controls to prevent support personnel from reaching client data.
- Similarly, a CSP must provide physical security to ensure a data center is difficult or impossible to penetrate. Reliable electrical power, for example, using uninterruptible power supplies (UPSs), and environmental controls are also important to ensure against data loss.
- In any secure system, data at rest should be encrypted. This is especially important in a cloud.
- Access controls should be controlled only by the cloud client or data owner. It's highly desirable that the CSP support fine-grained user privileges so that access can be finely matched to individual users. The CSP should provide and maintain logging of all access attempts.
- The CSP must have an adequate plan for data backup and recovery and for business continuity.
- Encryption is important to protect cloud transactions over open networks, for data protection and for security of user data, logs, and other sensitive information. Advanced encryption algorithms used in cloud environments include various types of Attribute-Based Encryption (ABE) in which a set of attributes of the ciphertext and of the user's secret key must match before decryption can be performed.
- Compliance with applicable laws and regulations may be different in a cloud environment than with a system controlled by the using organization. An example is the approach used for data locating. A cloud services client must be aware of any such differences and take steps to ensure compliance.

10.8 Secure Architecture Modeling

Representing security features and functions in an architecture model is not fundamentally different from any other architecting task. There is not yet a UML/SysML profile for security analogous to SoaML, but stereotypes, design patterns, and other elements tailored to security can be used in a very similar fashion to SOA modeling as discussed in Chap. 7.

10.8.1 Security in the MBSAP Methodology

The following are important aspects of embedding security in a system architecture, following the MBSAP sequence of model development.

Requirements Security requirements, as discussed earlier in this chapter, are part of the system requirements baseline and may include FRs such as specific security controls and associated processes and NFRs such as a requirement for single sign-on or for strong password enforcement.

Operational Viewpoint System-level security features and functions are modeled first in the OV.

- Major groupings of security resources may be modeled as separate Domains. An example would be a BSS like the one shown in Fig. 10.7. Alternatively, they can be included as Subdomains of a Domain like Computing Infrastructure. Responsibilities, Values, and Operations should be captured as for any other structural element.
- A secure system should have Use Cases (and may profitably use Abuse Cases) with specifications and behavioral diagrams for security-related processes such as malware detection and user authentication. It may be useful to create an SMD showing security states and the transitions among them.
- Foundation Classes in the CDM commonly have Values and Operations associated with security; e.g., defining a Sensitivity or Classification Value ensures that all System Classes inheriting from the Foundation Class will have this key attribute.
- Any system and enterprise security services used should be documented in the services taxonomy and modeled as described in Chap. 7.
- The Contextual Perspective is a logical place to file information about the cybersecurity environment, RI/RA results, and other factors bearing on the secure solution.

Logical/Functional Viewpoint As with any other architecture aspects, the high-level structures, behaviors, data, and services from the OV are detailed to create a functional design in the LV.

- Security design details and processes are modeled in structural and behavioral diagrams just like any other. Figure 10.7 showed a structural example, and

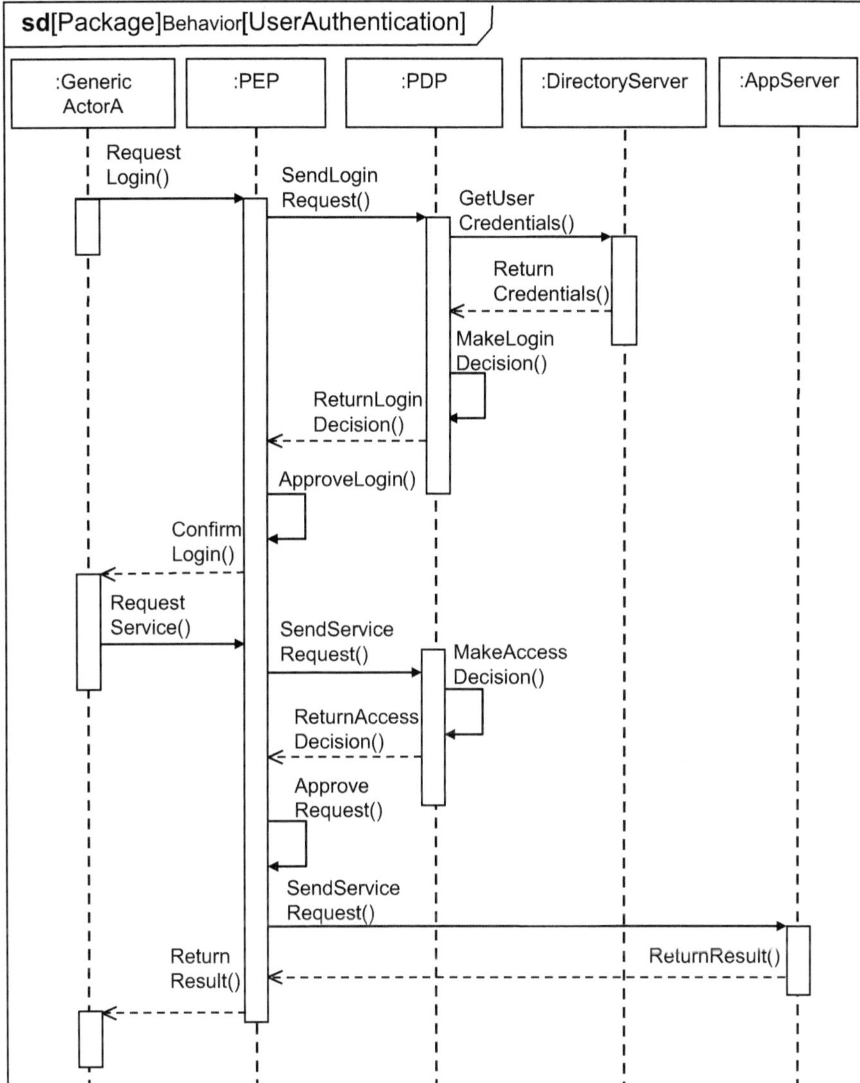

Fig. 10.10 Sequence diagram showing a user login and access to services

Fig. 10.10 shows a typical behavior, in this case a user login and access to an application. The user's login request is processed by the PEP based on current policy and the permissions and other attributes provided by a Directory Server (e.g., Active Directory), followed by a request to the application server and the return of a result. Many security controls are associated with interfaces and should be captured in interface specifications.

- Security requirements are flowed down from the OV and allocated to system elements. Security analysis and testing to demonstrate that the system achieves an acceptable risk level can be modeled using <<VerifiedBy>> Dependencies.
- The LDM may contain specific security data entities, and other data may have security-related Values and Operations.
- Domain-level security services from the OV should be further decomposed as appropriate and mapped to system elements such as processors, network devices, storage, and interfaces.
- Tailored artifacts such as a ports and protocols matrix are typically needed for system certification and may be generated from the model.
- The Contextual Perspective might well contain documentation of the rationale for the security design, special considerations of C&A or other authorization to operate processes, and anything else that will assist with long term-security maintenance and upgrading.

The participants in the behavior diagrammed in Fig. 10.7 all have security characteristics. Users have attributes and privileges that determine their ability to access the system and its content. The PEP and PDP are part of the BSS and have been described earlier. The Directory Server has the data and functions to securely manage user accounts. Finally, the Application Server has both Endpoint and Application Layer security controls as described under Defense in Depth.

Physical Viewpoint As with all other system elements, the PV documents the security products used and other implementation details. This could well include EALs for critical components.

Others Diagram annotations, linked files, and other documentation are often simple and effective ways to model and communicate a security design. Particular attention should be paid to information of interest to cybersecurity stakeholders, including accreditation or other approval authorities and the SYSADS/NETADs who are responsible for maintaining the system's security posture. Much of this belongs in a Security-Focused Viewpoint.

10.8.2 Security-Focused Viewpoint

A Security-Focused Viewpoint is essential for any system that deals with sensitive information, especially if a formal process such as C&A is required to obtain authority to operate. This is likely to be a large set of information, including graphics showing the placement of protective measures, a System Security Architecture document, documents defining security rules and procedures, and so forth. One approach is to take the evidence package prepared to support testing and accreditation and supplement it with any materials needed to convey the security design to program managers, other decision-makers, and system users. It is also useful to annotate the primary artifacts of the PV to show where security functions, security service interfaces, enclave boundaries and cross-domain information exchanges, encryption, trusted software, and other design elements are implemented. The Security Architect or Security Team Lead is normally responsible for compiling and maintaining this Viewpoint.

10.9 Secure Software Development

Reported data on successful cyberattacks shows overwhelmingly the consequences of software vulnerabilities. Many of these can be traced to poor coding practices and failure to implement well-known protective features and functions. System architects and engineers dealing with secure systems have a critical responsibility to ensure that software, whether purchased, newly developed, or modified, has been built and tested using best practices to eliminate security flaws.

10.9.1 Secure Software Development Life Cycle (SSLDC)

Figure 10.11 shows schematically how the phases of a software development effort must be matched with security processes to create a Secure Software Development Life Cycle (SSDLC). The figure also lists some typical activities such as choice of security controls and other risk treatments. Both testing during software development and system testing are key elements of the SSDLC. Ultimately, the maximum feasible level of protection is achieved by a combination of:

- Deploying software that is inherently hard to attack because it is free of exploitable flaws and has any discovered vulnerabilities promptly patched
- Deploying software in a robust layered defense environment that minimizes the chances of an attacker gaining access to it

A representative list of secure software development principles would include the following. The goals are better code, less vulnerable code, an overall smaller system attack surface for improved cybersecurity, and lower total cost. Systems engineers and architects should ensure that these or their equivalents are implemented in the software process.

- *Include Abuse Cases in SW Requirements/Use Cases* – modeling the behaviors associated with potential attacks and system malfunctions helps ensure high-quality RI/RA and define an effective set of risk treatments.
- *Include security risk analysis in requirements, architecture, and design* – this is an integral part of the principle of building in security from day one of a software effort.
- *Include testable security functions and features in requirements* – this is the fundamental principle of ensuring that security requirements are part of a comprehensive and balanced security baseline.

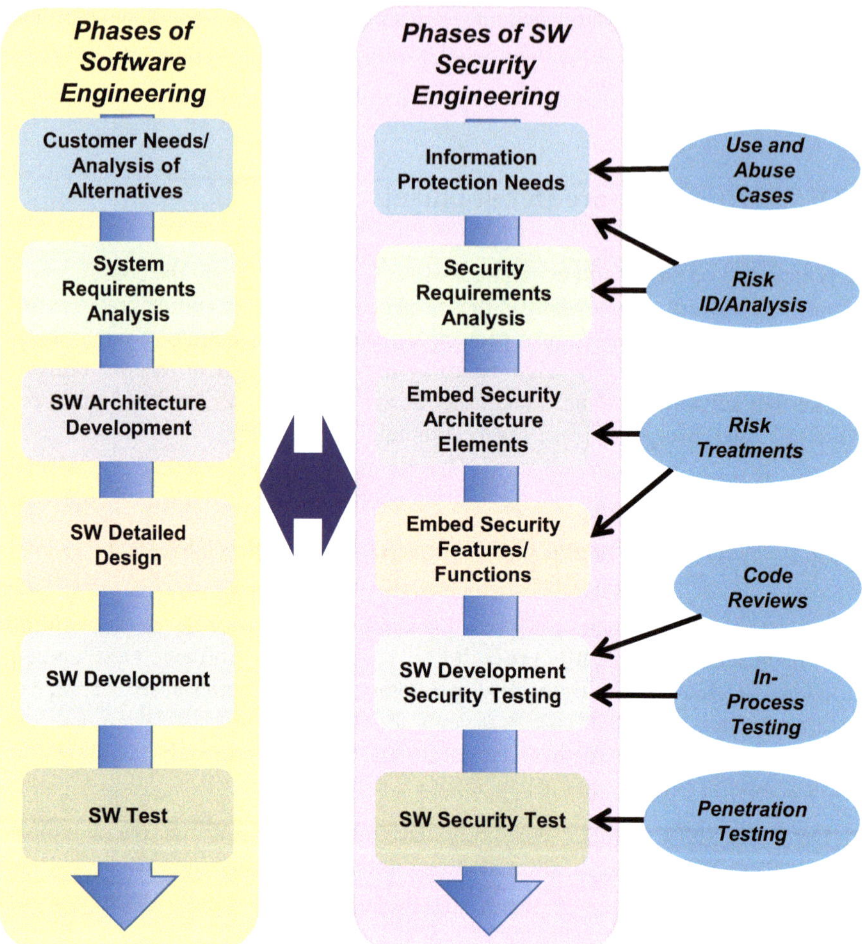

Fig. 10.11 Parallel phases of software development and secure software engineering

- *Include security functions and features in the design* – this is the remainder of the parallel secure software flow in Fig. 10.11; examples of these features are strong input checking, strong authentication, audit logs (which should be signed), code that is free of known vulnerabilities and signed to prevent tampering, privacy, and prevention of stack and buffer overflows.
- *Include vulnerability detection in design and code reviews* – the key point is to make software security evaluation an integral part of the development process.
- *Include risk-based security vulnerability testing and scanning during development and in system test* – it's important to use tools to scan for errors and vulnerabilities, and a wide variety of effective testing tools, many of them free, is available; see the next section for more details.

- *Perform penetration testing during system test and operations* – despite the best efforts to find and fix flaws during software development, it's vital that the final design be thoroughly tested to catch the vulnerabilities that are likely to remain; this continues through the system's operational life.
- *Implement software on secure, hardened servers with no access by programmers to the production environment* – an important aspect of the software development and test environment is that it be completely isolated from the operational system on which the software will be deployed.
- *Retest regularly for security and patch any vulnerabilities discovered or reported* – this has been stressed several times; the reality is that software development (or maintenance) and vulnerability testing never end.

The IEEE Computer Society Center for Secure Design (CSD) publishes a list of Top 10 Security Flaws that can serve as a useful checklist for a secure software process. These get to a level of software engineering detail that is beyond the scope of this book, but they indicate the kind of concrete coding practices that are essential to achieve trustworthy software. Appendix H gives a brief summary of the current list, which is updated periodically. Known best coding practices can eliminate common vulnerabilities to attacks based on injection of spurious user inputs, buffer overflows, and others as described in Appendix G. For example, the attack surface associated with the user access behavior in Fig. 10.10 could be quite large if secure software development is not applied to the various system components involved in it.

10.9.2 Security Testing

A variety of testing methods has been developed for assessing the effectiveness of security controls and other safeguards and for probing a system or its individual components for flaws and vulnerabilities. This is a matter of importance both to the engineers and architects responsible for a secure system solution and to the engineering team in general. In a system development or modification program, security testing must be integrated with the overall system test effort. Once a system is delivered and in operation, the need for ongoing assessment as part of governance means that some level of security testing will be a routine occurrence.

In the most common situation where system hardware is primarily purchased off the shelf, security testing is largely a vendor responsibility aimed at establishing required levels of trust through CC testing or some equivalent process. EALs are commonly published as a key element of product specifications. System developers must employ trusted components in a fashion that does not compromise their secure attributes, and testing of the complete, integrated system should confirm that no hardware flaws have been introduced.

The main focus of security testing is on software because that's where most vulnerabilities arise. Testing should be an integral part of software development, whether in an Agile methodology such as Sprint or in a more traditional approach. This simply means that security testing is incorporated along with all the other testing that is a natural element of software development. The software development team should include, or have easy access to, security software testing expertise. Appendix I summarizes the most important categories of software security testing and lists some commonly employed tools.

10.10 Cybersecurity for the Smart Microgrid

We can illustrate the application of the concepts and techniques covered in this chapter using our Community Smart Microgrid (SMG-A) example. We will make some assumptions about the security characteristics and requirements of the SMG and consider how a tailored cybersecurity solution can be defined that maintains system performance and reliability while providing adequate protection against cyberattacks. Electric power is a core element of national critical infrastructure and is of great and growing concern as a target of malicious actors. As in other uses of the E-X and SMG examples in this book, we will not attempt a comprehensive treatment, but will illustrate some key aspects of achieving an acceptable level of security risk.

First, we need to identify the critical assets of SMG-A that require protection. Such a list, in approximate order of sensitivity, would include:

- Equipment control processes and data – this is the most critical asset category since the most serious threat involves interrupting electric service by seizing control of power resources or causing damage.
- Current operational and status data – this includes levels and directions of power flows, technical data exchanges with a Primary Grid (PG), equipment status, and similar data.
- Customer data – this includes exploitable information such as names and addresses and credit cards used to pay bills.
- Business data – this includes data exchanged with PG service providers.

The first two asset categories are typically associated with the functions of a System Control and Data Acquisition (SCADA) system. It has been widely reported that existing SCADAs, which are ubiquitous in the power industry and were designed with little or no consideration of cyber threats, are especially vulnerable to attack. A related hazard is that an attacker who gains control of the SMG may cause unsafe operating conditions that threaten the lives and safety of the system's legitimate users.

Next, we must consider possible threat agents, i.e., attackers. We will address three kinds:

- Hackers and thrill seekers who, as noted earlier, are after perverse satisfaction simply by making mischief

- Terrorists and the Advanced Persistent Threat (APT) who are attempting to cause damage to the nation and its citizens
- Insiders who may be either malicious or simply careless and poorly trained

The vulnerabilities threat agents may seek to exploit could include the following:

- Software containing unpatched vulnerabilities, which could enable unauthorized actions, injection of false commands, exposure of sensitive data, and other attacks
- Inadequate malware protection, potentially allowing an attacker to:
  - Corrupt or steal ("exfiltrate") SMG data and software.
  - Gain privileged access to the system, even to the extent of taking control.
  - Infect the Primary Grid and other external entities.
  - Commit other attacks such as ransomware.
- Inadequate access control, which allows an attacker to penetrate the system to inflict damage, steal information, and command unsafe equipment operation
- Inadequate physical security that allows an attacker to access system hardware to alter configurations and settings, replace trusted components with malicious ones, override or disable security controls, and do many other things

Given this threat environment, Table 10.4 contains some representative content of a security policy or SECOPS for the SMG along with corresponding security

Table 10.4 Examples of security policy and requirements for the Smart Microgrid

Policy statements	Requirements
The system shall implement strong access controls to ensure Confidentiality of system content	• The system shall implement Mandatory Access Control • The system shall enforce multifactor authentication • The system shall enforce a strong password policy with automated strength checking • The system shall enforce strong user account management • The system shall implement Role-Based Access Control with fine-grained access to resources (Principle of Least Privilege)
The system shall implement a layered defense to maximize Confidentiality and Integrity of system content	• The system shall implement security controls at the system periphery, Microgrid LAN, Supervisory Control, application software, database, and Microgrid Device levels • [Specific controls at each level should be specified]
The system shall protect data at rest, in transit, and in use	• The system shall encrypt all data in storage • The system shall implement VPNs for all external data communications, including with the PG • The system shall employ application software that prevents unauthorized exposure of data

requirements. Both policy and requirements should be supported by a rigorous risk analysis as well as a clear statement of security policy endorsed by the management of the SMG and coordinated with the PG. For example, the SECOPS might specify overarching security policy such as:

- No more than one data breach per year of any kind and no more than one breach of critical operational content every 10 years
- Service outages not exceeding 4.4 h/year (operational availability of 99.95%) whether due to hostile actions or natural occurrences

Specific policies and requirements like those in Table 10.4 would then be defined to support these overall security outcomes.

The SMG is a relatively small system, and careful design is essential to meet the stringent security policy with acceptable cost and operational impact. While some of the requirements cited in Table 10.4 might appear excessive, the importance of resilient cybersecurity for the SMG means that they should be satisfied to the maximum possible extent. There are two information processing enclaves: the Supervisory Control computer and the controllers and other processing embedded in SMG devices. These correspond to the Supervisory and Device Levels in Fig. 5.6. Most security controls apply to the Supervisory Control computer, but the lower level must also be accounted for. The system users are the operational, technical, and business members of the SMG staff, and selected PG personnel also need access to certain SMG information. A representative set of security controls would include the following specific safeguards, and Appendix F gives further detail.

- A Unified Threat Management (UTM) Firewall protecting the Supervisory Control computer and providing packet inspection, IDS/IPS, DLP, and possibly other perimeter defense functions. A Web Application Firewall may be required, depending on what functionality is exposed via the Web Port. Use of one or more advanced anti-malware products that use techniques such as deep learning (artificial intelligence) is mandatory.
- A whitelist of approved external IP addresses allowed to interact with the system via the Web Port.
- Monitored input/output, including traffic logging and detection of suspicious behavior.
- Secure software procedures, including:
  - Use of software products with appropriate trust certifications, including both system software (operating system, utilities, and database management) and applications
  - Continuous monitoring of published software security patches with immediate installation
  - Periodic penetration testing of the SMG network
- VPNs for all external data communications.
- Prohibition of the use of vulnerable devices such as wireless connections and USB memory.

- An encrypted SMG LAN with dedicated IP addresses for all connected devices (static LAN topology).
- Access controls, including multifactor authentication and enforced strong passwords; since both the number of users and the number of protected processes and databases are small, an effective access control solution consists of:
  - A basic system-level account manager requiring two-factor authentication (e.g., password and physical token or biometric) to log on to the system
  - Logging of all successful and unsuccessful log-on attempts
  - Automatic password strength checking plus forced password changes every 30 days
  - Secondary password-protected access to individual functions and data records
- Logging and weekly analysis of system activities; when feasible, this can include additional information such as machine identifiers, equipment and user locations derived from GPS, and many other details that can be important in detecting malicious activities and in analyzing cybersecurity incidents.
- Annual security refresher training for all personnel with system access.
- Physical security, including a perimeter fence around the central facility, locking doors, security cameras, and locking enclosures for equipment. At the device level, all sensitive items should have anti-tamper enclosures.
- A plan for incident response (both natural events and cyberattacks) with provisions to restore or maintain electrical service with minimum outage time.

10.11 Summary

This chapter has briefly surveyed an exceptionally broad and complex topic. Both private and public sector systems and the information environments that support them are under serious and growing attack from many quarters, including personnel wrongly assumed to be loyal and honest and national entities interested in stealing national secrets and intellectual property or even causing harm to a population. Every aspect of architecture practice is implicated in achieving secure, safe, and reliable systems, from initial threat assessment and security requirements to the provision of all required protective mechanisms, to product selections and system security testing. Much of the detail of security architecture and design is best entrusted to experts, but the system architect requires a working knowledge of the problem, the available solutions, the state of the technology, and the policies and best practices involved.

Exercises

1. List some typical security threats to information-intensive systems such as (a) a major bank or other financial institution, (b) an online retailer, and (c) a transportation company such as a railroad.
2. Define the following elements of a cyberattack by a foreign criminal syndicate against a domestic manufacturer: Threat Agent, Threat, Asset, Vulnerability, Security Control, and Exposure.

3. List some of the major factors in a successful system security solution for a school district that must protect a wide range of personal information, test materials, and other sensitive data.
4. List the three categories of cybersecurity safeguards or preventive measures and give examples of each.
5. How would ABAC and RBAC apply to the behavior diagrammed in Fig. 10.10?
6. What are the two fundamental elements of a resilient and affordable cybersecurity solution?
7. Describe the basic steps of risk identification (RI) and risk analysis (RA) for a national telecommunications network.
8. Describe the primary differences between traditional Certification and Accreditation and the more recent Risk Management Framework, and identify the advantages of the latter.
9. Considering the secure architecture development flow in Fig. 10.5, identify some aspects of Governance that would apply to the Secure Architecture, Implementation, and Monitoring, Maintenance, and Updating phases of the process.
10. Why would security controls such as firewalls and intrusion detection systems be applied at multiple levels of a Defense-in-Depth structure?
11. What is the basic use of an Evaluation Assurance Level (EAL)?
12. Considering the Boundary Security System in Fig. 10.7 and the Domain structure of the E-X example in Figs. 4.5 and 4.6, where would the various components of the BSS be implemented in the E-X?
13. What are the factors that would lead to the choice of a secret key or public key encryption scheme for a particular system or enterprise?
14. List some user attributes that can be used in an Attribute-Based Access Control scheme.
15. What are some factors that make securing a cloud environment more challenging?
16. Considering the phases of a Secure Software Development Life Cycle as shown in Fig. 10.11, list some specific elements of Software Security Engineering that might be applied at each phase of the Software Engineering Flow.

Student Projects

Students should add cybersecurity content to their architecture models based on the nature of each system using the concepts and techniques presented in this chapter, including risk identification and analysis, application of layered defense, and considerations of initial and recurring system testing and approval to operate.

References

1. Verizon Enterprise Services (2017) Data Breach Investigations Report (DBIR). http://www.verizonenterprise.com/verizon-insights-lab/dbir/2017/. Accessed 30 May 2017
2. Veris Community (2014) VERIS framework. http://veriscommunity.net/veris-overview.html. Accessed 31 May 2017

3. Whitmore JJ (2001) A method for designing secure solutions. IBM Syst J 40(3):747–768
4. Hershey P, Silo C (2012) Procedure for detection of and response to distributed denial of service cyber attacks on complex enterprise systems. In Proceedings of 6th Annual International Systems Conference, Vancouver, 19–22 March 2012, p 85–90
5. NIST Computer Security Division (2006) Guide for developing security plans for federal information systems, NIST SP 800-18. rev 1. http://nvlpubs.nist.gov/nistpubs/Legacy/SP/nistspecialpublication800-18r1.pdf. Accessed 31 My 2017
6. NIST Computer Security Division (2006) Minimum security requirements for federal information and information systems, FIPS 200. http://nvlpubs.nist.gov/nistpubs/FIPS/NIST.FIPS.200.pdf. Accessed 31 May 2017
7. Snyder D et al (2015) Improving the cybersecurity of U.S. air force military systems throughout their life cycles. Rand Corporation Research Report, Santa Monica, CA
8. SANS (2016) The CIS critical security controls for effective cyber defense ver 6.1. https://www.sans.org/critical-security-controls, http://www.tenable.com/solutions/council-on-cyber-security-critical-security-controls. Accessed 30 May 2017
9. NIST Computer Security Division (2012) Guide for conducting risk assessments, SP 800-30. http://nvlpubs.nist.gov/nistpubs/Legacy/SP/nistspecialpublication800-30r1.pdf. Accessed 31 May 2017
10. Microsoft (2014) Threat modeling tool. https://www.microsoft.com/en-us/download/details.aspx?id=42518. Accessed 31 May 2017
11. MITRE Corp (2017) Common weakness enumeration. https://cwe.mitre.org. Accessed 31 May 2017
12. Chew E et al (2008) Performance measurement guide for information security, NIST SP 800-55. National Institute for Standards and Technology, Gaithersburg
13. Bowen P et al (2006) Information security handbook: a guide for managers, NIST SP 800-100. National Institute for Standards and Technology, Gaithersburg
14. Rozanski N, Woods E (2005) Software systems architecture: working with stakeholders using viewpoints and perspectives. Addison-Wesley, New York
15. Kindervag J (2013) Market overview: network segmentation gateways, Q4 2013. Forrester Research, Cambridge
16. Jones R, Horowitz B (2012) A system-aware cyber security architecture. Syst Eng 15(2):225–240
17. Trusted Computing Group (2014) TPM library specification. https://trustedcomputinggroup.org/tpm-library-specification/. Accessed 31 May 2017
18. Neuman C, et al. (2005) The Kerberos network authentication service (V5). https://tools.ietf.org/html/rfc4120. Accessed 31 May 2017
19. NIST Computer Security Division (2001) Specification for the Advanced Encryption Standard (AES), FIPS 197. http://nvlpubs.nist.gov/nistpubs/FIPS/NIST.FIPS.197.pdf. Accessed 31 May 2017
20. Rivest R, Shamir A, Adleman L (1978) a method for obtaining digital signatures and public-key cryptosystems. Commun ACM 21(2):120–126
21. NIST Computer Security Division (2001) Security requirements for cryptographic modules, FIPS 140-2. http://nvlpubs.nist.gov/nistpubs/FIPS/NIST.FIPS.140-2.pdf. Accessed 31 May 2017
22. Priebe T et al (2007) Supporting attribute-based access control in authorization and authentication infrastructures with ontologies. J Software 2(1):27–38
23. McGraw RW (2014) Risk Adaptable Access Control (RAdAC). http://csrc.nist.gov/news_events/privilege-management-workshop/radac-Paper0001.pdf. Accessed 31 May 2017
24. OASIS (2004) SAML 2.0 profile of XACML. http://docs.oasis-open.org/xacml/access_control-xacml-2.0-saml_profile-spec-cd-02.pdf. Accessed 31 May 2017
25. Keegan W (2014) Separation kernels enable rapid development of trustworthy systems. COTS J 16(2):26–29
26. Kaplan F (2016) Dark territory: the secret history of cyber war. Simon and Schuster, New York

27. National Computer Security Center (1987) A guide to understanding audit in trusted systems. https://fas.org/irp/nsa/rainbow/tg001.htm. Accessed 31 May 2017
28. Scarfone K, et al. (2008) Technical guide to information security testing and assessment, NIST SP 800-115. http://nvlpubs.nist.gov/nistpubs/Legacy/SP/nistspecialpublication800-115.pdf. Accessed 31 May 2017
29. National Vulnerability Database (2017) Guide for assessing security controls in federal information systems and organizations, NIST SP 800-53, rev 4. https://nvd.nist.gov/800-53/Rev4. Accessed 31 May 2017
30. NIST Computer Security Division (2014) Guide for applying the risk management framework to federal information systems, a security life cycle approach. http://csrc.nist.gov/publications/nistpubs/800-37-rev1/sp800-37-rev1-final.pdf. Accessed 31 May 2017
31. NIST Computer Security Division (2014) Federal Information Processing Standards (FIPS). http://csrc.nist.gov. Accessed 31 May 2017
32. Stoneburner G, Hayden C, Feringa A (2004) Engineering principles for information technology security (A baseline for achieving security), rev A. http://nvlpubs.nist.gov/nistpubs/Legacy/SP/nistspecialpublication800-27ra.pdf. Accessed 1 June 2017
33. Common Criteria Recognition Agreement (2017) Common Criteria, ver 3.1, rel 5. http://www.commoncriteriaportal.org/cc/. Accessed 31 May 2017
34. DISA (2004) Determining the appropriate evaluation assurance level for COTS cybersecurity and cybersecurity-enable products (white paper). Defense Information Systems Agency, Ft Meade
35. Yang K (2016) A 'demonically clever' backdoor. Michigan Engineer, Fall 2016
36. Cloud Security Alliance (2017) Cloud security research reports (multiple). https://cloudsecurityalliance.org/. Accessed 31 May 2017

Chapter 11
Using Prototypes, Verification, and Validation to Evaluate and Enhance System Architecture

11.1 Architecture and System Evaluation

In Chaps. 1–10, we have developed the MBSAP methodology and explored extensions to deal with a variety of system types. We now turn our attention to the vital question of establishing that an architecture, and a physical system that instantiates it, will satisfy the needs and expectations of system stakeholders. There are multiple methods for doing this. Testing begins as part of system design, (referred to in this chapter as Development Test, or DT) and continues with testing of the system itself in its intended operational and support environment (referred to as Operational Test, or OT). These and other methods of assessing compliance with requirements are discussed in Sect. 11.3. Our focus here is on the central role of prototyping, both physical and virtual, in system assessment and refinement. We treat all of these as a single, integrated topic, in which the various assessment methods are complementary and each method makes a particular contribution.

Most system engineers think about this subject in terms of verification and validation (V&V) [1]. We will discuss V&V in detail, but for the moment, we simply note that verification deals with proving that a system complies with its requirements, sometimes expressed as "Did we build the system right?" Then validation deals with whether those requirements, and thus a system that implements them, represent an acceptable solution to stakeholder needs and expectations, sometimes expressed as "Did we build the right system?"

In the SE Vee diagram, Fig. 3.2, horizontal links connect the left and right sides of the Vee at several levels. These represent the conventional concept of V&V in which each stage of system integration and test (I&T), the right leg of the Vee, provides measured and analytical data to evaluate the corresponding level of the system design in the left leg. With MBSE, the architecture model and virtual prototype allow an earlier start on V&V while the details of the system design are evolving, and subsequent data from I&T activities then serve to refine and confirm the modeled results.

J. M. Borky, T. H. Bradley, *Effective Model-Based Systems Engineering*,
https://doi.org/10.1007/978-3-319-95669-5_11

Before getting into the particulars of V&V, we need to expand our treatment of prototyping to a fuller consideration of the nature and roles of prototypes in the MBSAP methodology. Chapter 3 showed the MBSAP cycle with a prototyping environment as the central element that ties the various Viewpoints together. We described this in terms of physical and virtual prototypes. A well-designed and properly resourced prototyping environment knits together the various aspects of architecture and system evaluation. Importantly, this includes early use of modeling, simulation, and analysis (MS&A) to explore alternative system concepts and to refine and validate requirements and a system concept with stakeholders.

As architecture development proceeds through the Operational, Logical/Functional, and Physical Viewpoints (OV, LV, and PV), more and more detailed MS&A, which is the content of a system virtual prototype, supports a continuing stakeholder dialog to ensure the emerging system solution is correct and satisfactory. As a physical prototype becomes available, measured data allow both the architecture team and the stakeholders to confirm MS&A results, refine and calibrate the virtual prototype, and build confidence in the system design. At the same time, the outputs of these prototypes are invaluable in identifying problems, developing corrective actions, and feeding the results back into the requirements and the system architecture. This feedback enables a kind of "continuous V&V" under which early problem correction avoids larger cost and schedule impacts when problems surface in later phases of a program. We will return to these ideas in Sect. 11.3.

11.2 Prototyping in MBSAP

Given the importance of prototyping in architecture assessment, we now add some important practical considerations that underlie an effective MBSAP prototyping environment. High-technology systems have a lot of moving parts, meaning that processes, data, external events, computing resources, networking, machine and vehicle control, and other things interact in complex ways that can be hard to predict on the basis of the individual parts.[1] SysML modeling in the MBSAP methodology is intended to help tame that complexity, but even with the best of models, there remains a risk that undetected design flaws, unpredictable sequences of behavior, unsatisfactory system states, and other problems may occur. Well-designed architecture diagrams can describe a system clearly and in detail, and a skilled architect can use them to make judgments about important behavioral issues. However, diagrams are static and must be brought to life before the dynamics of system behavior can really be explored. Moreover, design mistakes and anomalous behaviors often show up only when real software runs on real hardware and processes real data derived from real equipment, in other words in the actual system or at least a prototype of it.

[1] This is the essence of "emergent behavior."

An adequate prototyping capability is a critical element of a complete architecture and SE strategy, and a program's chances of success improve dramatically when one is provided and properly used. For example, the specialized discipline of aircraft flight control engineering has for decades insisted on what is usually called an "iron bird," especially in developing a control system for an advanced aircraft with highly dynamic flight characteristics. The iron bird is specialized prototype that provides an integration test bed for electrical, hydraulic, flight control, and other vital aircraft systems. Among other things, it is used to refine and validate the parameters ("gains") of the control laws in the flight control system. Actual flight test is still required, and gains are adjusted until the required stability and flying qualities are achieved, but without the iron bird, the aircraft would require much more expensive flight testing and might not be initially safe to leave the ground. Perhaps less dramatic, but equally important, benefits of prototyping apply in any complex system.

11.2.1 Physical Prototypes

The physical part of a prototyping environment, especially for systems with significant software and data content, is often called a System Integration Laboratory (SIL). This is where many of the activities in the right leg of the Engineering Vee of Fig. 3.2 take place. It may actually involve a hierarchy of SILs used by component developers or vendors, subsystem and system developers and integrators, and enterprise integrators. These SILs are used to build up, test, and verify parts, components, subsystems, and ultimately a complete deliverable system. In an evolutionary spiral program, a progressively more complete prototype results from successive spirals, ultimately becoming the actual system. There is inevitable tension between the desire to make a given prototype as complete as possible and the reality that the necessary hardware, software, and instrumentation can be expensive, especially when basic design decisions such as product selections are not final, creating a risk that costly products and tools will be bought but ultimately not used. Partly due to this, provisioning a prototype and the SIL or other environment in which it is built has historically been a popular area for program management to be penny wise and pound foolish. The architect, as part of overall responsibility for the technical integrity of the design, should have a prominent voice in defining and justifying the necessary prototyping resources.

Following the general principle that the prototype is matched to the system, an integration facility reflects the content and priorities of the program it supports. For example, a program in which both hardware and software are being developed, especially when advanced technology is employed, will necessarily assemble the prototype in a sequence that is driven by the availability of components and subsystems. In another common situation, where a program is developing or integrating software on an off-the-shelf computing platform, the prototyping strategy will typically focus on instrumentation for diagnosing software defects and for measuring performance parameters such as thread execution times. To deliver best results, the integration environment needs comprehensive diagnostic and performance measurement resources for both hardware and software to enable problems to be traced to their root causes. This key aspect is the subject of Sect. 11.2.3.

A physical prototype is often needed to find and fix subtle problems such as:

- Design flaws in hardware and software components that only appear when those components operate together in realistic conditions
- Inconsistent or erroneous design specifications
- Problems with timing and synchronization, including those due to improperly designed cabling
- Inconsistent or erroneous specification and implementation of interfaces
- Unpredicted interference caused by electrical noise
- Flaws in the design of power and cooling

Regardless of the specifics of a given system or development program, the following paragraphs summarize some overall considerations in defining and using physical prototypes.

Incremental System Builds

A specific set of hardware, software, and other elements that make up all or part of a system in a prototype is commonly called a "build." It's highly useful to provide for both major "system builds" and smaller, more frequent "development builds" as a system is progressively integrated and tested. The details depend on a specific system development cycle, e.g., an agile SE methodology vs. a traditional waterfall strategy. The general principle is to provide for prototyping experiments on a scale from small, involving only one or a few elements of the system, all the way up to the full system. A major system build might constitute completion of a development spiral, the integration of a significant subsystem, or some other primary integration milestone. Development builds are smaller in scope, less formal, less expensive to plan and execute, and suitable for approval at lower levels of the engineering organization. They add relatively small capability increments, verify problem fixes, and support ongoing engineering activities such as verifying the results of trade studies. A series of development builds generally leads up to a system build.

Integration Hierarchy

Integration and test proceed up the right leg of the SE Vee, starting with components and small assemblies and moving on to subsystems and the full system. Effective system integration and prototyping require that lower-level integration facilities, e.g., at subcontractor or vendor facilities, also be adequately resourced. Here, the key principle is that lower level prototype must be able to progressively build, integrate, and test system elements in order to minimize the danger that undiscovered or unresolved problems will propagate to later stages of integration where they are much more expensive to correct. Similarly, it's important to ensure that integration facilities at all levels are adequately instrumented so that test conditions and data collection are sufficiently comprehensive to allow efficient I&T and to diagnose problems as early as possible.

Realistic Prototyping

When using test equipment such as sensor and communications device emulators to stimulate a configuration under test with data and signals, it's important to ensure that there is adequate fidelity in reproducing the outputs of the real components or subsystems. This is particularly challenging when those system elements are being developed in parallel with the resources undergoing I&T and may call for emulation of worst case data rates, timing, file sizes, and so forth.

Prototype Configuration Management

The SE process should impose configuration management on even early prototypes and should enforce adequate test planning and documentation of results for both system and development builds. The prototype configuration must faithfully track the currently approved design baseline, which should be documented authoritatively in the architecture model and supporting artifacts. For example, the racks and cabling used to install and connect system hardware in the integration environment should exactly match the actual design to ensure that things like signal delays and voltage drops are accurately reproduced.

Long-Term Prototype Use

Finally, an effective program should provide for the prototype, which often represents a significant investment, to be planned with a long-term view, so that it will be useful over the system's lifetime to support modifications, software maintenance, hardware problem fixes, technology refreshment, and other activities. Thus, for example, it should not be placed in a space likely to be claimed for another purpose after the system is initially delivered and should have facilities for engineering workstations, data archiving, detailed test setups, and other aspects of a stable and productive design, integration, and test environment.

A simple SIL or other integration environment might resemble the laboratory sketched in Fig. 11.1. Depending on the prototype, the resources typically include:

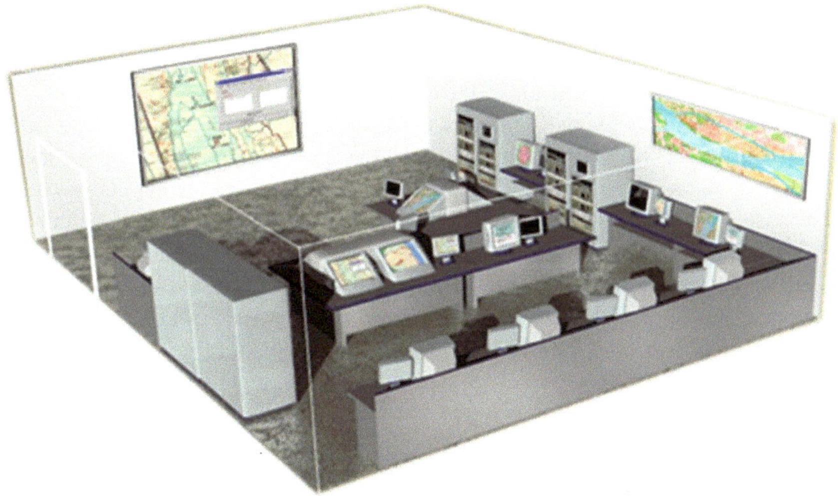

Fig. 11.1 Basic System Integration, Laboratory (SIL)

- One or more areas where operating mission equipment such as sensors, actuators, electrical controls, radios, navigation systems, and operator stations can be assembled, interconnected, and exercised; this is often referred to as a "spread bench" or "mock-up."
- Equipment racks with computing infrastructure, power supplies, audio communications, hardware instrumentation such as bus monitors and test data recorders, and other equipment.
- Networks and buses for interconnection, generally with extra ports to accommodate instrumentation; as was noted earlier, the cabling and other wiring that connects system elements must accurately match the design of the real system.
- Stations at which software engineers can access, modify, and load software in conjunction with working on the prototype.
- Additional computing resources for simulators, network and processor monitors, external network connections, test controllers, and utilities such as user account management.
- Large format displays to allow groups to view scenarios as they are executed and to review prototype data and status.

11.2.2 Virtual Prototypes and Executable Architectures

A simulation derived from an architecture model is sometimes called an "executable architecture." In its most general sense, this includes any method of animating a modeled representation of the architecture, at any level of abstraction, to dynamically reproduce one or more aspects of its behavior. Executable architectures have in common that their output includes observable changes over time in such system characteristics as platform location and velocity, system state and operating mode, parameter values, data flows, operator interactions, task queues, resource loadings, and many others. This definition actually covers physical as well as virtual prototypes, but the term is usually applied to the latter, and that is the focus of the discussion. For example, McQuay [2] has described an initiative that seeks to make virtual prototyping the central element of a collaborative engineering environment aimed at reducing cost, risk, and schedule in the development of complex avionics systems, from early concept and requirements analysis, through technology development, and into system design and implementation.

Figure 3.8, reproduced here for convenience as Fig. 11.2, introduced the notion of MS&A at four levels of abstraction corresponding to the Axis of Abstraction of the architecture taxonomy in Chap. 1. By definition, a physical prototype resides at the lowest level. In MBSAP, a virtual prototype is the assembly of whatever simulations at whatever levels of abstraction and with whatever degrees of fidelity are judged useful in developing, analyzing, refining, and communicating the architecture. Accordingly, a system in which requirements and their validation depend heavily on operational scenarios will need attention to the operational level, while one that has significant performance and cost issues associated with defining the

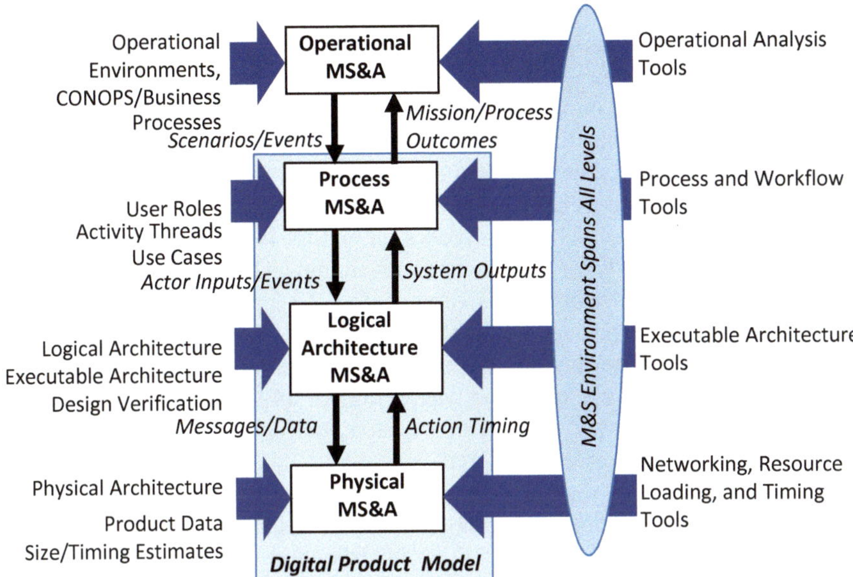

Fig. 11.2 Levels of abstraction matched with levels of modeling, simulation, and analysis to deal with architecture and design issues, using corresponding tools and interacting with the layers above and below

computing infrastructure and allocating software and data to it would probably want to do significant physical modeling. In the literature and in common practice, the term "executable architecture" is generally taken to refer to the logical and sometimes the physical level in terms of animating a UML/SysML model, but we use it more generally.

The tools that are selected for a given program strongly affect the ways in which prototypes are constructed and used, especially the virtual kind built up from layers of MS&A. In this book, we stress general principles and remain as nearly as possible agnostic with respect to tools, except for the examples later in this chapter, which necessarily use specific tools. This is partly due to the fact that any tool-specific methodology will rapidly become obsolete because these products constantly evolve and partly from a desire to be independent and objective. Consequently, this will be a relatively brief discussion that will provide overall guidelines, assuming that architects will, in the normal course of their practice, do tool trade-offs and define a prototyping environment that meets their needs. The following paragraphs expand on the preliminary description of virtual prototyping that we gave in Chap. 3.

Operational MS&A

A great many tools are available for representing an operational environment. For example, a space system might use a simulation that includes things like orbital dynamics, space weather, ground tracks, sensor field of view, and communications links, as well as high-level system behaviors. A commercial aviation analysis might look at statistics of on-time departures and arrivals, aircraft utilization, cargo delivery, and required aircrews. Operational models commonly allow events to be introduced and might even model the decision logic employed by synthetic human players. For an industrial or other private sector system, such tools are less likely to exist, and the architect may find it necessary to build enough of a simulation of the external environment to provide inputs for both virtual and physical prototyping. Typical Measures of Performance and Effectiveness (MOPs/MOEs) involve things like probability of successful business process or mission accomplishment, resource consumption, execution times and latencies, maximum task loading, and interoperability within an enterprise or other environment.

In MBSAP, Operational modeling can be used to support the mapping of the Capabilities Database to the OV by yielding insight into the makeup, interactions, information requirements, and capabilities of an enterprise and its nodes and systems. The customer for a given development effort may have preferred OA tools, standard scenarios, and predefined MOPs/MOEs. Provided the tools and models of choice are mature and properly documented, and especially if they conform to MS&A standards like the High-Level Architecture (HLA) [3] and Distributed Interactive Simulation (DIS) [4], they can be readily used in an MS&A framework.

Process MS&A

Operational MS&A provides a macroscopic view of a system's environment and operational scenarios. Next, this picture is refined to be specific to the activities of the enterprise or system being architected through modeling of business processes, mission threads, workflows, or other high-level behaviors. This is sometimes called Business Process Modeling (BPM). These models are typically built in a general-purpose process modeling tool. They are associated with Use Cases and are used to animate and instrument scenarios. For example, a process model could represent the individual activities involved in receiving, analyzing, scheduling, and carrying out a task. In execution, the model would have blocks for actions, processing steps, and human decision times at each stage and would connect these blocks with links that are parameterized with the data rates and latencies of the communications channels to be used. This same sequence of activities is the basis for constructing a workflow as described in Chap. 7 to be managed by a software component in the infrastructure of a system or node. Once again, a variety of tools are available, and both continuous and discrete event simulations may be used, depending on the nature of the problem.

Operational and process analyses are both primarily associated with the OV and have significant interaction, including exchanges between the two layers as suggested in Fig. 11.2. In a typical example, a scenario generator in the operational MS&A suite is used to create a time-tagged set of events that are then used to stimulate a process model. Conversely, the results of process and workflow MS&A may be used both to calibrate the steps of a flow in the more abstract operational MS&A environment, especially in terms of timing, and in exploring the details of critical events and behaviors. Modeling at this level of abstraction is important to ensure that behaviors in the OV are fully understood. It is also very useful in performing top-level performance and timing estimates, allocating deadlines to various workflow steps, balancing task loads among system operators, and analyzing various other aspects of the architecture.

In a Service-Oriented Architecture (SOA, Chap. 7), an alternative approach to process modeling is the use of BPEL, short for the Web Services Business Process Execution Language [5]. BPEL is an XML-based language standard for specifying Web Service interactions and can therefore be used to model SOA processes in general and orchestrations in particular, especially those involving service invocations on external entities. BPEL deals with messages and their timing, variables, decisions, and other aspects of a Web Service-based process. As described in Chap. 7, an orchestration can be controlled by a central management component such as a workflow engine to implement the process in a system. If a graphical process description is desired, e.g., to support up-front process definition, the Business Process Modeling and Notation (BPMN) specification [6] defines flowchart-like graphical semantics. Some tool vendors are working to achieve an integrated BPEL/BPMN modeling environment in which graphical process design leads to an executable process model.

Logical Architecture MS&A

This level is concerned with what the system architecture community most commonly refers to as executable architecture. As the functional domains of the OV are realized using Blocks in the LV, the diagrams of the SysML model can be simulated/animated and analyzed. In current architectural practice, this stage of modeling commonly falls into either or both of two basic categories.

- *Simulation of Model Dynamics* – the simplest approach, putting Behavior Diagrams, typically Sequence, Activity, and State Machine Diagrams, into a tool that can simulate the modeled behaviors to explore timing (e.g., by calibrating process steps with estimated execution times), parallelism (e.g., by verifying that synchronization steps occur correctly), and responses to external events, but without trying to generate real software. Increasingly, mainstream SysML modeling tools provide this capability, and this is the primary type of executable architecture of interest to system engineers and architects.
- *Executable UML* – used when dealing with an architecture that primarily concerns software, such that UML is the language of choice. This approach is commonly called Executable UML (xUML). Tools are available that can start with a properly annotated UML model and generate code (generically called autocoding), either as a simulation of the system or as actual system software. These tools require both that the UML model be complete and correct and that action semantics that fully define behavior be provided. In terms of the Object Management Group's Model-Driven Architecture (MDA), xUML is the primary method of autocoding used to transform a Platform-Independent Model (PIM) into a Platform-Specific Model (PSM), which can then be run on the target system computing platform.

Once the commitment to this level of MS&A is made, it's a natural step to treat the LV as the primary design representation, because it's the point of departure for implementation. Another useful feature is that the process of animating a diagram in a tool involves rigorous testing of the completeness and correctness of the model (basically, conformity to the SysML language standard) since if the model contains errors, the tool will fail to produce the animation. This is essentially the same process that tools employ in performing model validation. Still another benefit involves timing analysis. Earlier chapters described the initial stages of timing analysis with static allocations of deadlines to Sequence Diagrams and sometimes with Timing Diagrams. An LV executable or simulation adds important fidelity by incorporating more granular timing, synchronization, balking conditions (where a process is halted while waiting for an input, message, event, or other information), and sensitivities to variables such as data rates and task loadings. These results can be invaluable in focusing specific timing measurements in a physical prototype and in defining timing experiments in the physical level of MS&A based on stressing conditions such as deadlines that are exceeded or are met with inadequate margins.

Autocoding is growing in popularity as tools grow in power, efficiency, and generality. For some time, it has been used mainly in areas such as control systems and signal processing which are well defined and reasonably limited in scope so that algorithms, data sets, constraints, and similar factors are manageable. Eventually, autocoding can be expected to become very widespread and to largely supplant traditional source code written by programmers. That progress should broaden the applicability of this type of Logical Architecture executables. For now, the only general rule is that an architecture team should evaluate such tools and make a determination if one or more are available that meet the needs of the program and produce cost, quality, and efficiency advantages that warrant adoption.

Physical MS&A

Some aspects of design correctness and system performance can only be addressed by high-fidelity modeling of processor throughput, network data rate and latency, dynamics of controlled devices and equipment, file access times, routing overhead, sensor and communication device physics, and other physical phenomena. This stage of analysis is focused on specific system resources and design issues, because physical prototypes are expensive and a complete physical model of the system would *be* the system. Typically, physical MS&A looks at things like real or potential process bottlenecks, resource load balancing, and statistical patterns of behavior. The results can be used to calibrate architecture and process models to improve their accuracy. Experience on many programs has shown the power of properly scoped physical MS&A as a way to spot problems with infrastructure, connectivity, and margins early in the development when they are easiest to correct. Accurate physical models can subsequently be linked to I&T activities to extend the range of operating conditions beyond what is affordable with a physical prototype as well as using test results to calibrate and refine the models.

11.2.3 Instrumentation and Control

If the results of prototyping, whether physical or virtual, are to be accurate and useful, attention must be paid to the ways in which stimuli are applied, results are measured, and internal behavior is tracked. Fundamentally, a prototype at any level of the system structural hierarchy must be exercised in a sufficiently realistic representation of its actual operational environment, the right behaviors must be invoked, and the correct functioning and compliance with requirements of the design must be assessed. This sounds obvious, but there are subtleties in properly instrumenting and controlling a prototype that can impede effective use and even lead to false or misleading results. Mittal et al. [7] discuss this in terms of an Experimental Frame that includes stimulus generation, transducers that receive prototype outputs and transform them as necessary, and the resources needed to monitor an experiment and assess the results against criteria for success. Those criteria typically include MOPs/MOEs derived from operational analysis.

The following are some of the forms inputs to a prototype can take, depending on the nature of the system and the status of prototype development.

- *Input Data Stream* – either recorded or artificially generated data can be fed to prototype input points. Examples of electronic stimuli include digitized sensor data (which might be recorded off an actual sensor), a message sequence representing information from a network or data link, and discrete signals indicating events or the status of controlled equipment.
- *Additional Inputs* – a prototype might also need to experience mechanical loads, environmental stresses such as temperature and vibration, or actions of a human operator.
- *Equipment Emulator* – one step-up in fidelity from recorded data is the use of an emulator that acts like a real sensor, terminal, or other equipment item and produces, ideally, the same output as the real thing. Emulators are often available from equipment vendors and have the advantage that they can be configured or scripted to produce specific outputs that would arise in a particular scenario.
- *Physical Equipment* – the ultimate source of realistic inputs to a prototype is an actual mission equipment item, subsystem, or system. Depending on the content of the prototype and the specifics of an evaluation event, such equipment might be another part of the system itself or something in the external environment with which the system interacts. For example, a real communications terminal might be connected to a prototype of a system's data processing resources to verify the interface between them. This is typically the most expensive approach, especially if the equipment must be operated in a real environment, for example, by installing a sensor outdoors and piping the data to the SIL. Prototype testing at this level of fidelity is likely to happen toward the end of the I&T process when the system configuration is well established and the actual equipment is available. The goal is to ensure that various products work together correctly (a key aspect of verification) and that system processes produce the required results

(a similarly key element of validation) before committing to installation in an aircraft, satellite, or other platform.

Additional instrumentation is typically needed for both hardware and software components. Hardware instrumentation can include signal monitors, timing references, bus monitors, recording equipment connected to various internal nodes, built-in diagnostics of the equipment itself, and anything else needed to ensure behavior is within specified tolerances and to help isolate failures. Software instrumentation starts with familiar debugging techniques such as checkpointing but can include displays of workflow execution, recording of software exceptions, data checking against limits (e.g., by analyzing files from hardware instrumentation), and use of utilities that are provided by the products that make up the software infrastructure. A well-equipped integration environment also includes computing resources to control its operation, automate test activities, and allow centralized monitoring and display. A certain amount of judgment is required to balance the expense of instrumentation against the need to find and fix problems early in development, integration and test; to support effective system test planning; and to generate data for V&V. Once again, the architect, as part of the responsibility for ensuring the technical integrity of the system design, should be a strong advocate for adequate prototyping facilities and equipment. Properly managed, this investment can be amortized across multiple programs to minimize the expense to individual programs and users.

11.3 Verification and Validation

We have laid the groundwork for the primary topic of this chapter, which is assessment of architecture and systems against stakeholder needs. We will consider this as an integrated approach in which test, analysis, prototyping, and other methods are used where each is most effective.

11.3.1 Verification and Validation Methods

Verification and Validation (V&V) of system designs is a core SE responsibility. Although they are distinct processes, they are closely related and draw upon many of the same evaluation methods. The differences among types of requirements as discussed in Chap. 4 mean that no single assessment method is appropriate for all of them. Four principal methods are commonly accepted for evaluating compliance with requirements:

- *Test* – the most rigorous method places the system in a situation where specific parameters can be measured and compared to required values.

- *Demonstration* – when quantitative measurements are not appropriate or possible, the system can be exercised or manipulated so that its behavior can be observed to determine whether required functions or characteristics are exhibited.
- *Inspection* – nondestructive examination of the system, typically to determine if certain features or functions are implemented.
- *Analysis* – use of models, computations, theoretical analysis methods, etc., using appropriate supporting data, to evaluate or predict some aspect of system behavior.

However, there are additional sources of knowledge about system characteristics and behaviors, including:

- *Experimentation* – specific investigations of technologies, products, and even concepts relevant to a system to establish their properties
- *Certification* – data from events used to certify a system against criteria such as safety and security, which can also support V&V more generally
- *Modeling and simulation* – the exploitation of virtual prototyping, as further discussed in following sections
- *Comparison* – using the fact that a system, subsystem, or component is identical, or nearly so, to another that is known to be satisfactory as evidence of its compliance with requirements
- *Implication* – in other cases, taking the results of a V&V method as implying that additional requirements have been met even if they are not separately evaluated

Regardless of the methods selected, V&V is most effective when it is integrated with system development and flows into system integration and test. This can be thought of as "seamless verification," and it has multiple benefits in a system development program. It minimizes the seams between program participants, especially the acquiring organization and the contractor or contractors, who design, build, integrate, and test the system. It applies integrated testing techniques to promote a smooth transition from developmental testing to operational system testing. Most importantly, it facilitates early detection and correction of design flaws, poorly formulated requirements, and other problems. It also allows an early start on architecture validation, reducing the risk that defects in the system's foundation will lead to problems during development.

11.3.2 Verification

MBSAP supports these methods. Requirements verification begins with traceable flowdown of requirements to both structural and behavioral elements in the architecture model. This supports verification activities at each level of the system hierarchy based on the resources and behaviors associated with each requirement. These are modeled as <<testCase>> Blocks and linked to <<requirement>> Blocks using

<<verify>> Dependencies. Figure 11.3 shows generically what this looks like. As the figure indicates, the details of a test case can be embodied in a separate document that is linked to the <<testCase>> Block or, in the case of a relatively simple test, simply placed in the Block specification. Note that although we use the term Test Case, the verification event can include any of the methods defined in Sect. 11.3.1. With these relationships established, an architecture tool like those discussed in earlier chapters can generate a report showing how each requirement will be verified. This is the content of a Verification Cross Reference Matrix (VCRM), which is essential content for a test plan.

Whenever feasible, the preferred way to verify that a system satisfies its requirements is realistic and statistically valid testing. An architecture model supports testing in a variety of ways, especially when combined with physical and virtual prototypes.

- *Test Planning* – a common concern with complex, high-technology systems is that exhaustive testing of every system behavior under every operating condition is prohibitive in cost and schedule. Effective test planning is therefore based on identifying stressing test cases whose measured outcomes yield adequate confidence that the full requirements space has been evaluated. Important insight into this challenge can be obtained by analysis of behavioral diagrams to identify conditions such as maximum loading of a system resource or a critical timing condition. As shown in Fig. 11.4, the testing process is often iterative, with

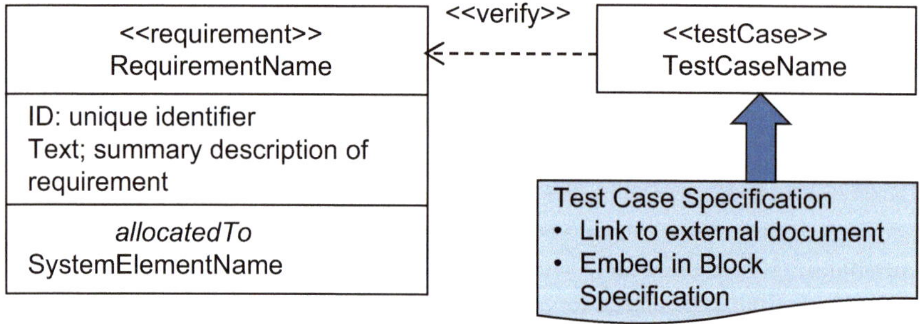

Fig. 11.3 Elements of a requirements diagram linking a test case to a requirement

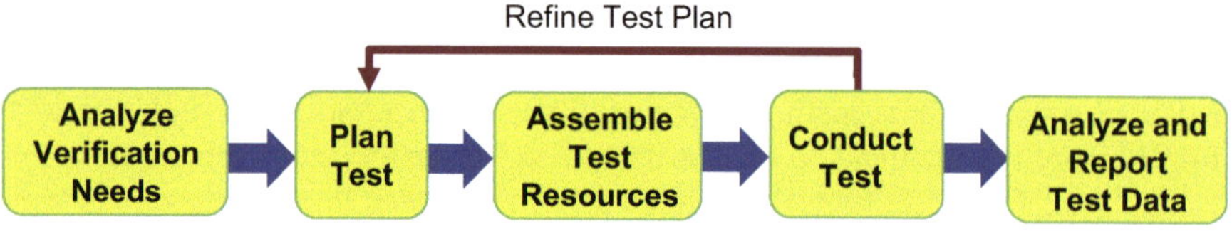

Fig. 11.4 Example of using feedback from test results and other verification events to refine pest plans

feedback from earlier tests, as well as from prototyping, analysis, and other verification methods, serving to refine test plans.

- *Predictions* – analysis of behavioral diagrams, as well as MS&A results, can be used to predict the outcome of a test, which can then be compared to required performance parameters.
- *Test Analysis* – in the non-uncommon situation where test results are anomalous or unexpected, the model is often a valuable tool for diagnosing the measured data and determining underlying causes.
- *Test Documentation*—the model, especially the specifications for model elements, provides a convenient repository for test results, e.g., by linking in test reports for easy retrieval. One of many ways this can be helpful is to integrate test results with other analyses such as a Failure, Reporting, Analysis, and Corrective Action System (FRACAS) [8].

In Sects. 11.6 and 11.7, we show simulations of stressing behaviors for the E-X and Smart Microgrid (SMG) examples. In each case, MS&A produces predictions of test results when performance is measured in a physical prototype. When that data is obtained, we can explore differences from the predicted values by varying model parameters and examining the sensitivities of computed outcomes to those variations. This can both refine the simulations and yield insight into changes in system design or configuration to deal with any measured performance shortfalls. For requirements verification by demonstration or inspection, the situation is similar except that the test case defines a demonstration or inspection event rather than a test point.

Finally, when analysis is the appropriate verification method, the computations can often directly address satisfaction of specific requirements. For example, reliability assessment in terms of a mean time between failures is often carried out using a dedicated reliability modeling tool. However, in cases where the analysis depends on the values of system parameters, the SysML Parametric Diagram as we introduced in Chap. 2 is often the most convenient approach. The necessary equations are captured in Constraint Blocks, and a Parametric Diagram then produces the computed result.

11.3.3 Validation

With MBSE, validation must be considered in two senses: validation of the architecture model itself and validation of the system that implements the architecture. Kossiakoff and Sweet recommend an initial validation step in which the justification for a system, together with customer requirements, technical feasibility, affordability, and other factors, is evaluated analytically against stakeholder needs, desires, and expectations [9]. This concerns the source materials of a system development and is independent of the SE methodology to be employed. Subsequent phases of validation apply to the selected system concept and to the system's design and the capabilities it delivers. The concept is validated analytically against the same stakeholder concerns as the initial analysis, while design is validated against more detailed requirements to determine the system's ability to achieve them. These include specific functions and behaviors needed by a system's users, overall characteristics such as reliability and maintainability, internal and external interfaces and interactions, and so forth. For example, a system that is required to function as an element of a larger networked enterprise has an important validation issue based on establishing that all necessary enterprise processes, services, interfaces, data definitions, and networking mechanisms are properly accounted for.

MBSAP support to validation begins with using the OV as a tool to communicate the architectural foundation to stakeholders, especially using operational simulation. This interaction seeks to confirm that the system concept and overall characteristics are satisfactory or if necessary to correct any deficiencies. A specific example is to demonstrate that the Use Cases are correct and complete in defining all the system behaviors spelled out in a Concept of Operations (CONOPS). Another is to use high-level Domain Diagrams to validate User Roles and their associations with major system areas or subsystems. If stakeholders have preferred operational analysis tools and scenarios, it is important to show how the conceptual system functions when modeled and exercised in that environment. Figure 11.5 shows early, iterative refinement and validation of requirements and architecture, and stakeholder interaction is central to this process.

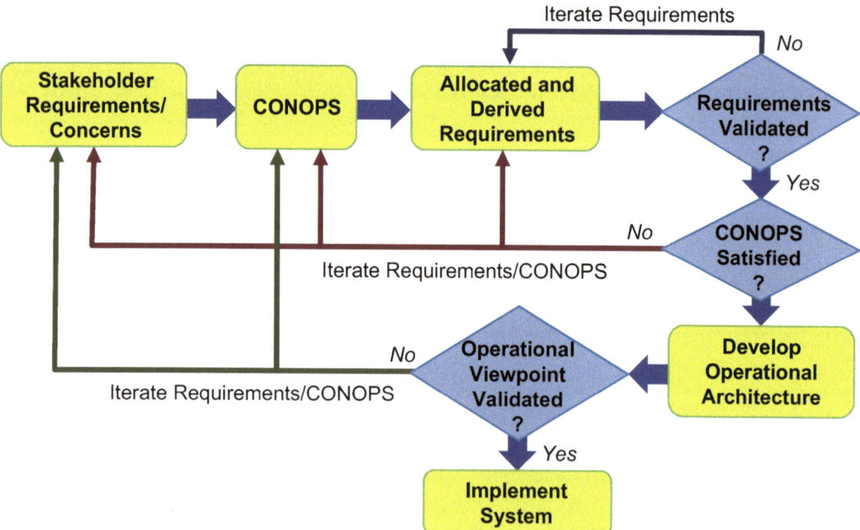

Fig. 11.5 Early validation and refinement of requirements and operational architecture

Operational analysis supports determination of the effectiveness of alternative system designs and employment tactics. It can be helpful for a number of purposes, especially early in a program. These include:

- Defining and refining requirements
- Supporting rapid prototyping to illustrate alternative solutions to operational personnel
- Exploring mixes of organizations, resources, policies, and procedures or tactics
- Identifying issues associated with system operations in a larger enterprise
- Developing and validating operational architectures

As the requirements baseline is defined, refined, and documented, the rigorous flowdown and allocation of individual requirements to system structure and behaviors is a key validation activity. Once the requirements baseline has been approved as fully representing stakeholder needs, validation involves analysis to demonstrate that the architecture provides the necessary system resources and functions. Realistically, the requirements baseline will often evolve as inconsistences and omissions are corrected, new or different stakeholder needs are identified, technology evolves, and other unforeseen factors come into play. This means that validation is likely to continue to be an iterative process, and the architecture model provides a powerful tool for maintaining the requirements baseline and tracking the status and impacts of any changes. Stakeholder dialog continues as progressively more detailed and complete MS&A results from the virtual prototype are communicated and reviewed to ensure the emerging system design remains aligned with their needs and expectations.

The overall system validation process uses information from multiple sources. The results of requirements verification as discussed in Sect. 11.3.2 establish that those requirements have been met, once any deficiencies have been corrected. In most cases, this equates to a system that successfully delivers what its stakeholders need. However, there is more to validation than meeting requirements, which, despite the best efforts of an SE team, may fail to capture everything that is important to operational users and other stakeholders. Accordingly, operational system testing goes beyond requirements verification to confirm that the delivered system has all the features and functions needed to make it acceptable to its users. This testing must confirm that the completed system, in its intended operational and support environment, satisfies critical operational interests, has effective human-machine interfaces, is interoperable within a larger enterprise, meets reliability and maintainability parameters, handles worst-case conditions, and meets any other vital stakeholder concerns.

11.4 Additional Roles for Prototypes

The preceding sections have dealt with the role of prototyping in V&V, complementing test, analysis, demonstration, and other sources of system knowledge. In addition to their use in system evaluation, prototypes can support system development in a number of other useful ways. The following are examples of such areas that can be important in some development programs.

- *Spiral Development* – as noted earlier, in the usual definition of an evolutionary spiral development and integration program, each spiral yields a prototype, ideally both physical and virtual. This was originally conceived as a way to deal with situations where:
  - Requirements cannot be fully defined in advance and must evolve and mature in conjunction with system development as the limits of technical feasibility, cost, operational suitability, customer feedback, and other factors are established during design and test efforts.
  - The system is too complex to be developed in a single pass.
  - Customer needs to dictate early delivery of selected high-priority capabilities.
- The flow of MBSAP supports a spiral paradigm. There are many variations, but a typical sequence for a spiral would be (a) planning and analysis (conducted in conjunction with developing the OV), (b) assessment of alternatives and design (part of the LV), (c) implementation (PV), (d) I&T, and (e) feedback to refine system requirements and select the increment of capabilities to be implemented in the next spiral. It would be rare in dealing with a nontrivial system for even a single spiral to run in a smooth sequence through these steps, and, once again, a key role of the prototyping environment is to provide effective feedback paths by which results of later stages can be used to correct problems in earlier ones.

- *Executable Specifications* – unlike natural language documents, a well-formed model is a clear and unambiguous description of a system. The history of complex system development is replete with examples where inconsistent interpretation of timing relationships, data formats, control flows, and other details created major problems, and no one has found a way to write a conventional specification that two parties cannot manage to interpret differently. It is therefore a logical extension of an architecture-centric methodology to treat the model and its artifacts as the authoritative specification of the system. Such an "executable specification," in addition to the role of the system model in the SE process, can be used to communicate across language barriers, to move design information among tools, to quantitatively compare predicted and measured performance, and to support system modifications, upgrades, and technology refreshment.
- *Software Testing* – extending the overall role of a system prototype in testing, additional considerations may apply to a software system or product. There is a formal discipline for software testing that has come to be called Model-Based Testing [10]. This starts with a model of the system, as opposed to just the software source code, and applies a range of methods to derive a suite of test cases. The approach is supported by a variety of commercial tools. Clearly, the existence of an accurate model from the core development methodology makes the application of Model-Based Testing more feasible.
- *Trade Studies* – virtual prototypes at various levels of abstraction are frequently an effective way to quantify the alternatives in a trade study. Operational MS&A

can be used with requirements trades for such purposes as computing the value of a performance requirement at which the cost to satisfy it becomes disproportionate to the benefit (i.e., the knee in the cost/benefit curve). MS&A at more concrete levels can often be helpful in conjunction with design and product selection trades.

- *Human-in-the-Loop* – often an important component of stakeholder dialog, used before a physical prototype is available, is to allow operational uses and others to experience system aspects such as a human-machine interface by working with a simulation. Typically, this provides an early opportunity to identify things that are important to users but not obvious to system designers. This is another way in which early prototyping can reduce the risk that expensive design changes will be called for later in system development.

11.5 Petri Nets

A related technique for executable architecture modeling deserves a brief mention. Petri Nets, and more recently Colored Petri Nets, can be thought of as an extension of graph theory, which was briefly summarized in Chap. 2, to the modeling and analysis of discrete, distributed, and concurrent systems [11, 12]. The model contains nodes, or "places," that usually represent things like resources and data, connected by edges, or "transitions," that correspond to things like processes, events, or tasks. Places can exchange tokens that indicate conditions such as the completion of a task or the availability of a resource. Colored Petri Nets allow for multiple token types (colors) and provide other enrichment of the semantics. Because this method allows rigorous mathematical analysis of behavior, e.g., to compute behavioral invariants and bounds, it appeals to many architecture theorists. Importantly, in UML 2 and SysML, Activity modeling has been converged with the fundamental semantics of Petri Nets to achieve a more powerful and unambiguous representation of behavior. MBSAP does not explicitly employ Petri Nets because SysML Activity Diagrams provide the necessary capability for most system architectures. However, the mathematical foundation of Petri Nets is useful, particularly when dealing with architecture challenges such as quantitative assessment of complexity, defining behavioral invariants, and establishing provable limits on performance and possible system states.

11.6 Smart Microgrid Simulation

Our first illustration of the use of virtual prototyping involves its use to develop, analyze, refine, and communicate the Smart Microgrid (SMG) architecture. Specifically, we seek to prototype and assess an important aspect of the SMG logical architecture. The particular functionality of interest is the behavior that responds to a requirement that:

> "The SMG shall control internal energy flows to meet any component-communicated constraints."

We will use this model to build test cases for verification of the performance of the SMG against this requirement. As shown in Fig. 11.2, a simulation of the logical architecture of a system is often constructed to interact with the physical architecture models (the next lower modeling level) and process architecture models (the next level up). Our example follows this basic principle. We will examine an especially relevant Use Case from Fig. 4.20 entitled "Control MG Energy Storage." The central importance of this UC to the function of the microgrid is reinforced by its definition as the main control loop in Fig. 4.21. This simulation further develops the "Compute Generation Commands" action defined in Fig. 4.21.

To create this simulation of the SMG logical architecture, our model must build on simulations of the physical behavior of key components. We have used estimates of key component specifications to construct physical simulations that will interact with behaviors in the logical architecture. For this executable architecture example, the simulation includes physical models of the photovoltaic (PV) system, AC Loads, electrical connections, and the battery. For the battery model, we have used estimates of battery type, sizing, and communication protocols to construct a physical-level simulation. By modeling the battery at the physical level, the timing of battery charge/discharge cycling and the action of battery charge/discharge power cutoff can be simulated with adequate fidelity.

A relatively simple battery model suffices to meet these modeling objectives. Conceptually, it consists of an open circuit potential that is a function of state of charge (SOC), a linear internal resistance, a 100% coulombically efficient SOC model, and a set of contactors that connect and disconnect the battery from the other components of the SMG. Thermal considerations, cell-level electrochemistry, geometry, and other considerations are not simulated, as they are not relevant to the architectural requirements being assessed; any or all of these could be important in dealing with other requirements. Figure 11.6 illustrates this battery model as implemented in the chosen MATLAB™, SIMULINK™, and STATEFLOW™ modeling environment.[2] Input to this model is the physical power that is going to or from the inverter and a command to close contactors. The output is a set of communication signals to the battery controller.

The battery model and the other physical models are composed into an SMG system-level simulation in MATLAB/STATEFLOW as shown in Fig. 11.7. Physical electrical connections are modeled using single-lined arrows, and communication connections are represented as three-lined arrows. As described in the SMG logical architecture discussion in Chap. 5, the PV System (through its maximum power point tracker (MPPT) controller) and Battery (through its battery control unit) com-

[2] These are not SysML diagrams, although they could be readily converted to SysML for inclusion in the architecture model.

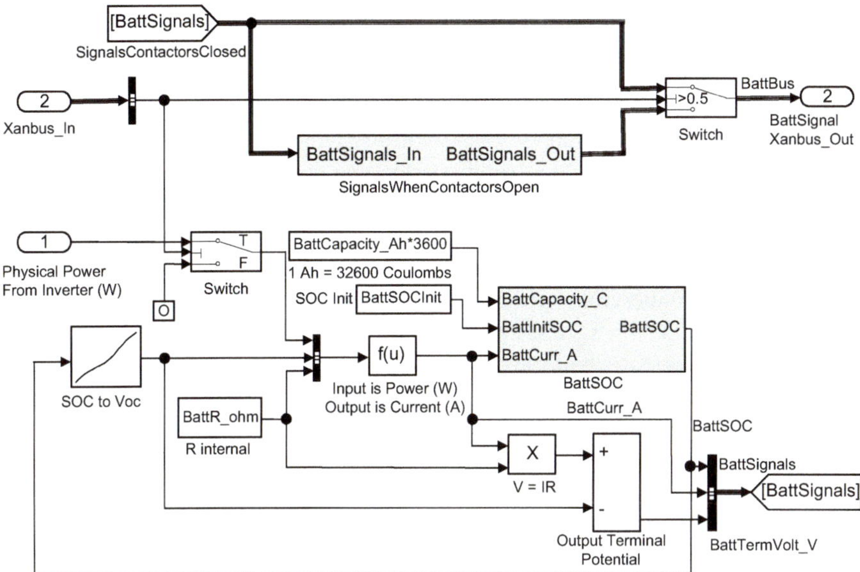

Fig. 11.6 Simulation of battery physics for the smart microgrid

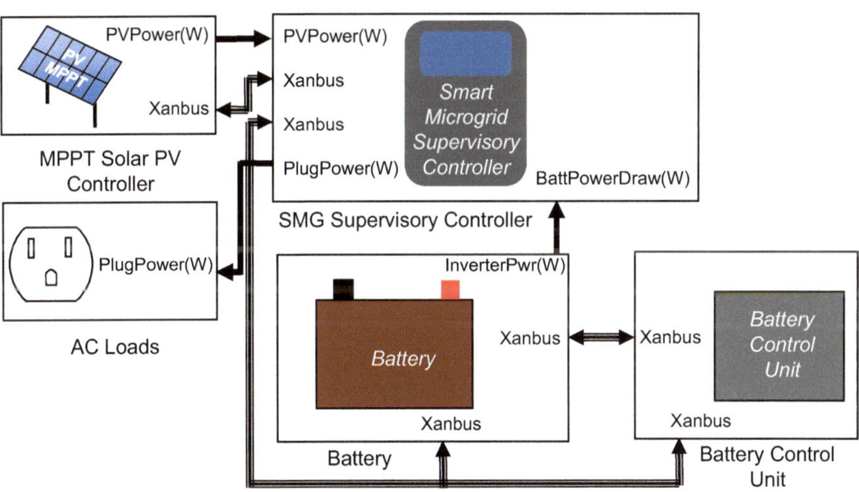

Fig. 11.7 Simulation of SMG components and interactions

municate with the Supervisory Controller via the Xanbus™ communication protocol. This is the physical implementation of the Microgrid Network identified with the <<MGNetwork>> stereotype in Fig. 5.30. The Xanbus System is commonly used in electrical power systems to create a smart network for integrating, monitoring, and controlling a variety of devices. Our model assumes no data communications between AC loads and the SMG Supervisory Controller.

Inside the SMG Supervisory Controller/Inverter is a model of the physical characteristics of the inverter and a model of the logical properties of the "Compute Generation Commands" activity of the SMG Supervisory Controller. In order to meet its requirements, the system executes what is known as an "enhanced AC support" function, where the Supervisory Controller will meet plug loads (power consumed by devices connected to the microgrid), and use any excess energy to charge the battery, while satisfying battery current constraints that are a function of battery SOC. The logic that implements this behavior is captured in the form of a StateFlow™ executable State Machine Diagram (SMD) in Fig. 11.8. This SMD describes a set of states including BatteryIdle (default), DischargeMode, and three charging modes corresponding to the three phases of battery charging that are supported by this controller. Battery Current Command (issued to the inverter) is defined in each of the states and is limited to 0.8 of battery capacity, C, in bulk charging mode, 0.1 C in absorption charging mode, and 0.1 C under float charging mode.

Inputting a repeating random signal for PV power and a constant demand for AC plug loads, we can see the behavior of the SMG system over a number of hours of operation. Figure 11.9 plots representative data. While the battery is at a low SOC,

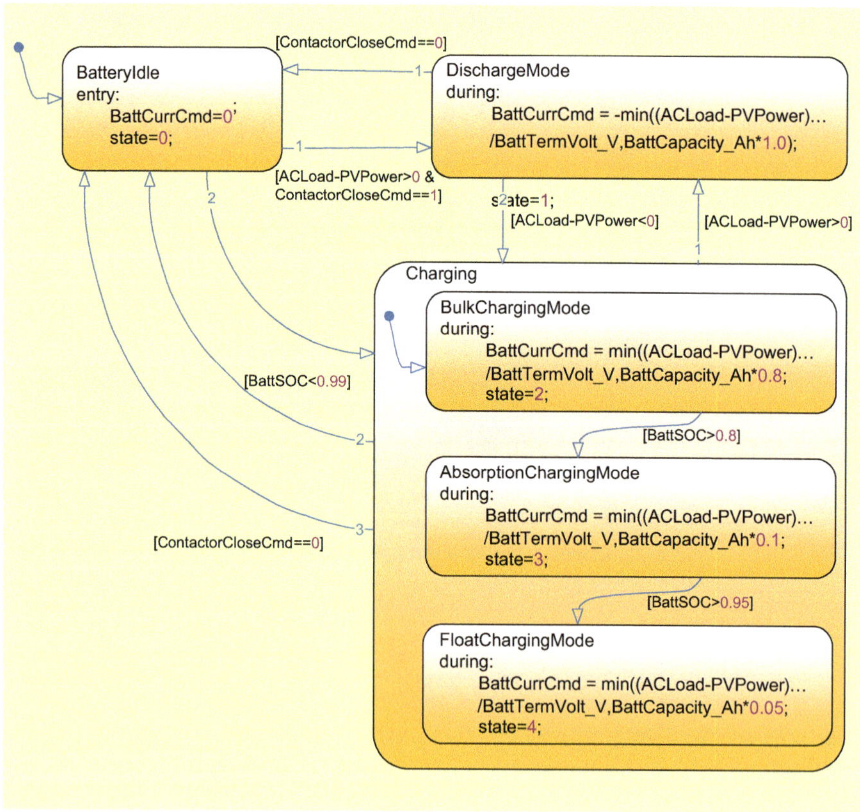

Fig. 11.8 Executable state machine diagram of the SMG supervisory Controller

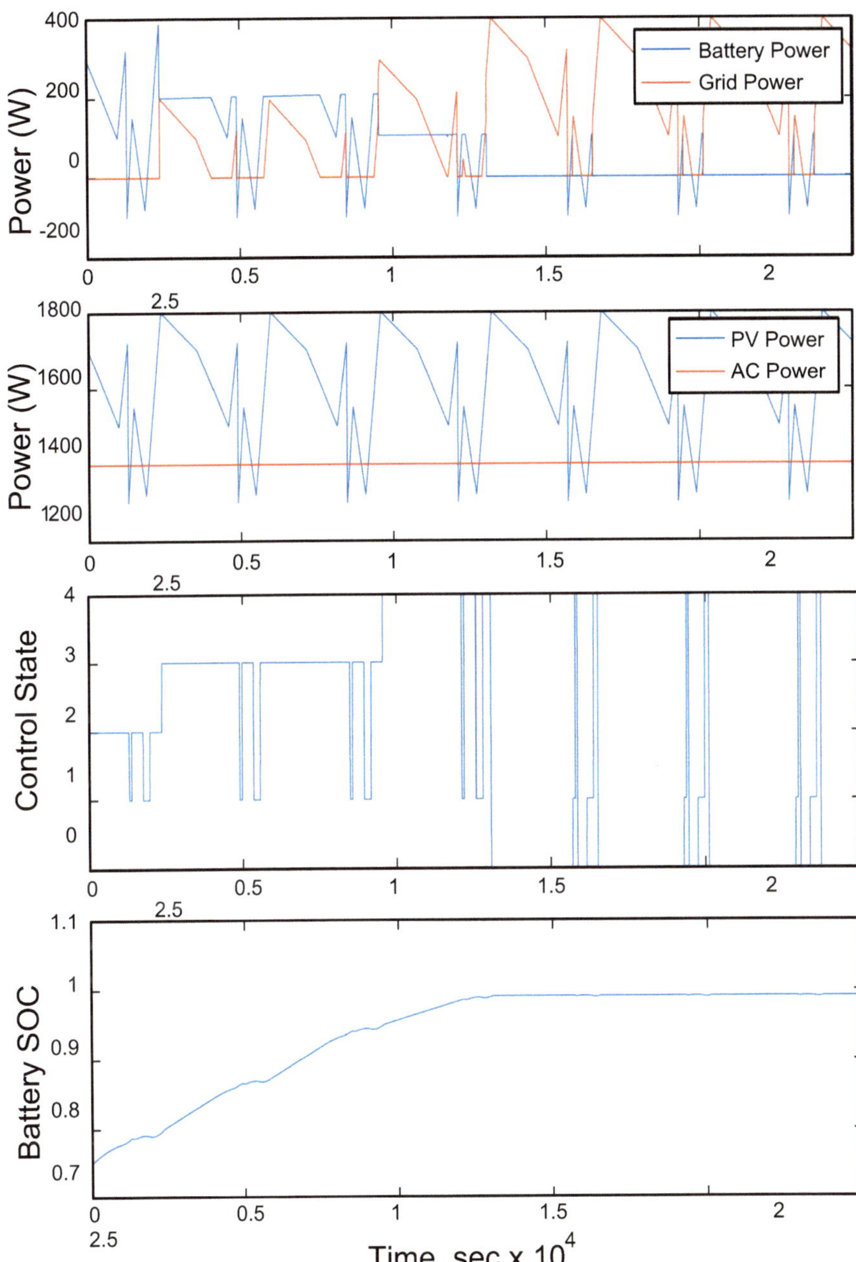

Fig. 11.9 Results of the SMG simulation during a model daytime period, in terms of the component and system power flows, control states, and battery state of charge

the power that is generated by the PV panel (PV Power) in excess of the plug loads (AC Power) is used to charge the battery. When PV output power is less than the plug loads, the battery discharges to support the microgrid. When the excess power is higher than the battery can use for charging, the remainder is transmitted (sold) to the primary electric grid. When the battery SOC is at its upper limit (~99%), the battery only provides power to the microgrid when loads cannot be met by PV power alone.

On the basis of this type of simulation, the architect can begin to assess some of the behavior and timing required of the "Compute Generation Commands" activity of the SMG Supervisory Controller. For example, as the update frequency for the Supervisory Controller goes up, the tolerance on the current control to or from the primary grid improves. In general, a tolerance of less 1 A on current to and from the electric grid is acceptable in practice. We can measure the error in grid power, when 0 W is commanded, as a function of the frequency with which "Compute Generation Commands" activity occurs. Figure 11.10 shows that the tolerance on power transferred to the grid improves from approximately 125 W at an update frequency of 1 Hz, to approximately 10 W at an update frequency of 100 Hz. In addition, this type of model can be used in MS&A at the operational and process levels of Fig. 11.2. For example, it could be integrated with process or operational MS&A to assess the system's capability to meet the battery's life targets or sustainability

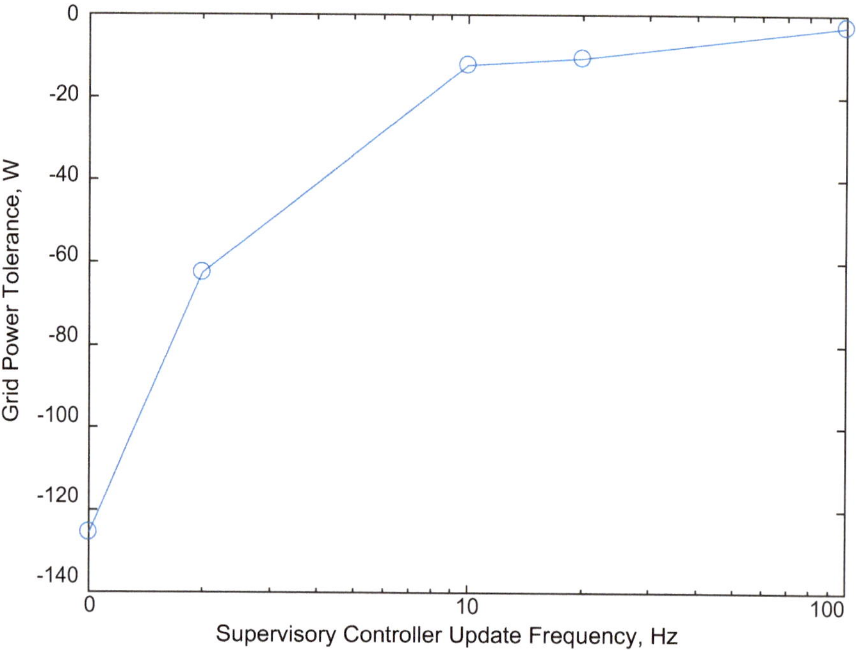

Fig. 11.10 Results of the SMG simulation demonstrating system-level performance management trade-off against logical architecture characteristics

requirements that might require prototyping as part of applying the MBSAP methodology.

MS&A results like those in Figs. 11.9 and 11.10 apply directly to V&V for the smart microgrid system. First, the computed data shows that the requirement being analyzed is feasible and that the associated numerical parameters can be "tuned" to optimize system performance. Next, it validates the system design in terms of delivering the required capability to meet stakeholder needs for an efficient and resilient microgrid. An analyst skilled in using the simulation tools and familiar with the underlying technologies and products used by a microgrid like this can build and exercise a virtual prototype in a short period of time. The computed results can then inform physical prototyping efforts to confirm the computed results and refine the simulation. The resulting smart microgrid model has long-term utility in developing specific smart microgrid solutions to meet customer needs.

11.7 E-X Process Simulation

Next, we will illustrate simulation at the process level in Fig. 11.2 by examining the primary flow of activities of the E-X example when executing a collection plan and creating ISR reports. Figure 11.11 shows a simple process model built in a tool called Vensim® from Ventana Systems. The model consists of "levels" shown as

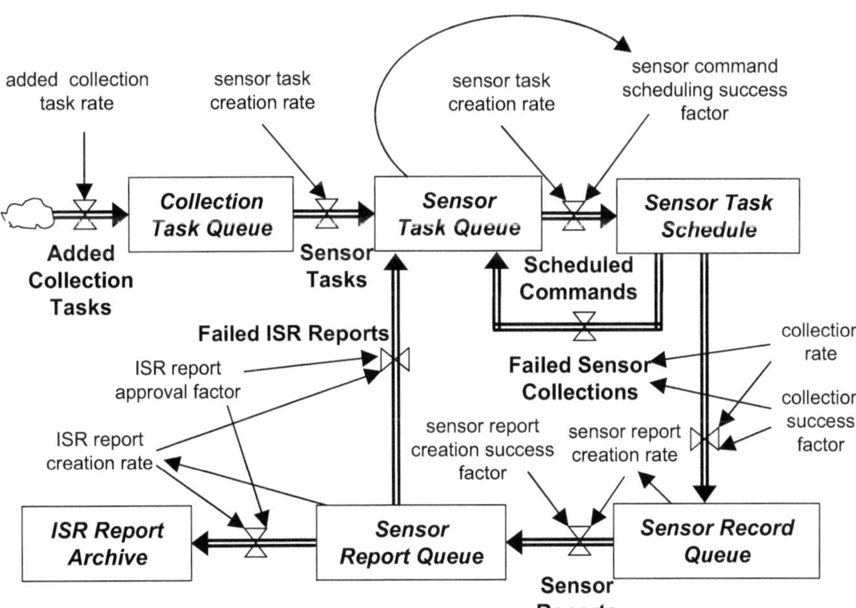

Fig. 11.11 Basic process model for the E-X with intermediate stages of collection task planning modeled as queues with queue servicing rates

boxes that are connected by flows. The boxes, in this case, represent queues of tasks and reports, and the flows or "rates" model the inputs and outputs of the various queues. Additional variables, called "exogenous variables" in Vensim, are shown by their names and connected to the queues and flows by arrows. Each variable in the model is governed by an equation, which may simply be a fixed value. Tasks, Commands, Records, and Reports are associated with the E-X Mission Management, ISR Management and Sensor Management Domains.

This structure corresponds to the design pattern introduced in Chap. 4 with flows of tasking downward and reports upward among these Domains.

The process starts with an initial set of Collection Tasks from a Collection Plan that is owned by the Mission Management Domain. We assume an initial set of 1000 such tasks, to which additional tasks are added at a constant (average) rate to simulate "pop-up" tasking sent to the E-X during a mission. We assume an average rate of one new task every 10 min. We further assume an on-station mission duration of 6 h (3600 min) during which the platform attempts to perform the original 1000 tasks plus the 360 new ones. In the model, the Collection Task Queue accumulates these and shows the number of tasks remaining to be executed versus time.

The flow out of the Collection Task Queue is governed by a Sensor Task Creation Rate variable that models the transformation of Collection Tasks to Sensor Tasks by the ISR Management Domain, accumulated in a Sensor Task Queue. Sensor Tasks are processed by the Task/Sensor Pairing and Scheduling logic of the Sensor Management Domain, resulting in a queue of scheduled sensor tasks to be sent to various sensors. As a way of exploring the sensitivity of system performance to the effectiveness of the pairing and scheduling process, the Scheduled Commands flow is parameterized with both a "normal" Sensor Command Creation Rate and a "success factor" that allows the architecture analyst to show how this rate may be affected by various circumstances such as whether a sensor has the target for a requested collection task within its field of view.

The flow from the Sensor Task Schedule queue to the Sensor Record Queue collects all the activities and external influences involved in sending a task to a sensor, attempting to collect the desired information, and returning that information as a Sensor Record. Once again, there is a normal Collection Rate tied to the current size of the Sensor Task Schedule queue and the capabilities of E-X sensors along with a success factor that can be manipulated in various ways, e.g., to reflect the time for a collection by a given sensor. There is a feedback flow that models Failed Sensor Collections being returned to the Sensor Task Queue for rescheduling.

Sensor Records are processed by the Sensor Management Domain to create Sensor Reports that are sent to the ISR Management Domain to be accumulated in a Sensor Report Queue. This flow has a normal Sensor Report Creation Rate, modulated by yet another success factor. An ISR Operator evaluates Sensor Reports, and those that do not meet mission criteria, labeled Failed ISR Reports, are sent back to the Sensor Task Queue. As with the earlier flows, there is an ISR Report Creation Rate and an approval factor. The final flow of approved ISR Reports accumulates in an ISR Report Archive. An obvious MOE for the modeled system performance is the ratio at the end of a mission of the ratio of records in the ISR Report Archive to the total number of Collection Tasks, which we call a Mission Success Parameter (MSP).

To establish a model baseline, we first set all the exogenous variables to constant values and adjust these to obtain as nearly as possible ideal system behavior. This is shown by an empty Collection Task Queue at the end of the 6 h mission, an MSP very close to unity, and consistent low levels in the intermediate queues, representing efficient queue servicing. For additional reality, we set success factors to less than 100%. A typical result of this initial model manipulation results in the values shown in Table 11.1, and the queue sizes are shown in Fig. 11.12. This results in an MSP of 0.91, which would be considered excellent in a real-world situation.

We can now extend and modify the basic model to examine multiple aspects of E-X design and operation. Valuable insights can be gained by simply varying parameters and looking at the effects on MSP and queue levels. The following are a few of the possible model refinements.

- Assumed values for exogenous variables can be made more realistic using data from existing systems, detailed physical simulations of data processing, or simple human-in-the-loop experiments to assess operator decision times. The model also allows a quick examination of the benefits of scaling system capacity through things like multiple simultaneous processing threads and multiple operators assigned to various User Roles.
- Simple constant values for variables, especially success factors, can be replaced by functions that account for various real-life circumstances such as task saturation.
- The modeling of sensor collections can be made more realistic by making the collection success factor depend on things like target ranges, weather obscuration, and intervening terrain features.

Table 11.1 Parameters for the baseline E-X process model configuration

Model parameter	Value
Mission duration	3600 min
Initial collection tasks	1000
Additional collection task rate	0.1 tasks/min
Total collection tasks for mission	1360
Sensor task creation rate	0.375 tasks/min
Sensor command creation rate	0.475 tasks/min
Sensor command scheduling success factor	0.95
Collection rate	0.45 records/min
Collection success factor	0.8
Sensor report creation rate	0.375 records/min
Sensor report creation success factor	0.95
ISR report creation rate	0.36 records/min
ISR report approval factor	0.95

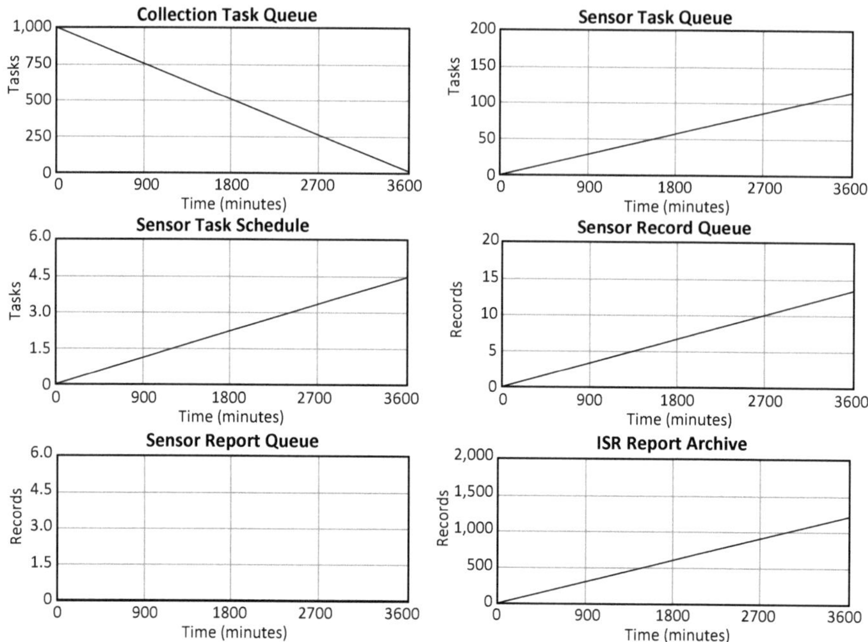

Fig. 11.12 Simulation results with the baseline E-X process model

As a simple example, we next examine the possibility that the effectiveness of the Task/Sensor Pairing and Scheduling logic falls off as the size of the Sensor Task Queue approaches the limit for which the scheduler was designed. To do this, we replace the fixed value of the Sensor Command Scheduling Success Factor variable with a graphical relation (called a "lookup" in Vensim) that starts at the nominal baseline value of 0.95, stays relatively constant until the Sensor Task Queue grows to a size of 75 tasks, and then drops off at an increasing rate with further queue growth. Figure 11.13 shows the resulting queue contents, and the MSP drops to 0.83. This result is reasonable for the assumed loss of efficiency in a critical step of the process.

There are subtle effects on the interior queues that might be important to examine more closely as part of optimizing the design of the affected system elements. However, the overall behavior of the model is reasonable. In particular, in working with the parameters of the modeled task scheduling effectiveness curve, it quickly becomes apparent that relatively small changes have quite large impacts on the number of ISR reports created by the end of the mission. This results from the fact that as the rate at which Scheduled Commands are created drops off, the Sensor Task Queue grows, which further impairs the scheduling function. This is a positive feedback, and when carried to a certain level, it was observed to produce a growing oscillation in this part of the model. That effect is clearly nonphysical, but these results highlight both the fact that the scheduler has a significant impact on overall system performance and the importance of adequate fidelity in the model to match actual system behavior.

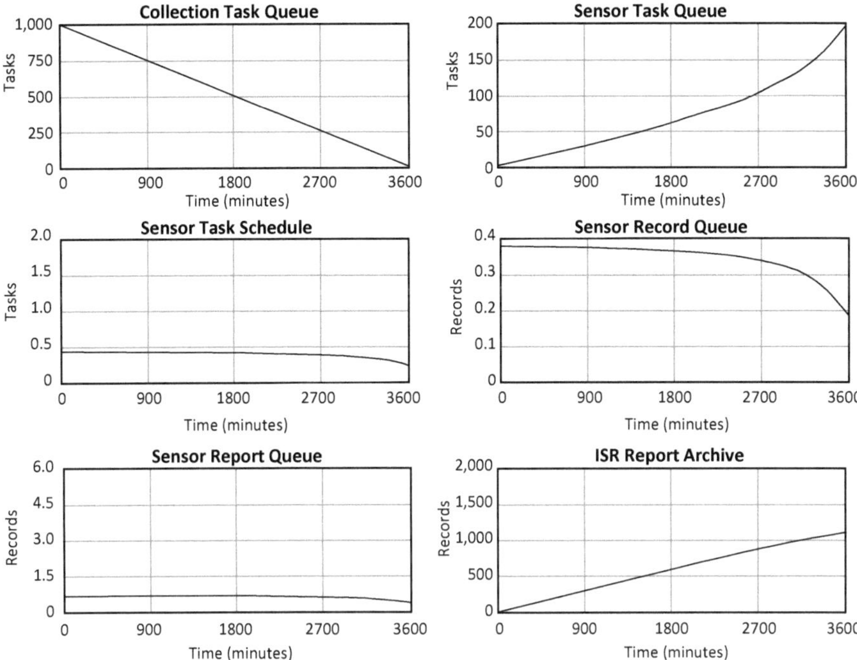

Fig. 11.13 Simulation results with degraded sensor task scheduling due to process saturation

Even a model this simple is valuable in assessing a system architecture, finding and optimizing design parameters that have high leverage on performance, and predicting and interpreting the results of physical prototyping and actual system testing. For example, we can set up a Monte Carlo experiment varying the model parameters within expected ranges to get a prediction of the system's ability to meet a performance requirement such as a minimum MSP. By identifying the sensitivity of system performance to various parameters such as success factors, we can gain insight into areas where design refinement (e.g., additional processing capacity) can make an important difference. Similarly, these results provide insight into the critical parameters that need to be examined during system test and evaluation. In the long term, in a fashion similar to the smart microgrid simulation, the model becomes a tool that can be used in succeeding system developments once it is validated with experimental data.

11.8 Summary

Prototyping, both physical and virtual, is an essential element of the MBSAP methodology. A SysML model is the starting point for MS&A in a virtual prototype and lays the foundation for one or more physical prototypes, which are aimed at evaluating specific system characteristics and functions. Working together, these prototypes provide early risk reduction, support optimization of system requirements and design, quantify performance, support the planning and execution of system integration and test, and play a crucial role in V&V. Readers interested in pursuing this more deeply may want to review a detailed example of a SysML model with executables for the Thirty Meter Telescope (TMT) being developed by the TMT Observatory Corporation [13]. In this chapter, we have provided overall guidelines and identified issues that frequently arise. Both the strategy and the implementation of a prototyping environment should be intimately related to the overall architecture methodology, and adequate resources should be provided. Prototyping is a powerful tool to maximize the success and minimize the total cost of a program that delivers a solution to meet stakeholder needs.

Exercises

1. Identify the preferred verification method for the following requirements:
   (a) E-X sensor performance
   (b) Operator interface displays and menus
   (c) Mean time between critical system failures
2. What is primary motivation for prototyping as part of a system development effort?
3. Describe the impact on system validation of a stakeholder action to add a significant new performance requirement.
4. Describe a comprehensive approach to V&V for Smart Microgrid System A (community smart grid).
5. List some specific ways in which a virtual and a physical prototype of a given system complement and refine each other.
6. Describe the primary elements of a System Integration Lab (SIL) for the E-X airborne sensor platform.
7. Consider a large information management system used to process orders and control deliveries by an online sales company. Describe an executable architecture for this system.
8. For the E-X airborne sensor platform, summarize the data that would be exchanged between:
   (a) The Operational and Process levels of MS&A
   (b) The Logical Architecture and Physical levels
9. Describe the use of an architecture model as an executable specification and summarize the essential content.
10. Identify some additional ways in which the process model in Fig. 11.9 could be modified and used to explore important aspects of system behavior.

Student Projects

Students should define an appropriate prototyping environment for their project systems and list the ways in which physical and virtual prototypes will be used in support of design activities and V&V. If a laboratory is available with suitable simulation tools, MS&A experiments similar to those described in this chapter can be added.

References

1. Project Management Institute (2013) A guide to the project management body of knowledge, 5th edn. https://www.pmi.org/pmbok-guide-standards/foundational/pmbok. Accessed 29 Apr 2018
2. McQuay WK (2011) Collaborative environment for 21st century avionics. IEEE A&E Syst Mag 26:4–11
3. IEEE Std 11516-2010 IEEE standard for Modeling and Simulation (M&S) High Level Architecture (HLA) – framework and rules. https://standards.ieee.org/findstds/standard/1516-2010.html. Accessed 27 May 2017
4. IEEE Std 1278.1-2012 IEEE standard for distributed interactive simulation – application protocols. https://standards.ieee.org/findstds/standard/1278.1-2012.html. Accessed 26 May 2017
5. OASIS (2017) Web Services Business Process Execution Language (WSBPEL) TC. https://www.oasis-open.org/committees/tc_home.php?wg_abbrev=wsbpel. Accessed 26 May 2017
6. Object Management Group (2016) Business Process Model and Notation (BPMN)
7. Mittal S, Ziegler BP, Risco Martin JL, Sahin F, Jamshidi M (2009) Modeling and simulation for system of systems engineering. In: Jamshidi M (ed) System of systems engineering: innovations for the 21st century. Wiley, Hoboken, pp 101–149
8. Weibull, FRACAS: an essential ingredient for reliability success. http://www.weibull.com/hotwire/issue122/relbasics122.htm. Accessed 7 May 2018
9. Kossiakoff A, Sweet W (2003) Systems engineering: principles and practice. Wiley, New York http://www.bpmn.org/. Accessed 26 May 2017
10. Utting M, Legeard B (2007) Practical model-based testing: a tools approach. Morgan-Kaufmann, New York
11. Petri Nets Steering Committee (2017) Petri nets world. http://www.informatik.uni-hamburg.de/TGI/PetriNets/index.php. Accessed 26 May 2017
12. Zurawski R, Zhou M (1994) Petri nets and industrial applications: a tutorial. IEEE Trans Ind Electron 41(6):567–583
13. GitHub – Open-MBSE (2017) TMT-SysML-Model. https://github.com/Open-MBEE/TMT-SysML-Model. Accessed 27 May 2017

Chapter 12
Using Reference Architectures and Frameworks

12.1 The Nature and Uses of Reference Architectures

Chapter 5 stressed the use of design patterns as a way to achieve design reuse and to benefit from proven solutions to common problems. This chapter expands the concept to architecture as a whole. Just as a familiar software design pattern is an abstraction of a set of concrete designs that can be tailored to a new situation while retaining the proven underlying principles, a reference architecture (RA) abstracts the structure, behavior, and rules of one or more successful real-world architectures to create a template. The fundamental motivation is to save effort and reduce risk in developing a new system or enterprise, and a number of approaches to this have been tried in recent years. Ellis et al. [1] define an RA in terms of the highest-level concept of a system, and an early architecture reference model developed by the Department of Defense called for an RA to provide the essential objects and processes of a system concept [2]. The ISO/IEC/IEEE standard 42010:2011 [3] also provides guidance for the structure and use of RAs. In this chapter, we apply the MBSAP methodology to RA creation and use.

12.1.1 Reference Architecture Terminology

Overlapping and conflicting terminology generates a fair amount of confusion in discussions of RAs. For example, the International Council on Systems Engineering (INCOSE) uses the term to describe an early architectural construct used during what is called the Exploratory Research Stage of a system development to support analysis of user needs, technology risks, and early cost and schedule estimates, but makes no mention of using a template derived from earlier architectures [4]. In this chapter, we define a set of terms, based on community consensus where it exists, to at least avoid adding to the problem.

J. M. Borky, T. H. Bradley, *Effective Model-Based Systems Engineering*,
https://doi.org/10.1007/978-3-319-95669-5_12

First, we use the term RA for an architecture that generalizes or abstracts the characteristics of a group of actual architectures. We further require that the content and organization of an RA follow those of MBSAP, dealing with structure, behavior, and rules. By its nature, an RA includes primarily Operational and Logical/Functional Viewpoints and is less specific and detailed than an actual system architecture. Broadly applicable physical detail can be included, but the essence of an RA is a functional design ready to be tailored in accordance with particular customer requirements.

It's useful to think about RAs in the context of the architecture taxonomy of Chap. 1. In terms of the Axis of Abstraction of the architecture taxonomy in Fig. 1.3, an RA does not contain the kind of point design data that would be documented at the Physical level for a specific system or enterprise, while the Operational, Process/Workflow, and Logical/Functional levels are relevant. RAs can be defined at any point on the Axis of Organization, but we focus on the System level. Occasionally, an RA is defined for an enterprise (system-of-systems), but the result is basically similar to a system RA. At the enterprise level, the architecture template must include features to deal with geographical dispersion of nodes and systems, including wide area networking and interoperability among resources using heterogeneous technologies. We refer to these as Enterprise Reference Architectures because, as discussed below, the term Enterprise Architecture (EA) is widely used in a somewhat different sense.

Companies, Government agencies, and other communities may define RAs and other forms of architecture guidance and may make the application of an RA mandatory. The motivation is a desire to capitalize on previous architecture investments, enhance architecture quality and compatibility, and profit from lessons learned [5]. Most major Information Technology (IT) vendors publish RAs for the system categories and processes their products support. Many constructs that are labeled RAs adopt a methodological focus, seeking to prescribe the way in which architectures should be developed and documented. A more appropriate term for this category of architecture guidance is architecture framework, another topic that is discussed later in this chapter.

Still another term often encountered in this context is Business Reference Model, which is a systematic approach to capture and analyze the operations and processes of an organization. This can be reflected in the Behavioral Perspective of an Operational Viewpoint rather than in a separate model so that operations can be mapped to organizations and resources within a common model along with rigorous tracing of functional allocations to the elements of a solution.

12.1.2 Reference Architecture Uses

RAs have a number of basic uses that drive the way in which they are constructed and employed in an architecture process. The following paragraphs describe some of these. It's important to note that, to be effective, an RA approach must be made an integral part of an organization's SE strategy and process, so that the reuse and guidance it offers can be exploited at every stage of architecture development and refinement.

Providing an Advanced Starting Point An RA can reduce the up-front effort in developing a new system or enterprise architecture. The resources needed to build and then maintain an RA can be amortized across multiple programs and yield significant aggregate savings, especially by reducing the initial architecture and SE effort and by avoiding false starts and dead ends that have been discovered in earlier programs. For example, an RA frequently includes a requirements template that documents both the overall requirements structure and the definitions of requirements that apply to all systems in the category governed by the RA. In these ways, an RA can save very substantial amounts of SE labor in building and validating requirements for a new system.

Promoting a Consistent Architecture and Functional Design An RA can facilitate commonality across systems in an application domain. This is the idea of using large scale design patterns as the basis for design reuse. In addition to further savings in system development effort and cost, an RA inherently facilitates interoperability and cooperative functions across systems because each will have the same fundamental interfaces, functions, and data and will use the same standards and design approaches in its implementation.

Facilitating Product Reuse Modularity, loose coupling, and other open architecture principles are central to enabling reuse of components across systems. When an RA helps achieve common partitioning of the structure, common interface definitions, common functions and control mechanisms, common data models, and other elements, the feasibility of product reuse is even greater. Among other benefits, this means that many system components will be products with which the developers are already familiar. This strategy creates opportunities for strategic vendor relationships that may produce both advantageous pricing and insight into future product development plans.

Facilitating a High-Quality SE Process Given an architecture-centric methodology like MBSAP, the use of a shared architectural foundation helps ensure that mature, well-controlled SE processes will emerge and be applied in a development organization, helping to ensure consistent quality. This manifests itself in many ways such as common configuration management structure, predefined Quality Attributes, and reuse of consistent and proven integration and test methods.

Achieving and Proving Policy Compliance As many Government programs, and a growing number of commercial customers, require the implementation of RAs, reference models, architecture frameworks, and other overall design standards, a body of architecture artifacts and materials that have been shown compliant and can be reused in new programs reduces the risk and effort in demonstrating that this requirement has been satisfied.

12.1.3 Reference Architecture Hierarchy

Experience has shown that the RA construct is inherently hierarchical, progressing from very general principles and guidance down to a fully specified architecture for an actual system. Figure 12.1 suggests such a structure. A number of such hierarchies have been proposed, but they tend to be similar to this one. The various levels are summarized in the following paragraphs.

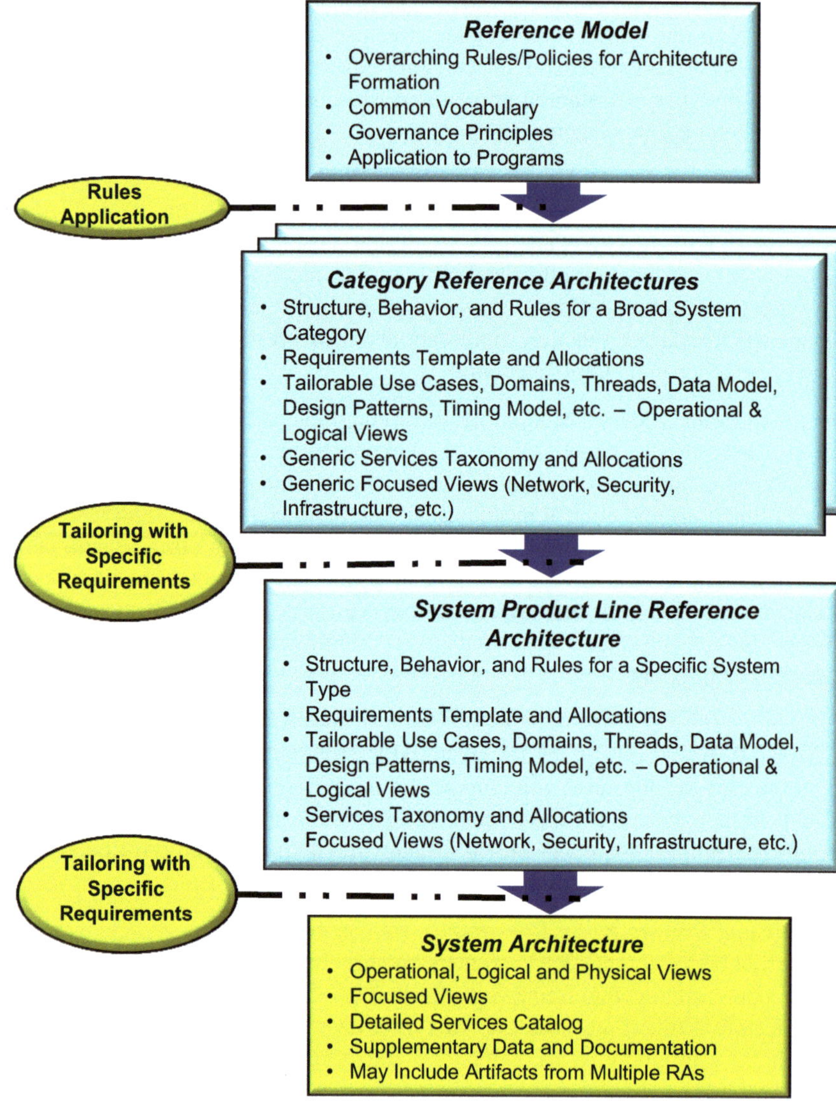

Fig. 12.1 Typical reference architecture hierarchy, from the most general to the most concrete

Reference Model (RM) This is not an architecture per se, but rather a set of rules and guidelines for constructing architectures. Such a model establishes overarching principles with the goal of converging the efforts of multiple architecture developments toward a consistent and interoperable portfolio of operational systems. There are a number of important aspects to structuring an effective RM.

- *Overall rules and policies* – including broad guidance for the architectures covered by the RM. An example might be that all architectures conforming to the model must support networked interoperability, policy-managed functionality, or other overall attributes. In particular, an RM at the system level may specify that systems will conform to a larger EA (defined below), which would then need to be fully described. This part of the RM could well include Quality Attributes prescribing overall characteristics such as modularity, loose coupling, and standards compliance, along with a default standards profile.
- *Common vocabulary* – a standard dictionary together with a requirement that these terms and definitions be used wherever they apply. Recognizing the pernicious effect of confusing terminology, a good RM includes terminology that becomes the core of the data model for all systems that implement the RM. A logical extension of this area would be an overall services taxonomy that identifies and defines the services that all participating systems will be expected to implement and use.
- *Governance approach* – provides system architects with sound, actionable guidance on the strategy to be applied to architecture governance, especially when an individual system is to be part of a larger enterprise that requires centralized control. This part of the RM should prescribe roles, responsibilities, processes and mechanisms to implement governance. Chapter 15 expands on this subject in more detail.
- *Application guidance* – describes how these RM rules and principles are to be applied in real programs. This might include guidance on the ways compliance will be verified, how waivers should be requested, and what sources of expertise are available. In Government programs, an RM is often the embodiment of a set of policies. A good RM should clearly describe the target systems to which it is intended to apply and the benefits of such application.

Category Reference Architectures Below the RM in Fig. 12.1 is a set of architectures corresponding to the Axis of Categories in Fig. 1.3. In some situations, these will be organized by the Hard Real-Time, Decision/Process Real-Time, and Management Information System Categories defined in Chap. 1, but any set of fundamental system categories of interest to an organization can be the basis for such a family of RAs. Basic differences in performance, technology, organization, mission criticality, geographic distribution, security, safety, exportability, or other factors could influence the selection of system categories that will be implemented in multiple real systems and thus are candidates for top-level RAs. As Fig. 12.1 indicates, the development of these RAs involves the application of rules and other guidance from one or more applicable RMs.

System Product Line Reference Architectures The next most specific type of RA deals with product lines that can be defined at any organizational tier, usually to describe systems and components that have many features in common and are used for similar purposes. A product line might be defined for a category of sensors or controllers that perform generally similar functions but are different in their specific features and capabilities. In terms of using RAs for the purposes described above, the primary emphasis is on the system level, although component product lines are briefly discussed later in this chapter. At this level of the hierarchy, content is selected and tailored from one or more Category RAs to develop more specific System Product Line RAs. A properly defined product line allows the RA to come very close to the architectures of actual systems or products. This potentially leads to the major source of time and cost savings in the overall RA approach because a modest amount of tailoring can replace a much larger total architecture development that lacks this advanced starting point. Depending on the diversity of its product offerings, a development organization may find it necessary to invest in a number of RAs at this level.

It's entirely likely that more than one Category RA will feed a system product line because most real-world systems will have features and requirements that cut across any manageable set of categories. Returning to the E-X example as an illustration, the product line is airborne, multisensor ISR systems, and the architecture includes both Hard Real-Time and Process/Decision Real-Time elements. This was highlighted in the Domain structure in Figs. 4.6 and 4.7 where the Aircraft Systems, Sensors, and External Communications Domains have deadlines driven by control and data processing for things like flight control, sensor transmit/receive modes, and data links, while the other domains have human-in-the-loop timelines. Chapter 8 dealt with the architecture differences encountered in RT applications. As long as the repository used to store and access RA artifacts and the architecture-centric development environment are built on a common methodology and tool set, selecting, modifying, and combining architecture model artifacts in this way are straightforward.

System Architecture A final stage of architecture tailoring is based on specific customer needs and transforms a Product Line RA into the design of an actual system. This includes the full set of MBSAP Viewpoints, Perspectives, and artifacts, including a PV. Translating a product Line RA into a system architecture entails primarily two kinds of activities:

- Tailoring the artifacts of the Operational and Logical Views to include only the content appropriate to the system being architected, which may require both deletions and additions of content
- Implementing the LV in a PV as described in Chap. 6

Several other RA types should be added to those in Fig. 12.1 to create a complete taxonomy.

Target System Also called an Objective System, this is the final goal of applying RAs in a specific development program. It is the ultimate deliverable system intended to satisfy a set of customer requirements and to function in a prescribed operational and support environment. Here is where the real payoff of an investment in developing RAs is realized. Speed to market is improved because much of the architecture and system engineering work to craft a solution for a specific customer need has already been done. Cost, reflected in the price presented to a customer, is lower for the same reason. Preferred products and components to implement the architecture, along with their vendors, have been established so that subcontracting and procurement actions are accelerated. Integration and test are reduced in scope and can leverage I&T plans from previous projects. Training and field support, maintenance and repair, accreditation of classified systems, and many other aspects of an individual program benefit in similar ways from the leverage that an RA provides.

Enterprise Architecture (EA) In MBSAP, the term enterprise is reserved for the top tier of the Axis of Organization, but the concept of Enterprise Architecture is widely used in a different sense. In this alternative context, the "enterprise" is simply a major organization, and the purpose of an EA is to define that organization's overarching processes, system strategy, data, rules, and other information-related aspects. Most such EAs are composed of a set of models, and these are, unfortunately, often called "reference models," creating yet another term with multiple conflicting definitions. These reference models typically include:

- A business model to describe the processes and products of the organization
- A technical model offering guidance such as mandatory standards for systems that implement those processes
- An enterprise data model
- Various others such as a metrics model for assessing program status and verifying requirements

The overlap of this kind of EA with both RMs as defined earlier and frameworks, defined next, is obvious, and the architect can usually treat this sort of EA as a kind of RM that provides guidance and imposes constraints on architecture development.

Architecture Framework Architecture frameworks are usually described as an approach to creating an EA and can be applied at the enterprise, node, or system level of the Axis of Organization. Such a framework defines an orderly sequence of architecture development, including the Viewpoints (or simply "Views") and artifacts to be used to define the architecture, a step-by-step methodology, and supporting information such as architecture team roles, responsibilities, and required expertise. However, frameworks differ widely in their organization and content, reflecting various architecting styles, company products, and the needs of sponsoring organizations. A sound architecture-centric system engineering process can usually be mapped to the Views of any mainstream framework, which is important when different customers prefer different frameworks. MBSAP is, in fact, an architecture framework in its own right and was initially optimized for aerospace and defense system and enterprise architecting. Over time, the methodology has been refined and extended to deal with almost any architecture category. A later section of this chapter discusses some typical frameworks.

12.2 Reference Architecture Process

The next subject is how to put an RA to work in a program. Figure 12.2 summarizes the overall process. Because RAs are primarily about design reuse, the starting point is one or more systems that are considered successful and operationally effective in the category to which the RA is intended to apply. Since Category RAs are the most

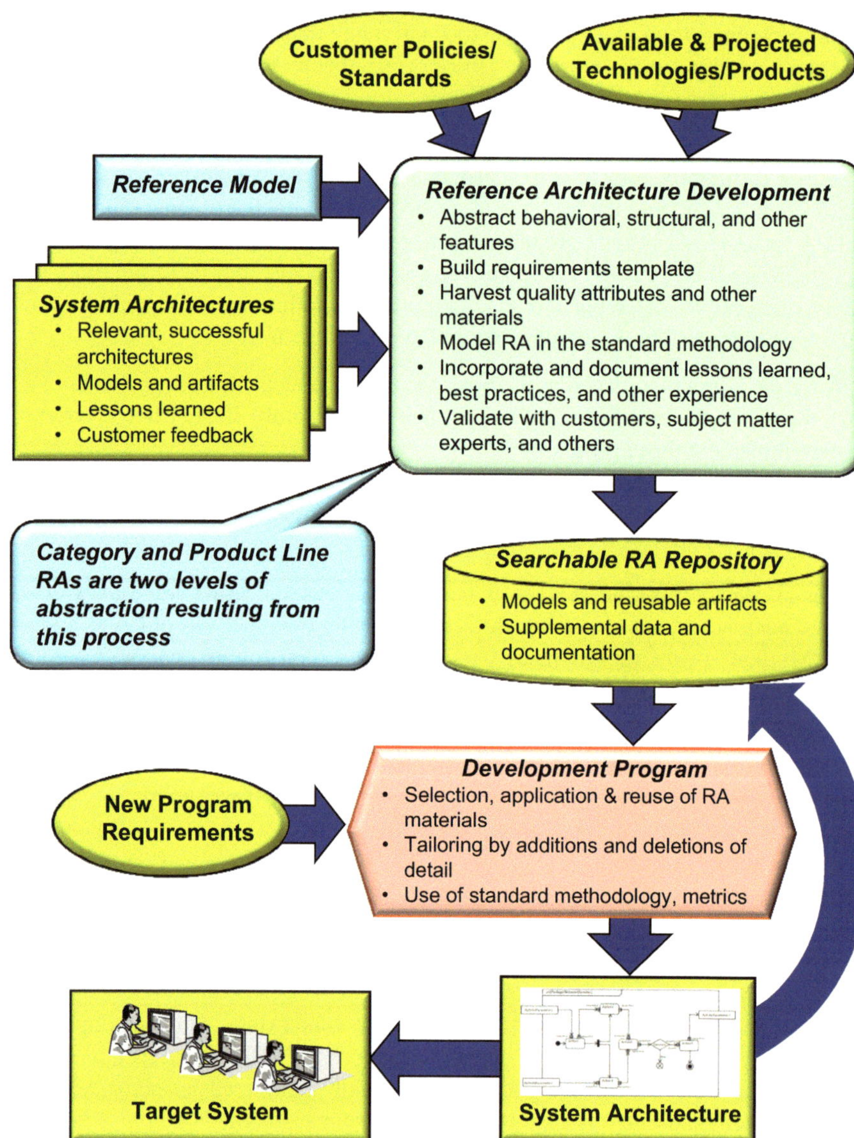

Fig. 12.2 Overall reference architecture process and methodology

general, spanning the largest range of target systems, their inputs must be similarly broad and representative of the overall design space, and their content will be rather general. However, this is the best tier to embody the RM, sometimes by directly importing things like a data dictionary and sometimes by shaping the RA model to conform to the applicable rules and policies. This content will then be inherited by System RAs and on down to the actual system architectures. The data for a System Product Line RA comes from successful architectures and system designs in the area of interest and is correspondingly more detailed, although the RA team must take care not to overspecify content based on things like personal familiarity with individual predecessor systems.

In addition to RM content, the process of building an RA should consider two other general kinds of input. The first results from an assessment of the situation within which system development and acquisition take place. A variety of factors impact the creation of new systems that will be built on the RA, including the operational environment in which these systems will operate; policies, priorities, and strategic plans of the acquiring organization; and constraints imposed by schedule (e.g., time to market) and resources. This kind of current program guidance is especially relevant when dealing with Government customers. Factors like these are different from an RM, which is a technical construct aimed at guiding architecture development. They deal instead with other aspects of target systems, the process through which they will be acquired, and the budget and schedule constraints that influence the feasible level of performance and sophistication. For example, an RM might require that architectures support a networked enterprise, while current guidance might include a roadmap showing which external systems must be interfaced to at various points in time and which standards are projected for those interfaces.

The second general input to RA development is an assessment of the current and emerging technologies and products that will be available for system implementation. A generation of information technology (IT) can be as short as 18 months, and even foundational standards go through updates every few years and are quite likely to be superseded within the lifetime of a system. The performance, size, weight, and cost of components and equipment items improve continuously, and these trends inevitably influence system architectures. Emerging technologies create new and better options for the architect. Non-IT technologies also change with time, albeit usually less quickly. Sensors improve in sensitivity and resolution, communications devices grow in range and bandwidth, equipment of all sorts requires increasingly sophisticated controls, and similar evolutionary trends affect virtually every system area. Dealing with these realities is an imperfect process at best, but it is incumbent on an RA team to devote time and resources to surveillance of the technology landscape, to maintain relationships with key suppliers, and to try to develop RA structures that are robust in the sense of recognizing the existing state-of-the-art and accommodating likely future trends.

Figure 12.2 lists some typical RA development activities, and the next section expands on these. If the investment in building RAs is to yield full benefits, their content must be easy to find, access, and use. The preferred approach is a searchable repository that helps a user find content of interest based on a variety of criteria.

The repository might support access to one or more entire RAs in a particular category or system product line with content as detailed as a specific Use Case, data object, interface definition, or design pattern. There are many good ways to accomplish this. An architecture modeling tool can provide some level of content search through its built-in model browser, and a set of RA models can be captured in a database with appropriate metadata. In fact, a well-structured architecture model is inherently navigable and searchable and can contain embedded files or links to supplementary information, so that the repository function amounts to finding and providing (potentially as a service) the right model. The point is that the RA job is not finished until the results are available and usable across the supported organizations and their architecture teams.

A key principle of the recommended approach is that projects that use and refine an RA have an obligation to refresh the contents of the repository by correcting or refining existing artifacts and adding new ones. If everyone is using the same modeling environment, making updates is mechanically simple. However, the repository must be subject to governance or random changes will quickly destroy the integrity of the RA models. Chapter 15 discusses governance in the broad sense, but for now it's important to note that there should be an organization-level Architecture Board, or equivalent, as recommended by the Open Group [6], that rules on recommended RA changes and maintains strict configuration control.

The payoff comes when a program needing to develop an architecture imports the appropriate content from the repository and then uses a methodology like MBSAP to transform this starting point into a physical solution that meets specific system or enterprise requirements. Use of RA does not invalidate the methodology, rather it accelerates it by, in effect, doing in advance much of the work of the early phases of the architecting process. Tailoring can involve deleting unneeded elements of the RA, adding new ones when repository content does not address particular requirements, and tailoring the characteristics of system elements. In general, "tailoring down" by deleting unnecessary content is easier than "tailoring up" by creating new architecture elements, and for this reason, System Product Line RAs should be inclusive in the sense of covering as many design alternatives as possible. Even so, it's common for the architecture team using an RA to discover the need to add new Domains, Use Cases, Mission Threads or Business Processes, Blocks, Behavior and Timing Diagrams, Services, Interfaces, and other elements. These become candidates for addition to the RA through the controlled updating process of RA governance.

Finally, it will always be the case that RA artifacts require additional specificity to be added in the course of transforming the RA to fully define an actual system architecture. This can involve adding and refining Block Properties and Operations, adding details to interface and service definitions, modeling and analyzing system-specific timing relationships, modifying user roles to conform to customer requirements and practices, and many, many others. However, with a good RA starting point, the new architecture will still require less time and effort than achieving the same result by starting with a clean sheet of paper. A critical point that bears repetition is that this whole RA strategy depends for effectiveness and efficiency on using a standard methodology and tool set across the organization for both development and application of RAs.

12.3 Building a Reference Architecture

Applying MBSAP to develop an RA is in many ways the same as the start-from-scratch approach we described in earlier chapters, but there are some important differences. We have already noted that since an RA is an abstraction, the Operational and Logical/Functional Viewpoints contain the primary content. This is sometimes accompanied by templates or other guidance for Focused Viewpoints and by any selected physical data that applies to every system covered by the RA. The OV and LV Perspectives and their artifacts are much the same as we presented in Chaps. 4 and 5 but kept at a level of generality appropriate to an RA. Other activities such as selecting and defining Quality Attributes that can be used in any application of the RA can also add great value.

RA Viewpoints There are two significant differences in developing an RA. The first is that starting with existing architectures, rather than with a set of system requirements, changes somewhat the development of the OV. Ideally, the RA team can select structural, behavioral, and other artifacts from the existing architectures that have been chosen as inputs. More commonly, it will be necessary to synthesize a set of general architecture elements by combining content from several predecessors. Often, an existing system that is considered successful and desired as an RA input will date to an earlier era and have an "architecture," if one exists at all, that is not much more than a requirements database, a collection of design documentation and perhaps some analysis results. In that case, the RA team will be faced with deriving architecture content by reverse engineering the legacy design to extract its features and using these to build a real architecture model.

For example, assessment of traditional block diagrams and organization charts can be the basis for discovering the de facto Domain structure, while documents such as operator manuals and troubleshooting guides may be the best source of insight into Use Cases and business processes or mission threads. If existing systems being used as inputs to an RA have specialized Use Cases for activities like strategy development, operational planning, operational oversight and assessment, process control, and process measurement and reporting, they can be copied in the RA. The RA team will typically have to analyze these raw materials to discover and describe universal behaviors such as strategy, planning, process execution, communications, system monitoring, and reporting that become the top-level Use Cases of the RA. These Use Cases can then be fleshed out with content such as pre- and postconditions, associated User Roles and data objects, and scenarios. It's often helpful during this process to arrange for reviews with subject matter experts (SMEs), especially people with direct knowledge of the predecessor systems. Activity modeling to explore the subtle differences among Use Cases of the existing systems can be another powerful technique here. Similar analysis of domain structures, data models, services taxonomies, and other content of the source architectures is required to develop the other Perspectives of the RA's OV.

For the LV, provided the existing system's design documentation is adequate, it will usually be less difficult to discover the design patterns, interfaces, component and system functions, and even the primary data objects that contribute to an LDM. Product data sheets, detailed physical block diagrams, software documentation, and other design data can be used to work out a functional design that applies across the architectures being used as RA sources. If an existing system is built on a SOA, or at least uses services, there will generally be documentation such as WSDL files that can be reused or updated.

RA Validation The second significant difference has to do with RA validation. Whereas a system architecture is validated against the requirements, policies, operational concepts and other aspects of the development program, an RA must be validated against the range of applications to which it is directed. To get the desired generality, coverage, and objectivity in RA content, assessment by a group of experts with a range of backgrounds and interests is usually important. SMEs in each of the real-world systems addressed by a Category or System Product Line RA should be brought in to apply their experience and insight as to whether the content is valid and addresses the right things. Another technique is to reverse the flow of RA development and try applying an RA to one or more of the systems used as sources. If the RA is reasonably easy to specialize to one or more actual system design(s), it is probably valid. When customer RMs and other policies and mandates are an issue, it will be essential to validate the RA against them, point by point. If the organization has an Architecture Board, a formal review by these senior architects can be an invaluable check on the extent to which the RA is correct, well formed, and complete. Basically, the RA team needs to attach high priority to validation and to make use of any available resources to accomplish it.

RA System Categories A Category RA seeks to identify and model a set of characteristics that fundamentally distinguish a class of architectures based on design approaches, product types, special features, and shared requirements that are not found in other categories. Chapter 1 suggested real-time performance as one way to separate the universe of architecture into three categories. Other categorization schemes might use:

- *Security* – systems with the special features and products required to protect data and processes at various levels of sensitivity
- *Logistics* – systems that perform data collection and processing associated with maintenance and supply activities, often with links among factories, depots or warehouses, transportation, and field support sites
- *Manufacturing* – integrated functions associated with material handling, machine control, inventory management, and other aspects of a factory
- *Finance* – systems that perform functions such as transaction processing, global fund transfers, financial market participation, and account management, with special needs for security and integrity
- *Management* – systems with tailored functionality for organizational and program management, with requirements to link multiple organizations and layers of control, again with special needs for security
- *Manned versus Unmanned Systems* – automated versus manually operated equipment and machinery or piloted versus unpiloted vehicles and aircraft

12.4 Category Reference Architecture for the E-X

An organization that adopts the RA strategy will be able to choose categories most appropriate to its activities and products. To illustrate the key features of a Category RA using material already presented in this book, we can consider a Category RA that covers the Process/Decision Real-Time domains of the E-X example. The E-X system is concerned with executing airborne data collection, processing, and dissemination, and the RA will have this flavor. With this approach, this section shows the organization and basic content of a Category RA. The reader may want to refer to the E-X architecture diagrams in earlier chapters and in Appendix C to get a sense of how the specific E-X architecture is being generalized in the RA. The RA diagrams will closely resemble those for the E-X with some detail removed and with suitable adjustments to model element names. Model element specifications are revised to reflect the common characteristics of all systems in the RA category. The Profile created in the E-X architecture to hold stereotypes can be imported directly into the RA.

Requirements A requirements template can be created using Requirements Diagrams that lay out the functional and nonfunctional requirements applicable to any system addressed by the RA. Nonfunctional requirements should have recommended Quality Attributes, with Openness being especially important to the E-X category. With some careful editing, the E-X requirements baseline can be extended serve the same purpose in the RA since any system in this category will have very similar characteristics and functions. For example, specific requirements of the E-X are converted to generalized requirements for sensors, communications, report generation, and other functions to accommodate the fact that various airborne ISR platforms will employ different mixes of these resources and processes.

Operational Viewpoint We first create the content of the Perspectives of the OV.

- *Structural Perspective:*
  - A representative set of generalized Domains includes Executive Management/Command, Planning, Process Execution and Control, Operational Support, Infrastructure and Information Management, and Networking and Communications Management.
  - Blocks modeling Domains have Ports typed by appropriate Interfaces that model typical internal and external interaction points.
  - User Roles are nearly identical to those of E-X. They include Executive Manager/Mission Commander, Planner, Operations Specialist, Technical Specialist, Support Specialist, System Specialist, and Communications Specialist (the correspondence of these Roles to the Domains is obvious and not accidental).
- *Behavioral Perspective:*
  - Use Cases in the RA include Perform Planning, Perform Coordination (with higher authorities, collateral organizations and subordinate organizations),

Perform and Control Operations, Assess and Replan Processes/Operations, Perform Support Operations (dealing with logistics, training, facilities, etc.), Perform Information Management, Perform Resource Management, and Perform Communications Management.

- Each of these Use Cases should have a generalized specification and Activity Diagram showing the flow of the behavior. The examples in Chaps. 3 and 4 show what these look like. Table 12.1 suggests content for a generic Perform Planning Use Case.
- The RA also includes an overarching Mission Thread or primary Business Process (called a "Day-in-the-Life" in Chap. 4), which is also modeled in an Activity Diagram. This diagram shows a generic operational sequence from planning and approval through execution and assessment, with dynamic replanning and support activities incorporated. As always, this diagram includes Domain swim lanes.

- *Data Perspective:*
  - The Conceptual Data Model (CDM) for this Category RA includes Task, Plan, Report, Message, Resource, Reference Data, Visualization, and perhaps other Foundation Classes.
  - If an RM is applied, it may include additional content for the CDM.

Table 12.1 Typical category reference architecture use case specification for the perform planning use case

Use case details	Definition
General description	This Use Case describes the activities associated with developing, approving, documenting, and promulgating a plan for system/enterprise operations
Preconditions	The system/enterprise has been tasked to execute operations; planning guidance information has been received
Postconditions	An approved planning package has been published
User Roles	Executive/Mission Commander, Planner
Primary Data	Operational Tasking, Planning Guidance, Operational Plan(s), Support Plan(s), Communication Plan(s)
Primary Scenario	The organization receives operational tasking and planning guidance. The organization develops a planning package, including Operational, Support, and Communications Plans The organization coordinates draft plans with collateral and subordinate organizations and makes appropriate modifications The organization submits the planning package for approval and makes modifications in response to additional guidance, concerns and decisions Higher authorities approve the planning package The organization publishes the planning package
Secondary scenario	During mission operations, the organization modifies the planning package in response to revised tasking and guidance, mission assessment, and other factors
Tailoring guidance	This Use Case should be tailored to include the structure (Domains), User Roles and activities (Mission Threads), data content, etc. of the system category being modeled

- *Services Perspective* – a Category RA includes the services that are implicit in its behaviors. There are Data Services for the objects in the CDM. The functions of the various Domains can be declared as services when external access is required. The E-X services taxonomy provides a solid starting point for this part of the RA.
- *Contextual Perspective:*
  - An Operational Context Diagram is very important to help an RA user understand the intended applicability and overall content of the model.
  - Supplementary data can include information about the application domain, customer policies and mandates that apply to the RA, and any other factors that influence RA usage.
  - Format samples for artifacts and documents prescribed by frameworks likely to be required by customers for systems derived from the RA are valuable.

Logical/Functional Viewpoint We populate the Perspectives of the LV in a way similar to the OV.

- *Structural Perspective:*
  - This starts with a catalog of design patterns for systems in the category. Examples from the architectures used as sources in developing the RA would also be useful. Multiple examples of design patterns are defined in Chap. 5.
  - BDDs and IBDs initially created for the E-X and other systems are generalized in much the same way as we did for Domains in the OV. Typical Blocks are Sensor, Communications Device, Workstation, LAN, Server, and so forth. E-X Block specifications are also generalized. Figure 12.3 illustrates the process of generalization or abstraction. This figure starts with Fig. 5.27, the IBD that models the E-X ISR Analysis and Reporting Subdomain. In the RA, the sensors are abstracted as Sensor 1, 2, and 3, and the Block Values and Operations indicate the kind of content to be defined in applying the RA to a specific system with a specific set of sensor resources.
  - General descriptions of internal and external interfaces are valuable as starting points for a system architecture that instantiates the RA.
- *Behavioral Perspective:*
  - Generalized versions of E-X Sequence and State Machine Diagrams showing the behavior of individual Classes/Blocks and of design patterns complement the reusable structural materials.
  - Diagrams for complex activities in Activity Diagrams modeling business processes or mission threads include general timing annotations.
  - If the fully mature E-X system architecture, and perhaps the architectures of other systems used as RA sources, includes Parametric Diagrams for commonly required computations, these are added to the RA.
- *Data Perspective* – this is a Logical Data Model with reusable data objects and schemas, largely copied from the E-X architecture with extensions as needed.

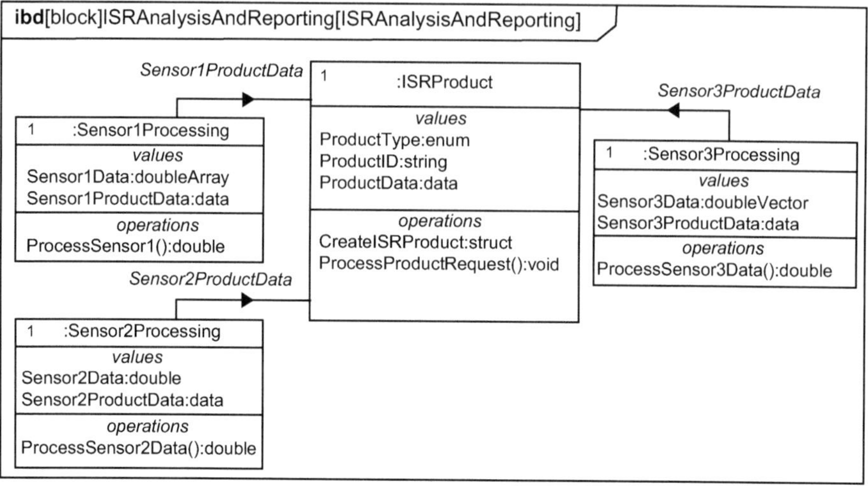

Fig. 12.3 Generalized ISR analysis and reporting subdomain

Additional valuable RA content includes Flow Items, Quantity Kinds, Units, and Value Types.

- *Services Perspective* – continuing the Services Taxonomy created in the OV, the LV for the RA includes a Services Catalog and functional definitions of services and their interfaces. We expect systems in this category to use services extensively, and the services defined for the E-X and other source architectures provide the basis for this content.
- *Contextual Perspective* – this might include nonfunctional requirements and Quality Attributes allocated to Blocks and interfaces, policies and other constraints that apply to functional design, supplementary data associated with elements of the LV, and format samples for artifacts and documents prescribed by frameworks likely to be required by customers for systems derived from the RA.

We have described a Category RA, derived from systems like the E-X, which could be transformed into a System Product Line RA that would apply to many airborne, multisensor systems. This would involve the addition of appropriate detail, such as functional models of the full range of sensors and communications devices that might be employed by such platforms. Such an RA would contain the same list of OV and LV artifacts in the same formats as shown for the E-X in Appendix C. In addition, this type of RA can provide a great deal of value by including good templates for Focused Viewpoints, even though these are primarily associated with the PV. For example, a generic Security Viewpoint for a product line could define and illustrate the full set of protective measures, security devices, policies, services, and other elements of a security architecture that area likely to be needed by systems or other products in the product line. This would reduce both the amount of effort to develop the actual Viewpoint and the need for often-scarce specialized expertise. A system architecture team could start with the RA template for the Viewpoint as a tool for ensuring security requirements are complete and correct and for developing a basic security design. A security specialist could then come in and work efficiently to validate the requirements, assist with detailed design for the specific system, select products to be used in the implementation, plan certification and accreditation actions, and so on. Similar payoffs are possible using RA Focused Viewpoints for networking, infrastructure, and other areas.

12.5 Profiles and Libraries for Reference Architecture

Two powerful techniques for customizing a modeling language, specifically SysML, can make an RA strategy more efficient and productive by tailoring it for a particular system type or domain [7]. These are the use of Model Libraries and the construction of category- or application domain-specific Profiles. The models within an RA repository can include a special type of package called a Model Library containing model elements such as Blocks and Activities that are intended for reuse. Most often, these elements are imported from an actual system model that is being used as an RA input. Such a package is identified with the keyword <<modelLibrary>>. Just as SysML is a profile of UML 2 that specializes the language for use in SE, it can itself be further specialized with Profiles for specific system categories. A Profile normally consists of stereotypes that extend the language and can contain properties and constraints. Once such a Profile is defined and then applied to a particular system model, values for the stereotype properties and stereotype constraints are applied as well. An organization that uses RAs to achieve efficiencies and improved system quality will maximize these payoffs by investing in the skills and content to create and apply Libraries and Profiles.

12.6 Architecture Frameworks

An RA can be thought of as a complement to an architecture framework. A framework is essentially a structured approach to architecture development and documentation. If the organization that uses one or more RAs has also decided on a preferred framework, the RA(s) should be developed and applied within that structure. Although this book, with the exception of Chap. 2 and some of the Appendices, emphasizes the practical aspects of system architecting over the theoretical foundation, we make an exception here in introducing architecture frameworks by placing them in their proper context of fundamental principles.

The following simple relationship shows the underlying concept.

Ontology ⇒ *Framework* ⇒ *Viewpoints* ⇒ *Artifacts*

The basis for a framework is an *ontology* that defines the elements out of which architectures will be developed within a particular category of systems or enterprises, along with the associated terminology. This creates a dictionary for the architecture category in question. Next, a *framework* prescribes the organizing principles to be followed in developing architectures from the elements defined by the ontology. A framework may define a methodology, usually at a fairly high level, and normally establishes a set of *viewpoints* (sometimes called models or architectures) for describing an architecture and the specific *artifacts* within each viewpoint. Clearly, in developing the MBSAP methodology, we have been following this progression from abstract concepts to the details of real system architectures. The artifacts are the ultimate realization of the ontology in a particular architecture space. In the following paragraphs, we take a brief look at a number of frameworks that an architect may encounter. These are often described in conjunction with EAs. We emphasize the point that frameworks and EAs are numerous, overlapping, and very different in the guidance they offer. Chapter 13 addresses EAs more fully.

One of the oldest and best known is the Zachman Framework™ that was cited in Chap. 3. It provides an architecture ontology, but does not include a methodology beyond a suggested sequence of development. Accordingly, it is widely used in the early stages of architecture development to organize thinking about an architecture, to ask the right questions that will bring together the necessary information, and to understand the data and the artifacts that will be produced as an architecture development progresses through levels of abstraction from overall scope and context down to physical implementation. It can be helpful in focusing effort on the most useful artifacts for a particular system. The tiers of the Zachman Framework correspond to the flow of MBSAP as shown in Table 12.2; there is not a perfect on-to-one equivalence, but the progression from abstract to concrete is very similar.

The Open Group, a consortium of organizations from industry, Government, and universities, has published and continuously refined an architecture framework (eponymously labeled The Open Group Architecture Framework, or TOGAF) [8] that is aimed at supporting the overall goal of promoting open systems. This defines a detailed and rigorous architecture development process built on an Architecture Development Method (ADM) supporting Business, Data, Application, and Technology Architectures. ADM is an alternative to MBSAP and is optimized for commercial systems and enterprises where applications are selected and integrated to implement business processes. It begins by establishing an architecture vision and iterates a sequence of Business Architecture, Information Systems Architecture,

Table 12.2 Relationships between the tiers of the Zachman Framework™ and MBSAP

Zachman level	Associated MBSAP content
Scope	System boundary, customer requirements, legacy assets, rules and policies, planned evolution to target system
Business Model	Requirements database, Operational Viewpoint
System Model	Logical/Functional Viewpoint
Technology Model	Physical Viewpoint
Detailed Representation	Models, artifacts, and documentation

and Technology Architecture development phases, followed by implementation, governance, and change management. Table 12.3 shows a basic mapping of the four TOGAF architecture categories to typical content of the MBSAP Viewpoints. An introduction to the latest version, TOGAF 9.2, can be downloaded from the Open Group Web site, and the full TOGAF document is available for purchase.

Another popular commercial framework that is methodology-oriented and tied to the Rational™ family of tools is called the Rational Unified Process (RUP) [9], which is explicitly targeted to software development using UML. RUP defines a well-known "4 + 1 Views" construct where the basic architecture views are:

- *Logical/Design View* – Class, Communication, Activity, and Statechart Diagrams describing software structure, design mechanisms, and functions
- *Process View* – processes and threads, including Active Classes that "own" threads, with runtime decomposition and performance/scalability parameters
- *Implementation View* – content defining software implementation mechanisms and diagrams documenting the point design
- *Deployment View* – infrastructure and hardware/software mapping
- *Use Case View* – the "plus 1" View that accounts for requirements

In MBSAP, these software Views have been incorporated into the more general Operational, Logical/Functional, and Physical Viewpoints.

An alternative commercial framework is called Rapid Object-Oriented Process for Embedded System (ROPES) [10] and was defined in conjunction with early versions of the Rhapsody™ tool. ROPES is also a software architecture process using UML with emphasis on embedded processing for real-time systems. It is an evolutionary spiral that iterates Analysis, Design, Translation, and Testing phases, with incremental buildup of a system prototype using the results from each spiral. There are many others; any architecture tool vendor is likely to have defined a process framework based on the features of the tools they offer.

The Department of Defense Architecture Framework (DoDAF) [11] is widely used by the aerospace and defense community and has influenced the evolution of other frameworks. DoDAF was instrumental in establishing a structure (Operational,

Table 12.3 TOGAF architecture categories mapped to MBSAP viewpoints

TOGAF architecture category	Operational viewpoint	Logical viewpoint	Physical viewpoint
Business	Use Cases, Mission Threads		
Data/Information		Logical Data Model	Physical Data Model
Applications/ Systems		Functional Blocks, Behavior Diagrams, Services	Products/Physical Blocks, Timing, Service Specs
Technical/ Infrastructure		Functional Blocks, Behavior Diagrams, Services	Products/Physical Blocks, Timing, Service Specs

System and Technical Views) and a vocabulary for describing system and system-of-systems architectures. With DoDAF Version 2, a more elaborate set of Viewpoints, each containing Views and Models, has been defined, and emphasis has shifted from diagrams and tables to a rigorous architecture data modeling foundation that supports visualization of various aspects to support a wide range of development activities and decisions. Importantly, DoDAF 2 incorporates a number of extensions and refinements to the original that were incorporated in the United Kingdom's Ministry of Defence Architecture Framework (MoDAF). Yet another important development is the Unified Profile for DoDAF and MoDAF (UPDM), [12] sponsored by the Object Management Group, which aims to move these key frameworks firmly into the world of object orientation and to define UML representations of their artifacts. The two primary points to be made here are that an architect working in Aerospace and Defense needs to be fluent in DoDAF and that the methodology used must efficiently support creation of DoDAF Views and Models.

In a Defense program, compliance with DoDAF typically calls for design models that are ancillary to MBSAP. Many of the models and views called for in the DoDAF System, Services, Standards, and Data and Information Viewpoints are generated from the MBSAP Logical/Functional and Physical Viewpoints. UPDM defines a scheme for constructing the full set of DoDAF models and views in a variety of standard languages, including SysML. DoDAF has a major weakness in its failure to distinguish between product-independent and product-dependent stages in the evolution of an architecture. It is also true that many commonly required DoDAF Views and Models have little or no value in actually architecting a system. In practice, no program would be expected to require (or be willing to pay for) the full range of DoDAF artifacts. There is considerable redundancy of content among these Views and Models, and many of them reflect a legacy view of SE and have limited value in a modern MBSE process. Table 12.4 presents a representative set of DoDAF artifacts based on a SysML model, with appropriate formats, that will serve the needs of most programs.

In response to a public law called the Clinger-Cohen Act, which sought to force major improvements in the use of information technology in the Federal Government, the Office of Management and Budget (OMB) issued a Federal Enterprise Architecture Framework (FEAF) [13, 14] and mandated tailored implementations of the framework according to the missions and processes of individual agencies. Today, this is simply referred to as the Federal Enterprise Architecture. It is structured like TOGAF with Business, Data, Applications, and Technology Architectures and is built on a set of six Reference Models (Performance, Business, Service Component, Data, Technical, and Security). The Reference models were updated to Version 3.1 in 2013, and a guide to FEA implementation is now mandatory by policy for use by all Executive Branch agencies and their suppliers. Under this policy, Architecture Frameworks and EAs have been developed by the National Aeronautics and Space Administration (NASA), the DoD Military Departments, the intelligence community, the North Atlantic Treaty Organization (NATO), and many others. Any architecture team whose work is subject to one or more of these will have to become familiar with its structure and products and to ensure the methodology being used creates the required artifacts.

Table 12.4 Summary of typical DoDAF artifacts and formats

Architecture artifacts		Formats
All Viewpoint		
AV-1	Architecture Overview and Summary Information	Text, Linked Document
AV-2	Integrated Dictionary	Spreadsheet, Text, Linked Document
Operational Viewpoint		
OV-1	High-Level Operational Concept Graphic	Graphic, Linked Document
OV-2	Operational Node Connectivity Diagram	Block Definition Diagram (BDD), Internal Block Diagram (IBD)
OV-4	Organizational/Command Relationships Chart	Use Case Actors Diagram
OV-5	Operational Activity Model	Use Case and Activity Diagrams, Use Case Specifications
OV-6b	State Transition Description	State Machine Diagram
System Viewpoint		
SV-1	Systems interface Description	IBD
SV-2	Systems Resource Flow Description	IBD, Model Element Specifications
SV-4	Systems Functionality Description	BDD, Block Specifications
SV-10c	Systems Event-Trace Description	Sequence Diagrams
Data and Information Viewpoint		
DIV-1	Conceptual Data Model	BDD
DIV-2	Logical Data Model	BDD, XML Schemas
DIV-3	Physical Data Model	Design Documentation, Linked Documents
Standards Viewpoint		
StdV-1	Standards Profile	Spreadsheet, Linked Document
StdV-2	Standards Forecast	Spreadsheet, Linked Document
Capability Viewpoint		
CV-1	Capability Vision	Text, Linked Document
CV-2	Capability Taxonomy	BDD
CV-6	Capability to Operational Activities Mapping	Model Element Specifications
CV-7	Capability to Services Mapping	Model Element Specifications
Project Viewpoint		
Developed by Government program office with contractor support		
Services Viewpoint		
SvcV-1	Services Interface Description (Services Taxonomy)	BDD

Model other Services Viewpoint content using annotations and model element specifications

One further EA that may influence information technology in DoD programs and systems is the Defense Information Enterprise Architecture (DIEA), now issued in Version 2.0 [15]. This pulls together a number of related information management (IM) and information technology (IT) policy and guidance documents to define a set of core principles and rules that can serve as criteria for IT investments to improve effectiveness, efficiency, and interoperability. DIEA offers both general and specific guidance on implementing a large, heterogeneous information enterprise. While some of the content is unique to DoD, the basic principles, such as decoupling data and services from the applications and systems that provide them, are excellent rules and best practices that can profit any high technology system or enterprise architecture.

A search of the system architecture literature will turn up other frameworks and methodological approaches. A system development organization, or even an individual architecture team, may find it worthwhile to do a trade study to select a method and tool set, especially if it is possible to invest in hands-on evaluation of the primary alternatives against the type of architectures to be developed. Provided there is an underlying commitment to employ a rigorous architecture-centric system engineering and design process, supported by mainstream tools and controlled by objective metrics, many approaches can be successful.

12.7 Product Line Considerations

In earlier generations of systems and enterprises, performance was often the overwhelming priority, and it was assumed that virtually every element of every system would be a new design based on optimizing performance. Today, the soaring cost and complexity of these systems and the reality of development times that routinely span a decade or more have motivated a different philosophy, one based on commonality and reuse. Even in industrial and commercial markets, the consequences of increasing complexity and cost are prompting customers to seek returns on development investments across multiple products and applications.

Product lines offer one possible strategy for reaching these goals. Clements [16] defines a product line in terms of a set of products associated with a particular mission, process area, or market segment and emphasizes the importance of using them in an architecture-based development environment with an organization that understands and is committed to the product line philosophy. Clements is concerned with software, but the same fundamentals apply to hardware products. These can range from an electronic board or module that can be mixed with other modules in a chassis up to entire subsystems or systems intended for use on multiple platforms. Companies have been selling products lines like these for many years. What is relatively new is the idea of using RAs as a tool in such a strategy.

Our discussion of RAs in this chapter has emphasized design reuse, primarily at the system level, with the goal of saving time and reducing the nonrecurring engineering effort in a development program. Successful product lines have been created in areas as diverse as information processing, industrial controls, telecommunications, health care, and aerospace. There are cases where a system point design effectively satisfies the requirements of different customers with different business goals or missions. Familiar examples include computers and networks, robots and other industrial machinery, commercial aircraft, and satellites.

In many cases, a basic product must be customized for specific uses. Operational, logistic, economic, and even cultural differences among system users routinely require significant changes in capabilities and therefore variations in system design. For example, a communications satellite bus can be fitted with a specific mix of transponders and antennas to deliver specified services to specified geographic areas. Furthermore, a successful product line can be expected to undergo technology and design upgrades over time to remain commercially viable. This is where a System Product Line RA yields its second major benefit, beyond the efficiencies cited earlier, by providing a basis for efficiently tailoring a basic product design to accommodate such differences. At best, however, this is still design reuse rather than true commonality.

The greatest payoffs can be realized when component and subsystem product lines are combined with RAs. Common components can produce savings through larger procurement quantities, reduced qualification testing, and reduced risk resulting from product familiarity. Further savings are possible in system integration because known products produce fewer surprises when employed in new configurations. Software technologies have made great strides in enabling true components, i.e., software packages that are encapsulated and autonomous with well-defined interfaces so that they can execute correctly in multiple system environments. This is what enables the mix-and-match integration methods discussed in Chap. 6.

Similarly, the pervasive use of open standards for COTS IT products allows an infrastructure development team to assemble and integrate a computing infrastructure, including the software that creates an execution platform. Hardware components intended for multiple applications often have built-in interfaces for a variety of network and bus standards, use standard power, and sometimes offer a variety of packaging options to accommodate different levels of environmental stress. On the software side, many product lines deal with infrastructure components such as operating systems and utilities and with application components that accomplish widely needed processing tasks. Such software product lines include an enormous range of products for business management, document production, data management, planning, equipment control, analysis and visualization, communications, and many others. Component-based software design methods and compliance with mainstream information technology standards and protocols are obviously very important in making such products lines broadly usable and easy to integrate.

Architects and System Engineers can benefit from maintaining product databases, vendor relationships, and lessons learned from prior product uses to allow sound and timely decisions about product adoption. Conversely, RA development uses this information to ensure that design patterns, requirements allocations, interface definitions, and other content are as compatible as possible with available and projected products in order to enjoy the greatest benefits in their markets and in real programs. The product line concept is perhaps not as new and radical as some proponents suggest, but it is nonetheless an important element of a successful and competitive architecture-based system development and integration process.

12.8 Summary

This chapter has looked at the power of architecture methods to reduce cost, schedule, and risk in programs by enabling design reuse and commonality. Customers who want these savings, and who often have other high priorities for interoperability, networking, compatible operational concepts, common logistics support, and other things, frequently seek to achieve them by imposing architecture rules and mandates. These often take the form of RMs, RAs, and EAs that prescribe structured approaches to developing and documenting architecture as well as dictating standards.

A comprehensive architecture-centric SE strategy, such as MBSAP, gives a development organization and its programs the key tools to deal with this environment. The same process is used to develop and maintain RAs at various levels and to tailor and apply them in programs. The architecture taxonomy helps sort out confusing terminology and keep technical discussions focused on the correct issues at any given stage of system development. Clear separation of functional from physical architecture is essential to RA use and is enforced by MBSAP. The RA process itself closes the loop to ensure that the repository is constantly refreshed based on real-world outcomes. Predefined and validated requirements templates and Quality Attributes reduce the effort to provide what's needed in specific programs. The core methodology can produce artifacts that conform to various customer architecture guidelines, so that the need for multiple tools and methods is minimized and supports proof of compliance. The result is improved probability that programs will succeed and will deliver what system customers need.

Exercises

1. For the Smart Microgrid A system example described in earlier chapters, list the content of a Category Reference Architecture (RA) similar to that in Sect. 12.4.
2. For the RA in Question 1, list some content of a Reference Model that could be used as a starting point for the RA in order to improve its quality and consistency.
3. How do Reference Architectures relate to the larger subject of Design Patterns?
4. For the RA in Question 1, list the sources of information that are important in creating the RA.
5. Consider a development program for a new communications satellite. Spell out the ways an RA can reduce the time and cost to design and build the satellite.
6. Assume a customer has ordered a new communications satellite in the category to which the RA in Question 1 applies. Further assume the customer requires a specific set of capabilities in terms of data and voice communications to various areas on the Earth's surface. List the sequence of activities to instantiate the RA to meet these requirements with minimum time and cost.
7. For the RA in Question 1, list the ways in which it can be validated. Similarly, describe a process for maintaining and updating the RA.

8. How would the Category RA in Question 1 be converted to a Product Line RA?
9. How can libraries and profiles be used in constructing and using Reference Architectures?
10. What does a typical Architecture Framework deal with?

Student Projects

Students should expand their projects by creating a new model for a Reference Architecture derived from their project system and perhaps others in the same category. This new model should contain representative examples of Requirements, OV and LV artifacts, and other relevant content to demonstrate the process of RA creation.

References

1. Ellis WJ, et al. (1996) Toward a recommended practice for architectural description. Paper presented at the 2nd International Conference on Engineering of Complex Computer Systems, Montreal, 21–25 October 1996
2. Dept of Defense (2006) The Net Centric Objective and Operations Reference Model (NCOW RM). https://dars1.army.mil/IER2/. Accessed 1 June 2017
3. ISO/IEC/IEEE (2011) Standard 42010:2011, Systems and software engineering – architecture description. http://www.iso-architecture.org/ieee-1471/. Accessed 2 June 2017
4. INCOSE (2017) Systems engineering handbook: a guide for system life cycle processes and activities, ver 4. http://www.incose.org/ProductsPublications/sehandbook. Accessed 1 June 2017
5. Reed P (2002) Reference Architecture: the best of best practices. https://www.ibm.com/developerworks/rational/library/2774.html, http://www.ibm.com/developerworks/rational/library/2774.html#N100E0. Accessed 1 June 2017
6. The Open Group (2009) Architecture board. http://www.opengroup.org/architecture/togaf7-doc/arch/p4/board/ab.htm. Accessed 1 June 2017
7. Friedenthal S, Moore A, Steiner R (2015) A practical guide to SysML, 3rd edn. Morgan Kaufman/Elsevier, New York
8. The Open Group (2017) The Open Group Architecture Framework (TOGAF). http://www.opengroup.org/subjectareas/enterprise/togaf. Accessed 5 May 2018
9. IBM (2001) Rational unified process: best practices for software development teams. https://www.ibm.com/developerworks/rational/library/content/03July/1000/1251/1251_bestpractices_TP026B.pdf. Accessed 1 June 2017
10. Douglass BP (2007) ROPES: Rapid Object-oriented Process for Embedded Systems. http://www.techonline.com/electrical-engineers/education-training/tech-papers/4124793/ROPES-Rapid-Object-oriented-Process-for-Embedded-Systems. Accessed 2 June 2017
11. Department of Defense (2010) The DoDAF architecture framework, ver 2.02. http://dodcio.defense.gov/Library/DoD-Architecture-Framework/. Accessed 2 June 2017
12. OMG (2013) Unified profile for the department of defense architecture framework (DoDAF) and the ministry of defense architecture framework (MODAF). http://www.omg.org/spec/UPDM/Current. Accessed 2 June 2017
13. US Government (2013) Federal enterprise architecture framework, ver 2. https://obamawhitehouse.archives.gov/sites/default/files/omb/assets/egov_docs/fea_v2.pdf. Accessed 1 June 2017

14. US Government (2012) The common approach to federal enterprise architecture. https://obamawhitehouse.archives.gov/sites/default/files/omb/assets/egov_docs/common_approach_to_federal_ea.pdf. Accessed 1 June 2017
15. US Government (2012) DoD information enterprise architecture, ver 2. http://dodcio.defense.gov/IntheNews/DoDInformationEnterpriseArchitecture.aspx. Accessed 1 June 2017
16. Clements P (1999) Software product lines: a new paradigm for the new century. https://pdfs.semanticscholar.org/f815/594fd0f8bb500c4895856edc72c62768a7a3.pdf/. Accessed 1 June 2017

Chapter 13
Architecting the Enterprise

13.1 Enterprise Architectures

Chapter 1 introduced an Axis of Organization with an overarching Enterprise level of organization at the top. Organizations from the worlds of industry, commerce, academia, and government have progressively exploited the growing power and availability of the Internet and private networks to connect locations, systems, and users, often geographically dispersed, in ways that improve the capability, speed, efficiency, and accuracy of their operations. Indeed, the term Enterprise Architecture (EA) has evolved to embrace much more than the technical topics of this book and now includes everything from organizations, business processes, and policies to socioeconomic factors and corporate culture. The Enterprise Architect is becoming recognized as a specialist who has the broad technical and organizational skills to deal with the many dimensions of EA. Before undertaking a discussion of applying MBSAP at the level of an enterprise, it's appropriate to introduce the subject in this broader context.

The Institute for Enterprise Architecture Developments (IFEAD) [1], an international organization that is one of many seeking to standardize and improve EA practice, defines an EA as a master plan dealing with the goals, strategies, governance, business processes, organizations, data, infrastructure, and other aspects of an enterprise. IFEAD approaches EA in terms of people, business objectives, processes, and technology, to which might be added data, organization, and other dimensions. Organizational and staff development, strategic planning, process distribution and synchronization, organizational metrics, and many other matters of concern to senior management can and do become part of an EA project. The IFEAD Web site offers a rich assortment of standards and source materials.

In this chapter, we define an enterprise as the *organizational* entity whose resources and processes are to be captured in a formal architecture. We then take the term system-of-systems (SoS) to mean a *technical* construct in which a number of independent systems are integrated to create or compose higher-level capabilities.

J. M. Borky, T. H. Bradley, *Effective Model-Based Systems Engineering*,
https://doi.org/10.1007/978-3-319-95669-5_13

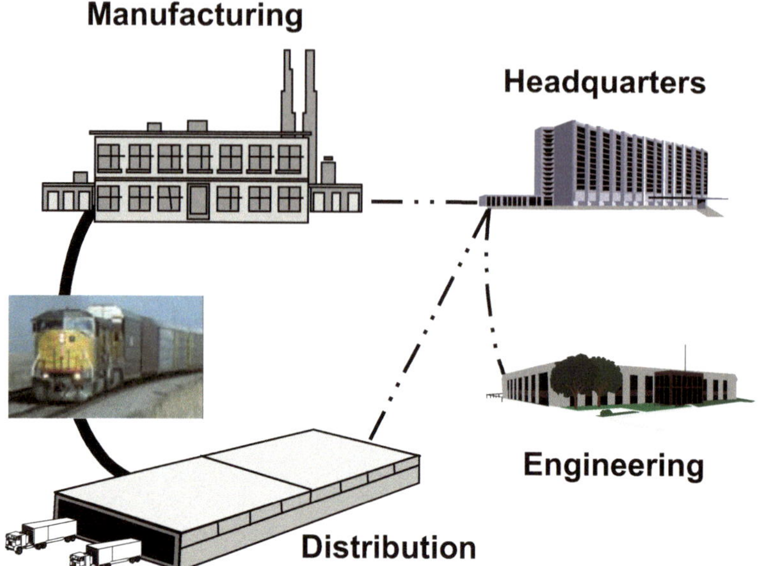

Fig. 13.1 Notional industrial example of the enterprise level of architecture

The two terms are often used interchangeably. We focus on technical aspects of EA, and we extend the concepts of service-oriented architecture (SOA) from Chap. 7. It is now generally the case that an EA will be a SOA for the reasons discussed at length in Chap. 7, and, in fact, the whole field of SOA arose largely in response to the need to integrate diverse and distributed resources within enterprises. The enterprise architect will be most interested in a SOA's ability to implement agile processes and to deal with heterogeneous technologies and dissimilar local processes across the enterprise.

Figure 13.1 sketches a very simple notional industrial enterprise in which a company headquarters, an engineering center, a factory, and a warehouse are networked together. This allows projects, orders, material management, shipping, and other activities to be centrally managed, implemented through workflows, and supported with exchanges. Many real-world enterprises look something like this and have similar needs and issues.

Initially, the elements of the enterprise are likely to have been developed as stand-alone entities. They typically have different information technology (IT) baselines, they were likely originally designed to operate autonomously, and they tend to have many local processes, rules, and conventions that complicate the task of integrating them into a coherent whole. Often, the systems that comprise an enterprise are owned and controlled by individual organizations whose priority is to maximize their performance and efficiency independent of the interests of the enterprise as a whole. These circumstances are what set enterprise architectures apart from the system architectures we have examined in previous chapters. Even so, the fundamental concepts and activities are the same, provided the methodology is extended to deal with the added challenges created by an enterprise.

13.2 The Nature of a System-of-Systems

Just as the discipline of Systems Engineering (SE) emerged from the extraordinary technological and engineering advances of World War II, the advent of an increasingly networked world has created a demand for rigorous methods to deal with larger enterprises, resulting in the still-evolving discipline of System-of-Systems Engineering (SoSE). Computing centers, production and logistics facilities, vehicles, ground and airborne sensors, individual mobile devices, and many others are now routinely connected by various communication paths and required to function as a collaborative whole. Calling such an ensemble a "system-of-systems" suggests that individual systems must become, in effect, the subsystems of the next higher level of the hierarchy. An SoS can be implemented at a single node or location as well as across an enterprise of geographically distributed nodes, and similar architecture issues and approaches apply to both. In fact, a sufficiently complex enterprise might be decomposed into additional levels of SoS, each including entities that are composed of higher-order entities in higher layers, but no new fundamental principles are involved. Figure 13.2 suggests the general nature of an SoS whose interactions are implemented as information exchanges via networks that create an "enterprise interconnect."

13.2.1 Common SoS Characteristics

Earlier chapters have discussed a number of important aspects of an SoS, including the power of SOA in enterprise integration. Chapter 6 dealt with techniques to host heterogeneous products from multiple sources on a shared infrastructure. Chapter 7 explored SOA implementation, and Chap. 9 discussed networking. This chapter is

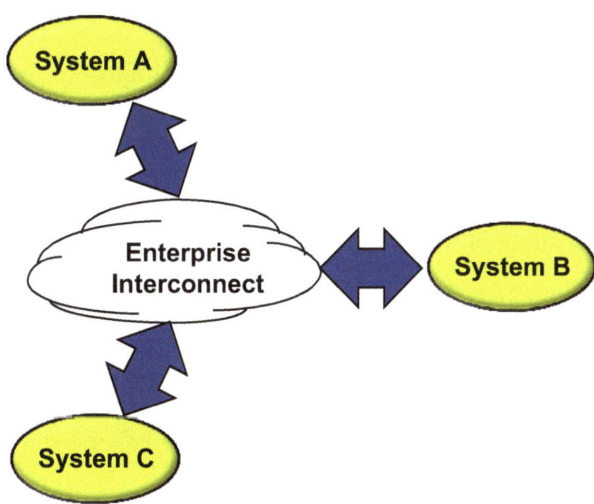

Fig. 13.2 An enterprise composed of participating systems interacting through networks or other interconnections

concerned with some additional issues and architecture principles that arise when integrating a group of independently developed systems to create a whole that is greater than the sum of its parts. Many of the topics in this chapter are described in greater detail in a compilation of SoS material edited by Jamshidi [2].

There are many definitions of an SoS, but in the present discussion, the emphasis is on the integration of a set of independently developed and operated systems with the goal of creating different, usually greater, capabilities than the individual participants can achieve. Jamshidi describes this as a complex assembly of individually complex, independent, and heterogeneous systems. The Department of Defense (DoD), which increasingly relies on networked enterprises and information superiority for operational advantage, defines an SoS as a situation in which independent and useful systems are integrated into a larger system that delivers unique capabilities [3].

Wells and Sage, in Chap. 3 of Ref. [3], identify five commonly cited characteristics of an SoS that were originally articulated by Maier [4]:

- *Operational independence of the individual systems* – each performs useful functions independent of others.
- *Managerial independence of the individual systems* – each is separately provided and independently operated and sustained.
- *Geographic distribution* – individual systems may be widely separated, even globally, and as a result may have constraints on their interactions.
- *Emergent behavio*r – the ensemble may perform functions that are not unique to any participating system and may be difficult to describe in advance of composing the SoS to create the higher-order behaviors that it makes possible.
- *Evolutionary and adaptive development* – the SoS evolves and adapts over time and in response to experience as well as changes in its environment and functions.

In addition to increasing the sheer complexity to be dealt with by architects and designers, an SoS frequently presents two basic challenges over and above those raised by an individual system:

- Some or all of the constituent systems are developed, evolved, and even operated by independent organizations that are not under the control of the SoS owner and often have competing requirements, timelines, and priorities.
- Individual systems are likely to have various technology baselines, interfaces, standards profiles, and other technical details that complicate their integration into a smoothly functioning SoS (see Heterogeneity below).

Maier and others distinguish SoS categories based on the ways in which they come into existence, including:

- *Directed* – a centrally managed SoS built and operated to fulfill specific purposes; to varying degrees, the central authority may be able to modify, configure, and operate the constituent systems to maximize goal achievement for the enterprise.

- *Collaborative* – an SoS without strong central management in which systems voluntarily cooperate to achieve enterprise purposes; such an enterprise may be composed temporarily for a particular time period or operational contingency and then dissolved.
- *Virtual* – an "accidental enterprise" with no central control that functions based on convenience and beneficial outcomes as perceived by the participants.

The Department of Defense adds a fourth category, *Acknowledged*, which is intermediate between Directed and Collaborative and that has central management and resources for the SoS while participating systems retain their particular ownership, objectives, resources, and approaches. This corresponds to many real-world SoS situations and depends for success on the commitment of the participants to the goals of the SoS and to the cooperation and compromises necessary to achieve them.

An enterprise can require a variety of exchanges among its participating systems. For example, a manufacturing business like that in Fig. 13.1 might need to move raw materials from sources to factories, finished goods from factories to warehouses, and individual orders from warehouses to customers. However, most writers on SoS topics restrict their interactions to information, and that is our focus in this chapter. That, in turn, means that SoS implementation primarily deals with the interfaces through which systems communicate and the protocols, data models, control methods, security mechanisms, and other elements of these communications. Maier [4] emphasizes that the enterprise architect must express an overall structure primarily through the specification of communication standards. Dahmann, Lane, and Lowry provide a useful comparison of equivalent artifacts in SoSE and SE contexts [5].

Early approaches to SoS tended to assume that universally mandated enterprise standards would solve problems with interoperability and cooperative functioning, and this simplistic view often yielded unsatisfactory results. Standards by themselves cannot resolve underlying incompatibilities in processes, organizations, priorities, and other factors that often vary considerably across the participants in an enterprise. The outstanding exception is the Internet, which is an enormous SoS and a mainstay of modern society. To a considerable extent, modern architecture strategies based on robust open system principles and technologies such as SOA are derivatives and refinements of Internet principles.

A successful SoS will generally be found to incorporate open architecture on a large scale. However, even the Internet has widely recognized shortcomings in areas such as security and time-sensitive performance, both of which are matters of vital interest to many systems and enterprises. Accordingly, in this chapter we concentrate on some important aspects of applying the overall MBSAP methodology to improve the outcome when an SoS is assembled from systems that were not originally developed with enterprise integration in mind. The basic methodology remains the same, but there are important subtleties in its application to an SoS.

13.2.2 SoS Challenges

We have noted several times that an SoS is generally created on the basis of achieving capabilities greater than those of the participating systems acting alone. The following are some typical categories of enterprise-level requirements that result from this.

- *Interoperability* – this obvious but frequently hard to satisfy requirement involves the degree and efficiency of system-to-system interactions that are necessary for the SoS to function correctly, especially when the participants are technically and organizationally heterogeneous; this is discussed in more detail in a later section.
- *Coordination toward common goals* – the enterprise requires adequate means to plan, prioritize, synchronize, and adjust individual system activities so that the overall objectives of the SoS or enterprise are satisfied; this requirement is often in conflict with Operational and Managerial Independence.
- *Consistent service to users* – by definition, an enterprise exists to satisfy user needs and should ideally present its capabilities to those users as consistent services based on mutually supported functions that are as nearly seamless as possible.
- *Enterprise optimization* – it will almost always be the case that an enterprise that is optimum from cost, performance, and supportability viewpoints will have participating systems that are suboptimized, e.g., enterprises that surrender some capabilities to other systems in order to eliminate duplication or achieve more efficient implementation. Such compromises may be possible in a Directed SoS but can be hard to achieve in the general case of highly independent systems whose owners have vested interests in maximizing their stand-alone capabilities.

The following paragraphs describe some of the challenges standing in the way of achieving these overall enterprise goals.

Heterogeneity An SoS will almost always incorporate systems that have important differences, including:

- Design based on varying technology baselines and design principles
- Operations and updates that are managed on different timelines
- Dissimilar and sometimes conflicting priorities
- Differing operational concepts and procedures

This is the technical consequence of Operational and Managerial Independence. Basic things like systems using different database engines and operating systems can complicate enterprise integration (and provide powerful motivation for a SOA approach), while incompatible processes, data models, external interfaces, and governance schemes only make the challenge greater. Every information system undergoes a continuous process of version upgrades, service pack installations, security patches, and problem fixes, any of which can break the interfaces to other systems if open architecture principles have not been enforced in system design. The enterprise architect needs a well-stocked toolkit of abstraction and adaptation methods to make fundamentally incompatible systems work well together.

Complexity. The range of functions, data, interfaces, constraints, and technologies involved in a high-end enterprise can greatly exceed those of an individual system. Indeed, an SoS can rise to a level of complexity where emergent behavior becomes a significant factor in ensuring satisfactory performance for users and where adaptive behavior is essential to maintain satisfactory operation in the face of dynamic environments and tasking. Such an enterprise is inherently nonlinear because the output is something more than a proportional combination of the inputs. This is a rich area of current research, which is briefly summarized later in this chapter. In general, a major challenge in enterprise architecture is the reality that such entities are inherently more dynamic and unpredictable in both structure and function than a typical single system.

Extended SE Methods. Many writers have observed that established SE tools and methods are inadequate to deal with the complexity of enterprises like those considering in this chapter [6]. A generally accepted SoSE methodology has yet to arise, although some pragmatic approaches have been defined. This is also the subject of a later section.

Requirements, Definition, and Validation. A particularly difficult aspect of extending SE to SoSE involves requirements because:

- The ultimate capabilities of an SoS can be hard to identify in advance.
- There are usually multiple stakeholders associated with various participating systems who often have competing priorities and needs.
- Traditional tools and methods of requirements analysis are likely to be inadequate.

Again, a later section expands on some practical approaches to this challenge.

Interface Definition and Management. An SoS poses special challenges in identifying, defining, and controlling interfaces as a result of its size and complexity, together with the technical and functional differences between systems on either side of any given interface. It's particularly important to adequately analyze and document both functional and physical interfaces between participating systems. This is coupled to requirements analysis to ensure appropriate allocation of requirements to interfaces, e.g., so that an interface shall support all data formats, data rates, error detection and correction methods, and communication protocols involved in system-to-system interactions that use it. Major interfaces need full documentation such as Interface Control Documents (ICDs), at least at Domain boundaries and system-to-system interfaces, and these can be difficult to construct when they involve reverse engineering existing systems to extract the required details. It is also very important to select and enforce standards at these interfaces, including requirements that individual systems either implement the standards directly or provide adapters to transform their native interfaces to be standards-compliant.

Enterprise Evolvability. An architecture and SE team may find ways to compose an enterprise that meets its requirements at an initial point in time only to have the inevitable changes in the participating systems and in the environment and goals of the enterprise itself create breakdowns as time passes. For example, in a secure system, even simple changes in participating systems may necessitate an expensive exercise to maintain the enterprise's authority to operate in order to continue to handling sensitive information. Some of the ways in which this kind of fragility can be at least partly mitigated include:

- Maintaining open architecture and SE discipline in areas such as preserving loose coupling and generous margins
- Enforcing rigorous change control and configuration management
- Ensuring that reliability and availability requirements continue to be met
- Ensuring that human-system interfaces continue to function as designed

The level of cooperation among organizations that control individual systems in forecasting changes, along with their willingness to be proactive in minimizing the consequences of those changes, can spell success or failure in preserving the viability of the enterprise as it evolves.

Governance. Governance of an SoS is inherently more complicated than that of a single system. There are many dimensions to the challenge of governance of an assembly of largely independent entities. A Directed SoS will have governance as part of the central authority, but a Collaborative or Acknowledged SoS needs voluntary compliance on the part of the participating systems, and a Virtual SoS may not have much more than community consensus. SoS governance is focused on defining and enforcing standards for communications and services (including Quality of Service), enterprise data models, SoS processes, configuration management at the SoS level, and management of the deployment of successive versions and updates. Standards bodies, by regulating the process of updating key enabling standards and working through community agreement on key issues, can be a powerful agency in enterprise governance. The organization employing an SoS will need an Architecture Board or equivalent coordination and control authority. Chapter 15 gives more details.

13.2.3 Adaptive and Emergent Behavior

As noted earlier, one consequence of the level of complexity reached by high-end information-intensive enterprises is emergent behavior, usually defined along the lines of behavior that is not uniquely associated with any individual system element but is instead a global characteristic of the ensemble [7]. Keating, in Chap. 7 of Ref. [3], notes that behavioral patterns and other properties of a complex system will be observed in operation but can't be anticipated in advance. This complicates the task of defining, architecting, and implementing an SoS because the behaviors resulting from composing a set of systems may not be predictable from the characteristics of those systems when standing alone and may or may not meet user needs and expectations. Another possible source of complexity is the desire in some enterprises for adaptive behavior, which changes in response to operational situations, patterns of events, predictions of future occurrences, or other stimuli.

Such complex behaviors are conditioned by the context of the enterprise, which may involve the operational situation, perceptual biases of operators, governing policies and priorities, and even things like severe weather and civil disturbances in addition to the logic embedded in the participating systems. Some are benign, for example, the fact that a given task from a queue may be assigned to any one of a set of pooled resources for execution without knowing in advance which one will be chosen is probably unimportant. On the other hand, an enterprise that makes unpredictable decisions about triggering a power outage or blowing up a satellite launch vehicle would create a very large problem.

The SoS literature contains a number of examples of both emergent and adaptive behaviors, and many researchers have suggested ways to cope with them, although no universally accepted approach has yet emerged. Kilicay and Dagli [8] recommend something very much like MBSAP with emphasis on detailed executables as key tools for behavior analysis. Bar-Yam [9] notes that the complexity of interfaces in an SoS is typically much greater than that of the individual resources behind the interfaces and advocates an evolutionary approach supported by an environment in which the ensemble evolves from continuous incremental changes in the constituent parts rather than from focused efforts to design the enterprise as a whole. This is a logical extension of the physical and virtual prototyping environment. Recommended approaches fall into three broad categories:

- *Build a little, try a little* – incrementally assemble the SoS, and exercise it to determine its behaviors and responses at each stage, with provisions to correct or limit unacceptable behaviors before adding the next system(s) or set of capabilities.
- *Modeling and simulation (M&S)* – build high-fidelity simulations or emulations of systems, networks, operators (including human-in-the-loop simulation), and environments, and explore the largest possible combination of tasks, policies, operational circumstances, and other factors to try to establish the bounds on behavior and the sequences of system actions that lead to them.
- *Design for stability* – incorporate generous margins, stress loose coupling among systems, and seek to establish stable operating points to which the SoS reverts after excursions, such as a set of system states whose outputs and allowed transitions are well understood and known to represent allowable enterprise behavior.

An SoS with adaptive and emergent behaviors may also exhibit self-organization, that is, the ability to select and combine resources to perform a given function without operator intervention. A simple example from Chap. 9 is a MANET whose distributed control algorithms allow it to manage entering and departing participants, do dynamic message routing based on the current network geometry, and manage traffic around outages or interference. Another example is an enterprise of sensors and effectors, using either a centralized control authority or distributed control logic, which can select the optimum sensor(s) to detect and measure an event or condition in the enterprise operational environment based on the configuration of sensors and the location of the event, and then select the best effector to perform the indicated action. An industrial example of adaptive behavior would be a manufacturing facility that can automatically route a shipment of incoming raw materials to an appropriate inspection line, schedule the initial processing steps based on the results of that inspection, and then send the processed materials to a fabrication line. As the complexity of the enterprise and the scope of its available behaviors increase, self-organization can be both powerful and hard to predict. This is an area where high-fidelity M&S can be very useful.

There are many variations on this complex behavioral theme. Systems may implement polymorphic services whose behavior when invoked depends on the specifics of the service request and the accompanying parameters. As rule- and policy-based processes and workflows become more common, the possibility of unpredictable combinations of these arising through interactions of systems in an enterprise is a likely source of overall unpredictable behavior. For example, if different systems have different rules for dealing with faults and contingencies, they may try to take conflicting actions when such a situation occurs during a cooperative activity. This brief discussion has only hinted at the challenges of complex behavior in an SoS, and an enterprise architect is advised to investigate this rapidly evolving field of theory and practice more fully. We address the related subject of dynamic enterprises later in this chapter.

13.3 Enterprise Architecture Methodology

We recommend in this book that enterprise architecting be approached through the same basic MBSAP methodology that we have been applying individual systems, with additions and special emphasis to cope with the challenges described in the previous section. Figure 13.3 annotates the stages of MBSAP to highlight some of these differences.

Organization, Roles, and Responsibilities. The independence and diversity of systems in an SoS have important implications for the structure and responsibilities of the architect and the architecture team. The architect's role is likely to have a large component of coordination, negotiation, and compromise in trying to reconcile the capabilities and priorities of the participating systems, organizations, and programs. The architecture team should have representation across the participating organizations and will be most effective if it is empowered by all concerned stakeholders to pursue the best solution possible under the limitations inherent in an SoS. This is especially true when there is a requirement to use existing systems that reduces the flexibility to optimize the enterprise as a whole. The architecture team must control the enterprise-level architecture and its model and will need resources to carry out analysis, prototyping (especially virtual), and other activities under MBSAP (or whatever methodology is adopted).

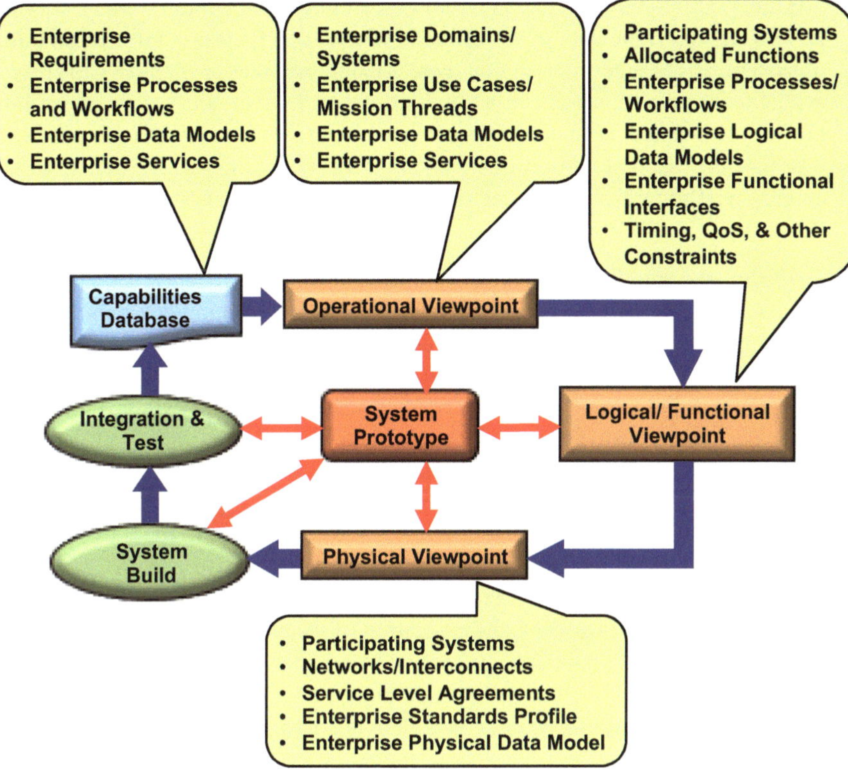

Fig. 13.3 Special areas of emphasis in applying MBSAP to an SoS architecture

Although this book deals with architecture rather than program management, the enterprise architect has a vital interest in an effective structure for planning, executing, and tracking a project. Specifically, the architect needs the ability to ensure that the program structure includes essential architecture efforts and supports the development and maintenance of an SoS solution with high technical integrity. Conversely, an astute program manager will appreciate the power of architecture in supporting both technical and programmatic decisions and will work with the architecture team to that end. The following are some specific examples.

- The overall plan and schedule, which might be an Integrated Master Plan and Schedule (IMP/IMS), should have tasks, milestones, and criteria corresponding to SoS architecture development and evaluation.
- The IMP, or equivalent, should include architecture processes, metrics, decision processes, program review criteria, and other content that enable a robust architecture program; a uniform architecture methodology applied across the SoS is essential to success, but it must also accommodate established tools, methods, and products from individual system development efforts.

- SoS integration and test (I&T) should be carefully defined and sequenced with system and development builds (Chap. 11) for an orderly and controlled addition of systems to the enterprise so that there are opportunities to identify and correct problems at minimum cost. In a common situation, individual systems operate on their own schedules and deliver products, upgrades, and problem fixes whenever they are ready or to meet other users' needs, rather than for the benefit of the enterprise. The I&T process needs extensive and proactive coordination plus the flexibility to accommodate these asynchronous timelines.

Capabilities and Requirements. As discussed under SoS challenges, the primary emphasis in establishing the requirements foundation of the architecture must be on capturing enterprise-level content. This includes requirements to be satisfied by the enterprise, presumably exceeding those of the individual systems, along with processes and workflows that create functionality for users. In other words, enterprise-level requirements must include higher-level behaviors that are not available from the participating systems acting alone. It is also essential to ensure that the needs and expectations of all stakeholders are identified and either incorporated in requirements or adjudicated and resolved if they are not going to be met. These overarching issues are often best dealt with in the context of an enterprise or SoS concept of operations (CONOPS) that synthesizes all the capabilities to be delivered and how they will be employed to achieve the objectives of the enterprise.

If enterprise-level data models and service taxonomies are defined, they should be incorporated in the baseline as the starting point for more complete data and services definitions that will emerge as the architecture is developed and analyzed. A classic error in enterprise architecting is to allow incompatible allocations of responsibility for data or services to participating systems, creating built-in conflicts and errors. To an even greater extent than with a system architecture, the capabilities baseline will be a living document that is updated as the behaviors, interactions, user functions, and other characteristics of the SoS are discovered. Another important early consideration is how enterprise requirements will be validated; this is likely to require extensive use of Operational and Process MS&A as discussed in Chap. 11.

Operational Viewpoint. The top-level partitioning of an SoS into Domains normally starts with the participating systems, which are then in turn partitioned as usual under MBSAP. The interfaces among these systems are crucial to the success of the SoS; they need to be clearly identified and associated with the Business Processes or Mission Threads and other behaviors they support. Each interface must be carefully mapped to the system resources that will be responsible for realizing and using it. Behaviors are still captured in Use Cases and Mission Threads. They are modeled with Activity Diagrams but with swim lanes devoted primarily to systems and users to show the flow of behavior and data exchanges among systems.

A Conceptual Data Model and Services Taxonomy, which may include a priori content from the Capabilities Database, must be completed for the full SoS. Few artifacts are more valuable to an SoS project than an accurate and comprehensive Enterprise Dictionary, especially when inconsistent terms and definitions used by various systems must be harmonized. Another important part of building the OV is research to understand the capabilities, processes, policies and procedures, organization and user roles, and other important characteristics of the participants. Both commercial and Government systems are often developed or procured for a variety of operational users, and it is critical to identify any requirements or priorities arising from other applications of a system that may conflict with the needs of the SoS that is being architected.

The Services Perspective of the OV starts with compiling Mission and Data Services of the individual systems. Then we add a set of Enterprise Services that provide essential glue to hold the SoS together. Enterprise Services typically include categories such as:

- *Security* – services for enterprise-wide user authentication, data confidentiality, event logging and auditing, and other functions like those discussed in Chap. 10
- *Backup and Recovery* – enterprise-level data backup to a secure storage facility plus services for restoring data and system states following an outage
- *Data Dissemination and Synchronization* – services for providing current and consistent versions of enterprise data and for resolving conflicts between competing versions
- *Policy* – services for maintaining and updating policies governing enterprise operations such as task priorities, communications, and access to sensitive information
- *Configuration Management* – services for auditing and reporting system configurations and for consistently pushing updates out to individual sites and systems
- *Health and Status* – services for reporting the condition of the enterprise and its constituent systems to a central monitoring point
- *Workflow, Choreography, and Orchestration* – enterprise-level SOA mechanisms for composing and coordinating services
- *Networking* – enterprise-level communications, including QoS

Like the Enterprise Dictionary, a Services Registry is an indispensable architecture artifact, both to support SoS integration and during operations to allow services to be discovered and invoked.

Logical/Functional Viewpoint. SoS architecture design consists largely of building the most complete functional model of the participating systems, their interfaces, and their interactions that available data permits. The model can be expected to evolve significantly over the course of a project as more is learned and design decisions are made. SysML Blocks with Part Properties and Ports provide a natural way to declare systems and their internal partitioning, define interfaces associated with subsystems or other resources, and represent associations and flows between systems. Figure 13.4 is a simple example of one such format, using a BDD with systems interacting via declared interfaces and an IBD showing the internal Domains of System A modeled as Parts with a `<<Domain>>` stereotype. Port 1 is delegated to Domain 2 for realization. Assuming the enterprise is implemented using a SOA, the primary content of interface definitions will be Service-Level Agreements

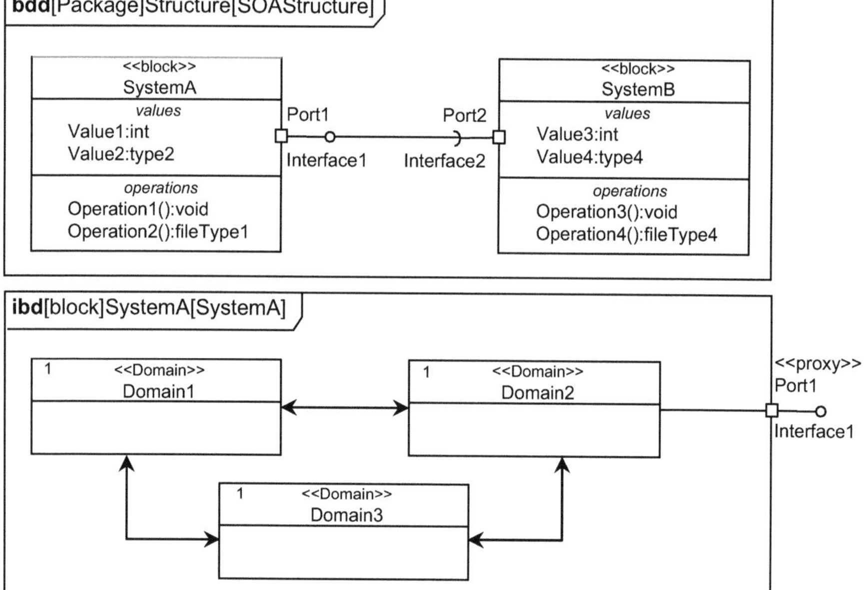

Fig. 13.4 BDD and IBD showing one approach to model an SoS structure

(SLAs) or some equivalent for the Services that are exposed at various Ports. There are many other ways to model SoS structure following the general principles of MBSAP discussed in earlier chapters.

The preferred mode of interaction within most SoS is message-based communications. Sequence Diagrams are therefore the usual means of modeling behavior, but in addition individual systems are frequently modeled using State Machines to show responses to events and signals from other systems and the external environment and to account for timing and synchronization. Loose coupling, implemented through SOA with messaging, is likely to be vital in taming the complexity of an SoS.

As is the case for a single-system architecture, an important element of the LV of an SoS is an Allocated Baseline that flows requirements down to Systems, Subsystems, and lower-level Parts. Whenever an SoS is being composed from existing systems, difficult decisions are often involved in deciding where certain requirements should be allocated, and thus where verification testing will be performed, when system capabilities overlap. A requirement that is to be met by cooperative functioning of multiple systems must be parsed and correctly allocated to each. When no existing capability satisfies a requirement, further difficult decisions are in order as to which system or systems should take responsibility and how the necessary modifications or enhancements will be implemented. Any such decision will generally need support from trade studies, M&S, cost and risk assessments, and other SoSE activities. Like most aspects of an SoS architecture, the Allocated

Baseline is very likely to change over time as the design matures and supporting analyses are completed.

SoS behaviors may be "hard wired" in the sense that systems are programmed to execute specified functions and produce outputs in response to a fixed sequence of control inputs, procedure calls, timed events, or other predetermined stimuli. However, in the preferred approach where systems are service-enabled and integrated using SOA, cooperative behavior will normally be defined using workflows and orchestrations. This starts with functional definitions in the LV. A Logical Data Model, derived from the Conceptual Data Model, is intimately related to these processes and workflows. In fact, many architects prefer to start with data flow modeling as the basis for understanding enterprise behaviors that cross system boundaries and then flesh out these flows by adding the activities that create, transform, and use the data.

In many enterprises, the independence of the individual systems requires special attention to matters of timing, networking, results checking, resource management, information assurance, safety, fault detection and recovery, and other aspects of correct operation. These become, in effect, constraints on the enterprise architecture and must be dealt with in the functional design of the LV. Some examples include:

- An SoS that is primarily interconnected by one or more networks whose performance is significant in meeting process deadlines requires Quality of Service (QoS) mechanisms as discussed in Chap. 9 to ensure highest priority traffic is handled with minimum delays and errors.
- An SoS with requirements for security and safety requires coordination of the affected functions and features of the individual systems to ensure criteria are met at the enterprise level; this typically creates requirements on individual systems to use enterprise-defined services.
- An SoS with real-time requirements such as process deadlines, especially when the systems comprising the enterprise are collocated, typically needs a master timing service or clock to ensure all participants adhere to a common definition of time.

Physical Viewpoint. The SoS architecture should include the most complete set of physical data possible on the participating systems, recognizing that in some cases details of internal design will not be available for proprietary, security, or other reasons. As long as physical interfaces are fully defined, this may not be a concern. SoS physical design emphasizes the networks or other interconnects and details of system interfaces such as ICDs and SLAs. In some cases, it's important to capture physical interfaces such as connectors and to define required flows such as power and cooling. Individual systems have likely been designed using different methods and conventions for their human-machine interfaces (HMIs), and it's important to have these definitions to support operator training, e.g., to allow operators, who may initially be familiar with only a subset of the SoS, to access full enterprise capabilities.

The PV also has enterprise-level content. Important details include networks and other connectivity, messaging schemes, and additional interaction models such as events, alerts, and information dissemination mechanisms. The configuration of an Enterprise Service Bus, if one is used, must be verified and fully documented. There will be an enterprise standards profile, largely dealing with interfaces, communications, data, security, and HMIs. Finally, the Logical Data Model may be implemented in a Physical Data Model, at least for data resources that are shared across the enterprise. If the participating systems expose their information content via data services that comply with enterprise standards, the details of their database schemas are much less important. If the enterprise has real-time requirements, the PV will need timing analysis using the appropriate diagrams and simulations at the enterprise level. It's obviously essential to establish that the participating systems can each complete their functions as part of an enterprise behavior such as a workflow within allocated deadlines. This analysis may result in the design of enterprise-level clocks or other timing services, as identified above under the LV, to synchronize and control cooperative actions among enterprise participants.

Prototype. Since cost, geographical dispersion, and other factors are likely to make a full physical prototype of an SoS impractical, virtual prototyping takes on added importance. A common strategy is to network together existing simulators for various participating systems to build up at least a partial representation of the enterprise that can be used to exercise specific functions, conduct training, validate interfaces, and determine whether modifications to individual systems will cause enterprise processes to fail. A later section of this chapter focuses on enterprise M&S. Eventually, the physical configuration must be assembled and tested. Establishing network connections among the locations of participating systems often does this. An initial prototype may be built up as participating systems provide their assets and ultimately evolve into the actual enterprise.

13.4 System of Systems Engineering (SoSE)

We are now in a position to present an SoSE approach that deals with many of the shortfalls in earlier SoS methodologies. The principles and methods of SE in general and MBSAP in particular apply to an SoS, but with additional issues and activities to deal with the challenges discussed in this chapter. DoD has published an SoSE guide [10] that focuses on military enterprises but brings together a wide range of research and offers a good overall approach. As with any program dealing with complex, information-intensive entities, the SoSE Team should document the methodology and related materials in a Systems Engineering Plan (SEP) or equivalent document so that all participants know how the enterprise will be managed from a technical viewpoint. In addition to the development of Operational, Logical/Functional, and Physical Viewpoints as discussed above, a sound SoSE strategy includes the topics defined in the following paragraphs.

Requirements Analysis and Management. Any complex system needs decomposition, flowdown, and allocation of requirements. In an SoS, there is a special need to work on the relationship between enterprise- and system-level requirements. The latter are often pre-determined, and significant changes may entail cost and risk. It's therefore especially important to analyze the capabilities that are created by the SoS and to ensure that enterprise requirements are achievable within the available resources and schedule and that traceability to system requirements is clearly defined and understood by all participants. Once an enterprise requirements baseline is established, continuous coordination with participating system programs and organizations is vital to anticipate changes and to manage (ideally, to minimize) the resulting impacts on the enterprise.

Enterprise User Focus. Requirements analysis should retain an enterprise user perspective and should ensure that the overall capabilities that motivate creation of the SoS are fully documented. In many situations, the SoS will present itself to an operator performing a given User Role through a specific system, and that system's user interface must handle both the system's native functions (with which the operator is likely to be already familiar) and the additional capabilities contributed by the SoS (which may be new to the operator). This is important to ensuring that the SoS meets operational user needs and expectations as well as for training, updates to procedures, and preparing for effective operator participation in enterprise processes and workflows. A strong focus on CONOPS in the OV is important in this regard.

Capability Evolution. An earlier section noted that in addition to technical differences, the systems in an SoS usually have their own schedules that may or may not support the desired enterprise timeline. As a practical matter, the SoSE process becomes one of continuous evolution, integrating systems and upgrades as they become available. Figure 13.5 gives a very simple view of this situation. Each participating system follows the engineering "Vee" from Chap. 3 and produces results ready for integration into successive builds of the SOS on some staggered schedule. Assuming the SoS has limited ability to influence participating system schedules, the SoSE process must include an integration and test methodology that efficiently accommodates the sequence and timing of these deliveries. A well-provisioned prototyping environment is key to making this possible.

A Master Evolutionary Plan (MEP) can be another invaluable tool that provides a systematic mechanism for forecasting changes in the requirements and environment of the enterprise as well as planned changes to the participating systems. For example, the MEP has milestones for the deployment of significant enterprise capabilities over time. Such a "big picture" overview of the evolutionary future can help the SoSE team spot both potential problems and opportunities far enough in advance to take appropriate action. It also provides participants a vehicle to communicate individual system plans for modifications, upgrades, and eventual retirement. The MEP should also deal with technical planning, including forecasting and preparing for product obsolescence and replacement, and identifying emerging technologies and products of interest to the enterprise. It's a good idea to develop and prominently display a master schedule that puts all enterprise participants on a common

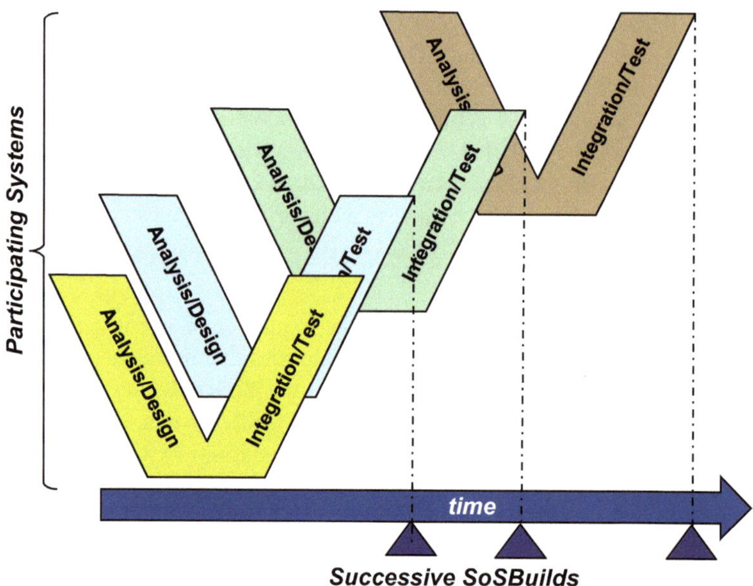

Fig. 13.5 Asynchronous schedules of individual programs resulting in an evolutionary sequence of SoS builds

timeline with milestones such as planned upgrade and fielding events, plus dates for system retirement ("deprecation") and for introducing new participants.

Open, Loosely Coupled Architecture. As important as these tenets are in architecting any information-intensive system, they are even more vital in the context of an SoS. An attempt to simply connect a group of systems together using their native interfaces and relying on their built-in business and control logic to support enterprise-level processes would be high risk at best. The SoSE process should be architecture-centric and should define a framework of open, standardized interfaces through which message-based transactions and event- or data-driven process control can be implemented. This is based on understanding the participating systems and their relationships as both a foundation for and a constraint on the architecture. It's important to emphasize again the vital importance of an enterprise data model and associated services for any information that must be exchanged among systems and with users and the external environment. Negotiation may be required to secure compliance by the participating systems, and resources will be needed if they must make modifications to implement the enterprise architecture, e.g., by transforming their native data formats to those of the enterprise. The hallmarks of an effective enterprise architecture are that it creates an effective framework for implementing the SoS and that it persists over multiple iterations of the constituent systems, lending essential SoSE stability to cope with the reality of continuous change.

Metrics. Technical assessment of an SoS depends on a manageable set of enterprise-level measures. These may include Quality Attributes (QAts), Technical Performance Measures (TPMs), Key Performance Parameters (KPPs), Measures of Performance and Effectiveness (MOPs/MOEs), or other metrics that can be quantified, tracked over time, and compared to required values. The difficulty comes in selecting from the enormous range of things that could be measured and tracked in a complex enterprise. What's needed is a relatively small set of metrics that yields high insight into the quality and effectiveness of the design without creating an excessive overhead burden for data collection, analysis, and presentation. Openness is an essential QAt. When the enterprise has clear overarching operational objectives such as primary business processes or missions whose success or failure can be established, these are likely to lead to good MOEs. When key technical parameters such as thread timing, control errors, or failure rates are important, they may support good TPMs. There is no universal answer, but early attention to an effective metrics framework will pay large dividends in assuring the quality and integrity of an evolving SoS.

Decision Analysis. In a general SE methodology, decision analysis is a set of formal techniques such as trade studies, analysis of alternatives, risk assessments, and performance assessments that provide a factual basis for selecting among alternatives. In an SoSE methodology, the fundamentals are the same, but there can be significantly more variables in identifying the optimum alternative, and the analysis must assess consequences at the enterprise level. As in other areas, a high-quality virtual prototyping environment can be the most powerful and affordable tool for SoS decision analysis. Metrics are an important part of decision analysis because it is often critical to assess the impact of various alternatives on the overall measures of goodness of the design in order to guard against expedient short-term decisions that compromise the long-term integrity of the enterprise. A commonly encountered example of such short-term thinking that produces long-term consequences is to be too lenient in granting waivers to enterprise standards. This may spare participating systems the initial expense of fully integrating into the enterprise while having the effect of either foreclosing options for achieving full enterprise capabilities or raising the cost to achieve them later.

Cybersecurity. If the enterprise is required to handle classified, proprietary, or other sensitive information, yet another dimension of coordination among systems is created. Although the SoS in this case will have to undergo cybersecurity assessment and authorization as discussed in Chap. 10, that process is based on, first, accreditation of the individual systems and then on evidence that interactions within the enterprise do not compromise their security attributes. Fielding an initial or upgraded configuration of an enterprise can be delayed, with attendant cost and operational impacts, if even a single participating system fails to complete its approval process or is late in delivering the required documentation for enterprise accreditation. Furthermore, a change to any system in the enterprise typically requires re-accreditation of the whole.

Performance Measurement and Requirements Verification. Both development and operational testings, as described in Chap. 11, are required at the enterprise level, first to determine what the SoS is actually capable of doing and then to verify that enterprise-level requirements are satisfied. This can be a difficult SoSE task as a result of the complexity of the enterprise being tested, the fact that it is often impossible to literally test every function under every possible condition, and, in most cases, the need for human-in the-loop testing with operational personnel to determine if the enterprise meets the needs of its users. When participating systems are already fielded, data and operator involvement from their operations may be useful in enterprise testing. The SoSE Team, with the backing of the governance structure, has a critical responsibility to ensure that adequate analysis and testing are accomplished to ensure that the enterprise is operationally effective.

Configuration Management. CM of the systems within an enterprise usually remains the responsibility of the owning organizations. However, the SoSE Team needs full access to current baselines in support of enterprise-level planning, analysis, and design. Additionally, requirements and design baselines for the enterprise as a whole are a core SoSE activity. As with a single system, a well-maintained architecture model creates an ideal framework for CM of a complex entity like an SoS.

Kaplan [11] provides an alternative perspective. He offers an overview of SoSE with a set of recommendations based on establishing program management and technical authorities empowered to enforce a disciplined and comprehensive approach, which would be associated with a Directed SoS. His approach is broadly similar to that defined in this chapter and includes:

- Establish a System-of-Systems Engineer reporting to the System-of-Systems Authority (SoSA, which could well be an enterprise architecture board).
- Preserve the integrity of participating systems/programs, and do not set up competing governance.
- Charter the SoS Engineer to help deliver the best SoS, create the support environment, and coordinate technically across programs; the SoS Engineer pursues enhanced information availability; seeks to optimize the SoS in terms of performance, cost, risk and agility; and provides overall analysis and support.
- Employ the SoS Engineer in a classical Systems Engineering role for the SoS, creating an environment in which individual program System Engineers can work together in a common context.

13.5 Interoperability

We use the term interoperability for the overarching characteristic of a successful SoS that achieves the full required set of capabilities to communicate, cooperate, and integrate functions and information. Like many architectural concepts, this appears at first to be rather simple and intuitive and then proves in practice to be quite otherwise. Meaningful interoperability goes far beyond networking or message exchanges. It is the heart blood of any enterprise or SoS because entities that cannot interact correctly and consistently cannot coordinate their actions to produce higher capabilities. A reasonable approach to the subject is to build up from the physical basis for information exchange to the information-enabled processes and the shared understanding that are the ultimate test of interoperability. Along the way, we will discover characteristics that the participants in an enterprise must exhibit to make full and robust interoperability possible.

Interoperability has been given many definitions, but in MBSAP terminology, it is the ability of a group of systems and their users to exchange information in ways that enable common understanding, collaborative decision-making, and coordinated action to achieve operational goals. It is closely related to concepts like service-orientation and robust networking, which can contribute powerfully to its attainment. However, this discussion deals with the essence of interoperability, which comes down to sharing information to enable cooperative action.

Figure 13.6 lays out a hierarchy of interoperability levels. This construct traces in large part to the work of Julian Ranger of the former Stasys, Ltd. (now part of Lockheed Martin), who described the highest level as "brain-to-brain." It builds on

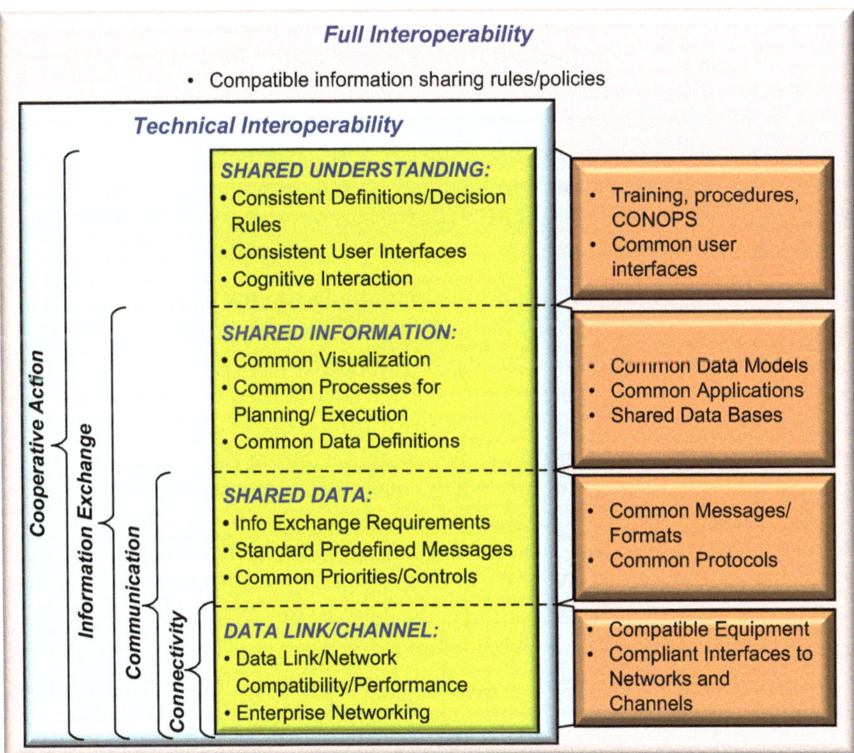

Fig. 13.6 Levels of interoperability leading up to cooperative activity among enterprise participants

the layered structures of the OSI model and the Internet protocol stack as presented in Table 9.1. The following are some of the key points in understanding what facilitates and what impedes interoperability, especially in the context of an SoS.

Data Link/Channel. The lowest level establishes physical connectivity such as a network or point-to-point channel over which participants can send energy that carries information. It corresponds to the bottom of the Internet layer structure. Interoperability here amounts to establishing a communications medium and using compatible devices to send and receive. In the early days of telephony and radio, this was the only interoperability consideration, and it is still essential, but only as the foundation for higher levels. The enterprise architect must typically cope with some combination of existing and new communications and networking resources and must ensure that data rates, latencies, error handling, security, and reliability meet performance requirements with adequate margins. Both physical M&S and adequate communications testing under real-world conditions are essential.

Shared Data. The next level in Fig. 13.6 deals with compatible messages, protocols, data formats, and similar standards or conventions used to exchange comprehensible content. This is, essentially, the Transport and Network layers of the Internet and creates the basis for a communication or networking session. An enterprise-level messaging model typically establishes protocols, formats, control and traffic management methods, session management, security mechanisms, policies, and priorities. For example, if the network implements QoS, all enterprise participants should use a uniform policy or rule set. A Communications Management Domain like that introduced in Chap. 4 for the E-X can be readily extended to optimize the management and use of individual system communications resources as well as those of an enterprise in order to do the best possible job of communications among participants under overall constraints, rules, and priorities.

Shared Information. As shown in Fig. 13.6, the Data Link/Channel and Shared Data levels of the interoperability hierarchy establish a basic communications capability for exchanging messages and data. However, it is all too common for messages to be incomplete or misleading and for data to be interpreted in different ways. The next level brings in improved ways of processing, displaying, and using data to create true information sharing. Data becomes information as it is transformed by processing, combined with other data, interpreted to elucidate its significance, and presented so that it can support decisions. This level in Fig. 13.6 involves a number of aspects of the Internet model Application layer.

This level of interoperability starts with exploiting a standard enterprise data model so that all participants use the same data definitions and structures. Next, interoperability is materially improved if all participants use common applications, or at least the same algorithms, logic, and rules, in processing the data to create information. Then, standard ways of visualizing the information, for example, displays with common formats and symbology, to generate common pictures of the operational situation across an enterprise, help all participants to use the information consistently to achieve enterprise goals through cooperative action.

Shared Understanding. With the capabilities of the first three levels, the enterprise can have robust information exchange. Traditional definitions and testing of interoperability have gone no further, and traditional network models have no higher levels to represent shared understanding. Nevertheless, experience with information-rich enterprises has shown that a still higher level of interoperability is both possible and required for effective enterprise operations, especially in executing critical or time-sensitive processes. The basic premise is that any operator or other user in the enterprise, when presented with a given set of information, should understand and interpret it in the same way, resulting in timely, consistent, and correct collaborative decisions. When enterprise operations depend on quick action and timelines do not allow operators to discuss information and debate its meaning, this kind of shared understanding can spell success or failure.

Shared understanding begins with operational concepts, business process definitions, trainings, and procedures applied consistently across enterprise participants so that all have the same background for understanding and acting on information. The common operational visualization discussed at the Information Sharing level must be expanded to consistent, if not identical, HMIs so that all operators are working in the same environment with the same available actions and choices. Ultimately, and usually over a period of time, the enterprise seeks to support cognitive interaction, meaning that operators working with each other, even at geographically dispersed sites, come to know how their colleagues think and achieve the highest level of interoperability. By instinctively making good shared decisions and invoking mutually supportive activities, decision-makers significantly improve the chances of success. This is the essence of "brain-to-brain" interoperability.

Full Interoperability. The four levels just described deal with the technical basis for interoperability. Figure 13.6 indicates that another dimension is required for full interoperability. Here, the concerns are policies and rules that restrict or prohibit information exchange based on a variety of sensitivities. An organization or individual manager may determine that certain information is not to be released, or not released to certain recipients, even when it is technically feasible to do so. The enterprise architect will generally be unable to do much to overcome such artificial barriers but needs full awareness of their existence so that any resulting impairment in cooperative action among enterprise participants can be accounted for. It may be important to model the effects of restricted interactions in virtual prototyping so that the impact on achieving enterprise requirements can be understood and displayed to higher levels of management.

13.6 Enterprise Modeling and Simulation

Throughout this chapter, we have identified virtual prototyping as a key element of SoSE, but some specific considerations warrant further attention. One is the fact that just as an enterprise is often geographically dispersed and composed of independent systems, M&S facilities associated with various nodes and systems are commonly distributed and controlled by their owning organizations. If some or all of the constituent systems of the enterprise have existing simulators for training, SE, test support, or other purposes, the most affordable and highest-fidelity enterprise virtual prototype will probably be achieved by networking these together like the enterprise itself. However, constructing a network of existing simulations that use various modeling methods and tools has been historically a very difficult challenge.

In an increasingly networked world, this has become a significant issue and has motivated research, standards development, and products to support distributed M&S. An important example is the High Level Architecture (HLA), originally sponsored by the Defense Modeling and Simulation Office (DMSO) and now governed by IEEE Standard 1516 [12]. HLA seeks to allow the definition and use of federations of simulations by providing an overall architecture, interfaces to a runtime infrastructure that controls interaction among distributed simulations, a template for the information exchanged among simulations, and a set of rules that participating simulations must follow in order to interact. As noted above, this is an exceptionally hard problem because the developers of individual simulations generally seek to optimize their stand-alone performance, resulting in unique and incompatible designs. As a result, the early success of HLA has been limited, but a number of organizations continue to refine the HLA standards, and they are being used in a growing number of programs.

A more fine-grained standard for distributed M&S is Distributed Interactive Simulation (DIS), governed by IEEE Standard 1278 [13]. This is based on Protocol Data Units (PDUs) that codify messages, events, data entities, and other exchanges among participating simulations. Like HLA, the DIS standard has had varying degrees of success and continues to evolve. Both HLA and DIS have been applied to simulation middleware whose purpose is to facilitate composition and operation of distributed federations of heterogeneous simulations [14].

Because distributed M&S generally involves exchanges of discrete content among largely autonomous participating simulations, modeling approaches based on token passing, such as Petri Nets (Chap. 11), or on generation and reception of events, such as state machines, are widely used. Because Petri Nets apply graph theory to model an interacting ensemble as a set of nodes or places connected by edges or links, they are a natural way to model an SoS connected by networks and channels. Tokens can represent messages, tasks, or events passed from one node to another and can have guards and other logic controlling the exchanges. State machines can abstract detailed behaviors of participants into high-level actions that are triggered on transitions or when entering, occupying, or exiting a state, especially to generate events or other interactions. To be useful, distributed enterprise M&S must be able to account for timing, including thread executions at nodes and interactions among nodes. Popular modeling tools provide features such as clocks and statistical distributions of task generation and execution times.

13.7 Robust Networking

The enterprises we discuss in this chapter, almost without exception, depend on reliable, high-performance networking for cooperative functioning of participating systems. A typical definition of this concept of network-enabled capability is given by Albers et al. [15]. Their focus is on a "net-centric" enterprise built on a framework that supports both human and machine interactions. The ideal is a networking framework that shares information when, where, and to whom it is needed with high reliability and cybersecurity. Just as an SoS is intended to provide greater capabilities as a whole than those of its separate parts, a networked enterprise is meant to exploit information for operational advantage in ways individual information systems cannot attain when acting alone.

There are three foundational concepts.

- *Robust Networking* – enables information exchange with high performance and reliability under all operational circumstances; by preference, information is pulled by the party needing it rather than pushed to all participants, many of whom will not need it, and networks are intelligently managed to allocate capacity in accordance with policies and priorities and are secured to ensure confidentiality and integrity of information.
- *Information Sharing* – creates higher-quality information from all available sources and uses it to create common awareness of the operational situation among all participants; this is the Shared Information level of interoperability from Fig. 13.6.
- *Shared Situational Awareness* – enables collaborative decisions, self-organization and synchronization of activities, mutual support and sustainability, and speed and correctness of decision-making; this relies on the Shared Understanding level of interoperability.

This kind of robust networking is relevant to any enterprise where common understanding, agility in applying resources, and good collaborative decisions are important. The referenced publications agree that the characteristics of an enterprise with robust net-centric capabilities include:

- *Reach* – access to information across the enterprise in minimum time, regardless of geographic dispersion
- *Richness* – ability to access and employ any expertise, information, or resources of the enterprise when responding to any task or contingency
- *Agility* – ability to rapidly react and adapt to any situation by assessing needed actions, taking correct decisions, and applying resources; this is one important way in which service-oriented architecture facilitates net-centricity through agile processes
- *Assurance* – high confidence that the enterprise will identify and implement a correct and effective response when challenged, including protecting sensitive information from a competitor or adversary

13.8 Dynamic Enterprises

Chapter 7 discussed composing lower-level services to create more complex capabilities. More generally, many organizations use the familiar concept of a team or task force that assembles specific expertise and resources to accomplish a particular task. As enterprises incorporate more kinds of resources and provide them with increasingly capable interconnections, these ideas can be generalized in the form of dynamic enterprises whose participating nodes and systems are designated, linked into a temporary enterprise, tasked, and employed for a specific purpose and then disaggregated and recomposed to create new assemblies for new tasks. With truly robust networking, an enterprise may exhibit a measure of self-synchronization or self-organization in which a dynamic enterprise can arise spontaneously within it, even if the dynamic behavior is limited to rapid coordination of action among existing organizations. There can be two basic approaches to creating and controlling such an enterprise:

- *Central Control* – a central authority makes decisions about composing, tasking, and controlling a dynamic assembly of resources.
- *Autonomous Control* – nodes and systems have embedded logic, controlled by policies and rules, which enables them to respond to operational situations, high-level tasking, or other situations by combining and cooperating to accomplish a task.

In such a dynamic enterprise, a combination of sensors, effectors, processors, communications, platforms, and other elements would be selected, composed, and tasked for a specific purpose based on their capabilities, locations, and other tasking that is currently in effect. This same idea can be applied in a wide range of industrial and commercial situations. An example is a space enterprise consisting of satellites, communication links, and ground stations that can be linked and configured to execute a specific set of communications or data collection tasks. Another example is a completely agile manufacturing process that can be managed on a production lot-by-lot basis by selecting machines, part conveyors, inspection and packing stations, and other elements of an end-to-end flow based on considerations of optimum use of capital resources and minimum processing time for the highest-priority orders.

13.9 Summary

This chapter has extended the concepts of system architecting and the MBSAP methodology to a higher level of enterprise and SoS architectures. The overarching goal is to combine the capabilities of an ensemble of systems synergistically to achieve operational outcomes greater than their separate functioning can deliver. The key differences are greater levels of complexity, the reality that participating systems in an enterprise are usually independent and have competing priorities and requirements, and the need to harmonize systems with diverse technologies and design approaches. We have summarized the challenges and suggested some pragmatic approaches to deal with them. The field of SoSE is still emerging, and agreement on terminology, technical approaches, measures of effectiveness, and principles of governance can be hard to come by. However, the very nature of an increasingly networked and automated world is such that SoS approaches are often inescapable. The enterprise architect needs all the skills of a system architect plus a generous measure of coordination, negotiation, and leadership ability to get the parties to an enterprise to work together effectively. Open architecture, prototyping, configuration management, and other fundamental disciplines are, if anything, more important in this context than when dealing with individual systems. SoSE is evolving rapidly in theory and practice and will become an increasingly important area for the application of modern architecture practices.

Exercises

1. Referring to Fig. 4.2, sketch a BDD showing the E-X aircraft as a participant in an enterprise involving the other ground, air, and space nodes with which it interacts. Include a Block modeling the enterprise as a whole.
2. For the enterprise-level Block in your answer to Question 1, sketch an IBD showing interactions among the E-X and the other nodes.
3. Considering the enterprise shown in Fig. 13.1, list some factors that would be involved in a decision to implement a Service-Oriented Architecture to achieve the required enterprise-level functionality.
4. Considering the Smart Microgrid A system example from Chap. 4, describe how the commonly recognized characteristics of a system of systems in Sect. 13.2.1 would apply.
5. Continuing with Smart Microgrid A, describe how some commonly encountered SoS challenges would be likely to arise.
6. List some additional responsibilities of an Enterprise Architect, dealing with multiple systems, compared to a System Architect dealing with only one.
7. For the enterprise in Fig. 13.1, describe a prototyping strategy to evaluate important aspects of an SoS architecture. Include the possible applications of the DIS and HLA standards.
8. Consider an airport control tower operating a node in a larger air traffic control enterprise. Describe how it might exhibit adaptive behavior in response to (a) an aircraft emergency requiring an immediate landing and (b) an unexpected severe storm.
9. Continuing with the control tower from Question 8, describe the nature of the Shared Information and Shared Understanding levels of interoperability as shown in Fig. 13.6.

Student Projects

Students should create a basic enterprise architecture OV placing their project systems in the context of external entities. This should include structural, behavioral, data, and, if appropriate, services Perspectives and should include functional descriptions of enterprise interfaces.

References

1. IFEAD (2017) Institute for Enterprise Architecture Development (IFEAD) home. http://www.enterprise-architecture.info/index.html. Accessed 2 June 2017
2. Jamshidi M (2009) System of systems engineering: innovations for the 21st century. Wiley, Hoboken
3. Department of Defense (2017) Defense acquisition guidebook. https://www.dau.mil/tools/t/Defense-Acquisition-Guidebook. Accessed 2 June 2017
4. Maier MW (1998) Architecting principles for systems-of-systems. Syst Eng 1(4):267–284
5. Dahmann J, Lane JA, Lowry R (2011) System engineering artifacts for SoS. IEEE A&E Syst Mag 26(2):22–28
6. Software Engineering Institute (2017) System-of-Systems engineering. http://www.sei.cmu.edu/sos/research/sosengineering/. Accessed 3 June 2017
7. Bar-Yam Y (1997) Dynamics of complex systems. Addison-Wesley, Reading
8. Kilicay N, Dagli C (2007) Methodologies for understanding behavior of system of systems. In: Paper presented at the conference on systems engineering research. Stevens Institute of Technology, Hoboken
9. Bar-Yam Y (2003) Engineering complex systems: multiscale analysis and evolutionary engineering. New England Complex Systems Institute Briefing, Cambridge
10. ODUSD(A&T)SSE (2008) Systems engineering guide for systems of systems, version 1.0. http://www.acq.osd.mil/se/docs/SE-Guide-for-SoS.pdf. Accessed 3 June 2017
11. Kaplan JM (2006) A new conceptual framework for net-centric, enterprise-wide system-of-systems engineering. http://ctnsp.dodlive.mil/2006/07/01/dtp-030-a-new-conceptual-framework-for-net-centric-enterprise-wide-system-of-systems-engineering/. Accessed 3 June 2017
12. IEEE (2010) Std 1516-2010 – IEEE Standard for Modeling and Simulation (M&S) High Level Architecture (HLA) – framework and rules. http://ieeexplore.ieee.org/document/5553440/. Accessed 3 June 2017
13. IEEE Standards Organization (2015) Std 1278.2-2015 – IEEE Standard for Distributed Interactive Simulation (DIS) – Communication services and profiles. https://standards.ieee.org/findstds/standard/1278.2-2015.html. Accessed 3 June 2017
14. Riley P (2003) MPADES: Middleware for Parallel Agent Discrete Event Simulation. In: Kaminka GA, Lima PU, Rojas R (eds) Lecture notes in artificial intelligence. Springer Verlag, Berlin, pp 162–178
15. Albers D, Garstka J, Hayes R, Signori D (2001) Understanding information age warfare. http://www.dodccrp.org/files/Alberts_UIAW.pdf. Accessed 34 June 2017

Chapter 14
Applying Advanced Concepts

14.1 Advanced Concepts and Trends

Like many books that attempt to compactly survey a large and diverse subject, this one has ended up with a collection of topics that may be of interest to various readers but do not conveniently fit into the major themes. Several are collected here in an attempt to make the discussion of information-intensive system architectures as complete as possible. Many other subjects could be added to this list, from quantum computing to robotics, but we have selected the contents of this chapter on the basis of emerging technologies and system concepts with near-term impacts. The treatments are brief and introductory, and as always, the reader is invited to pursue further any of these concepts that appear interesting and relevant.

A number of trends that concern the ways in which organizations obtain and use information technology (IT) create both opportunities and challenges for system architects, and several have already been discussed. Chapter 7 examines the growing importance and sophistication of service-oriented architectures, Chaps. 9 and 13 address the challenges of networking geographically distributed enterprises to enable them to work together to deliver enhanced information processing capabilities, and Chap. 10 surveys the essential trends in cybersecurity. Some others, including virtualization, clustering, and distributed computing, are covered in the paragraphs that follow. These trends are not independent. A networked enterprise commonly requires a SOA foundation in order to operate successfully. Distributed computing is only possible with robust, high capacity networking. Collectively, these concepts establish a new paradigm for employing advanced information and communication technologies to meet the needs of systems and enterprises in many communities.

J. M. Borky, T. H. Bradley, *Effective Model-Based Systems Engineering*,
https://doi.org/10.1007/978-3-319-95669-5_14

14.2 Virtualization

In the most general sense, virtualization involves the use of software to emulate or abstract computing resources. This is a mainstream IT strategy that is of great interest to system architects because of its potential to improve performance, reliability, and affordability in complex systems. It is increasingly likely that effective use of virtualization will make it feasible to satisfy both functional and nonfunctional requirements to degrees that would otherwise entail unacceptable cost and risk.

Virtualization is commonly used (a) to make a collection of computing resources, which may be dissimilar and look like a pool that can be allocated based on current processing or storage needs; (b) to make a single resource, such as a server, emulate multiple resources; or (c) sometimes, to combine both approaches. Hardware virtualization creates a Virtual Machine (VM) that can then be used to host and execute software as if it were the physical machine that is being emulated. The component that runs directly on the hardware to create a virtualization environment is often called a hypervisor.

Fischer and Mitasch [1] group virtualization techniques into three categories:

- *Hardware Virtualization* – creates virtual machines; this is by far the most frequently encountered.
- *Para-Virtualization* – reduces overhead by only partially emulating the underlying hardware and requiring changes to operating systems (OS) that are to be hosted.
- *OS Virtualization* – uses a single OS kernel running on the hardware that supports both host and guest systems, again to improve speed and achieve higher resource utilization at the cost of limiting the flexibility of the solution.

Some specific capability virtualization offers to a system architecture include:

- Allowing an operating system (OS) to run on a non-native hardware architecture or an application to run on a non-native computing platform (non-native signifies an environment different from the one for which the software was designed).
- Allowing a computer to host a variety of OS.
- Allowing one physical server (or group of servers) to create a variety of virtual servers.
- Allowing a collection of disks to function as a storage pool that can be allocated dynamically as storage needs change.
- Allowing a workstation desktop to be remotely accessed on another machine; this is one approach to creating thin clients by putting client-side functions on a server under central management and providing images of approved workstation configurations to users; connection to a remote desktop often uses a VPN for security (see Chap. 10).
- Allowing network virtualization, often referred to as segmentation or zoning.
- Allowing remote access to a secure network, again usually through a VPN.

In the overall IT community, virtualization is popular as a way to make more efficient use of a set of physical resources, thereby saving money and energy. Some major VM advantages, especially in the context of cloud computing, include: [2]

- Server consolidation to reduce required hardware, input/output, networking, and other resources
- Isolation among applications running in individual OS environments
- Rapid scaling, deployment, and configuration of virtual machines to keep pace with changing requirements

Virtualization products can also provide functions such as resource management and automated backup and recovery. Applied across multiple servers, virtualization can provide redundancy to help maintain high operational availability (see Clustering below). Figure 14.1 shows the basic features of server consolidation with multiple VMs creating execution environments for heterogeneous applications, including the appropriate OS for each. The underlying hardware can be made redundant to maintain high availability while reducing the number of discrete hardware servers. This is often referred to as "hosted virtualization," and it combines server emulation and OS virtualization. The base, or host, OS creates the environment on top of which the virtualization layer (or hypervisor) creates a set of virtualized servers. Figure 14.1 shows this both using the layered architecture graphic convention from Chap. 5 and as a segment of a SysML BDD with downward dependencies among the software Blocks and allocations to the server that hosts them.

Network virtualization is generally implemented at the switch level and on the host computers connected to the network. Virtual local area networks (VLANs) provide isolation and security as discussed in Chaps. 9 and 10. The sender of a

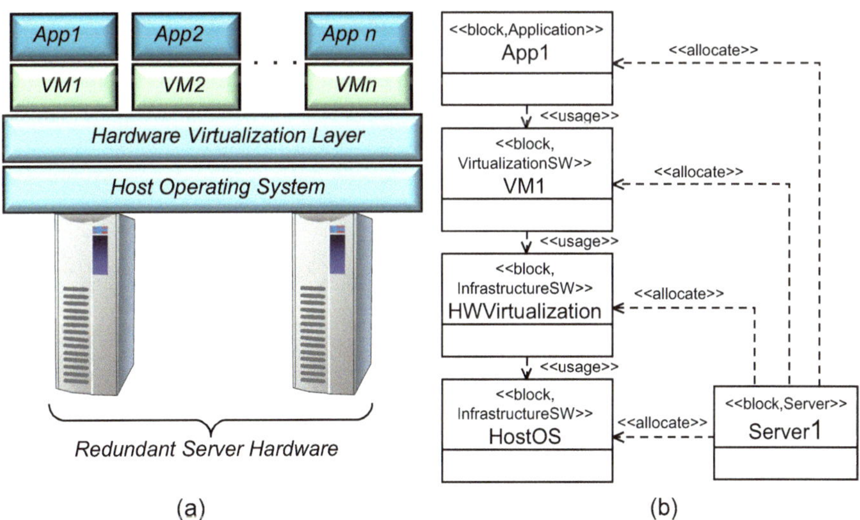

Fig. 14.1 Basic virtualization involving server consolidation to host multiple virtual machines (VMs) on a single box or set of boxes, shown as (**a**) graphic, (**b**) block diagram

message does not know the actual IP address of the final destination. Properly done, this set of virtualization techniques can provide security, state monitoring, load balancing, and transparent failover to redundant resources in the event of a failure. Some of the security benefits of network virtualization, especially in a cloud environment as discussed below, are the following:

- Isolation of virtual networks and environments from each other and from the physical platform, e.g., to separate a development/test environment from a production environment.
- Segmentation to control traffic between network segments and tiers.
- Advanced security service deployment and management.
- Consistency in physical and virtual security models.

Virtualization adds processing overhead and therefore results in a performance penalty. This is acceptable, even negligible, in many systems and continues to improve as products become more flexible and efficient. However, systems with the most demanding and time-sensitive requirements are not good candidates for virtualization. Some writers address virtualization of services, which simply means applying the principles presented in Chap. 7 to expose service implementations through a stable interface while providing governance to ensure that the service-level agreement (SLA) is honored.

Nested virtualization is a recent innovation that has certain advantages, especially in connection with cloud computing, which is addressed in a later section. Essentially, it involves running one or more VMs inside another VM, and this can be carried to several levels of nesting. Because the higher-level VMs can be created, deleted, or modified without shutting down the underlying infrastructure and can be moved to any host, the cloud environment seen by users is highly elastic to respond to changing demands. Nested virtualization facilitates the creation of a virtual data center in the cloud space that exactly mirrors a user's private data center. It makes the cloud environment seen by the user transparent to resource updates implemented by the cloud provider. There are other benefits as well, including improved backup and recovery and the ability to create a mirrored user environment where vulnerability testing, evaluation of new or upgraded products, training, and other activities can be accomplished without impacting the user's production environment.

14.3 Clustering

Clustering is the common term for virtualization in which two or more resources, especially servers, are unified to appear as a single resource. Various clustering patterns have been defined, including high availability or failover schemes such as a standby server that can assume the role of any other server in the cluster after a failure. Fischer and Mitasch [1] describe a variety of schemes for combining virtualization and clustering. Primary motivations typically include the following:

- *Reliability and Availability* – this starts with minimizing single point failures from virtualized resources such as servers; this is likely to be the main objective of clustering in a high availability system, especially if a single computer is emulating multiple servers so that a hardware failure could take down multiple key resources.
- *Load Sharing/Load Balancing* – which usually involves trying to keep all computers comparably busy to optimize collective performance and to avoid resource bottlenecks.
- *Scaling* – to efficiently support varying numbers of users and levels of resource usage.

Perhaps the most relevant clustering for mission-critical systems is called high availability or failover. Here, a standby server backs up one or more primary servers and monitors their health and status, often using a "heartbeat" signal that is generated periodically to show that the software on a primary computer is running normally. Upon the loss of the heartbeat or receipt of some other fault indication, the backup takes over the processing tasks of the primary server, ideally with minimum downtime and no loss of data or transactions. Figure 14.2 gives a basic illustration, including redundant resources for the LAN, shared storage, and connections, once again depicted both in a graphic and as a SysML diagram.

Effective clustering generally depends on networks with high performance and reliability, especially a LAN connecting the computers in a local cluster. Clustering

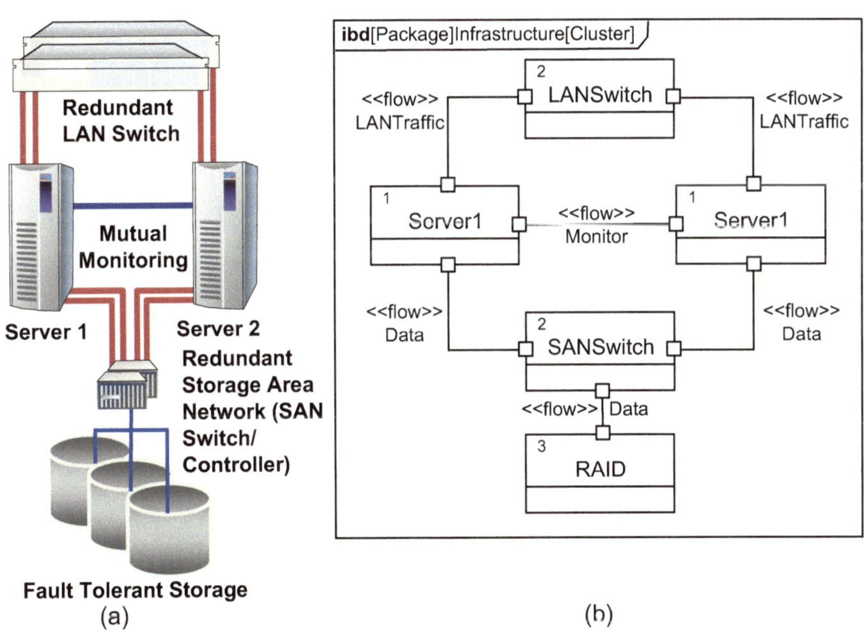

Fig. 14.2 Typical high availability clustering approach with redundant resources and connections, shown as (**a**) a graphic, (**b**) an internal block diagram

schemes that are optimized for aggregate computing performance have many points in common with the distributed computing approaches described in the next section. Like other varieties of virtualization, clustering offers the architect and the implementation team some powerful ways to optimize a system design for performance, reliability, and cost.

14.4 Distributed Computing

Distributed computing describes a variety of schemes in which an ensemble of computing resources that are physically separated and often located in multiple places can be employed as a single capability. Such an approach can be motivated by such considerations as follows:

- The desire to attack large, computationally intensive problems by harnessing a number of relatively low-level or low-cost computers such as Personal Computers (PCs) and spreading the work among them, creating the equivalent of a more expensive, higher-capacity computer
- The desire to make an organization's computing assets, which in some cases have low utilization rates for significant periods of time, more productive (and more profitable) by selling spare capacity to a variety of users
- The desire to guard against natural or man-made disasters by spreading computing nodes geographically

Depending on the specific situation and task, various distributed computing schemes imply conditions over and above the obvious need for reliable network connectivity with adequate data rates. For example:

- Algorithms and standards are required to divide a large complex computing task into smaller pieces that can be allocated to participating machines for processing and then recombined to create the overall result.
- The owners and operators of independent computers participating in a distributed computing enterprise must often commit to remain enrolled, executing assigned computations as long as a customer job that they support remains active.

Liu et al. [3] describe a spectrum of distributed computing approaches. These are summarized in the following paragraphs.

Clustering. This is the concept just discussed, with the assumption that the computers involved are distributed and loosely coupled, working under a Cluster Operating System that creates a Single System Image (SSI). On this, a layered architecture is erected, consisting of a guest OS, any additional required infrastructure software, and applications.

Peer-to-Peer. This approach applies chiefly to mobile devices and is closely related to the Mobile Ad Hoc Networking (MANET) idea in Chap. 12. Participants dynamically enroll and disenroll, and the network does not aim for high-end computing but is essentially an autonomous, self-organizing, distributed system.

Grid Computing. This is an established approach to creating a powerful virtual platform from a pool of participating computing resources [4]. Foster, Kesselman, and Tuecke [5] define the "grid problem" as an approach to flexible, secure, and coordinate sharing of resources among dynamic collections of resources and individuals, considered to be "virtual organizations." For the basic grid construct to work, the distributed problem must be capable of being "parallelized" such that it can be broken into essentially autonomous tasks whose results can be efficiently assembled into the larger output. Then a relatively simple layer of grid "middleware" can handle the registration of participating computers, dispatch and reception of tasks, formatting of the result, and tracking progress. The Globus Alliance [6] develops open-source grid software in the form of a toolkit for building grid systems and applications. The Globus Open Grid Service Architecture (OGSA) [7] defines extensions to Web Services for grid computing and provides a standard grid programming model. Connectivity in a grid often involves VPNs.

Cloud Computing. The latest, and certainly the most widely used, variety of distributed computing today is an extension of the Web Service concept in which resources that are hosted anywhere on the network can be accessed anywhere else. The National Institute of Standards and Technology (NIST) [8] defines cloud computing as a model for enabling ubiquitous, convenient, on-demand network access to a shared pool of configurable computing resources (e.g., networks, servers, storage, applications, and services) that can be rapidly provisioned and released with minimal management effort or service provider interaction. It should always be borne in mind that cloud computing raises significant cybersecurity concerns, which were addressed in Chap. 10.

The motivation is for the cloud to relieve users of the need to maintain computing resources, applications, and even expertise by allowing them to rely on capabilities provided by others on the network and to pay less than they would by owning them. Essentially, everything becomes a service that a Cloud Service Provider (CSP) will make available, reducing the costs to service consumers while bringing in significant revenue. Virtualization, including nested virtualization as described earlier, is the foundation of cloud architectures using VMs that expose the various categories of services. A commonly cited list of cloud attributes includes the following:

- On-demand service with quality and usage metrics
- Rapid elasticity (scaling) to adjust resources to user needs
- Resource pooling to share and minimize costs
- Wide network access (ubiquitous computing)

The rapid growth in popularity of cloud computing has led to the phrase Everything as a Service (XaaS). Specific varieties include the following:

- *Software as a Service* (*SaaS*) – a flexible licensing scheme in which users can employ applications hosted in a cloud only when needed, typically using a thin client or browser; this is sometimes referred to as Service as a Service.
- *Infrastructure as a Service* (*IaaS*) – allows users to provision processing, storage, networking, and other resources to meet specific needs and to adjust as necessary.

- *Platform as a Service* (*PaaS*) – closely related to IaaS but focused on allowing users to develop and run their own applications on a CSP-provided platform.
- *Many others* – such as Communications as a Service (CaaS), Database as a Service (DBaaS), Desktop as a Service (DaaS), Security as a Service (SECaaS), and Identity Management as a Service (IMaaS).

Cloud users can employ a number of deployment models, including:

- *Private* – created for and used by a single organization
- *Public* – accessed by anyone over an open network
- *Hybrid* – composed of two or more clouds using multiple deployment models
- *Community* – used by multiple organizations having common interests and needs

In general, distributed computing schemes are extensions of networked service-oriented architecture (SOA, Chap. 7). The distributed resources are encapsulated and made available as services, and the protocols that implement a given distributed computing model are generally built on basic Web Services. The specialization associated with creating and using a grid, cloud, or other computing model is often added to the application layer.

14.5 Agents

We now turn to some primarily software technologies that have important implications for architectures of modern, information-intensive systems. In parallel with the development of the Internet and of modern software engineering, a variety of methods have been explored for extending the basic concepts of software to automate system functions and interactions. Chapter 7 looked at workflows, orchestrations, and choreographies as ways to automate service invocations to implement a business process. Another approach, which has an extraordinarily diverse set of flavors and variations, is the use of agents. A software agent is a specialized program whose purpose is to automate the execution of an action or succession of actions that a system user could otherwise invoke manually. A classic survey paper by Nwana [9] lays out the history and categories of agents, and the field continues to grow and proliferate. Murch and Johnson [10] present a representative treatment. As with the other topics in this chapter, we discuss agents to make the reader aware of their potential significance in developing and optimizing system and enterprise architectures.

Universally accepted terms and definitions for agents are hard to come by, but a typical description would be "a resource, which may be software, hardware or both, that acts to accomplish a task or tasks on behalf of a user or client." An agent can be as simple as a piece of software that automates a common task like checking for messages and as complex as an adaptive search service or an agent that monitors and controls systems and machinery. Most agents of interest for the system architectures considered in this book are deployed via the Internet or an organizational network to perform information gathering and coordination tasks. Agents that collaborate in such activities, especially if they use services provided by external systems and other resources, need a common language and shared semantics, typically metadata schemas, for the information objects on which they act. The Simple Agent Communication Protocol (SACP)/Agent Communication Transfer Protocol (ACTP) is one approach to these interactions [11].

Many agent taxonomies have been suggested, using such categories as intelligent agents, various kinds of "bots," fixed vs. mobile agents, and so forth [12]. An agent is likely to be tailored to an individual or community for maximum effectiveness. A mobile agent literally consists of executable code that migrates over the net and executes on a machine that can host it and make the necessary resources available. This raises obvious concerns in any situation involving secure computing and networking, and most secure systems explicitly forbid mobile code. The same functionality can often be achieved through messaging or remote procedure calls to invoke the desired process on the remote system. Regardless of such specifics, some commonly accepted characteristics include the following:

- *Autonomy/Proactivity* – the agent decides when to act and acts without direct human intervention.
- *Cooperation/Collaboration* – agents act together to accomplish their tasks.
- *Learning/Adaptation/Reactivity* – agents interact with their context and the environment and modify their behavior as a result.

Agents can do a variety of useful things in information-intensive systems. Perhaps most obviously, automating the discovery, retrieval and exchange of information within a system and across an enterprise can be an important contributor to correctness and timeliness in executing business processes or mission threads. Agents can monitor equipment, mission status, and various aspects of the operational environment such as weather and external events, message traffic, and a host of other things and either take action or alert operators to conditions requiring attention. Under the Simple Network Management Protocol (SNMP), agents are deployed on network-connected devices and report back to a central monitoring server. Agents can also be used in a cybersecurity context to detect, report, and even respond to activities or conditions associated with security threats.

An agent never gets tired, bored, or distracted and is likely to be much more reliable than a human in detecting critical but rare occurrences separated by long periods of inactivity. An effective agent will generally require the investment of development resources, although tools and design patterns are available. Even so, in the right situations, agents can be a useful architecture tool.

14.6 Proxies

A proxy is another common IT design technique that the architect can often use to advantage regardless of the domain of interest. In this context, the word has a meaning very much like the basic dictionary definition of "one who acts for another." A proxy server functions as an intermediary between a client and a service provider to abstract the location and implementation details of the service implementation and simplify the client's action in requesting and receiving that service. The proxy design may include both client-side and server-side components. A proxy interface running in the client environment makes the service invocation appear to be a purely local transaction, with the client-side proxy component taking care of formatting, messaging, and other mechanics.

This can be thought of as taking to a higher level the basic principle of exposing services through an interface that hides the details and remains stable as the hardware and software that create the service evolve. The proxy offers a number of useful functions over and above those of a basic service interface, including:

- Filtering service requests and rejecting those that fail to meet certain criteria, so that server resources are not wasted in processing them.
- Caching and prioritizing service requests in situations where a server is too busy to handle them as they are received; the proxy server may even be able to store client information in order to speed the processing of service requests.
- Supporting logging and analysis of service requests as an input to resource management and planning.
- Various security and policy enforcement functions such as user authentication and content filtering to protect the resources of a critical server and ensure that clients conform to established rules; a component called a Reverse Proxy may be used in a Demilitarized Zone (DMZ) as described in Chap. 10 to provide these protective measures.

Proxies come in multiple types for various purposes. In addition to the basic task of managing client access to servers, they may be optimized for World Wide Web traffic, secure transactions, and other applications. The basic idea of a proxy can be used in conjunction with a broker that mediates the connection of client requests to available services. A proxy server is often in a different address space than the main server, especially when that address space may change with various server deployments. Proxies can be valuable architecture elements in many practical situations.

14.7 Advanced Computing Algorithms

Although the selection and implementation of specific computational methods is often the task of a specialist, or even defined by the customer, the system architect needs a general familiarity with available algorithms because they may be important in determining the technical feasibility and cost of achieving requirements and may influence the system design. Journals such as the *IEEE Transactions on Aerospace and Electronic Systems* and the *AIAA Journal of Aerospace Computing, Information and Communication* publish the ever-expanding body of research in this area. The subject would fill a large book by itself, and so this discussion is limited to illustrating advanced algorithms with a few examples.

Artificial Neural Networks (ANNs). The term artificial intelligence (AI) is applied to a variety of computing schemes, including rule bases that are called "expert systems" and artificial neural networks. Today, ANNs are the primary area of interest, both in terms of research and in practical products. Scientists have long wondered at the ability of the human brain, with neurons that "switch" at far slower speeds than modern electronics, to achieve computational power exceeding that of even the largest computers. Whatever else may be involved, it is clear that massive parallelism and concurrency among billions of neurons with very complex connections and interactions are key factors. ANNs attempt to mimic this, especially in situations where multiple dissimilar input data types must be transformed into outputs such as control signals, facial identification, and other kinds of pattern recognition [13, 14]. Strictly speaking, an ANN is a nonlinear statistical data modeling technique. It can be thought of as a tool of reinforcement learning that can be used to match the input/output response of a real-world model, referred to as a Contributing Analysis (CA). An ANN can handle discontinuities, nonlinearities, and other mathematically intractable conditions in the observed behavior of complex systems.

Figure 14.3 shows the basic structure of an ANN of the most common feed-forward type. Part (a) of the figure shows a basic three-layer ANN that is matched to a CA with three inputs (X) and two outputs (Y). More generally, the ANN has i inputs, j hidden nodes or neurons, and k outputs. Part (b) of the figure shows typical computations in each of the ANN layers. The input layer neurons simply apply a

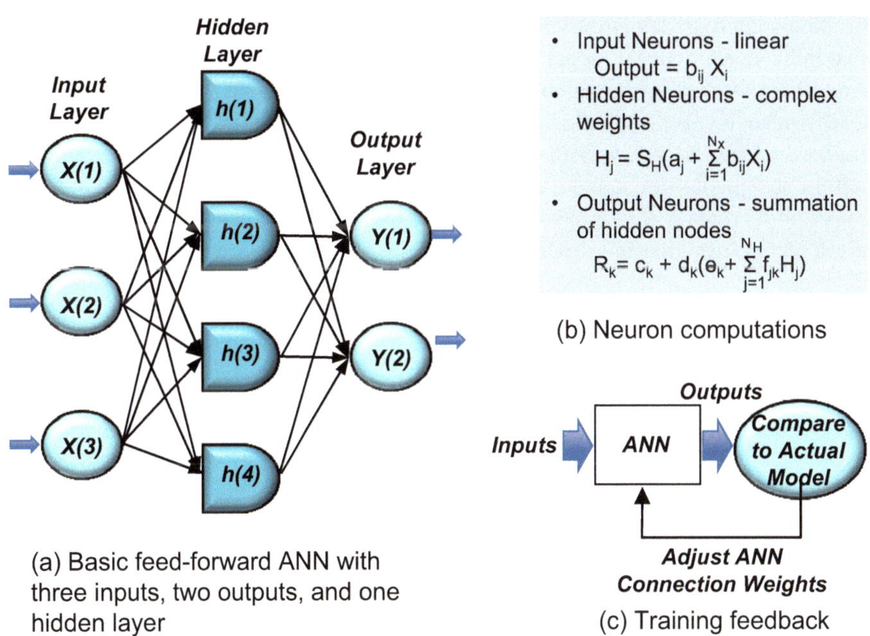

Fig. 14.3 Structure and features of a simple feed-forward artificial neural network (ANN). (**a**) Basic feed-forward ANN with three inputs, two outputs, and one hidden layer, (**b**) neuron computations, and (**c**) training feedback

linear weighting to the input values. The hidden layer neurons compute a summation of the weighted input values, with an adjustable parameter, a_j, for each hidden node and a typically nonlinear "logistic function," S_H. Finally, the output neurons sum the values from the hidden layer with additional parameters and weights. As shown in part (c) of the figure, the ANN must be "trained" using one or more of a variety of methods to obtain values for the parameters that yield the best match to the CA.

ANNs have been used in a variety of complex systems and are potentially applicable to architectures when modeling complex behavior, representing relationships induced from data, and creating surrogate nonlinear models for computationally costly parametric solutions. The most common are various kinds of pattern recognition or object classification, especially when different sensor or intelligence data streams must be quickly processed to yield an estimate of an object or an operational situation, a problem known as data fusion. In some cases, an ANN is used as a preprocessor that transforms complex input data sets into a set of variables that are normalized in some sense such as having common dimensions, after which more conventional computing methods can be used to complete the task and produce the desired final result.

A particular example is the application of ANNs to the AI area called deep learning. Here, an ANN with multiple hidden layers is trained to extract new insights and information from a set of data. These designs are referred to as Deep Neural Networks (DNNs). Using sophisticated (often vendor-proprietary) learning algorithms to train the DNN, such an approach shows great promise in dealing with problems such as detection of sophisticated malware and of patterns of behavior that indicate a threatened or actual cyberattack. Another variety is the Convolutional Neural Network (CNN), which has shown promise in machine image recognition. CNNs also involve multiple layers of neurons that perform filtering and feed-forward of results in a way that keeps the total number of required computations within acceptable limits and can be trained to remarkably high accuracy in distinguishing features within images. For a system architect, the key point is that ANNs are a technology that is growing in capability and likely to become an important enabler of many complex systems.

Fuzzy Logic. While ANNs deal with situations where complex data sets must be transformed into an optimal value or estimate, fuzzy logic applies to analysis and control tasks where the input data is noisy, imprecise, or unreliable [15]. Just as neural nets are based on emulating features of brain organization, fuzzy logic tries to emulate the way humans reason when confronted with this kind of decision situation. Fuzzy sets can therefore have value in modeling human decision-making aspects of an architecture.

In contrast to the familiar control system approach of computing an output algebraically from a set of control inputs, error signals, and their derivatives and integrals, fuzzy logic uses IF-THEN-ELSE rules applied to the inputs available at any given time. Such rules typically require only that an input be known to lie in a range of values and an output would typically be something like "Increase Power." As with neural networks, these rules, especially the "if" and "then" conditions and the actions commanded in response to various combinations of rule outputs, can be trained to optimize performance in the real world. The definitive reference on fuzzy logic is Zadeh [16].

Genetic Algorithms. The final topic in this section is a brief introduction to yet another class of algorithms inspired by the study of living organisms, this time based on evolution. Whitley [17] summarizes the approach using algorithms that mimic a chromosome-like data structure and applies recombination operators in such a fashion that critical information is preserved, much as biological reproduction conserves the beneficial characteristics of organisms. Genetic algorithms are often used to optimize functions but have been applied to a broad range of problems.

Chromosomes that "reproduce" correspond to better solutions, and the population at any given epoch of the analysis identifies the current best solution. Further, problems to which genetic algorithms are applied are often nonlinear and have dependencies among their variables, making traditional optimization approaches hard to apply, and have ranges of possible variable values large enough to make simple enumeration of alternatives impractical. Finally, these algorithms do not use gradient methods and so can be used with functions that cannot be differentiated. They do represent a global search strategy applied as the population evolves in response to some mathematical assessment of the "fitness" of individuals. Whitley summarizes the mathematical arguments that these algorithms implement effective searches of the solution space. As usual, there are numerous variations on the original algorithm.

To the system architect, genetic algorithms are of interest chiefly as an optimization technique for highly complex systems, especially for systems that have the characteristics of nonlinearity, coupling among the functions that describe them, and large ranges of possible outcomes. However, the computational load, especially when large numbers of complex fitness functions must be evaluated, can force the use of approximations that weaken the case for convergence to a truly optimal solution; there is often convergence to a local rather than a global optimum. There may be utility in hybrid approaches that combine optimization methods. In any case, the architect confronted with a problem to which a genetic algorithm may offer an answer needs to understand the strengths and weakness of the approach well enough to interact with an expert analyst and to understand the significance of the result.

14.8 Mechanical Issues in Architecture

The focus of this book, and the great preponderance of its content, has been on the information content of a system or enterprise and on the resources that ingest, use, create, store, and export information. However, many systems are installed in facilities or on platforms that have physical constraints. Even fixed sites may require the ability to relocate and may have limitations on things like available power and environmental control. A design that looks attractive when viewed from the standpoint of system functional requirements may have to be modified to live within such physical boundaries. This section highlights a few of the mechanical problems a system architect or engineer may confront in arriving at an operationally suitable solution.

Size, Weight, and Power-Cooling (SWAP-C). For any weight- and power-constrained system, the mission equipment suite will have a SWAP-C allocation because even the largest platforms have limitations. Even systems installed at ground sites or in vehicles are routinely required to stay within the values prescribed in an allocated SWAP-C budget. Together with requirements for environmental hardening, reliability, security, safety, resource margins, and other attributes, SWAP-C can routinely be a challenge in arriving at a physical architecture, as well as a factor in trade studies to select products and set levels of performance.

In many organizations and programs, the system-level SWAP-C budget is held by a System Engineering Integrated Product Team (SEIPT),[1] which suballocates to the various subsystems and components. The E-X example can illustrate some practical matters involved in dealing with limited SWAP-C. First, it is conventional in airplane design to distinguish between the A-Kit, which includes items more or less permanently installed in the airframe, and the B-Kit, which counts removable equipment. The racks holding the equipment and the operator seat and any power wiring, cooling ducts, and other mechanical support would be part of the A-Kit, while the plugged-in electronics would be in the B-Kit. They are clearly not independent. For example, the electronics dictates the rack layout, and there is a tradeoff between the mechanical ruggedness of the electronics and the shock and vibration isolation provided by the racks, with one or the other incurring the associated weight and cost.

This is the kind of situation where the SEIPT justifies its existence in finding the best solution for the overall system. The mission system team will have trades in areas like choosing components that meet required performance specifications while remaining within weight, power, and cost targets, e.g., selecting a workstation computer with embedded storage vs. an external storage peripheral.

Environmental Specifications. Unless intended for a benign environment such as a permanent building with air conditioning and high-quality electrical power, a mission equipment suite will have an environmental specification to meet. Equipment will require qualification against these requirements, and there may well be associated weight and cost penalties. Depending on the situation, the environmental requirements may include the following:

- *Power Quality* – even a system installed in a building is likely to have a specification for the range of electrical power parameters it can tolerate.
- *Loads* – the maximum loads, measured in multiples of the force of gravity, G, that equipment must sustain in all axes without damage.

[1] Sometimes System Engineering, Integration, and Test (SEIT)

- *Temperature and Humidity* – the values of temperature and relative humidity that the equipment must survive without damage or malfunction while operating and in storage.
- *Vibration* – the levels of vibration the equipment must survive without damage or malfunction, often expressed as a power density spectrum in G^2/Hz as a function a frequency.
- *Shock* – the level of shock the equipment must survive, including magnitude, waveform, and duration.
- *Pressure* – for airborne and spaceborne systems, the ambient pressure under which the equipment must operate correctly, including cabin altitude for aircraft and (typically) vacuum for spacecraft.
- *Other* – factors such as resistance to fungus and corrosion (including salt spray for maritime environments), sand and dust penetration, and exposure to lightning and perhaps other electromagnetic stresses.

Safety and Security. A further set of mechanical constraints often arises in conjunction with meeting safety and security requirements. Both are usually governed by standards. Safety can include anything from interlocks on equipment access panels to disconnect high voltage and thus minimize the risk of electrocution to protection for sharp metal edges. Physical security concerns include locking enclosures and accreditable means to zeroize (erase) sensitive data in electronic equipment. These often require coordination with platform engineers to ensure all required design features are included and with test personnel to make certain the required data to demonstrate requirements compliance are compiled. For example, system operators might require controls that allow them to shut down nonessential onboard systems in an emergency. Especially if the architecture is accorded its proper place as the central organizing mechanism of a program, the architecture team takes on responsibilities to ensure all these considerations are identified and captured in the requirement baseline, and appropriate actions are defined to deal with them.

14.9 Enhancing Performance and Scalability

Performance and scalability are two basic and important architecture tenets. However, achieving them within constraints of budget, schedule, and competing stakeholder interests is a challenge that system architects endlessly confront. There is no magical solution, but the following are some time-tested methods to improve the ability of a solution to meet its requirements initially and over time.

Quantify and Validate Functional Requirements. It is all too common for a stakeholder community to impose requirements which are (a) vague and ambiguous, or (b) extreme, in the sense of being disproportionately expensive or risky such that they unbalance the overall system. This is often described as being "past the knee of the curve" that relates benefits to costs and shows that beyond a certain point, a small improvement in performance, reliability, or some other desired characteristic entails a large increase in cost. When such a situation arises, the best recourse of the architect is to (a) force the stakeholders to quantify the requirement or do the analysis that yields the required values and (b) put the total system impact of the requirements in front of the all stakeholders and argue for something more realistic. This often relaxes a burden on specific system resources and their attributes in ways that promote a more flexible, robust, and evolvable result.

Use Modeling, Simulation, and Analysis. The virtual prototype can support experiments in which the architecture is stressed with various scenarios involving increased performance and scalability in order to identify limiting factors. These insights can then support adjustments to make the architecture more robust for the long term.

Do Realistic Prototyping and Testing. No amount of analysis and modeling can fully predict how a system will perform under stress. Test data from prototypes and early system instances can be far more effective in finding bottlenecks, critical timing relationships, and other factors that limit performance and scalability. As with analytical results, these can then enable effective improvements to the architecture.

Optimize High Rate Functions. A rule of thumb says that 20% of the functions of a system account for 80% of the execution time (the 80/20 rule). When time and money to refine the system architecture are limited, an obvious tactic is to identify the key 20%, either software, hardware, or both, and focus efforts to improve the design of these.

Reduce Resource Contention. A similar rule of thumb holds that simultaneous threads or processes that compete for a limited set of key resources can be delayed or "baulked" by the competition. Remedies may include adding resources (e.g., multiple instances of a key application on a server), balancing loads across a set of resources, and tuning thread control algorithms to deconflict demands on the key resource and mitigate such competition.

Prioritize Tasks. One simple approach when a constrained resource cannot be multiplied is to schedule access to it in accordance with thread or process priorities. Chapter 8 talks about important aspects of contention and schedulability.

Partition and Parallelize Processes. This amounts to breaking down complex processes into smaller units that can be assigned in parallel to resources. A simple example is to allocate processes to foreground and background threads, where the latter are lower in priority and execute during slack periods.

14.10 Summary

This chapter has introduced some additional considerations that may impact the task of the system architect, especially when dealing with extreme requirements for performance, efficiency, system installation and operation, and environmental stresses. Rapid progress in computer science and the trend toward universal network access create alternatives to traditional dedicated information systems. Advanced algorithms offer faster, higher-quality results in selected situations. Mechanical and environmental design is essential in dealing with physical constraints. Each of these is the subject of a particular engineering specialty, and the architecture team will need to avail themselves of the appropriate expertise. Factors such as these are architecturally significant because they impact the feasibility and cost of achieving functional and nonfunctional requirements and because they may constrain design choices, especially in the Physical Viewpoint. As the ultimate technical generalist, the architect needs sufficient familiarity with these aspects of design to account for these and to deal effectively with specialists.

Exercises

1. Consider an information system that integrates four legacy systems, each of which was originally created in a different programming language and designed to execute on a different operating system. List some ways virtualization can help deal with this situation.
2. For the system in Question 1, list some benefits that could motivate the use of clustering.
3. Consider a scientific information processing task involving very large numbers of computations and very large data sets. List some factors that influence a decision to employ Grid Computing or Cloud Computing to accomplish this task.
4. List some specific functions that the Hardware Virtualization Layer (Hypervisor) in Fig. 14.1 must provide to make the system work.
5. What is the difference between an agent and a proxy?
6. Consider a security product like those mentioned in Chap. 10 that are designed to scan incoming messages and files and to detect malicious content such as embedded code that can implant malware in the receiving system. List some factors that would influence a decision to employ an Artificial Neural Network.
7. How might a SWAP-C budget for a system under design be managed to optimize the overall system?
8. Consider a mobile system that is to be installed in a trailer and used to collect and process environmental data in desert locations. List some environmental stresses that such a system might be required to withstand.

Student Projects

As an optional enhancement to their architecture models, students can review the content of this chapter and incorporate examples where appropriate to the nature and functions of their systems.

References

1. Fischer W, Mitasch C (2006) High availability clustering of virtual machines – possibilities and pitfalls. Paper presented at the 12th Linuxtag, Berlin 23–26 May 2006
2. Smith J, Nair R (2005) The architecture of virtual machines. Comp IEEE Comp Soc 38(5):32–38

3. Liu B, Chen Y, Hadiks A, Blasch E, Aved A, Shen D, Chen G (2014) Information fusion in a cloud computing era: a systems-level perspective. IEEE A&E Syst Mag 29(10):16–24
4. Srinivasan L, Treadwell J (2005) An overview of service-oriented architecture, web services and grid computing. HP Software Global Business Unit, p 2
5. Foster I, Kesselman C, Tuecke S (2001) The anatomy of the grid: enabling scalable virtual organizations. Int J Supercomput Appl High Perform Comput 14(9):1–24
6. GLOBUS (2017) GLOBUS for research providers. http://www.globus.org/providers/. Accessed 4 June 2017
7. GLOBUS (2017) Open Services Grid Architecture (OGSA). http://www.globus.org/ogsa/. Accessed 4 June 2017
8. Mell P, Grance T (2011) The NIST definition of cloud computing. NIST Special Publication, Gaithersburg, MD, pp 800–145 http://nvlpubs.nist.gov/nistpubs/Legacy/SP/nistspecialpublication800-145.pdf. Accessed 4 June 2017
9. Nwana HS (1996) Software agents: an overview. Knowl Eng Rev 11(3):1–40
10. Murch R, Johnson T (1998) Intelligent software agents. Prentice-Hall, Englewood Cliffs
11. Artikis A, Pitt J, Stergiou C (2000) Agent communication transfer protocol. Paper presented at the 4th International Conference on Autonomous Agents, Barcelona, 3–7 June 2000
12. Hector A, Narasimhan V (2005) A new classification scheme for software agents. Paper presented at the 3rd International Conference on Information Technology and Applications (ICITA'05), Sidney, 4–7 July 2005
13. Bar-Yam Y (1997) Dynamics of complex systems. Addison-Wesley, Reading
14. Gurney K (1997) An introduction to neural networks. Routledge, Abingdon
15. Lee C (1990) Fuzzy logic in control systems. IEEE Trans Syst Man Cybernetics SMC 20(2):404–435
16. Zadeh LA (1998) Fuzzy logic. Computer 21(4):83–93
17. Goldberg DE (1989) Genetic algorithms in search, optimization and machine learning. Addison-Wesley, New York

Chapter 15
Implementing Governance to Measure and Maintain Architecture Quality

15.1 Ensuring Quality in an Architecture

In our treatment of architecture development, we have stressed the importance of fundamental architectural tenets and best practices in achieving results that meet customer needs, both initially and over the long term. Unfortunately, our experience has been that systems and enterprises often, over their operational lifetimes, fall short in enforcement of these rules. Common causes are shortsighted expedience ("we can't afford to track architecture quality"), conflicting priorities ("nothing matters but performance"), or lack of architecture skills and discipline ("it's really just about software running on hardware"). The result has been architectures that are either flawed from the outset or allowed to deteriorate with the passage of time.

We conclude our discussion of MBSE and MBSAP by focusing on two practical and effective ways to protect against this loss of architecture integrity: metrics and governance. The approach is based on two premises: that it is both possible and necessary to objectively assess the inherent quality of an architecture and that it is possible and necessary to put in place mechanisms that enforce the rules and practices through which that quality is achieved and maintained. Unless these basic principles are applied, no architecture methodology can ensure that a system will not eventually have serious problems.

Recent publications, conferences, and organizations that focus on architecture assessment and governance show that these matters are becoming recognized as essential. As with many other topics in this book, we can present only an introduction to the basics, and the interested reader is invited to pursue areas of interest in more detail. Perhaps surprisingly, there is good agreement in the architecture community on the fundamentals of metrics and governance, but, inevitably, organizations and individuals have come up with rather different terms and methods. The following discussion describes approaches that are consistent with MBSAP and representative of published results and recommendations. It's important to recognize that assessment and governance need to be carefully tailored to the specific

J. M. Borky, T. H. Bradley, *Effective Model-Based Systems Engineering*,
https://doi.org/10.1007/978-3-319-95669-5_15

content and goals of individual systems and enterprises if they are to achieve their intended payoffs without imposing excessive overhead or unduly constraining system design.

Figure 15.1 provides a high-level view of the interaction of metrics and governance with a system over the course of its development and operational lifetime. It begins with the familiar MBSAP flow, including provision for an iterative spiral strategy if appropriate, and continues with fielding and operation. As the figure indicates, we assume there will be periodic upgrades, modifications, and technology refreshment that trigger new development cycles. Metrics and governance interact with the program and with each other at every stage. Even before development begins, in conjunction with requirements analysis, the SE team should pay close attention to defining a small set of metrics (typically five to ten) that are feasible to quantify and analyze and that yield high insight into the status and quality of the architecture. At the same time, a governance strategy, also crafted to meet the immediate and long-term needs of the program, should be devised, coordinated with all stakeholders, and implemented with authority from senior management. During development, metrics are both assessed at regular intervals and used to support design decisions with the goal of maintaining the desired system attributes. Governance at this stage has to do with ensuring that the architecture rules, organizational roles and responsibilities, applicable policies, and other elements of the

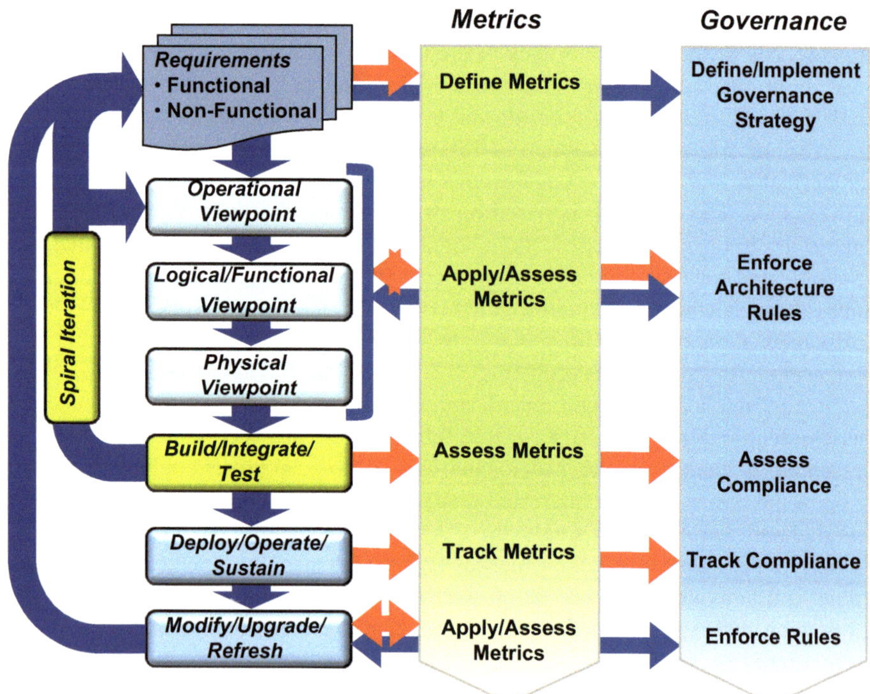

Fig. 15.1 Architecture metrics and governance applied to the entire life cycle of a system

strategy are being adhered to. Test data from a prototype or from the actual system is assessed against both metrics and the rules of the governance strategy.

Once the system is fielded, it's essential that tracking of metrics and rules continue. This is the time, usually much longer than the development period, when inattention can result in poor decisions that degrade architecture quality, often with the consequences going undetected until after the damage is done. These changes may involve system configuration, operational processes and procedures, technology and product upgrades, and other departures from the original system baseline. Decisions about such system changes should be subject to the same assessment and enforcement processes as those applied during development. In some cases, the customer dictates governance requirements and constraints, e.g., by mandating configuration management and change control processes.

Reference architectures and frameworks (Chap. 12) can add value by including guidance on governance rules and guidelines as well as example requirements for performance and technical aspects of the system that influence the choice and use of metrics. Again, some customers have established enterprise architectures, reference models, governance organizations and processes, and other overarching requirements that must be harmonized with the governance and assessment strategy of a particular system. There is no perfect solution to architecture assessment and enforcement, but the principles laid out in this chapter will help the architect craft an effective approach that tracks the quality of the architecture and supports good design and management decisions to maintain that quality.

15.2 Defining Metrics and Quality Attributes

Table 1.2, suggested some overall architecture tenets that should guide and inform decisions during system development, fielding, and sustainment. We now go a step further, moving from general guidance to a defined architecture assessment strategy. We need to consider both functional and nonfunctional requirements (FRs and NFRs) in selecting assessment metrics. FRs commonly have defined quantitative values that are measured in system testing as discussed in Chap. 11. There can also be nonquantitative FRs that are associated with things the system must do to but for which numerical measures are hard to define, e.g., a set of data formats that it must be able to ingest or a set of algorithms it must employ, which are verified in terms of correct implementation of the functions. For NFRs, Chap. 4 introduced the idea of Quality Attributes (QAts) as the preferred method for assessing compliance. Most of the measures of goodness listed in Table 1.2 fall into this category, involving overall architecture characteristics, and are associated with QAts.

A well-known formal method for defining architecture metrics and using them in assessment is the Architecture Tradeoff Analysis Method (ATAM™) developed by the Software Engineering Institute (SEI) [1]. Although ATAM is explicitly intended for software architectures, the general principles apply to system architecture in general. Fundamentally, ATAM is a structured approach to defining architecture goals and preferred approaches based on determining stakeholder interests and priorities, analyzing driving requirements, assessing risks, and selecting a consensus set of QAts and other metrics. The ATAM approach recommended by the SEI begins with a large-scale workshop involving system stakeholders and facilitated by trained staff, usually consultants. Even if the full ATAM methodology is not used, a structured and properly documented process is important, and a workshop, or series of workshops, is an efficient way to get the needed answers and to ensure that all interested parties have their concerns and priorities recognized and appropriately treated, while recognizing that it's probably not possible to give everyone all that they want.

Architecture metrics are important content in a reference architecture as discussed in Chap. 12 because much of the effort involved in defining metrics can be shared among multiple systems in a given category. The key, as in all reference architecture matters, is that functions, system and enterprise structures, user roles, operational concepts, and priorities are similar if not identical. If the data and analysis leading to a set of metrics are properly documented, an expedited process can be employed to tailor the results to a different application that shares the basic characteristics of the system or systems for which the metrics were originally derived.

15.2.1 Metrics for Functional Requirements

The first and somewhat simpler type of architecture metrics applies to FRs. A straightforward analysis of the requirements baseline yields the list of quantitative and functional performance items, and these become, in effect, a checklist for this part of architecture assessment. Depending on the customer's preferences and applicable acquisition rules, FRs may be described in various ways. Quantitative requirements are often called Technical Performance Measures (TPMs) and may aggregate groups of related detailed requirements into a smaller number of major ones. Functional requirements may be called Key System Attributes (KSAs). In Government programs, there will usually be a number of Key Performance Parameters (KPPs) that are intended to ensure that the highest priority requirements are highlighted and tracked. Failure to satisfy a KPP may be grounds for program cancellation, whereas other requirements may be negotiable to less challenging values, especially if they can be shown to cost more than can be justified by the capability they deliver.[1]

The metrics challenge results from the reality that a major system might have hundreds or even thousands of FRs, each of which must have an approved verification method such as an analysis or test procedure. For reporting and tracking purposes, these might be rolled up into an aggregate metric such as percentage of FRs

[1] This is the rationale behind a method called Cost as an Independent Variable (CAIV) in which requirements are included in trade studies along with cost, risk, and other factors in an attempt to guard against excessive requirements (beyond the knee of a cost/benefit curve) that disproportionately drive program cost.

satisfied (or projected by analysis to be satisfied), plus, perhaps, a separate report detailing the requirements that are not being met. In any case, a system development program needs a formal system for tracking and reporting FRs. The architecture team must determine the features and functions required to support FRs and use the tracking system as an important tool in making design decisions. The following paragraphs give some examples of the architecture issues associated with FRs.

Completeness System architects and designers must ensure that the full set of components or products needed to satisfy requirements has been identified and accounted for in the architecture model. An allocated baseline mapped to the Logical/Functional Viewpoint is a vital tool for verifying completeness and correctness. Anything from a major system component that performs a function like controlling a machine tool to a supporting utility like a chat tool for operators could be needed to fully meet the requirements of a particular system.

Resources A related concern is to ensure that the architecture provides adequate resources to satisfy quantitative requirements such as control response time, simultaneous task loading, communications and network throughput, service orchestrations and workflows, and other capabilities. A sound architecture methodology includes guidelines for margins on all such resources for a robust design, meaning one that can deal with changes in system requirements and processes and with variability in operating conditions. Both analysis and modeling are likely to be needed to determine performance requirements for infrastructure and other system components based on the FRs. As discussed in Chap. 8, a real-time architecture will have especially demanding performance requirements requiring very thorough analysis.

Integration of Existing Components The architecture must support integration of all legacy and third-party components and products that are dictated by the customer or determined to be needed to satisfy FRs.

Interoperability The architecture must support interoperability with the customer-specified external environment, including all systems, networks, and other resources with which the system is required to interact. Although standards compliance is treated as an NFR, there may be specific functional and performance aspects of external interfaces that must be accounted for on the FR list such as the ability to access particular data bases hosted on external nodes and systems within allowable times and the ability to perform remote system backups.

As a simple example, we refer to the examples of high-level functional requirements for the E-X System in Table 4.1. Table 15.1 shows how some of these might be transformed into performance metrics. Some, such as radar range, are quantitative, while others dictate functionality that must be present, for example, to perform sensor fusion, for which quantitative measures cannot be prescribed. In effect, the performance specification becomes the overall performance metric for the architecture.

Table 15.1 Examples of performance metrics for the E-X system

Requirements area	High-level requirements	Performance metrics
Sensing	• The Multimode Radar shall perform surface surveillance using synthetic aperture radar (SAR) and ground moving target indicator (GMTI) modes • The ELINT sensor shall collect and analyze RF emissions in specified bands • The EO sensor shall perform imaging of surface targets in specified bands	• Detect radar targets above a specified size at a specified range • Detect ELINT signals in a specified frequency range with a specified sensitivity • Collect EO imagery with a specified resolution and image quality
Sensor Exploitation	• The system shall detect objects and events of interest in sensor data • The system shall perform tracking, geolocation, and identification of objects of interest • The system shall manage and update tracks. • The system shall fuse sensor reports from onboard and off-board sensors to create estimates of track state • The system shall declare and report targets and target updates based on track parameters	• Detect and classify specified objects and events in calibrated sensor data • Perform tracking, geolocation, and identification with specified accuracy • Fuse sensor reports from specified sensors using specified algorithms to create estimates with specified accuracies • Create target reports based on specified criteria and with specified characteristics

15.2.2 *Quality Attributes for Nonfunctional Requirements*

The central concern in developing and maintaining an architecture that will exhibit high quality, i.e., will meet the needs of stakeholders throughout the operating lifetime of the system, is the definition and use of overall measures of goodness that we call Quality Attributes (QAts).[2] QAts can deal with many system attributes, including application of good design practices, robustness (tolerance for change), reliability, and many others, depending on the nature of a system and the priorities of its stakeholders. Reliability is an example of a quantitative NFR since it has numerical parameters such as mean time between failures. Many QAts, however, are hard to measure and are assessed using knowledge gained through expert judgment, conformity to best practices, checklists, and comparison to other systems. Dozens of candidate QAts have been proposed. A disciplined approach is essential in selecting a manageable set of QAts that are most effective for a given system and a given set of stakeholder concerns.

Because various stakeholders in a system typically care about different things, the goals they espouse are often in competition. A widely cited example concerns a

[2] The somewhat contrived QAt acronym is used because QA is so commonly taken to mean Quality Assurance. These are also sometimes called Quality Attribute Parameters (QAPs).

goal of "ease of use" that appeals to system operators and trainers; because it can require additional system features and functions, it conflicts with a goal of "efficiency" that lowers the cost of meeting requirements. Another example is a goal of "high reliability" that requires a level of redundancy and component testing that conflicts with "affordability." In addition, goals can be "hard" in the sense that they are mandatory or "soft" in the sense that they represent desirable characteristics. Largely as a result of the inevitable compromises that result from trying to balance conflicting interests, soft goals are often satisfied to various degrees, and the criterion for success may be more in the nature of "good enough" than of absolute achievement of a desirable characteristic. In short, QAts can be expected to be:

- Hard to quantify and measure, sometimes requiring decomposition into more detailed characteristics that are easier to assess
- In conflict with each other and hard to prioritize when attempting to select the most useful ones while meeting the needs of all system stakeholders
- Intimately bound up with design changes that can, individually or as a cumulative result, affect QAts for better or worse

The last of these is especially important in using QAts to maintain architectural integrity. System design decisions can be driven by a wide variety of considerations, including correcting performance shortfalls, reducing cost to stay within budget, and making product selections based on vendor relationships or customer preferences. An architecture-centric SE process should require that impacts on QAts be a factor in such decisions, and the QAts themselves can often provide valuable data to assess the alternatives. This is the central concept in using QAts to ensure that decisions take into account effects on overall system technical quality since otherwise this may not even be considered in a focused analysis that deals with a specific issue.

Cleland-Huang et al. [2] propose a goal-centric traceability method for managing NFRs. In this approach, a Softgoal Interdependency Graph (SIG) is applied as the basis for a sequence of goal modeling, impact detection, goal analysis and decision-making. The objective is to provide system developers a means of assessing the impact of design decisions on NFRs, as measured by QAts, by tracing the impact of a change to a specific goal through the entire goal structure to determine its total impact. This is done by drawing a goal decomposition tree; adding dependencies to show which higher-level goals are impacted, positively or negatively, by which lower-level ones; injecting a change associated with a design decision; and tracing how the impacts ripple through the structure and ultimately affect high-level goals. Other published strategies in this area are also based on the concept of goals that an architecture should satisfy [3]. Svahnberg and Henningson [4] propose a way to understand the relationships among QAts. Subramanian, Chung, and Song [5] describe an NFR framework and a way to establish traceability between system and enterprise architectures. A paper by Suppakul and Chung [6], which was first cited in Chap. 4, on formally incorporating NFRs in UML models includes associating soft goals with Use Cases.

A further general characteristic of QAts is that they are hierarchical. Overarching system characteristics are achieved through a combination of more specific

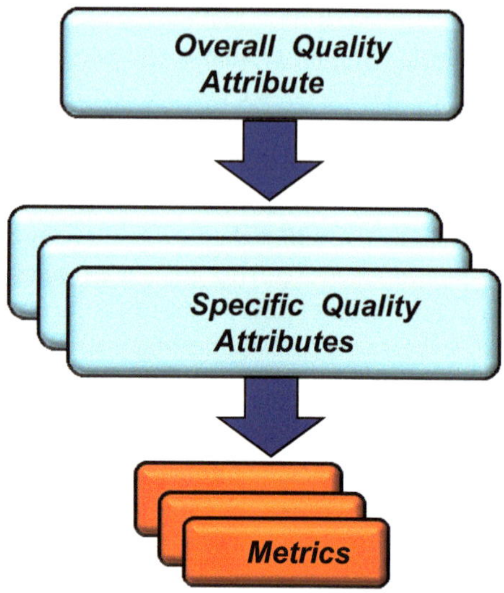

Fig. 15.2 Basic quality attribute hierarchy

characteristics. For example the Software Engineering Institute (SEI) has developed a hierarchy with QAts as the top level, followed by Strategies that represent design approaches and finally Tactics, which are specific design features within a Strategy. Lower-level QAts are often easier to assess with effective metrics. Our experience has shown that the basic structure in Fig. 15.2 is appropriate in many situations. Overall QAts are made up of Specific QAts that can be assessed with objective metrics. Note that metrics may be associated with one or more than one of the Specific QAts. Depending on their nature, these metrics can be scored either quantitatively or on a scale such as Missing, Partially Achieved, Fully Achieved, and Exceeded.

As an example, the important overall QAt of Modularity is achieved through Specific QAts such as Modular Partitioning, Interface Definition and Control, Loose Coupling, and Localization of Change Effects. Some candidate metrics would then be:

- Fraction of system elements or resources that have been assigned to modular groupings that minimize dependencies across module boundaries
- Fraction of modular groupings that have fully defined and stable interfaces that are independent of the internal details of the module
- Fraction of modules that have established change confinement such that regression testing after a design change is limited to the affected module.

It is no accident whatever that object orientation and service orientation offer powerful support to achieving this QAt.

Barbacci et al. [7], at the SEI, published one of the early and influential treatments of QAts in which they proposed a basic taxonomy, again focused on software, based on:

- *Concerns* – what a given QAt deals with and what problem or problems it seeks to solve
- *Factors* – barriers that impede achieving the characteristics measured by a QAt

- *Methods* – possible ways in which those barriers can be overcome to yield the desired result

We have extended this approach for MBSAP. We define a QAt in terms of:

- *Definition* – a clear description of the QAt that is accepted by the stakeholders.
- *Objectives* – the outcomes, benefits, or payoffs in terms of the delivered system that the QAt seeks to ensure or enhance; this can be anything from aspects of operational effectiveness to competitive advantage in the marketplace.
- *Challenges* – the practical considerations standing in the way of achieving the QAt.
- *Methods* – analysis and design approaches, procurement techniques, or any other practical avenue to achieving the desired attribute; for example, Commonality and Interoperability QAts might be enhanced by application of an appropriate Reference Architecture.
- *Metrics* – a description of the approach to assess achievement of the QAt, which may evolve as experience is gained.
- *Issues* – a running tally of matters that must be dealt with, such as nonconcurrence in a QAt on the part of a key stakeholder or an open technical risk that must be resolved before the desired design can be adopted.

Continuing with the earlier Modularity example, the following is the content of the six elements of the QAt.

- *Definition* – the system is partitioned into subsystems and components that localize interactions and change effects, promote loose coupling, and have stable, fully defined, and appropriately standardized interfaces that abstract the internal details.
- *Objectives* – the system is scalable and evolvable at a fine-grained modular level, confines the effects of design changes and failures within known boundaries to minimize regression testing, and can be composed using best available products from multiple sources.
- *Challenges* – the system must deal with heterogeneity in technologies and designs among modules and must achieve performance requirements in the presence of overheads associated with interactions among modules.
- *Methods* – the system design is based on an object-oriented and service-oriented architecture, using an architecture-centric system engineering methodology.
- *Metrics* – the three candidate metrics for Modularity that were listed in an earlier paragraph will serve as a starting point that can be refined with experience.
- *Issues* – to be determined, but might include things like the necessity of getting all parties to the system development to implement a common methodology and set of metrics.

There can be no single, universal list of candidate QAts; every program will need a carefully chosen and tailored set. Table 15.2 gives some examples that may help with the selection. In one sense, this is an elaboration and codification of the thoughts about architecture tenets that were first expressed in Chap. 1. The table follows the structure of Fig. 15.2 and adds some ideas about the nature of various metrics.

Table 15.2 Examples of possible quality attributes

Overall quality attribute	Definition	Specific quality attribute	Definition	Candidate metric
Performance	Architecture has all characteristics required to satisfy performance requirements	Functional Requirements Support	All architecture features required to meet FRs are incorporated	Verification that FRs are satisfied
		Adaptability	System can add or modify behaviors in response to operational changes	Expert judgment that architecture supports timely, affordable changes
		Resource Allocation and Margins	Required resources for worst case loading with adequate margins	Analysis and testing of resource use under worst case loading conditions
Openness	Architecture has characteristics that promote competitive acquisition and upgrading, long- term operational suitability, design reuse, interoperability, and other benefits	Standards Compliance	Applicable standards applied to interfaces, components, and services	Analysis of design vs. approved standards profile
		Scalability	Capacity can be adjusted by addition or deletion of fine-grained components; modules can be selected and combined to span a range of system complexities	Analysis and testing of capacities as modular resources are added; analysis and testing of system variants spanning a range of sizes and functional complexities
		Evolvability	Modifications, upgrades, and technology refreshment at component level	Expert judgment that architecture supports component replacements
		Composability	Architecture supports integration of diverse products from multiple sources	Design analysis; results of system integration and test

Modularity	Architecture employs modular design principles to enable tailoring, robustness, system evolution, and other benefits.	Modular Partitioning for Loose Coupling	Resources are grouped to minimize dependencies and interactions across module boundaries	Expert judgment: modular groupings effectively localize interactions and minimize external dependencies
		Interface Definition and Control	Modules have stable, fully defined interfaces that abstract internal details	Design analysis to determine fraction of interfaces with required definition
		Localization of Changes	Modules employ encapsulation such that internal design changes do not induce changes in other modules (also called change confinement)	Design analysis to establish change confinement boundaries; regression testing limited to changed modules
Operational Suitability	Architecture supports correct and reliable operation when performing system functions in the intended operational and support environment	Interoperability	All architecture features required to interact with specified external parties to share information and allow collaborative actions	Testing of external interactions For a SOA, analysis of functions and data exposed as standards-compliant services
		Commonality	Architecture supports use of common designs, components, and subsystems	Design analysis of applicable common items employed
		Environmental Compatibility	Architecture features enable system to function correctly in the operational and support environment	Design analysis, testing, and field experience to establish system performance and supportability
		Mobility	Architecture features support system movement and correct operation in various locations	Design analysis, testing, and field experience to establish system movement and correct operation
Ease of Use	Architecture incorporates features that promote user effectiveness and minimize unique training requirements	Role-Based Human-System Interfaces	Architecture enables displays and operator functions tailored to efficiently support specific operator roles	Expert judgment and operational user feedback on HSI effectiveness
		Embedded User Support	Architecture enables automation, help, alerts, error checks, and other functions that promote user effectiveness	Expert judgment and operational user feedback on effectiveness of embedded support functions
		Ergonomic Design	Architecture incorporates ergonomic design standards and practices	Expert judgment of ergonomic design effectiveness

(continued)

Table 15.2 (continued)

Overall quality attribute	Definition	Specific quality attribute	Definition	Candidate metric
Availability	Architecture incorporates features that minimize the occurrence and impact of system faults and failures to maximize the ratio of time the system is ready for use to total time	Intrinsic Reliability	Architecture incorporates components with high reliability in the environment	Component analysis and reliability testing; Mean Time Between Failures (MTBF)
		Redundancy	Architecture incorporates redundancy to minimize high occurrence single point failures	System reliability analysis and testing; Mean Time Between Critical Failures (MTBCF)
		Fault Management	Architecture features support containment of and recovery from faults	Design analysis and testing of fault management features
		Maintainability	Architecture features support rapid system repair and restoration	Analysis and testing of repair actions; Mean Time to Repair (MTTR)
		Health Management	Architecture supports analysis and projection of failures	Expert judgment of prognostic and health management functions
Security	See detailed discussion in Chap. 9			
Affordability	Architecture enables a solution compatible with the customer budget	Acquisition Cost	Architecture features minimize initial cost to acquire the system	Acquisition cost vs. budget
		Operations and Maintenance Cost	Architecture features minimize cost to operate, maintain, and upgrade system	Life Cycle Cost vs. budget and projections

Additional QAts that are more related to programmatic concerns and might be important in some situations include:

- *Technology Maturity and Risk* – a system that exploits advanced technology might need an overall measure of the maturity (proven feasibility and affordability) and availability of products incorporating that technology; the inverse of maturity is risk.
- *Exportability* – a system intended for sale to overseas customers might want to track the degree to which performance, technology, and other sensitive aspects are consistent with current export regulations.
- *Schedule* – if a system is urgently needed, it might be important to explicitly tie design decisions to the program schedule, usually in terms of projected impact to the critical path.

15.3 Establishing Effective Governance

Just as the basic purpose of any governance regime is to promote and enforce long-term strategic goals of the governed entity, architecture governance is concerned with preserving overall architecture integrity through the life cycle of a system or enterprise. With any but the simplest systems, the investment made to develop a good architecture, especially through architecture-centric Systems Engineering, will yield a much higher return if it is complemented by a good governance strategy.

Earlier in this chapter, we noted that an architecture that starts out with high quality can degrade over time as the cumulative effects of decisions about system evolution and use conflict with, and sometimes overcome, the original QAts. Loose coupling might be impaired by the creation of interactions among modules that create new dependencies on the grounds that this is a faster and cheaper way to implement a design change than to modify the internals of the modules. Margins on system resources are often gradually consumed as functionality is added until the ability to deal with surges in task loading or to quickly add resources to scale the system to a higher level of performance is lost. Reliability may be compromised by the selection of lower-quality replacement parts to reduce support costs. These and similar decisions may actually be justified when there is no better alternative, but a solid governance process will at least ensure the decision-maker knows what is being given up in terms of the long-term effectiveness, supportability, and cost of the system.

Other motivations for governance include ensuring that only authorized changes are made to a system and only authorized users or customers have access to system products and resources. Complex systems and enterprises typically have overarching goals and long-term strategic plans, and governance processes are essential to ensure that any additions, deletions, or changes are aligned with those larger objectives. Changes that may look appealing from a purely technology or cost point of view might have unforeseen and deleterious effects on business processes that a system or enterprise supports; again, good governance provides insurance against such an outcome.

15.3.1 General Architecture Governance

Today, the term "architecture governance" is often associated with service-oriented and enterprise architectures (SOA and EA), but it's important to start with a more general discussion. Any governance process involves a common set of basic steps, as sketched in Fig. 15.3:

- Define goals and objectives for the entity in question and its implementing program.
- Select a strategy to achieve those objectives.
- Identify and obtain concurrence on rules and guidelines for implementing the strategy, generally in the form of policies.
- Establish an organizational context in which the strategy can be carried out along with a framework for making decisions.
- Establish processes for implementing the strategy and for applying policies.
- Enforce policies in both the near- and far-terms.
- Measure achievement of objectives and make any necessary corrections to policies, processes, and organizations to deal with problems as they arise.

The Open Group (TOG) [8] publishes extensive guidance on architecture governance, specifically, in Part IV of The Open Group Architecture Framework (TOGAF), Phase G of the Architecture Development Methodology. TOGAF calls for architecture governance at an enterprise-wide level to manage and control architectures at all levels of an organization. This makes an important point: governance does not operate for the benefit of a single system, but rather to achieve the goals of the organization that employs the system, and therefore is carried out within a context of larger, overarching strategies, policies, constraints, and processes. TOG also emphasizes that governance must include an overall approach, a set of processes, a supportive organizational culture, and clear responsibilities if the goal of maintaining the integrity and effectiveness of an organization's architectures is to be achieved. In a similar vein, Winter and Schelp [9] note that consistent and timely enterprise architecture results depend on clear and effective guidance.

The underlying assumption is that corporate architecture will be implemented through a series of projects that achieve the desired changes, either by introducing new capabilities or by upgrading legacy architectures, and this is usually planned in increments that are small enough for the organization to absorb and cope with. These projects involve individual systems and systems-of-systems whose architecture is the subject of this book but with the added dimension of conformity to and support for larger corporate or enterprise architecture goals and policies. Any information-intensive system will be a participant in a larger enterprise and must conform to the policies of that enterprise. Indeed, one of the most important reasons for the existence of Reference Models as described in Chap. 12 is to establish a foundation for governance of the systems to which they apply, typically expressed in terms of Enterprise Architectures (EAs).

As with QAts, there is a considerable level of consensus among researchers and practitioners about the elements of architecture governance. Figure 15.1 provided a

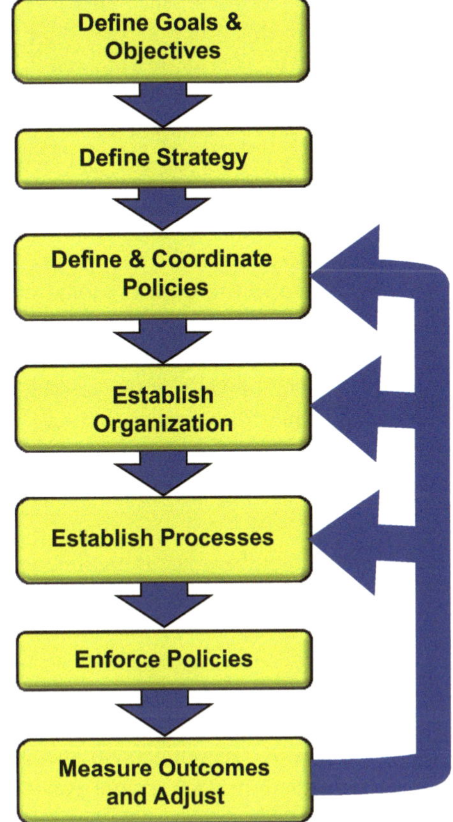

Fig. 15.3 Basic steps of a governance process

high-level view of the relationship of governance activities to a system life cycle and associated metrics. Architecture governance can be thought of as one element of overall corporate or organizational governance, focusing on the processes by which technology is implemented, managed, and used to achieve overall goals. Just as corporate governance is the responsibility of a board of directors, TOG recommends establishment of an architecture board that is responsible for architecture strategy and management and is accountable to higher governance authorities and stakeholders. Indeed, many writers about architecture governance stress the complementary aspects of policy enforcement and accountability. Some other important points TOG makes in regard to an architecture governance framework include:

- Develop and apply governance processes for gathering, auditing, and disseminating information about architecture implementation and employment in areas such as policy management, compliance enforcement, product selection, etc.
- Keep processes clearly distinct from governance content (rules, standards, existing architectures, etc.).

- Ensure required governance authority is defined and flowed down from senior management; this might be the Chief Information Officer or Chief Technology Officer (CIO/CTO) with empowerment from the board of directors, Chief Executive Officer (CEO), or other executive authority.
- Assess compliance continuously against established rules, standards, interface contracts, business processes, and architecture strategies; it is especially important to monitor the impact of architecture deployment on the efficiency, integrity, reliability and other key aspects of organization's processes that the architecture is intended to enhance;
- Establish an effective organizational structure with clearly stated responsibilities for the development, implementation, and deployment of the overall architecture and of individual system architectures within it.
- Establish a comprehensive set of architecture principles, traceable to the strategy, to guide architecture implementation.
- Deal with the realities of culture change and organizational politics that arise as a result of the changes associated with architecture implementation.

In an organization that uses IT as an important enabler of business or mission processes and, therefore, needs architecture governance for that technology, the primary functional areas likely to be involved in governance are business or operations and the IT staff, but many other functional organizations may deserve a role in architecture decisions. Figure 15.4 gives a very simple sketch of a governance organizational structure. The importance of full empowerment from the senior management of the organization cannot be overstated.

Following TOG recommendations, Fig. 15.4 shows an Architecture Board (sometimes called an Architecture Review Board) as the primary decision body. This board includes the organization's technical and business architecture leaders and has participation from whatever functional areas are impacted by its decisions or have roles in carrying them out. These might include Finance to deal with budget

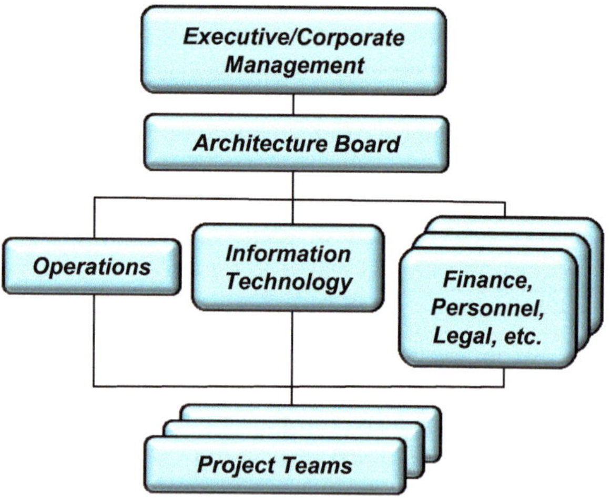

Fig. 15.4 Typical organizational structure for architecture governance

allocations and the impact of architecture on overall organizational financial performance, Personnel/Human Resources to deal with the required staff skills and training, and even the Legal staff if the operations affected by technology have implications for corporate liability, intellectual property, and other such matters. This board needs comprehensive expertise in the technologies that underpin the architecture, including IT, incorporated products, process management, data management, quality assurance, and security. Finally, either standing or ad hoc project teams are commonly set up to actually do the work of implementing, supporting, and upgrading the organization's technology resources and processes.

15.3.2 Governance of Service-Oriented Architectures

Much of the literature on architecture governance derives from work on SOA, and it is undeniable that such architectures pose additional governance challenges. For one thing, most writers and researchers working in this area assume that they are dealing with projects in which an organization has decided to migrate the technology foundation of a legacy system or enterprise to a SOA. This brings up a host of issues with changing processes and organizations, indeed, cultures, in addition to the purely technical aspects. In Chap. 7, we addressed the very real challenges that can become significant in a SOA implementation when it changes the ways in which people are accustomed to think and act. In many organizations, the fundamental concepts of enterprise architecture and of services as the basis for transforming traditional ways of doing things are unfamiliar and may be controversial. Some of this was suggested in the preceding section, and this one delves into it a little more deeply. When a new system is to be developed with the opportunity to be "born service-oriented," the tasks involved in a transition to SOA are reduced or eliminated, but there are often still problems to solve in integrating the new capability into an enterprise whose other participants are likely to include both non-SOA systems and systems based on a variety of SOA technologies and policies. This section draws heavily on the work of Marks and Bell [10].

Anticipating and planning for such culture change is an important part of a SOA project and is a key part of establishing a long-term governance strategy. If the affected personnel understand the goals and benefits of a new way of working, they will be less likely to resist the strictures imposed by governance. Large organizations are likely to have far more elaborate governance structures, especially when public law or policy affects the use of technology and when multiple agencies are involved. But the basic principles of top-down support, empowered decision-making with involvement of all stakeholders, continuing focus on alignment with policy and strategic objectives, and clear direction to the organizations responsible for implementing such projects are universal.

Chapter 7 dealt with the basic nature and implementation of a SOA, but the treatment of governance has been deferred to this final chapter because it is a subtopic of architecture quality and its management. Marks and Bell stress the following key points:

- SOA is achieved incrementally over a period of time, largely at the product level, by continuously defining and enforcing standards and policies that create a conceptual architecture, on which the solution will be based.
- When implemented, these policies help an organization achieve its SOA vision and business goals.
- Finally, these policies become the basis for a SOA governance model.

As with general architecture governance, a framework is needed in which the affected organizations and functions collaborate with IT to provide leadership, set goals and strategy, establish roles and responsibilities (often set forth in formal charters), allocate budgets, and track progress and outcomes. Governance enforces overall SOA strategy and controls implementation, management, and use of technologies and services. The following are some additional governance challenges that SOA brings to the organization.

- *More extensive and profound impacts of SOA on organizations and processes compared to traditional IT strategies* – transitioning to SOA can, indeed should, result in basic changes in the way the organization does business. This may well require new rules for behavior to realize the full benefits of those changes. SOA governance typically entails more coordination and issues resolution across organizational elements and their activities precisely because SOA promotes a higher level of integration, automation, and measurement of the organization's processes. It is therefore essential that the strategy for a SOA implementation be comprehensive and holistic, accounting for all organizational areas and activities that will be affected.
- *Additional technologies and standards with the resulting need for additional and sometimes more complex policies* – for example, a policy might be needed to control dynamic composition of business processes to take advantage of SOA's agility while maintaining quality of products or services and avoiding disruption of manufacturing, transportation, supply chain, and other affected areas.
- *Additional security measures resulting from greater information sharing and access to resources* – the existing cybersecurity structures and processes of the organization will require at least a major updating and sometimes wholesale changes.
- *The need to ensure discipline in the development, deployment, support, and upgrading of services to avoid disruption of service-enabled processes due to poorly conceived or incompletely analyzed changes* – for example, there should be policy covering service publication and access, messaging and networking, standard data models, and security with which any proposed change must comply.

The governance framework presented earlier undergoes some subtle alterations to deal with a SOA. The following paragraphs describe some important considerations.

- There is an urgent need for careful analysis of the processes that will be created or modified as services are introduced to ensure that goals, policies, and strategy are based on clear understanding of what is going to change, what benefits are sought, and how success will be measured. This is an essential element of the up-front analysis and planning that must precede the definition of services as well as their development and deployment.

- Similar analysis of the organizational elements and functions that will be affected and the ways in which roles and responsibilities, staff skills and training, and even reporting relationships will change is also essential. Introducing SOA can result in anything from redefining some individual staff members' duties up to full-blown culture change, and many writers stress the importance of attention to behavioral rules, training, and management emphasis to help the organization and its staff adapt to the new situation.
- A successful SOA implementation requires selection of the appropriate standards, guidelines for selecting or developing implementing technologies and products, rules and procedures for using services and composing business processes, and any other necessary guidance so that complete, clear, and effective policies can be written.
- The initial analysis leads to identification and definition of services that can achieve the desired process transformation, followed by rigorous change control and service configuration management. Analysis of SOA projects that failed outright or failed to achieve the desired results frequently cite insufficient attention to this step and to the analyses that must precede it as root causes. Conversely, effective governance results in services that are reusable, composable, and evolvable by enforcing design rules and best practices and ensuring that they are derived from business processes/mission threads so that they deliver truly useful capabilities. Often, adopting a standard for service-level agreements (SLAs) promotes these desired outcomes.
- Another powerful support to SOA project success is thoughtful selection of initial or pilot projects to validate the strategy and demonstrate early payoffs. Pilot projects are an opportunity to try out the SOA governance strategy and to refine it where necessary. Pilot projects create the foundation for defining and executing the ensuing projects that complete the SOA implementation and then evolve and refine it using metrics that track the results. For example, a pilot project is generally focused on a specific process or subprocess where a limited set of services can support the desired changes and demonstrate measurable improvements. Long-term SOA evolution should be guided by feedback from metrics data, responses to changes in the organizational strategic plan, forecasts of technology evolution, actual or predicted product obsolescence, changing customer requirements, and a host of other factors.
- As with any governance process, definition and tracking of metrics that clearly establish with data the ways in which services improve outcomes such as process time, resource utilization, quality, and cost are essential.
- The implementation, integration, and management of services are typically more complex than the introduction or upgrading of basic IT resources and require a broader range of skills and experience.

- Planning, allocating, and tracking of resources employed in a SOA project may be different compared to traditional IT, especially if some services are outsourced to vendors or provided by another organization or agency. Resource management processes from traditional IT often need extensions and modifications to support SOA. For example, configuration management, system administration, load balancing, and security management take on additional content when services must be managed in addition to hardware and software.
- Assuming the system being transitioned to SOA is a participant in a larger service-enabled enterprise, support for and use of enterprise services must be accounted for. This is where governance links to enterprise architecture as discussed in Chap. 14.

Portier [11], in addressing SOA governance, puts the emphasis on the human dimension, specifically the need for clear chains of responsibility and authority, good organizational communications, and mechanisms for architecture measurement, policy enforcement, and control. To cite a rather different perspective, the National Association of State Chief Information Officers (NASCIO) [12] has published a number of studies on SOA in terms of its benefits to and impacts on state governments. NASCIO places great emphasis on SOA governance, noting that without it, there is a risk of unnecessary duplication of capabilities, failure to comply with regulations, and excessive costs. The NASCIO SOA governance model includes organizational design, funding, management of staff development (knowledge, skills, and experience), principles, standards, operational processes and tools, change management, risk management, and compliance and performance metrics, which is a formula very similar to the other approaches described in this section.

The implementation of a typical SOA governance process can be laid out in five phases as summarized in the following paragraphs.

- *Service Identification* – at the outset, the functions to be implemented as services should be identified and modeled in the context of the overall architecture structure and behaviors. A service inventory as discussed in Chaps. 4–7 should be documented and assessed for key characteristics, including reusability, composability (e.g., in workflows), and relevance based on operational need. This is an element of completing the Operational and Logical/Functional Viewpoints.
- *Service Definition* – in the transition to the Physical Viewpoint, fully detailed service contracts (Service-Level Agreements) with provisions for interoperability, composability, and discoverability are essential. Definitions should also establish that services are stateless whenever possible since this facilitates loose coupling and service composition.
- *Service Implementation* – as part of creating the software objects that will implement services, it's important to verify that the attributes identified during service definition are achieved in the code. Many of the standards and specifications summarized in Chap. 7 are essential elements of service implementation.
- *Service Integration* – in this phase, services are tested to verify the required characteristics, deployed on the SOA infrastructure, and tested in a realistic operational context, including interactions with other services and composition in higher-level services and workflows.

- *Service Management* – in this ongoing phase, services are monitored for performance, reliability, and usage. As service-enabled processes evolve, dynamic service composition, new workflows, service modifications and upgrades, and other management activities are accomplished.

We conclude this discussion of architecture governance with some final points the SOA architect should consider. A widely accepted best practice involves separating or decoupling policy from service definition, essentially by binding policy to a service endpoint rather than to the service behind it. However, some SOA architects choose to embed policy assertions in a WSDL document defining a service. We argue that this should only be done when necessary. Another practical matter is the growing availability of SOA governance tools, often included in application servers, Web servers, and various middleware products such as enterprise service buses and explicit web service management (WSM) products. Such tools reduce the effort required to manage services much as other resource management tools do for system and network administrators. Today, architecture governance in general, and especially for a SOA, is understood, supported by technology, and essential to success.

15.4 Summary

This chapter brings the discussion of implementing architectures and model-based system engineering methods for complex, information-intensive systems to a close with the vital subject of measuring and enforcing architecture quality and technical integrity through rigorous governance. The discussion has shown that objective, ideally quantitative, measurement of a well-chosen set of architecture attributes is feasible and affordable and that it provides invaluable insight into characteristics that greatly affect the operational effectiveness, supportability, and total cost of a system or enterprise. Tracking architecture metrics over the life cycle of development, deployment, operational support, and upgrading supports making good decisions and avoiding bad ones about initial design and long-term evolution of the system. Metrics data is then a key input to architecture governance which establishes the formal policies, controls, and procedures needed to keep complex systems aligned with organizational goals; to enforce essential rules for access, operation, and support; and to plan and manage changes to deal with inevitable evolution of requirements and environments over long system lifetimes. Governance is vital for any complex entity and doubly so for SOA. Perhaps the ultimate architectural tragedy is for an initially well-designed system to lose focus on technical integrity so that it becomes unnecessarily limited in performance and expensive to own and operate through lack of sustained governance.

Exercises

1. Consider a system that is used by a maritime shipping company to plan and manage the operations of a fleet of cargo vessels. List five Quality Attributes that

would help the company establish and maintain an effective enterprise architecture, with a brief rationale for your selections.
2. Assume that a company whose business involves collection, storage, processing, and protection of large amounts of sensitive data has decided to contract with a software developer for a new and more capable information system. Summarize the elements of a plan to apply the Architecture Tradeoff Analysis Method (ATAM) to derive a set of architecture metrics that represent the key concerns of the system's stakeholders.
3. Assume that the organization that uses a system to automate a diverse set of business processes has identified open architecture as a key system attribute. Further assume that the organization's IT staff has identified an opportunity to reduce annual software licensing costs by signing an exclusive long-term contract with a specific software vendor. Discuss how a management decision on whether to accept the IT staff recommendation should include consideration of Quality Attributes such as Openness.
4. Referring to the Security Quality Attribute in Table 15.2 and to Chap. 10, write a definition of the QAt, identify and define three Specific QAts, and suggest metrics that could be used to assess them.
5. Referring to Fig. 15.1, discuss the differences between metrics and rules and the way they interact in a governance process.
6. Suggest some steps an organization planning a project to introduce SOA to a legacy business process could take to prepare the organization's personnel for the project.

Student Projects

Students should develop and document a governance strategy for their systems, with consideration of stakeholders, metrics for functional and nonfunctional requirements, organizational roles and responsibilities, and the essentials of a governance policy.

References

1. Barbacci MR, Clements PC, Lattanze AJ, Northrop LM, Wood WG (2003) Using the Architecture Tradeoff Analysis Method (ATAM) to evaluate the software architecture for a product line of avionics systems: a case study. http://resources.sei.cmu.edu/library/asset-view.cfm?assetid=6447. Accessed 5 June 2017
2. Cleland-Huang J, Settimi R, BenKhadra O, Berezhanskaya E, Christina S (2005) Goal-centric traceability for managing non-functional requirements. Paper presented at the 27th Intl. Conf. on Software Engineering (ICSE'05), St. Louis, 21–15 May 2005
3. Easterbrook S (2004) Goal modeling tutorial. http://www.cs.toronto.edu/~sme/CSC340F/2005/slides/goal_Modeling.pdf. Accessed 5 June 2017
4. Svahnberg M, Hennigsson K (2009) Consolidating different views of quality attribute relationships. Workshop presented at the 31st Intl. Conf. on Software Engineering (ICSE'09), Vancouver, British Columbia, 16–24 May 2009

5. Subramanian N, Chung L, Song Y (2006) An NFR-based framework for establishing traceability between enterprise architectures and system architectures. Paper presented at the 7th ACIS International Conference on Software Engineering, Artificial Intelligence, Networking, and Parallel/Distributed Computing, Las Vegas, NV, 19–20 June 2006
6. Suppakul S, Chung L (2005) A UML profile for goal-oriented and use case-driven representation of NFRs and FRs. Paper presented at the 3rd ACIS International Conference on Software Engineering Research, Management & Applications (SERA2005), Central Michigan University Mount Pleasant, Michigan, 11–13 August 2005
7. Barbacci M, Klein M, Longstaff T, Weinstock C (1995) Quality attributes. http://resources.sei.cmu.edu/library/asset-view.cfm?assetid=12433. Accessed 5 June 2017
8. The Open Group (2006) Architecture governance. http://www.opengroup.org/architecture/togaf8-doc/arch/chap26.html. Accessed 5 June 2017
9. Winter R, Schelp J (2008) Enterprise architecture governance: the need for a business-to-IT approach. Paper presented at 2008 ACM Symp. on Applied Computing, Fortaleza, Brazil, 16–20 March 2008
10. Marks E, Bell T (2006) Service-oriented architecture: a planning and implementation guide for business and technology. Wiley, Hoboken
11. Portier B SOA terminology overview, Part 1: Service, architecture, governance, and business terms. https://www.ibm.com/developerworks/webservices/library/ws-soa-term1/. Accessed 5 June 2017
12. NASCIO (2006) Service oriented architecture: an enabler of the agile enterprise in state government. https://www.nascio.org/Publications/ArtMID/485/ArticleID/229/Service-Oriented-Architecture-An-Enabler-of-the-Agile-Enterprise. Accessed 5 June 2017

Appendix A: Quick Reference to Object-Oriented Design (OOD) and the Unified Modeling Language (UML)

General

Fundamental Concepts of Object Orientation

- *Abstraction*—capture the essential, shared characteristics of an entity or group of entities.
- *Hierarchy*—rank or order abstractions to capture common and unique features among entities.
- *Modularity*—decompose complexity to enable understanding; localize effects of changes.
- *Generalization/Inheritance*—define elements such as Classes whose characteristics apply to multiple more specialized elements; child elements inherit the characteristics of the parent.
- *Encapsulation*—conceal information that does not need to be exposed to an outside entity; bundle-related information and functionality within an entity.
- *Interfaces*—clearly define interactions among entities and with the external environment; allow entities to interact by exchanging messages or other information across controlled boundaries.
- *Polymorphism*—hide details of implementation behind an interface; allow the same service invocation to yield different results depending on circumstances (e.g., the data type of a passed parameter).

J. M. Borky, T. H. Bradley, *Effective Model-Based Systems Engineering*,
https://doi.org/10.1007/978-3-319-95669-5

Dealing with Complexity Using OOD

- Abstraction focuses on the most important aspects of complex entities (Objects) and identifies common or shared characteristics; objects correspond to real-world entities that have defined "state" and "behavior" (data and functions or operations).
- OOD facilitates decomposition of complex entities into simpler ones that can be more readily specified, implemented, tested, and used.
- Interface definition is inherent and natural; captures real-world interactions of a system.
- Polymorphism allows tailored functionality with common interfaces.
- Facilitates open, loosely coupled, service-oriented, and component-based ("plug and play") architectures.
- Encapsulation facilitates localization and confinement of the effects of design changes or system failures.
- Inheritance minimizes the difficulty of handling variations among related information and actors.
- OOD facilitates understanding of system states and transitions via the attributes and associations of objects.

UML Basics

Classes and Objects

- An Object is a basic entity of a system, providing an identifying label and encapsulating the information and operations associated with that entity; it is the physical entity that can hold information, perform functions, and provide services to other Objects or external entities.
- An Object is the instantiation of a Class; the Class is the primary type of "classifier" that abstracts the shared characteristics of its Objects.
- An Association between Classes represents the ability to exchange messages and to request and deliver services; a Link is a semantic relationship between Objects and the instantiation of an Association.
- Class and Object Diagrams capture the identity and interactions of Classes and Objects.

UML Metamodel

- Type specifies a domain of values and a set of operations on those values.
- Class implements a Type by representing the Attributes and implementing the Operations.
- Type Class includes as subclasses:
  - Primitive Type—not implemented by a Class (e.g., integers, strings)
  - Class—implements the Type
  - Use Case—sequence of actions performed by a system and actors

- Class comprises as subclasses:
  - Active Class—anchors one or more execution flows; a Task Diagram is a Class Diagram showing only active Classes, used to describe multitasking.
  - Signal—named event.
  - Component—reusable element containing physical constituents of model elements.
  - Node—physical device on which artifacts are deployed to execute.

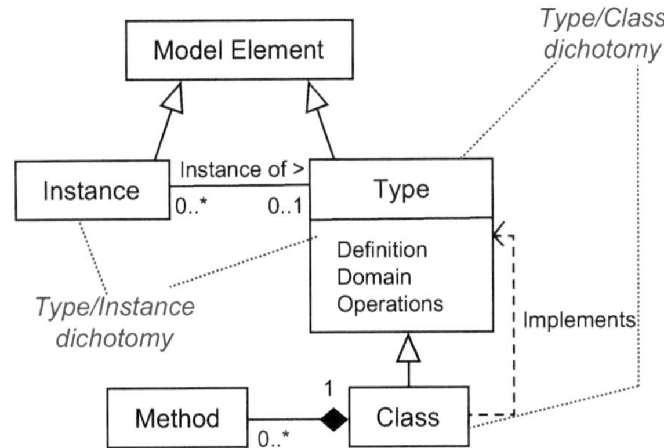

Fig. A.1 UML Metamodel, defining the Basic Constructs of the Language

UML Common Elements

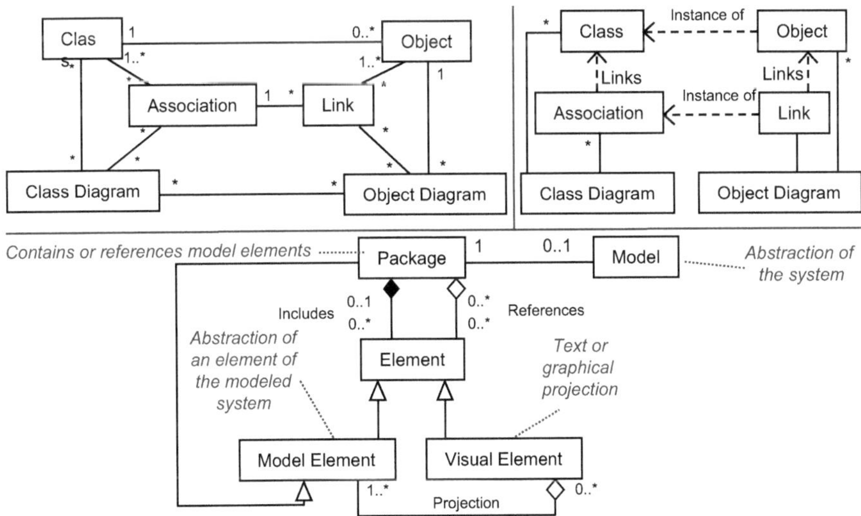

Fig. A.2 UML Common Elements

UML Structural Metamodel

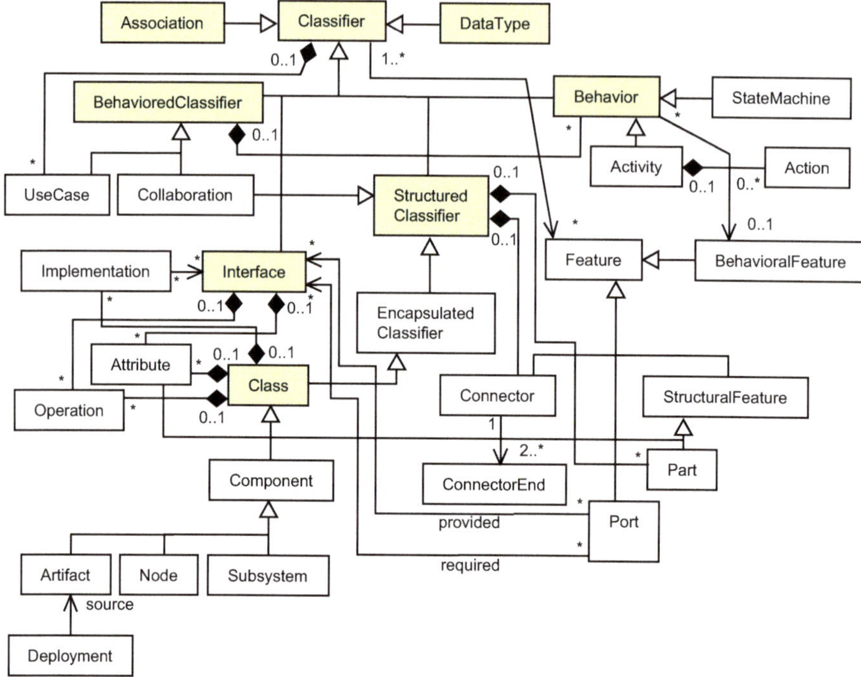

Fig. A.3 UML Structural Metamodel

UML Diagrams

UML consists of Structural and Behavioral specifications:

- *Structural Diagrams*—model the static aspects of a system or enterprise.
- *Behavioral Diagrams*—model the dynamic aspects, how the structural aspects change over time.

*Supplementary Diagrams and Artifacts**

Artifacts not defined in the UML standard are often very useful in communicating key aspects of architecture and design; examples include:

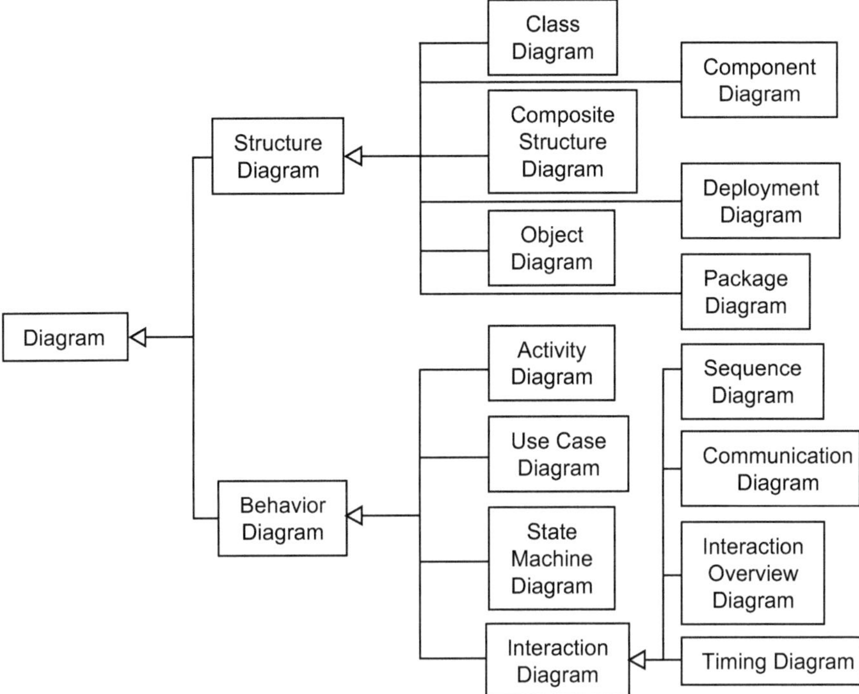

Fig. A.4 UML Diagrams

- *Hardware Block Diagrams*—the conventional way to show hardware items and interconnects
- *Entity-Relation Diagrams*—used to show data relationships that may not be obvious in Class Diagrams.
- *Decision Tables*—used to show concisely complicated logical conditions and decision alternatives.
- *Tabular Data and Taxonomies*—used to identify, define, and describe activities, services, information exchanges, interfaces, performance parameters, procedures, rules, and many other details; a common example is a Data Dictionary.

*NOTE: This does not refer to the UML Artifact model element that represents a physical element of a system.

Structural Diagram Elements

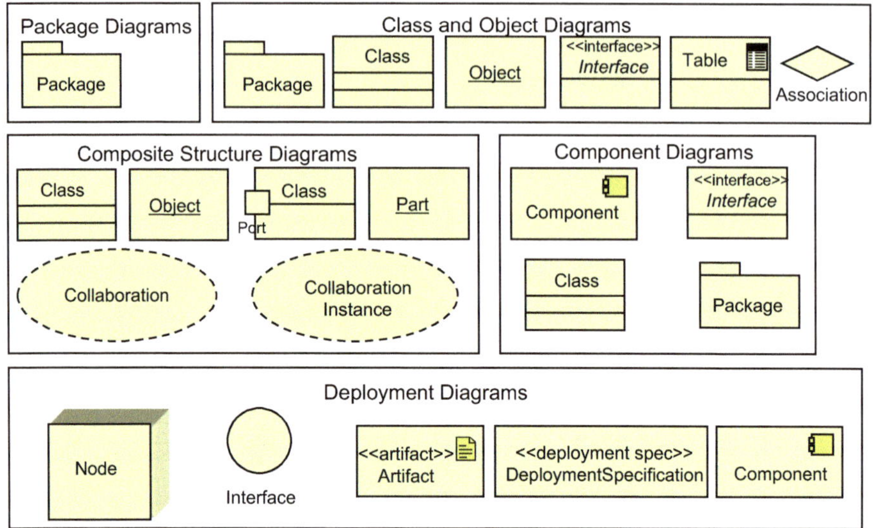

Fig. A.5 UML Structural Elements

UML Common Mechanisms

- Keyword—a predefined term used to categorize or describe a model element; notated as `<<import>>` or `{abstract}` depending on the term (generalizes and replaces the Stereotype construct from UML 1, but Stereotype is still frequently used as the generic term).
- *Stereotype*—a class, usually associated with a profile, that extends another class or stereotype and can have properties whose values become tagged values of the element to which the stereotype is applied; notated as `<<stereotype>>`.
- *Tagged Value*—an extension of a UML element that allows creation of new information in the element's specification; a (name, value) pair that describes a property of a model element.
- *Note*—a comment attached to one or more model elements; a stereotyped Note can be a Constraint.
- *Constraint*—any kind of semantic relationship between model elements; may be expressed in a formal language such as the Object Constraint Language (OCL); extends the semantics of an element to add or modify rules; semantics are:

```
<constraint> ::= '{' [ <name> ':' ] <Boolean-expression> ' }'
```

- *Dependency*—defines a unidirectional use relationship between model elements, referred to as source and target, client and server, etc.; a change to the dependent element may affect the semantics of the independent element.
- *Type/Instance and Type/Class Dichotomies*—the split between the essence or specification of an element (Type) and its realization/implementation (Instance or Class);

a Type is a stereotype of a Class that specifies a domain of objects with their Operations (but not Methods); Data types do not have Stereotypes, Tagged Values, or Constraints.

- *Responsibility*—a concise statement of what a model element, usually a class, must do for the system to function as designed; can be shown as comments on a Class Diagram, documented in a Class Specification, etc.
- *Realization*—the source implements/realizes the destination; shows traceability.

Behavioral Diagram Elements

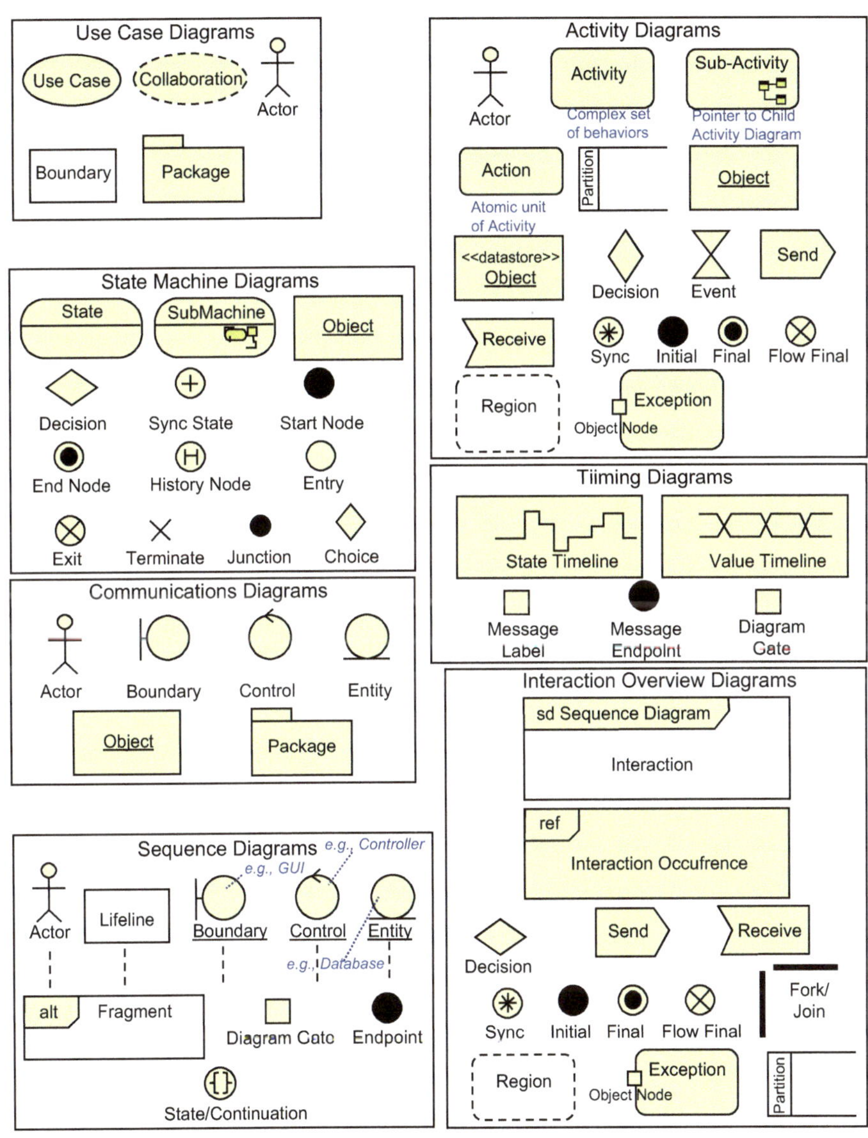

Fig. A.6 UML Behavioral Diagram Elements

Common Symbology

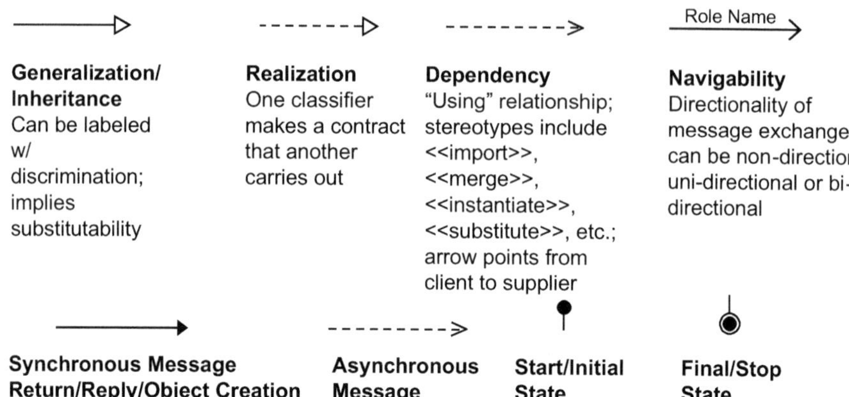

Fig. A.7 UML Common Symbology

Classification

The relation between an object and its type or classifier may be:

- *Single*—object belongs to a single type, which may inherit from multiple supertypes.
- *Multiple*—object may be described by several types that are not necessarily related by inheritance; the generalization arrow is labeled with a "Generalization Set" ("discriminator") that identifies the specific generalization relationship.
- *Static*—object cannot change classes within the subtyping structure (default relation).
- *Dynamic*—object can change classes, combining the concepts of state and type.

Classifiers

A Classifier is any model element or mechanism that can type another element by describing its structural and behavioral characteristics.

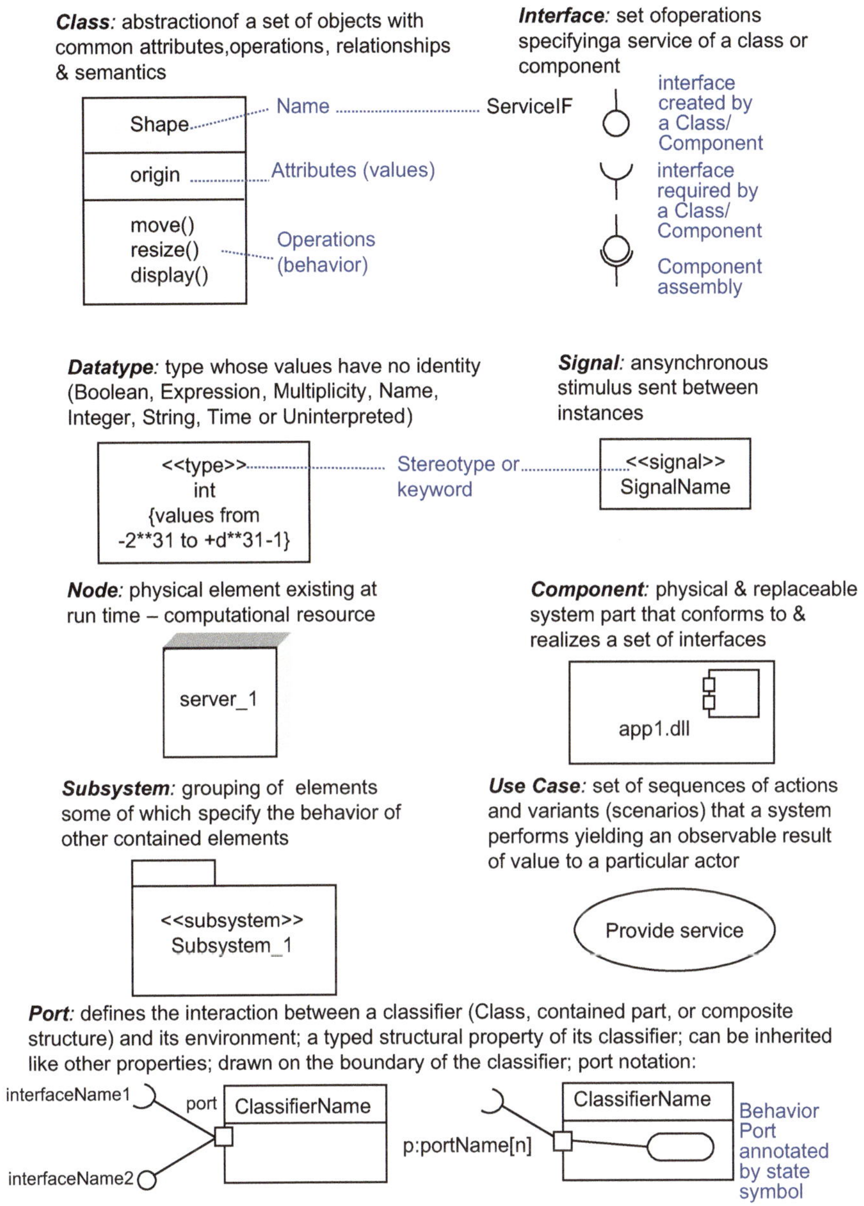

Class: abstractionof a set of objects with common attributes,operations, relationships & semantics

Shape — Name
origin — Attributes (values)
move()
resize()
display() — Operations (behavior)

Interface: set ofoperations specifyinga service of a class or component

ServiceIF — interface created by a Class/ Component
interface required by a Class/ Component
Component assembly

Datatype: type whose values have no identity (Boolean, Expression, Multiplicity, Name, Integer, String, Time or Uninterpreted)

<<type>>
int
{values from
-2**31 to +d**31-1}

Stereotype or keyword

Signal: ansynchronous stimulus sent between instances

<<signal>>
SignalName

Node: physical element existing at run time – computational resource

server_1

Component: physical & replaceable system part that conforms to & realizes a set of interfaces

app1.dll

Subsystem: grouping of elements some of which specify the behavior of other contained elements

<<subsystem>>
Subsystem_1

Use Case: set of sequences of actions and variants (scenarios) that a system performs yielding an observable result of value to a particular actor

Provide service

Port: defines the interaction between a classifier (Class, contained part, or composite structure) and its environment; a typed structural property of its classifier; can be inherited like other properties; drawn on the boundary of the classifier; port notation:

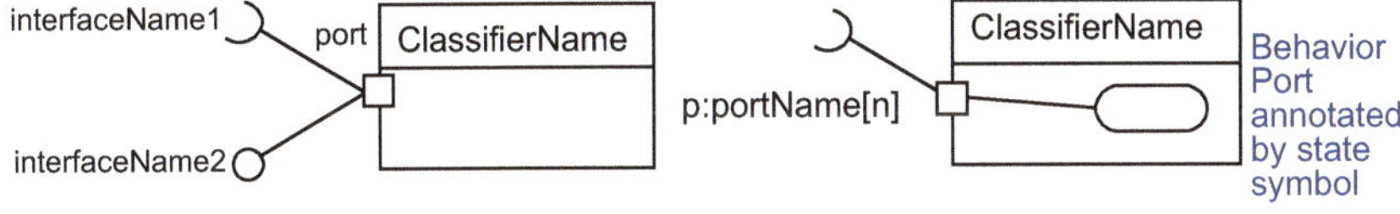

Fig. A.8 UML Classifiers

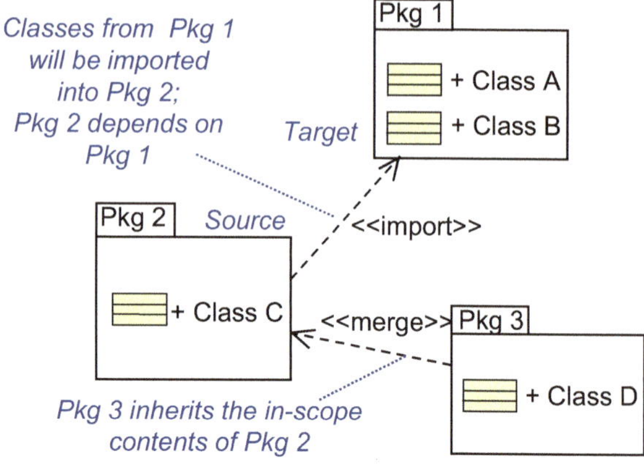

Fig. A.9 Packages with Dependencies

Packages

Used to partition and organize a model and to group (encapsulate) model elements:

- Common stereotypes are `<<category>>` (Logical/Functional Viewpoint) and `<<subsystem>>` (Implementation/Physical Viewpoint).
- Package defines a namespace and can contain a mix of packages and model elements; packages can be nested to any depth.
- A dependency exists between packages if at least one class of the Client Package uses the services of at least one Class in the Supplier/Server Package (Dependency points from the Client to the Provider).
- A Package has an interface and an implementation; contained elements can be Public or Implementation (Private); only Public classes appear in the interface.
- Importing from one Package to another and merging package contents are shown as stereotyped Dependencies.

Visibility

Specifies if a feature can be used by another classifier:

- *Public*—any outside classifier with visibility to the given classifier can use the feature; denoted +.
- *Protected*—any descendant of the classifier can use the feature; denoted #.
- *Private*—only the classifier itself can use the feature; denoted –.

Scope

Specifies if a feature appears in each instance or as a single instance of the feature for all instances of the classifier:

- *Instance*—each instance of the classifier holds its own value for the feature.
- *Classifier*—one value of the feature for all instances of the classifier; denoted by underlining the feature label.

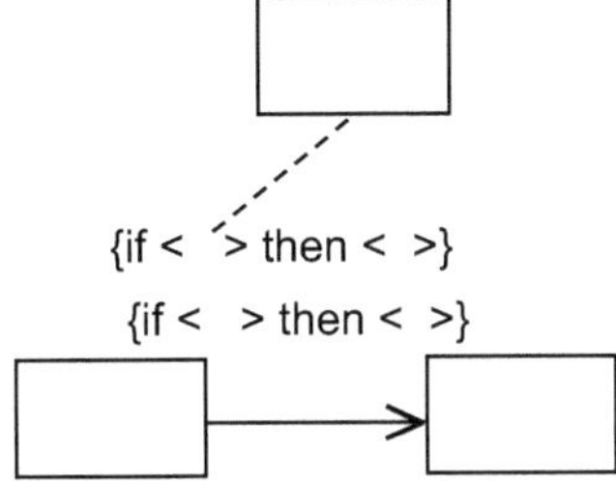

Fig. A.10 Methods of Annotating Constraints

UML Extensions

- `<<keyword>>` or `<<stereotype>>`—extends an existing model element with a project-specific element derived by specialization from any existing metaclass.
- *Tagged value*—`{attribute = value}`; new modeling Attribute added by a user to hold and convey additional meaning; can serve to parameterize a model element; can be attached to a stereotype.
- *Constraint*—`{constraint}` new semantics; an additional condition applied against a modeling element, often for "well-formedness."
- *Profile*—a coherent set of extensions that specializes UML for a vertical domain, environment, customer, etc.; may also hold reusable model elements (e.g., types) and sample diagrams.

Class/Object Diagrams

Class diagrams show the logical structure of a system; classes are the templates from which physical objects are instantiated.

Classes

- Classes can inherit and delegate attributes and behaviors.
- Class Properties can be:
  - Attributes—values held within a class or object; collectively, they define state.
  - Associations—declared interactions (ability to exchange messages) between "source" and "target" classes; attributes can be shown at the target ends of associations.
- Association ends *can* (usually should) be labeled with role names showing what each Class does via the Association.
- Multiplicity means the number of objects that can participate in the relationship (1, 0..1, m..n, * 0..*, 1..*, etc.).

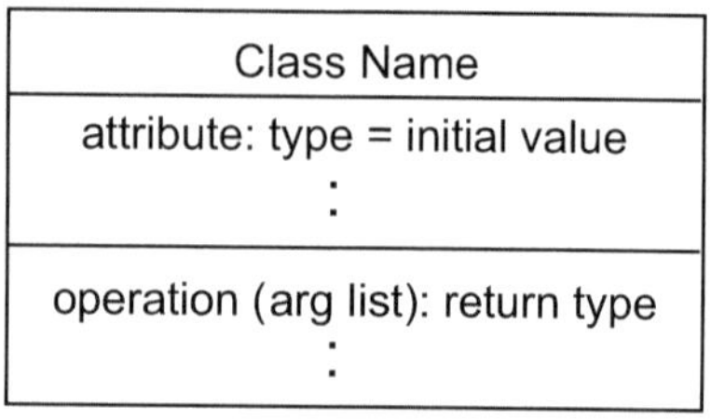

Fig. A.11 Basic Annotation of a Class

Class Name
attribute: type = initial value
:
operation (arg list): return type
:

- Navigability shows "responsibilities" or "pointers" and can be uni- or bidirectional.
- An Operation is the "signature" of a Method; Operations are the activities or behaviors a class can exhibit.
- Abstract Class:
  - Defines a Class *type*; never directly instantiated
  - Annotated with <<abstract>> keyword or *italicized* class name
  - Usually includes at least one abstract operation
- Class Keywords/Stereotypes:
  - <<signal>>—notable event that triggers a transaction within a state machine; asynchronous stimulus that evokes an asynchronous response in a receiver
  - <<interface>>—description of visible operations
  - <<metaclass>>—the class of a class
  - <<utility>>—class reduced to the concept of a module; cannot be instantiated; groups global variables and procedures in the form of a class declaration

Objects

- *Attributes*—data, values
- *Behavior*—operations, methods
- *State*—memory; a mode or condition of existence with specific events that can be processed and operations that can be performed
- *Identity*—unambiguous specification of which object is involved
- *Responsibilities*—roles taken by the object in the system; defined in annotations on a Class Diagram

Class Diagrams

- Properties:
  - Can be shown as either Attributes within Classes or as named Associations between Classes.

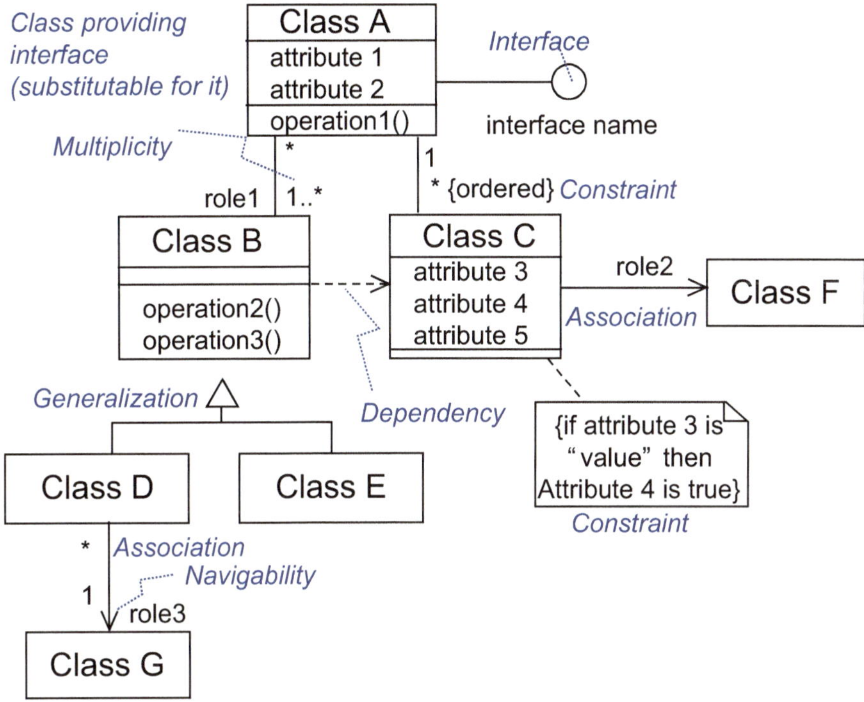

Fig. A.12 Basic Elements of a Class Diagram

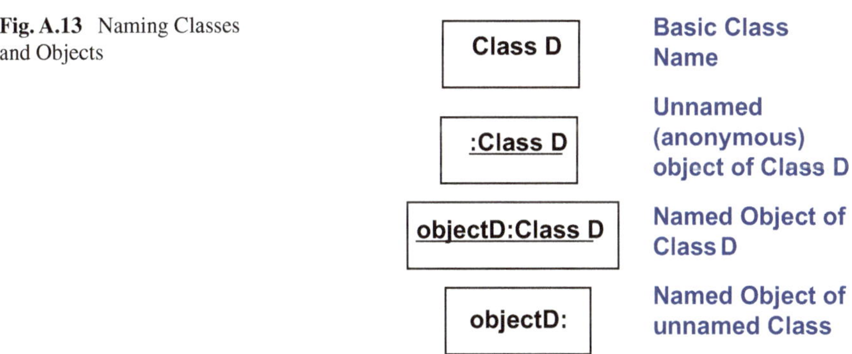

Fig. A.13 Naming Classes and Objects

  - Multiplicity of a Property shows how many Objects can fill the property; most common are 1, 0..1, *, or m..n.
  - Typical Constraints on multivalued Properties include `{ordered}` (indicates the order of objects has significance) and `{nonunique}` (indicates duplicates are allowed).

- Static Relationships include associations and subtypes.
- Associations:

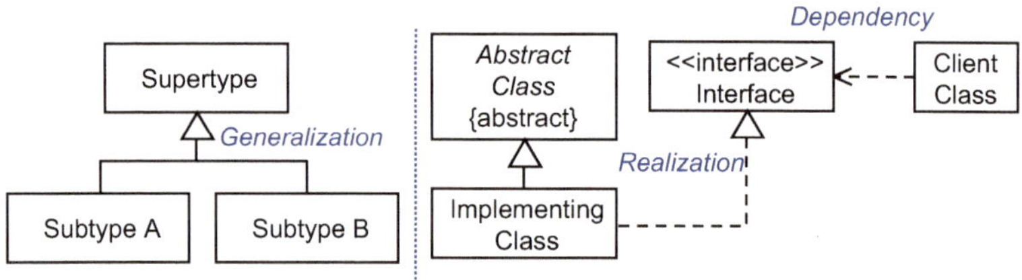

Fig. A.14 Examples of Class/Object Relationships

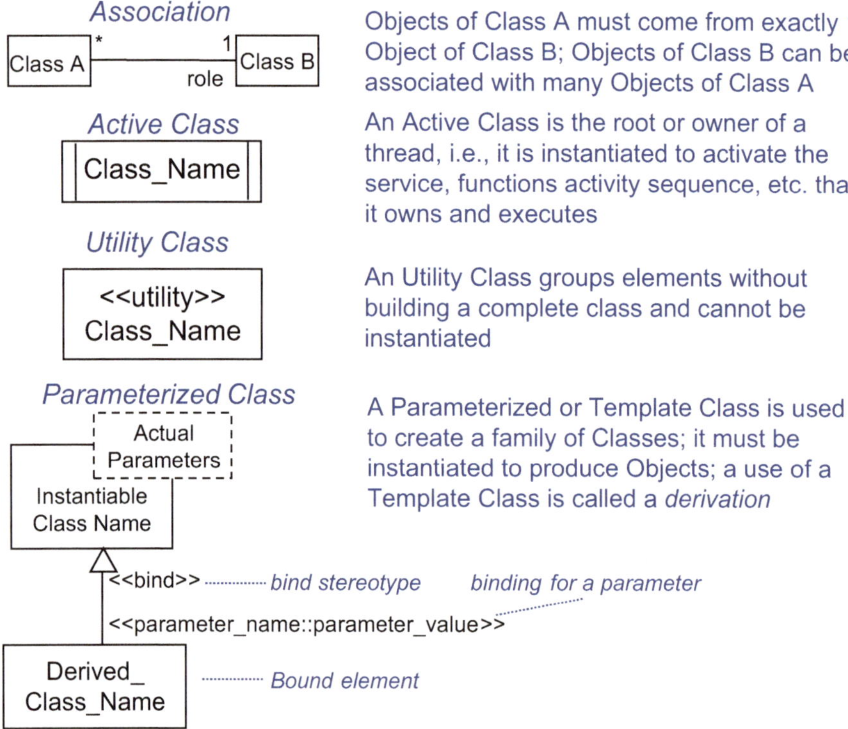

Fig. A.15 Additional Class/Object Semantics

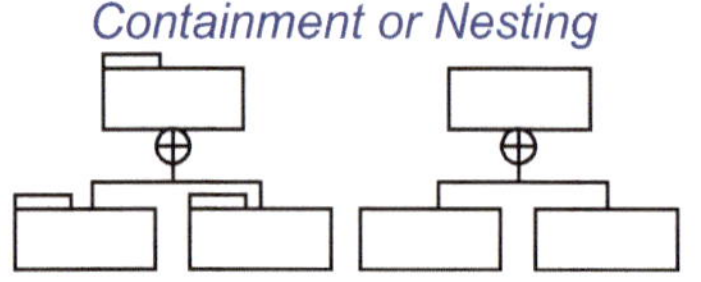

A Containment Hierarchy is a namespace hierarchy involving model elements and the containment relationships among them

Fig. A.16. Containment Hierarchy

Fig. A.17 Class Attributes

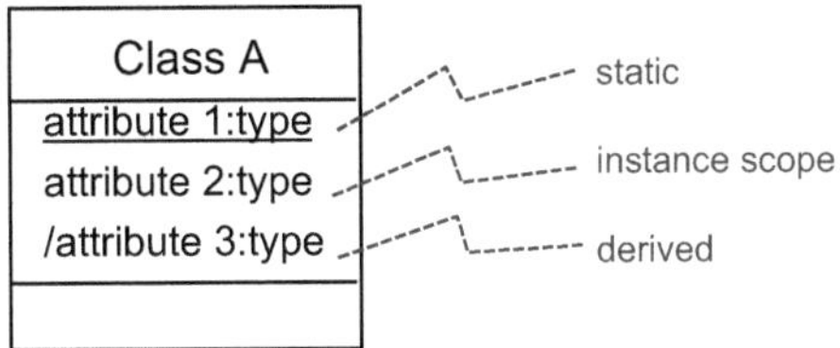

  - Show interactions between classes, meaning that messages, objects, values, etc., can be exchanged).
  - In code, associations are instantiated as target class variables:

    A bidirectional Association can be shown as: ——— or ⟷

  - Labeling Associations

    Property—shows Objects exchanged between Classes
    Verb Phrase—shows relationship between Classes; shows direction of the relationship with > or ►

- Interfaces:

  - Interface Class has no implementation; all features are abstract (see Java, C#, etc., interfaces).
  - Use keyword `<<interface>>`.
  - A Class provides an interface if it is substitutable for the interface.
  - A Class that requires an interface to function has a dependency on the interface.
  - Preconditions—guaranteed to be true by the object sending a message.
  - Postconditions—guaranteed to be true by the time the message is processed.
  - Signature—mechanism used for message transfer (function call, bus protocol, etc.).
  - Abstractly, an interface is a named collection of operations (NOT attributes); it is realized by either logical elements (classes and packages) or runtime artifacts (components).

- Attributes:

  - Attributes contain data, fields, etc., associated with the Class; Attribute values define the state of an Object.
  - Syntax—attribute visibility designator (default is public) followed by

    ```
    AttributeName: type [multiplicity] = defaultValue
    {property string}
    ```

    where only the Name is mandatory.
  - Static attribute—applies to a class, rather than an instance; underlined on a Class Diagram; annotated as:

`ClassName.attributeName.`

- Instance Scope attribute—applies to an object; not underlined on a Class Diagram.
- Derived attribute—computed from other values; annotated as

`/name:type=f(base attributes).`

- Visibility—can be + (public), # (protected), – (private), or ~ (package); exact definitions depend on the implementing software language.
- The Type of an Attribute restricts the Objects that can be placed in the Attribute.
- `defaultValue` is assigned to a newly created Object.
- Property String allows additional properties to be specified, e.g., `{ReadOnly}`.
- In Specification and Implementation phase diagrams, attributes imply navigability from the type to attribute and that type contains solely its own copy of the attribute object; any type used as an attribute has a value.
- Attributes correspond to Public fields in languages that support Properties and to Private fields in languages that do not.

- Operations:
  - An Operation specifies a service that can be requested from an Object of the Class; a Method is the implementation of a Operation
  - Syntax—`visibility OperationName (parameter list): return-type-expression {property string}.`
  - Static Operation—applies to a class, rather than an instance; underlined on a Class Diagram; annotated as:

`ClassName.OperationName().`

  - Visibility—+ (public), # (protected), – (private), or ~ (package); exact definitions depend on the implementing software language.
  - `OperationName` is a string.
  - Identifying the owning class: `ClassName::OperationName().`
  - `Parameter List` contains comma-delimited parameters with same syntax as attributes plus direction (in, out, inout).
  - `Property String` indicates values for the Operation; e.g.:

`+balanceOn (date:Date):Money`

  - Query—operation that gets a value without changing the state of the system.
  - Modifier—operation that changes the state.
  - Method—body of a procedure.

- Polymorphism—supertype with multiple subtypes; e.g., an Operation implemented with multiple methods depending on, e.g., the data type of an input parameter; an Operation can have multiple Methods (procedures) when Polymorphism is used.
- Abstract Operation—never directly implemented; shown in italics: *`operationName()`* often in an abstract SuperClass as a directive to implement in subclasses.
- Typical Property Strings:

  `{query}`—denotes an Operation that gets a value from a Class without changing system state
  `{modifier}` or `{command}`—denotes an Operation that changes state
  `Get()` and `Set()` Operations return or put a field value while doing nothing else

- *Constraints*:
  - Syntax—any form but enclosed in { }.
  - Pre- and postconditions apply to operations; specify checking responsibilities.
  - Exception—occurs when a precondition is met by an operation that can't meet a postcondition.
  - Invariant—1 an assertion about a class:

    Always true for all instances
    Added to pre- and postconditions
    Must be true at the end of an operation
    Can be used in polymorphism to ensure conformance
    Used only to increase subclass responsibilities:

      Weaken a precondition; strengthen an assertion/postcondition.
      Facilitate dynamic binding.
      Include in code defining the interface.

Additional Details of Relationships Among Classes and Objects

- *Association*—models a structural relationship between Classes of Objects governing exchanges of messages.
  - Physically instantiated by a Link.
  - Assumed to bilateral unless marked with a Navigability arrow < >.
  - Objects use/exchange services but do not own each other; at runtime, an Object of one Class may send a message to an Object of the other Class.
  - Associations can be named, and their ends can have Roles; `LinkName` is underlined in an Object Diagram.
  - Association Class adds Attributes and Operations to an Association.

- *Role*—name for the behavior of a Class in conjunction with an Association, i.e., with respect to the other Class.
  - Role Names:
    Can be a Class method, placed at the end of the association where the method is applied.
    Can be an attribute of the Class at the other end.
    Roles can be shown with visibility (same arrowheads as above).
    Typical role pair would be "#Uses" and "+IsUsedBy" on opposite ends of an association.
  - Role Attributes:
    Aggregation—see below.
    Changeability—True means association semantics are preserved when one instance replaces another.
    Ordering—True means instances are ordered.
- *Multiplicity*—prescribes how many Objects are instantiated and will participate in the Role of an Association at runtime; also called Cardinality.
- *Aggregation*—defines an "includes" or "is a part of" relationship where one Object is logically or physically contained in another.
  - A Class is part of another Class, attribute values propagate, and actions of one Class imply actions of another.
  - Objects of one Class are subordinates of Objects of another.
- *Composition*—strong aggregation; physical incorporation of attributes by value:
  - Owning Object is responsible for creating/destroying its Part objects.
  - Composition and attributes are semantically equivalent.
  - An instance can be a component of only one owner.
  - Navigability shows that any given instance of the target, Class B, can only be "owned" by an instance of one class, in this case Class A.
- *Generalization*—defines an "is a kind of" relationship; child/subclass Object is a specialization that inherits aspects of parent/superclass/supertype Object.
  - Subclass/subtype must be substitutable for superclass/supertype without changing semantics.
  - Subclass/subtype can inherit from multiple superclasses.
  - Subclass inherits a superclass interface—must include (conform to) the superclass interface (limited generalization).
  - Subclass inherits all superclass methods and fields (conventional generalization).
  - Subtypes can be implemented with subclasses or delegation.
  - Visibility can be + (public), # (protected), and – (private).
  - Constraints on Generalization relationships can be {disjoint}, {overlapping}, or {complete}.
  - Attributes of a generalizable element:

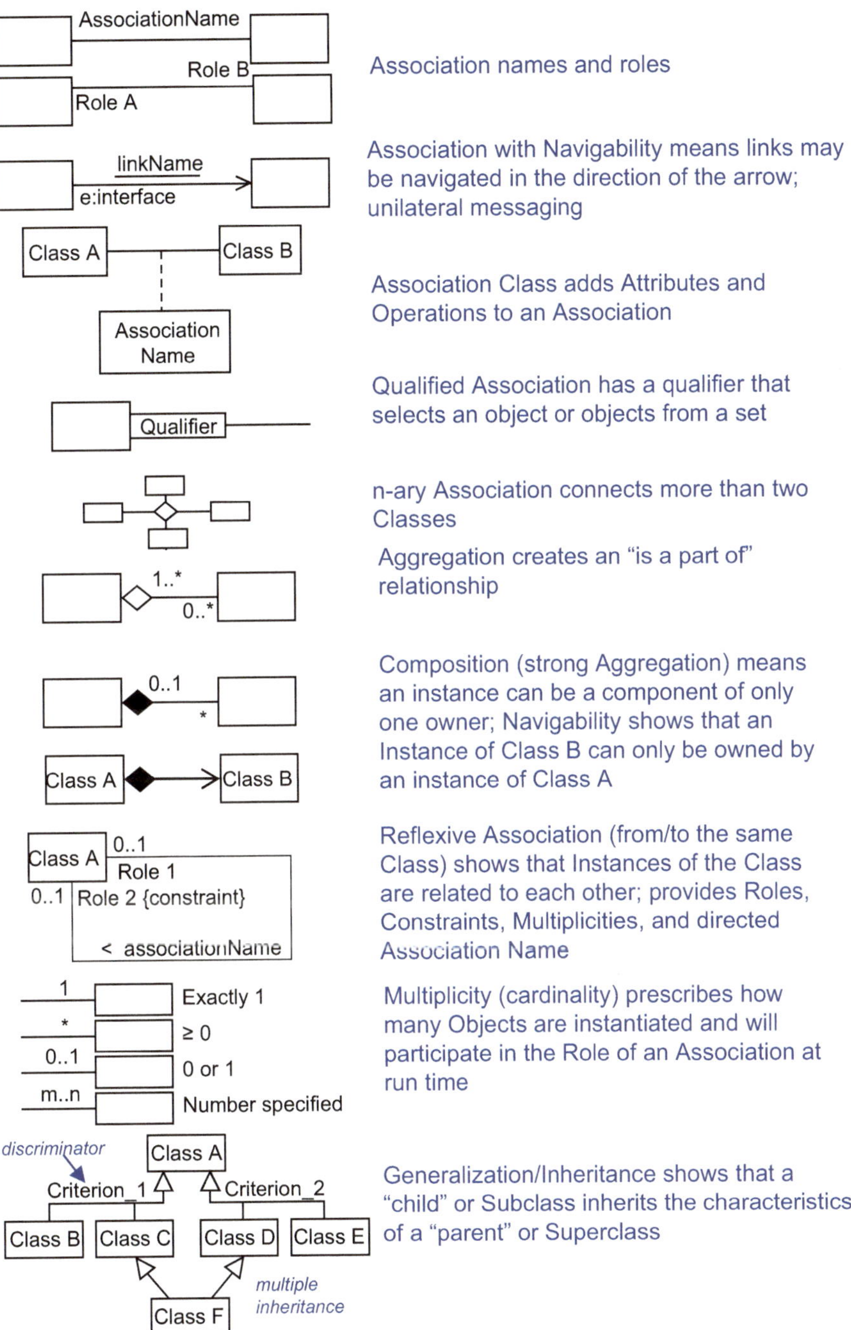

Fig. A.18 Class/Object Relationships

Abstract—True means element can't be directly instantiated.
Leaf—True means element has no subtypes.
Root—True means element has no supertypes.

- *Dependency*—defines a "using" relationship; implies changes to source or provider may cause changes to the client; includes multiple stereotypes:

<<call>>	Source calls an operation in Client
<<create>>	Source creates an instance of Target
<<derive>>	Source is derived from Target
<<permit>>	Target allows Source to access private features
<<realize>>	Source implements a Target specification or interface
<<refine>>	Shows relationship among semantic levels
<<substitute>>	Source is substitutable for Target
<<trace>>	Shows relationships among models/areas
<<use>>	Source requires Target for implementation

- *Abstract Class*—cannot be directly instantiated; indicated by name in *italics*

Structured Class

- Structure refers to a composition of interconnected elements collaborating over communication links to achieve some common objectives; key mechanisms include internal structures, ports, collaborations, structured classes, and actions.

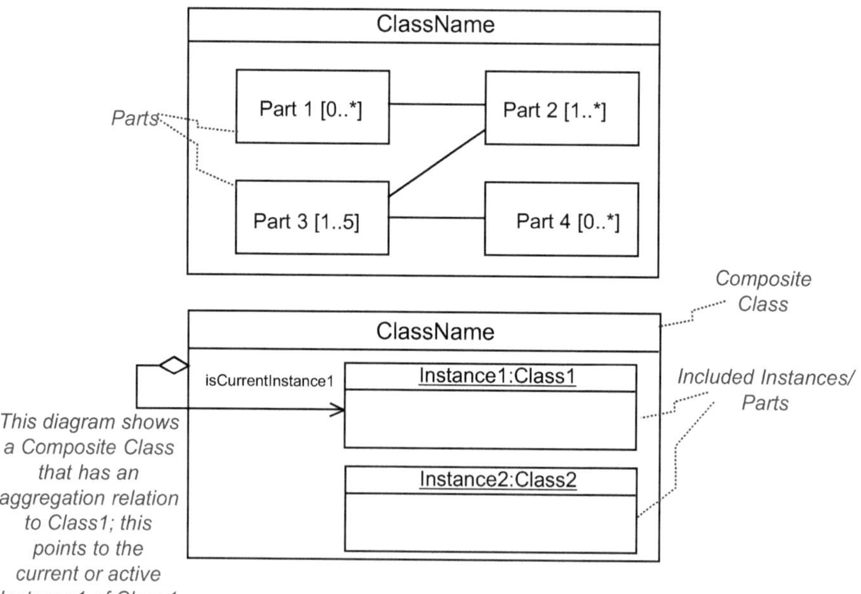

Fig. A.19 Structured Class

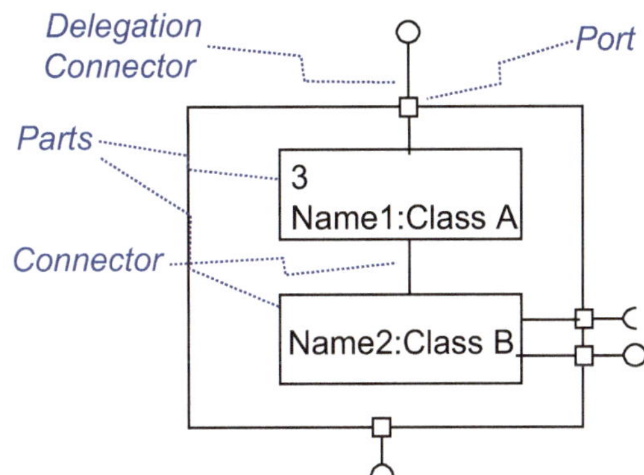

Fig. A.20 Composite Structure

- A structured classifier is an abstract metaclass that represents any classifier whose behavior can be fully or partly described by the collaboration of owned or referenced instances.
- A Structured Class typically shows an internal structure made up of Parts with Associations; Parts are properties of the abstract Class, and an instance of the abstract Class may include instances of the collaboration of Parts it contains; Parts can have multiplicity.

Composite Structure

Allows a complex Class to be decomposed into its Parts; shows collaboration of Classes, Interfaces, or Components to describe a functionality.

Use Cases

- *Use Case*—description of a sequence of actions, including variations, that a system performs that yields an observable result of value to an actor; a set of scenarios involving a common user goal.
- *Scenario*—an instance of a use case; a sequence of steps describing interaction between a user and a system.
- *Use Case Realization*—a stereotyped collaboration, represented by a set of diagrams showing classes/objects, their relationships, and their interactions that implement the Use Case functionality; each Use Case/Collaboration should have a Class diagram.
- *Use Case Model*:
  - *Actor*—Role a user or other external entity plays with respect to the system
  - *Include Dependency*—used with behavior that is common across multiple Use Cases; can be used to decompose a complex Use Case

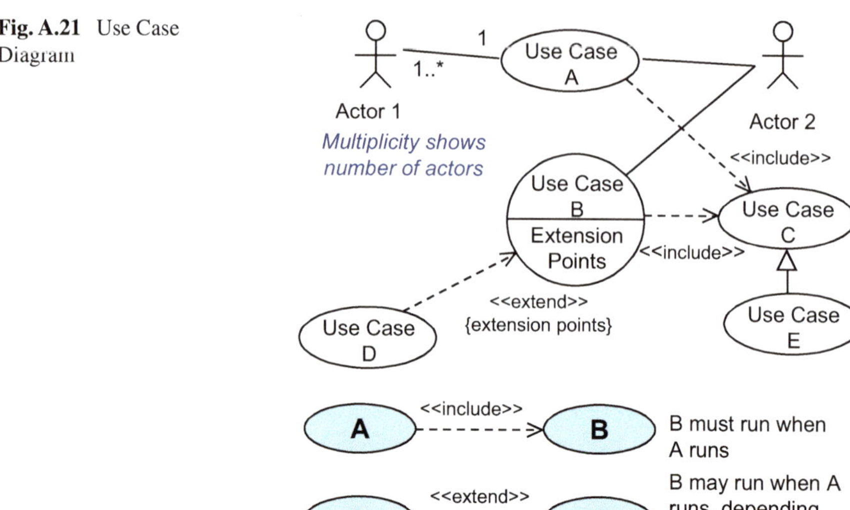

Fig. A.21 Use Case Diagram

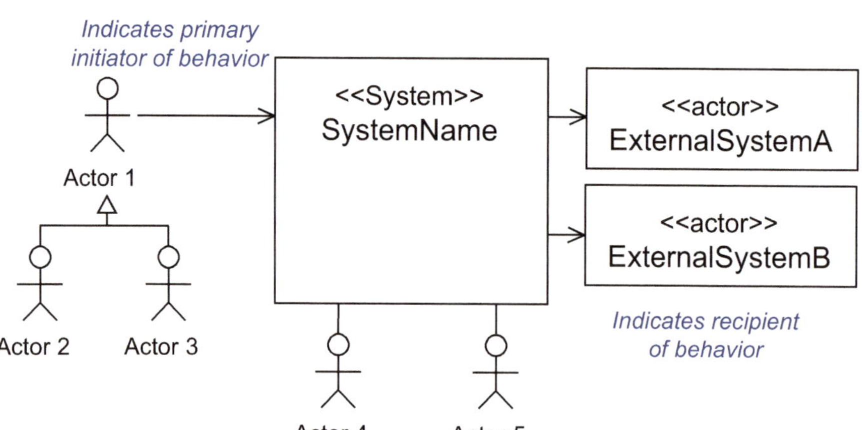

Fig. A.22 Use Case Context Diagram

- *Generalization*—used to describe a variation that is more detailed or specific
- *Extend Relationship*—used to rigorously describe a variation using extension points which define where the extending Use Case modifies behavior in the basic Use Case
- *Workflow/Scenario*—describes what the system must do to provide value to an Actor; can have main and alternative flows; described with Activity Diagram
- *Glossary*—defines key terms for consistent usage across system/project
- *Supplementary Specification—1* defines functional/nonfunctional requirements not captured by Use Cases, e.g., constraints, reliability, timing, etc.

- *Use Case Environment*:
  - *External Events Diagram*—spreadsheet of events received or issued by the system; includes events, data, timing, statistics, etc.
  - *Context Diagram*—idiomatic object diagram of the system, actors, and messages; places the system in the context of the environment, users, stakeholders, and others that will impact its design and behavior

- *Use Case Behavior*:
  - Informal textual description, e.g., problem/requirement/function, capability, and constraints.
  - Scenarios and variations—explore interesting or significant paths through the Use Case; since the scenario is an instance of the Use Case, it's labeled with the Use Case Name and underlined.

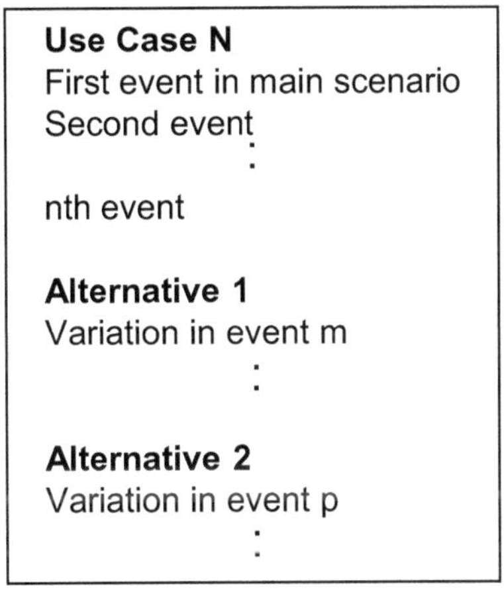

Fig. A.23 Informal Use Case Behavior Description

Use Case N
First event in main scenario
Second event
:
nth event

Alternative 1
Variation in event m
:

Alternative 2
Variation in event p
:

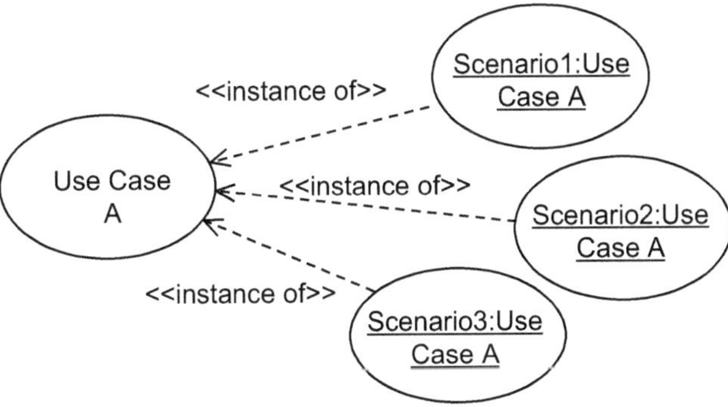

Fig. A.24 Use Case Scenarios

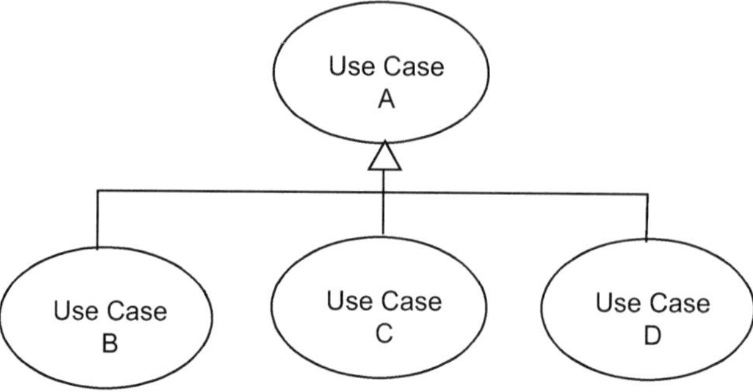

Fig. A.25 Use Case Generalization

- Sequence Diagrams—show sequence of messages among objects in a collaboration of classes implementing a Use Case
- Activity Diagrams—show sequence of actions involved in a Use Case scenario
- State Diagrams—show discrete-state behavior, transitions among states, events causing transitions, and actions

- *Generalization*—can be used to specialize particular behaviors from a general behavior

Collaboration

A collaboration defines a set of cooperating Roles and their connectors to illustrate a specific functionality:

- A society of Classes, Interfaces, and other elements that work together to provide a cooperative behavior, e.g., of a Use Case.
- The specification of how an element is realized by a set of classifiers and associations.
- An instance of a (design) pattern.
- Structure—usually represented by Class Diagrams.
- Behavior—represented by Behavior Diagrams.
- Mechanism—instance/realization of a collaboration; a design pattern that applies to a society of classes; a Use Case is realized by a Collaboration of Classes.
- A Classifier represented by a Collaboration is shown with an `<<occurrence>>` stereotyped dependency.
- A Classifier using a Collaboration is shown with a `<<represents>>` stereotyped dependency; Fig. A.26 shows that the Class uses the Collaboration of Objects in its implementation; such Collaborations can be design patterns.

Fig. A.26 Collaborations

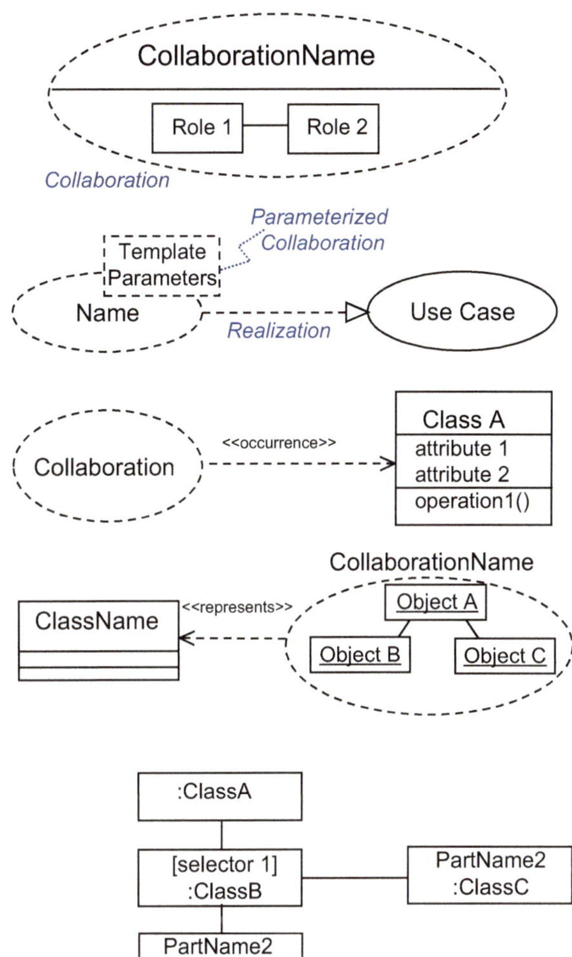

Fig. A.27 Collaboration of Objects/Parts

- A Collaboration can be developed from either an Object Diagram or a Composite Structure Diagram; the elements are Objects or Parts, labeled:

```
ReferenceName [selectors] : ClassName
```

  - `ReferenceName` is an optional label for the Part.
  - `Selectors` is an optional field that selects a particular Object from a set; can be a qualifier, an indexing subscript, or a Boolean expression (not needed if the multiplicity of the object is 1).

- Class/Object Collaboration—used to show interactions across domain boundaries.

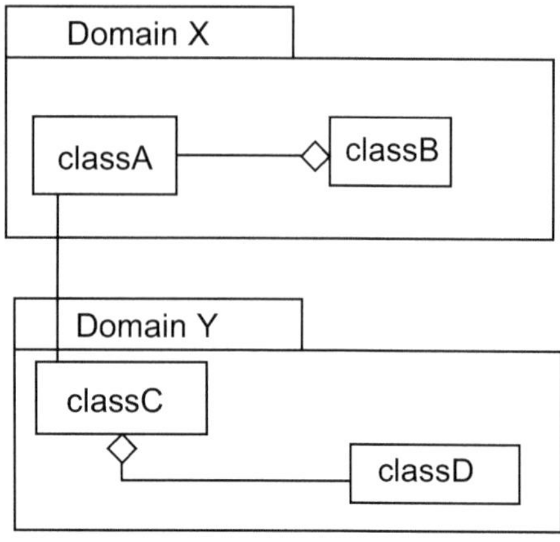

Fig. A.28 Class/Object Collaboration

Fig. A.29 Signal used in a Flow of Control

Flows of Control

- Process—a "heavy" flow that can execute concurrently with other processes
- Thread—a "light" flow that can execute concurrently within a process (sometimes treated synonymously with Process
- Signal—a named object dispatched by one object and received by another, e.g., an exception

Object Constraint Language (OCL)

- A formal language for defining constraints that reference classes, attributes, rules, and operations in a Class Diagram—usually attached to operations.
- Syntax—Figure A.30 illustrates OCL syntax.
- Keywords—pre, post, inv identify the constraint as a precondition, postcondition, or invariant.
- An optional `ConstraintName` can be used to allow the constraint to be referred to again.
- OCL Expression—logical expression evaluating to True of False that can be used in a constraint.

Operation or Use Case to which the constraint applies

Defines constraint applicability

Type of value returned by the behavior

Context Type::behaviorName(parameter1:type, parameter2:type . . .):returnType

System, Subsystem, Class or Type that is the subject of this behavior

Parameter list of the behavior

Fig. A.30 Object Constraint Language Statement Syntax

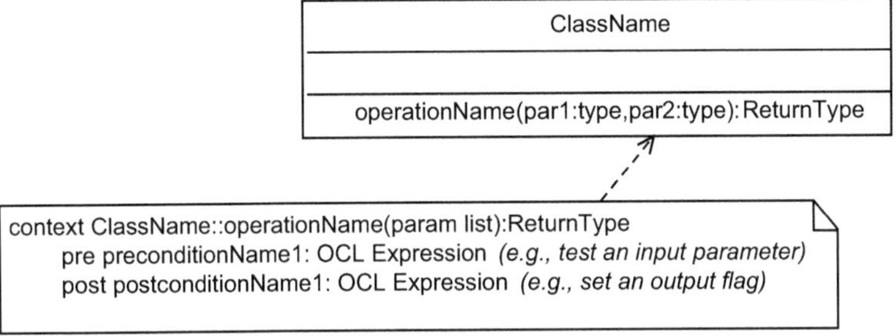

Fig. A.31 Example of OCL to define a Class Operation

- OCL Dot Operator, A.B:

`object.attribute`	The Attribute B of object A
`object.queryOperation`	Result of calling query Operation B on Object A
`Class.staticAttribute`	The static Attribute B of Class A
`object.rolename`	Set of Objects playing role B across an Association from A
`EnumeratedType.literal`	Value of literal B of the enumerated data type A

- OCL example—Figure A.31 shows a typical use of OCL.

UML Behavior Models

- Specify how the structural aspects of a system change over time (dynamic aspects).
- Three behavior models:
  - Activities focus on the sequence, conditions, and inputs and outputs for invoking other behaviors.
  - State Machines show how events cause changes of object state and invoke other behaviors.
  - Interactions describe message passing between objects that causes invocation of other behaviors.

- Each kind of behavior model emphasizes a different aspect of system dynamics, making one or the other more suitable for a particular application, or stage of application development.
- Any behavior can be parameterized (activities, state machines, interactions, etc.) and can be either invoked separately or treated as methods on objects.
- User-defined behaviors are also Classes, executed as runtime instances.

Sequence Diagram

- Message:
  - Associated with a `SendEvent` at the sending Object and a `ReceiveEvent` at the receiving Object.
  - Usually an Operation Call from the sender—operationName(parameter list).
  - Can use several kinds of signaling:
    Procedural/Synchronous Operation Call
    Return from a call/asynchronous Message
    Generic/Asynchronous Call, Event
    Object Creation (can use the `<<create>>` stereotype)
    Object Destruction (can use the `<<destroy>>` stereotype).

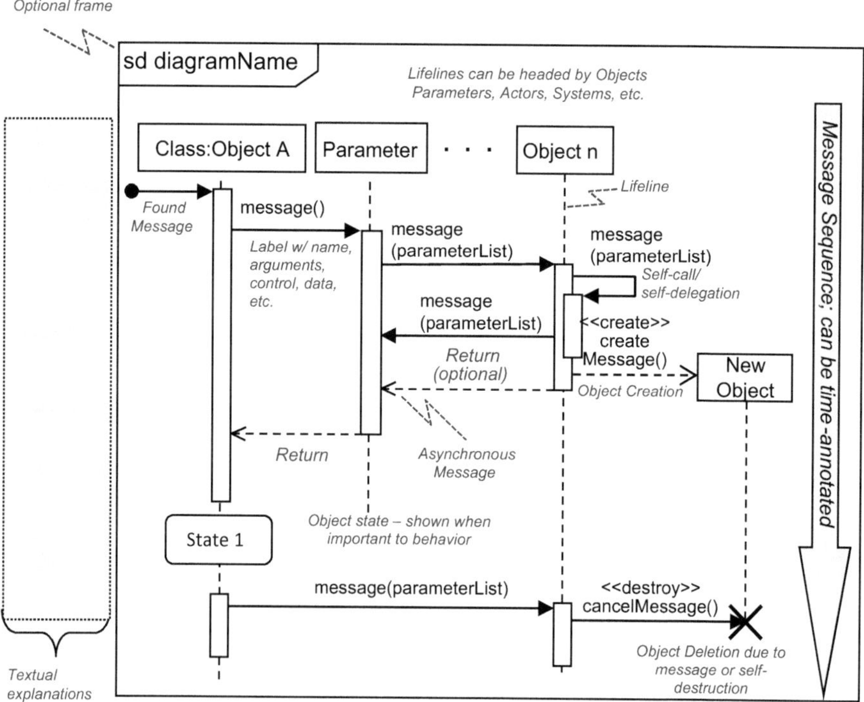

Fig. A.32 Sequence Diagram

- Composing complex, reused, and alternative behaviors:
  - Frame—defines and names an interaction or part of an interaction.
  - Combined Fragment—reflects a piece or pieces of interaction (called interaction operands), separated by dashed lines, controlled by an interaction operator whose Boolean conditions are called interaction constraints.
  - Iteration Frames—mark off part of a Sequence Diagram for looping, conditional behavior, etc.; frames can be nested as required.
  - Interaction Occurrence—a reference to a reusable piece of an interaction that is defined elsewhere; identified by the `ref` operator.
- Commonly Used Operators:

`alt`	Alternative fragments; fragment with guard = True executes
`opt`	Optional; fragment executes if guard = True
`par`	Parallel; fragments run in parallel
`loop`	Fragment loops based on guard condition
`region`	Critical region; fragment has one thread executing at once
`neg`	Negative; fragment shows invalid interaction
`ref`	Reference; refers to an interaction on another diagram or `interaction occurrence; can have parameters and return values`
`sd`	Sequence diagram; surrounds entire diagram

Others include `ignore`, `consider`, `strict`, `assert`, and `critical`

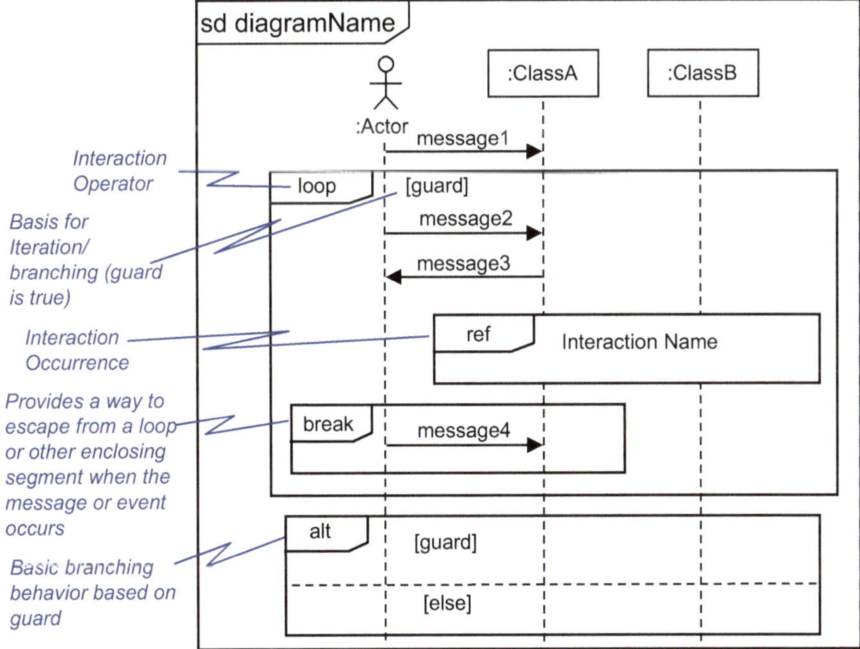

Fig. A.33 Sequence Diagram with Multiple Interaction Fragments

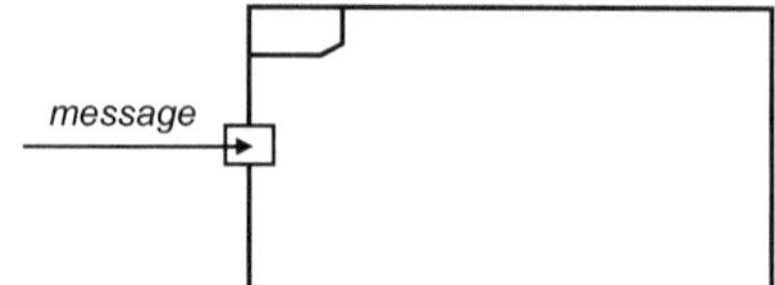

Fig. A.34 Diagram Gate

- Diagram Gate—shows the point at which messages can be transmitted into/out of an interactive fragment

Communication Diagram

- Formerly called a "Collaboration Diagram."
- Thread—show objects as icons; show message sequence by numbering.
- Nested message numbers—show more precise message sequencing: 1.1 is the first activity after 1, 1.1.1 is the next subordinate message within the scope of message 1, etc.; this is called *outline numbering* of messages.
- Appending an asterisk * to a message number indicates that the message may be sent multiple times (similar to Class multiplicity).
- Arrows show the direction of a message.
- Message can consist of:
  - Procedure call—returnVariable=operationName(parameter list)—receiving class must have this operation; arguments and types must match; objects must be connected by links.
  - Signal, e.g., an exception—receiving class must have this signal reception.
- Looping—to repeat a message a specific number of times, the syntax is: `SequenceNumber*[iteration clause]`
- Concurrent Threads—use same sequence numbers for concurrent messages but with thread identifiers:

```
3.4thread1:message1(), 3.4thread2:message2(),
3.4thread3:message3()
```

Activity Modeling

- Activity—the state of doing something, executing a process; decomposes into Actions which are atomic units of behavior:
  - An Activity models user-defined behavior; it defines a virtual machine that can be implemented many ways.

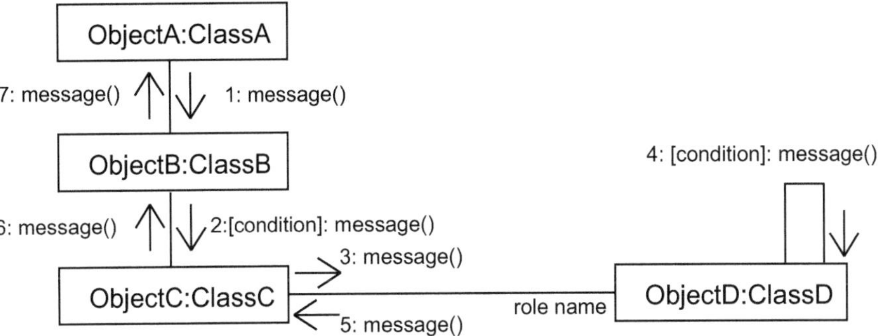

Fig. A.35 Communication Diagram

- An Activity can be an operation on a Class/Object and can also be invoked directly (e.g., by a Call); it can be initiated by invocation Actions and can exchange data with invokers via parameters.
- Activities contain nodes connected by edges, forming complete flow graphs (token flow semantics, similar to Petri nets):

  Action nodes operate on control and data values that they receive and provide control and data to other actions.
  Control nodes route control and data tokens through the graph, including decision points, forks, etc.
  Object nodes hold data tokens temporarily as they wait to move through the graph.

- An Activity can have pins to connect actions in a flow; pins support data and Object flows to/from an Action.
- Two kinds of directed nodes:

  Control flow edges connect actions to indicate that the action at the target end of the edge (the arrowhead) cannot start until the source action finishes.
  Object flow edges connect object nodes to provide inputs to actions.

- Activity Package dependencies:
  - Action—an atomic increment of behavior; an Activity decomposes into Actions.
  - Actions are the only UML elements that can query objects, have a persistent effect on them, invoke operations on them, and invoke behaviors directly.
  - Actions are defined by the UML standard rather than by users.
  - Actions are contained as nodes in activities.
  - Actions execute when all input data and controls are available.
  - Actions use Attribute values for static information inputs.
  - Actions include CallBehaviorAction, CallOperationAction, SendSignalAction, BroadCastSignalAction, and SendObjectAction.

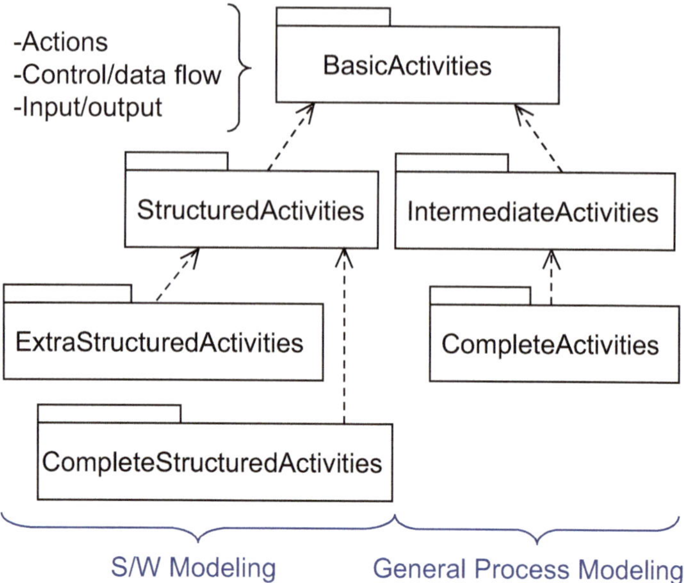

Fig. A.36 Hierarchy of Activity Types

Activity Diagram

- Fork denotes the start of parallel activities, some of which may be guarded.
- Decision (formerly "Branch") denotes selection of one of a set of guarded transitions.
- Join synchronizes parallel threads; in general, all threads must complete before the join transition.
- Merge ends conditional behavior from a branch.
- Expansion Region shows multiple concurrent executions of an activity or iteration without constructing a loop.
- Activities can be sorted vertically or horizontally into "swim lanes," e.g., by domains or resources; formally, these are "Activity Partitions."
- Diagram can include data objects to show when they are created and where they are sent.
- Activities can be decomposed into Subactivity diagrams.
- Activities can be organized vertically into *swim lanes* (labeled as Domains in this example) that identify the parties responsible for sets of actions.
- A pin is an object node for data input to/output from actions; distinguishes value flow from control.

- Object Node Exchange or Flow between Actions:

  – Activity modeling uses Token Flow semantics—"tokens" for control and data flow through a graph of nodes connected by edges; this is equivalent to the Action, Control, and Object Nodes described earlier.

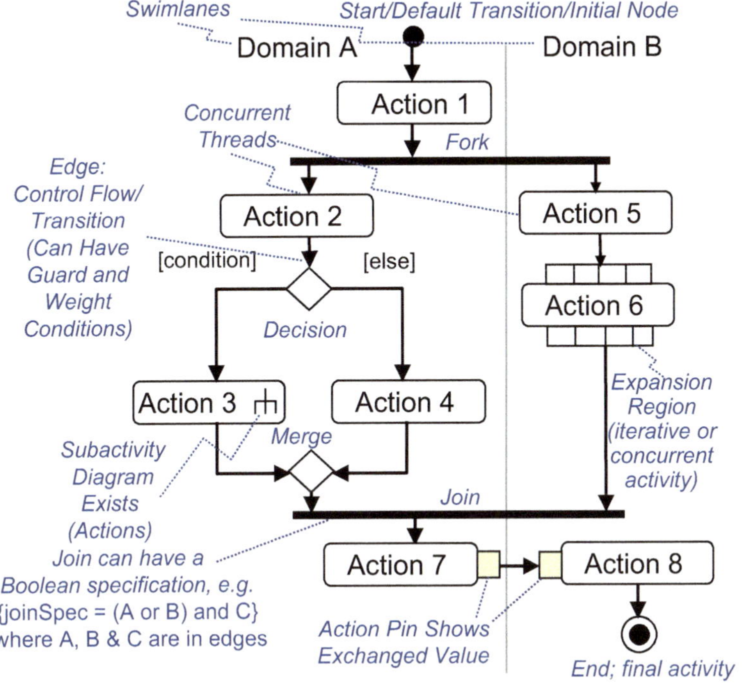

Fig. A.37 Basic Activity Diagram

- The basic semantics for Object exchanges or flows are based on Pins:

  Modeled as small squares on the boundaries of Actions.
  Input pins bind input parameters to local variables.
  Output pins bind output parameters to output variables.
  Action starts when input data appears on all input pins.
  Action completes when data is present on all output pins.

- An alternative representation uses Object Nodes in the diagram, connected by Object Flows.
- Streaming Parameter—annotated with {stream} or a filled-in Pin.
- Exception parameter—annotated with a triangle.
- Connector—provides a way to link flows among diagrams.

• Show pre- and postconditions for an activity (can be those of a Use Case that the Activity Diagram illustrates):

  `<<precondition>>` expression evaluates to True to begin activity.
  `<<postcondition>>` expression evaluates to True at end of activity.

• Local pre-/postconditions only apply to a specific activity; annotated as note with:

  `<<localPrecondition>>`

• Show intermediate activity termination with an abort/termination node:

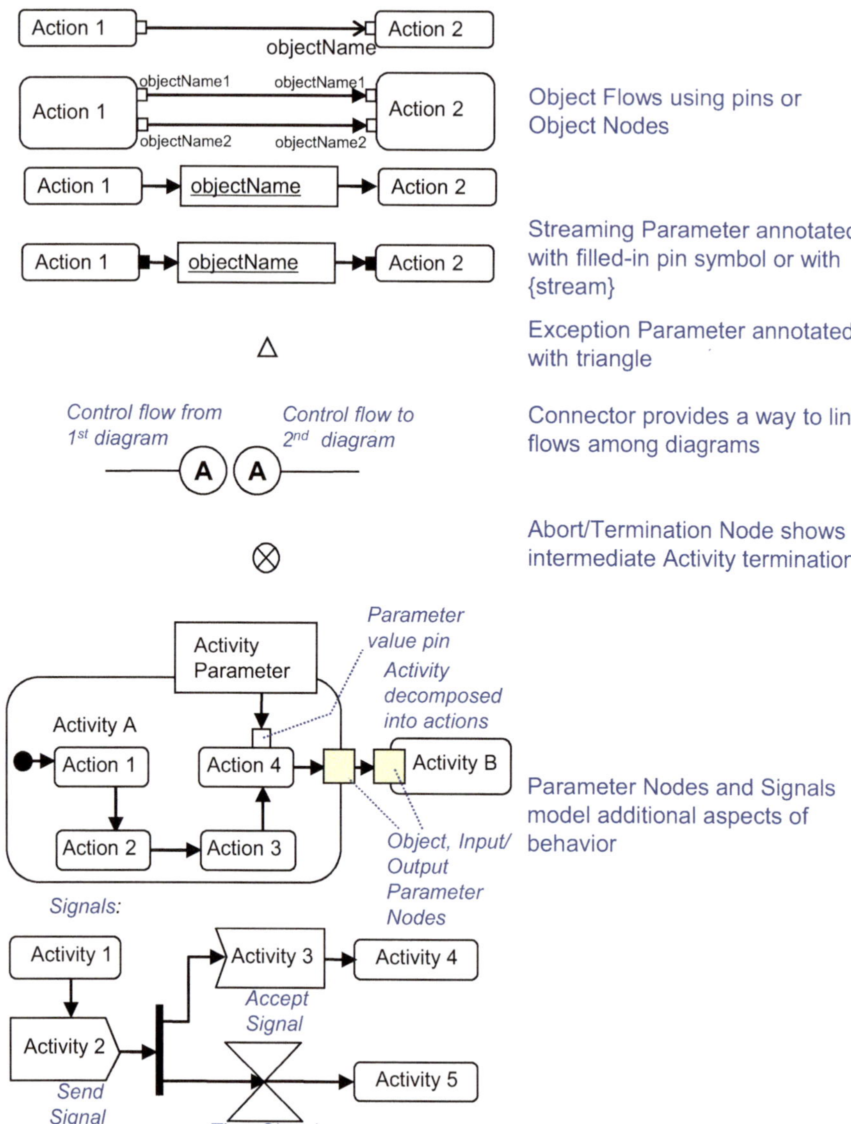

Fig. A.38 Additional Activity Diagram Syntax

Interaction Overview Diagram

Combines Activity and Sequence diagrams; drawn as an Activity diagram with embedded Interaction (usually Sequence) diagrams instead of activities.

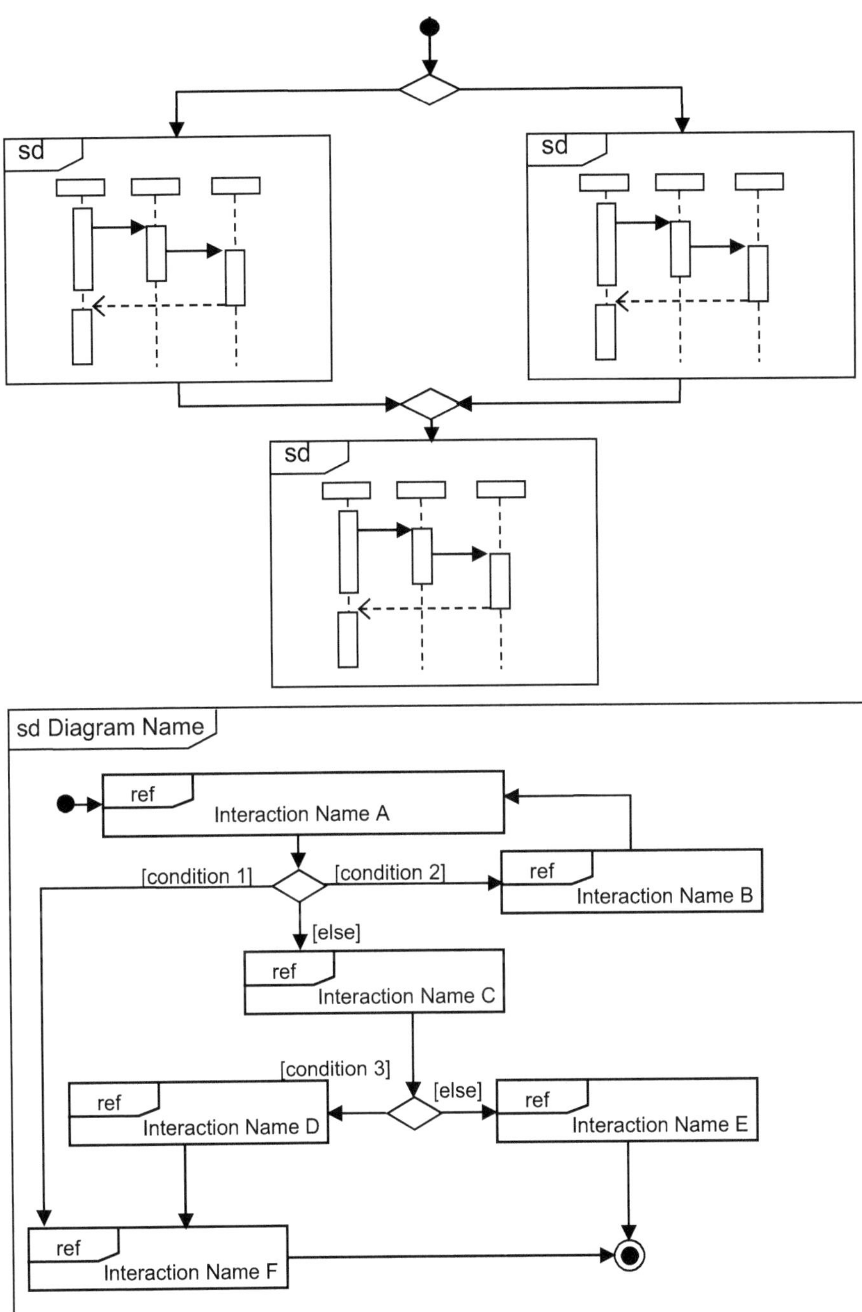

Fig. A.39 Interaction Diagrams

State Machine Diagram/Statechart

Models stateful behavior, i.e., the entity can be in one or more States that represent behavioral configurations (parameter values, possible Actions, etc.) and depend on prior behavior.

- States have unique identifiers and can have a variety of Activities/Actions:
  - `entry/<action>`—action performed upon entry into the State.
  - `do/<action>`—action performed while in the State.
  - `exit/<action>`—action performed upon leaving the State.
  - `internalEventName/<action>`—action triggered by an internal event.
  - `Defer/<action>`—action may be performed if a blocking condition is removed.
- Transitions can have triggering events, parameters, guard conditions, and actions; all are optional; an unguarded transition happens as soon as the flow reaches it.
- A guard condition is Boolean; if it evaluates to True, the transition is enabled.
- An event that cannot be handled can be deferred (blocked) until it can be handled.
- Concurrent States:
  - Used when an Object has sets if independent behaviors.
  - Object has only a single State when exiting the concurrent States.
  - OR State means the Object must be in exactly one State of the Superstate.
  - AND states communicate via broadcast Events, propagated Events, Guards (`[isIn(stateName)]`), and Attributes of the Object.
- States can have `entry/, do/, exit/ internalEvent/`, or `deferred/` Actions/Activities
- Time Triggers generate time events to model the expiration of a deadline or the arrival of a specified instant of absolute time:
  - Relative Time Trigger measures time from State entry or similar condition, e.g., `after(n seconds).`
  - Absolute Time Trigger uses an expression that evaluates to True when an absolute time value is reached, e.g., by a clock.
- Events can be of several types:
  - Signal—asynchronous signal received from elsewhere in the model
  - Call—Event generated by an Operation Call
  - Change—Event generated by a change in a value
  - Time—as above, an Event generated by arrival of a specified instant `(when(timeCondition))` or elapse of a relative deadline `(after(timeInterval))`

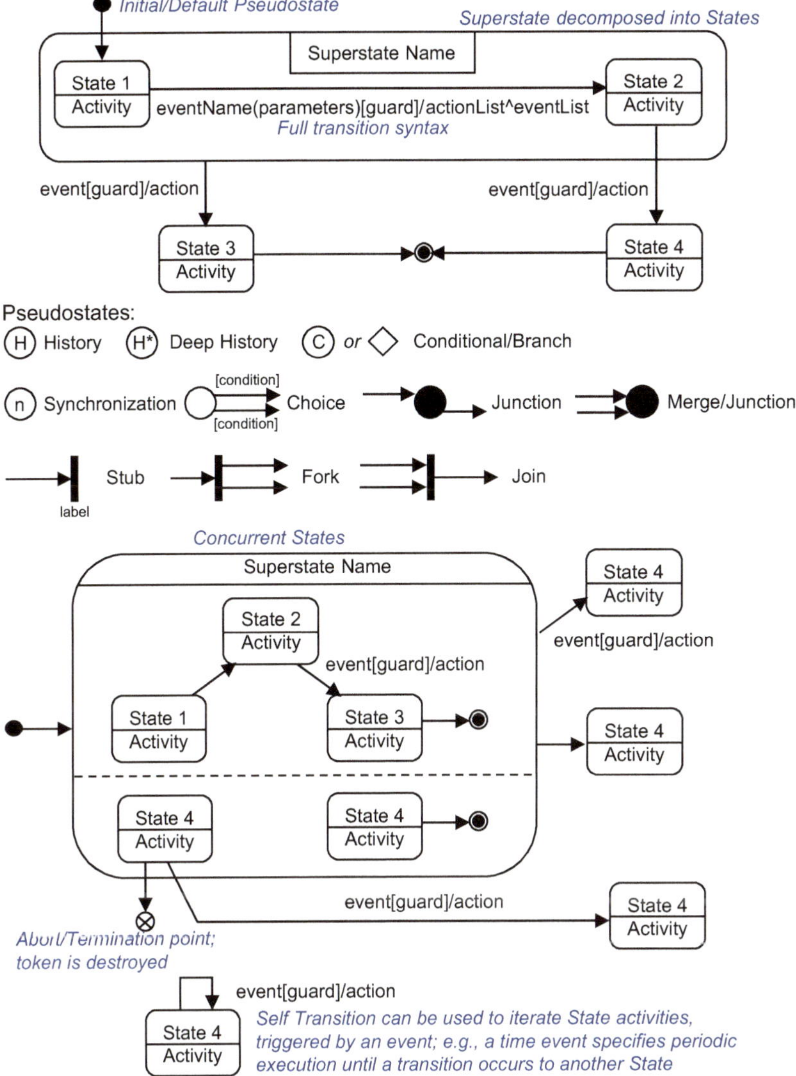

Fig. A.40 State Machine Diagrams/Statecharts

Physical Diagrams

Used to model physical elements of a system and their relationships, including nodes (computational resources, usually hardware) and connections (paths over which nodes interact)

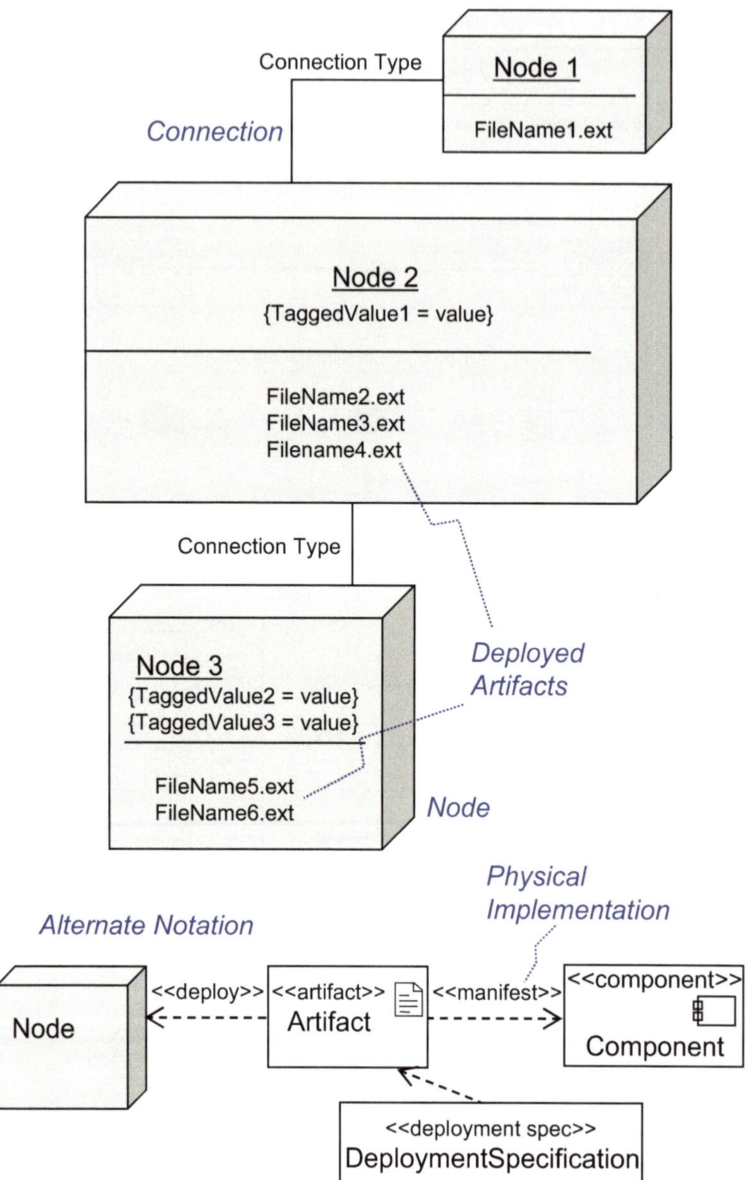

Fig. A.41 Deployment Diagram

Deployment Diagram

Shows physical relationships among software and hardware components—what software runs on what machines, how components and objects are routed in a distributed system

- Dependencies—show how changes to some components cause other components to change; show communication among components.
- Tagged Values—define node characteristics such as Operating System, Processor Type.
- Typical stereotypes for nodes:
  - `<<device>>`—node with processing capability
  - `<<application server>>`—provides remote service for an application
  - `<<mobile device>>`—device with wireless connectivity
  - `<<embedded device>>`—real-time embedded processor
  - `<<execution environment>>`—virtual node providing the environment for program execution
  - `<<container>>`—node that holds components in a Java environment
- Typical stereotypes for communication paths:
  - `<<serial>>`—serial port connection between nodes
  - `<<parallel>>`—parallel port connection between nodes
  - `<<USB>>, <<Ethernet>>, etc.`—specific connection types/protocols
  - `<<LAN>>, <<WAN>>, etc.`—for local or wide area networks
  - `<<Internet>>`—Internet connection, e.g., for Web services.
- Hardware items can also be stereotyped with model numbers or other common characteristics.
- Artifact—mode physical implementation of a Component or Subsystem, hence, dependent on that component or subsystem; common stereotypes:
  - `<<executable>>`—can be run as a program on a computer
  - `<<library>>`—file that is a static or dynamic link library (DLL)
  - `<<script>>`—source code files that are interpreted at runtime
  - `<<page>>`—a single html page
  - `<<file>>`—generic file artifact

Component Diagram

- Component—a modular unit with well-defined interfaces that is replaceable within its environment:

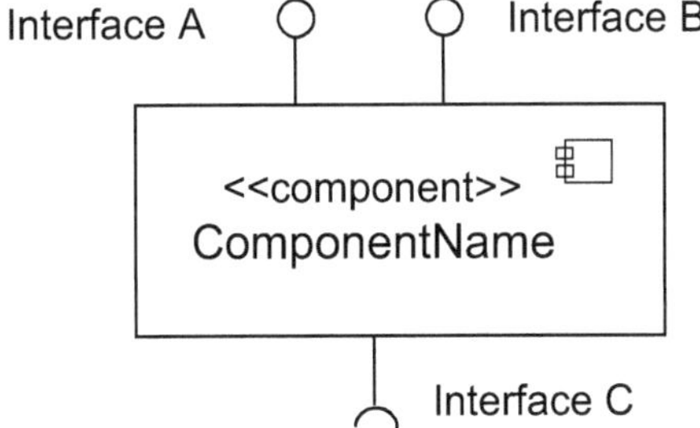

Fig. A.42 Component

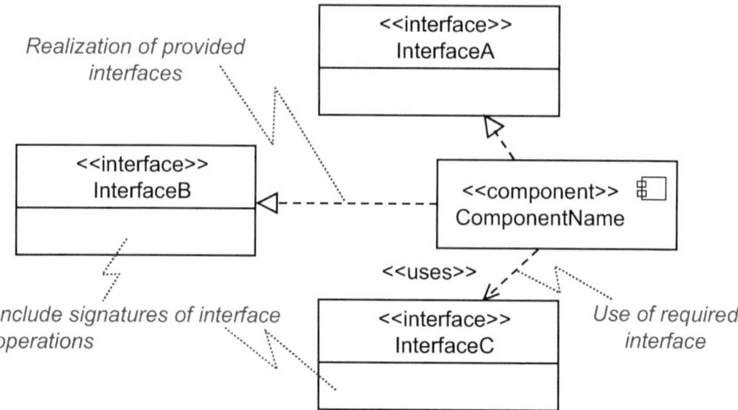

Fig. A.43 Component Interfaces

- Can be a `<<subsystem>>` with internal classes that work together to realize the interfaces
- Can be logical or physical

- Behavior is specified in terms of provided and required interfaces.
- Connectors can be Assembly or Delegation; the latter links a component contract at an external interface to the internal elements that realize the behavior.

Composite Structure Diagram with Internal Component Structure

Typical stereotypes for internal parts include:

- `<<focus>>`—executes some or all of the business logic of the enclosing component
- `<<process>>`—executes a transaction and ensures completion
- `<<service>>`—stateless part that computes a value
- `<<entity>>`—part whose values, behavior, and state persist after runtime
- `<<auxiliary>>`—part that assists the focus part

Fig. A.44 Internal Components

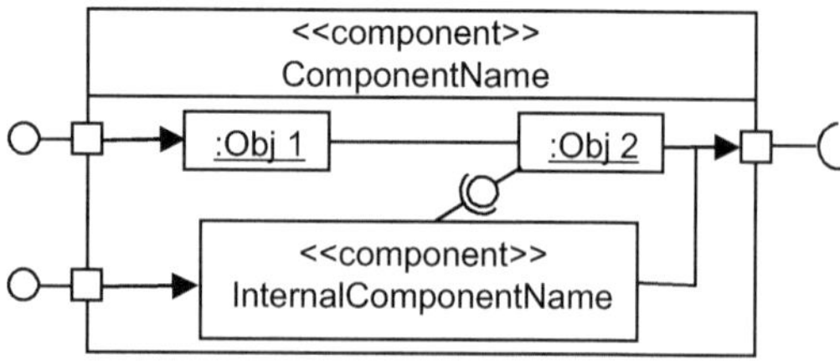

Component Diagram with Associated Active Objects

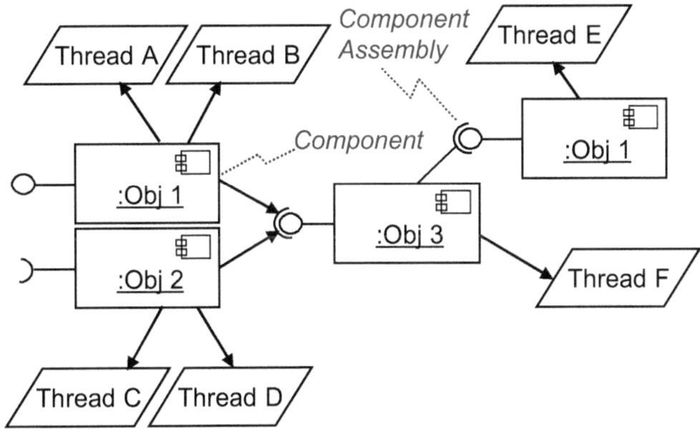

Fig. A.45 Components with Active Objects

Timing Diagrams

- Show explicit changes of state along a linear time axis
- Specific type of Interaction Diagram with focus on time
- Task or Thread—sequence of operations that executes independently of other tasks/threads (see also Flows of Control)
- Deadline—time point or interval by which a system action must occur

Semantics of Real-Time Systems

A real-time (RT) system has deadlines associated with functions or threads with system or mission failure ("hard" RT) or degradation ("soft" RT) resulting when deadlines are exceeded. Typical attributes include:

- *Schedulability*—functions and operating system threads can be deterministically scheduled; criteria such as Rate Monotonic Analysis (RMA) or Deadline Monotonic Analysis (DMA) are supported.
- *Preemption*—higher-priority functions can preempt lower-priority functions.
- *Memory Protection*—mechanisms are employed to prevent one process from overwriting or corrupting another.
- *Fault Detection and Recovery*—mechanisms are employed to detect failures, exceptions, suspended threads, etc. and to restore normal operation or initiate backup modes.

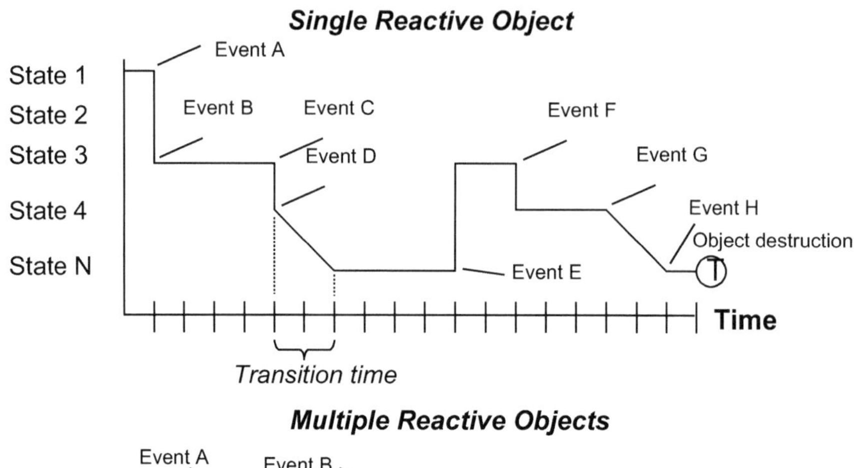

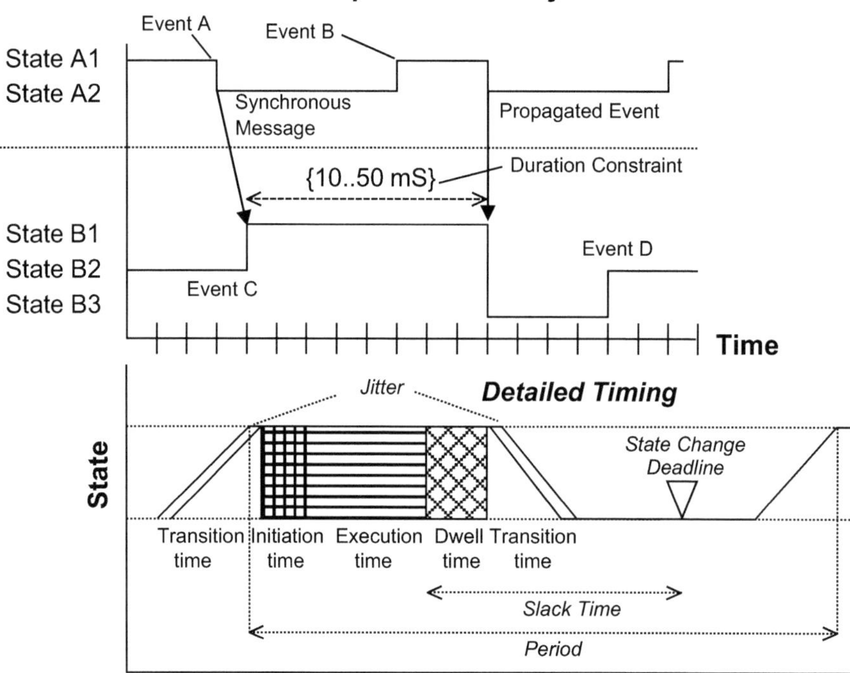

Fig. A.46 Timing Diagrams

Keywords/Stereotypes for Real-Time Systems

Class	<<active>>	Class is the root of an OS thread
Message	<<synchronous>>	Association will be realized as a simple function/ method call
	<<blocking-local>>	Association will cross thread boundaries; caller will baulk until receiver returns a value
	<<asynchronous-local>>	Association will cross thread boundaries by placing msg in target thread's incoming queue
	<<waiting-local>>	Sender will wait a fixed period or until receiver responds
	<<synchronous-remote>>	Association crosses processor boundary; sender will baulk until explicit return from receiver
	<<asynchronous-remote>>	Association crosses processor boundary; sender will continue immediately
	<<periodic>>	Message is sent periodically
	<<episodic>>	Message is sent based on event of interest
	<<epiperiodic>>	Message is sent both periodically and episodically
Operation Semaphore	<<guarded>>	Execution is guarded by a mutual exclusion ("thread-reliable")
Node	<<processor>	Device that executes designated software
	<<device>>	Device does not execute SW but has an interface
	<<sensor>>	Device monitors external world or another device's operation
	<<actuator>>	Device affects external world or another device's operation
Operation	<<display>>	Device displays information to user
	<<knob>>	Interface control knob
	<<button>>	Interface control button
	<<switch>>	Interface control switch
	<<watchdog>>	Sensor causes fail-safe behavior after timeout

Appendix B: Quick Reference to the System Modeling Language (SysML) Profile of the Unified Modeling Language (UML)

General

- SysML is a UML 2 profile and reuses a subset of the basic UML modeling elements, labeled UML4SysML; unneeded UML elements are excluded from the profile, and additional/modified diagrams and modeling elements are added.
- SysML is focused on common System Engineering tasks and deals with:
  - Requirements
  - System structure
  - Functional behavior
  - Allocations
  - Basic testing
  - Basic trade studies during the specification and design phases of Systems Engineering

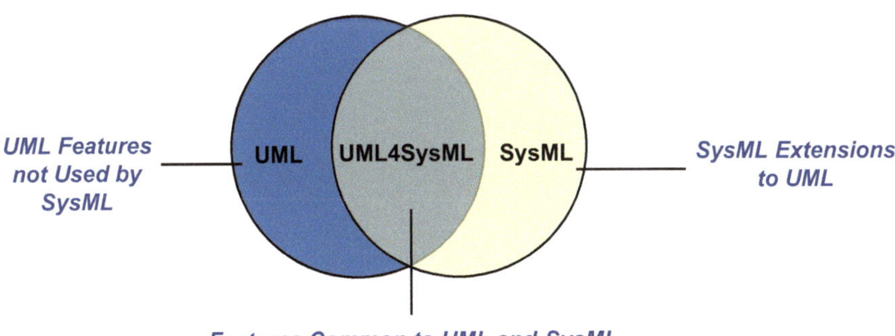

Fig. B.1 UML/SysML Scope and Overlap

J. M. Borky, T. H. Bradley, *Effective Model-Based Systems Engineering*,
https://doi.org/10.1007/978-3-319-95669-5

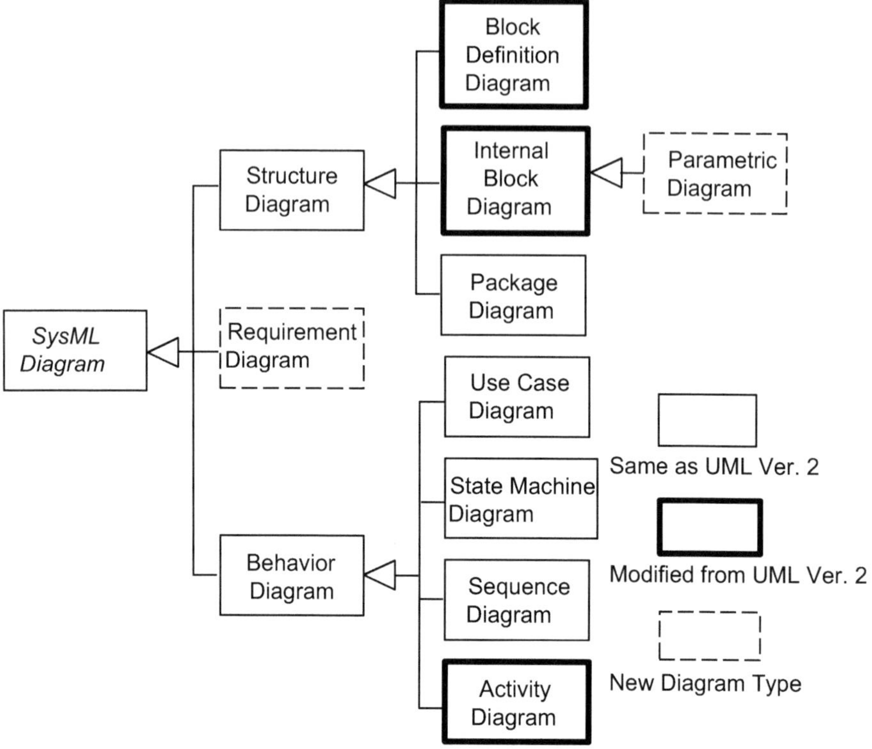

Fig. B.2 SysML Diagrams

SysML Diagrams

SysML eliminates the following UML diagrams: Object Diagram (replaced by Instance Specification), Deployment Diagram (replaced by Allocations on Block Diagrams), Timing Diagram (replaced by timing parameters in Parametric Diagrams), Communication Diagram, Component Diagram, Interaction Overview Diagram, and Profile Diagram.

SysML Extensions of UML 2

Includes stereotypes, diagram extensions, and model library elements

SysML Packages

Shows the basic structure of the SysML Profile

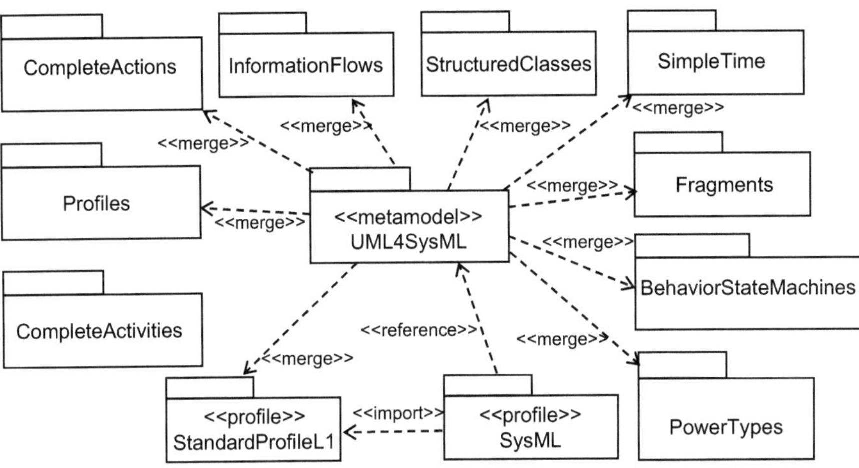

Fig. B.3 SysML Extensions of UML 2

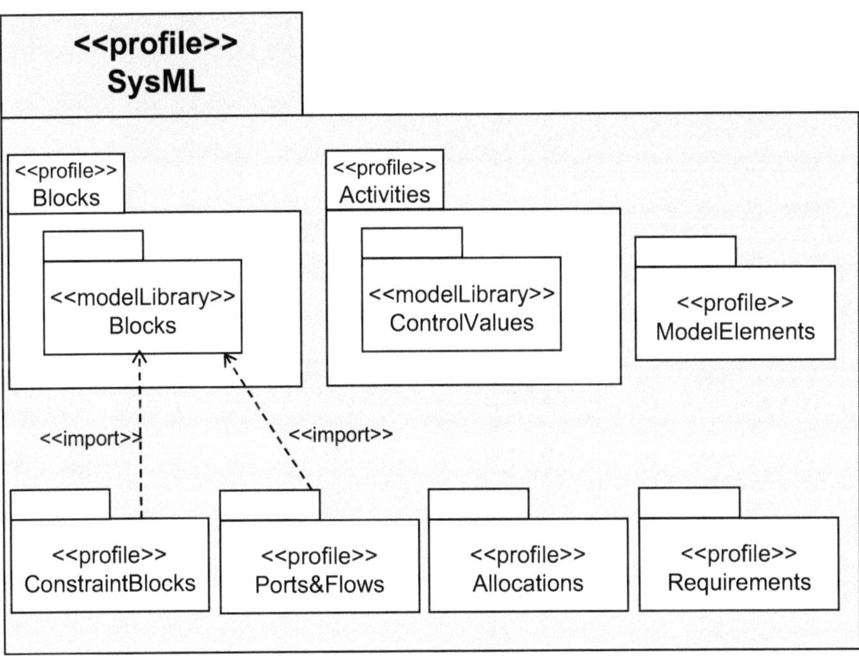

Fig. B.4 SysML Packages

Diagram Frames

Frames are mandatory for all SysML diagrams; a Note on the frame gives information about the diagram.

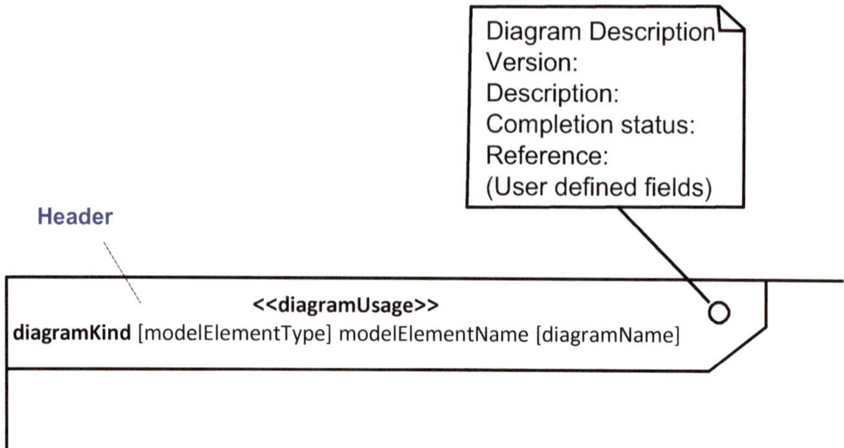

Fig. B.5 Diagram Frame with Descriptive Note

View/Viewpoint Diagram

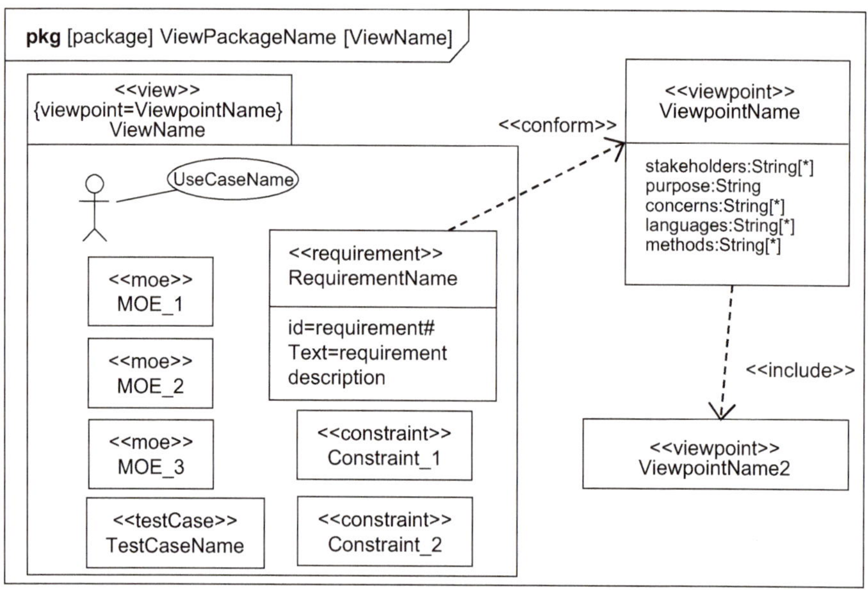

Fig. B.6 View or Viewpoint Diagram

Stereotypes Defined by the SysML Model Elements Package

- A `<<conform>>` relationship is a dependency between a supplier viewpoint and a client view that fulfills the requirement.
- A `<<problem>>` documents a deficiency, limitation, or failure of one or more model elements to satisfy a requirement or need, or other undesired outcome.
- A `<<rationale>>` can be attached to any element and is used to document analysis results and requirements and design decisions.
- A `<<view>>` is a representation of a whole system from the perspective of a single stakeholder viewpoint.
- A `<<viewpoint>>` is a specification of the conventions and rules for constructing and using a view for the purpose of addressing a set of stakeholder concerns.

Classifiers

- Model elements that can type a set of lower-level elements by specifying their common characteristics.
- General Classifiers, or Superclasses, and type specialized Classifiers, or Subclasses.
- Classifiers include Blocks, Value Types, Interfaces, and Signals.

Crosscutting Constructs

- *Allocations*—the organized cross-association (mapping) of elements within the various structures or hierarchies of a user model:
  - Allocate—a dependency based on UML::abstraction, used to associate elements of different types, or in different hierarchies, at an abstract level; a stereotype of a UML4SysML::Abstraction; an `<<allocate>>` relationship between model elements is a precursor to a more concrete relationship between the elements, their properties, Operations, Attributes, and subclasses.
  - Functional Allocation—used to separate form (structure) from function (behavior).

    Allocation of Usage—allocation between "elements of usage" such as parts, actions, and connectors; only allocates specific usage, not general definition, so doesn't apply to other elements that are typed by the same Block
    Allocation of Definition—allocation between Blocks, Activities, Associations, etc.; applies to every property that is typed by the elements of definition, e.g., allocates an Action in an Activity Diagram to a Part in an Internal Block Diagram or a Flow in an Activity Diagram to a Connector in an Internal Block Diagram

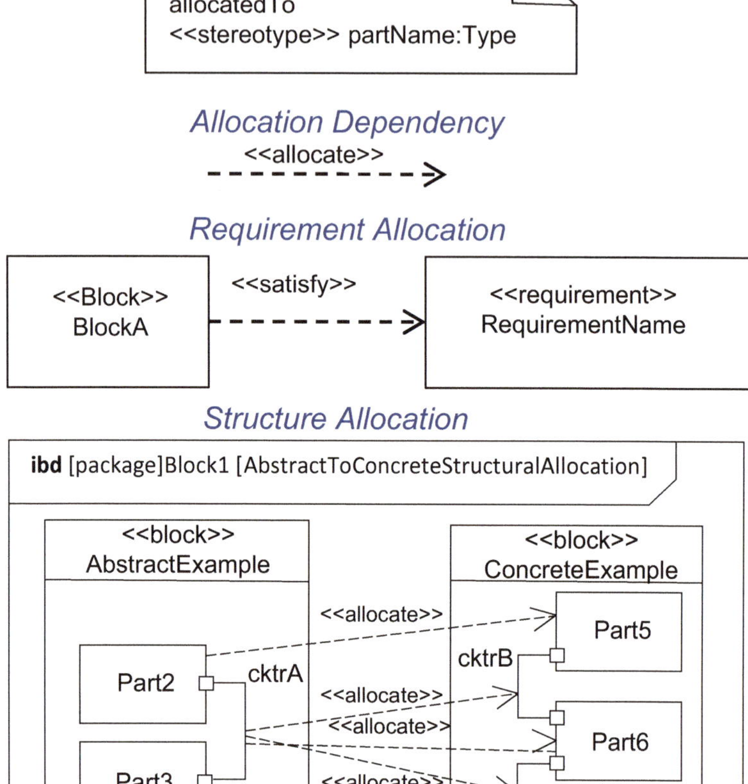

Fig. B.7 Allocations

– On a Block Definition Diagram, show an Allocation Dependency between an Activity and a Block.
– Flow Allocation—specifically maps flows in functional system representations to flows in structural system representations; can be shown using a Note containing Allocated To or Allocated From with the name of the Allocation source or target, anchored to the Object Flow or Connector involved in the Allocation; can also be shown in an `allocatedTo` or `allocatedFrom` compartment on a Block.
– Structure Allocation—used to separate "logical" and "physical" representations of a system.

- *Requirements Allocation*—used to associate a requirement with a structural or behavioral element; does NOT use `<<allocate>>`, but DOES use `<<satisfy>>`, `<<verify>>`, `<<refine>>`, `<<deriveReqt>>`, `<<copy>>`, `<<trace>>`, etc.

    NOTE: allocate arrow points to Requirement because Block design depends on the requirement.

    - Structure Allocation—used to separate "logical" and "physical" representations of a system.
    - Software-Hardware Allocation—UML Deployment diagram is replaced by the use of allocation in Internal Block Diagrams and Block Definition Diagrams.
    - Properties Allocation—allocate performance, SWAP-C, etc., to Property Values of components.

- *Profiles and Model Libraries*—mechanisms that allow metaclasses from existing metamodels to be extended to adapt them for different purposes; can also be used to restrict the language by identifying the subset of the base metamodel that is required for the specific domain.
- *Stereotype*—the primary mechanism used to create profiles to extend the metamodel; defined by extending a metaclass and then applied to the relevant model elements in the user model.

Four "Pillars" of SysML

- Structure
- Behavior
- Requirements
- Parametrics

Structure

Blocks

- Blocks are the basic units of structure in a SysML model; they have structural and behavioral features, including Properties, Values, Operations, and Constraints; they define the types of systems, system elements/components, flow items, etc.
- Blocks can be instantiated, with instances that share the Blocks' characteristics; this can be modeled using Generalization with the addition of instance-specific characteristics.

- A Block can have internal structure and can be decomposed into lower-level Blocks; these become Part Properties ("Parts") of the composite Block that owns them; a Part is typed by the Block of which it is an Instance.
- Block Properties:
  - *Part Properties (Parts)*—describe internal compositional structure of a Block; always typed by a Block and used to show an instance of that Block in the context of the composite Block that owns the Part; these create a structural hierarchy.
  - *Value Properties (Values)*—define quantitative characteristics of a Block.
  - *Reference Properties (References)*—indicate a non-compositional relationship between Blocks that may belong to separate compositional hierarchies; these create a logical hierarchy; e.g., a container may reference its contents, but each is in a distinct compositional hierarchy; these create a functional hierarchy.
  - *Constraint Properties ("Constraints")*—logical or quantitative condition which a Block must satisfy.
- Blocks can have allocated behavior and can be connected into composite structures.
- Blocks can have Ports that model points at which they interact with other Blocks, as well as connect Parts within a Block:
  - In SysML up to Version 1.2, there are two kinds:

    Standard Port—supports general interactions such as provided and required services; often used for communications between software components

    Flow Port (In, Out, or InOut)—supports quantities that flow in or out of a Block or Part, defined by Flow Items; often used for communications between processes
  - From Version 1.3, these change to:

    Full Port—a Part on a Block boundary that handles flow, non-flow, and behavioral interactions without exposing any of the Block's internal details; it's typed by a Block and used to model a physical part of the system.

    Proxy Port—not a Part but a window exposing the Block's inner workings; flow, non-flow, and behavioral features are provided or used by something within the Block; typed by an Interface Block and used when the interaction does not involve an actual physical part.

- Flow Ports and Flow Specifications are deprecated; Flow Properties and Flow Items remain and are used to type flow features in ports, along with direction.
- These Ports facilitate definition of physical connections.

• On diagrams, Blocks can optionally show characteristics in a variety of Compartments:

- *Parts*—lists the Parts defined by Part Properties, shown as:
  `partName:blockName[Multiplicity]`
  where `blockName` identifies the Block that types the Part.
- *References*—lists the references defined by Reference Properties.
- *Values*—lists the values defined by Value Properties, which are defined using Value Types.
- *Operations*—lists the Operations that have been defined for a Block, shown as:
  `operationName (parameter list) : returnType`
- *Receptions*—lists the Signal Receptions that have been defined for a Block, shown as:
  `<<signal>> receptionName (attributeList)`
  and the attribute list has entries as:
  `attributeName : attributeType`
- *Ports*—lists the Ports defined on a Block.
- *Constraints*—lists constraints and constraint properties owned by the block.
- *Requirements*—identifies requirements satisfied by the block.
- *Namespace*—shows blocks that are defined in the namespace of the block.
- *Structure*—shows connectors and other structure elements of the block.
- Other compartments, e.g., Attributes, may be supported by SysML tools.
- Block Diagrams can include Dependencies, Realizations, Generalizations, and other elements of UML.

Block Definition Diagram (BDD)

• Derived from the UML Class Diagram, used to declare Blocks and their compositional, logical, and Generalization/Inheritance relationships.
• Used to define Blocks for use in other diagrams.
• UML Classes and Objects become Blocks and their instances.
• Parts can be tied together by Connectors; a physical connection between instances of Parts is a Link; Connectors can be typed by Associations that can be defined by Association Blocks.

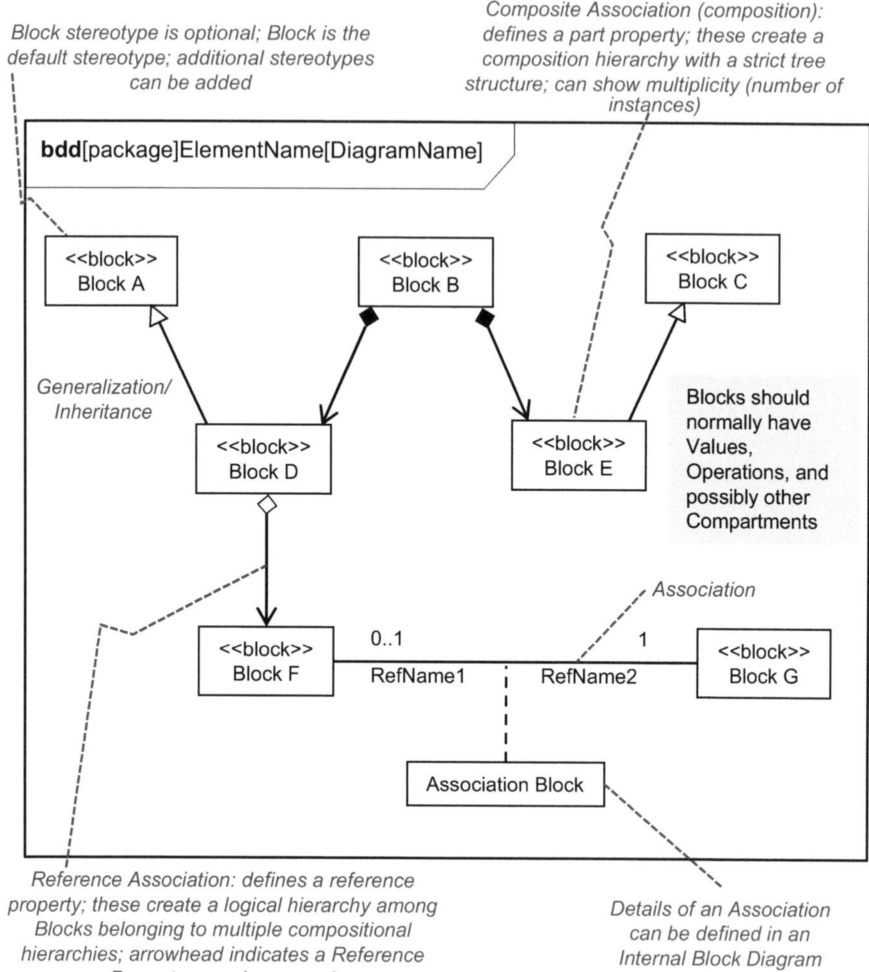

Fig. B.8 Block Definition Diagram

Internal Block Diagram (IBD)

- Derived from the UML Composite Structure Diagram, used to show connections and interactions among Parts within a Block; the diagram frame represents the composite Block.
- Ports on Blocks and Parts can be connected by Connectors or by Item Flows.
- Reference Properties are shown as Parts within a Block but with dashed boundaries; connections are the same as with Parts.

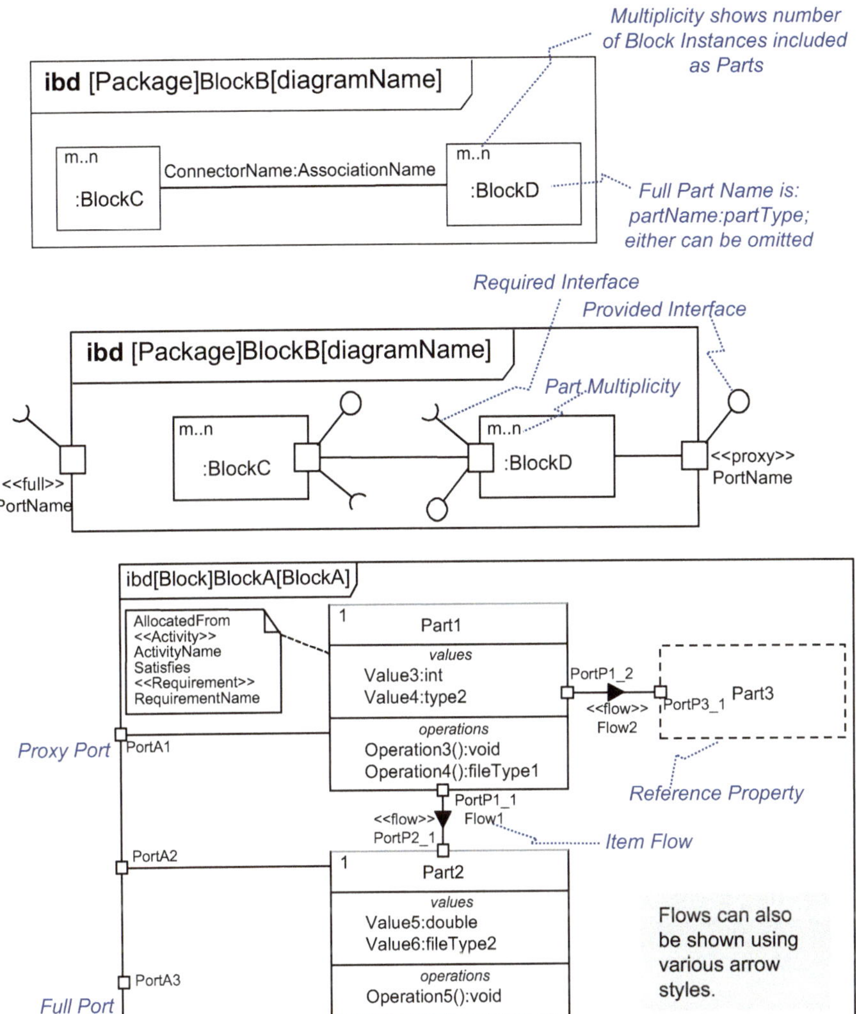

Fig. B.9 Internal Block Diagrams (IBDs)

Associations

- *Composite Association*—shows physical composition, part-whole relationships with a black diamond at the whole end; these declare subordinate Blocks as Parts of higher-level Blocks.
- *Directed Composition* (composition with navigability arrowhead)—shows the Part is uniquely owned by the higher-level Block.
- *Reference Association*—shows that the block on one end can be referenced by the block on the other end with a hollow diamond:

  - Can specify a reference property on the blocks on one or both ends.
  - Can be used to create logical aggregations of parts from various physical composition hierarchies that act together to create an entity (the owner of their reference properties) and accomplish some function.
  - Has an open arrowhead on the end toward the type of a reference property and away from the owner of the reference property (used when there is a reference property on only one end).
  - Can use a white diamond on the owner end (optional).
  - Reference Properties can be listed in a *references* compartment of a block.

- Associations can have Role Names when useful to clarify the function of the part or reference property.
- Associations can have multiplicities on one or both ends; default is 1.
- Associations typically have open arrowheads on the Part or type end.

Connectors

- Used on an IBD (not a BDD) to show connections between Parts of a Block
- Allows Parts to interact without specifying the interaction, which may be:
  - Flows of inputs/outputs among parts
  - Invocation of services
  - Exchange of messages
  - Constraints among properties of parts
- Can connect Ports
- Can be depicted using item flows
- Can have multiplicities defining the number of instances that can be connected
- Can be typed by an Association to show more detailed characteristics, including an Association Block whose internals are shown on a separate IBD

Flows

- Model exchanges of tangible quantities such as liquid, power, or even a message stream over Connectors between Blocks, Parts, and the external environment.
- Can be uni- or bidirectional.
- See Flow Properties and Item Flows below.

Ports

- Interaction points owned by Blocks that allow item flows, service invocations, and other interactions; annotated using small squares on the Block boundary with Port names, stereotypes, associated Interfaces, etc.

- Current version of SysML (ver. 1.3 and higher; current version is 1.4) replaces the original Standard and Flow Ports with Full and Proxy Ports:
  - *Full Port*—a Part Property of the owning Block; it handles flow, non-flow, and behavioral interactions without exposing any of the Block's internal details; typed by a Block, of which it is an instance, and used to model physical resources of the Block that create and support the interaction point; defining Block can include one or more Interfaces.
  - *Proxy Port*—associated with (delegated to) one or more Parts within the Block that provide the supporting resources; typed by an Interface Block; Interface can be provided, required, or both; handles all the interactions that the implementing Part or Parts can support.
- Flow Properties and Item Flows are retained in ver. 1.3.
  - *Flow Property* of a Port describes *possible* Item Flows between a Block and its environments, including data, material, energy, etc.
  - *Item Flow* describes an *actual* flow between Blocks or Parts; can be allocated from object nodes in activity diagrams or signals from state machines.

Value Properties

- Model the quantitative characteristics of a block; represent values not entities:
  - Can have initial values and probability distributions of values.
  - Based on Value Types that describe the values for quantities, including data structures and allowed ranges; types include:

    Primitives—e.g., integer, string, Boolean
    Enumeration—sets of named values (literals)
    Structured—data structures with multiple elements, each having a value property
- Value types have units and dimensions.
- Value properties are shown in a *values* compartment of a block.
- Show quantities with probability distributions as:

```
<<DistributionName>>[parameter string]PropertyName:TypeName
```

- Show Value Properties in a values compartment of a Block.

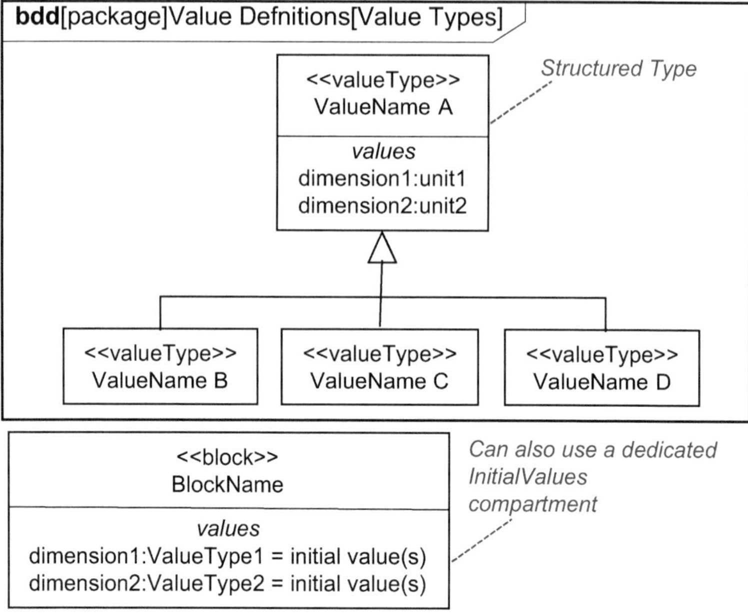

Fig. B.10 Value Definitions and Properties

Behavior

- SysML models behavior using:
  - Activities—includes additions to the basic Activity Diagram from UML; shows the flow of Actions and Objects that make up an Activity.
  - State Machines—Harel Statechart shows stateful, event-driven behavior.
  - Interactions—Sequence Diagrams show time-ordered interactions among model elements.
- Behavioral features of a Block define the requests to which the Block can respond:
  - *Receptions*—associated with Signals (messages with attributes) that transmit asynchronous requests; Signal is defined by a rectangle with the `<<signal>>` stereotype (or keyword) and a compartment for attributes, annotated as:

    ```
    attributeName:attributeType[multiplicity]
    ```

  - *Operations*—associated with synchronous requests; have parameters that define arguments passed in and out and a type for any returned value; Operation Signature is:

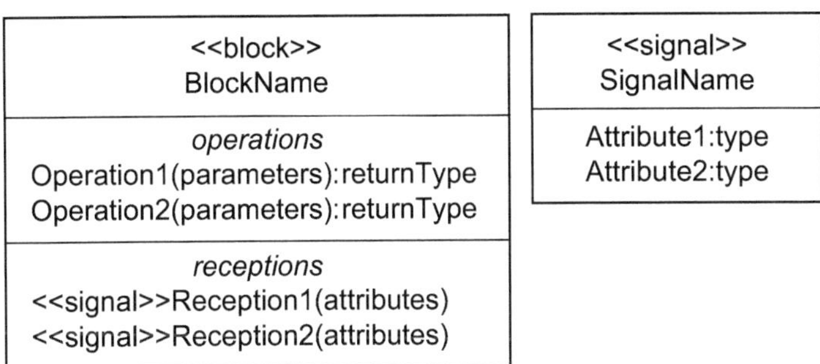

Fig. B.11 Block Behaviors

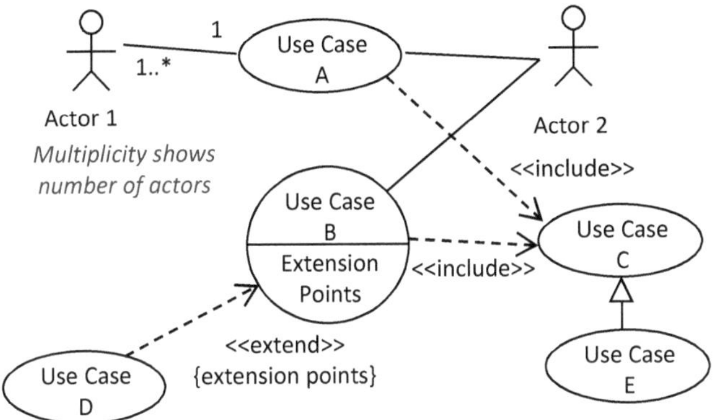

Fig. B.12 Use Cases

```
operationName(parameterList):returnType
```

and Reception is annotated as:

```
<<signal>>ReceptionName(attributes)
```

Use Cases

- Identical to UML; used to model primary behaviors or functions with associated Actors.
- Use Case specification defines behavior, pre- and postconditions, and other key aspects.
- Use Cases may be elaborated via other behavioral diagrams and scenarios, especially Activity Diagrams.
- `<<extend>>` and `<<include>>` Dependencies model optional and common behaviors; `<<include>>` can be used to decompose higher- to lower-level Use Cases.

Activity

- Emphasizes the inputs, outputs, sequences, and conditions for coordinating other behaviors (see UML 2 for details of Activity modeling).
- Extends UML to allow activities to be stopped while executing, as well as started.
- Control Value—an input or output of a control operator, which is how control acts as data.
- Continuous Systems—SysML allows both continuous and discrete flows of material, energy, and information; objects can update and discard values to maintain current values and handle transients.
- Activities can be modeled as Blocks in Block Definition Diagrams with the `<<Activity>>` stereotype can have composition relationships when one activity invokes other activities.
- Instantiating/destroying an instance of an Activity causes the Activity to initiate/terminate.

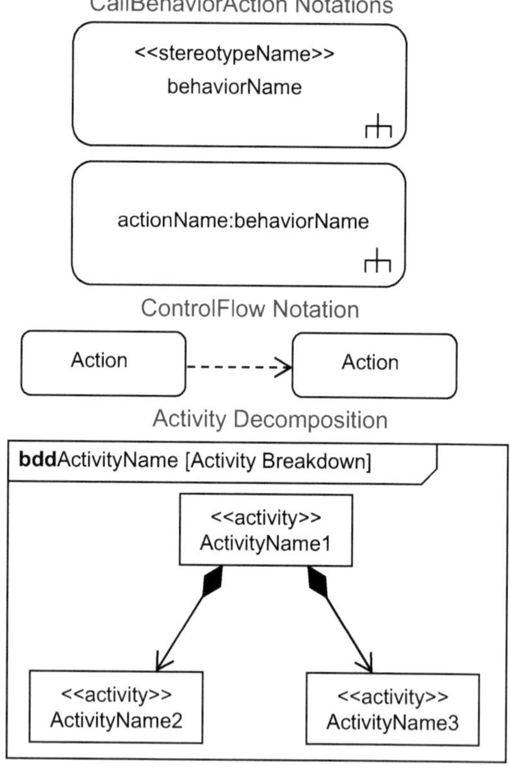

Fig. B.13 Activity Notations

Activity Diagram

- Shows the flow of activities; see UML 2 for basic activity diagram syntax.
- SysML also allows the use of Enhanced Functional Flow Block Diagram (EFFBD).
- Swim lanes ("Activity Partitions") and allocations tie activities to other model elements, especially to structural units that perform Actions and own data.
- Activity Parameter Nodes provide a clear way to input and output values.
- Alternative Object Flow notation.

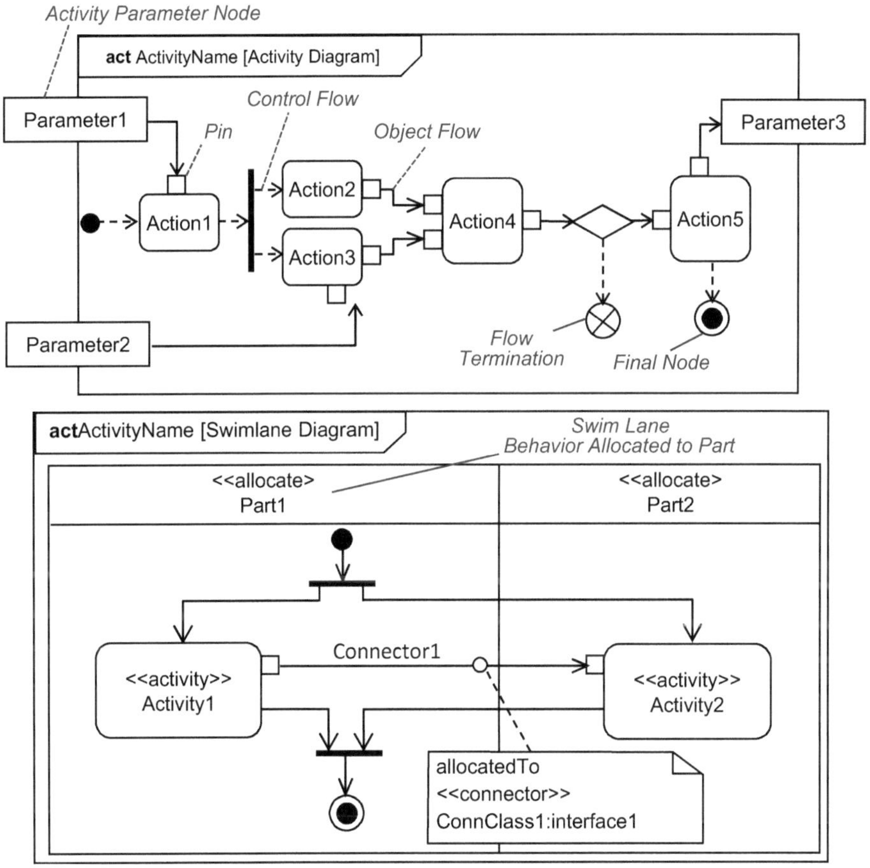

Fig. B.14 Activity Diagrams

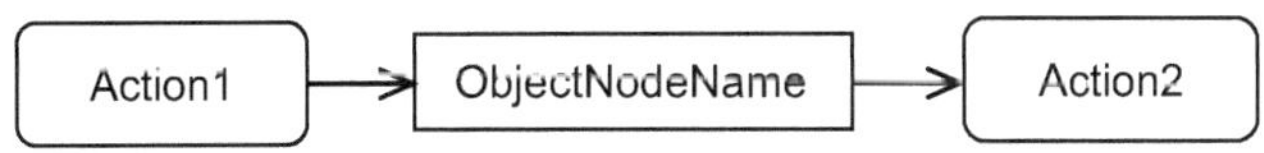

Fig. B.15 Alternative Object Flow Notation

Stereotypes Defined in the SysML Activity Package

- `<<Continuous>>`—type of flow rate with time increment approaching zero.
- `<<Control Operator>>`—a behavior that is intended to represent an arbitrarily complex logical operator that can be used to enable and disable other actions.
- `<<Discrete>>`—flow rate where the time between items is non-zero.
- `<<NoBuffer>>`—arriving tokens are discarded if they are refused.
- `<<Overwrite>>`—tokens arriving at a full node replace those already there.
- `<<Optional>>`—lower multiplicity limit of a parameter with this stereotype must be zero.
- `<<Probability>>`—provides an expression for the probability that an edge from a decision or object node will be traversed.
- `<<Rate>>`—specifies the number of items and values that traverse an edge per unit time.

State Machine

- Shows the life cycle of a block.
- Models discrete, typically event-based behavior.
- See UML 2 for basic diagraming.

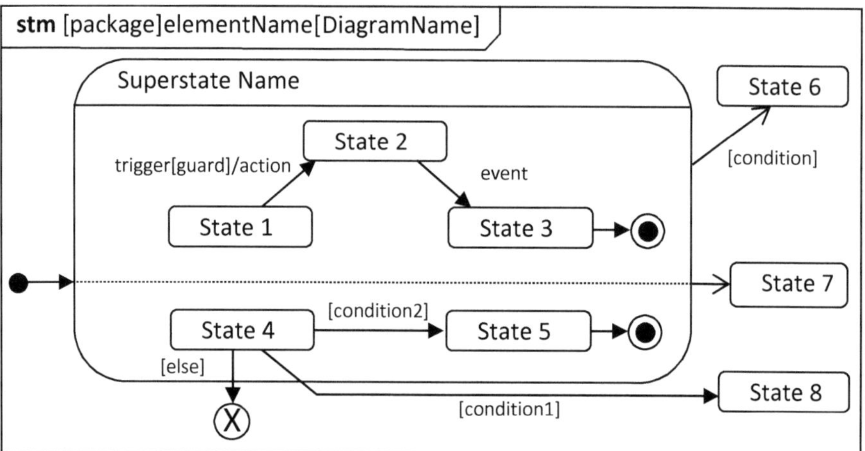

Fig. B.16 State Machine Diagram

Interaction

- Sequence Diagrams describe interactions among entities.
- Communication and Interaction Overview Diagrams (UML 2) are excluded from SysML.
- *Black Box Interaction*—composes reference frames to show high-level interactions in complex scenarios.
- *White Box Interaction*—see UML 2 Sequence Diagram.

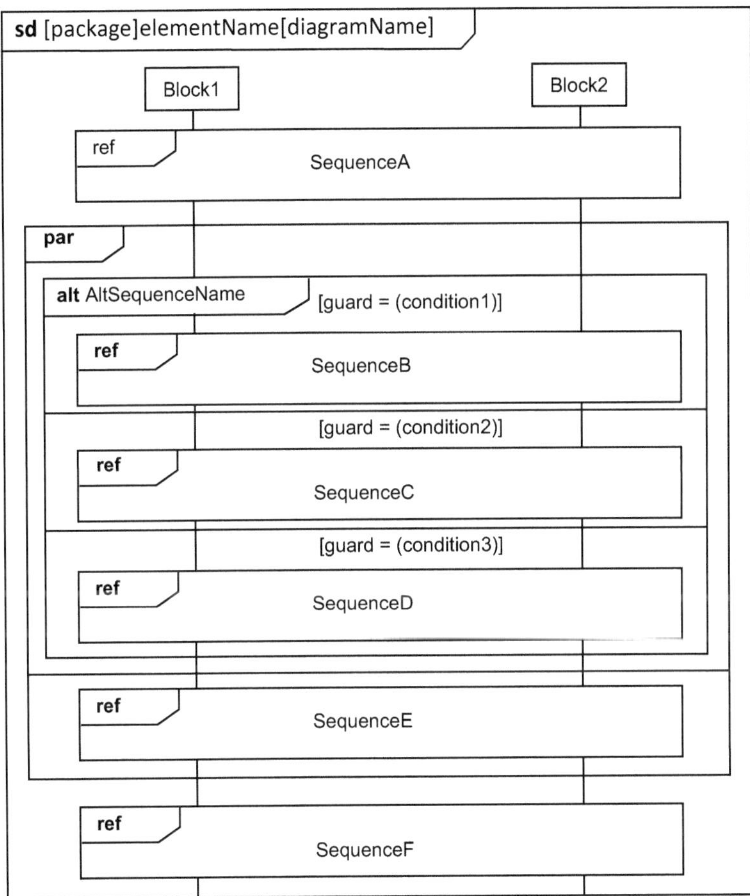

Fig. B.17 Sequence Diagram modeling a "Black Box" Interaction Behavior

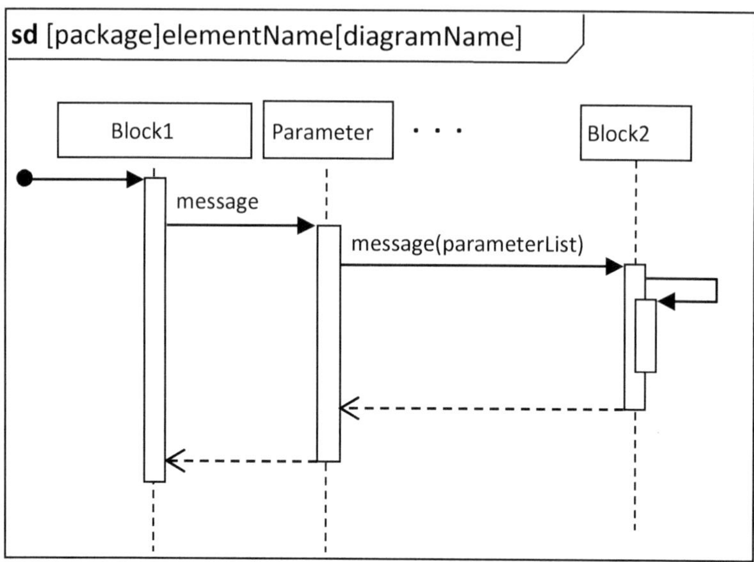

Fig. B.18 Sequence Diagram modeling a "White Box" Behavior

Requirements

- A requirement is a capability or condition that must (or should) be satisfied; defined in a Block with the <<requirement>> stereotype.
- May specify a function that a system must perform or a performance condition a system must achieve.
- Composite Requirement—can contain sub-requirements in terms of a requirements hierarchy, specified using the UML namespace containment mechanism.
- Slave Requirement—a requirement whose text property is a read-only copy of the text property of a master requirement; permits requirements reuse.
- Derive Requirement Relationship—relates a derived requirement to its source requirement; annotated as a <<deriveReqt>> Dependency.
- Satisfy Relationship—describes how a design or implementation model satisfies one or more requirements; annotated as a <<satisfy>> Dependency.
- Verify Relationship—defines how a test case verifies a requirement; annotated as a <<verify>> Dependency.
- Refine Requirement Relationship—can be used to describe how a model element or set of elements can be used to further refine a requirement; annotated as a <<refine>> Dependency.
- Generic Trace Requirement—relationship provides a general-purpose relationship between a requirement and any other model element; annotated as a <<trace>> Dependency.

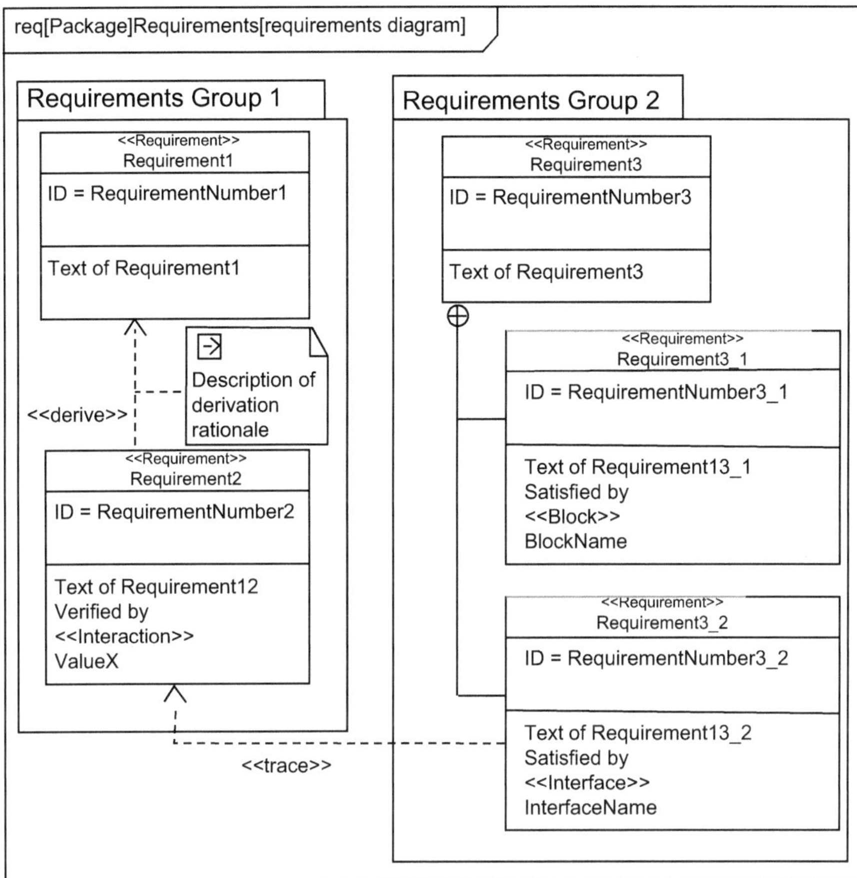

Fig. B.19 Requirements Diagram

- Copy Relationship—a `<<copy>>` Dependency between a supplier requirement and a client requirement that specifies that the text of the client requirement is a read-only copy of the text of the text of the supplier requirement.
- Requirements can be attached to any diagram/element.
- Requirements can be expressed in diagrams or in tables showing requirements satisfaction.

Parametrics

- A mechanism to model constraints/relations (equations) among value properties
- Integrates engineering analysis (performance, reliability, etc.) with SysML Blocks

Constraint Block

- Captures equations in formal or informal language.
- Defined by the `<<constraint>>` keyword applied to a Block definition.
- Includes the constraint (e.g., an equation) and its parameters.

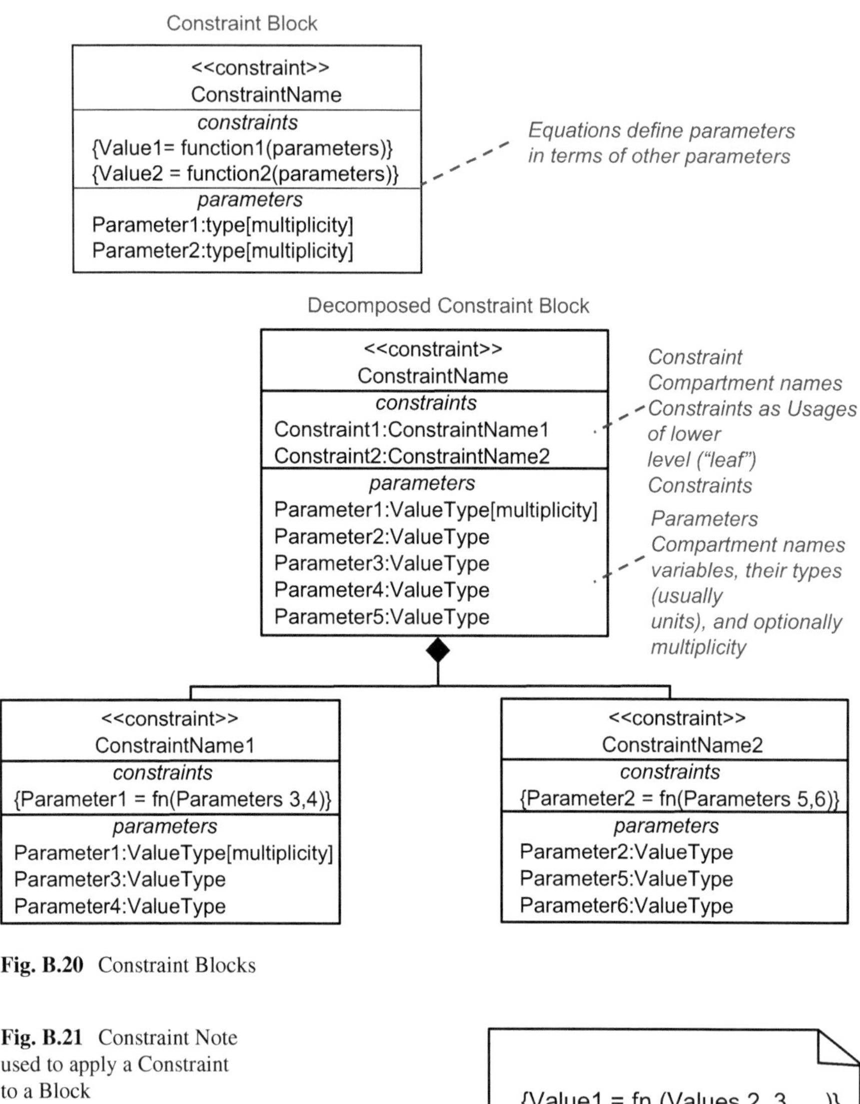

Fig. B.20 Constraint Blocks

Fig. B.21 Constraint Note used to apply a Constraint to a Block

{Value1 = fn (Values 2, 3, …)}

Note is anchored to the Block it constrains

- Can be decomposed into lower-level Constraint Blocks using Composition Associations in a BDD.
- Constraint expression is between curly brackets ({ }).
- Can be used on Parametric diagrams or IBDs to constrain the properties of another Block.
- Parameter Values can be used to specify system state.
- Can be used to specify a function used to perform trades with Parametric diagram.
- A usage of a Constraint Block is called a Constraint Property.
- The frame of a Parametric diagram represents a Constraint Block for which it shows bindings among parameters of its Constraint Properties using binding connectors.
- Constraint Blocks can constrain the Item Properties of Item Flows by using a Parametric diagram with constraints that relate the items.

Constraints on Blocks

- Created either in a Constraints compartment or by attaching a Constraint Note.
- Curly brackets indicate a Constraint relation.

Parametric Diagram

- Creates a system of equations by "wiring up" constraints with values created and used in the system.
- The frame represents a Block or Constraint Block.

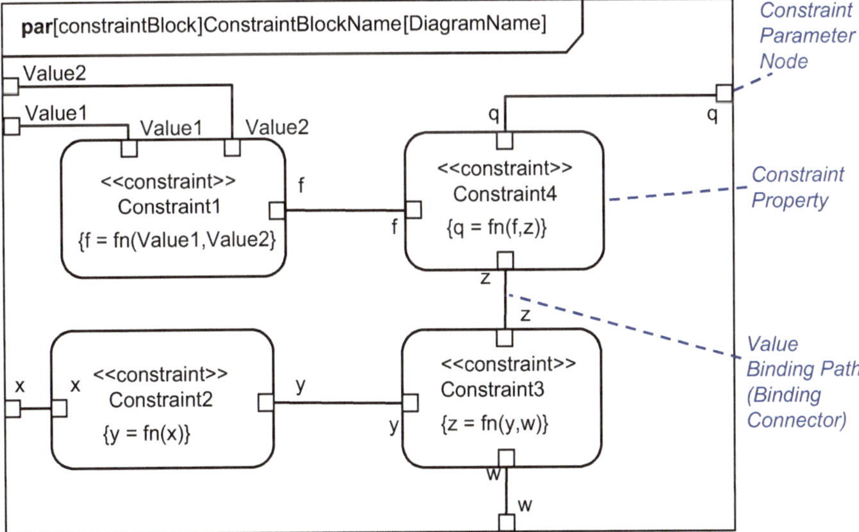

Fig. B.22 Parametric Diagram

- A Value Property can be included in Parametric Diagrams as a rectangle containing the name, type, and multiplicity and then bound to the appropriate Constraint Parameter Node with a Value Binding Path.
- Constraints involving time-varying parameters can be modeled by including a Time Value Property in the Parametric Diagram.
- Parametric Diagrams can be created for Item Flows that bind flow parameters to Constraint parameters.

Relationships Among Diagrams

Examples of connecting relationships among the four SysML "Pillar" diagrams.

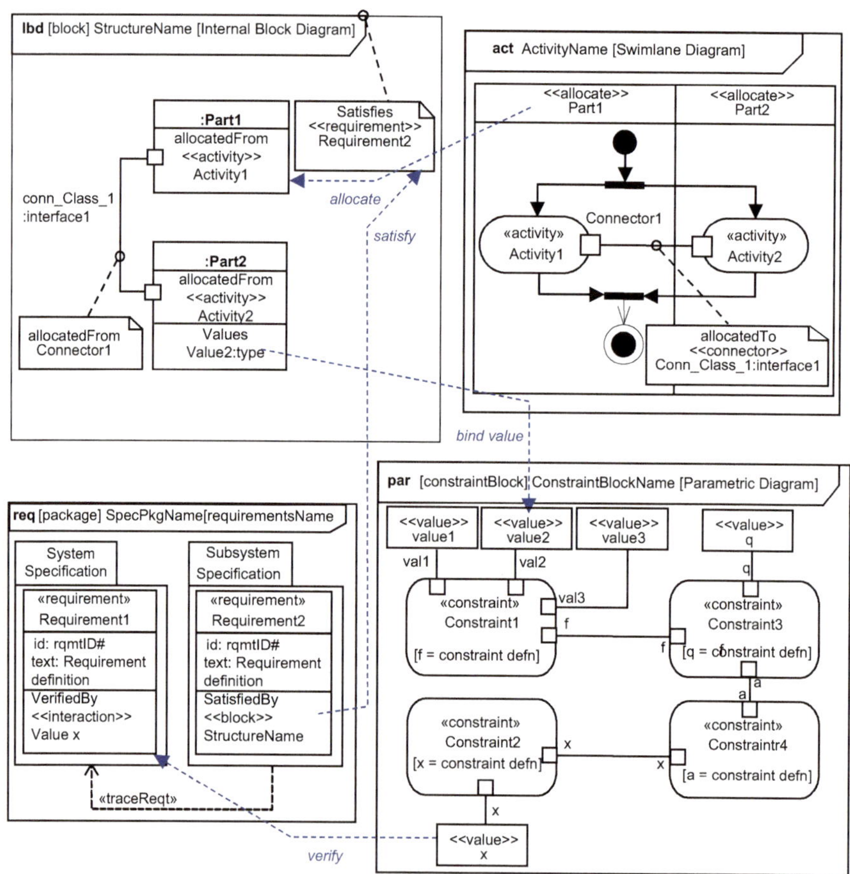

Fig. B.23 Linkages within a SysML Model

Appendix C: System Architecture Example: E-X Airborne Multi-Sensor Platform

Introduction

This appendix presents example content of a system architecture for a notional system called the E-X Airborne Sensor Platform, developed using the Model-Based System Architecture Process (MBSAP) methodology that is presented in the main body of this book. Excerpts from this architecture are used throughout the book to illustrate various points, but those artifacts are collected here to provide context and additional detail. Although this architecture is far from complete and does not pretend to represent a real system, these diagrams, text, and other materials are representative of a real system architecture and are intended to be used as format samples. MBSAP stresses the importance of documenting the elements of an architecture model, and this appendix has a number of examples of specifications for model elements. However, due to space limitations, many model elements are given only brief summary descriptions, while in a real architecture they would have complete specifications. The Physical Viewpoint, representing a point design that implements the architecture at a specific time, is limited to summaries of the kinds of content and architecture analyses that would be required in a real system since an impractical volume of physical detail would be required to support actual product trade studies, standards selections, and other point design content.

This appendix follows the layout of a typical system architecture report. The E-X architecture was developed using the System Modeling Language (SysML) profile of the Unified Modeling Language (UML). The architecture structure follows the Viewpoints and Perspectives of MBSAP, but, as in a real-world project, these artifacts were developed iteratively and not necessarily in the sequence of this report. In particular, key definitional diagrams were completed first so that their elements could be consistently applied in other diagrams. Attachment 1 describes the notional E-X system and summarizes the MBSAP process and products. It provides an example of an Architecture Summary and Overview document as described in Chap. 3. Attachment 2 illustrates an Integrated Dictionary, which is an essential document for a complex system architecture. Note that some of the

J. M. Borky, T. H. Bradley, *Effective Model-Based Systems Engineering*,
https://doi.org/10.1007/978-3-319-95669-5

diagrams in this document use arrows that are variants of the graphics conventions of the SysML Language Standard.

Model Organization

This appendix follows the model structure recommended by MBSAP that has proved effective in a number of programs. The organization of an architecture model has a great impact on its usability. Depending on the modeling tool used in a given architecture development, there will be variations, but the fundamental structure is the same in virtually any situation. The key points of an effective model organization include the following.

Top-Level Structure: The top-level structure of an architecture model, which is normally established in the browser of the selected tool, includes the following directories or packages. These primarily reflect the Perspectives within the Operational, Logical/Functional, and Physical Viewpoints (OV, LV, and PV) of the MBSAP methodology.

(a) *Use Cases*—contains Use Cases, Actors, and associated diagrams; this includes Activity Diagrams that model the scenarios of individual Use Cases, which are created under their respective Use Cases in the model.
(b) *Structure*—contains Block Diagrams; Blocks are progressively decomposed in more detailed Block Diagrams created under the respective Blocks in the model. Similarly, behaviors that are limited to a single Block or other element (referred to as Block Behaviors in the SysML language standard) are created under the respective Blocks; these are typically State Machine Diagrams showing stateful behavior of individual Blocks.
(c) *Behavior*—contains Sequence Diagrams, State Machine Diagrams, and occasionally Activity Diagrams that model higher-level behaviors involving more than one Block.
(d) *Data*—contains Block Definition Diagrams for the Conceptual and Logical Data Models. Physical Data Model details, e.g., relational database schemas, when available, may be linked to this directory.
(e) *Context*—contains material on the external environment and interactions that doesn't fit in the other Perspectives.
(f) *General Information*—contains the Architecture Summary and Overview, Integrated Dictionary, and any other related material, typically as linked or archived files.
(g) *Requirements*—contains Requirements Diagrams and related documents when these are created as part of the architecture model.

The most important benefit of this organization is that it facilitates browsing in the model to find information of interest, typically by navigating down from a Domain or Block in the Structure package or from an Activity in the Behavior package to the desired level of decomposition.

Operational Viewpoint: The OV is assembled from model content drawn from the Requirements package, Use Case package, top-level diagrams from the Structural and Behavioral package, the Conceptual Data Model (CDM) from the Data package, and much of the content of the Context and General Information Packages.

Logical/Functional Viewpoint (LV): The LV decomposes the high-level elements of the OV to create a functional architecture. The Logical Data Model (LDM) defines the System Data Blocks with inheritance from the Conceptual Data Model.

Physical Viewpoints: The PV is created primarily by adding physical (point design) detail to the Blocks of the LV. The Physical Data Model (PDM) similarly defines the physical details of data storage and management for the Blocks of the LDM.

Focused Viewpoints: These may employ model content directly or may involve other graphics, text, tables, etc., depending on the stakeholder needs being addressed. An example is that a ports and protocols matrix for the Local Area Network, an important element of a Networking Viewpoint, could be built as a spreadsheet and imported or linked to the specification for the LAN Block.

Requirements

The starting point for the E-X architecture is a requirements baseline that has been derived from available sources, captured in a structure such as a database or model, coordinated with system stakeholders, and placed under change control. Both functional and nonfunctional requirements are included. This appendix assumes that E-X requirements are modeled in Requirements Diagrams; they may also be documented in a separate requirements management tool.

Figure C.1 is an example of a Requirements Diagram for the E-X that covers top-level functional requirements. Figure C.2 shows how requirements are decomposed, in this case for the E-X Multimode Radar. In a full system architecture, this is carried to the necessary level of detail to allow allocation, satisfaction, and other linkages of requirements to system structure and behavior. The Requirements Blocks have "Allocated to" compartments to show one way of linking requirements to system elements. Figure C.2 illustrates requirements decomposition, specifically for the E-X Multimode Radar. Block specifications or linkages to external requirements documents allow the full detail of each requirement to be included.

Under MBSAP, nonfunctional requirements are dealt with using Quality Attributes (QAts). Table C.1 shows representative high-level QAts that could be applied to the E-X.

Operational Viewpoint Artifacts

The OV is largely concerned with transforming system requirements into an architectural context and describing the overall concept of the system or enterprise that will satisfy those requirements.

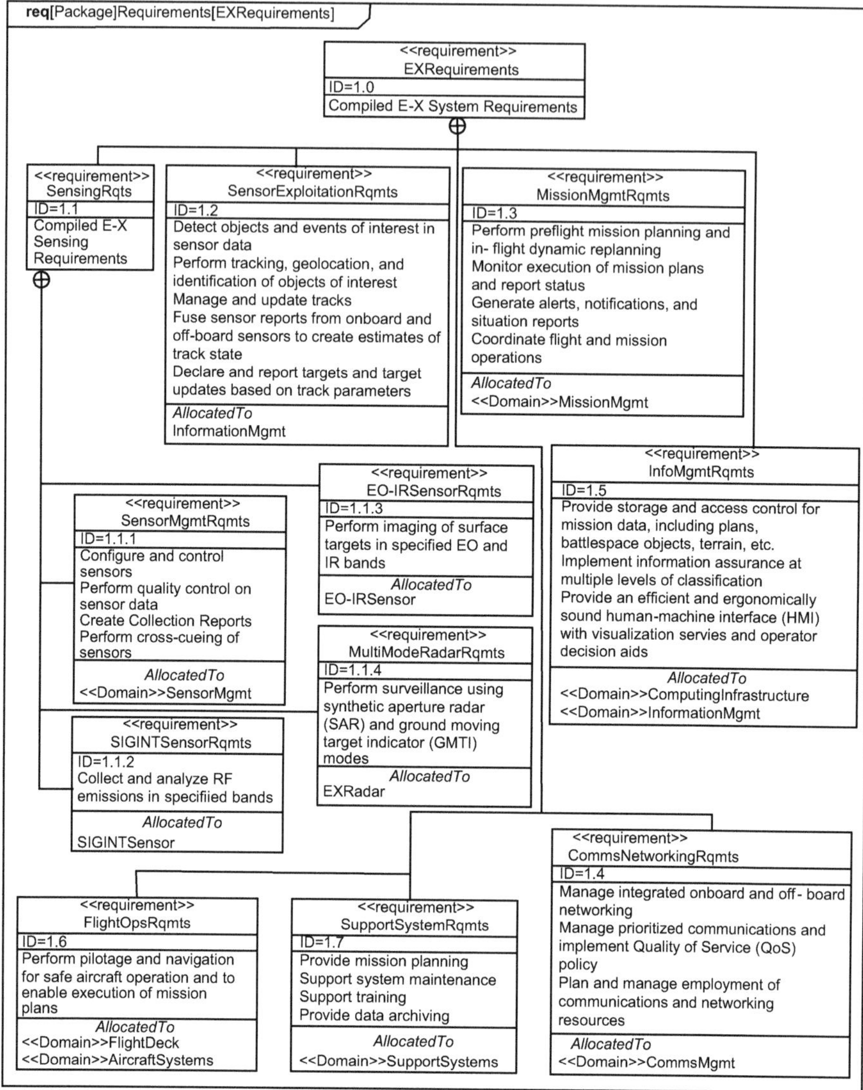

Fig. C.1 High-Level Requirements Diagram for the E-X

Structural Perspective

Figure C.3 defines the Domains that represent the top-level partitioning of the E-X architecture and show Associations with external Actors, primarily User Roles using a Block Definition Diagram (BDD). Domains collect the functions and resources associated with particular User Roles. Diagrams are followed by example specifications for Domains, Subdomains, Use Cases, User Roles, and other

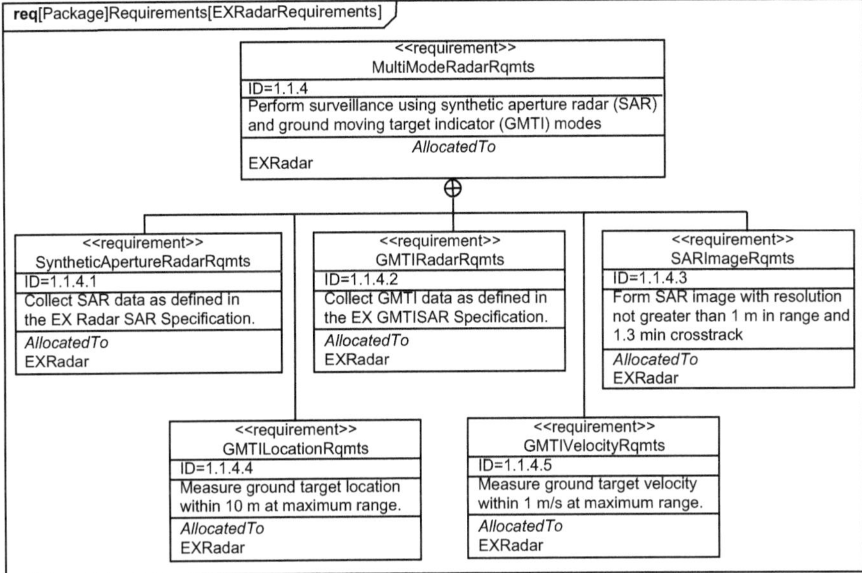

Fig. C.2 Requirements Decomposition for the E-X Multimode Radar

Table C.1 Quality Attributes for the E-X

Quality attribute	Definition
Openness	System can be integrated and evolved using products from multiple suppliers; system conforms to standards in the operational environment
Modularity	System can be integrated and upgraded at a modular level; system can deal with changes in design and operations at a modular level
Operational Suitability	System can perform its missions and interoperate with external systems
Security	System can be accredited to collect, process, and disseminate sensitive data
Affordability	System can be procured in sufficient numbers and kept effective and available over its lifetime within available budgets

elements of the OV. Figure C.3 uses several graphical conventions that are employed throughout this architecture example. The Block representing the E-X system has a patterned background to distinguish it from subordinate Blocks. Several Domains have gray backgrounds, denoting the fact that they exhibit real-time behavior. Actors modeling off-board User Roles also have gray shading. User Roles are defined in Fig. C.5 and associated with Domains in Fig. C.3.

The following are examples of model element specifications for the Blocks in Fig. C.3.

System: E-X

Description: The E-X is a manned airborne multi-sensor platform that collects information and creates and disseminates Intelligence, Surveillance, and Reconnaissance (ISR) products in coordination with a Ground Station and various supported and supporting organizations. There are obvious military applications of a system like this, and military terminology such as ISR is used for consistency. However, aerial imaging and other sensing on both manned and unmanned aircraft are growing in importance in civilian applications ranging from agriculture to fire fighting, so this example is relevant to many civil activities.

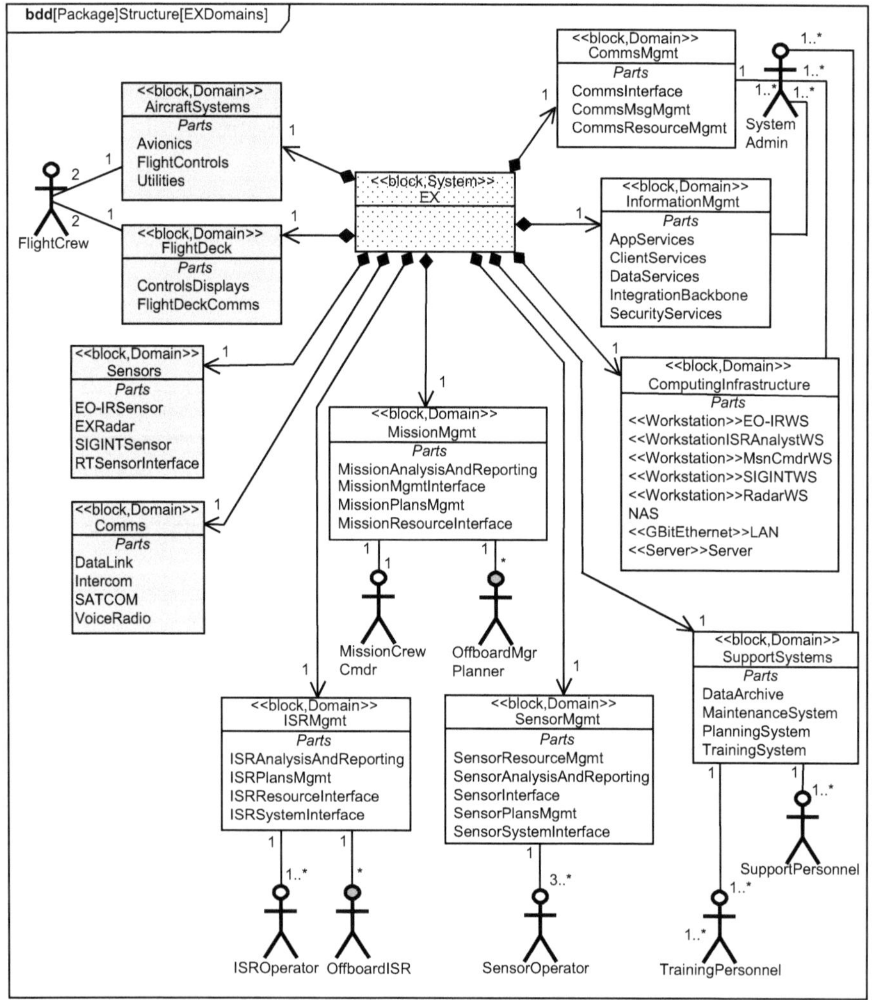

Fig. C.3 E-X Domains

Domain: Aircraft Systems

General: This Domain contains the systems, equipment, and resources that are part of the E-X aircraft and interact with Mission Equipment and the Mission Crew.
Owner: aircraft manufacturer.
Definitions: see Integrated Dictionary.
Operations:

- Flight Controls
- Propulsion Controls
- Navigation
- Aircraft Communications
- Prime Power Generation and Control
- Airframe Equipment Control
- Aircrew Life Support

Data:

- Flight and Aircraft Data
- Control Data

Interfaces:

- Flight Crew (Flight Deck)
- Mission Equipment

Allocated Requirements: *1.6 Flight Operations Requirements* associated with the aircraft and its systems, e.g., range, airspeed, altitude, and endurance.

Domain: Flight Deck

General: This Domain contains the controls, displays, and interfaces that the Flight Crew employs to conduct flight operations and coordinate with the ISR Crew.
Owner: aircraft manufacturer.
Definitions: see integrated dictionary.
Operations:

- Flight Controls
- Propulsion Controls
- Navigation
- Aircraft Communications
- Prime Power Control
- Airframe Equipment Control

Data:

- Flight Plan and Supporting Data
- Communications Plan

- Flight and Aircraft Data
- Control Data

Interfaces:

- Aircraft Systems
- Mission Equipment

Allocated Requirements: *1.6 Flight Operations Requirements* associated with the Flight Crew controls and displays and interfaces to Mission Equipment.

Domain: Mission Management

General: This Domain contains the resources and functions used by the Mission Crew Commander and any supporting staff to exercise overall control of mission execution.
Owner: Mission Equipment supplier under direction from the E-X organization
Definitions: see integrated dictionary
Operations:

- Mission Plan monitoring, execution, and replanning
- Coordination with Flight Crew, Ground Environment, and other external organizations
- Supervision of ISR operations
- Review, approval, and dissemination of ISR products

Data:

- Mission Plan and Supporting Data
- ISR Reports
- Mission Plan Revision Data

Interfaces:

- ISR Management
- Flight Deck
- Information Management
- Off-board ISR Manager/Planner

Allocated Requirements: *1.3 Mission Management Requirements* associated with the overall execution, monitoring, replanning, and reporting of ISR mission operations, including coordination with Ground Environment and other external organizations.

Domain: ISR Management

General: This Domain contains the resources and functions used by the ISR Operators and any supporting staff to exercise overall control of mission execution.

Owner: Mission Equipment supplier under direction from the E-X organization
Definitions: see integrated dictionary.
Operations:

- Collection Plan monitoring, execution, and replanning
- Coordination with Sensor Operators, Mission Crew Commander, and Off-board ISR personnel
- Supervision of sensor operations
- Creation and review of ISR products

Data:

- Mission Plan and Supporting Data
- ISR Task
- ISR Report
- Supporting Data

Interfaces:

- Mission Management
- Sensor Management
- Information Management
- Off-board ISR Personnel

Allocated Requirements: *1.2 Sensor Exploitation Requirements* associated with analysis of sensor data and Collection Reports, creation of ISR Reports, development of Sensor Tasks, coordination with Off-board ISR personnel, and other ISR Management activities.

Domain: Sensor Management

General: This Domain contains the resources and functions used by the Sensor Operators to configure and control sensors and process collected data.
Owner: Mission Equipment supplier under direction from the E-X organization
Definitions: see integrated dictionary.
Operations:

- Sensor Task execution and monitoring
- Resource Management of sensors
- Sensor Data quality control
- Creation and review of Collection Reports

Data:

- Mission Plan and Supporting Plan Data
- Sensor Task
- Collection Report
- Supporting Sensor Data

Interfaces:

- ISR Management
- Sensors
- Information Management

Allocated Requirements: *1.1.1 Sensor Management Requirements* associated with control of sensors and reporting of collected data.

Domain: Sensors

General: This Domain contains the E-X sensors and the interface between sensors and Sensor Management. Physically, the interface is implemented via the Computing Infrastructure LAN.
Owner: Sensor supplier(s) under direction of Mission System Integrator.
Definitions: see integrated dictionary.
Operations:

- Multimode Radar Sensing
- EO/IR Imaging/Video Sensing
- SIGINT Sensing

Data:

- Sensor Task
- Sensor Data
- Sensor Status

Interfaces:

- Sensor Management
- Computing Infrastructure

Allocated Requirements:

- *1.1.2 SIGINT Sensor Requirements*
- *1.1.3 EO/IR Sensor Requirements*
- *1.1.4 Multimode Radar Requirements*

Domain: Information Management

General: This Domain contains the resources and functions that integrate and support E-X information processes, including access to onboard and off-board data, data storage and retrieval, access to external communications, and security services. It is implemented as a layered service-oriented architecture (SOA) that hosts mission applications from the primary mission domains and provides messaging and file sharing among them.

Owner: Mission System Integrator.
Definitions: see integrated dictionary.
Operations:

- Internal data access, retrieval, and storage
- External data and communications access
- Messaging
- Mission applications support
- System services
- Security services
- Data: all Mission Systems data

Interfaces:

- Mission Management
- ISR Management
- Sensor Management
- Computing Infrastructure
- Communications Management
- Mission Applications

Allocated Requirements: *1.5 Information Management Requirements*

Domain: Computing Infrastructure

General: This Domain contains the computers, networks, and storage that execute Mission System software and allow User interaction with the system. It forms the lowest tier of the layered architecture.
Owner: Mission System Integrator.
Definitions: see integrated dictionary.
Operations:

- User workstation functions
- Server functions
- Local Area Network (LAN) functions
- Network-Attached Storage (NAS)

Data: all Mission Systems data.
Interfaces: Information Management.
Allocated Requirements:

- Requirements are primarily derived from the functionality to support E-X information processes.
- Some requirements may be set by the E-X organization, e.g., required margins and storage capacity, as part of *1.5 Information Management Requirements*.

Domain: Communications Management

General: This Domain contains the resources and functions that control external communications and networking and provide the interface between the onboard information processes and the external communications environment.
Owner: Mission System Integrator.
Definitions: see integrated dictionary.
Operations:

- Monitor and control communications devices.
- Process external communications traffic (voice and data).
- Process external network interactions, including Quality of Service (QoS) functions.
- Control communications security functions.

Data:

- Message (external)
- Report (external)

Interfaces:

- Information Management
- External communications and networks (via communication devices)

Allocated Requirements: *1.5 Information Management Requirements*, associated with control and employment of external communications.

Domain: Communications

General: This Domain contains the radios and other resources and functions that implement external communications and networking.
Owner: Communications equipment provider under the direction of the Mission System Integrator and the E-X Organization.
Definitions: see integrated dictionary.
Operations:

- Satellite communications (voice and data)
- Line of Sight (LoS) communications (voice and data)

Data: Message (external, including data links)
Interfaces:

- Communications Management
- External communications and networks

Allocated Requirements: *1.5 Information Management Requirements*, associated with external communications functionality, performance, protocols and standards, etc.

Domain: Support Systems

General: This Domain contains the resources and functions used to plan and support E-X operations.
Owner: Support System Suppliers and the E-X Organization (some systems may be customer-provided).
Definitions: see integrated dictionary.
Operations:

- Mission Planning
- System Maintenance and Servicing
- Training
- Data Archiving

Data:

- All Mission System Data
- Training Data
- Mission Plan
- Maintenance

Interfaces:

- Flight Deck (data transfer)
- Mission Management (data transfer)
- Support Personnel
- System Administrator
- Training Personnel

Allocated Requirements: *1.7 Support System Requirements*

Figure C.4 is the corresponding Domain Interaction Diagram using an Internal Block Diagram (IBD). Ports and Flows show the primary internal and external interactions of the system. Figure C.5 defines User Roles representing onboard and off-board personnel who operate and support the E-X system. These are used in multiple diagrams. General or abstract User Roles are specialized by Inheritance to actual User Roles. Specifications for these Roles follow. Visibilities and return types for Values and Operations are indicated.

Actor: Mission Crew

General: This is an abstract Actor that models all User Roles associated with flying the E-X aircraft and operating its systems.
Qualifications:

- General airmanship and survival training current
- General E-X crew qualification current

Attributes:

Aircrew Qualification Status	Public	String
Clearance	Public	String
Flight Status	Public	String
Grade	Public	String
ID Number	Public	Int
Unit of Assignment	Public	String

Operations:

Accomplish Training	Public	Void
Debrief Mission	Public	Void
Perform Aerial Mission	Public	Void
Perform Mission Planning	Public	Mission Plan

Actor: ISR Crew

General: This is an `<<abstract>>` User Role that distinguishes the crew members who execute ISR functions from the Flight Crew.
Actor: Mission Crew Commander.
General: This Actor models the senior Mission Crew member responsible for mission execution, including replanning in response to operational tasks and circumstances, direction of the Mission Crew, and coordination with external entities.
Qualifications: Mission Crew Commander qualification current.
Responsibilities:

- Executive control and coordination of E-X mission operations.
- Supervision and direction of ISR and Sensor Management personnel.
- Development and approval of Mission Plan changes.

Attributes:

Mission Commander Qualification Status	Public	String

Operations:

Monitor Mission Plan Execution	Public	Void
Perform Replanning	Public	Mission Plan Change
Issue ISR Tasking	Public	ISR Task
Perform Coordination	Public	Void
Review and approve Products	Public	Void

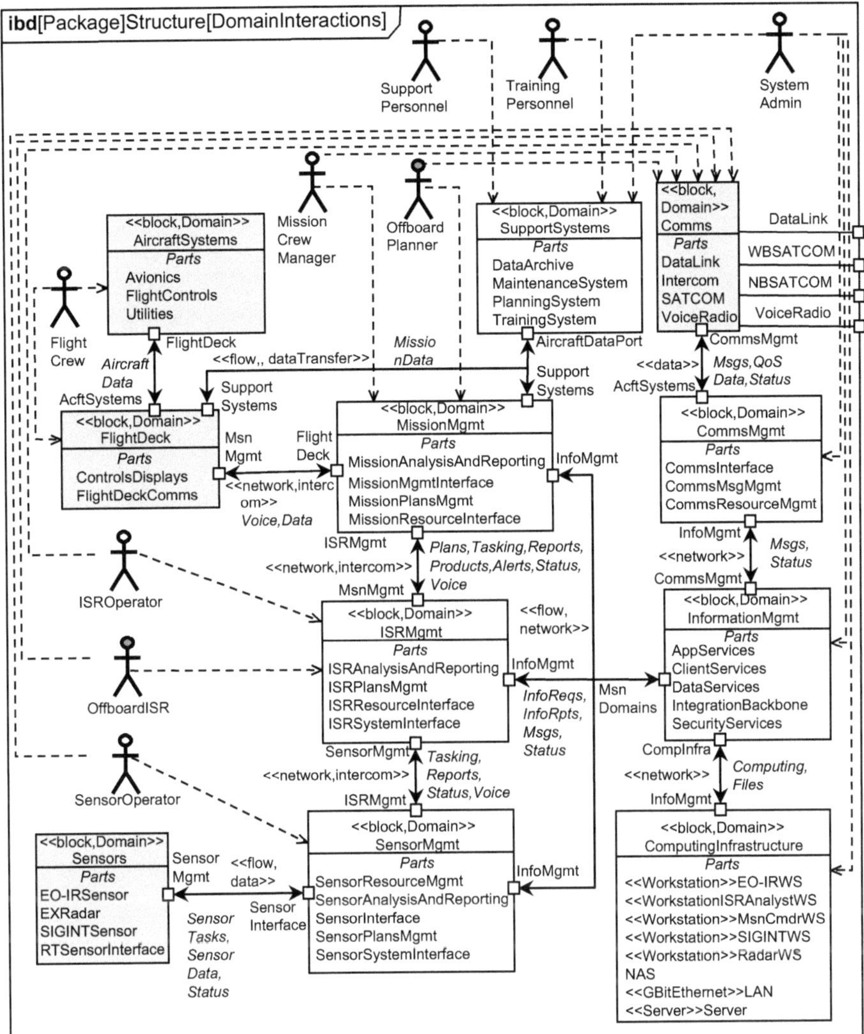

Fig. C.4 E-X Domain Interaction Diagram

Actor: Flight Crew.

General: This Actor models the User Role of the pilot and copilot who are responsible for planning and executing flight operations in coordination with the other Mission Crew members, air traffic control, and others.

Qualifications:

- Pilot rating, instrument, multiengine
- E-X operational qualification

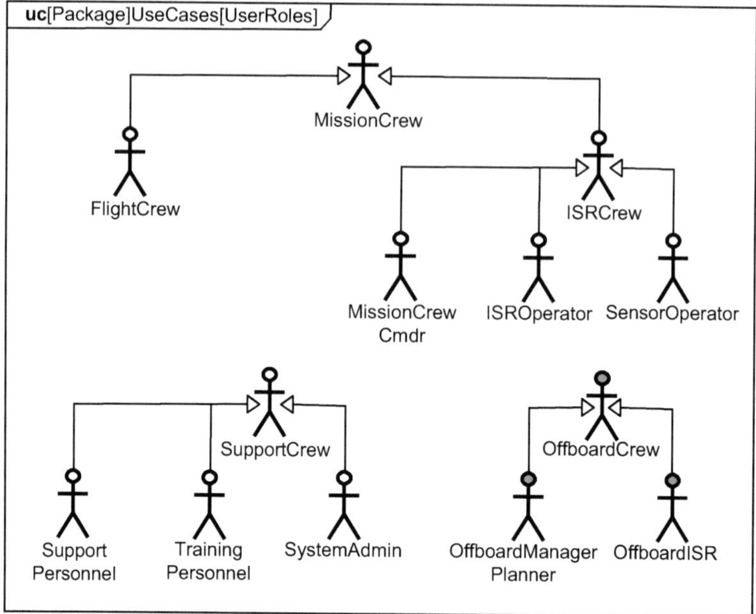

Fig. C.5 E-X User Roles

Responsibilities:

- Plan and execute flight operations.
- Serve as Pilot in Command.
- Coordinate mission operations with other E-X crew members and external organizations.

Attributes:

Pilot Qualifications Status	Public	Int

Operations:

Perform Flight Planning	Public	Flight Plan
Perform Aircraft Operations	Public	Void

Actor: ISR Operator.

General: This Actor models the User Role associated with using the resources of the ISR Management Domain to transform ISR Plans into Sensor Tasks, monitor plan execution, assess Sensor Products, create ISR Products, coordinate mission activities, and perform other assigned tasks.

Qualifications: ISR Operator qualification current.
Responsibilities:

- Manage ISR operations, including issuing collection tasking and analyzing collection reports.
- Coordinate ISR operations with other E-X crew members.

Attributes:

ISR Operator Qualifications Status	Public	String

Operations:

Perform ISR Planning	Public	Mission Plan
Perform ISR Management	Public	Void
Issue Collection Tasking	Public	Collection Task
Perform ISR Analysis	Public	Analysis Report

Actor: Sensor Operator

General: This Actor models the Mission Crew User Role that directly interacts with E-X sensors to control their operation, monitor and assess sensor data, and create and distribute Collection Reports and other sensor data.
Qualifications: Sensor Operator qualification current
Responsibilities:

- Configure and control sensors.
- Quality check sensor data and issue Collection Report.
- Coordinate sensor operations with other E-X crew members.

Attributes:

Sensor Operator Qualifications Status	Public	String

Operations:

Monitor and control Sensors	Public	Void
Create Collection Report	Public	Collection Report

Actor: Support Crew

General: This Actor models the abstract User Role that includes various personnel required to support the E-X system and its operations.
Qualifications:

- General airmanship and survival training current
- General E-X crew qualification current

Attributes:

System Support Qualification Status	Public	String
Clearance	Public	String
Grade	Public	String
ID Number	Public	Int
Unit of Assignment	Public	String

Actor: Support Personnel

General: This Actor models the User Role at the E-X operating location that performs system maintenance and servicing, pre- and postflight inspections and problem correction, aircraft movement and marshaling, supply functions, and other support tasks associated with achieving system operational ready status and supporting sortie generation and recovery.
Qualifications: System Support qualification current per assigned tasks.
Responsibilities:

- Perform and record system inspection, fault diagnosis, maintenance, and servicing.
- Certify operational readiness status.

Operations:

Perform System Maintenance	Public	Void
Perform Preflight	Public	Void
Perform Postflight	Public	Void
Perform Logistic Support	Public	Void

Actor: Training Personnel

General: This Actor models the User Role that plans, delivers, and tracks training for E-X personnel, specifically concerned with system operations and maintenance, and including both classroom and on-the-job training.

Qualifications: E-X Trainer qualification current per assigned tasks.
Responsibilities:

- Provide initial and recurring skills training and certification.
- Maintain training records.
- Maintain training equipment and materials.

Operations:

Perform Training Planning	Public	Void
Perform Training	Public	Void
Maintain Training Records	Public	Void

Actor: System Administrator (SYSAD)

General: This Actor models the User Role at the E-X operating location and on the aircraft responsible for the configuration and maintenance of onboard and off-board information systems, including user accounts, security controls, data archiving and restoration, hardware and software updates, and other system administration tasks.
Qualification: E-X SYSAD qualification current per assigned tasks.
Responsibilities:

- Support and ensure correct content and functioning of IT systems and data, both onboard and off-board.
- Provide secure long-term data storage/archiving.
- Maintain system IT configuration.
- Maintain and analyze usage and event logs, including security.

Operations:

Perform Information System Maintenance	Public	Void
Perform Data Archiving	Public	Void

Actor: Off-board Crew

General: This is an abstract Actor that models the characteristics of Ground Station personnel.
Qualifications:

- E-X Ground Station qualification
- General ISR systems and functions qualifications, per assigned tasks

Attributes:

E-X Ground Station Qualification Status	Public	String
Clearance	Public	String
Grade	Public	String
ID Number	Public	Int
Unit of Assignment	Public	String

Actor: Off-board Manager/Planner

General: This Actor models the User Role that plans and directs the operations of a Ground Station in coordination with the E-X Mission Crew and external entities.
Qualification: E-X Ground Station Supervisor qualification current.
Responsibilities:

- Participate in E-X mission planning and develop Ground Station plans
- Oversee and direct Ground Station operations and personnel
- Coordinate E-X mission operations with Mission Crew Commander and with external organizations and systems

Operations:

Perform Ground Station Planning	Public	Void
Direct Ground Station Operations	Public	Void
Coordinate Mission Operations	Public	Void
Assess and disseminate Products	Public	Void

Actor: Off-board ISR

General: This Actor models the User Role within a Ground Station that complements the E-X onboard ISR Operator(s) by performing additional analysis and ISR Product creation using downlinked Sensor Reports.
Qualification: E-X Ground Station Operator qualification current.
Responsibilities: Perform analysis and report generation using downlinked information
Operations:

Perform ISR Analysis	Public	Void
Coordinate Operations	Public	Void

Behavioral Perspective

Use Cases: Figures C.6 and C.7 show the Use Cases associated with E-X missions and support activities, with a level of decomposition and User Role associations. The diagrams are followed by example specifications for the principal Use Cases.

Use Case: Conduct Aircraft Operations

General: This Use Case models behaviors associated with preparing the E-X aircraft for flight, conducting flight operations, and recovering the aircraft after flight. This includes pre- and postflight inspections and servicing as well as ground and air operations.

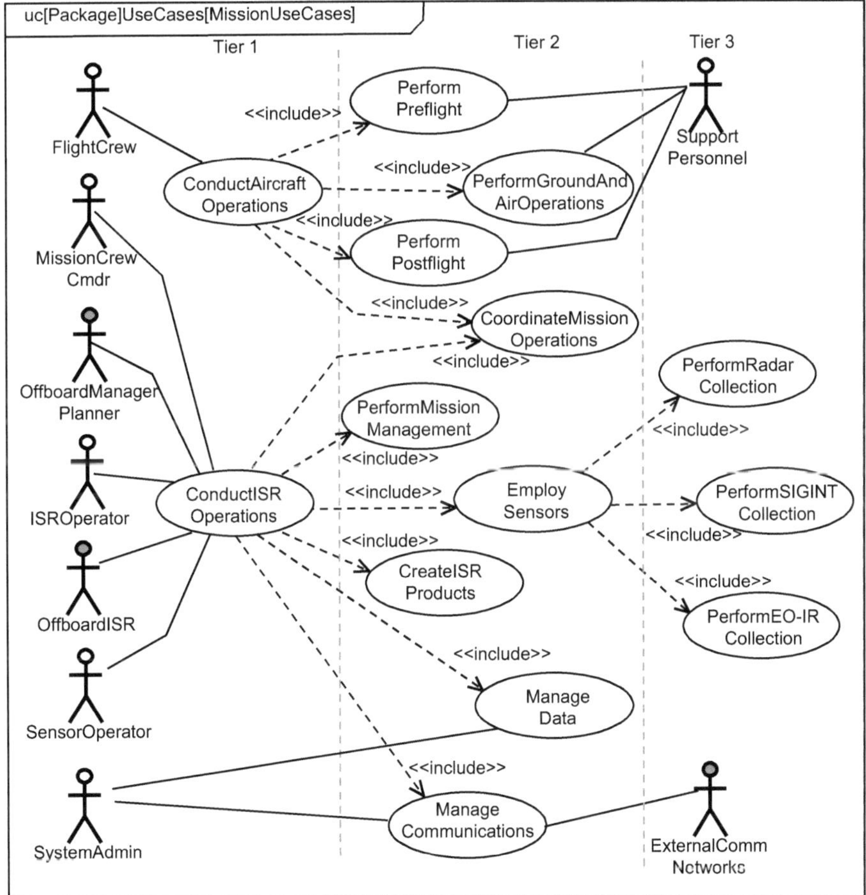

Fig. C.6 E-X Mission Use Cases

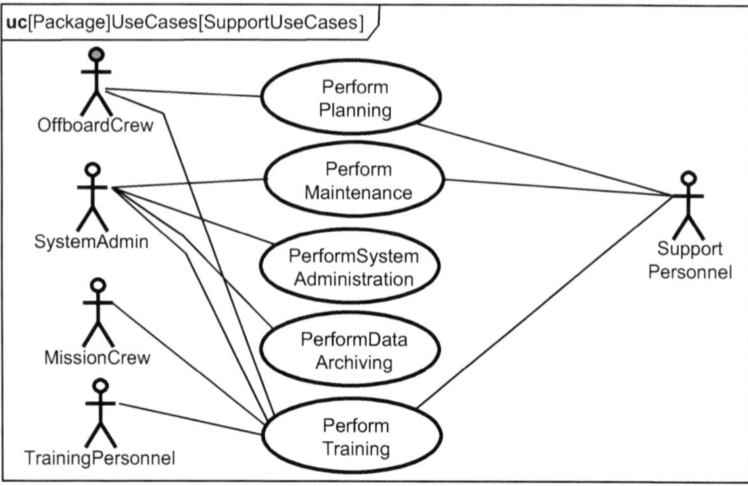

Fig. C.7 E-X Support Use Cases

Preconditions: Aircraft and Flight Crew assigned; Mission Plan published.
Postcondition: Aircraft recovered and secured.
User Roles: Flight Crew, Support Personnel.
Primary Data Objects:

- Flight Plan
- Maintenance Report

Scenarios:

Primary—Normal Mission:

- Support Personnel position aircraft, conduct preflight inspection, and perform any required servicing.
- Flight Crew performs walk-around and loads flight plan data.
- Any correctable faults identified during preflight are corrected.
- Other Mission Crew perform preflight equipment checks and configure mission equipment.
- Flight Crew obtains taxi clearance, conducts ground operations, and takes off.
- Flight Crew executes Flight Plan in coordination with Mission Crew Commander, Air Traffic Control, and other external entities.
- Flight Crew executes return to base, landing and ground operations, parking in designated ramp position.
- Support Personnel conduct postflight inspection and servicing.
- Mission Crew debriefs mission with Support Personnel.
- Any required maintenance actions are logged and reported to Maintenance.

Secondary—Ground Abort:

- Non-correctable fault(s) is identified during preflight.

- Ground abort is declared.
- Alternative aircraft is assigned or mission is cancelled.

Secondary—Air Abort:

- Mission-impacting fault occurs in flight.
- Mission is terminated, and Flight Crew executes return to base and other actions as in Primary Scenario.

Allocated Requirements: TBD, associated with Flight Operations

Use Case: Conduct ISR Operations

General: This Use Case describes the use of E-X mission equipment to gather, analyze, store, and communicate ISR information. The Mission Crew executes a Collection Plan that defines how sensors are to be employed and how collected information is to be treated. The preplanned Collection Plan may be modified during a mission based on circumstances such as urgent requests for information. Sensors may be employed individually or cooperatively, e.g., using cross-cueing.
Preconditions: The E-X system is in position to conduct ISR operations, and the mission equipment is powered up and operationally ready (OR). [NOTE: degraded operations may be possible after failure or degradation of individual mission equipment items.]
Postconditions: The Collection Plan has been competed, time on station limit has been reached, the E-X has left the operational area, or the mission equipment has been shut down or has suffered a nonrecoverable failure (any of the above is sufficient to terminate this Use Case).
Actors: Mission Crew Commander, ISR Operator, Sensor Operator, Off-board Manager/Planner, Off-board ISR Operator
Primary Data Objects:

- Collection Plan
- Flight Plan
- Communications Plan
- Information Management Plan
- Mission Equipment Health and Status
- ISR Task
- Sensor Task
- Mission report
- ISR Report
- Sensor Data
- Message
- Supporting Data

Scenarios:

Primary—Preplanned Mission

- Mission crew performs equipment power-up and checkout; correctable problems are fixed.
- Mission crew loads plans and other mission data.
- Mission crew employs sensors per tasks in Collection Plan, including cross-cueing.
- Mission crew exploits collected ISR data, develops reports, performs onboard archiving, and disseminates reports per Information Management Plan.
- Off-board Manager/Planner and Off-board ISR Operator collaborate with Mission Crew and perform supplemental tasks such as data exploitation, reporting, and dissemination.
- Upon completion of Collection Plan, expiration of time on station, or other termination event, Mission Crew secures mission equipment for return to base.
- Mission Crew performs data download and supports off-board data archiving and post-mission processing

  Secondary—Dynamic Replanning
  Same as Primary except that during mission execution, the Collection Plan or individual collection tasks are replanned in response to dynamic tasks such as urgent information request from supported organization
  Secondary—Mission Abort
  Initially same as Primary, but nonrecoverable equipment failure during mission causes Mission Crew to secure mission equipment for unscheduled return to base

Allocated Requirements

- Mission Management Requirements
- ISR Management Requirements
- Sensing Requirements; specific Multimode Radar, EO Sensor, and SIGINT Sensor requirements allocated to perform Radar Collection, perform EO/IR Collection, and perform SIGINT Collection Use Cases
- Information Management and Infrastructure Requirements allocated to manage Data Use Case
- Communications/Networking Requirements Allocated to manage Communications Use Case

Use Case: Perform Planning

General: This Use Case describes the activities associated with developing and publishing operational and support plans.
Preconditions: Mission tasking has been issued, aircrew has been assigned, and mission support data is available.
Postcondition: A Mission or Support Plan has been published.
User Roles: Mission Crew, Off-board Crew
Primary Data Objects:

- Mission Task
- Mission Plan

- Collection Plan
- Flight Plan
- Comms Plan
- Info Management Plan
- Support Plan

Scenarios:
Mission Plan

- Mission Crew receives tasking, compiles mission data, and develops Flight, Collection, Comms, and Information Management Plans.
- Mission Plan is reviewed with higher authority, revised as necessary, approved, and published.

Support Plan

- Support Personnel compile maintenance and support data.
- Support Personnel schedule maintenance activities, manage supply actions, and perform other support tasks.
- Support Personnel develop Support Plan.
- Support Plan is reviewed with higher authority, revised as necessary, approved, and published.

Allocated Requirements: The system shall have the ability to develop Mission and Support Plans.

Use Case: Perform Maintenance

General: This Use Case models the activities associated with correcting system failures or faults and restoring the E-X aircraft and mission equipment to Operationally Ready (OR) status.
Preconditions: Fault and condition data is available; aircraft is available for maintenance.
Postcondition: Maintenance action(s) completed.
User Roles: Support Personnel
Primary Data Objects:

- Maintenance Report
- Support Plan

Scenarios:
Primary Maintenance Action

- Support Personnel receive system logs, Mission Crew debriefing, and any other sources of equipment status and failure data.
- Maintenance Control schedule maintenance actions and assigns personnel.
- Maintenance personnel perform diagnostic and corrective actions to correct failures and perform required equipment servicing
- Maintenance personnel record maintenance actions and report aircraft status.

Additional Maintenance

- Same as primary, except maintenance personnel determine that one or more failures cannot be corrected with available resources.
- Maintenance personnel and Maintenance Control define and schedule additional maintenance activities.
- Upon completion of failure correction, maintenance actions are recorded, and aircraft status is reported.

Allocated Requirements: The system shall support maintenance actions to restore functionality after faults or failures.

Use Case: Perform System Administration
General: This Use Case describes the activities associated with maintaining and supporting the Information Technology (IT) resources of the E-X, including hardware and software configuration management and updating, user account management, information security management, and associated record keeping and reporting.
Preconditions: Appropriate system resources are available for System Administrator (SYSAD) actions.
Postcondition: SYSAD action(s) completed.
User Roles: System Admin (SYSAD).
Primary Data Objects:

- User Account Data
- Hardware Configuration Logs
- Software Configuration Logs

Scenario:

- SYSADs create, cancel, or modify user accounts for E-X Mission and Support Systems in response to approved requests.
- When tasked, SYSADs perform configuration audits, install upgrades, and do other tasks associated with maintaining currency in E-X information systems.
- SYSADs support maintenance actions involving E-X information systems.
- Allocated Requirements: The system shall support system administration actions to maintain system configuration, security, user access, event and condition logging and analysis, reporting, and related functions.

Use Case: Perform Data Archiving

General: This Use Case models the activities associated with transferring mission data from an E-X aircraft to an information processing and archiving environment for archival storage and post-mission analysis.

Preconditions: Mission data has been removed on portable media and provided to the System Administration staff.
Postcondition: Data storage completed.
User Roles: System Admin (SYSAD).
Primary Data Objects: Mission Data (multiple categories).
Scenario:

- Following Postflight actions and crew debriefing, removable media with mission data are transported to the Support Environment.
- SYSAD verifies data integrity and transfers files to the Data Archive.
- Archived data is made available to qualified users for post-mission analysis and reporting.

Allocated Requirements: The system shall support downloading, storage, backup, retrieval, and safe long-term retention of system and operational data.

Use Case: Perform Training

General: This Use Case models the activities associated with providing training to E-X personnel; training is specialized by User Role.
Preconditions: Training materials are available, personnel to be trained are available, and training events are scheduled.
Postcondition: Training event completed.
User Roles:

- Training Personnel
- All other User Roles as trainees

Primary Data Objects: Training Data
Scenario:

- E-X organization establishes training requirements and schedule.
- Training Personnel conduct training, supervise self-study, conduct evaluations, and maintain training records.

Allocated Requirements: The system shall support initial and recurring training and qualification of mission and support personnel.

A Use Case specification is normally complemented by an Activity Diagram (AD) that makes the scenarios of the behavior explicit and easy to follow. As a single example, Fig. C.8 is an AD for the Create ISR Product Use Case. Larger behaviors can be modeled in an AD like Fig. C.9, which shows the elements of a typical E-X mission. This starts with tasking from higher authorities and proceeds through the planning and execution of the mission, reporting of collected information, and recovery and servicing of the aircraft. Many of the Actions in this diagram represent individual Use Cases as defined earlier.

A diagram like Fig. C.9 is sometimes referred to as a "Day in the Life" and serves to provide context for the more specific behaviors associated with Use Cases and more detailed Mission Threads. A Mission Thread or Business Process models a major end-to-end function and may be composed as a sequence of Use Cases, Operator actions, external interactions, and other elements. Examples are shown in section "Logical/Functional Viewpoint Artifacts"

Another important aspect of system-level behavior in the Operational Viewpoint involves the States in which the E-X system can be, along with the transitions, among them. Figure C.10 illustrates a system State Machine Diagram that includes a completely powered-down condition, various States associated with preparing for and executing and recovering from operations, and system maintenance.

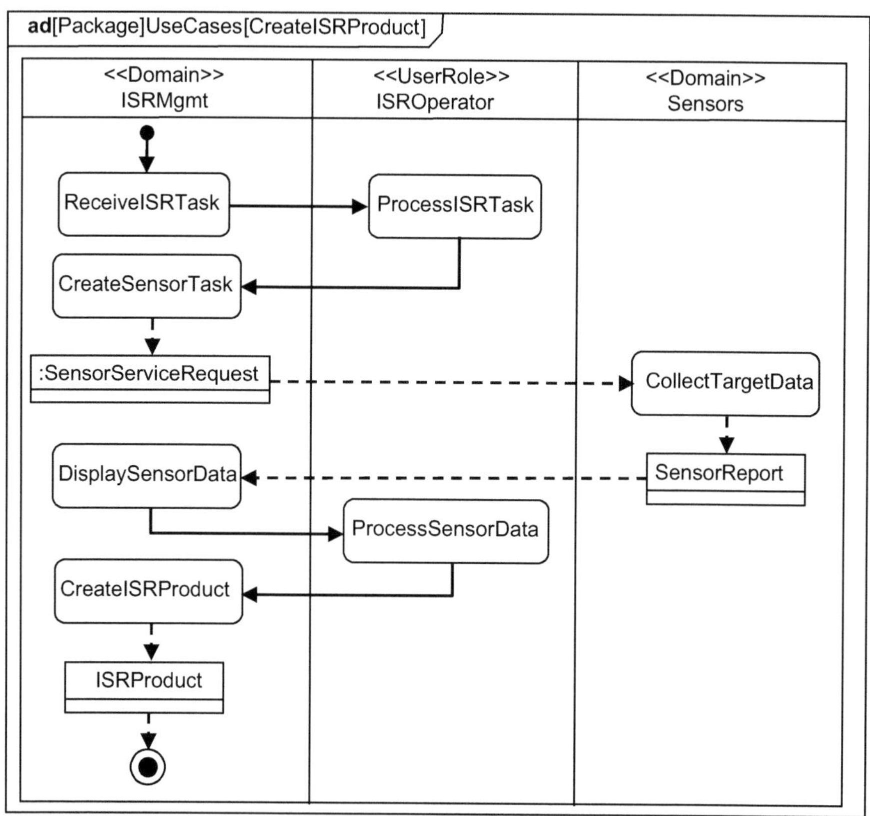

Fig. C.8 Activity Diagram for the Create ISR Product Use Case

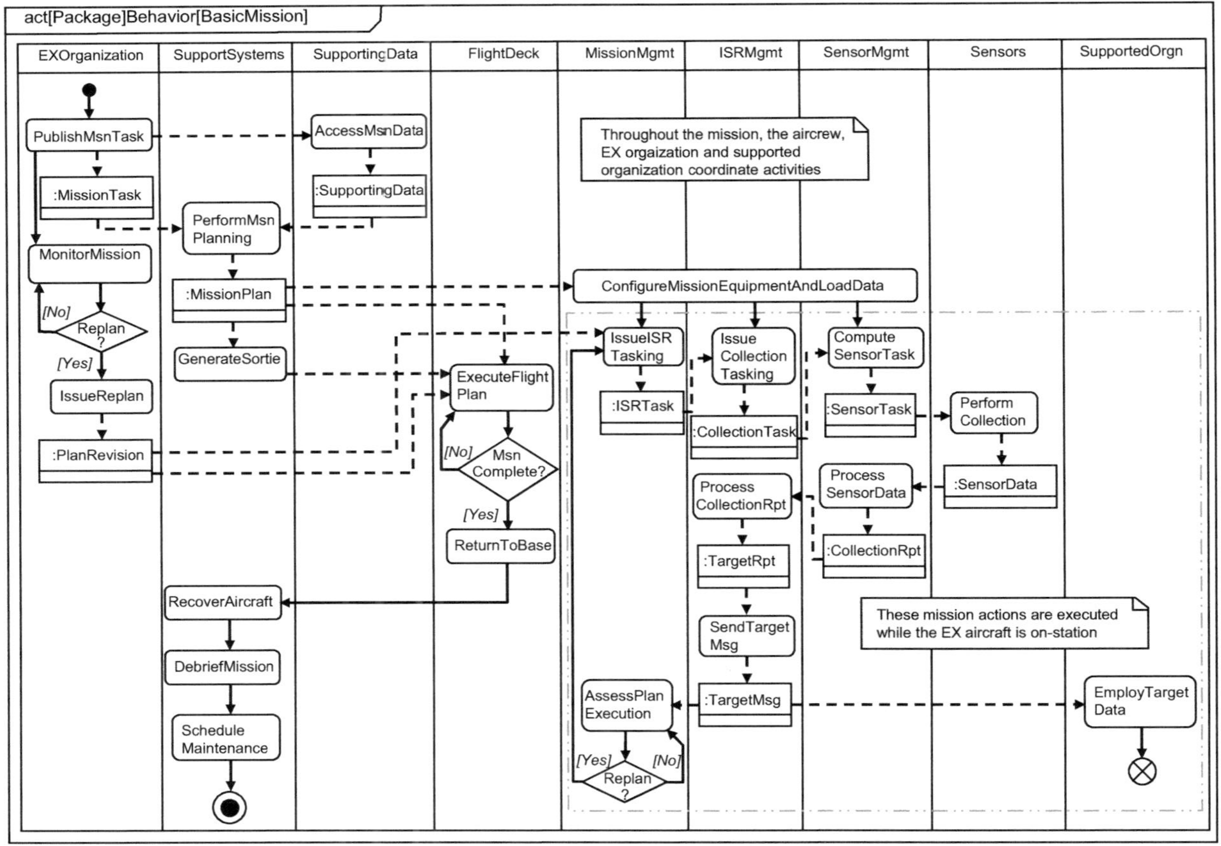

Fig. C.9 E-X Basic Mission or “Day-in-the-Life” of the System

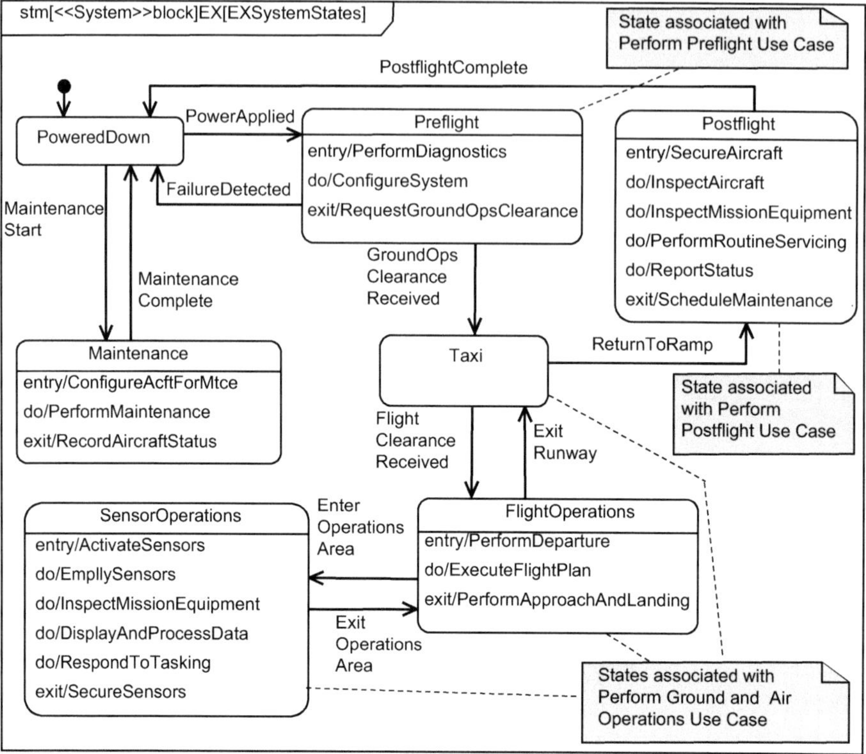

Fig. C.10 E-X State Machine Diagram

Data Perspective

Figures C.11, C.12, C.13, and C.14 show the initial structure and concept of a Conceptual Data Model for the E-X, with emphasis on Foundation Classes. Note that these are modeled using SysML Blocks, but the term Foundation Class is retained because it is widely used and understood for high-level entities in a system architecture. Note also that both data and resource categories are modeled as Foundation Classes and that they are stereotyped as both `<<abstract>>` and either `<<Info Element>>` or `<<Resource>>`. The following text gives short descriptions of CDM data Blocks, and the diagrams provide Values and Operations. Collectively, this creates a basic specification for an `<<abstract>>` CDM Block.

- *Aircraft Systems*—Foundation Class for resources that are integral to the E-X aircraft and distinct from Mission Equipment; provides abstraction of the content of the Aircraft Systems Domain
- *Alert*—Data Item involved in urgent notification of an event or situation
- *Analysis Data*—Information content of an Analysis Report

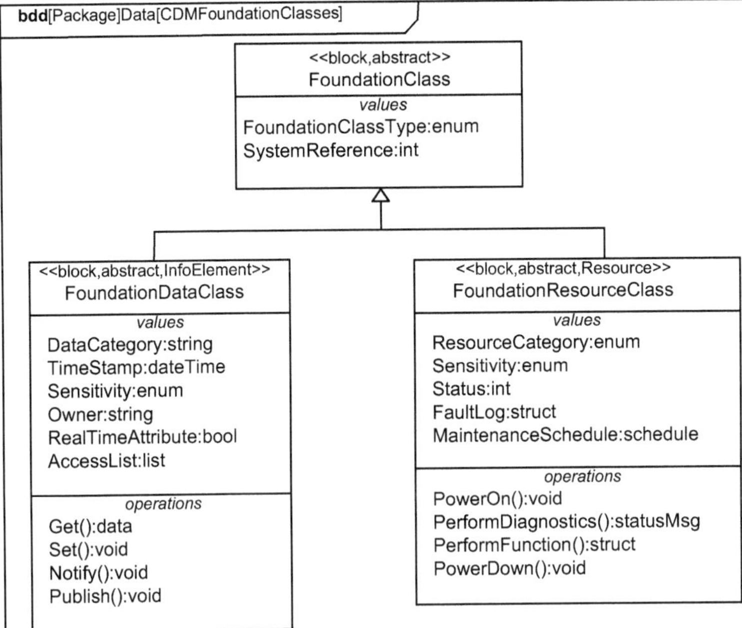

Fig. C.11 Overall E-X Foundation Classes

- *Communications Device*—Foundation Class for external communications resources; provides abstraction of the content of the Communications Domain
- *Data Item*—Foundation Class providing abstraction of system data categories
- *Foundation Class*—Highest level of abstraction of system content, including data and resources; used to establish overall patterns that are specialized through inheritance to create system Classes and Blocks. [*NOTE*: the terminology Class is retained in this SysML model both because of common usage of the term Foundation Class and to emphasize that this is an `<<abstract>>` entity that is never directly instantiated, as in a Block.]
- *ISR Task*—Task category for tasking issued to the ISR Management Domain.
- *Maintenance Data*—Data Item for information associated with maintenance actions or status.
- *Plan*—Foundation Class for content of plans in various categories.
- *Message*—Foundation Class for content of information exchanges via messaging.
- *PNT*—Data Item for various types of Position, Navigation, and Timing (PNT) data.
- *Report*—Foundation Class for information formatted in Reports in various categories.

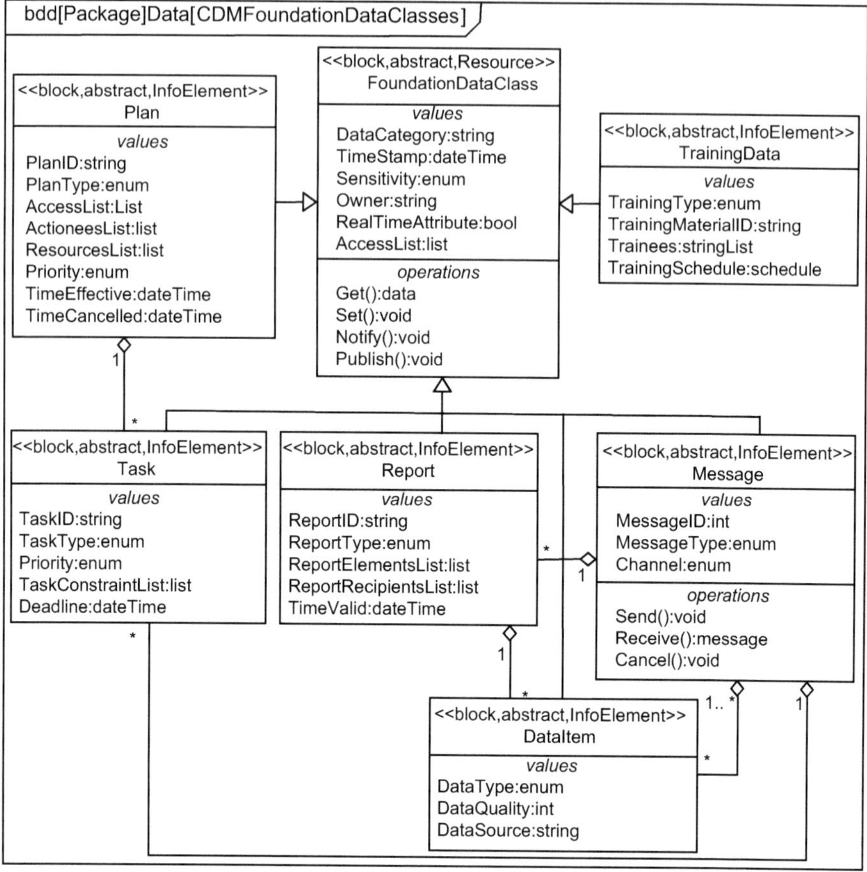

Fig. C.12 Data Foundation Classes

- *Sensors*—Foundation Class for E-X sensor resources; provides abstraction of the content of the Sensors Domain.
- *Sensor Data*—Data Item for data from E-X sensors.
- *Task*—Foundation Class for tasking issued to system resources or personnel.
- *Training Data*—Data Item for information associated with training, including training materials, records, training device data, and others.
- *Visualization*—Data Item for information used to create screens on user workstations and on training device displays.

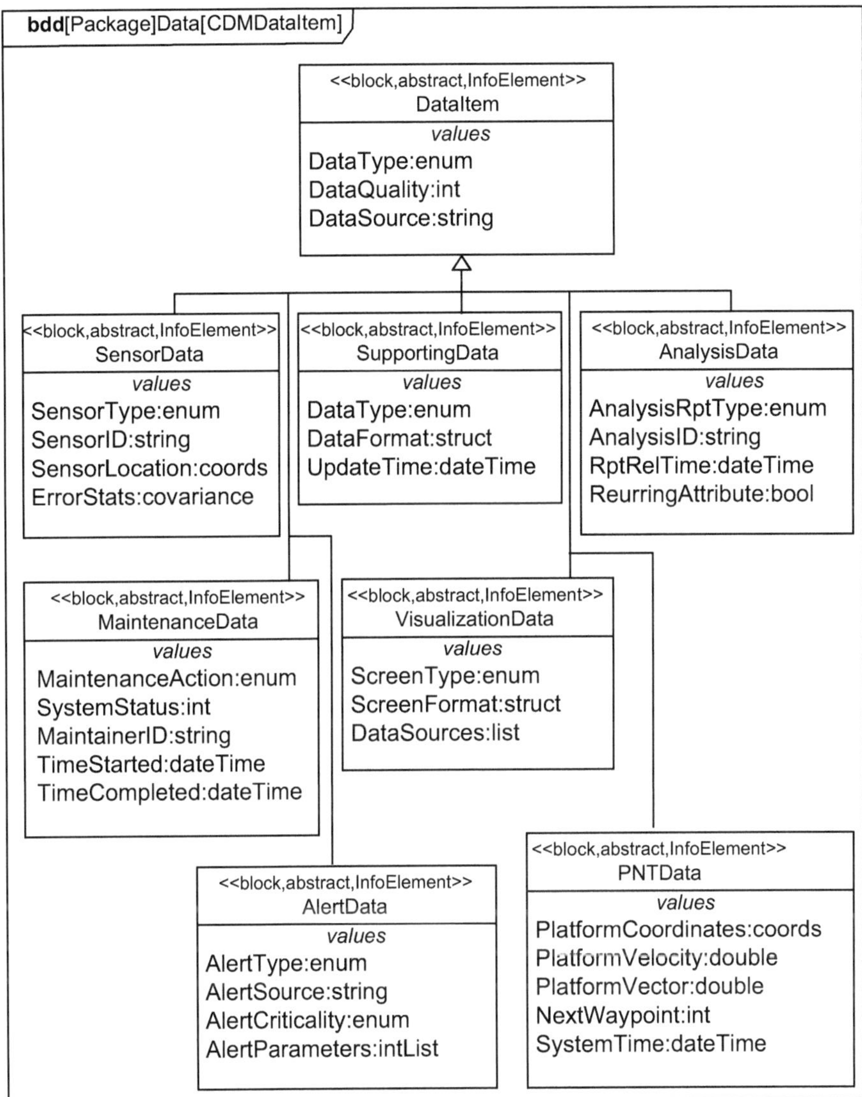

Fig. C.13 Data Item Foundation Classes

Contextual Perspective

Figure C.15 shows the operational context of the E-X. These external entities would, in the real world, be documented by the appropriate organizations. Figure C.16 provides another element of system context in the form of a high-level organizational diagram showing how the unit that operates the E-X system is structured. The following are brief descriptions of the external entities in Fig. C.15 and the organizational entities in Fig. C.16.

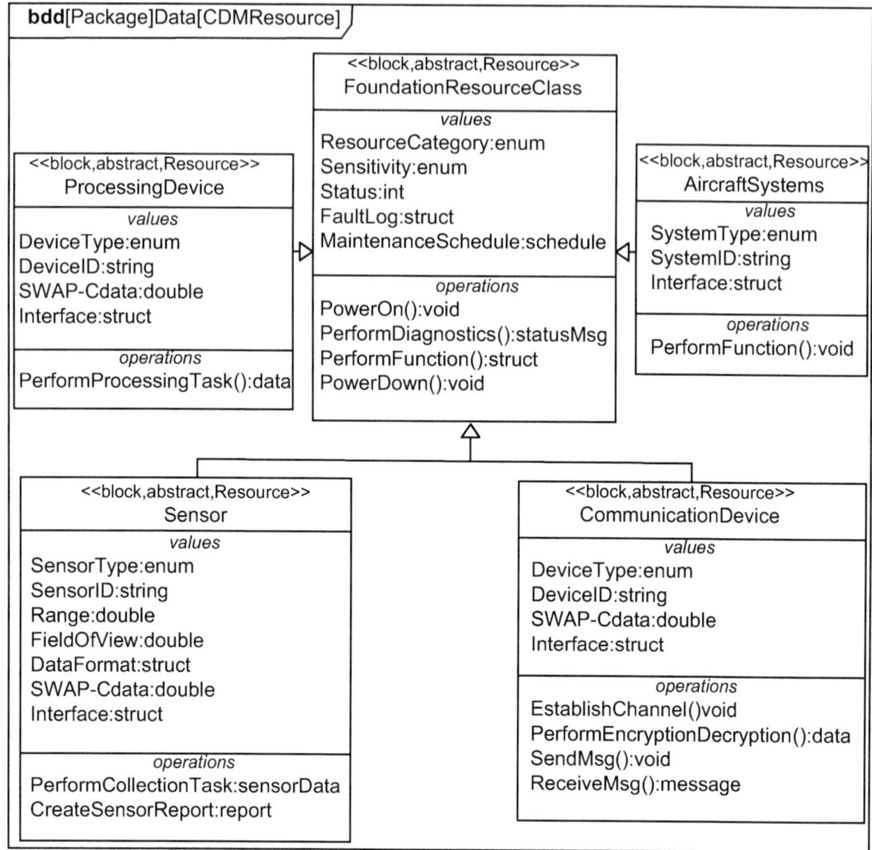

Fig. C.14 Resource Foundation Classes

Narrowband Satellite Communications (NBSATCOM)—Models the satellite constellation and associated ground environment used by E-X for narrowband SATCOM communications.

Wideband Satellite Communications (SATCOM)—Models the satellite constellation and associated ground environment used by E-X for wideband SATCOM communications.

Global Positioning System (GPS)—Models the GPS constellation that provides position, navigation, and timing (PNT) data for precision E-X navigation.

Command and Control (C2) Platform—Models external platforms (e.g., aircraft) that provide supervisory and coordination services with which the E-X interacts, e.g., coordinating flight operations of multiple aircraft.

Target—Models entities in the external environment on which the E-X collects information using sensors; includes locations and tracks for point targets and imagery and video collected using SAR and EO/IR sensor modes.

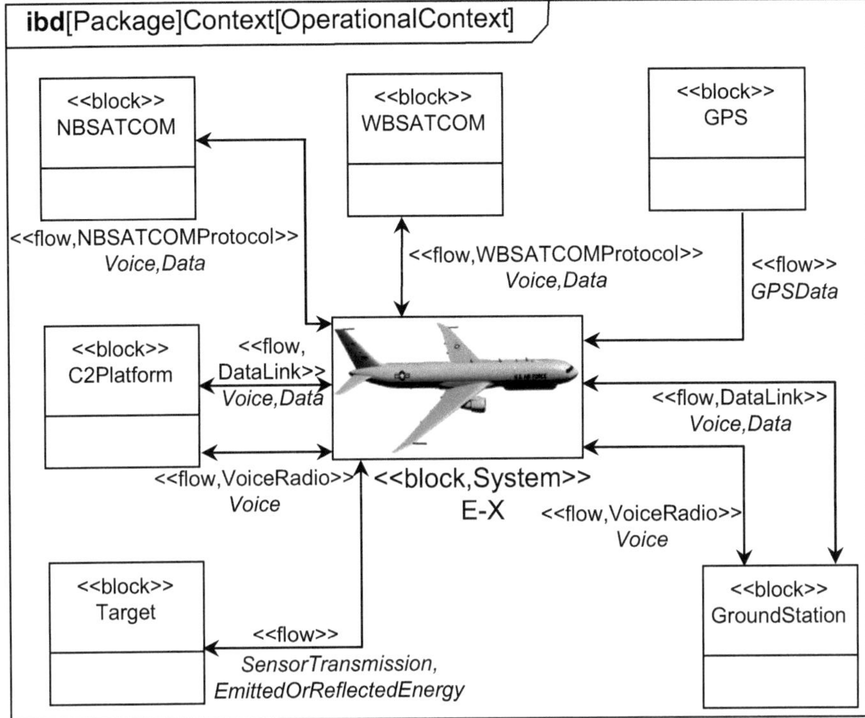

Fig. C.15 E-X Operational Context

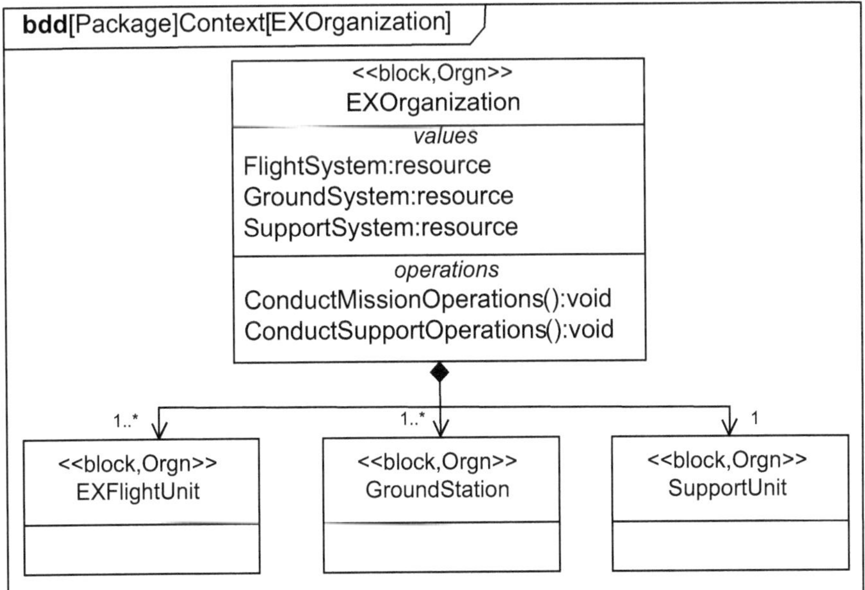

Fig. C.16 E-X Organizational Structure

Ground Station—A terrestrial location hosting communications and information processing resources that participates in E-X missions to supplement onboard capabilities for analysis, reporting, and mission coordination; this is the Unit of Assignment for Off-board Crews.
E-X Organization—Models the organization that owns, operates, and supports the E-X platform and associated Ground Station; details are shown in Fig. 3.15.
E-X Flight Unit—Unit of Assignment for E-X Mission Crews.
Support Unit—Unit of assignment for E-X Support Crews.

Logical/Functional Viewpoint Artifacts

Structural Perspective

Figures C.17, C.18, and C.21 show the decomposition of the primary Mission Domains into their major Parts, each of which can be treated as a Subdomain and further decomposed. Details such as dividing these major components into client-side and server-side components residing on workstations and within the Information Management Domain, respectively, will be settled in detailed design and implementation and documented in the Physical Viewpoint. The ports of each Domain are delegated to Subdomains/Parts. As illustrations, Figs. C.19 and C.20 show decomposition of the ISR System Interface and ISR Analysis and Reporting Subdomains from Fig. C.18.

As was done for the OV, the diagrams of the LV are followed by text descriptions of elements of the architecture. Together with Attributes, Operations, Interface Definitions, and other details, these create the functional specifications for the E-X design from which the actual system will be realized in the PV.

Mission Management Domain: The following text gives short descriptions of the Subdomains of the Mission Management Domain.

- *Mission Management Interface*—this Subdomain provides the interfaces of the Mission Management Domain to the Information Management Domain and the Flight Deck. It provides the User Interface for the Mission Crew Commander.
- *Mission Plans Management*—this Subdomain processes the Mission Plan and provides a prioritized queue of ISR Tasks to be sent to the ISR Management Domain. It also maintains the status of task execution and supports replanning as required to deal with mission situations.
- *Mission Analysis and Reporting*—this Subdomain processes ISR Reports to support Mission Crew Commander situational awareness and decision-making and to develop Mission Reports for dissemination and archiving.
- *Mission Resource Interface*—this Subdomain provides the functional interface to the ISR Management Domain and to Mission Equipment resources, activities, and Status. The physical link is implemented via the LAN.

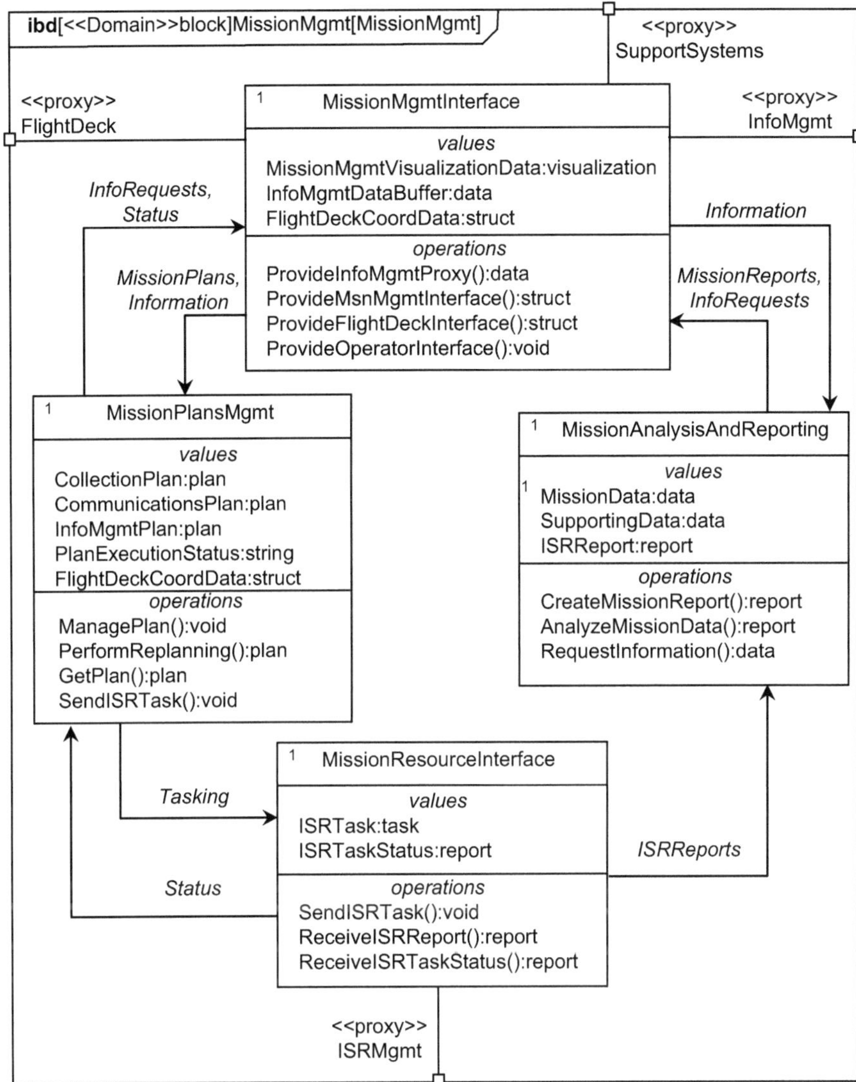

Fig. C.17 Mission Management Domain Decomposition

ISR Management Domain: The following text gives short descriptions of the Subdomains of the ISR Management Domain.

- *ISR System Interface*—this Subdomain provides the interfaces of the ISR Management Domain to the Mission Management and Information Management Domains. It provides the User Interface for the ISR Operator. Figure C.19 shows the next level of decomposition to a point where the Parts are candidates to be implemented as software components. ISR Visualization implements the HMI for

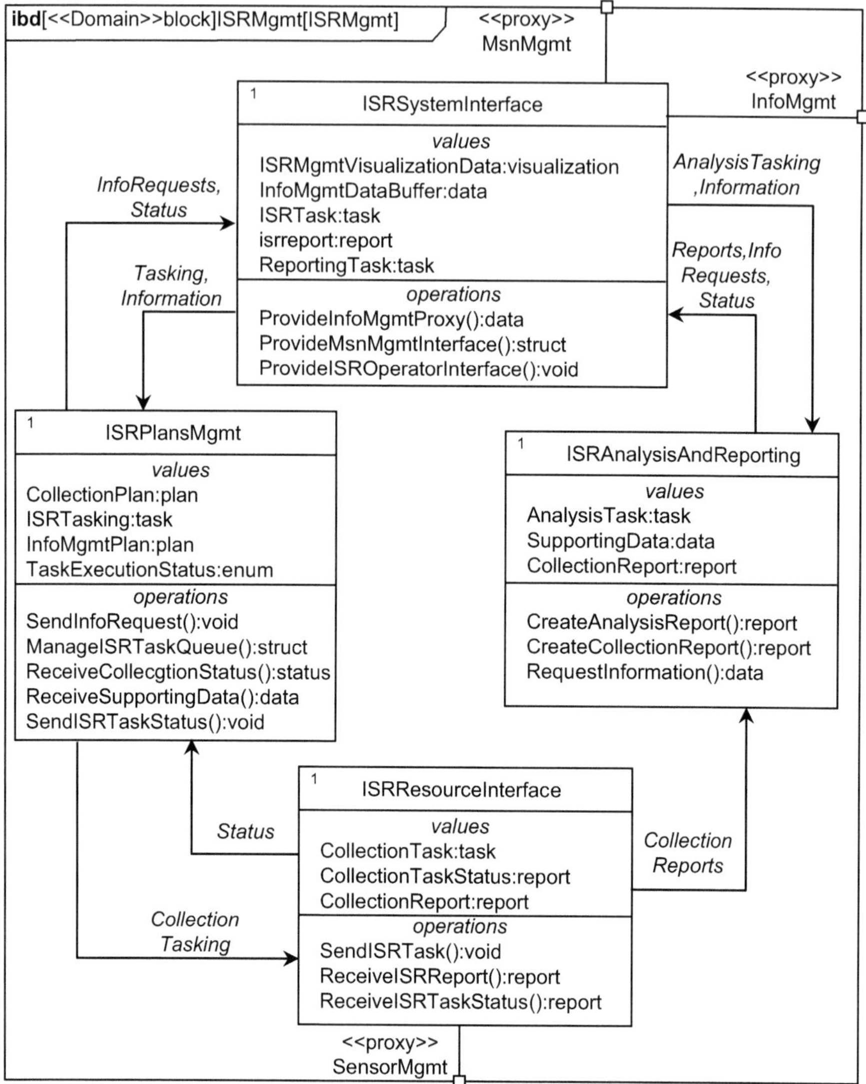

Fig. C.18 ISR Management Domain Decomposition

onboard and off-board ISR Operators. ISR Information Manager implements the interface to the Mission Management Domain, and an Information Management Proxy creates a local endpoint for the services of the Information Management Domain that can be efficiently used by the entire ISR Management Domain.

- *ISR Plans Management*—this Subdomain processes ISR Tasks and provides a prioritized queue of Collection Tasks to be sent to the Sensor Management Domain and Analysis Tasks to be sent to the ISR Analysis and Reporting Domain.

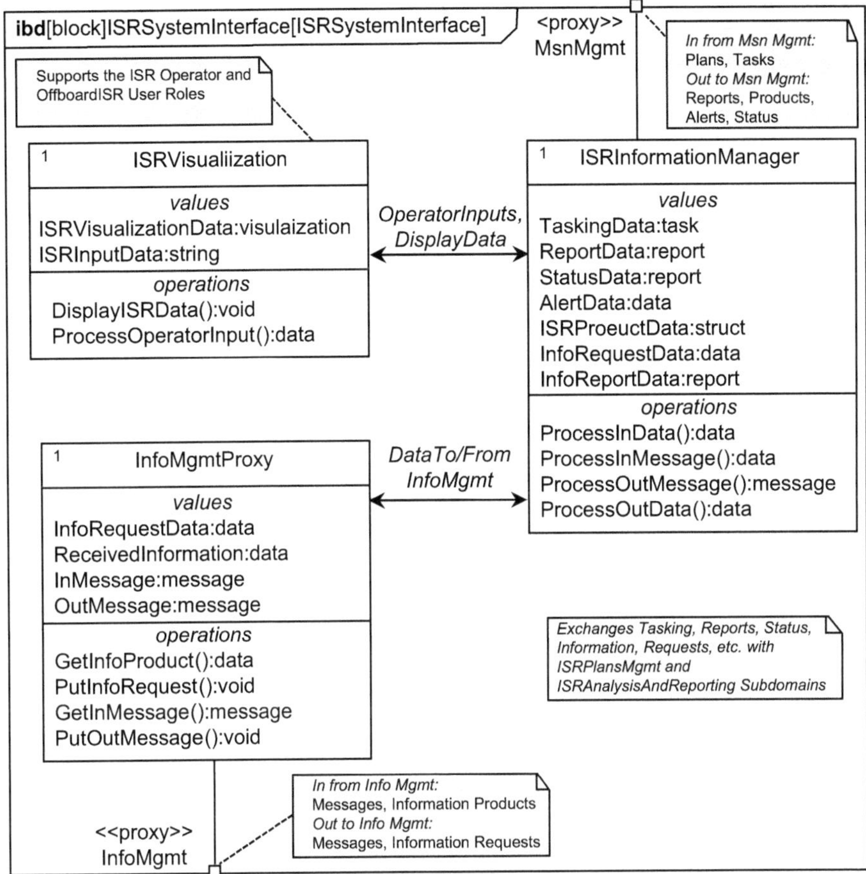

Fig. C.19 ISR System Interface Subdomain Decomposition

It also maintains task status to support task queue optimization and maintenance.

- *ISR Analysis and Reporting*—this Subdomain processes data from Collection Reports along with supporting Technical Data to create ISR Reports. It also supports various ISR Operator analysis and reporting tasks. Figure C.20 shows the components that handle data from the various E-X sensors.
- *ISR Resource Interface*—this Subdomain provides the functional interface to the Sensor Management Domain. The physical link is implemented via the LAN.

Sensor Management Domain: The following text gives short descriptions of the Subdomains of the Sensor Management Domain (Fig. C.21).

- *Sensor System Interface*—this Subdomain provides the interfaces to ISR Management and Information Management. It provides the User Interface for the Sensor Operator.

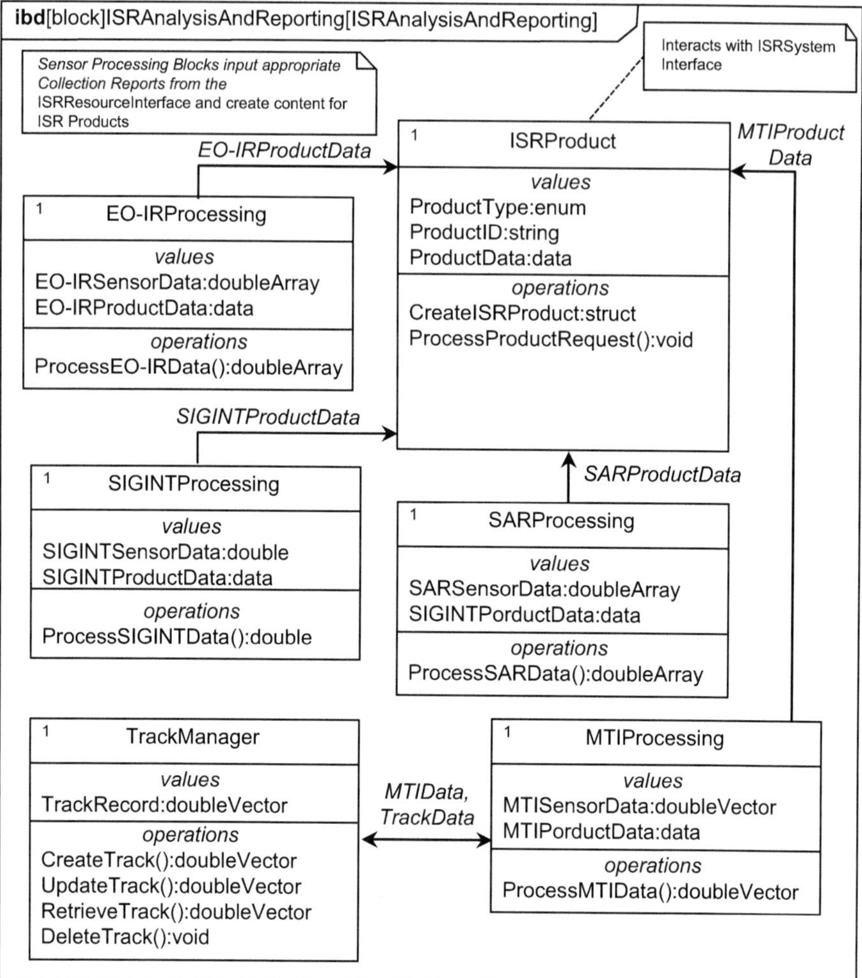

Fig. C.20 ISR Analysis and Reporting Subdomain Decomposition

- *Sensor Plans Management*—this Subdomain processes Collection Tasks and provides a prioritized queue to the Sensor Resource Manager. It also maintains task and sensor status to support task queue optimization and maintenance.
- *Sensor Resource Management*—this Subdomain is responsible for pairing Collection Tasks with Sensor Assets and scheduling collection actions. Sensor Tasks are sent to individual Sensors, and task status feedback is used in computing subsequent task schedules. Figure C.22 shows the decomposition of the Subdomain, following the overall Resource Management design pattern and showing interactions with other components and data sources. The behavior of this Subdomain is further described in the LV Behavioral Perspective.

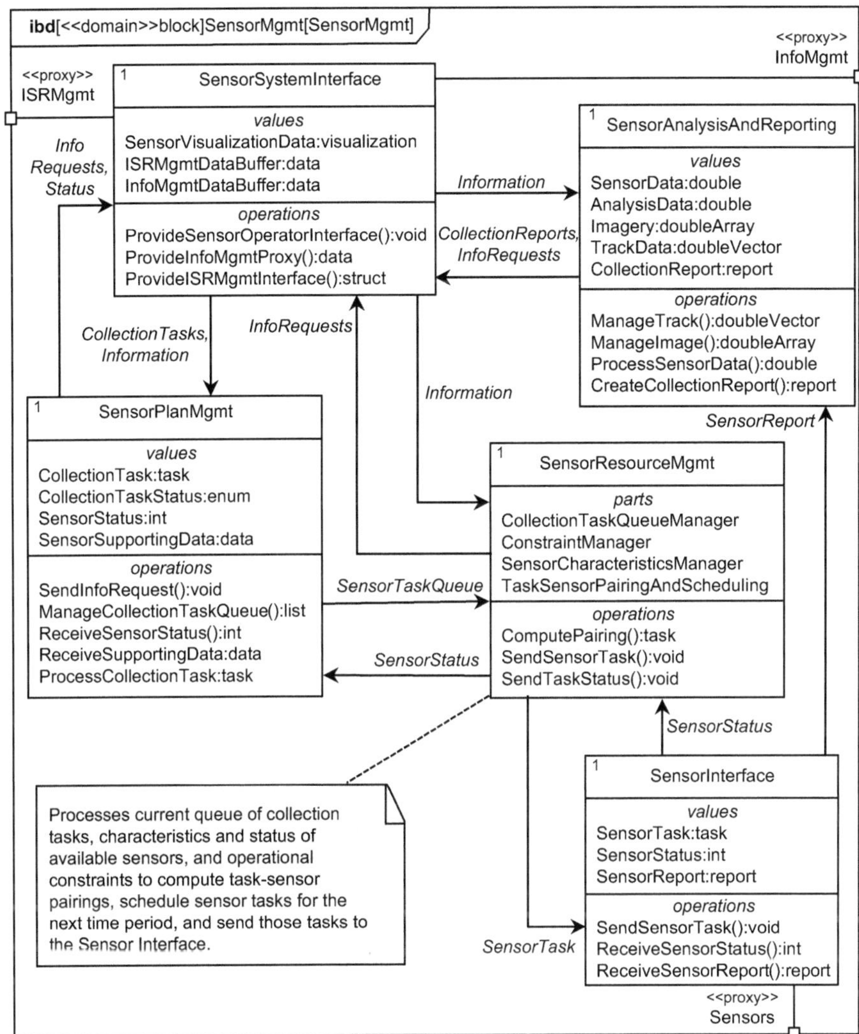

Fig. C.21 Sensor Management Domain Decomposition

- *Sensor Interface*—this Subdomain provides the functional interface to the Sensor Real-Time Interface component in the Sensors Domain. The physical link is implemented via the LAN.
- *Sensor Analysis and Reporting*—this software component processes data from Sensor Reports to create Collection Reports. It also supports various Sensor Operator analysis tasks and quality control of collected data.

Sensors Domain: Figure C.23 shows the decomposition of the Sensors Domain, including the radar, EO/IR and SIGINT sensors, and the Real-Time Sensor Interface, that enables interaction between the hard real-time environment of the sensors, with

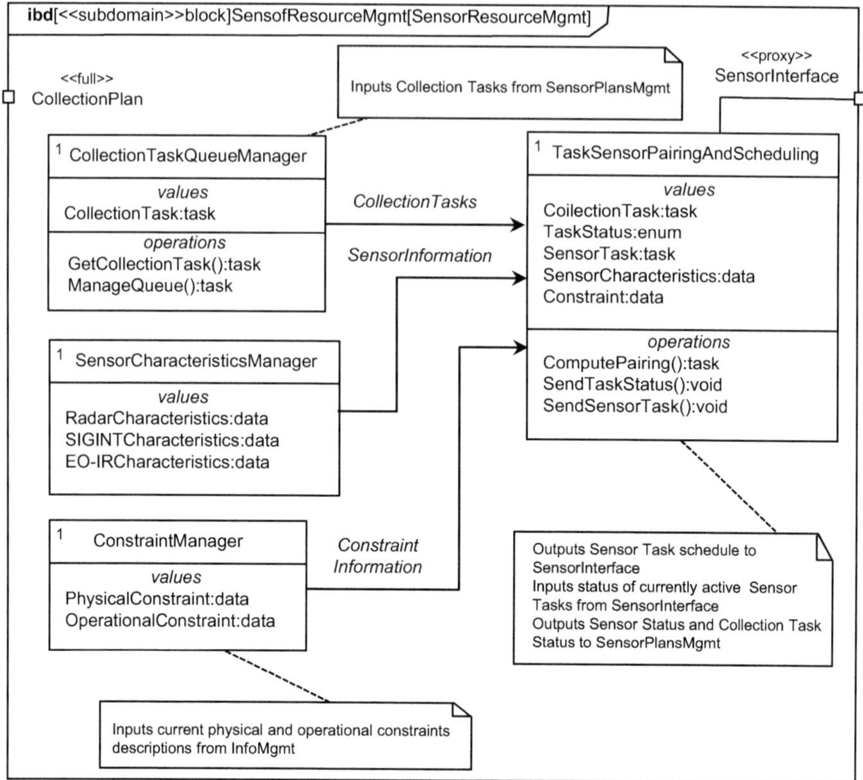

Fig. C.22 Sensor Resource Management Decomposition

their embedded processors, and the soft real-time environment of the mission domains and the crew members. The ports have associated provided and required interfaces that will be defined in the Physical Viewpoint with the specifics of timing, message formats, error detection and correction, electrical and mechanical design, and so forth. The following text gives brief descriptions of the Subdomains of the Sensors Domain.

- *Real-Time Sensor Interface*: The Sensor Interface allows sensors, with embedded real-time processing, to exchange information with the Sensor Management Domain by providing data buffering, signal translation, data formatting and conversion, and other functions.
- *Multimode Radar*: The Multimode Radar detects, tracks, and identifies targets, including a Ground Moving Target Indicator (GMTI) mode and collects imagery using a synthetic Aperture Radar (SAR) mode.
- *Signals Intelligence (SIGINT) Sensor*: This is a spectrum surveillance sensor that uses a variety of antennas and receivers to detect, analyze, and identify RF signals.

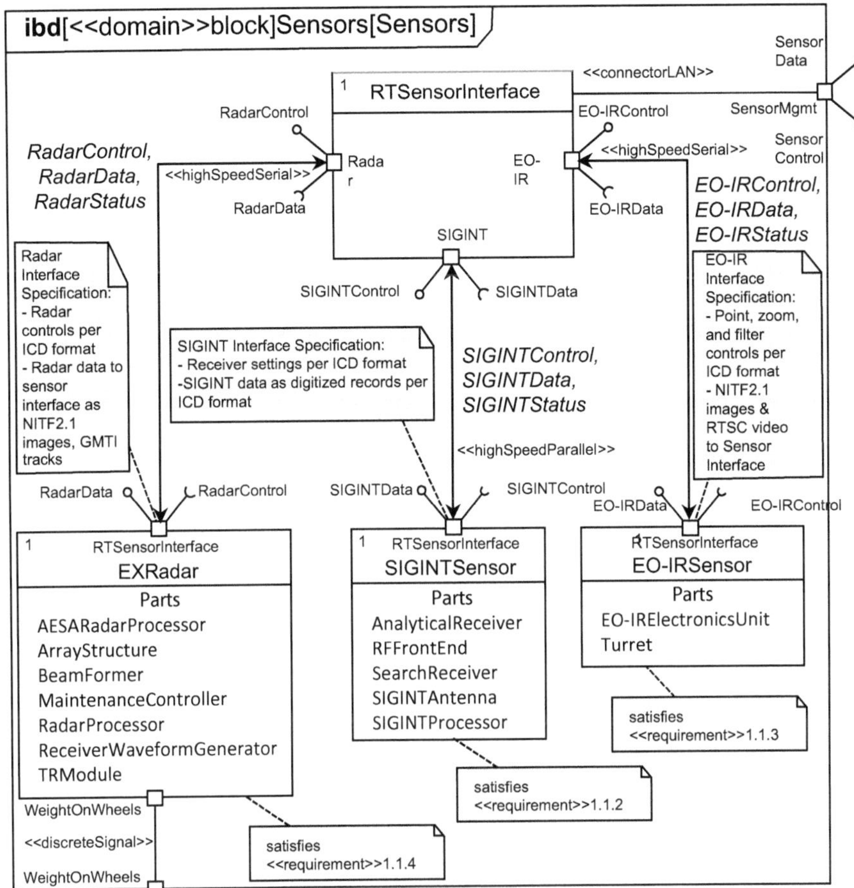

Fig. C.23 Sensors Domain Decomposition

- *EO/IR Sensor*: The EO/IR Sensor uses a steerable turret with optics and detectors in multiple spectral bands to collect imagery and video.

Radar Design Pattern: As part of documenting the rationale for the design of the E-X radar, Fig. C.24 is a design pattern expressed in a BDD that illustrates the inheritance of a system resource, the Multimode Radar, from a more general or `<<abstract>>` entity, a general Sensor (defined below as a Resource Foundation Class), to an RF Sensor, to an Active RF Sensor, and, finally, to a Radar. The pattern divides the Radar into an RF Aperture and a Processor and allows for multiple aperture types. Figure 4.10 shows this design pattern specialized to the design of the Multimode Radar of the E-X platform. The E-X uses an active electronically scanned antenna (AESA), commonly called a phased array, as shown in Fig. C.25. The design pattern in Fig. C.24 can also be realized in a radar that employs a mechanically steerable antenna with a single high-power transmitter to generate and point the beam.

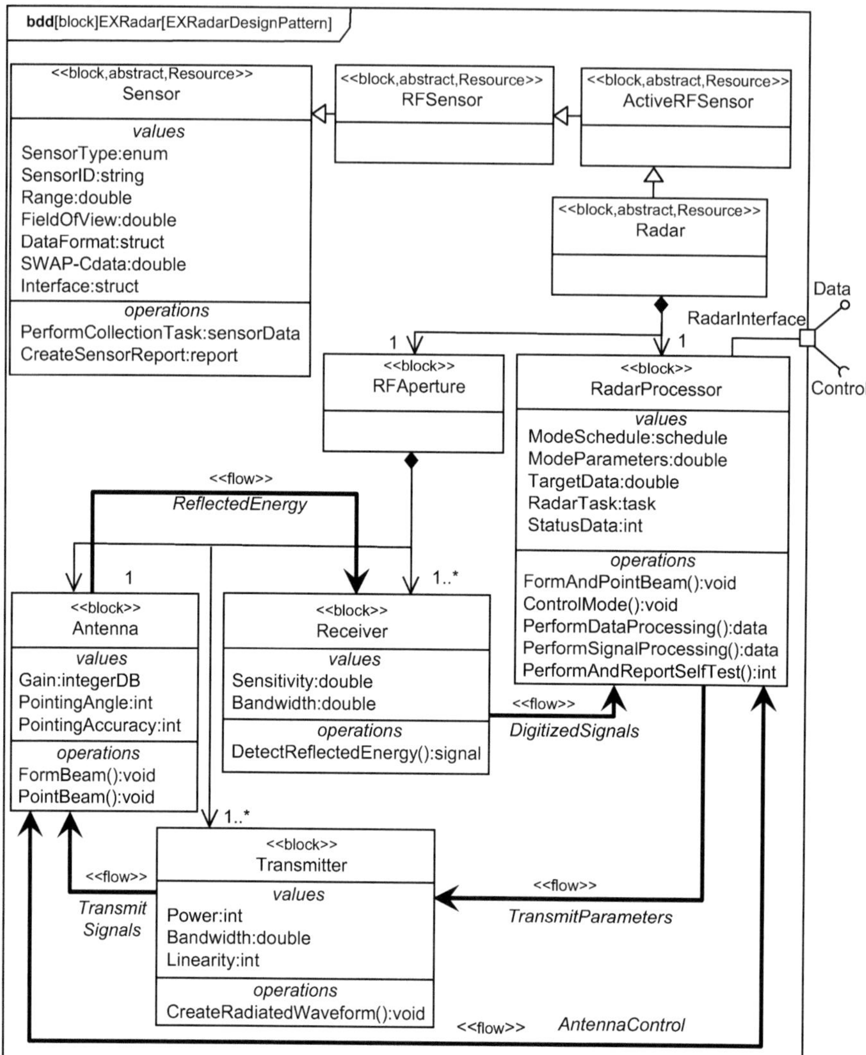

Fig. C.24 Design Pattern for a Radar, with Inheritance from the Sensor Resource Foundation Class

Radar Processor: Figure C.26 is a very simple design for the Radar Processor, showing Signal Processing, general-purpose Data Processor, and Radar Controller, each with delegated ports. The specific functions and interactions of these sub-components (Parts) will be spelled out in detailed design in the Physical Viewpoint.

EO-IR Sensor: Figure C.27 is a BDD showing the composition of the EO/IR Sensor, which collects both still images and video in several spectral bands. Figure C.28 is the corresponding IBD showing interaction among the sensor parts. The following text describes the main components of the sensor.

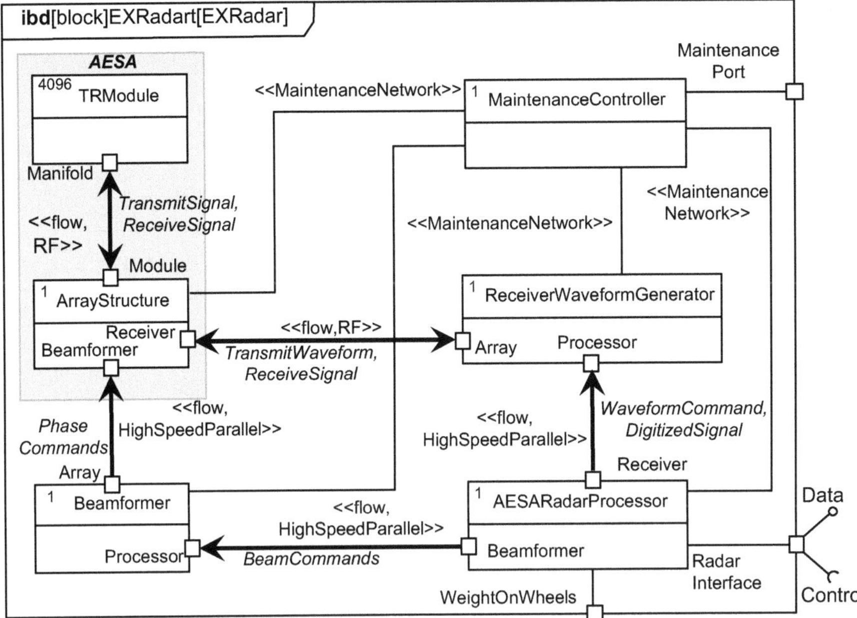

Fig. C.25 E-X Multimode Radar, using an Active Electronically Steered Array (AESA)

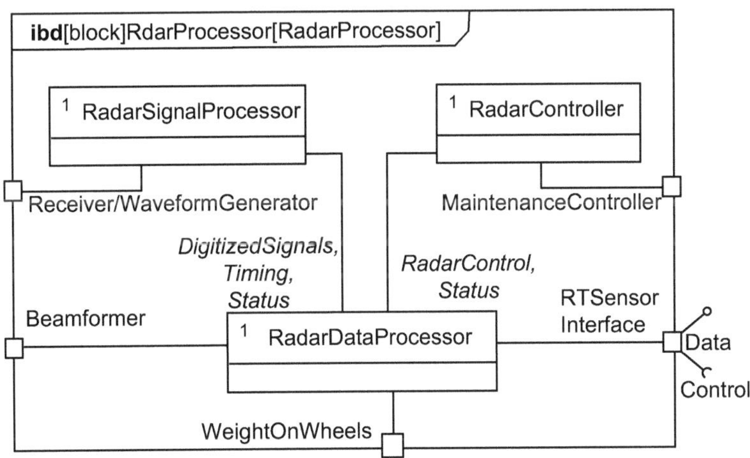

Fig. C.26 Top-Level Partitioning of the Radar Processor

- *Turret*: The EO/IR Sensor gimbaled turret houses the optics (telescopes) for EO and IR spectral bands, EO and IR focal planes mounted on cryogenic coolers for maximum sensitivity, the drive mechanisms for pointing in azimuth and elevation, focus and magnification controls, and other components.

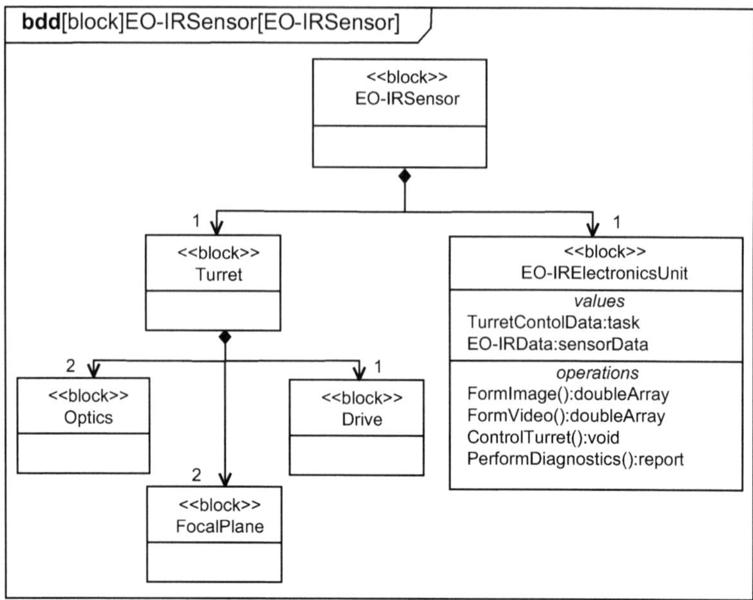

Fig. C.27 Composition of the EO/IR Sensor

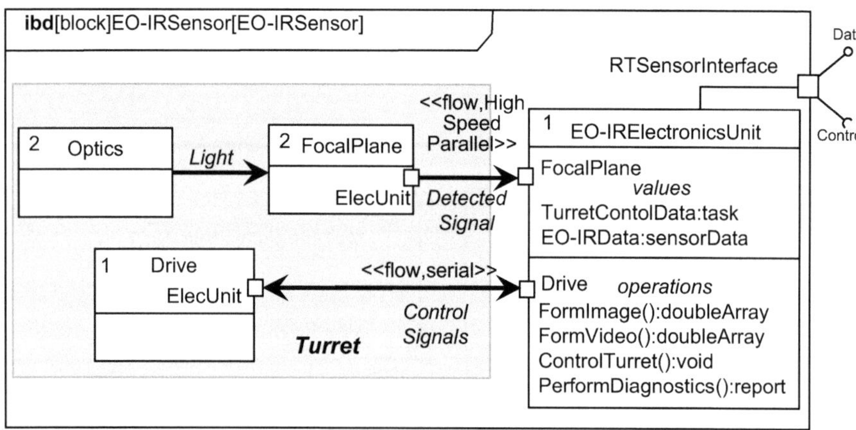

Fig. C.28 Interaction Diagram for the EO/IR Sensor

- *EO/IR Electronics Unit*: The EO/IR Electronics Unit controls the Turret and processes detected signals to create images and video.

Signals Intelligence (SIGINT) Sensor: The third E-X sensor gathers RF energy across a broad spectrum to detect and analyze signals. Figure C.29 shows the structure and interactions of the SIGINT Sensor. The Parts are defined in a BDD similar to Fig. C.27, which is omitted to conserve space.

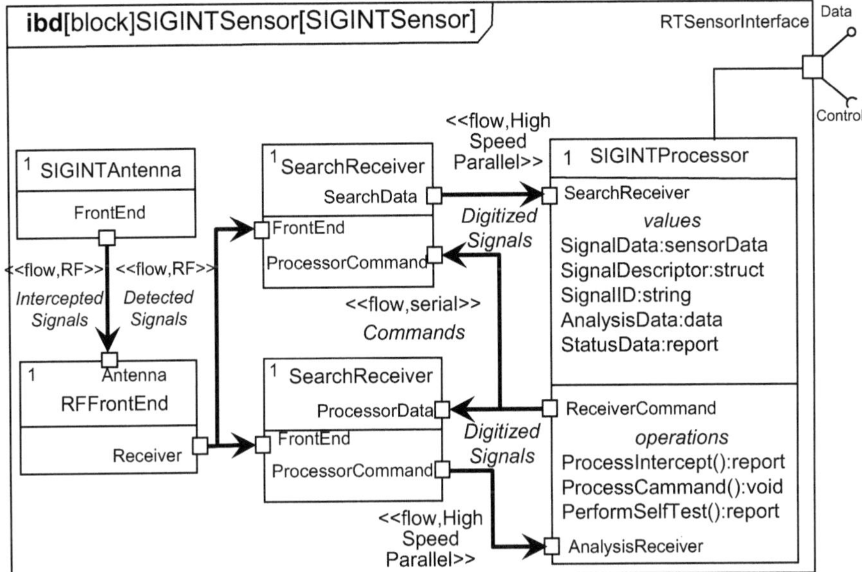

Fig. C.29 Interaction Diagram for the SIGINT Sensor

Information Management Domain: Figure C.30 begins the design of the system infrastructure, starting with the decomposition of the Information Management Domain. This Domain is built on a layered Service-Oriented Architecture (SOA), with an Integration Backbone (IB) that exposes infrastructure services to Mission Applications and system operators, a middleware layer that supports system functions, and a System Services layer that controls and provides interfaces to system computational and other resources. The InfoMgmt (Information Management) Port on the IB models the complex interface through which Mission Applications access the resources and functions of the infrastructure. The CommsMgmt (Communications Management) Port handles the flow of messages and status to and from the Communications Management Domain, and the CompInfra (Computing Infrastructure) Port supports the interface to the underlying computing infrastructure and resources. Connections among the Parts show how highly interconnected the Domain is.

Middleware: Middleware provides a wide variety of services that support the hosting and execution of Mission Applications and the functions of the Mission Crew. These include:

- *Workflow and Orchestration*—used to automate system functions and improve operator productivity
- *Common Utilities*—functions such as network time transaction management that are used by Mission Applications
- *Data Services*—support storage, retrieval, and management of information
- *Security Services*—implement required protective mechanisms for Information Assurance

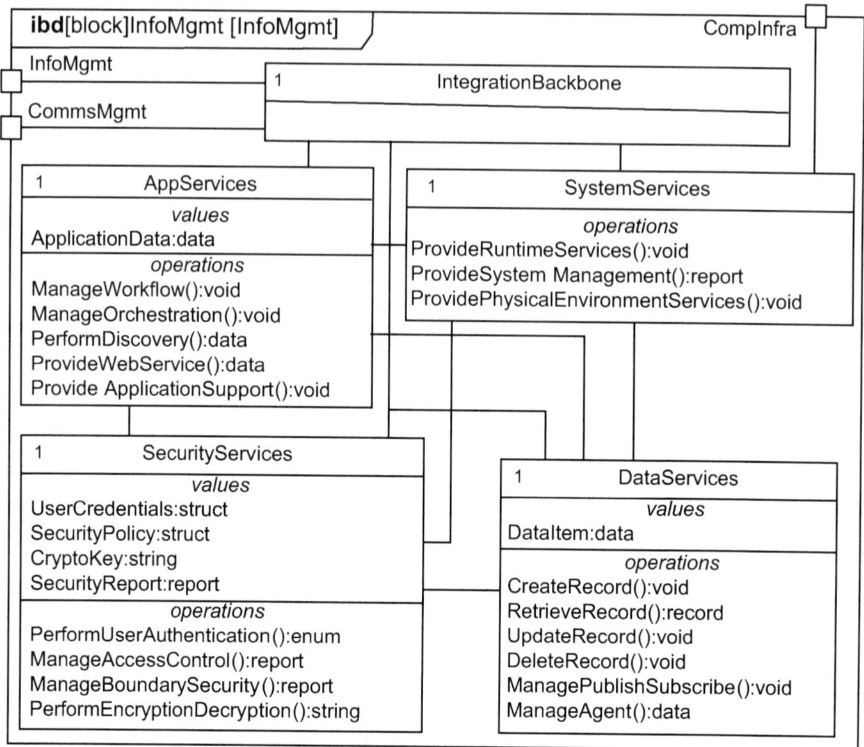

Fig. C.30 Information Management Domain Decomposition

In the PV, each of these will be implemented with one or more software components running on the Computing Infrastructure to provide the indicated functionality. As before, `<<proxy>>` Ports are delegated to the appropriate components. Client side components of System Services and Mission Applications are hosted on Operator Workstations.

Integration Backbone: Figure C.31 shows the primary components of the IB Subdomain, which is the central hub of E-X information processes and exchanges. It creates the environment in which Mission Applications are hosted and executed and provides access to the information services required by other Domains. It provides internal messaging and interacts with the Communications Management Domain to implement external messaging, including services provided by a Routing component. A Mediation component resolves conflicts over competing values for a given data item. This Subdomain uses Security Services, Application Support Services, System Services, Data Services, and other capabilities of the Information Management Domain to realize these functions. It may be implemented using an Enterprise Service Bus (ESB) or equivalent components. The Integration Backbone also supports the interface to external communications as well as internal communications among Mission Applications and other system elements.

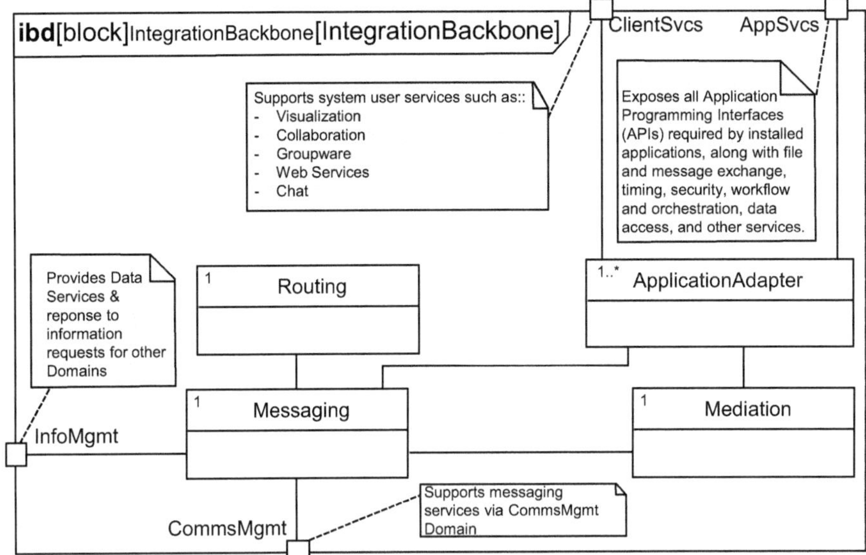

Fig. C.31 Decomposition of Integration Backbone

- *Application Adapter*—this software component creates the Application Programming Interface (API) required by each Mission Application to execute on the Integration Backbone and to interact with other applications and system resources.
- *Routing*—this software component supports the messaging component by determining addresses and paths for message transmission.
- *Mediation*—this software component resolves conflicts over differing data item versions.
- *Messaging*—this software component exchanges messages among Mission Applications and other system resources and manages message traffic with the external environment via the Communications Management Domain.

Application Services: The Application Services Subdomain provides a variety of functions used by Mission Applications. Workflow and Orchestration are used to automate mission processes through sequential invocation of Mission Applications and User interactions. Remotely hosted functions are accessed via a Discovery service.

Data Services: The Data Services Subdomain provides access to internal and external data resources. It supports Publish/Subscribe and Agent modes for information discovery and dissemination. It manages E-X databases.

System Services: The System Services Subdomain creates the underlying environment for E-X information processes and resources. It abstracts higher-level services and Mission Applications from the Computing Infrastructure. It includes components such as Operating Systems, Virtualization, File Handling, System Time, Diagnostics Management, Configuration Management, and Fault Handling.

Computing Infrastructure: Figure C.32 models the computing environment on which the Mission Domains and supporting services are hosted and executed. These model elements are specializations of the Computing Device Resource Foundation Class. This is a classic n-tier client-server design with workstations supporting various User Roles, a server (possibly multiple computers) and Network-Attached Storage (NAS) supporting shared functions and interactions, and a network interface to the real-time Sensor Domain, all connected by a dual switched Ethernet Local Area Network. Network ports are given "pseudonames" that will be replaced by actual port numbers and identifiers in the PV. The primary components are stereotyped as <<Workstation>>, <<Server>>, <<NAS>>, etc., and these will have Instance Specifications and other physical product data in the PV. The following text gives brief descriptions of the principal Blocks in Fig. C.32.

- *Mission Commander Workstation*—the Mission Commander Workstation provides the Mission Commander interface to the system and uses Mission Applications on the Server and local applications on the workstation to provide the required functions.
- *ISR Analysis Workstation*—this workstation provides the system interface for the ISR Operator and uses Mission Applications on the Server and local applications on the workstation to provide the required functions.
- *EO/IR Workstation*—this workstation provides the system interface for the EO/IR Sensor Operator and uses Mission Applications on the Server and local applications on the workstation to provide the required functions.
- *SIGINT Workstation*—this workstation provides the system interface for the SIGINT Operator and uses Mission Applications on the Server and local applications on the workstation to provide the required functions.
- *Radar Workstation*—this workstation provides the system interface for the Radar Operator and uses Mission Applications on the Server and local applications on the workstation to provide the required functions.
- *Local Area Network (LAN)*—this is a switched, dual, GBit Ethernet Local Area Network connecting the Computing Infrastructure.
- *Server*—the server hosts the primary Mission Applications and databases, system utilities such as built-in diagnostics and security, Communications Management software, and other system functions.
- *Network-Attached Storage (NAS)*—the NAS device provides high-speed, high-capacity, nonvolatile shared storage for databases and files.

Communications Management Domain: Figure C.33 shows the primary components within the Communications Management Domain. A Communications Message Management component creates the interface to the rest of the system and processes inbound and outbound messages. A Communications Interface component creates the required set of interfaces to the various communications devices. A Communications Resource Management component matches message traffic to communications channels and schedules external transactions.

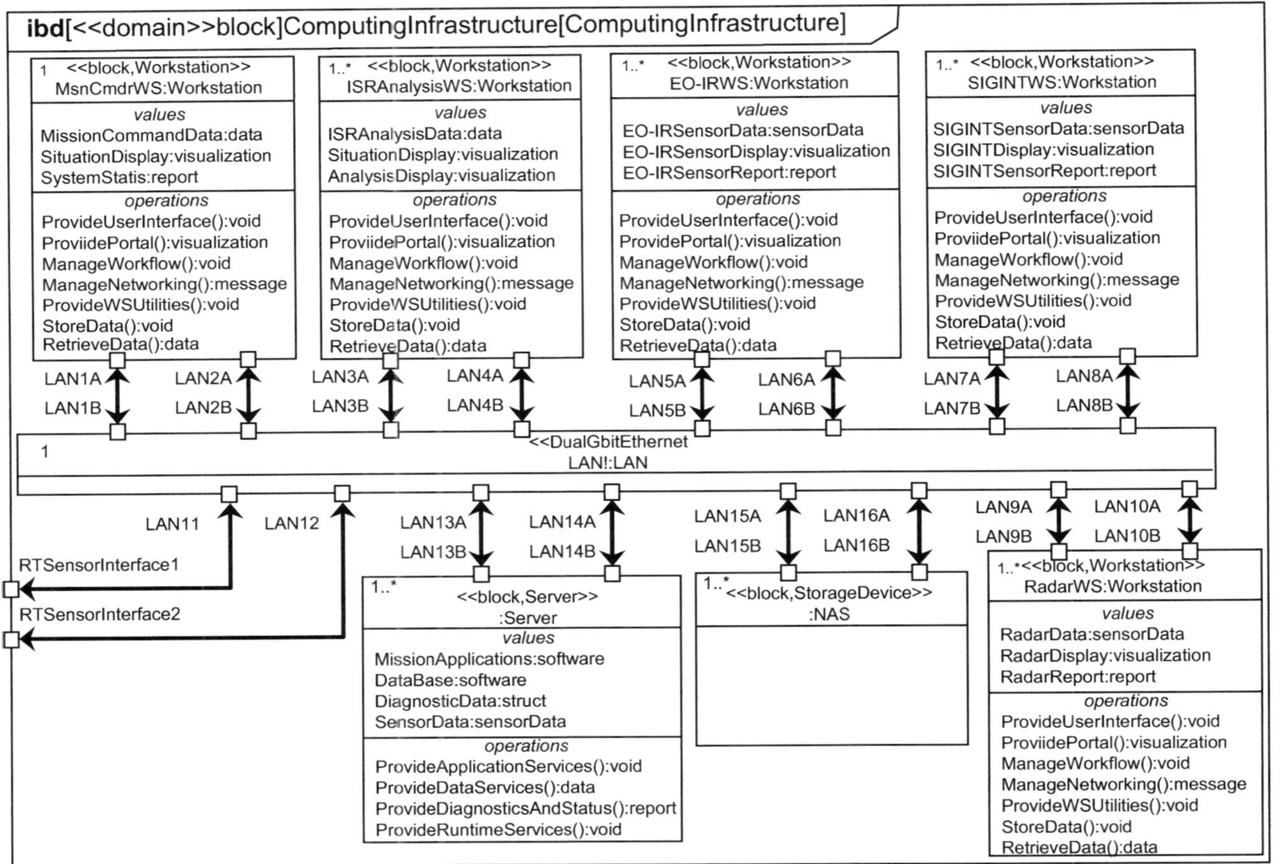

Fig. C.32 Computing Infrastructure Decomposition

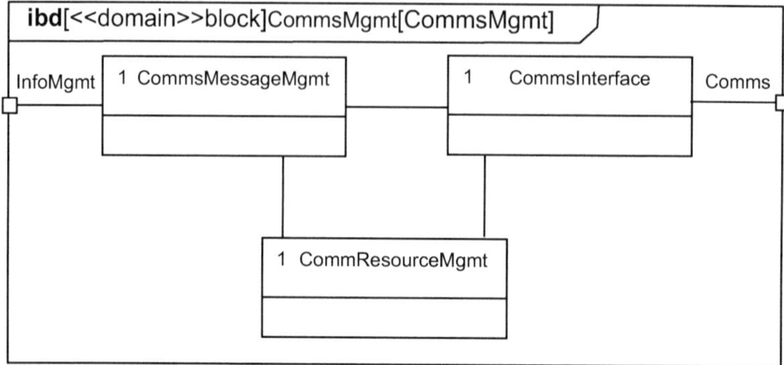

Fig. C.33 Communications Management Domain Decomposition

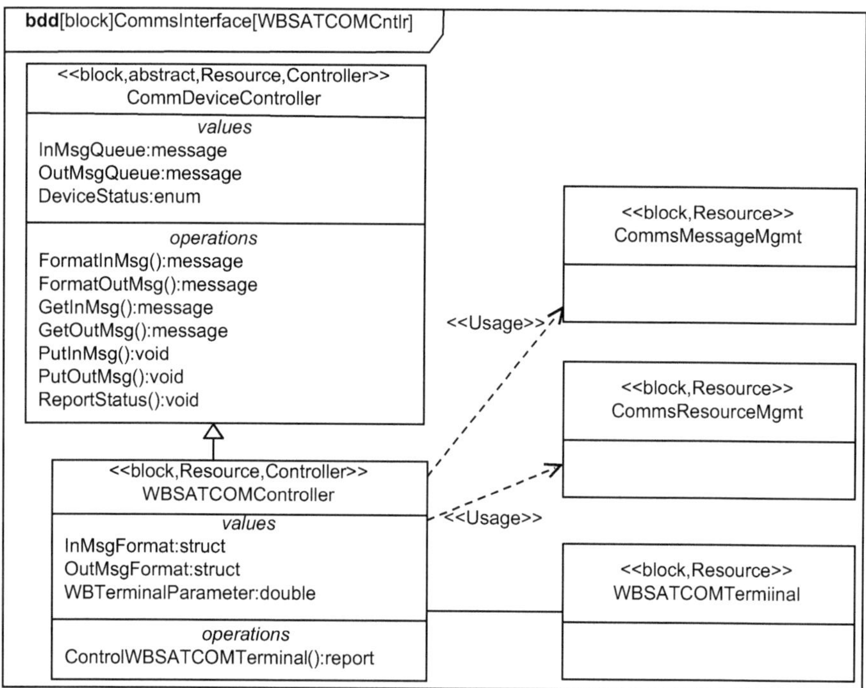

Fig. C.34 Wideband Satellite Communications Controller

Wideband Satellite Communications Controller: Figure C.34 illustrates a specific device controller within the Communications Interface component, in particular for the Wideband Satellite Communications channel, that inherits from a generic Communications Device Controller block. It uses the functionality of the Communications Message Manager and Resource Manager and interacts directly with the Wideband SATCOM terminal.

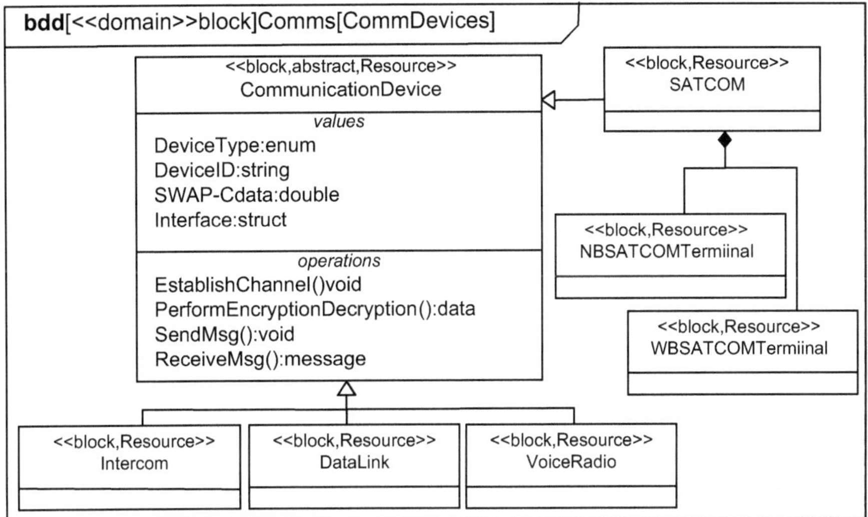

Fig. C.35 Resources of the Communications Domain

Communications Domain: Figure C.35 shows the content of the Communications Domain, with each device shown as a specialization of the Communications Device Resource Foundation Class.

Behavioral Perspective

Figures C.36, C.37, C.38, C.39, C.40, C.41, C.42, C.43, C.44, and C.45 show behavioral aspects of employing the E-X sensors. A real system architecture model would have many more, but these examples illustrate the nature and diversity of LV behavioral modeling. Again in a real model, each diagram would have, at a minimum, a model element specification describing the modeled behavior and giving the rational for the selected implementation.

Figure C.36 models a mission thread in which the Sensor Operator uses the Radar and SIGINT sensors cooperatively to detect, track, and characterize a target. The following text gives typical content of a specification for this diagram.

Mission Thread: Perform RF Sensing

Description: This diagram models the cooperative functioning of the E-X Multimode Radar and SIGINT Sensor to create and update target tracks. A track is the location and velocity of an object on the surface. This thread uses the Perform Radar Collection and Perform SIGINT Collection Use Cases and employs resources of the Sensors, Sensor Management, ISR Management, Information Management, and Communications Management Domains.

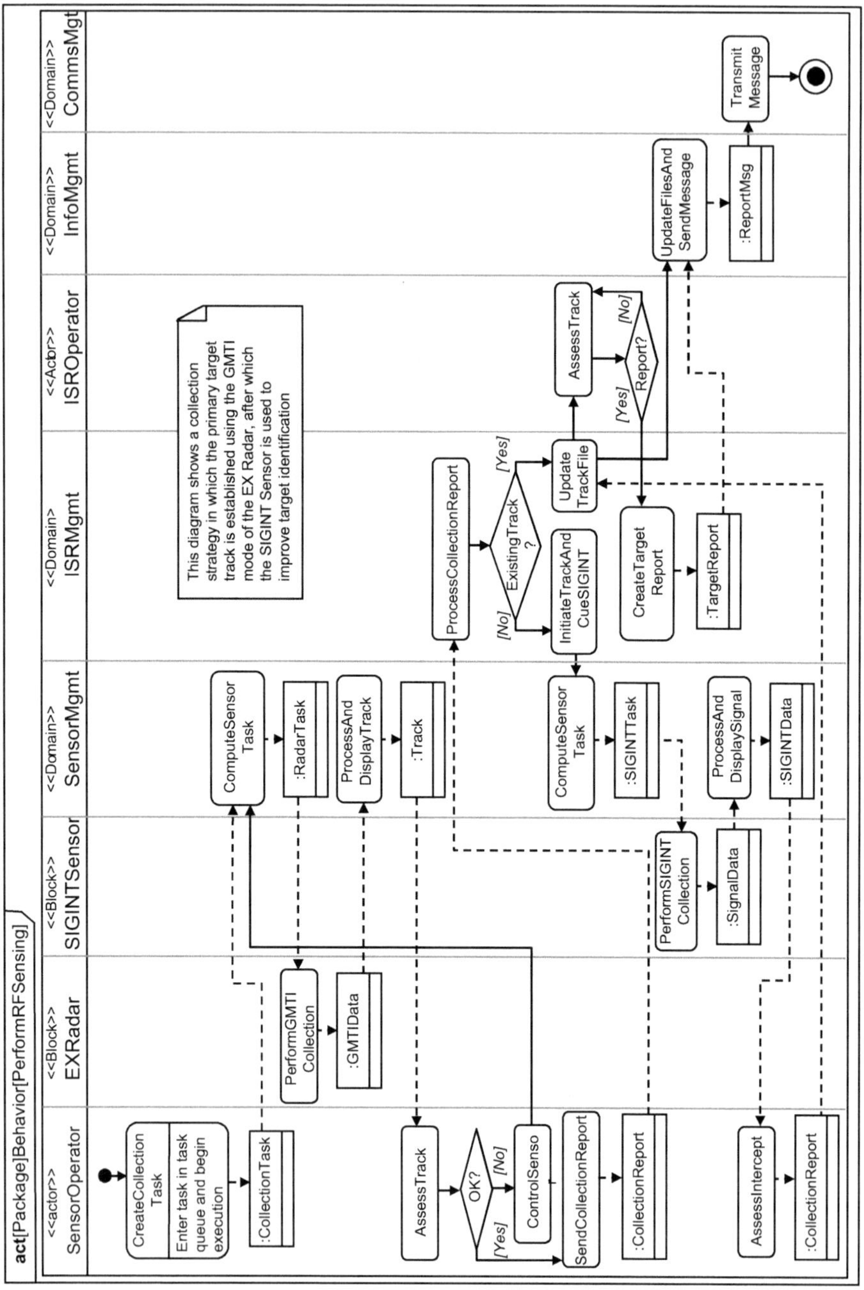

Fig. C.36 Mission Thread Activity Diagram for RF Sensing

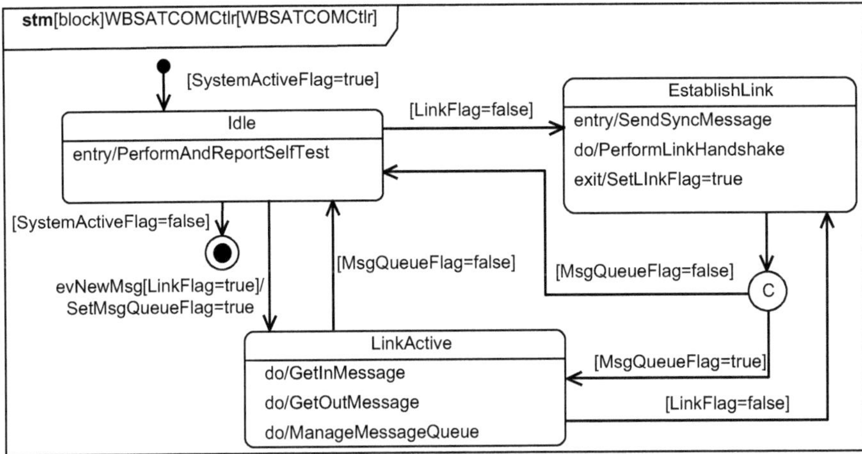

Fig. C.37 State Machine for the Wideband Satellite Controller

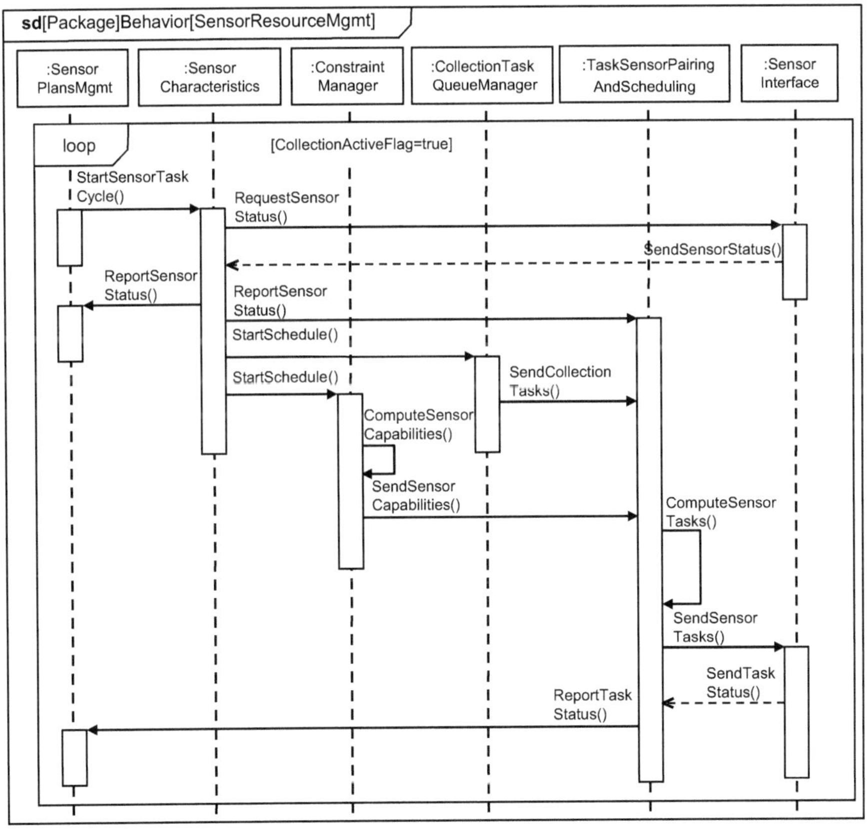

Fig. C.38 Sequence Diagram for the Sensor Resource Manager

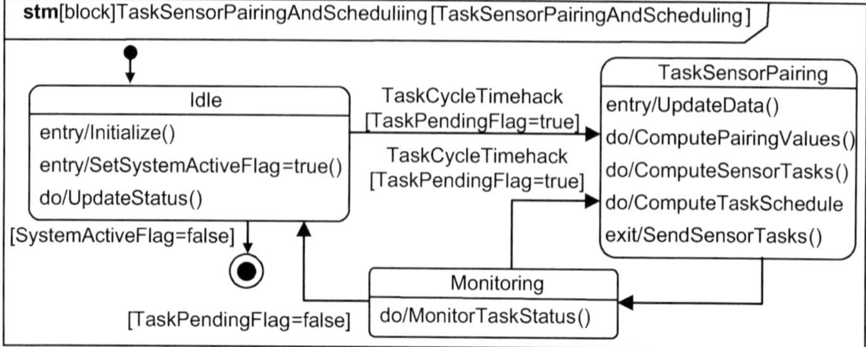

Fig. C.39 SMD for Sensor Task Pairing and Scheduling Component

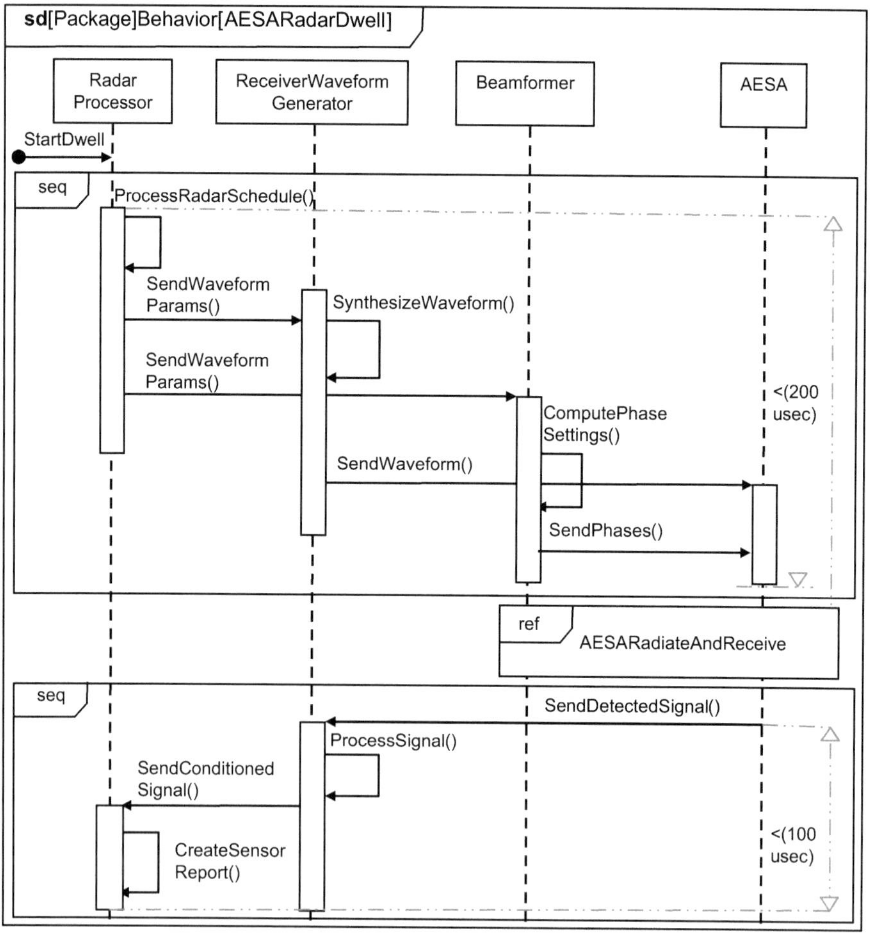

sd[Package]Behavior[AESARadarDwell]

Radar Processor

ReceiverWaveform Generator

Beamformer

AESA

StartDwell

seq

ProcessRadarSchedule()

SendWaveform Params()

SynthesizeWaveform()

SendWaveform Params()

ComputePhase Settings()

<(200 usec)

SendWaveform()

SendPhases()

ref

AESARadiateAndReceive

seq

SendDetectedSignal()

ProcessSignal()

SendConditioned Signal()

CreateSensor Report()

<(100 usec)

Fig. C.40 Sequence Diagram for an AESA Radar Dwell

Fig. C.41 Detailed Sequence Diagram for AESA Radar Collection of Target Data referenced in Fig. 4.24

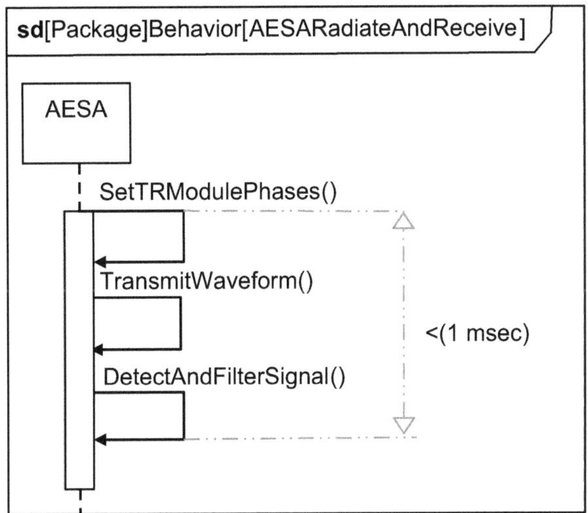

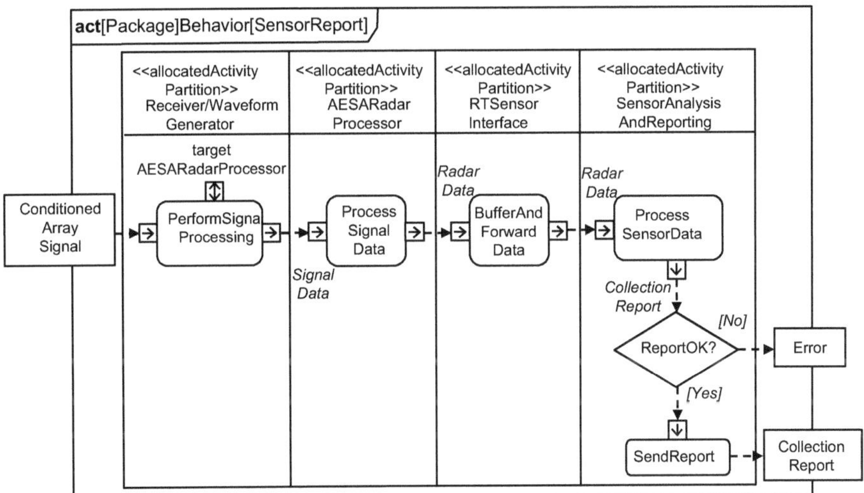

Fig. C.42 Activity Diagram showing Processing of Sensor Data into a Collection Report

The behavior begins when the Sensor Operator enters a Collection Task in the Sensor Plan Management Subdomain. The system develops a Radar Task causing Ground Moving Target Indication (GMTI) collection to be performed. The resulting collection report is assessed by the Radar Sensor Operator, and the collection is repeated if necessary. An approved Collection Report is forwarded to the ISR Management Domain where it is compared to the current Track File. If no match is found, the ISR Management Domain creates a SIGINT Task to collect identification data on the target. The SIGINT collection is assessed by the SIGINT Sensor Operator and, if approved, forwarded to ISR Management. The two sensor collec-

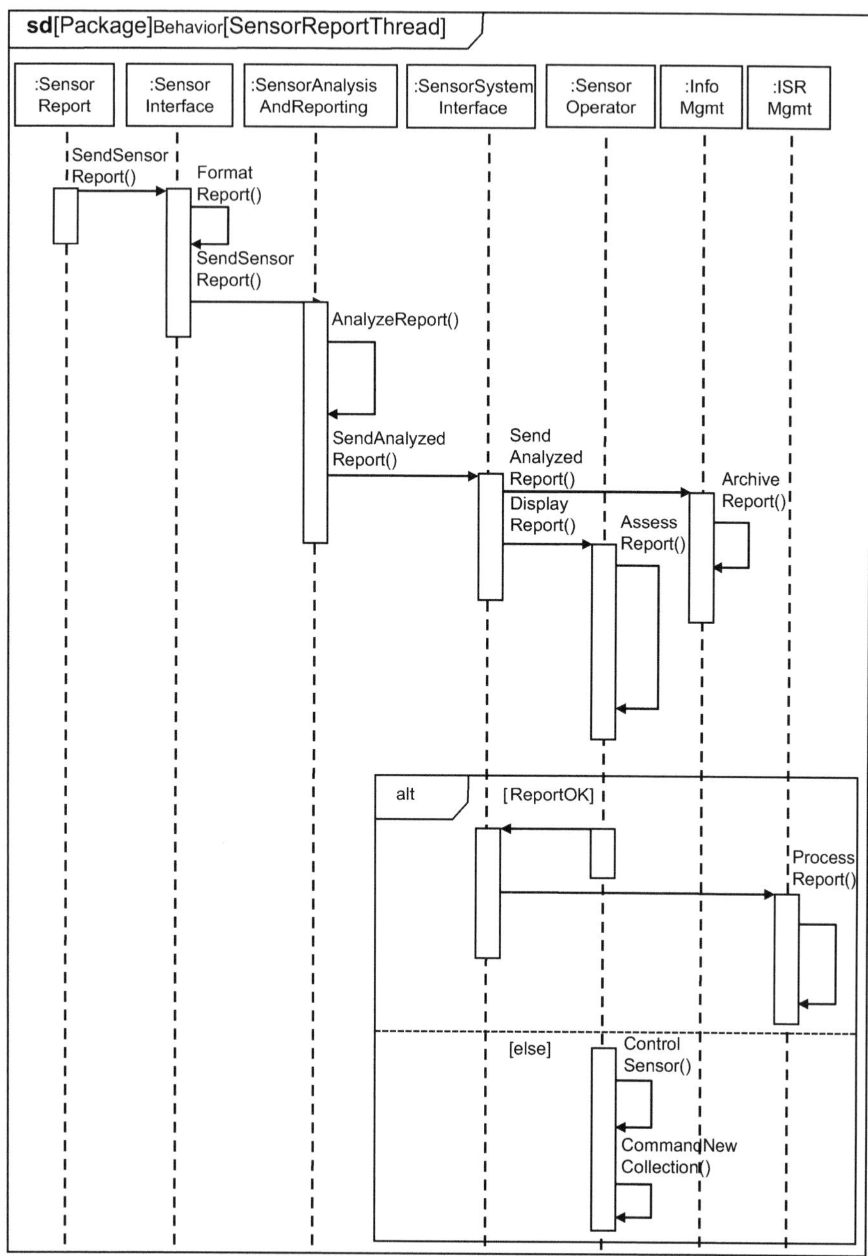

Fig. C.43 Sequence Diagram showing Creation, Assessment, and Forwarding of a Collection Report

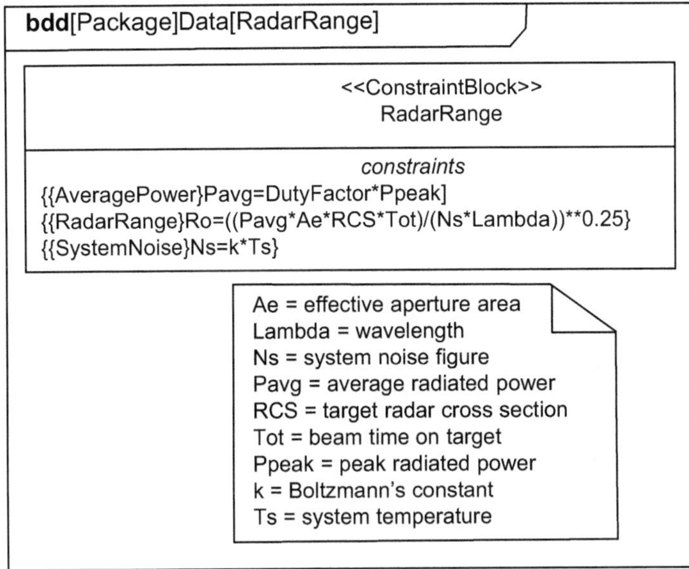

Fig. C.44 Constraint Block defining the equations for radar range

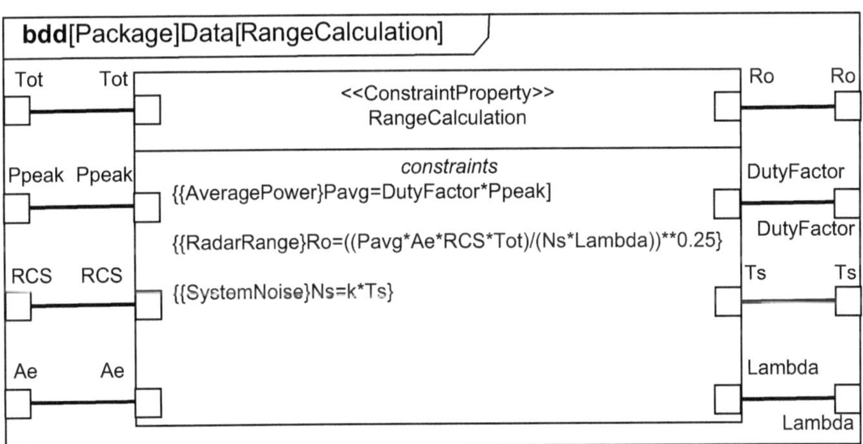

Fig. C.45 Parametric Diagram for the Radar Range Equation

tions are then used to update the Track File with a new track. If the new track correlates with a target in the Track File, the Track File is updated to refine the track. The ISR Operator assesses the new or updated track and, if it is approved, causes a Target Report to be created and sent to the Information Management Domain for processing and transmission to system customers.

Design Rationale: This thread employs E-X resources as intended in order to develop the most complete and accurate Track File possible. It does this by fusing multiple target detections, fusing track and SIGINT data to establish a target identification, and providing for Operator quality control. Track data is disseminated promptly to meet customer needs via the system infrastructure, using the best available communications channel.

Figure C.37 is a State Machine Diagram showing the primary functions of the Wideband SATCOM Controller, especially in establishing and maintaining the satellite link. Figure C.38 is a Sequence Diagram showing the behavior of the Sensor Resource Manager Subdomain. Expanding on this diagram, Fig. C.39 is an SMD showing aspects of the behavior of the Task/Sensor Pairing and Scheduling Component of the Subdomain. Figure C.40 shows the behavior of several components of the Radar to collect information on targets by radiating a waveform and receiving and processing the reflected signal during a scheduled operating period called a dwell. The diagram includes examples of time interval annotation. The diagram also illustrates the use of a Reference (ref) Interaction Occurrence to simplify a diagram by moving some detail to a subordinate diagram as shown in Fig. C.41.

Figure C.42 illustrates Activity Parameter Nodes as a way to model inputs and outputs. It uses Pin notation to show data object exchanges among Actions. The Perform Data Processing operation of the Radar Data Processor includes a Call Operation Action, modeled by the Pin on the top of the Action icon labeled targetRdrSignalProc. This invokes a supporting Operation in the form of methods provided by the Block identified by the "target" pin, in this case the Radar Signal Processor. Figure C.43 shows the mission thread that analyzes and approves sensor reports, including a decision by a Sensor Operator that is modeled by an `alt` fragment.

Finally, Figs. C.44 and C.45 illustrate the use of Parametric Diagrams to incorporate mathematical relationships in a model. Figure 4.28 shows a Constraint Block that defines the basic equations governing radar range performance, and Fig. 4.29 embeds the Block in a Parametric Diagram with connections to input and output values.

Data Perspective

Figures C.46, C.47, C.48, C.49, C.50, and C.51 illustrate the Logical Data Model by specializing Foundation Classes into System Classes; these are modeled in SysML as Blocks. Like all Blocks, these require element model specifications giving a description, owner, and any other details needed to fully define each information entity.

Physical Viewpoint Artifacts

Because the E-X is a notional example to illustrate an architecture model, the details of a point design that would constitute the Physical Viewpoint for a real system do not exist. The following sections summarize the nature and content of the artifacts in the various Perspectives of the PV.

Design Perspective

The Design Perspective of the E-X architecture PV is populated by adding physical details to the elements of the LV along with supplementary information needed to complete the system definition. The following examples illustrate this content.

Rationale for Product Selection—System elements that are defined by functional/performance specifications and in some cases constrained by Nonfunctional Requirements are physically implemented by either:

- Tasking a design team to develop, integrate, test, and qualify a hardware/software solution that satisfies allocated requirements
- Selecting an existing product, possibly with modifications or other tailoring to match requirements

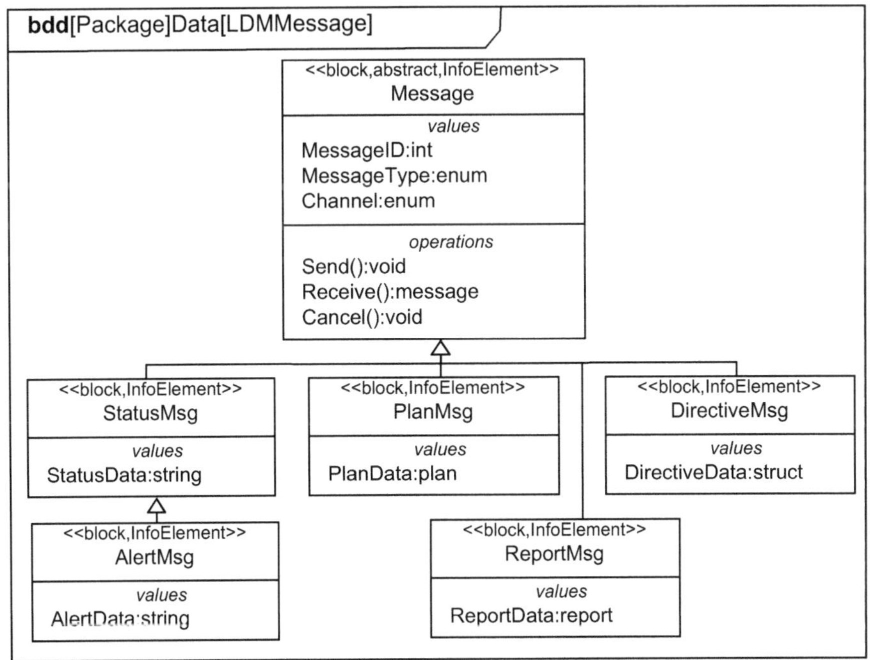

Fig. C.46 Logical Data Model for the Message Foundation Class

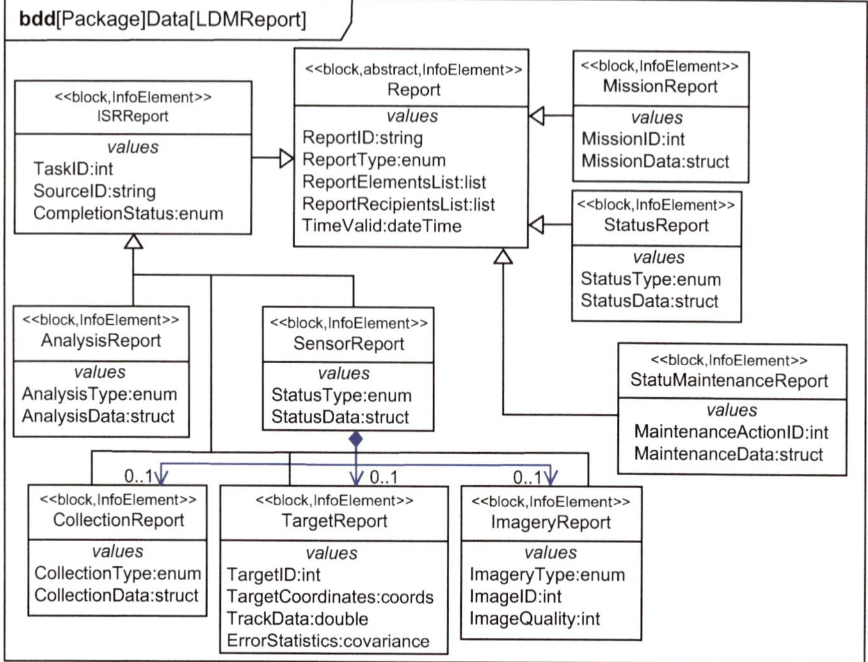

Fig. C.47 Logical Data Model for the Report Foundation Class

When product selections or design decisions are based on trade studies, they should be documented in the model or by reference to a design archive to ensure this rationale remains readily available. As an example, the following is a typical sequence of point design decisions associated with the E-X Multimode Radar:

- Derive by analysis the performance specifications required to meet system-level requirements.
- Collect and analyze data on existing radar products that are candidates, based on size, weight, performance, cost, etc., for use on the E-X.
- Perform a trade study to select the preferred alternative among existing products vs. a new radar development.
- Based on the trade study outcome, select an existing product, and existing product with modifications, or a new development.
- Incorporate a procurement or plan and execute a new radar development.
- Incorporate radar physical data in the architecture model or link to external documentation.

Model Element Specification: The following is typical of the data that would be added to the element specification in the model, either directly in the model tool window for the element, by providing links or references to external sources, or by incorporating documents in the model. This information represents the primary support provided by the architecture to system configuration management.

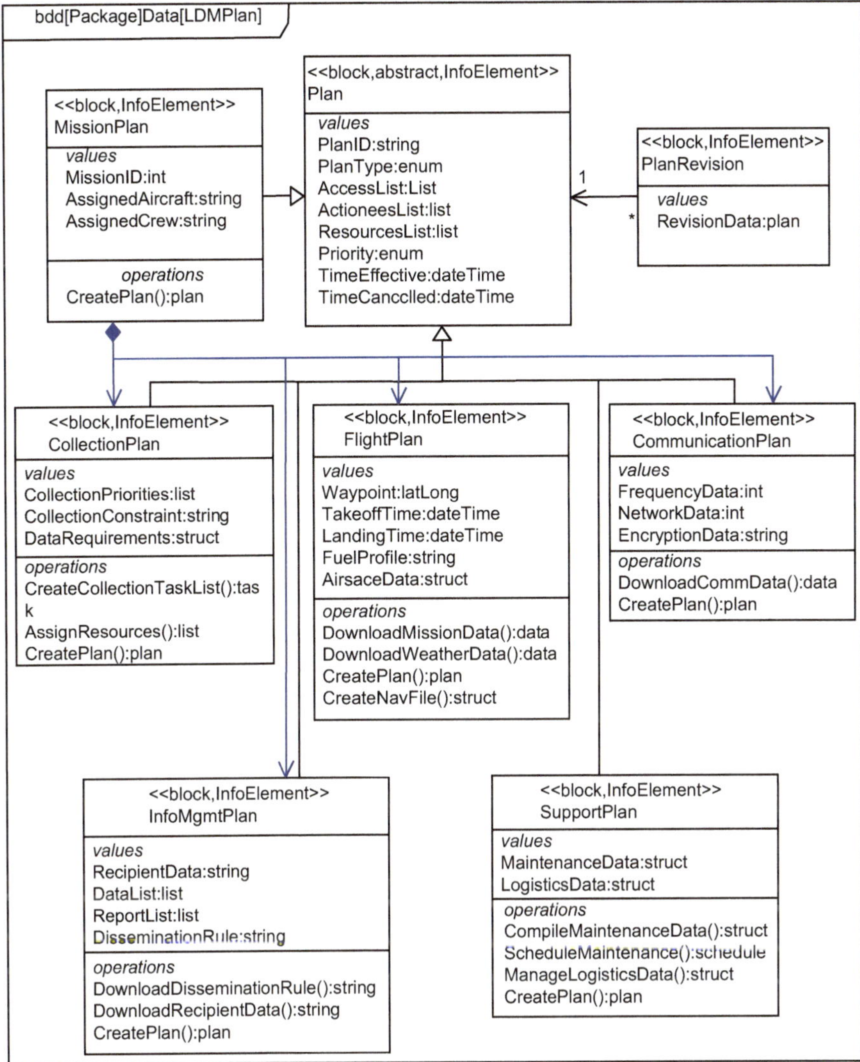

Fig. C.48 Logical Data Model for the Plan Foundation Class

- *Nomenclature*: product model number and name, revision number and date
- *Description*: summary of features and functions, product family, intended usage, implemented standards and practices, etc.
- *Drawings and specifications*
- *Physical characteristics*: dimensions, weight, power consumption, installation, cables and connectors, etc.
- Software documentation
- *Environment*: temperature range, shock and vibration, power quality, etc., in which the component is guaranteed to perform correctly and achieve specified reliability

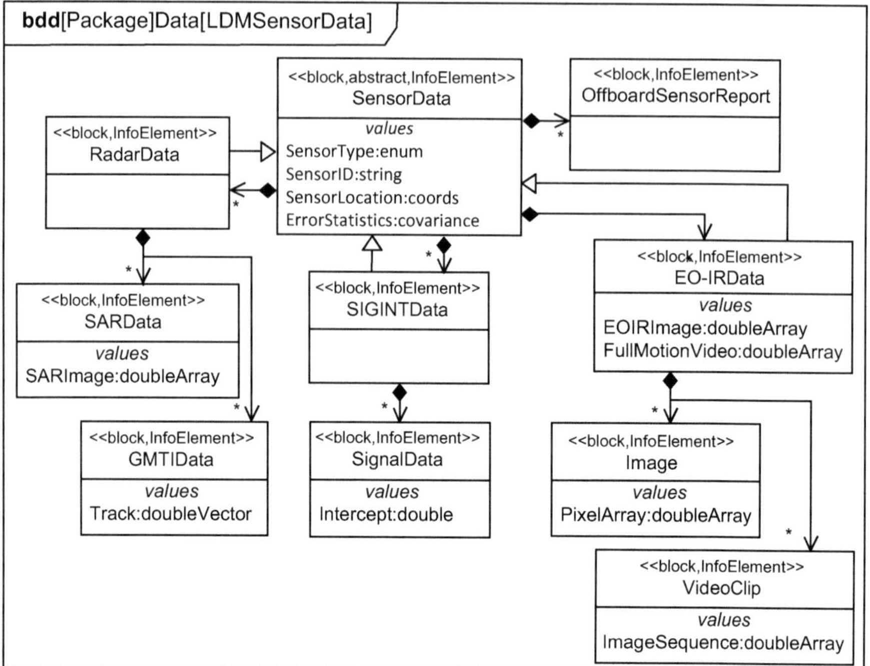

Fig. C.49 Logical Data Model for the Sensor Data Foundation Class

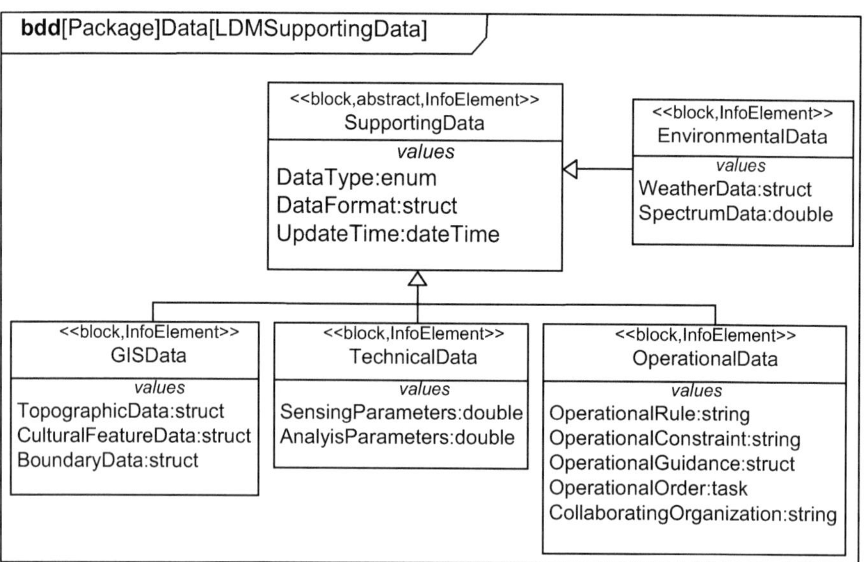

Fig. C.50 Logical Data Model for the Supporting Data Foundation Class

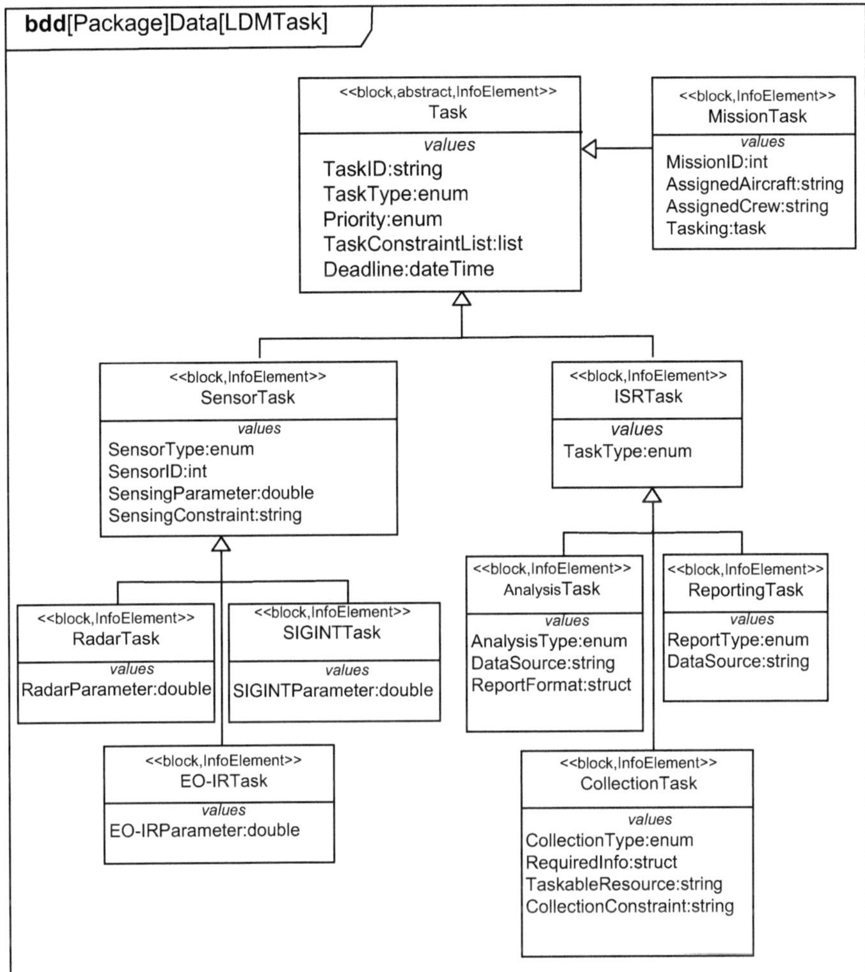

Fig. C.51 Logical Data Model for the Task Foundation Class

- *Maintenance, servicing, and repair data*
- *Supplemental Data*: any other information required to install, integrate, operate, upgrade, support, and potentially replace the component

Interface Specification: Full details of all internal and external interfaces are among the most important contents of the PV and can be captured in the model using the same mechanisms as Component Specifications applied to the Description and Contract tabs in the Features window of a Port that models the occurrence of an interface. The information may be contained in an Interface Control Document (ICD) or other formal documents. Both internal interfaces among system components and external interfaces to other systems or networks must be documented. ICD content typically includes:

- *Nomenclature and description*
- *Physical*: connectors, mounting hardware, vibration isolation, safety provisions, etc.
- *Electrical*: voltage and current, power quality parameters, etc.
- *Functional*: timing, data/message formats, electromechanical device characteristics, fluid pressures and flow rates, controls, standards, flow definitions, etc.

Another type of interface documentation applicable to an IP network is a Ports and Protocols Matrix that spells out for each network port such details as:

- Port Name
- Transport Protocol supported by the port
- Application Protocol supported by the port
- Allocation (static or dynamic)
- TCP/IP Port Number or range of numbers
- Source—server name, client name, or other using process
- Destination—same
- Direction—in, out, or bidirectional
- Rule—e.g., "Use default http proxy"
- Usage—time period when active, indefinite, etc.
- Description—brief statement of functions supported

System Design Data: Anything required to fully describe the entire system and not captured in component and interface specifications should be incorporated. This may include:

- System-level timing and performance
- Workflow and Orchestration descriptions and models
- User interface definitions and training materials
- System maintenance, servicing, calibration, troubleshooting, etc., including requirements for depot maintenance
- Security and information assurance
- System safety and hazard analysis

Standards Perspective

A table or spreadsheet identifying the standards implemented in the system should be included as an imported document or as a data table directly in the model. For the E-X, standards would include Information Technology standards for the Computing Infrastructure and User Interfaces, external communications and networking interface standards and protocols, equipment installation and power, security, safety, and others.

Physical Data Model

The E-X employs various databases and requires data and memory management in the Real-Time domains. The PDM would therefore include both database design documentation and details such as memory maps for embedded processing.

Attachment 1 to Appendix C: E-X Architecture Overview and Summary Information

Executive Summary

This document illustrates the organization and content of a typical Architecture Overview and Summary Information document, referred to in this attachment as the Architecture Overview. Content that describes the E-X architecture as if it were a real system is in plain text. [Content that is applicable to the E-X but is not defined by this architecture example is enclosed in square brackets.] *[Comments and other content that describes how this document would be modified for another architecture or situation are in italics and enclosed in square brackets.]*

This attachment is part of the example documentation based on the architecture of a notional airborne sensor platform referred to as the E-X. The E-X architecture is intended for use as a worked example of an architecture methodology called the Model-Based System Architecture Process (MBSAP) in a variety of educational situations. MBSAP is defined and illustrated in the main body of this book. The Architecture Overview summarizes the E-X architecture and the artifacts and supporting documents that define it. It follows the general outline of an All-Viewpoint 1 (AV-1) of the US Department of Defense Architecture Framework (DoDAF). *[Like other elements of the E-X example, it can be used as a format sample for preparing the Architecture Overview of an actual system architecture.]*

The following sections describe the E-X system; the scope, purpose, and perspective of the architecture; contextual information; schedule and resources for E-X architecture development; and the uses for which the architecture is intended. An important use of an Architecture Overview is to establish a de facto contract between program and technical leadership and the System Architect or Architecture Team, based on establishing bilateral commitments and understanding of the role of the architecture effort within the overall program context. *[These agreements, including empowerment of the Architecture Team to perform its proper functions in the program, should be agreed to by all parties and documented in the Architecture Overview.]*

The Architecture Overview is a living document that is updated as the architecture effort progresses. This example is written on the assumption that the E-X is nearing completion of functional design and preparing to enter implementation to develop an initial point design. *[In a real program, maintenance of a current Architecture Overview is an important responsibility of the System Architect; updates would typically be required after major design reviews, program milestones, or similar events. A final version accompanying the completed architecture model and summarizing Findings and Recommendations is an important element of development program documentation.]*

Architecture Identification and Description

General

The E-X is a notional system invented for educational purposes as an example of the application of the MBSAP methodology. It contains many of the features typically encountered in complex systems that relay on processing, display, and exchange of large amounts of information to accomplish business processes or mission functions. It therefore provides opportunities to illustrate the modeling and analysis of typical architecture content and artifacts. This Architecture Overview is part of an E-X architecture model and can be used as a format sample for an equivalent document in a real system architecture. [Specific program information includes:

- Formal title (nomenclature) of the system or enterprise being architected, in this example the E-X Airborne Sensor Platform
- Name(s) and Organization(s) of the Architect(s)
- Approving authority
- Number and completion date of the latest (current) architecture version]

Assumptions and Constraints

This section documents the principal assumptions underlying the architecture development and any constraints that impact the range of alternatives and the development effort. [The following typical examples would be tailored to the circumstances of an actual program.]

Assumptions: The E-X architecture has been developed based on assumptions like the following:

- The E-X system will be a primary operational resource for collecting, processing, and disseminating information in real- or near real-time about an area of interest using both data from individual sensors and integrated information derived from multiple sensors.
- The E-X will operate within a networked enterprise of sensors, effectors, and control nodes and will conform to enterprise standards and support enterprise services.
- The E-X will be required to evolve over time to accommodate new functions, technologies, and missions in order to remain operationally effective and supportable.

Constraints: Typical constraints on the E-X architecture are described in the following paragraphs:

- The E-X shall be based on an open architecture as described by [specific open architecture definitions, requirements, and assessment criteria]. [Many public and private sector organizations define and mandate open architecture policies and definitions.]

- The E-X Mission System (sensors, communications, onboard processing, etc.) shall be physically compatible with [specified aircraft platform], including maximum allowable size, weight, power and cooling, antennas and apertures, mission crew size, etc.
- The E-X Mission System shall conform to [applicable standards for safety, reliability, maintainability, etc.].
- The E-X Mission System shall be interoperable within [specified enterprise network] and shall employ and support [specified enterprise services].
- The E-X shall implement and comply with [specified customer architecture policies, including governance].
- The E-X shall operate and be supportable in the [specified operational environment].
- The E-X shall implement security requirements specified in [system requirements, System Security Plan, and other documents].

Architecture Development Resources

The E-X architecture is developed and maintained through the Model-Based System Architecture Process (MBSAP) implemented in [the architecture modeling tool or tools selected for the development program]. The estimated architecture development schedule, level of effort, and other required resources are defined in [specific program plans and budgets or provided here as a summary estimate].

Scope

The E-X architecture is baselined in a model conforming to the System Modeling Language (SysML) Profile of the Unified Modeling Language (UML) and includes artifacts in accordance with the MBSAP methodology that define Operational, Logical/Functional, and Physical Viewpoints (OV, LV, and PV). The SysML model is supplemented by Focused Viewpoints that describe specific aspects of the architecture of concern to various stakeholders. The model supports visualization of architecture data in various formats as appropriate for technical and programmatic reviews and decisions and to meet specific stakeholder needs. The architecture is documented and disseminated in [specified file or document formats]. The E-X architecture model is the foundation of a Model-Based System Engineering (MBSE) methodology that integrates tools, processes, and products for an efficient and effective program of development, integration, test, deployment, and support to realize system requirements.

Architecture Model and Artifacts

The architecture starts with a requirements baseline, which is refined and updated based on the results of architecture analysis. It proceeds through Operational, Functional, and Physical Viewpoints, each of which is organized into Perspectives that group artifacts according to the basic aspects of the architecture such as behavior, structure, data, and services. The Operational Viewpoint transforms requirements into an architectural context and establishes the overall organization of the architecture. The Functional Viewpoint fleshes out the architecture into a system design, with elements that represent each part of the system solution but without defining specific technologies, products, or standards. Finally, the Physical Viewpoint defines the implementation of the system by adding point design detail. The system then proceeds through build, integration, and test stages, and in a spiral development, the outcome of each architecture cycle is used to refine the requirements baseline and establish the functionality to be implemented in the succeeding spiral. Both a physical prototype and a virtual prototype composed of various system models and simulations tie the process together and support design and analysis at each stage. The process is fully reentrant and iterative; problems or opportunities arising at any point can be carried back to an earlier stage to refine the architecture. *[This section should include an MBSAP summary graphic such as Fig. 3.4.]*

Primary Artifacts

The following are the primary artifacts to be developed and used to define the E-X architecture. This list may be adjusted as required in response to program and customer needs. For each artifact, the format to be used is indicated; many artifacts are directly exported from the architecture model, and others are created and incorporated as text, graphics, and tables. These artifacts can be mapped to the content of standard architecture frameworks such as The Open Group Architecture Framework (TOGAF), the Federal Enterprise Architecture Framework (FEAF), and DoDAF.

Requirements: The E-X requirements baseline is defined and managed in [define here the process and artifacts to be used to manage requirements for the system]. *[A Requirements Perspective is optional under MBSAP if a separate requirements management tool is employed; however, including Requirements Diagrams in the architecture allows valuable allocation and traceability content to be explicitly captured and portrayed.]*

Operational Viewpoint: The OV includes the following Perspectives.

- *Contextual Perspective*:
  - Operational Context Diagram—a graphic with a high-level representation of the system, its environment, and interactions with external entities is a valuable artifact for helping users of the architecture understand scope and content.
  - Architecture Overview and Summary Information—this document.

- Integrated Dictionary—table of terms, definitions, and acronyms; Appendix C, Attachment 2 gives an example for the E-X.
- Additional Information—this includes any content that is required or useful in describing the system and its architecture. [Examples may include legal or policy documents, operational concept documents, descriptions of operational and support environments in which the system will function, etc. This section could be illustrated with a graphic depicting the operational context of the system.]

- *Structural Perspective*:
  - Domain Composition Diagram—Block Definition Diagram (BDD) showing top-level partitioning of the architecture into Domains. Blocks have compartments for Values, Operations, Parts, etc., as appropriate.
  - Domain Interaction Diagram—Internal Block Diagram (IBD) showing domains with ports and flows as well as associated User Roles.
  - Specifications—specifications for Blocks, Interfaces, Flow Items, Value Types, and other model elements, following formats prescribed by MBSAP.
- *Behavioral Perspective*:
  - Use Case Diagrams—showing primary Mission and Support behaviors with decomposition and associated User Roles
  - User Roles Diagram—specialized Use Case Diagram showing User Roles and the generalized role categories to which they belong
  - Activity Diagrams—a combination of Activity Diagrams for basic behaviors such as Use Cases and more complex diagrams that show the flow of Mission Threads across Domains
  - State Machine Diagrams—showing the states in which the E-X system can be and the transitions among those states, as well as any stateful behavior associated with Domains
  - Specifications—specifications for Use Cases, User Roles, etc.
- *Data Perspective*—a Conceptual Data Model (CDM) using BDDs with emphasis on Foundation Classes/Blocks representing the basic categories of information content; Blocks have compartments for Values and Operations and may require additional specification content for a full and unambiguous definition.
- *Services Perspective*:
  - Domains and their interfaces, including Ports, can be annotated to describe Services that are supported and exposed.
  - Artifacts such as a Services Taxonomy may be developed to identify internal and external (including Enterprise) Services and to allocate them to Domains.
  - Behavior diagrams can also be created or annotated to show the specific behaviors (e.g., Actions) that are associated with Services.
  - Specifications—high-level service definitions, which may include initial Service-Level Agreements.

Logical/Functional Viewpoint: The LV includes the following Perspectives.

- *Structural Perspective*—BDDs and IBDs showing decomposition of Domains, interactions among system elements and with external entities, and associated User Roles, as well as specifications; Ports and Item Flows are normally important content of this Perspective.
- *Behavioral Perspective*:
  - Sequence Diagrams—showing how groups of Blocks, User Roles, external entities, and others collaborate to realize behaviors, including timing annotations for time-sensitive behaviors.
  - State Machine Diagrams—showing stateful behavior of individual Blocks.
  - In some cases, Activity Diagrams may be useful, especially in modeling behavior involving external interactions.
- *Data Perspective*—a Logical Data Model (LDM) using BDDs to decompose the CDM to define system data entities with Inheritance from Foundation Classes in the CDM. Content may also include XML schemas or other formal data descriptions.
- *Services Perspective*:
  - Blocks and their interfaces, including Ports, can be annotated to describe Services that are supported and exposed.
  - Artifacts such as a Services Taxonomy may be decomposed and expanded to define Services in more detail and to allocate them to Blocks, Flows, and Interfaces.
  - Behavior diagrams can also be created or annotated to show the specific behaviors that are associated with Services.
- *Contextual Perspective*—a wide range of supplementary information and documents may be assembled and can be referenced by or incorporated into the architecture model.

Physical Viewpoint: The PV includes the following Perspectives.

- *Design Perspective*—adding detail such as product data to the elements of the Functional Viewpoint to document architecture implementation in a point design.
- *Standards Perspective*—a Standards Profile compiled in table format. If appropriate, projected future standards may be identified.
- *Data Perspective*—a Physical Data Model (PDM) showing detailed design of databases and other storage constructs.
- *Services Perspective*—adding detail to service interface specifications and, if available, service implementations, often complying with appropriate standards, to the Services Taxonomy.
- *Contextual Perspective*—a wide range of supplementary information and documents, including material such as training and maintenance manuals, may be assembled and can be referenced by or incorporated into the architecture model.

Focused Viewpoints: The following Focused Viewpoints address the concerns of specific system stakeholders. *[These are not illustrated in the E-X architecture example in Appendix C. Appropriate Focused Viewpoints should be selected and defined in accordance with Chaps. 3 and 6.]*

- *Runtime Environment Viewpoint*—representation of the software infrastructure and products used by executing mission applications
- *Hardware Installation Viewpoint*—traditional drawings and other documents for the hardware configuration of the system
- *Network Viewpoint*—similar drawings and documents focused on internal networks and interfaces to external networks
- *Communications Viewpoint*—collection of information on the communications equipment of the system and its configuration and operation
- *Security Viewpoint*—documentation of the mechanisms, processes, and procedures employed to implement a security policy and achieve required protection of sensitive information
- *Messaging Viewpoint*—representation of the messaging modes, protocols, policies, and other aspects of message-oriented information exchanges, internally and externally
- *System Management Viewpoint*—documentation of the products and procedures used to manage and control system resources, detect and recover from faults, perform information backup, and do other tasks essential to proper system operation

Architecture Timeline and Evolution

The E-X architecture describes the system at a point in time, assumed for this example to be completion of Functional Design, which would typically be approved at a Preliminary Design Review (PDR). The E-X will pass through an extended lifetime in operational service and is planned to undergo modifications and upgrades over time. Considerations of technology risk and available resources may dictate that the initially fielded system have only a portion of the capabilities that are ultimately intended. Considerations such as these mean that there is a time dimension to the architecture that includes the following.

- [Specific program phases and dates to which the current architecture applies, e.g., the date at which the system is planned for initial fielding.
- A timeline for future evolution of the system and its architecture, e.g., planning dates for major upgrades. Regardless of whether specific planning dates are available, it may be useful to designate successive system configurations and corresponding architecture baselines in terms of "current" or "initial" system, "interim" or "transitional" system, block upgrades, and an ultimate "target" or "objective" system.
- Identification of the process to be used to plan and execute system and architecture evolution over time.]

Organizations Involved

Participants in the architecture effort include [the system development organization, the customer, and any policy or regulatory bodies that establish requirements, standards, or other constraints on the architecture, e.g., for safety certification].

Purpose and Viewpoint

Architecture Purpose and Uses

The purpose of the architecture is defined in terms of specific stakeholder needs to be met. This content is essential to ensure a shared understanding between the Architect(s) and architecture users. Specific purposes include:

- Support to technical and programmatic decisions based on the impact of decision alternatives on overall system technical integrity and achievement of objectives
- Requirements analysis and refinement, including dialog with stakeholders, especially system users and support organizations
- Development of performance, design, and procurement specifications
- Performance, design, and other trade studies
- System design and analysis
- Standards selection and application
- Services and service-oriented architecture definition
- Integration and test planning
- Configuration management
- Support to system certifications such as safety and security
- Other system engineering processes

Architecture Analysis

A variety of analyses have been or are planned to be used in architecture development and validation. Details depend heavily on the specific nature and capabilities of a system. For the E-X under the MBSAP methodology, these include Modeling and Simulation, the basis for the Virtual Prototype, and a variety of specialized analyses. *[Architecture analysis tends to be system- and program-specific and requires tailoring and adaptation to meet the needs of a real program.]*
Modeling, Simulation, and Analysis (MS&A): MBSAP incorporates four levels of MS&A that comprise the Virtual Prototype. Each supports specific decisions in the development and implementation of the architecture. *[This section should include a graphic summarizing the MS&A environment such as Fig. 3.8.]*

- *Operational*—places a high-level functional representation of the system in one or more operational contexts and computes system-level Measures of Effectiveness (MoEs). These are used to refine and validate system-level performance requirements, including establishing the feasibility of meeting customer requirements.
- *Process*—creates an executable model of the overall processes involved in the Operational Viewpoint to allow analysis of activity threads, operator roles and task loading, key decision points, and other aspects of system behavior.
- *Architecture*—animates the system design as represented in the Logical/Functional Viewpoint to support validation of model completeness and correctness as well as initial timing and performance analysis, refinement of external interactions, and other design issues.
- *Physical*—uses a variety of tools for high fidelity analysis of specific issues such as resource loading, network throughput and latency, power and cooling, antenna and aperture parameters, electromagnetic compatibility, and many others.

The lower three levels of the MS&A hierarchy create an executable representation of the system and can be used as a Digital Product Model.

Specialty Analysis: Specific analysis tools and methods are required for the E-X. These include:

- *Reliability, Maintainability, and Availability (RMA)*—the architecture is being analyzed to ensure all required features to meet RMA requirements are present and correctly allocated to elements of the design.
- *Safety and Security*—the architecture is being analyzed as part of overall system safety and security assessments and may produce information to support certification and accreditation.
- *Human Factors*—the architecture is being analyzed to ensure all required features to satisfy human factors criteria are present and correctly allocated to elements of the design.
- *Interoperability*—the architecture is being analyzed to ensure all required features and functions required to interact with other systems and to participate in a networked enterprise are present, including appropriate standards, and correctly allocated to elements of the design.

Stakeholder Perspectives

The architecture is developed from the perspectives of the primary system stakeholders, including the development/acquisition organization, the system users, and the system maintenance and support community. Specific stakeholder concerns include [list here any specific stakeholder issues].

Context

Mission/Business Objectives

[The primary purpose for developing an Aerospace system is generally support for accomplishing specified missions. The equivalent in commercial architecture development is normally support for achieving business objectives. This section summarizes the purpose for which the system is developed and used.]

The mission of the E-X is to support operations in an area of interest with information about the operational environment and its content. This includes collecting multi-sensor data on terrain and targets, processing that data to produce usable information products, disseminating those products, and interacting with a ground station and other systems and organizations to accomplish operational objectives. *[In a real program, this section would typically also describe*:

- *An overall vision for the system*
- *Applicable operational doctrine*
- *Mission or business process scenarios, including any constraints, threats, or environmental conditions affecting the architecture*
- *Applicable standards, rules, criteria, and conventions, as well as their sources*
- *Linkages to other systems in use or in development, including a description of the enterprise in which the system being architected will participate*
- *Policies governing the system and its development or acquisition*
- *Information security and other aspects of the mission or business context]*

Architecture Status

This E-X Architecture Overview is written on the assumption that the system is in the late stages of Functional Design, ready to enter the Implementation phase. The architecture framework has been established; Operational and Functional Views have been modeled; and design detail is being added and refined based on the results of ongoing trade studies and detailed design efforts to produce a Physical Viewpoint. Coordination is underway with the organization and program responsible for the Ground Station as well as other stakeholders, especially the customers for the information products produced by the E-X. Mechanisms for this coordination include [specific Working Groups that meet periodically to share status, synchronize activities, and identify and resolve issues.]

The E-X architecture is being created using the MBSAP methodology in a SysML model and artifacts as detailed in section "Architecture Model and Artifacts." The architecture is being validated through a variety of analyses, including:

- Flowdown of requirements to ensure all are accounted for and have appropriate means of verification.
- Results of modeling and simulation at various levels to ensure that the system has the required behaviors and that predicted performance meets requirements.

- Dialog with stakeholders, especially system users and support personnel, to ensure the system operates as expected and has the features and functions to support operations in the intended environment; this may include human-in-the-loop simulations.
- Assessment of Quality Attributes to ensure overall technical characteristics are maintained.
- Data from I&T on a physical prototype to refine and verify the virtual prototype and confirm satisfactory system performance.
- Audits of the architecture against applicable policies and standards to ensure compliance.
- Results of specialty engineering analysis and testing, including failure analysis, safety analysis, and security analysis, to ensure that the system satisfies applicable criteria in all these areas.

Tools and Formats

The primary architecture tool in use is [identify the tool or tools selected for the program and the formats to be used for architecture artifacts]. Other artifacts, including text, spreadsheets, graphics, and presentations, are developed using [identify the tools and formats to be used for these artifacts].

Organizational Context

The architecture effort is under the direction of a Chief Architect who manages an Architecture Team and reports to a Chief Engineer or to the Program Manager. The organizational context is shown in [an organization chart showing participants, relationships, and responsibilities.] *[In a real program, specifics of the program organization will depend on the nature of the program.]*

Rules and Criteria

General

The E-X is being developed in full compliance with applicable customer policies and directives, including architecture frameworks, enterprise architecture, safety and security directives, supportability in the intended operational environment, and any others applied to the system and the development program. The architecture is being developed, modeled, analyzed, and documented to satisfy the purposes identified in this document, including decision support and coordination among system stakeholders. *[This part of the Architecture Overview depends heavily on specific policies and rules of the organization sponsoring the development.]*

Quality Attributes

The architecture is continually assessed in terms of Quality Attributes to ensure overall technical integrity of the solution, support technical and programmatic decisions, and provide early identification of problems and issues for timely resolution. Quality Attributes and the associated process are defined in [specific documentation, which may include an appendix to this Architecture Overview]. Quality Attributes help assure that the following criteria for success in E-X architecture development are satisfied:

- Providing support for the purposes listed in section "Primary Artifacts"
- Demonstrating compliance with customer policies and directives
- Creating data items, presentations, and other deliverables and supporting stakeholder dialog

Governance

Architecture governance mechanisms are employed to control and enforce the development and application of the E-X architecture, including:

- An Architecture Team that resolves issues, makes decisions, and documents architecture analysis.
- Configuration management and change control applied to the architecture model and its artifacts to maintain a current approved baseline.
- Program-level mechanisms, including Quality Attributes, for applying the architecture and enforcing architecture principles. These should be included in program plans and other appropriate documents.

Findings

Analysis Results

The E-X architecture has been subjected to analysis as described in section "Architecture Model and Artifacts" in developing the Operational and Functional Viewpoints. Key results of that analysis include:

- An open architecture foundation, assessed through an Openness Quality Attribute, has been defined and provides the basis for initial system implementation and long-term system evolution.

- A design pattern based on Mission Management, Intelligence/Surveillance/Reconnaissance (ISR) Management, and Sensor Management, supported by shared Information Management and Communications Management, provides an effective partitioning of the system in accordance with open architecture principles.
- Real-time processes associated with sensors and communications are effectively interfaced to human-in-the-loop processes through appropriate interfaces.
- All functional and nonfunctional requirements are satisfied at the functional level, subject to verification of the implemented system through test, demonstration, analysis, and inspection.
- Analysis of mission threads using Process Modeling, Simulation and Analysis shows that high-level timing constraints, operator workloads, resource loading, and other system performance aspects are compatible with requirements.
- A preliminary physical budget for size, weight, and power-cooling (SwaP-C) is compatible with the capabilities of the selected aircraft platform, subject to verification for the implemented system.

[The Architecture Overview for a real system would include content like this but specific to the system and program.]

Recommendations

- The E-X system architecture should proceed to implementation, including detailed design, integration, and test, following the methodology defined in this document.
- Quality Attributes should be tracked, analyzed, and reported as the basis for ensuring overall technical integrity of the solution and satisfaction of all stakeholder needs.
- The architecture model and artifacts, as well as virtual and physical prototypes, should be placed under rigorous configuration and change control.

[The Findings and Recommendations content of this document will be specific to an individual program and will become significantly more detailed as implementation and operational support of the system proceed.]

Attachment 2 to Appendix C: E-X Architecture Integrated Dictionary

Term	Model element	Definition
Active RF Sensor	CDM	<<abstract>> Class that models RF sensors which operate by transmitting a radio-frequency signal and detecting a target echo
Adapter	Component	Software component that creates the Application Programming Interface (API) required by each Mission Application to execute on the Integration Backbone and to interact with other applications and system resources
Aircraft Operations	Use Case	Activities associated with preparing the E-X aircraft for flight, conducting flight operations, and recovering the aircraft after flight, including pre- and postflight inspections and servicing as well as ground and air operations
Aircraft Systems	Domain	Contains the systems, equipment, and resources that are part of the E-X aircraft and interact with Mission Equipment and the Mission Crew
Aircraft System	CDM	<<abstract>> Foundation Class for resources that are integral to the E-X aircraft and distinct from Mission Equipment; provides abstraction of the content of the Aircraft Systems Domain
Alert	CDM	<<abstract>> Information Element that models data involved in urgent notification of an event or situation
Analysis Data	CDM	<<abstract>> Information Element that models the information content of an Analysis Report
Antenna	Component	Mechanical structure of an RF Aperture; may be a passive radiator or provide for installation of an array of transmit/receive modules
Application Services	Subdomain	Provides a variety of functions used by Mission Applications; Workflow and Orchestration are used to automate mission processes through sequential invocation of Mission Applications and User interactions; remotely hosted functions are accessed via a Discovery service
C2 Platform	Context Block	External platforms (e.g., aircraft) that provide C2 services with which the E-X interacts
Communications	Domain	Contains the radios and other resources and functions that implement external communications and networking
Communications Device	CDM	<<abstract>> Foundation Class for external communications resources; provides abstraction of the content of the Communications Domain
Communications Device Controller	Component	<<abstract>> Component that models the resources used to interface communications devices to the system
Communications Management	Domain	Contains the resources and functions that control external communications and networking and provide the interface between the onboard information processes and the external communications environment

Term	Model element	Definition
Communications Message Management	Subdomain	Communications Management Domain functions that control inbound and outbound message flows
Communications Resource Management	Subdomain	Communications Management Domain functions that control message routing and scheduling for communications devices
Computing Device	CDM	Foundation Class for the computing, networking and storage resources of the E-X; provides abstraction of the content of the Computing Infrastructure Domain
Computing Infrastructure	Domain	Contains the computers, networks and storage that execute Mission System software and allow User interaction with the system; forms the lowest tier of the layered architecture; components include Mission Commander Workstation, ISR Analyst Workstation, EO/IR Workstation, SIGINT Workstation, Radar Workstation, Local Area Network, Real-Time Sensor Interface, Server, and Network-Attached Storage
Data Archiving	Use Case	Activities associated with transferring mission data from an E-X aircraft to an information processing and archiving environment for archival storage and post-mission analysis
Data Item	CDM	<<abstract>> Foundation Class that models generic characteristics of system data categories
Data Services	Subdomain	Provides access to internal and external data resources; supports Publish/Subscribe and Agent modes for information discovery and dissemination; manages E-X databases
EO/IR Electronics Unit	Component	Controls the Turret and processes detected symbols to create images and video
EO/IR Sensor	Component	E-X Sensor that provides information about the external environment using cameras operating in the visible (referred to as EO) and infrared (IR) bands; consists of a sterrable turret and an electronics unit
EO/IR Turret	Component	Provides azimuth and elevation pointing of the optical components of the EO/IR Sensor; includes a telescope with selectable zoom optics, focal planes for each band, and drive mechanisms for turret pointing
EO/IR Workstation	Component	Provides the system interface for the EO/IR Sensor Operator and uses Mission Applications on the Server and local applications on the workstation to provide the required functions
E-X	System	Manned airborne multi-sensor platform that collects information and creates and disseminates Intelligence, Surveillance and Reconnaissance (ISR) products in coordination with a Ground Station and various supported and supporting organizations
EX Flight Unit	Organization	Block that models the personnel, resources and relationships associated with providing and managing E-X mission crews

Term	Model element	Definition
EX Organization	Organization	Block that models the personnel, resources, and relationships associated with owning and operating the E-X system
External Comms/ Networks	External Resource	External Communications Systems and Networks with which the E-X aircraft interacts
Flight Crew	User Role	Personnel responsible for aircraft operations; pilot and copilot
Flight Deck	Domain	Contains the controls, displays, and interfaces that the Flight Crew employs to conduct flight operations and coordinate with the ISR Crew
Foundation Class	CDM	Highest level of abstraction of system content, including data and resources; used to establish overall patterns that are specialized through inheritance to create system Classes and Blocks.
Foundation Data Class	CDM	<<abstract>> Information Element that models information entities
GPS	External Resource	Global Positioning System constellation that provides PNT data for precision E-X navigation.
Ground Station	Organization	A terrestrial location hosting communications and information processing resources that participates in E-X missions to supplement onboard capabilities for analysis, reporting, and mission coordination.
Information Management	Domain	Contains the resources and functions that integrate and support E-X information processes, including access to onboard and off-board data, data storage and retrieval, access to external communications, and security services; implemented as a layered service-oriented architecture (SOA) that hosts mission applications from the primary mission domains and provides messaging and file sharing among them
Integration Backbone	Subdomain	Provides the central hub of E-X information processes and exchanges; creates the environment in which Mission Applications are hosted and executed and provides access to the information services required by other Domains; provides internal messaging and interacts with the Communications Management Domain to implement external messaging; uses Security Services, Application Support Services, System Services, Data Services, and other capabilities of the Information Management Domain to realize these functions
ISR Analysis and Reporting	Subdomain	Processes data from Collection Reports along with supporting Technical Data to create ISR Reports; also supports various ISR Operator analysis and reporting tasks
ISR Analysis Workstation	Component	Provides the system interface for the ISR Operator and uses Mission Applications on the Server and local applications on the workstation to provide the required functions
ISR Crew	User Role	<<abstract>> User Role that models characteristics of personnel who execute ISR functions

Term	Model element	Definition
ISR Management	Domain	Contains the resources and functions used by the ISR Operators and any supporting staff to exercise overall control of mission execution
ISR Operations	Use Case	Use of E-X mission equipment to gather, analyze, store, and communicate ISR information; the Mission Crew executes a Collection Plan that defines how sensors are to be employed and how collected information is to be treated; the preplanned Collection Plan may be modified during a mission based on circumstances such as urgent requests for information; sensors may be employed individually or cooperatively, e.g., using cross-cueing
ISR Operator	User Role	Mission Crew member that uses the resources of the ISR Management Domain to transform ISR Plans into Sensor Tasks, monitor plan execution, assess Sensor Products, create ISR Products, coordinate mission activities, and perform other assigned tasks
ISR Plans Management	Subdomain	Processes ISR Tasks and provides a prioritized queue of Collection Tasks to be sent to the Sensor Management Domain and Analysis Tasks for the ISR Analysis and Reporting Domain; also maintains task status to support task queue optimization and maintenance
ISR Resource Interface	Subdomain	Provides the functional interface to the Sensor Management Domain; the physical link is implemented via the LAN
ISR System Interface	Subdomain	Provides the interfaces to the Mission Management and Information Management domains; provides the User Interface for the ISR Operator
ISR Task	CDM	Task category for tasking issued to the ISR Management Domain
Local Area Network (LAN)	Component	Switched, dual, GBit Ethernet Local Area Network connecting the Computing Infrastructure
Maintenance	Use Case	Activities associated with correcting system failures or faults and restoring the E-X aircraft and mission equipment to Operationally Ready (OR) status
Maintenance Data	CDM	`<<abstract>>` Information element that models information associated with maintenance actions or status.
Mediation	Component	Software component that resolves conflicts over data item versions
Message	CDM	`<<abstract>>` Foundation Class for content of information exchanges via messaging; Information Elements inheriting from Message include Status Message, Alert Message, Plan Message, Report Message, and Directive Message
Messaging	Component	Software component that exchanges messages among Mission Applications and other system resources and manages message traffic with the external environment via the Communications Management Domain

Term	Model element	Definition
Mission Analysis and Reporting	Subdomain	Processes ISR Reports to support Mission Crew Commander situational awareness and decision-making and to develop Mission Reports for dissemination and archiving
Mission Commander Workstation	Component	Provides the Mission Commander interface to the system and uses Mission Applications on the Server and local applications on the workstation to provide the required functions
Mission Crew	User Role	Onboard personnel who employ E-X Mission System resources to execute ISR operations
Mission Crew Commander	User Role	Senior Mission Crew member responsible for mission execution, including replanning in response to operational tasks and circumstances, direction of the Mission Crew, and coordination with external entities
Mission Equipment	Descriptor	Hardware and software resources installed in the aircraft to create required mission capabilities and provide crew interfaces
Mission Management	Domain	Contains the resources and functions used by the Mission Crew Commander and any supporting staff to exercise overall control of mission execution
Mission Management Interface	Subdomain	Provides Mission Management interfaces to the Information Management Domain and the Flight Deck; provides the User Interface for the Mission Crew Commander
Mission Plans Management	Subdomain	Processes the Mission Plan and provides a prioritized queue of ISR Tasks to be sent to the ISR Management Domain; also maintains the status of task execution and supports replanning as required to deal with mission situations
Mission Resource Interface	Subdomain	Provides the functional interface to the ISR Management Domain and to Mission Equipment resources, activities, and Status; the physical link is implemented via the LAN
Multimode Radar	Component	E-X Sensor that provides information about the external environment and targets using Synthetic Aperture Radar (SAR) and Ground Moving Target Indicator (GMTI) modes; can be based on either an Active Electronically Scanned Array (AESA) or a mechanically scanned reflector RF aperture
NB SATCOM	Context Block	Satellite constellation and associated ground environment used by E-X for narrowband SATCOM communications
Network-Attached Storage (NAS)	Component	Provides high-speed, high-capacity, nonvolatile shared storage for databases and files
Off-board Crew	User Role	<<abstract>> User Role that models the characteristics of Ground Station personnel
Off-board ISR	User Role	Ground Station personnel that complement the E-X onboard ISR_Operator(s) by performing additional analysis and ISR Product creation using downlinked Sensor Reports
Off-board Manager/Planner	User Role	Ground Station personnel responsible for planning and directing the operations of a Ground Station in coordination with the E-X Mission Crew and external entities

Term	Model element	Definition
Plan	CDM	<<abstract>> Foundation Class that models content of plans in various categories; Information Elements inheriting from Plan include Mission Plan, Collection Plan, Flight Plan, Communications Plan, Information Management Plan, and Support Plan; Plans are updated via Plan Revision Information element
Position, Navigation, and Timing (PNT)	CDM	<<abstract>> Information Element that models various types of Position, Navigation, and Timing (PNT) data, especially derived from the Global Positioning System (GPS)
Processing Device	CDM	<<abstract>> Foundation Class that abstracts computing, storage, networking, and other resources involved in information processing
Radar	Component	Active RF sensor that produces information about the external environment using various target tracking and imaging modes
Radar Processor	Component	Embedded computing resource within a radar that provides data and digital signal processing; components include Radar Signal Processing, Radar Data Processing, and Radar Controller
Radar Range	Algorithm	Constraint Block that defines the computation of range to a minimum detectable target based on radar parameters
Radar Workstation	Component	Provides the system interface for the Radar Operator and uses Mission Applications on the Server and local applications on the workstation to provide the required functions
Receiver	Component	Radio-frequency electronic component that provides detection, filtering, etc. for signals from an antenna
Report	CDM	<<abstract>> Foundation Class for information formatted in Reports in various categories; Information Elements inheriting from Report include ISR Report, Analysis Report, Collection Report, Sensor Report, Detection Report, Mission Report, Maintenance Report, Status Report, Target Report, and Imagery Report
Resource	CDM	<<abstract>> Information Element that models physical entities
RF Aperture	Component	The element of an RF sensor that radiates and receives radio-frequency energy
RF Sensor	CDM	<<abstract>> Block that models sensors employing radio-frequency energy
Routing	Component	Software component that supports the messaging component by determining addresses and paths for message transmission
Sensor Analysis and Reporting	Subdomain	Processes data from Sensor Reports to create Collection Reports; also supports various Sensor Operator analysis tasks and quality control of collected data

Term	Model element	Definition
Sensor Data	CDM	<<abstract>> Information Element that models the shared characteristics of data from E-X sensors; Information Elements inheriting from Sensor Data include Radar Data, SIGINT Data, EO/IR Data, SAR Data, GMTI Data, Track, Signal Data, Image, Video Clip, and Off-board Sensor Report
Sensor Interface	Component	Software component that provides the functional interface to the Sensors Domain; the physical link is implemented via the LAN
Sensor Management	Domain	Contains the resources and functions used by the Sensor Operators to configure and control sensors and process collected data
Sensor Operator	User Role	Mission Crew member that directly interacts with E-X sensors to control their operation, monitor and assess sensor data, and create and distribute Sensor Reports
Sensor Plans Management	Subdomain	Processes Collection Tasks and provides a prioritized queue to the Sensor Resource Manager; also maintains task and sensor status to support task queue optimization and maintenance
Sensor Resource Management	Subdomain	Responsible for pairing Collection Tasks with Sensor Assets and scheduling collection actions; Sensor Tasks are sent to individual Sensors, and task status feedback is used in computing subsequent task schedules; included Components are Collection Task Queue, Sensor Characteristics, Constraint Manager, and Task/Sensor Pairing and Scheduling
Sensor System Interface	Subdomain	Provides the interfaces to ISR Management and Information Management. It provides the User Interface for the Sensor Operator
Sensors	Domain	Contains the E-X sensors and the interface between sensors and Sensor Management; physically, the interface is implemented via the Computing Infrastructure LAN
Sensor	CDM	<<abstract>> Foundation Class for E-X sensor resources; provides abstraction of the content of the Sensors Domain
SIGINT Sensor	Component	E-X Sensor that provides information about the external environment by detecting and analyzing RF signals from multiple sources; components include one or more antennas, RF Front End, Search Receiver(s), Analysis Receiver(s), and SIGINT Processor
SIGINT Workstation	Component	Provides the system interface for the SIGINT Operator and uses Mission Applications on the Server and local applications on the workstation to provide the required functions
Support Crew	User Role	<<abstract>> User Role that includes various personnel required to support the E-X system and its operations

Term	Model element	Definition
Support Personnel	User Role	Personnel at the E-X operating location that perform system maintenance and servicing, pre- and postflight inspections and problem correction, aircraft movement and marshaling, supply functions, and other support tasks associated with achieving system operational ready status and supporting sortie generation and recovery
Support Systems	Domain	Contains the resources and functions used to plan and support E-X operations
Support Unit	Organization	Block that models the personnel, resources, and relationships associated with maintenance, training, and other support activities for the E-X System
Supporting Data	CDM	`<<abstract>>` Information Element that models multiple categories of data used to support mission tasks, including intelligence, Geographical Information System (GIS), equipment characteristics, weather, airspace, and other; Information elements inheriting from Supporting Data include Environmental Data, GIS Data, Technical Data, and Operational Data
Synthetic Aperture Radar (SAR)	Descriptor	Radar mode that yields imagery of an area on the Earth's surface under all weather conditions
System Administrator (SYSAD)	User Role	Personnel at the E-X operating location responsible for the configuration and maintenance of onboard and off-board information systems, including user accounts, security controls, data archiving and restoration, hardware and software updates, and other system administration tasks
System Services	Subdomain	Creates the underlying environment for E-X information processes and resources; abstracts higher-level services and Mission Applications from the Computing Infrastructure; includes components such as Operating Systems, Virtualization, File Handling, System Time, Diagnostics Management, Configuration Management, and Fault Handling
Target	Context Block	Entity in the external environment on which the E-X collects information using sensors; includes locations and tracks for point targets and imagery and video collected using SAR and EO/IR sensor modes
Task	CDM	`<<abstract>>` Foundation Class for tasking issued to system resources or personnel; Information Elements inheriting from task include Mission Task, Sensor Task, Radar Task, SIGINT Task, EO/IR Task ISR Task, Analysis Task, Collection Task, and Reporting Task
Training	Use Case	Activities associated with providing training to E-X personnel; training is specialized by User Role
Training Data	CDM	`<<abstract>>` Foundation Class that models information associated with training, including training materials, records, training device data, and other

Term	Model element	Definition
Training Personnel	User Role	Personnel that plan, deliver, and track training for E-X personnel, specifically concerned with system operations and maintenance, including both classroom and on-the-job training
Transmitter	Component	Radio-frequency electronic component that generates a signal to be radiated by an antenna
Turret	Component	Gimbaled turret that houses the optics (telescopes) for EO and IR spectral bands, EO and IR focal planes mounted on cryogenic coolers for maximum sensitivity, the drive mechanisms for pointing in azimuth and elevation, focus and magnification controls, and other components
Visualization	CDM	`<<abstract>>` Information Element that models information used to create screens on user workstations and on training device displays.
WB SATCOM	Context Block	Satellite constellation and associated ground environment used by E-X for wideband SATCOM communications
WB SATCOM Controller	Component	Communications Management component that provides the interface to the WB SATCOM Terminal
WB SATCOM Terminal	Component	Communications component that transmits and receives information via communication satellites

Appendix D: Summary of the Modeling and Analysis of Real-Time and Embedded (MARTE) Systems Profile of UML

Introduction

This appendix gives a summary of the concepts and terminology employed in the Modeling and Analysis of Real-Time and Embedded Systems (MARTE) Profile of UML.[1] At this writing, the Profile has been issued in Version 1.1 by the Object Management Group. MARTE extends and is intended to replace the UML Profile for Performance, Schedulability, and Timing (PST), which has been for some years the standard approach to representing and analyzing real-time behavior in UML models.

The Notes that follow have the structure of the MARTE specification. They are intended to provide enough information to give an initial understanding of the MARTE approach and to include basic content, especially stereotypes, in models. The reader is encouraged to obtain and use the full specification, which contains much more detail and rationale, along with defining diagrams and examples. Reserved terms that are part of the Profile are *italicized*, model code is printed in `Courier New font`, and terms that are defined for important concepts but not directly used in models are underlined.

MARTE begins with a MARTE foundations package containing five sub-profiles:

- `<<profile>>` CoreElements
- `<<profile>>` Nonfunctional Properties (NFP)
- `<<profile>>` Time
- `<<profile>>` Generic Resource Modeling (GRM)
- `<<profile>>` Allocation (Alloc)

There is then a MARTE design model package containing four additional sub-profiles that depend on and extend those of the foundations package:

- `<<profile>>` Generic Component Model (GCM)

[1] http://www.omg.org/spec/MARTE/1.1.

J. M. Borky, T. H. Bradley, *Effective Model-Based Systems Engineering*,
https://doi.org/10.1007/978-3-319-95669-5

- `<<profile>>` High-Level Application Modeling (HLAM)
- `<<profile>>` Software Resource Modeling (SRM)
- `<<profile>>` Hardware Resource Modeling (HRM)

Finally, there is a MARTE analysis model package with three sub-profiles, also depending on the foundations package:

- `<<profile>>` Generic Quantitative Analysis Modeling (GQAM)
- `<<profile>>` Schedulability Analysis Modeling (SAM)
- `<<profile>>` Performance Analysis Modeling (PAM)

The full MARTE specification includes a number of annexes, including:

- `<<profile>>` Value Specification Language (VSL)
- `<<profile>>` Repetitive Structure Modeling (RSM)
- `<<modelLibrary>>` MARTE_Library

Fundamentally, MARTE is concerned with providing the architect with tools and concepts to fully and accurately represent real-time and embedded systems as UML models and then to analyze those models to determine if the execution of system processes can be scheduled (schedulability) or if those processes can complete within defined times under specified operational conditions (performance). These two measures of behavior are obviously closely related, but real-time design tends to emphasize one or the other, and analysis tools similarly seek to assess schedulability or performance.

Basic Concepts

The Core Elements package contains Foundations and Causality packages.

- The *Foundations Package* contains basic elements used to represent the dual descriptor instance nature of any modeling entity.
- The *Causality Package* contains basic elements necessary for behavioral modeling and their runtime semantics.
- A model is composed of *ModelElements* which can be *Classifiers* or *Instances*: MARTE uses the *Classifier* → *Instance* concept from UML; the *Classifier* is the *Type* of the *Instance.*
- A *ModelElement* can be specialized by Inheritance as a *MultiplicityElement* which has Upper and Lower Value limits; a *MultiplicityElement* can be further specialized as a *Property*, which can then be incorporated by Composition in a *Classifier.*
- The *Common Behavior Package* within Causality defines the *Model Elements* used to represent behavior, including:
  - *Behavior*—describes how a system or entity changes over time; describes the behavioral aspects of a *Behaviored Classifier*; can be specialized into:

    Action—the atomic unit of behavior
    CompositeBehavior—a set of Actions and other *CompositeBehaviors*

- *Event*—specification of a kind of change of state in the system or entity (may be timed)
- *Trigger*—specifies an *Event* that may initiate a *Behavior* (may be timed)
- *Mode*—an operational segment within a system execution characterized by a set of active resources and operational parameters
- *ModeTransition*—models the dynamics of switching between modes
- *ModeBehavior*—*Behavior* characterized by an operating phase or condition, a reconfigurable system state, or other distinguishable state of system execution; composed of one or more *Modes* and *ModeTransitions*
- *BehaviorExecution*—specifies the execution of a unit of behavior or action; can be specialized into *CompBehaviorExecution* and *ActionExecution*
- *EventOccurrence*—specifies occurrence of an *Event* that can initiate a *Behavior*
- *Invocation*—Package of *Events* and *Occurrences*, which are *Instances* of *Events*, associated with the starting and ending of a *Behavior*, including *StartOccurrence*, *StartEvent*, *TerminationOccurrence*, and *TerminationEvent*
- *Thread*—basic unit of execution concurrency; equivalent to *Task* or *Process*; commonly owned by an *Active Object*
- *Passive Object*—not associated with a concurrent unit of execution but can be owned and called by an *Active Object*
- *Message*—basic unit of communication among *Instances*:

  SenderInstance executes a behavior resulting in an *Invocation Occurrence*; this results in one or more *Requests* (e.g., in a broadcast).
  Request specifies the communication in transit to the receiver and results in a *ReceiverInstance.*
  ReceiverInstance executes a behavior, resulting in a *Receive Occurrence.*

MARTE Profile Domains define stereotypes that extend concrete UML Model Elements and map concepts from MARTE; can be applied to Classifiers and Instances; these include:

- *Configuration*
- *Mode*
- *ModeTransition*
- *ModeBehavior*

For example, in a State Machine Diagram/Statechart, States can be stereotyped as `<<mode>>` and a transition between them as `<<modeTransition>>` as a way to model modal behavior.

Modeling Nonfunctional Properties (NFPs)

The NFP modeling framework provides the capability to describe various kinds of values related to physical quantities, such as Time, Mass, Energy, Number of Participants, etc. "Value" is used as in SysML through value properties and value types. Functional properties deal with runtime purpose; NFPs deal with fitness for purpose, i.e., how well an element fulfills its purpose. Examples of NFPs are the total delay of a *Step* when executed (including queuing delays), the *Utilization* of a resource, the number of clients subscribed to a service, and the *response time* and *throughput* of a service.

NFP Framework Elements

- *AbstractNFP* can be:
  - *Qualitative*—inherent or distinctive characteristics that may not be measured directly
  - *Quantitative*—characterized by a set of *SampleRealizations* and *Measures.*
- *Annotated Model*—a model that is extended with *Annotated Elements* such as:
  - *Step*—a unit of execution
  - *Scenario*—a sequence of steps
  - *Resource*—an entity that offers one or more services
  - *Service*—something of value offered by a *resource* or by a component of some kind
- *Modeling Concern*—a viewpoint or special interest to be addressed by a model, e.g., through NFPs
- *NFP Value Annotation*—attached to a model element to describe it
- *ValueSpecification*—defines the value expressions associated with NFPs; may be included in an *NFP_Constraint.*
- *NFPConstraint*—Boolean condition on NFPs; assertion of restrictions that must be satisfied by a real-time/embedded (RTE) system, e.g., maximum execution latency; can support multiple levels for a given service, e.g., varying by system mode
- *Value Property*—any kind of physical quantity relevant to the design of the system; can be further specialized as an NFP
- *ValueType*—a *TupleType* of a *Value Property*; can carry a measurement unit (by means of *unitAttribute*) and additional qualifiers to values (*qualifierAttributes*); can be further specialized to *NFP_Type*, e.g., *statisticalQualifier*, direction, value source, and measurement precision.

Stereotypes in the NFP Domain include:

- *Unit*—extends UML Enumeration_Literal.
- *Nfp*—extends UML Property.
- *NfpConstraint*—extends UML Constraint.
- *Dimension*—extends UML Enumeration.
- *NfpType*—specializes VSL *TupleType.*
- *TupleType*—combines different types into a single composite type. The parts of a Tuple Type are described by its attributes, each having a name and a type.

- *ConstraintKind* is an enumeration type that defines literals used to specify the nature of constraint assertions by either required, offered, or contract nature.

Value-Unit notation for Value Specification having *NfpType* as data type consists of a pair of a value and a unit:

```
<nfp-value> ::= <value-specification> [' ' <unit-enumeration>]
```

The following are typical examples:

5 ms	Specifies a duration value
50 kHz	Specifies a frequency value

This is aimed at readability in a graphical model and follows common usage in engineering models in general.

NFP Types and Annotations

MARTE uses the Value Specification Language (VSL) that extends and formalizes (by means of a metamodel and its associated concrete syntax) some concepts supported by earlier models to annotate constant, variable, tuple, and expression values; it defines four kinds of composite data types (data types allowing attributes): IntervalType, CollectionType, ChoiceType, and TupleType.

MARTE includes a library of predefined data types supporting commonly used NFP annotations in RTE systems and provides for users to define additional data types. Examples include NFP_Boolean, NFP_Real, NFP_DateTime, NFP_Duration, NFP_Frequency, and NFP_Power. They can have probabilistic distributions.

NFP Types have qualifier attributes, including source, SourceKind [0..1]; precision, Real [0..1]; statQ, StatisticalQualifierKind [0..1]; and dir, DirectionKind [0..1].

Three kinds of NFP annotation mechanisms:

- *Tagged Value*—associated with Attributes of specific UML stereotypes.
- *Constraint*—natural language or machine-readable expression that defines some semantics of one or more model elements.
- *Instance Specification Slot*—NFP is declared at a Classifier level, and NFP values are specified within the related slots.

The following is an example using VSL:

```
<<nfpContraint>> {kind= contract }
{ procUtiliz > (90 , percent ) ? clockFreq==(60 , MHz ) : clockFreq== (20 , MHz) }
                ↑                          ↑                          ↑
            Condition              If true expression         If false expression
```

This expression is contained in a Note and linked to the model element to which it applies.

Modeling Time

The time modeling framework deals with the association of behavior with time; objects, behavior executions, and event occurrences may explicitly refer to *clocks* considered as accessors to the time structure. The following are the primary Time Framework Elements.

- Provides three types of temporal abstractions:
  - *Causal/temporal*—concerned only about instruction precedence/dependency
  - *Clocked/synchronous*—adds a notion of simultaneity and divides the time scale in a discrete succession of instants.
  - *Physical/real-time*—involves the precise accurate modeling of real-time duration values, for scheduling issues in critical systems.
- *Time Structure*—one or more *Time Bases*, *Instants*, and *Time Structure Relations*
  - *Time Base*—container of an ordered set of *Instants.*
  - The structure of time is specified by the *nature* attribute that takes its values from the enumeration *TimeNatureKind.*
  - *Time Bases* are a priori independent but may be linked by *Time Instant Relations*, which involve special *junction instants.*
  - The *TimeStructureRelation* abstract class is subclassed into *TimeBaseRelation* and *TimeInstantRelation* abstract classes and further subclassed into concrete classes:

    CoincidenceRelation—links *junction instances* at same place and time
    PrecedenceRelation—weaker relation between *junction instants* where one precedes or follows the other.
    TimeIntervalMembership—characterizes *junction instants* (members) that are either in a given time interval or are coincident with *junction instants* in this time interval.

  - A *TimeBaseRelation* specifies a set of *TimeInstantRelations.*
  - In dense time, for any given pair of *instants*, there always exists at least one *instant* between the two; a dense *Time Base* can also own *junction instants.*
  - A discrete *Time Base* owns only *junction instants*, which discretize dense time and are potentially observable.
- *Time Access*—includes the abstract *Clock* class which is subclassed into concrete Logical and Chronometric clocks:
  - A *Clock* has a *timeBase* that refers to a set of *junction instants*; can also refer to a dense time base through the *coveringTB* property.
  - A *Clock* accepts units (*acceptedUnits* property), one of which is the *defaultUnit* which is attached to the *currentTime* value. The *resolution* property specifies the readout granularity of the clock, expressed in some *defaultUnit* unit; default value is 1.

- A *maximalValue* attribute is the *currentTime* at which a *Clock* rolls over.
- A *Clock* may own an event called a *clockTick.*
- *LogicalClock* has a *definingEvent* property and ticks at each occurrence.
- *TimeValue* refers to a *Clock* via the *onClock* property; may have a *TimeValueSpecification*; always discrete with integer values.
- *InstantValue* is a *TimeValue* referring to 0, 1, or more *junction instants.*
- *TimeIntervalValue* is a pair of *min* and *max InstantValues* referring to 0, 1, or many (due to clock folding) time intervals and having Boolean attributes *isMinOpen* and *isMaxOpen.*
- *DurationValue* is the time distance between two instants, i.e., the difference modulo *maximalValue* between *min* and *max* values of the *TimeIntervalValue.*
- *DurationIntervalValue* is defined by a pair of duration values, which specifies an interval of values; in a specification/model, a *DurationIntervalValue* indicates any duration value in the interval.
- *ChronometricClocks* are bound to physical time (which is dense); non-ideal behavior is modeled by NFPs such as:

  Stability—the ability for a clock to report consistent intervals of time
  Skew—offset between two clocks, one generally a reference
  Drift—first derivative of skew

- *Current Time*—the current time value supplied by a clock

• *Time Usage*—modeling entities that are bound to time (*TimeRelatedEntities*)

  - *TimedElement*—abstract class generalization of all other timed concepts; associates a nonempty set of clocks with a model element
  - *ClockConstraint*—constraint applied to two or more clocks, such as a requirement that their rates be harmonic, expressed by a *ClockConstraintSpecification.*
  - *TimedObservation*—subclass of *TimedElement* and superclass of *TimedInstantObservation* and *TimedDurationObservation*; made at runtime via the *observationContext* property from one or more *Clocks*, possibly in response to a *Request.*
  - *TimedEventOccurrence*—subclass of *TimedElement* and *EventOccurrence*; at property specifies the instant value of this timed event occurrence on one of its clocks.
  - *SimultaneousOccurrenceSet*—models a group of event occurrences that collectively trigger some behavior.
  - *TimedEvent*—an event the occurrences (which can be multiple) of which are bound to clocks; Boolean attribute isRelative specifies whether the time value is relative (the when property is a time duration value) or absolute (the when property is a time instant value); time values are specified by Clocked Value Specification Language (CVSL) expressions.
  - *TimedConstraint*—a constraint imposed on the occurrence of an event (*TimedInstantConstraint*), or on the duration of some execution, or even on the temporal distance between two events (*TimedDurationConstraint*); specified by predicates (*InstantPredicate* for instants and *DurationPredicate* for durations).

- *TimedExecution*—a behavior that has *startInstant* and *finishInstant TimeValues* associated with its *StartOccurrence* and *Termination Occurrence*; may also have a *DurationValue.*
- *TimedProcessing*—a generic concept for modeling activities that have known start and finish times or a known duration.

Stereotypes in the Time Domain include:

- *Clock*
- *ClockType*
- *TimedDomain*
- *TimedElement*
- *TimedValueSpecification*
- *TimedConstraint*
- *ClockConstraint*
- *TimedInstantObservation*
- *TimedDurationObservation*
- *EventKind*
- *TimedEvent*
- *TimedProcessing*

Generic Resource Modeling (GRM)

The GRM Domain provides constructs for modeling hardware and software (e.g., Real-Time Operating System [RTOS]) that creates platforms for RTE systems.

GRM Elements

- Resource Types:
  - A *Resource* represents a physically or logically persistent entity (structural) that offers one or more *ResourceServices* (behavioral) that provide the means to perform the expected duties and/or satisfy system requirements; characterized by the Boolean Attributes:

    `isProtected`—requires arbitration of access to the resource or its services
    `isActive`—has an independent course of action

  - A *Resource* is instantiated as a *ResourceInstance*, can have or require NFPs, and can have *ownedElements* modeling internal structure, e.g., a logical or chronological *Clock*; A *ResourceService* is instantiated as a *ResourceServiceExecution* and can also have or require NFPs.

- *ResourceReference*—used when modeling the dynamic creation of resources is required.
- *ResourceAmount*—generic quantity of the "amount" provided by the resource; may be mapped to any significant quantification of the resource, like memory units, utilization, power, etc.
- A *Resource* described only by its services is a "black box"; one with internal (lower-level) resources declared is a "white box"; modeling options include:

  One black box resource (e.g., an RTOS), which abstracts the hardware, considered as internal elements; services may constitute an API or have a simpler representation.

  A collaboration between a software layer and a hardware layer.

  A collaboration between basically hardware elements; software features of the execution platform may be represented by overheads on raw hardware performance.

  Any combination of these previous approaches depending on the type of development and analysis method applied by the user.

- *StorageResource*—represents memory with capacity expressed in number of elements; the size of an individual element in bits must be given; reference *Clock* determines time to access an element.
- *TimingResource*—represents a hardware or software entity that is capable of following and evidencing the pace of time; defined as a kind of chronometric clock and may be subclassed as a *ClockResource* or a *TimerResource.*
- *SynchResource*—represents the protected resources that serve as the mechanisms used to arbitrate concurrent execution flows, especially mutually exclusive access to shared resources.
- *ComputingResource*—represents either virtual or physical processing devices capable of storing and executing program code; active and protected.
- *ConcurrencyResource*—resource that is capable of performing its associated flow of execution concurrently with others, all of which take their processing capacity from a potentially different protected active resource (eventually a *ComputingResource*); active and protected.
- *DeviceResource*—external device that may require specific services in the platform for its usage or management; not part of the core model; may be active.
- *CommunicationResources*:

  CommunicationMedia—resource such as a bus or protocol stack.

  CommunicationEndPoint—resource whose services include sending/ receiving data, providing notification, etc.

- *ProcessingResource*—generalizes the concepts of *CommunicationMedia*, *ComputingResource*, and active *DeviceResource.*
- Additional *ResourceServices* include *Acquire*, *Release*, *GetAmountAvailable*, and *Activate.*

- Resource Management:
  - *ResourceBroker*—resource that is responsible for allocation and de-allocation of a set of resource instances (or their services) to clients according to a specific *AccessControlPolicy.*
  - *ResourceManager*—responsible for creating, maintaining, and deleting resources according to a *ResourceControlPolicy.*
- Scheduling—arranging runtime behavior by choosing the order of threads for execution:
  - *Scheduler—ResourceBroker* that provides access to its brokered *ProcessingResource* or *Resources* following a certain *SchedulingPolicy* which belongs to one of an enumerated list of *SchedulingPolicyKind*
  - *SecondaryScheduler*—subordinate scheduler in a hierarchy; receives resources from the main *Scheduler*
  - *SchedulableResource—ConcurrencyResource* with logical concurrency such that it takes the processing capacity from another active protected resource, usually a *ProcessingResource*, and competes for it with others linked to the same scheduler in accordance with associated concrete *SchedulingParameters*
  - *MutualExclusionResource*—for a shared *Resource*, defines the identification of the common resource, its protecting mechanism, and the associated *MutualExclusionProtocol* with its associated *ProtectionParameters*
- Resource Usage:
  - *ResourceUsage*—models the consumption of part of the "amount" provided by a *Resource*; may be *StaticUsage* or *DynamicUsage*
  - *UsageDemand*—models the mechanism that requires the usage of a *Resource*, e.g., in responding to an *Event*
  - *UsageTypedAmount*—defines a group of specific usages such as memory consumption, network throughput, CPU time, etc.

Stereotypes in the GRM Domain include:

- *Resource*
- *CommunicationEndPoint*
- *StorageResource*
- *SynchronizationResource*
- *ConcurrencyResource*
- *Scheduler*
- *SecondaryScheduler*
- *SchedulableResource*
- *MutualExclusionResource*
- *ProcessingResource*
- *CommunicationMedia*
- *ComputingResource*
- *DeviceResource*

- *TimingResource*
- *TimerResource*
- *ClockResource*
- *GrService*—maps the *ResourceService* domain element; general concept that helps in the definition of generic resource models
- *Acquire*
- *Release*
- *ResourceUsage*

The following Basic Types are defined in the GRM profile:

- `<<enumeration>>` SchedPolicyKind
- `<<enumeration>>` ProtectProtocolKind
- `<<enumeration>>` PeriodicServerKind
- `<<dataType>> <<choiceType>>` SchedParameters
- `<<dataType>> <<choiceType>>` ScheduleSpecification
- `<<dataType>> <<tupleType>>` EDFParameters
- `<<dataType>> <<tupleType>>` FixedPriorityParameters
- `<<dataType>> <<tupleType>>` PeriodicServerParameters
- `<<dataType>> <<tupleType>>` PoolingParameters
- `<<dataType>> <<tupleType>>` TableDrivenSchedule
- `<<dataType>> <<tupleType>>` TableEntryType

Compatibility of Scheduling Policy and Scheduling Parameters

The policy used by the scheduler (host) must be compatible with the scheduling parameters (schedParams) of the schedulable resource.

Scheduling policy	choiceAttribute(s) of the scheduling parameters used
EarliestDeadlineFirst	`edf`
FixedPriority	`fp`, `polling`, or `server`
LeastLaxityFirst	`edf` (combined with)/plus `server`
TimeTableDriven	`tableEntry`

Allocation Modeling (Alloc)

The MARTE profile specification states that allocation of functional application elements onto the available resources (the execution platform) is the main concern of real-time embedded system design. The Alloc Domain provides constructs for modeling the allocation of application elements to available resources, including spatial distribution and temporal scheduling ("horizontal" association). A complementary aspect is abstraction/refinement within a model ("vertical" association from more abstract down to more concrete) to provide the necessary detail for allocation matches. It frequently requires description of relationships among clocks associated with applications and platform and from abstract to more refined levels of modeling, e.g., using `<<ClockConstraint>>`. The following are the primary Alloc Domain Elements.

- MARTE *Allocation*—an association between a MARTE application and a MARTE execution platform; allocation can be:
  - *Structural*—association between a group of structural elements and a group of resources.
  - *Behavioral*—association between a set of behavioral elements and a service provided by the execution platform; ultimately, allocate Actions to Resources and Services.
  - *Hybrid*—allowed when the nature of the allocation is clear from the context, for instance, when an implicit service is uniquely defined for a resource.
- An *Allocation* has one or more sources called *ApplicationAllocationEnd* and one or more targets called *ApplicationPlatformAllocationEnd*; it also has an associated *impliedConstraint* describing the constraints that apply when the allocation is performed and whose type is *NFP_Constraint.*
- At various levels of abstraction/refinement, *Actions* can be grouped into *Services* and *Services* into *Compound Services* to support atomic associations with applications.
- Spatial distribution—allocation of computations to processing elements, of data to memories, and of data/control dependencies to communication resources; structural allocation enforces the corresponding behavioral allocation of encapsulated behaviors, so that contained elements "inherit" the allocation of compound structures.
- Scheduling—temporal/behavioral ordering of the activities (computations, data storage movements, or communication) allocated to each resource; represented as a relation between the respective time bases of application and platform elements; schedule analysis may result in "back refinement" of structural allocations.
- Spatial and temporal allocations must be mutually and globally consistent; analysis techniques based on MARTE allocation descriptions addresses this requirement; e.g., *StartEvents* in the *TimeBases* of the Application and Platform must have a coincidence relationship.
- Key points:
  - An allocation mechanism creates a mapping from logical parts (application model elements) to physical parts (execution platform).
  - Multiple allocations may be possible with various costs for power, time, and other NFP budgets.
  - Allocations can be made for spatial distribution of artifacts onto resources or to schedule algorithmic parts onto available resources.

- Allocation can also be shown as a Dependency (dashed arrow) with an `<<Assign>>` stereotype pointing from the logical element being allocated to the physical/execution platform element that is the target of the allocation.

Stereotypes in the Allocation Domain include:

- *Allocated*—identifies what can be allocated and what can be a target
- *Allocate*—identifies what is allocated onto what and the reasons and constraints; includes Properties *AllocationNature* (spatial or time distribution) and *AllocationKind* (structural, behavioral, or hybrid)
- *Assign*—extends the UML Comment metaclass with "to/from" attributes for the ends of an Assignment
- *AllocateActivityGroup*—extends the UML Activity Partition metaclass
- *Refine*—identical with UML keyword

Generic Component Model (GCM)

The GCM Domain provides constructs that build on UML Structured Classes and SysML Blocks to provide a general model not tied to specific execution semantics and to which RTE characteristics can be added. The following are the primary element of the GCM Domain.

- *StructuredComponent*—subclass of *BehavioredClassifier* that defines a self-contained entity of a system, which may encapsulate structured data and behavior; properties can be *AssemblyPart*, *Attribute*, or *AssociationConnector End.*
- *AssemblyPart*—property of a *StructuredComponent* that can be used as a part in an internal component structure description.
- *InteractionPort*—specialized property of a *StructuredComponent* that defines an explicit interaction point through which components may be connected (linked) through an assembly connector and through which they can communicate via message passing; related ports must be compatible in properties, Flow Specifications, directions, etc.; has a Boolean *isAtomic* property:
  - *FlowPort*—see SysML; has a *FlowSpecification* composed of *Flow Properties* that include a *FlowDirectionKind* (in, out, inout).
  - *Client/ServerPort*—supports a request/reply communication where messages that flow across ports represent operation calls or signals; has a *ClientServerSpecification* composed of *ClientServerFeatures* that may be *Actions* or *Receptions* and have a *ClientServerKind* (provided, required, proreq):

    Action—a service that the owning structured component may provide and/or require via this port
    Reception—a signal that may be published or consumed via the port

  - *Data Communication Semantics*:

    Pull—based on Events that trigger retrieval of data from a store.
    Push—arrival of data triggers a behavior; data is not saved at the port.

  - An *InvocationAction* on an *InteractionPort* can be subclassed as *SendSignalAction*, *CallOperationAction*, or *SendDataAction*; the latter maps to a *FlowProperty*.

- *Connector*—property of a *StructuredComponent* that can interconnect Parts or connect Ports to Parts; has a *ConnectorKind* property that can be delegation or assembly; owns two or more *ConnectorEnds* that may map to Ports or Parts:

  - Ports propagate requests/messages between the internals of a Component and the external environment; commonly, a Port is delegated to one or more Parts which handle incoming and generate outgoing requests or other messages.
  - Absent a Port, a request can also be sent directly to a Part, which is then responsible for handling it.

Stereotypes in the GCM Domain include:

- *FlowPort*—maps to UML Port metaclass
- *ClientServerPort*—maps to UML Port metaclass
- *FlowProperty*—maps to UML Property metaclass
- *FlowSpecification*—maps to UML Interface metaclass
- *ClientServerSpecification*—maps to UML Interface metaclass
- *ClientServerFeature*—maps to UML BehavioralFeature metaclass
- *GCM Trigger*—maps to UML Trigger metaclass
- *GCMInvocationBehavior*—maps to UML Behavior metaclass
- *GCMInvocationAction*—maps to UML InvocationAction metaclass
- *DataEvent*—maps to UML AnyReceiveEvent metaclass
- *DataPool*—maps to UML Property metaclass; has a *DataPoolOrderingType* property (LIFO, FIFO, UserDefined)

  Icons used with Ports are essentially identical to those in SysML.

High-Level Application Modeling (HLAM)

The HLAM Domain provides high-level modeling concepts to deal with real-time and embedded features and includes quantitative features such as deadline and period and qualitative features that are related to behavior, communication, and concurrency. The following are the primary elements of the HLAM Domain.

- Concurrency:

  - *RtUnit*—similar to an Active Object:

    Owns one or more schedulable resources that may be dynamically created at runtime; otherwise uses a pool of schedulable resources and may wait for one to be released (*poolWaitingTime* attribute).

Manages concurrency and RT constraints attached to incoming messages using a concurrency and behavior controller for managing message constraints according to its current state and the concurrent execution constraints.
Can satisfy several requests from several real-time units at the same time, enabling intra-unit parallelism.
May own one or several behaviors; owns a specific behavior, called operational mode, that usually takes the form of a state-based behavior where states represent a configuration of the *RtUnit* and transitions denote reconfigurations of the unit.
Owns a single message queue for saving the messages it receives once its execution has started; messages can trigger behaviors, invoke services, etc.
An application owns at least one *RtUnit* which includes a *Main* that executes to completion, handles Event Receptions, and invokes appropriate Services.

- *ProtectedPassiveUnit (PpUnit)*—used to model shared information:

  Specifies concurrency policy either globally via concPolicy attribute or through a concPolicy attribute for an individual *RtService.*
  concPolicy type is *CallConcurrencyKind* that can be *sequential*, *guarded*, or *concurrent.*
  Has an *executionKind* that can be *immediateRemot*e or *deferred.*

- *InMsgQueue*—plays the role of the broker for its schedulable resources; composed of one or more computing resource policies (*CompResPolicy*), each of which has a *SchedulingPolicyKind* attribute (LIFO, FIFO, FixedPriority, EDF, ELF, RoundRobin, Synchronous, and others).

• Real-Time Service (*RtService*):

  - Provided by *RtUnits* and *PpUnits.*
  - May specify a *RealTimeFeature* with additional parameters defining the execution of the service.
  - Execution may be atomic as defined by the Boolean *isAtomic* attribute.
  - Has a concurrency policy defined by a *concPolicy:ConcurrencyKind* attribute (reader, writer, parallel).

• Real-Time Action (*RtAction*):

  - Subclasses *InvocationAction*; basic mechanism for communication among RT processes; specializes a UML Call Action with features appropriate to RT
  - Can specify:

    Real-time features such as a deadline or period (*ArrivalPattern* data type)
    Size of the message generated when executing or the kind of synchronization (synchKind attribute, synchronous, asynchronous, delayedSynchronous, rendezvous, and others)

  - Can be defined as Atomic (isAtomic attribute)

Stereotypes in the HLAM domain include:

- *RtUnit*
- *PpUnit*
- *RtFeature*—used to annotate model elements with real-time features according to set of *RtSpecification* parameters
- *RtSpecification*—used in a Constraint to specify RT parameters of a model element (occKind, value, priority, miss, concPolicy, exeKind, syncKind, etc.)

Detailed Resource Modeling (DRM)

The DRM Domain provides concepts for representing the software and hardware elements of a platform; MARTE defines multiple special icons for notating various model elements.

Software Resource Modeling (SRM)

The SRM Domain focuses on modeling of application programming interfaces of a software multitasking platform with an RTOS; concurrent *RtUnits* are scheduled and executed, e.g., in an ARINC 653-1 environment; it also models multitasking libraries and more generally multitasking framework APIs, e.g., RTE middleware and RTE virtual machines. The following are the primary elements of the SRM Domain.

- *SW_ResourceCore*—provides the basic software resource concepts:
  - Models API definition of types and operations.
  - Type is modeled as *SwResource* that subclasses *ResourceManager* and inherits a set of *ResourceServices*.
  - *SwAccessService*—accesses resources.
- *SW_Concurrency*—classifies concurrent execution contexts:
  - *SwConcurrentResource*—models an entity that competes for computing resources; provides execution context (stacks, interrupts enables/disables, and registers); includes interrupt and schedulable resources; has an entry point that may be reentrant.
  - *SwConcurrentResource* may interact with:

    SharedDataComResource
    MessageComResource
    SwMutualExclusionResource
    NotificationResource

  - *InterruptResource*—subclasses *SWConcurrentResource*; models both hardware interrupts and exceptions (software interrupts) such as faults, traps, and aborts.

- *SwSchedulableResource*—subclasses *SWConcurrentResource*; linked to a software scheduler that determines order and timing of execution; examples are a thread or a task.
- *MemoryPartition*—models a virtual address space such that a *SwConcurrentResource* can only modify its own memory space.

- *SW_Interaction*—sorts communication and synchronization resources:
  - *SwSynchronizationResources*—control execution flows:

    SwMutualExclusionResource—models resources commonly used to synchronize mutual access to shared data
    NotificationResource—supports control flow by notifying occurrences of conditions (typically signals and events) to awaiting concurrent resources
    MessageComResource—artifact used to communicate messages, e.g., queue or pipe
    SharedDataComResource—defines specific resources used to share the same area of memory among concurrent resources

  - *SwCommunicationResources*—manage data flows

- *SW_Brokering*—refers to hardware and software resources management; models resources that perform brokering actions such as resource allocation and hardware device access:
  - *DeviceBroker*—also called a driver; interfaces peripheral devices to the software execution support
  - *MemoryBroker*—models allocation, protection, and mapping of physical memory onto virtual address ranges for memory partitions

Stereotypes in the SRM Domain include:

- *SwResource*
- *SwConcurrentResource*
- *Memory Partition*
- *SwAccessService*
- *EntryPoint*
- *SwSchedulableResource*
- *Alarm*
- *SwInteractionResource*
- *SwCommunicationResource*
- *SharedDataComResource*
- *MessageComResource*
- *SwSynchronizationResource*
- *NotificationResource*
- *SwMutualExclusionResource*
- *DeviceBroker*
- *MemoryBroker*

Hardware Resource Modeling (HRM)

The HRM Domain focuses on modeling a hardware platform through different views and levels of detail; permits specification of performance and other design parameters to support RTE analysis, e.g., Schedulability; expands on the UML Deployment Package; and is primarily concerned with real-time execution platforms. The following are the primary elements of the HRM Domain.

- *HW_Resource*—generic hardware entity; may encapsulate owned HW resources (decomposition).
- *HW_ResourceService*—provided by a *HW_Resource*; collaborations of services characterize the execution platform.
- *Hardware Logical Model*—provides a functional classification of hardware entities (e.g., computing, storage, communication, timing, or device resources) mainly in terms of their services:
  - *HW_ComputingResource*—generic functional resource; Attributes define design and performance details; can be subclassed as Processor, ASIC, PLD, etc.
  - *HW_Memory*—generic storage resource; Attributes define design and performance details (e.g., storage capacity); can be subclassed as *HW_ProcessingMemory* or *HW_StorageMemory*, each of which can be further subclassed by specific storage types.
  - *HW_StorageManager*—generic memory management resource; can be subclassed as *HW_MMU* (Memory Management Unit) or *HW_DMA* (Direct Memory Access).
  - *HW_CommunicationResource*—generic communication resource; can be subclassed as *HW_Arbiter*, *HW_Media* (e.g., bus or bridge) or *HW_EndPoint*.
  - *HW_TimingResource*—generic timing resource; can be subclassed as *HW_Clock* or *HW_Timer*.
  - *HW_Device*—auxiliary resource that provides functionality required by the execution platform; can be subclassed as *HW_I/O* or *HW_Support*.
- *Hardware Physical Model*—represents hardware resources as physical components with details on their shape, size, position within platform, power consumption, heat dissipation, and many other physical properties.
  - *HW_Component*—generic physical component; can be decomposed into lower-level owned components; Attributes define physical details; can be subclassed into *HW_Chip*, *HW_Unit*, *HW_Channel*, *HW_Card*, or *HW_Port*
  - *HW_PowerDescriptor*—describes instantaneous power consumption/dissipation of a *HW_Component*
  - *HW_PowerSupply*—models power provided to a HW_Component
  - *HW_CoolingSupply*—models cooling provided to a HW_Component

Stereotypes in the HRM Domain include:

- *HwResource*
- *HwResourceService*
- *HwComputingResource*
- *HwProcessor*
- *HwASIC*
- *HwPLD*
- *HwBranchPredictor*
- *HwISA* (Instruction Set Architecture)
- *HwMemory*
- *HwCache*
- *HwRAM*
- *HwROM*
- *HwDrive*
- *HwStorageManager*
- *HwMMU*
- *HwDNA*
- *HwCommunicationResource*
- *HwArbiter*
- *HwMedia*
- *HwEndPoint*
- *HwBus*
- *HwBridge*
- *HwTimingResource*
- *HwClock*
- *HwTimer*
- *HwDevice*
- *HwI/O*
- *HwSupport*
- *HwComponent*
- *HwPowerSupply*
- *HwCoolingSupply*

Generic Quantitative Analysis Modeling (GQAM)

The GQAM Domain provides foundational concepts and NFPs for formal quantitative analysis of software performance using mathematical models; there are sub-profiles for Schedulability and Performance Analysis. The core concept is a description of how the system behavior uses resources. Results are values of "output NFPs" such as response times, deadline failures, resource utilizations, and queue sizes based on "input NFPs" such as request or trigger rates, execution demands, deadlines, or QoS targets. Analysis is based on modeling:

- System-level operations (services)
- Frequency of service requests
- Execution conditions (environment)

Time-Related NFPs

- Hard Deadline—process must complete within stated time.
- Soft Deadline—specified percentage of process instances must complete within stated time.
- Delay Cost Function—measure of delay consequences; must be within allowed value or minimized.
- Other Statistical Measures—execution time must satisfy an average value, probability distribution, etc.

Other NFPs

- Memory Usage—under specified conditions and points in process execution
- Power Usage—under specified conditions and points in process execution

GQAM Elements

- *AnalysisContext*—root of the Domain model; Attributes/Parameters define different cases being considered for analysis and may affect the parameters of behavior and resources.
- GQAM_Workload—package of elements dealing with workload and behavior:
  - *WorkloadBehavior*—workload is represented as a stream of triggering events; composed of one or more *WorkloadEvents* and associated *BehaviorScenarios*.
  - *WorkloadEvent*—triggering event that may be timed, have an arrival pattern (e.g., periodic), or be generated by a *WorkloadGenerator* or *EventTrace*.
  - *BehaviorScenario*—describes response to a *WorkloadEvent*:

    May be decomposed into *Steps*; captures any system-level behavior description or any operation in UML and attaches resource usage to it.

    Has Attributes for things like *HostDemand* (time interval or number of operations), throughput, utilization, etc.; e.g., *respTime* is the end-to-end delay of the scenario.

    Can be represented in an Interaction (Sequence), Statechart, or Activity diagram.
  - *Step*—suboperation of a *BehaviorScenario*:

    Inherits from *BehaviorScenario* and can also be a *BehaviorScenario*

    Uses an Operating System service (*SchedulableResource*)

    Can be primitive, i.e., executed by a host processor

    Can be subclassed as:

      AcquireStep—acquire a resource.

      ReleaseStep—release a resource.

      CommunicationStep—convey information between system elements.

      RequestedService—employ a service provided by a subsystem.

    Can be associated with other Steps by a *PrecedenceRelation*

  - *PrecedenceRelation*—defines the predecessor-successor relationship between steps; can be a sequence, branch, merge, fork, or join.
  - *Time Interval*—defined by events associated with one or more units of behavior.
  - *Service*—associated with an operation included in some interface of a system component; modeled as a subclass of *Step*.

- *GQAM_Observers*—package of elements associated with RTE measures:

  - *TimedObserver*—conceptual entity that defines requirements and predictions for measures defined on an interval between a pair of user-defined observed events, named *startObs* and *endObs*.
  - *LatencyObserver*—subclass of *TimedObserver* that specifies a duration observation between *startObs* and *endObs*, including a miss ratio assertion and a jitter constraint.

- *GQAM_Resources*—package of elements that model things like hardware devices, software servers, and logical resources (e.g., locks); a resource has a scheduling discipline and a multiplicity; resources may be acquired and released as part of a *BehaviorScenario*:

  - *ExecutionHost*—processor or other device that executes operations specified in the model with host relationship to processes and *Steps* that execute on it
  - *CommunicationHost*—hardware link between devices with host relationship to conveyed messages
  - *SchedulableResource*—schedulable service like a process or thread pool, which is a software resource managed by the OS
  - *CommunicationChannel*—middleware or protocol layer that conveys messages

Stereotypes in the GQAM Domain include:

- *GaWorkloadBehavior*
- *GaAnalysisContext*
- *GaResourcesPlatform*
- *GaEventTrace*
- *GaWorkloadEvent*
- *GaWorkloadGenerator*
- *GaScenario*—identifies a *BehaviorScenario*
- *GaStep*
- *GaCommStep*
- *GaAcqStep*
- *GaRelStep*
- *GaRequestedService*
- *GaTimedObs*
- *GaLatencyObs*
- *GaExecHost*
- *GaCommHost*
- *GaCommChannel*
- *GaPerformanceContext*

NFP Attributes for Analysis

NFP	For resource	For scenario and step	For workload event
repetitions: NFP_Real[*]	Repetitions, N/A	Number of times the Step is repeated, once triggered (default = 1)	N/A
probability: NFP_Real[*]	Probability, N/A	Probability that the Step is executed, following its predecessor (for conditions)	N/A
hostDemand: NFP_Duration[*], hostDemand:Ops: NFP_Real[*]	Composite demand across all services of the Resource, in terms of time and processor operations	For a Step, CPU demand on the host of the process that executes the Step; for a Scenario, the sum of all demands for all Steps	N/A
priority: NFP_Integer[*]	N/A	For a Step, priority on its Host	N/A
respTime: NFP_Duration[*]	Response time, composite average respond time across all services offered by the Resource	Total delay from the trigger event until completion of the Step or Scenario	Required value for the Scenario
execTime: NFP_Duration[*]	Execution time, N/A (same as hostDemand)	Response time less any initial scheduling delays	N/A
interOccTime: NFP_Duration[*]	Inter-occurrence time, interval between successive requests for services	Interval between initiations	Interval between trigger events
throughput: NFP_Frequency[*]	Frequency of requests for all services	Frequency of initiations	Frequency of trigger events
utilization:NFP_ Real[*]	Fraction of time the resource is active (has an active service); for a multiple resource, mean number of busy units	Fraction of time the BehaviorScenario is active (between its trigger event and its completion)	N/A
utilizationOnHost: NFP_Real[*]	N/A	Fraction of time the host is busy executing the Behavior Scenario; for multiple hosts, this is a set of values	N/A
blockingTime: NFP_Duration[*]	Blocking time, N/A	Pure delay waiting for passive resources to be available or an event controlled elsewhere (value is an output variable)	N/A
selfDelay: NFP_Duration[*]	Delay	Pure delay controlled or requested by the Step (value is an input variable)	N/A

Schedulability Analysis Modeling (SAM)

The SAM Domain provides concepts for dealing with schedulability (the ability of the system to meet certain temporal constraints, e.g., deadlines and miss rations) and sensitivity (determining ways the system can be improved) as well as common annotations for model-based schedulability analysis. It specializes the GQAM Analysis Context dealing with *WorkloadBehavior* and *ResourcesPlatform.* In a Model-Driven Architecture (MDA) context, *WorkloadBehavior* elements are associated with a Platform-Independent Model (PIM), and *ResourcesPlatform* elements are associated with a Platform-Dependent Model (PDM).

SAM Elements

- *EndToEndFlow*—SAM_Workload construct referring to a set of stimuli requesting computations; a stimulus is related to a *BehaviorScenario*, and recurring events are related to a *WorkloadEvent*; Attributes include timing requirements and predictions that can be modeled by a *TimedObserver*; characterized by a set of NFPs.
- *SaStep*—specializes the GQAM *Step* with additional Attributes associated with schedulability.
- *SaCommunicationsStep*—specializes the GQAM *CommunicationStep* with additional Attributes associated with schedulability.
- *SchedulingObserver*—specializes the GQAM *TimedObserver* with Attributes such as suspensions, blocking time, and overlaps.
- *SaExecutionHost*—SAM_Resources construct that specializes the GQAM *ExecutionHost* with additional Attributes associated with schedulability; owns shared resources such as I/O devices, DMA channels, critical sections, or network adapters.
- *SaCommunicationHost*—SAM_Resources construct that specializes the GQAM *CommunicationHost* with additional Attributes associated with schedulability.
- *SharedResource*—specializes the GRM *MutualExclusionResource* with additional Attributes associated with schedulability; dynamically allocated to schedulable resources by means of an access policy such as first in/first out (FIFO), priority ceiling protocol, highest locker, priority queue, and priority inheritance.

Stereotypes in the SAM Domain include:

- *SaAnalysisContext*—adds schedulability parameters *to GaAnalysisContext*
- *SaEndToEndFlow*
- *SaStep*
- *SaCommStep*
- *SaSchedObs*
- *SuSharedResource*
- *SaCommHost*
- *SaExecHost*

Model Analysis Methods

- *Static*—scheduling performance is determined in advance based on full set of possible behaviors and execution contexts; typically uses mathematical scheduling policies such as Rate Monotonic Analysis and Deadline Monotonic Analysis.
- *Dynamic*—scheduling decisions made on the basis of changing values of importance of schedulable resources, dynamically allocated to schedulable resources by means of an access policy within the dynamic context of execution; often called value-based or utility-based scheduling; e.g., Earliest Deadline First.

SAM Notation Application Examples

- *General*—apply SAM elements and stereotypes to diagrams describing the structure and behavior of an RTE system.
- *Activity Diagram*
  - Apply <<GaWorkloadBehavior>> stereotype to the activity diagram.
  - Use Activity Partitions (swim lanes) to model concurrency; each swim lane contains a parallel end-to-end flow; where appropriate, synchronize flows with events.
  - Label swim lane as:

```
<<SaEndToEndFlow>>
Flow Name
{Constraints}
```

  - Use Workload Events to trigger flows, labeled as:

```
<<GaWorkloadEvent>>
Trigger Event Name
{Constraints}
```

  - Use Behavior Scenarios to model flows; decompose as needed to show detail; label as:

```
<<GaScenario>>
{Constraints}
Scenario Name
```

- *Sequence Diagram*
  - Apply <<GaScenario>> stereotype to the sequence diagram.
  - Construct the diagram using MARTE elements and other system objects from the model; label as:

```
<<Stereotype>> (e.g., Allocated, SaSharedResource)
:ObjectName
{Constraints} (e.g., AllocatedTo)
```

- Apply MARTE notation to Messages in the diagram; label as:

  `<<Stereotype>>` (e.g., `Step`, `ExecStep`, `AcqStep`, `CommStep`)
  `MessageName()`
  `{Constraints}` (e.g., Max and Min Execution Time)

- *Class/Object Diagram*
  - Define Classes/Objects using MARTE elements as defined in previous sections.
  - Construct the diagram using normal semantics, e.g., use `<<allocate>>` Dependency to associate a *SchedulableResource* with an *SaExecutionHost*.

- *Analysis Context*—provide a set of Analysis Context objects, stereotyped `<<SaAnalysisContext>>`, with attributes that define the parameters of the analysis.

Performance Analysis Modeling (PAM)

The PAM Domain provides concepts for analysis of temporal properties of best-effort systems and soft real-time embedded systems.

- Analysis outputs are statistical, and inputs may be probabilistic, e.g., randomly occurring inputs or events.
- Analysis techniques include simulations, extended queueing models, and discrete-state models and may include sensitivity analysis or scalability and capacity analysis.
- Performance analysis assesses how the system uses resources, including hardware, *ScheduledResources*, and *Logical Resources*. It associates a *BehaviorScenario* with workload, request, service, and response.
- *AnalysisContext* defines the scope of the analysis and connects a system *BehaviorScenario* and *Resources* with one or more workloads (from GQAM).
- A system model in UML can be parameterized to yield multiple performance models, called "cases," to model various dynamic situations. Each case is defined by a group of variables. Additional variations in a case result from design details such as choice of RTOS. A performance model is constructed from:
  - System UML model (specification)
  - System/Platform configuration and case parameters
  - Input NFPs (parameter values)
- A performance model is fed to some kind of performance model "solver" to compute output NFPs (performance results).
- Parameterized NFPs use:
  - Variables to define ranges of values that are global for a given *AnalysisContext*, e.g., workload intensity, resource scaling, data record size, etc.
  - Variable names in place of numerical values for input properties
  - Functional dependencies of input properties on global variables

PAM Elements

- *Behavior*—defined by a *BehaviorScenario* (see GQAM_Workload) in a behavior diagram; includes *Steps* and demands on system *Resources.*
- *Workload*—*AnalysisContext* can have any number of workloads as defined in GQAM, each with a distinct mechanism for initiating requests, a load intensity, and Quality of Service (QoS) requirements.
- *Service*—normal way to describe performance using offered and demanded QoS:
  - Defined by a *BehaviorScenario*, including offered QoS.
  - Requests queue for service from a resource; may include demanded QoS.
  - Service is included in the analysis by:

    serviceDemand from a *Step* to a *RequestedService*
    behaviorDemand from a *Step* to directly invoke a *BehaviorScenario*
    extOpDemand from a *Step* to request an external service (defined outside the system UML model, e.g., message conveyance via an external network)

- *Resource*—model as servers:
  - Active resources are self-contained with characteristic service times for each service.
  - Passive resources are acquired and released during a *BehaviorScenario* with holding times determined by their behaviors.
  - *External Resource*—active resource that abstracts an external subsystem; External resource is not associated with the software design that affects performance; usage is modeled by *externalOpDemands* and *externalOpCount* parameters in a *PAMStep.*
  - *Component*—software entity with interfaces, each of which offers one or more of the services of the component, each of which responds to one or more interface methods; may also be a passive resource if its execution is restricted.
- *Communication Channel*—models the mechanism by which two objects exchange a message; each message with performance significance is modeled as a *CommunicationStep*:
  - Within a process—implemented by the language runtime
  - Between processes—implemented by the Operating System
  - Between nodes—implemented by a *CommunicationChannel* (middleware service or other infrastructure)
  - Modeled as:

    Language runtime and OS costs ignored by default; may be computed as a *hostDemand* if the cost per byte to a *ProcessingHost* is known.
    Between nodes, compute *hostDemand* and insert link latency.
    Alternatively, between nodes, treat as an external operation based on a model of the communication layer.

Communication layer may be modeled as a Structured Class with send and receive operations.
Complex communication protocol may be modeled as a *BehaviorScenario* not associated with a system component but describing a collaboration of the hosts.

- *PAMStep*—extends GQAM *Step* with more kinds of operation demands during a step and the possibility of an asynchronous (non-synchronizing) parallel operation.
- *ResourcePassStep*—identifies the passing of a resource (usually a *SharedResource*) from one process to another.
- *LogicalResource*—any entity to which the software requires access and for which the program may have to wait at some point.
- *RunTimeObjectInstance*—an alias for a process resource that can be used in behavior specifications such as lifelines and swim lanes.

Stereotypes in the SAM Domain include:

- *PaStep*
- *PaResPassStep*
- *PaRequestedService*
- *PaCommStep*
- *PaLogicalResource*
- *PaRunTInstance*

Input/Output Properties for Performance Analysis

- *Duration*—NFP_Duration—time for an operation to execute or respond
- *ForcedDuration*—NFP_Duration—time interval within an operation, e.g., blocking time
- *Frequency*—NFP_Frequency—rate of occurrence of some event; alternatively, throughput
- *Probability*—NFP_Real—associated with event occurrence or other circumstances
- *Repetition*—NFP_Real—associated with a loop or repeated operation
- *Message or Memory Size*—NFP_DataSize

Performance Analysis Methods

- *Queuing Models*—define Workloads that are captured in scenarios and execute specific aspects of the software; may be simple (define scenarios and total average demand on system devices) or complex (define statistical demands, allow passive resources, allow parallel executions, etc.); compute throughput, utilization, and response time.

- *Simulation Models*—define multiple logical tokens which execute the software, following the detailed *BehaviorScenario* structure and using execution time distributions for the operations of each step; can compute a variety of measures, including distributions.
- *Discrete-State Models*—also define tokens which execute the software, following the detailed *BehaviorScenario* structure; based on some form of graph theory with nodes/places and edges/transitions along which tokens are passed; includes Petri Nets, Colored Petri Nets, Markov and Semi-Markov Chains, Stochastic Process Algebras, and Stochastic Automata.

PAM Annotation Examples

- Stereotypes and NFPs can be applied to any model elements; e.g., notate a server node in a Deployment Diagram as `<<GaExecHost>>` and constraints for parameters such as Communications Receiver and Transmitter Overhead (*commRcvrOhead*, *commTxOhead*).
- Case/Context parameters can be attached to a diagram as Constraints stereotyped as `<<GaPerformanceContext>>`.
- *Sequence Diagram*—notate similarly to SAM example:
  - Specify `<<GaPerformanceContext>>` and provide input parameters.
  - Annotate messages with `<<PaStep>>`, `<<GaWorkloadEvent>>`, etc., and provide parameters as NFPs such as *hostDemand* and *msgSize*.
  - Apply `<<PaRunTInstance>>` stereotype to lifeline Objects, and identify the instance, e.g., {instance = server}.
- *Activity Diagram*—notate similarly to SAM example:
  - Annotate swim lanes as `<<PaRunTInstance>>`, and provide NFPs as constraints.
  - Annotate Actions as `<<PaStep>>`, `<<PaCommStep>>`, etc., with NFPs as constraints.
  - Model both physical and logical resources as swim lanes.
  - Model resource acquisition, release, and planning as Actions and Transitions, e.g., a Transition stereotyped as `<<PaResPassStep>>` with the resource ID and other parameters in Constraints.
- *State Machine Diagram*
  - Annotate States as `<<PaStep>>` with corresponding NFPs as Constraints.
  - Transitions can also be annotated as `<<PaStep>>` in order to specify the transition probability to the next state.
  - For a diagram modeling a set of operations, annotate the start point as driven by a *WorkloadEvent*.

MARTE Acronym List

Acronyms associated with sections of the MARTE Profile Specification are shown with the corresponding Clause or Subclause, where applicable.

Acronym	Definition	MARTE clause/ subclause
AADL	Architecture Analysis and Design Language	
AHB	AMBA High-performance Bus	
Alloc	Allocation Modeling	Clause 11
AMBA	Advanced Microcontroller Bus Architecture	
ARM	Advanced RISC Machines	
CAN	Controller Area Network	
CCM	CORBA Component Model	
CCSL	Clock Constraint Specification Language	
CHF	Clock Handling Facilities	Annex C
CORBA	Common Object Request Broker Architecture	
CPU	Central Processing Unit	
CVSL	Clocked Value Specification Language	
DMA	Direct Memory Access	
DPRAM	Double-Port RAM	
DRAM	Dynamic Random Access Memory	
EAST-ADL2	EAST Architecture Description Language 2	
EDF	Earliest Deadline First	
EQN	Extended Queueing Network	
FIFO	First In First Out	
FP	Fixed Priority	
GCM	Generic Component Model	Clause 12
GQAM	Generic Quantitative Analysis Modeling	Clause 15
GRM	Generic Resource Modeling	Clause 10 a
GUI	Graphical User Interface	
HLAM	High-Level Application Modeling	Clause 13
HRM	Hardware Resource Modeling	Sub clause 14.2
LQN	Layered Queueing Network	
Lw-CCM	Lightweight CCM	
MARTE	UML profile for Modeling and Analysis of Real-Time and Embedded systems	
MDA	Model-Driven Architecture	
MoCC	Model of Computation and Communication	
NFP	Nonfunctional Properties	Clause 8
OCL	Object Constraint Language	
OS	Operating System	
PAM	Performance Analysis Modeling	Clause 17
QN	Queueing Network	

Acronym	Definition	MARTE clause/ subclause
QoS	Quality of Service	
QoS&FT	UML Profile for Quality of Service and Fault Tolerance specification	
RISC	Reduced Instruction-Set Computer	
RMA	Rate Monotonic Analysis	
RSM	Repetitive Structure Modeling	Annex E
RTOS	Real-Time Operating System	
SAM	Schedulability Analysis Modeling	Clause 16
SI	Système International	
SPT	UML Profile for Schedulability, Performance, and Time specification	
SRM	Software Resource Modeling	Sub clause 14.1
SysML	Systems Modeling Language	
RTE	Real-Time/Embedded	
RTM	Real-Time objects Modeling (RTE MoCC)	Clause 13
TCP	Transmission Control Protocol	
TPC-W	Transaction Processing Council Web benchmark	
TVL	Tag Value Language	
UML	Unified Modeling Language	
Time	Enhanced Time Modeling	Clause 9
VSL	Value Specification Language	Annex B
WCET	Worst-Case Execution Time	

Appendix E: Listing of Information Technology (IT) Core Standards and Sources

Name	Standard	URL
Atom	IETF Atom v1.0	http://www.ietf.org/rfc/rfc4287.txt
FTP	File Transfer Protocol (FTP), (IETF Standard 9/RFC 959)	http://tools.ietf.org/html/rfc959
HTML	HyperText Markup Language (HTML), v4.01	http://www.w3.org/TR/html401/
HTTP	HyperText Transfer Protocol (HTTP v1.1), (IETF RFC 7230-7235)	https://www.w3.org/Protocols/
Java Enterprise Edition JEE	JSR 244: Java™ Platform, Enterprise Edition 5 (Java EE v5) Specification JSR 316: Java™ Platform, Enterprise Edition 6 (Java EE v6) Specification	http://www.jcp.org/en/jsr/detail?id=244 http://www.jcp.org/en/jsr/detail?id=316
JAX-WS	JSR 224: Java™ API for XML-Based Web Services (JAX-WS) v2.0	http://jcp.org/en/jsr/detail?id=224
JMS	Java Message Service (JMS) Specification—v1.1, (JSR 914)	http://java.sun.com/products/jms/docs.html
JSR-220 FR	Enterprise JavaBeans v3.0	http://jcp.org/aboutJava/communityprocess/final/jsr220/index.html
JSR-286	Java Portlet Specification 2.0 (JSR-286)	http://www.oracle.com/technetwork/server-storage/ts-4817-158962.pdf
KML	Open Geospatial Consortium (OGC) KML (formerly Keyhole Markup Language) v2.2.0	http://www.opengeospatial.org/standards/kml/
LDAP	Lightweight Directory Access Protocol (LDAP) v3: Technical Specification, (IETF RFC 4510)	http://www.ipa.go.jp/security/rfc/RFC4510EN.html
REST	JSR 311: JAX-RS: The Java™ API for RESTful Web Services	http://jcp.org/aboutJava/communityprocess/final/jsr311/index.html
SOAP	SOAP v1.2 Part 1: Messaging Framework (2nd Edition)	http://www.w3.org/TR/soap12-part1/

J. M. Borky, T. H. Bradley, *Effective Model-Based Systems Engineering*,
https://doi.org/10.1007/978-3-319-95669-5

Name	Standard	URL
SSL	The Secure Socket Layer (SSL) Protocol v3.0	http://www.mozilla.org/projects/security/pki/nss/ssl/draft302.txt
TLS	The Transport Layer Security (TLS) Protocol v1.1, (IETF RFC 4346)	https://tools.ietf.org/html/rfc5246
UDDI	Universal Description, Discovery, and Integration (UDDI) v3.0.2	http://www.oasis-open.org/committees/uddi-spec/doc/spec/v3/uddi-v3.0.2-20041019.htm
WADL	Web Application Description Language (WADL)	http://www.w3.org/Submission/wadl/
WFS	OpenGIS Web Feature Service (WFS) Implementation Specification v2.0	http://www.opengeospatial.org/standards/wfs
WMS	OpenGIS Web Map Service (WMS) Implementation Specification v1.3	http://www.opengeospatial.org/standards/wms
WS-Addressing	Web Services Addressing (WS-Addressing)	http://www.w3.org/2002/ws/addr/
WS-BPEL	Web Services Business Process Execution Language (WS-BPEL v2.0), OASIS	http://docs.oasis-open.org/wsbpel/2.0/OS/wsbpel-v2.0-OS.html
WSDL 1.1	Web Services Description Language (WSDL) v1.1	http://www.w3.org/TR/wsdl
WS-Eventing	Web Services Eventing	http://www.w3.org/Submission/WS-Eventing/
WSRP	OASIS Web Services for Remote Portlets Specification, Version 2, April 2008	http://docs.oasis-open.org/wsrp/v2/wsrp-2.0-spec-os-01.pdf
WS-Security	WS-Security v1.1	http://www.oasis-open.org/committees/download.php/16790/wss-v1.1-spec-os-SOAPMessageSecurity.pdf
XLink	XML Linking Language (XLink) Version 1.1	http://www.w3.org/TR/xlink11/
XML	Extensible Markup Language (XML) Fifth Edition v1.0	http://www.w3.org/TR/REC-xml/
XML Namespaces	XML Namespaces v1.1	http://www.w3.org/TR/REC-xml-names/
XML Schema	XML Schema (XSD) v1.0, (XML Schema Part 1:2004—Structures 2nd edition), (XML Schema Part 2:2004—Datatypes 2nd edition)	http://www.w3.org/TR/xmlschema-1/
XMPP	Extensible Message and Presence Protocol (XMPP), RFC 6120: Core, RFC 6121: Instant Message and Presence, RFC 6122: Address Format (March 2011)	http://xmpp.org/xmpp-protocols/rfcs/
XPath	XML Path Language (XPath) v1.0	http://www.w3.org/TR/xpath
XQuery	XML Query Language (XQuery) v1.0	http://www.w3.org/TR/xquery/
XSLT	Extensible Stylesheet Language Transformations (XSLT) v1.0	http://www.w3.org/TR/xslt

Appendix F: Defense in Depth (DiD) Definitions[2]

Perimeter Security Layer

A set of physical and technical security and programmatic policies and safeguards that provide levels of protection against remote malicious activity: set to protect the back-end systems from unauthorized access. When properly configured, the perimeter defense security model prevents, delays, absorbs, or detects attacks, thus reducing the risk to critical back-end systems.

Perimeter Firewall: A device or set of devices on the outer edge of the network or networked enterprise designed to permit or deny network transmissions based upon a set of rules; frequently used to protect networks from unauthorized traffic while permitting legitimate communications to pass. The perimeter firewall is usually the first device encountered when entering an enterprise via a boundary router or other external network connection.

Intrusion Detection System (IDS): A passive system (does not sit in-line with network traffic, also known as "out-of-band") that monitors the perimeter for malicious activities, analyzing traffic by comparing it to information in its database that contains patterns, called "signatures," associated with known threats. It does not stop the "bad" traffic but identifies and reports it.

Intrusion Prevention System (IPS): An active ("in-band") system that analyzes traffic by comparing it to signatures (and sometimes patterns of behavior) and then blocks traffic identified as malicious from entering the network.

Secure DMZ: A network enclave (distinctly bounded outside the enterprise) that sits on the edge of the enterprise exposing external services, such as Email, web servers, Honeypots, proxy servers, and other public facing devices to the external environment. DMZ resources use a separate set of IP addresses from those of the enterprise and are separated from the enterprise by single or dual firewalls. A DMZ prevents direct access to back-end resources by an external entity and prevents malicious content from entering (see Message Security and Network Access Control below).

[2] Used by permission, Barry Lyons IV, CISSP.

J. M. Borky, T. H. Bradley, *Effective Model-Based Systems Engineering*,
https://doi.org/10.1007/978-3-319-95669-5

Application Security/Gateway/Firewall: Applications that run on publically facing servers inside the DMZ and incorporate application security, that is, the applications have been scanned for vulnerabilities (such as Injection Flaws and Cross-Site Scripting) and flaws have been mitigated. The applications may also incorporate active file integrity monitoring software to alert if a file or system has experienced an unauthorized change. The file monitoring system may also revert the system back to its last known "good" state. Web Application Firewalls (WAF) perform deep packet inspection looking for packets crafted in such a way as to exploit known application vulnerabilities.

Message Security: Implemented by servers in the DMZ that identify and block messages with potentially malicious content (either in the message itself or in an attachment). They can also be set up to screen out frivolous messages ("spam").

Honeypot/Honeynet: A server or group of servers in a DMZ that imitate back-end servers to trick malicious intruders. They are used to observe and obtain surreptitious intruders' characteristics and to determine the tactics, techniques, and procedures (TTPs) used in an attack.

Malware Analysis: Implemented by servers in the DMZ that analyze malware (infected software designed to launch a virus, worm, rootkit, or other malicious code); usually coupled with a Message Security Server.

Zero-Day Analysis: A device or service that scans incoming packets, email, attachments, and other items looking for behavioral anomalies that could exploit a vulnerability that has not previously been identified and therefore has not been patched.

Advanced Sensor: Sensors that are specifically designed to address the Advanced Persistent Threat (APT) and many times purposefully built to handle a specific attack signature based on TTP reconnaissance. Advanced technologies such as application of "deep knowledge" algorithms are being developed to significantly improve defenses against the APT.

Data Loss Prevention or Data Leak Prevention (DLP): Systems that identify, monitor, and protect data in use (e.g., endpoint), data in motion (e.g., network), and data at rest (e.g., data storage) through deep content inspection, contextual security analysis of transactions, and a centralized management framework for known security vulnerabilities. It is specifically designed to observe and stop data from leaving the Enterprise. The perimeter DLP is the last defense to prevent data loss.

Network Security Layer

The layer that applies security controls to the internal network of the enterprise and partitions that network into enclaves; an enclave is a distinctly bounded area enclosed within a larger unit. Enclaves incorporate their own individual access controls and protection mechanisms.

Enclave Firewall: A device that protects the enclave and is designed to permit or deny network transmissions into the enclave based upon a set of rules; it is used to protect the enclave from unauthorized access while permitting legitimate communications to pass. Often employs a Unified Threat Management (UTM) firewall that adds functions such as intrusion protection, anti-malware, and addresses filtering to basic firewall message scanning.

Enterprise IDS/IPS: These are the same devices as described in the Perimeter Security Layer section; however they are deployed throughout the enterprise, usually in combination with an Enclave Firewall.

VoIP Protection: Provides Voice over Internet Protocol (also known as IP Phones) security protection; can include policies, procedures, and VoIP switch security (embedded software).

Virtual Network Security: Policies, procedures, hardware, and software that protect a virtual network, which consists of virtual network links rather than physical (wired or wireless) connections between two computing devices; they implement methods that incorporate network virtualization. The most common forms are protocol-based virtual networks (Virtual Local Area Networks, VLANs; Virtual Private Networks, or VPNs; and Virtual Private LAN Services, or VPLSs) and virtual networks that are based on virtual devices (such as a network connecting virtual machines).

Web Proxy Content Filtering: A device (usually a dedicated server) that acts as an intermediary for requests from clients seeking resources from external sources, such as the Internet; used to keep users/machines behind it anonymous. Applies access policies to network services or content to block undesired sites and scans outbound content for DLP.

Network Access Control (NAC): There are two types, Passive and Active. Passive devices authenticate users based on a predetermined profile and grant access to authorized areas within the enterprise. The Active type is also known as a NAC Proxy Server. This acts as an intermediary for external requests seeking internal resources. It keeps enterprise servers hidden, only fetches data the user is allowed to receive, and keeps detailed logs of requests and usage.

Enterprise Message Security: Same as Perimeter Message Security except that devices sit inside the Enterprise. They can be designed to scan outbound messages looking to thwart DLP.

Wireless/Mobile Protection: Policy, procedures, hardware, and software design to defend wireless networks from surreptitious actives and intrusion. For example, connection to the Enterprise wireless network may be restricted to use a VPN (encrypted tunnel), or a wireless router may be protected by a trusted secure processor (endpoint) that allows only authorized users, connecting devices and traffic.

Enterprise Remote Access: Policy, procedures, hardware, and software design to enforce secure remote access. This may include desktop/notebook firewalls, anti-malware, and other items on remotely connected devices to ensure they are protected before connecting with the enterprise. For example, a NAC will confirm that the connected device meets all the policy elements required to connect remotely with the enterprise.

Data Loss Prevention or Data Leak Prevention (DLP): Same type of device or service as in the Perimeter Security Layer. At the Network Layer, multiple DLP solutions may be deployed throughout various enclaves.

Labeled Messaging: A service that attaches a special header or label to a message to restrict receipt of network traffic to authorized network participants and to prevent or detect message corruption or tampering.

Endpoint Security Layer

Security protection mechanisms and controls that reside directly on an endpoint device connected to the enterprise network such as a computer, tablet, smartphone, peripheral, etc. Host-based security is the use of controls like personal firewalls and desktop intrusion prevention systems (IPS) on individual devices.

Desktop Firewall: Firewall software that resides in the desktop/laptop that performs firewall functions at the individual device level.

Host IDS/IPS: Performs IDS/IPS functions within a host. For example, the IDS/IPS might detect an attached document that has a behavior that is foreign to how the associated application should work and can then block it from opening and report accordingly.

Content Security: Keeping the endpoint device up-to-date with the latest antivirus and anti-malware patches along with any other security software dictated by policy.

Endpoint Security Enforcement: Works in conjunction with Enterprise Remote Access; if the endpoint device is not current under Content Security, the Enterprise Remote Access system may not allow the endpoint device to connect with the enterprise. This requires that Content Security must be up-to-date and correspond with the Enterprise Security Policy.

Endpoint Threat Detection and Response (ETDR): Complements antivirus and other safeguards with features like continuous, proactive, enterprise-wide monitoring, attack detection, and automated alerts and response actions.

US Government Configuration Baseline (USGCB) Compliance: If the policy mandates that the system must comply with the USGCB, then the system will not be allowed to connect with the enterprise until it is in compliance. Additionally, the system must be updated as the Baseline is updated.

Patch Management: Confirming that all endpoint devices are current with their respective patches and there is a record of all patches including patch level, when patched and by whom. Patch Management is an essential element of Configuration Management.

Data Loss Prevention or Data Leak Prevention (DLP): Host software that can identify, monitor, and protect data in use at the endpoint device while stopping data leakage if an unauthorized event should occur.

Application Security Layer

Security protection mechanisms and controls that are embedded within the applications residing on enterprise servers, the DMZ, and endpoint devices.

Static Application Testing/Code Reviews: A form of software testing where the software isn't actually used. It checks mainly for the sanity of the code, algorithm, or document. It is primarily syntax checking of the code or manually reviewing the code or document to find errors. Code Review is the systematic examination (often as a peer review) of computer source code. It is intended to find and fix mistakes overlooked in the initial development phase. Reviews are done in various forms such as pair programming, informal walkthroughs, and formal inspections.

Dynamic Application Testing: Examination of the physical response from the system to variables that are not constant and change with time. In dynamic testing the software must actually be compiled and run, giving input values and checking if the output is as expected. These are the Validation activities. Unit Tests, Integration Tests, System Tests, and Acceptance Tests are few of the Dynamic Testing methodologies. Web Application Vulnerability Scans perform Dynamic Application Testing yielding exact vulnerability locations in the code and instructions on how to mitigate the vulnerabilities.

Web Application Firewall (WAF): Also known as a *Deep Packet Inspection (DPI)* firewall. Unlike the traditional firewall used in the Perimeter and Network layers, the WAF performs *deep packet inspection* seeking hidden scripts that can exploit application vulnerabilities such as injection and cross-site scripting flaws. The WAF rejects any packets incorporating embedded exploit scripts.

XML Firewall: Extensible Markup Language (XML) firewall used to validate XML traffic, control access to XML-based resources, filter XML content, and apply rate limiting to requests to back-end applications exposed through XML-based interfaces. An XML Firewall can be stand-alone hardware or a virtual appliance.

Database Monitoring/Scanning: DB Monitoring is the same as file integrity monitoring described in the Application Security section under the Perimeter Security Layer. It is the software that generates alerts when a file or system has experienced an unauthorized change; the file monitoring system may also revert the system to its last known "good" state.

Database Secure Gateway (Shield): Also known as a *Database Firewall (DBF)*. Some perform deep packet inspection protecting against injection exploits. They may include real-time monitoring, alerting and blocking, pre-built security policies, and audit rules.

Data Security Layer

Security mechanisms and controls that safeguard enterprise data regardless of the data's state, whether in motion, at rest, or in use.

Public Key Infrastructure: A general Access Control method that leverages all three forms of encryption (symmetric, asymmetric, and hash-based) to provide and manage digital certificates, which are public keys signed with a digital signature. Digital certificates can be server-based (as in SSL Web Sites) or client-based (bound to a person). If the two are used together, they provide mutual authentication and encryption. The standard digital certificate format is X.509.

Data at Rest, in Motion, and in Use (DAR/DIM/DIU) Protection: Measures to ensure Confidentiality, Integrity, and Availability (CIA) of data, regardless of its state. To accomplish this, the data's location must be known at all times, and, when applicable, data should be encrypted.

Data Classification: Data is categorized by its type, location, level of access, and level of protection. It may also be described in terms of adherence to a specific compliance regulation (such as a level of classification by a Government agency). Data controls relate to both a Subject (active entity seeking access) and an Object (passive data file or other object). Specific controls include:

- *Discretionary Access Control (DAC)*—gives Subjects full control of Objects to which they have been granted access, including sharing with other Subjects
- *Mandatory Access Control (MAC)*—system-enforced access control based on the Subject's clearances of permission and the Object's labels of other attributes
- *Role-Based Access Control (RBAC)*—groups Subjects into roles and gives access permissions based on the role, not the individual.

Enterprise Rights Management: Also known as Enterprise Digital Rights Management (eDRM) or Information Rights management (IRM); technology which protects sensitive information from unauthorized access, alteration, or copying. Used extensively in the music and motion picture industry to protect the unauthorized copying of data such as songs and movies.

Federal Identity Credential and Access Management (FICAM): A Federal Government document that presents a common framework and implementation guidance needed to plan and execute Identity, Credential, and Access Management (ICAM) programs.

Data Integrity Monitoring: Also known as "file integrity monitoring." Software that generates alerts if data has experienced an unauthorized change; it may also be set up to revert the data file back to its last known "good" state.

Data/Drive Encryption: Encrypting plaintext data (using a cryptographic algorithm) to convert it to unreadable ciphertext.

Data Wiping/Cleansing: Destroying the data and many times the media holding the data. Degaussing is a common data wiping method that destroys the integrity of magnetic media (such as tapes or disk drives) by exposing them to a strong magnetic field.

Data Loss Prevention or Data Leak Prevention (DLP): Same protective approach described under Perimeter Security Layer. It is specifically designed to observe and stop data from leaving the Enterprise.

Additional Defense-In-Depth Concepts

Mission Critical Assets: Refers to systems and other resources that perform functions essential to maintaining reliable operation and data whose compromise would severely impact the mission or the organization's ability to function. NOTE: It is important to always know all dependencies among Mission Critical Assets dependencies.

Monitoring and Response: Tools and processes involved in constant observation of the enterprise to recognize incidents and events and respond accordingly in a timely manner. This includes:

- *Security Information and Event Management (SIEM)*—provides real-time analysis of security alerts generated by network hardware and applications; may consist of software, appliances, or managed services, which are also used to log security data and generate reports for compliance purposes.
- *Security Dashboard*—an executive information system user interface delivering security event information that is designed to be easily read and understood.
- *Security Operations Center/Network Operations Center (SOC/NOC) Monitoring (24/7)*—monitoring conducted 24 hours a day and 7 days a week. The SOC/NOC monitors security and network events, and its success depends on the quality and expertise of the individuals performing Detection, Protection, Response, and Sustainment security best practices.
- *Focused Operations*—surveillance and reconnaissance operations that are focused on a specific target; usually associated with a Honeypot/Honeynet used to obtain an attacker's Tactics, Techniques, and Procedures (TTPs).
- *Digital Forensics*—recovering and investigating material found in digital devices, usually relating to a digital crime or when DLP occurs. The digital forensic process covers the seizure, forensic imaging (acquisition), and analysis of digital media.
- *Escalation Management*—approved policies and procedures to manage an event's response and its associated escalation to higher levels of authority.
- *Computer Incident Response Team (CIRT)*—a team trained to coordinate and handle computer incidents. Detection includes Correct Event Detection, Information Gathering and Historical Review, Event Triage, Escalation, and Optimization and Tuning (to improve detection accuracy). Response encompasses Preliminary Incident Analysis, Incident Containment, Incident Analysis, Incident Eradication and Recovery, Post-Incident Process Improvement, and Lessons Learned.
- *Continuous Monitoring and Assessment*—general term for the activities that observe and assess what is transpiring across the enterprise. It includes all the elements of Detection, Protection, and Sustainment and requires experienced, knowledgeable individuals who fully comprehend and can correctly act upon the data presented by the monitoring tools.
- *Situational Awareness*—broad understanding of all the elements that comprise the network Enterprise and its security in relationship to time and other variables that can alter the current "picture" of the enterprise environment. It is a "layered" relational view that includes the following separate diagrams: Geographic, Physical, Logical, Device, and Personal. These diagrams are then rolled into a single comprehensive view.
- *Security Service-Level Agreement (SLA)/Service-Level Objectives (SLO) Reporting*—reports showing the metrics outlining how well the security monitoring and response are meeting a contractual SLA/SLO.

Prevention: Policies, procedures, training, threat modeling, risk assessment, penetration testing, and all other sustainment activities to establish a secure posture for the enterprise, including:

- *Information Technology (IT) Security Governance*—a governing authority that defines the IT Security policy by law; also a security compliance directive or other regulation.
- *Security Architecture and Design*—security features and functions added to or embedded in a networked enterprise architecture.
- *Threat Modeling*—developing threat scenarios to help determine the levels of security controls to deploy. Models include Attack Tree Models, Qualitative Information Assurance Models, Quantitative Information Assurance Models, Multiple Objective Decision Analysis Models, Multiple Objective Decision Analysis for Information Assurance Model, and the Mission-Oriented Risk and Design Analysis Model.
- *Cyber Threat Intelligence*—gathering and examining cyber threat data from sensors, organizations, and other threat intelligence sources.
- *Security Policies and Compliance*—establishment of security controls approved by senior management to ensure security enforcement. Many Government security policies are set by preestablished directives or regulations that mandate strict compliance.
- *Security Technology Evaluation*—testing and evaluating security technologies for conformity or proving the security product performs as advertised.
- *Continuous Certification and Accreditation (C&A)*—the process by which a secure system is assessed to prove an acceptable level of risk as a prerequisite for obtaining permission to operate. Traditional C&A based on initial Authority to Operate followed by periodic reaccreditation is giving way to continuous monitoring and assessment processes such as the Risk Management Framework (RMF).
- *Risk Management Framework (RMF)*—a six-step process developed by NIST (National Institute of Standards and Technology) that provides a disciplined and structured process that integrates information security and risk management activities into the system development life cycle (SDLC). Defined in NIST Special Publication 800-37, "Guide for Applying the Risk Management Framework to Federal Information Systems."
- *Penetration Testing*—a method of evaluating the strength of a computer systems security posture by attacking the system to discover and exploit security vulnerabilities.
- *Vulnerability Assessment*—the process of identifying, quantifying, and prioritizing vulnerabilities found in a system. It does not include attacking the system (as done in a penetration test); rather it only scans the system for vulnerabilities.
- *Security Awareness Training*—the process of teaching personnel better "cyber consciousness," especially how to protect data and the importance of security controls. Comprehensive Security Awareness Training includes *why* security is important, *who* wants the data, *how* the data can be stolen, and *what* can be done by individuals to thwart data breaches. A prominent example is training to recognize and avoid social engineering attacks such as phishing.

Appendix G: Common Cyberattack Methods

Attack	Category	Description	Mitigation
Aggregation	Software, Database	Combining multiple low sensitivity records to create high sensitivity; can be basis for identity theft	Control database access permissions to limit ability to assemble high sensitivity objects
Applets (ActiveX)	Software	Rogue developer writes malicious code in an applet; client trusts the server and gives the applet control	Block ActiveX applets at the firewall or proxy server; establish a policy limiting user trust of applets; prohibit mobile code
Birthday Attack	Cryptography	Exploit the probability of two messages producing the same digest with the same hash function; digital signatures can be vulnerable (Note 4)	Use hash function long enough to make cracking computationally infeasible (may be ~2× the length to defeat a brute force attack)
Bluejacking and Bluesnarfing	Network Security	Send anonymous, unsolicited messages to Bluetooth-enabled devices; stealing personal data from a Bluetooth phone	Prohibit Bluetooth use outside secure areas (depends on device range)
Brute Force/ Dictionary	Access Control	Attempts to crack passwords, passphrases, or PINs by repeated guessing	Protect security account databases and password files Require strong, frequently changed passwords/PINs
Brute Force Key Search	Cryptography	Try all possible keys to find a secret key	Use longer keys or better Public-Key Encryption algorithms
Buffer/Stack Overflow	Access Control, Web Security	Often used in Denial of Service; based on inducing corruption of buffer/stack data; send machine language instructions as parts of queries to trick web server into running them	Identify and patch vulnerabilities in system, network, and applications

J. M. Borky, T. H. Bradley, *Effective Model-Based Systems Engineering*,
https://doi.org/10.1007/978-3-319-95669-5

Attack	Category	Description	Mitigation
Chosen Text Attack (CTA)	Cryptography	Choose plaintext and obtain ciphertext; includes Chosen Plaintext attack (CPA), Adaptive CPA (ACPA), Chosen Ciphertext Attack (CCA), and Adaptive CCA (ACCA)	Use a cryptosystem which is probably secure under chosen-ciphertext attack, including RSA-OAEP, Cramer-Shoup, and many forms of authenticated symmetric encryption
Ciphertext-Only Attack (COA)	Cryptography	Search for repeating patterns, and do statistical analysis (e.g., frequency analysis) on multiple intercepted messages using the same encryption algorithm	Use advanced, fully-vetted encryption standards
Credential Theft	Access Control	Various means of tricking users, often by social engineering, intercepting sign-ons, etc. to obtain credentials	Apply strong (e.g., two-factor) authentication; train staff to protect credentials
Cross-Site Request Forgery (CSRF)	Software, Web Security	Trick a user into clicking a link that performs a malicious action (e.g., a script embedded in a clickable image)	Include "nonces" on important pages Use a secondary approval dialog for important transactions Train users not to click untrusted content
Cross-Site Scripting (XSS)	Software, Web Security	Inject client-side script into Web pages viewed by other intended victims; can be nonpersistent (trick the victim into clicking a malicious URL containing a script) or persistent (store malicious code on a blog, message board, etc.)	Employ defensive coding practices such as limiting character sets Apply XSS patches from vendors Disable or discourage browser scripting functionality except for trusted sites Use techniques such as whitelists and security zones Use Phishing Attack warning Retype addresses rather than clicking hyperlinks
Denial of Service (DoS)	Web Security	Send malicious query to crash a server; send huge volumes of queries to overwhelm a server; includes Distributed DOS attacks	Identify and patch vulnerabilities, especially in vulnerable services such as NTP and SSDP Identify and block attacker addresses, especially known botnet C2 servers
Dictionary Attack	Access Control	Steal a password or shadow file, and then run password-cracking software that guesses words and combinations of words and numbers	Require strong passwords and frequent changes

Attack	Category	Description	Mitigation
Escalation of Privilege	Software	Penetrate a system and increase the attacker's privilege level	Defeat all possible system penetrations
Fraggle	Network Security	Variant of Smurf; using UDP Echo packets (UDP Port 7)	Configure routers to disable TCP and UDP services used in Fraggle attacks
FTP Server Exploit	Web Security	Access sensitive information on an unprotected Web-facing FTP site	Protect FTP servers: strong access controls, encrypted content, etc.
ICMP Flood	Network Security	Send large numbers of ICMP packets (usually Echo Request) to consume bandwidth or other resources	Drop ICMP packets at router or filter them out at the firewall
Injection Attack	Software	Insert computer instructions via a program's input field to trigger unintended functions; includes SQL Injection to a database and Script Injection or Cross-Frame Scripting (XFS) to a browser/Web page to steal data from other active frames; related to Click-Jacking	Combine defensive coding practices with vulnerability detection and runtime attack detection See Notes 3, 5
Insider Threat/ Insider Misuse	Multiple	Actions by trusted individuals, from theft and sabotage to poor practices such as circumventing controls	Train staff in good security hygiene, including reporting possible bad actors Collect and analyze logs for actions or patterns indicating poor practices or malicious intent
Known Plaintext Attack	Cryptography	Exploit previous plaintext/ ciphertext message pairs to decipher new messages	Use advanced encryption such as AES
Maintenance Hook	Software	Undocumented code in a product that triggers Maintenance or Debug mode and exposes sensitive content	Use only trusted products Set alarms on Maintenance/ Debug modes
Malicious Code	Software	See Note 1	See Note 2
Man-in-the-Middle (MITM) Attack	Access Control, Cryptography	Exploits TCP/IP vulnerabilities; intercept and alter messages by injecting packets, e.g., to insert an attacker's public key	Security perimeter defense, server and network component hardening, robust patch management, and following best security practices
Meet-in-the-Middle Attack	Cryptography	Encrypt known plaintext with each possible key, decrypt the ciphertext with each possible key, and then compare results in the middle; this can exponentially reduce the number of attempts in a brute-force attack	Protect plaintext and use advanced cryptography algorithms and long keys to make the attacker's storage requirement prohibitive

Attack	Category	Description	Mitigation
Multiple	Access Control	Any attempt to defeat controls	Threat modeling, asset valuation, vulnerability analysis, access aggregation (e.g., Single Sign-On)
Packet/Password Sniffing or Eavesdropping	Access Control	Capture network packets, and analyze content which may include user names and passwords, patterns of application use, etc.	Use secure protocols such as https, SFTP, and SSH Use VPNs and packet sniffer detectors
Payment Card Skimming	Physical Tampering	Device is implanted in a POS, ATM, or other device to capture magnetic strip data from cards	Use tamper-resistant terminals, tamper-sensitive controls, and device surveillance Use trusted terminals, protect PINs
Pseudo-Flaw	Software	Send scareware message to trick a user into downloading a phony patch which is usually a Trojan	Educate users and enforce best practices
Remote Maintenance	Software	Vendor support person gains access to target to insert malware, steal data, etc.	Only permit remote maintenance by absolutely trusted vendors Require notification of departed or disaffected vendor employees
Replay Attack	Cryptography	Intercept a session key and use against a later session between the parties	Incorporate and use a time stamp in the session key
RSA/ElGamal Attacks	Cryptography	Faster techniques for than brute force for breaking PKE system	Use longer keys or better PKE algorithms
Session Hijacking	Access Control	Similar to MITM except attacker impersonates recipient	Same as MITM
Script Injection	Web Security	Insert scripting language commands in form fields on web pages	Vulnerability analysis and testing, patch software to eliminate vulnerabilities
Smurf	Network Security	Variant of ICMP flood where ICMP Echo Request packets are sent to a network broadcast address using a spoofed IP address so bounces overwhelm the network	Drop ICMP packets at the router
Social Engineering	Access Control	Phishing, dumpster diving, shoulder surfing, outright burglary, etc. to trick a user into revealing passwords, etc., or to load malicious code; also Pharming, Spear phishing, Whaling, and others	Educate users and enforce best practices

Attack	Category	Description	Mitigation
Spam	Software	Unwanted emails that clog the network and may contain malware	Block spam using a central appliance, a blocking service, spam-blocking software, etc.
Spoofing	Software, Network Security	Change attacker's network identity to trick the target into granting access; often part of a Denial-of-Service attack	Avoid authentication based only on IP addresses Use packet filtering to block packets from outside the network using source addresses inside the network Use hard-to-predict sequence numbering in TCP
State Attack	System	Steal session identifier, take over the session, steal or otherwise exploit sensitive content	Ensure Web applications use session establishment logic that protects against theft of identifiers
SYN Flood	Network Security	Send TCP packets with spoofed source address requesting a connection (SYN bit set) to the target network, and then don't respond to the SYN ACK reply to create a half-open connection	Use router features such as TCP Intercept or SYN Defender and Committed Access Rate (CAR)
Teardrop	Network Security	Variant of a DoS attack involving fragmented packets; modify Length and Fragmentation offset fields of sequential IP packets	Use latest operating systems; patch TCP/IP fragmentation reassembly bugs
Theft	System	Illegal taking of resources, especially computers another storage	Full hard drive encryption; strong physical protection and accounting
Traffic Analysis and Inference	Network Security	Analyze network traffic patterns and other data to infer exploitable events; often part of a larger attack strategy	Vary scheduling of important events such as system backups Apply defensive measures against other elements of the attack strategy
UDP Flood	Network Security	Variant of a DoS attack involving large numbers of UDP packets sent to random ports to consume bandwidth or other resources	Drop unnecessary UDP packets at the router; place firewalls at key network points to block unnecessary UDP packets; if attack uses a required UDP port, other countermeasures are needed
Web App Attack	Software/ Access Control	Various methods used to gain access usually by organized crime to commit theft; includes stolen credentials, injection, backdoors, XSS, etc.; Web Shells are a form of backdoor often installed via Content Management Systems (CMS)	Use strong (e.g., two-factor) authentication; patch known software vulnerabilities, especially CMS code and third party plugins; train staff not to compromise credentials; harden Remote File Inclusion (RFI) and file upload functions

Note 1

- Virus—code that self-replicates, spread by boot sectors, executables, macros, images, controls, etc.; may be designed to turn the victim into a zombie or bot.
- Worm—code that propagates via system weaknesses; assumes control to launch attacks or do damage (e.g., alter data).
- Rootkit—code that hides to avoid detection; may steal or alter data, intercept communications; change system behavior; types include HW, firmware, hypervisor kernel, and library.
- Bootkit—kernel-mode rootkit that attacks computers with full-disk encryption.
- Trojan Horse—often injected via email payload such as an executable or macro or as a pop-up; tricks user into executing the code to perform attacks; may be designed to turn the victim into a zombie or bot; Dridex is a leading Trojan aimed at banking, delivered via malicious document attachment.
- Ransomware—code that encrypts the victims databases; attackers demand payment to release the data; Locky is a leading example delivered via document attachment.
- Hoax—false warning message.
- Logic Bomb—software that does damage when an event occurs, such as a date/time or a program execution.
- Malicious Applet—especially ActiveX controls that do damage.
- Trap Door—logic bomb that performs an undocumented function in a program such as granting root access.
- Hidden Code—malicious code hidden in a program; similar to alteration of authorized code.
- Adware—pop-ups installed with freeware/shareware.
- Backdoor—enables attacker to bypass authentication.
- Spyware—collects information without user/system owner knowledge; may interfere with operations, e.g., redirecting a browser or installing malware.
- RAM (or memory) Scraping—spyware that infects a point-of-sale (POS) terminal to steal payment card information while it's in cleartext during a transaction.

Note 2

Defenses against malware:

- Install and frequently update antivirus (AV) software.
- Use strong authentication: two-factor authentication, strong passwords, etc.
- Track remote logins and detect non-normal attempts.
- Do frequent AV scans; use more than signature-matching AV tools.
- Scan incoming emails and attachments.
- Train users.
- Allow only registered applications to run.
- Detect data leakage, and listen for malware "calling home."
- Separate Point-of-Sale (POS) and other public devices from the corporate LAN; do NOT expose them to the Internet.

Note 3

Defense in depth vs. SQL injection:

- Prohibit or sanitize user-supplied input in queries, e.g., input validation against a whitelist of allowed patterns.
- Parameterized queries.
- Stored procedures.
- Least privilege, e.g., only "select" allowed for general users.
- Honeytokens to detect unauthorized usage.
- Application log auditing.
- Penetration testing.
- Defeat Cross-Frame Scripting/Script Injection by hardening browsers with techniques such as filtering and scraping.

Note 4

Birthday Attack gets its name from the >50% probability that in among 23 people, 2 will have the same birthday. For that to be a specific day, there must be ≥253 people.

Note 5

SQL injection tools include Havij SQL Injection, Safe3SI, BSQL (Blind SQL) Hacker, SQL Ninja, Pangolin, and The Mole.

Note 6

There are common security incident patterns or categories; many attacks combine multiple elements:

- Point-of-Sale (POS) Intrusions
- Payment Card Skimmers
- Crimeware (generalized Malware attacks; specific patterns include cyber-espionage and POS intrusions)
- Web App Attacks
- Denial-of-Service Attacks
- Physical Theft/Loss (NOTE: simple loss is 100× more common that theft but can be just as damaging; institute training and procedures to reduce losses by staff.)
- Insider Misuse
- Miscellaneous Errors
- Cyber-Espionage

Appendix H: IEEE Computer Society Center for Secure Design (CSD) List of Top 10 Security Flaws

1. *Earn or give, but never assume, trust*—validate all data from untrusted clients, assume data have been compromised, and avoid security controls and sensitive data in client code (authorization, access control, etc.).
2. *Use authentication mechanisms that can't be bypassed*—require reauthentication when user changes identity; ensure strong authentication; use time-outs; insist on "Something you are, Something you have, and Something you know"; avoid sharing IP numbers and MAC addresses; and avoid predictable tokens.
3. *Authorize after authentication*—make authorization an explicit check; reuse common infrastructure to conduct these checks; consider both a user's privileges and the context of the request; and ensure authorization is revoked to prevent a user from exercising an out-of-date authorization.
4. *Strictly separate data and control instructions*; *never process control instructions from untrusted sources*—mingling data and control creates vulnerability to injection attacks; use hardware to enforce separation of code and data; use appropriate compiler/linker security flags; expose methods/endpoints that consume structured types; protect against injection-prone APIs that are subject to Cross-Site Scripting (XSS), Structured Query Language (SQL) injection, shell injection, and similar attacks; and avoid eval commands in interpreted languages.
5. *Define an approach that ensures all data are explicitly validated*—enforce comprehensive validation; enable security review of the validation scheme; use a centralized validation mechanism and canonical data forms (avoid strings); check all assumptions about data; and avoid blacklisting; use whitelisting instead.
6. *Use cryptography correctly*—this is hard and should not be home-brewed; it requires support from real experts; use standard algorithms and libraries; centralize and reuse crypto schemes; design for crypto agility; identify and resolve key management issues; and avoid nonrandom "randomness."
7. *Identify sensitive data and define how they should be handled*—know where sensitive data exist, including context that affects sensitivity; classify data in categories; apply data controls to files, memory, databases, etc.; plan for data

J. M. Borky, T. H. Bradley, *Effective Model-Based Systems Engineering*,
https://doi.org/10.1007/978-3-319-95669-5

protection to evolve over time; pay attention to trust boundaries; and remember that data protection is more than confidentiality.

8. *Always consider system users*—provide for deployment, configuration, use, updating, etc.; recognize that security is an emergent property of a system; account for user culture, experience, biases, and other factors; don't assume users care about security; make things secure by default; impose adequate security without impairing system functionality and use; and don't let users make security decisions.
9. *Understand how integrating external components changes your attack surface*—test components for security, including external components and dependencies; isolate components; follow publicly available security information about components; minimize composition, which is dangerous; recognize that a system can inherit security risks from components and that open source is not secure; only trust components after applying and reviewing controls; and avoid or disable extra (unneeded) functionality.
10. *Be flexible when considering future changes to objects and actors*—design systems for change and evolvability, including security updates; use code signing and protection properly; allow isolation and toggling; have a plan for "secret compromise" recovery; ensure security is and remains robust, not fragile or brittle; and ensure system administration maintains security.

Appendix I: Software Testing Methods and Tools

Functional Test

Functional test is concerned with assessing the correctness of software operations and identifying coding errors that could introduce vulnerabilities. This testing should be an integral element of any Agile or other software development methodology, and software teams should include or have easy access to security testing expertise. The following are some typical test categories.

- *Unit testing*:
  - Done on small units of code to verify correctness/completeness vs. allocated requirements, perform early error detection, respond to requirements changes, and validate/improve embedded documentation.
  - Specific test for the unit under test, applied to production code; failures lead to immediate code revision.
  - Integral element of agile SW development during each sprint or other development increment.
  - Can be the basis for Test-Driven Development (TDD): start with a test that fails, create or refine code until it passes, and refactor the code to eliminate redundancies.
- *Smoke or Sanity testing*—quick functional check to establish basic correctness.
- *Integration testing*—performed as code units, components, etc. are being built up toward a full system, and may be done top-down or bottom-up.
- *Interface and Usability testing*—focus on human-machine interfaces to assess the ability of users to work with the system.
- *System testing*—all-up testing of the complete system.
- *Regression testing*—retest of previously passed code following a design change to verify correctness or find any induced problems.
- *Pre-User Acceptance testing*—alpha and beta testing to get an initial assessment of software behavior in the real world.

J. M. Borky, T. H. Bradley, *Effective Model-Based Systems Engineering*,
https://doi.org/10.1007/978-3-319-95669-5

- *White/Black Box testing*—testing that depends on whether internal design details are known to the tester.
- *Globalization (Internationalization)/Localization testing*—the former is testing to ensure software works correctly in any location or culture globally; the latter is testing for capability and correctness in a specific location and may include translation.

Nonfunctional Test

Nonfunctional testing assesses overall characteristics and suitability of software, often expressed in terms of nonfunctional requirements. A large part of cybersecurity testing falls in this category. The following are some typical test categories.

- *Load/Performance/Stress/Volume testing*—verify correct behavior under maximum task loading, data rates, or other stressing conditions.
- *Ergonomics* testing—assess human suitability; may be done in conjunction with Interface/Usability testing.
- *Compatibility and Migration testing*—verify a system, subsystem, component, etc. and work correctly in a given environment and with other software that it must interact with.
- *Data Conversion testing*—verify that static or dynamic data conversion process operate correctly.
- *Security/Penetration testing*—focused testing based on known or potential vulnerabilities and threats to assess effectiveness of countermeasures/safeguards; may include testing applications security, network security, system security, etc., in a Defense-in-Depth environment.
- *Operational Readiness testing*—pre-deployment or delivery testing to verify correct behavior and capability in the intended operational environment.
- *Installation testing*—testing performed as part of an installation process to ensure software loads and runs correctly.
- *Fuzz testing (Fuzzing)*—inserting invalid, unexpected, or random data in the inputs to a system:
  - Varieties include mutation-based and generation-based; can be used in white, gray, or black box tests.
  - Look for crashes, memory leaks, faulty exception handling, etc.; can be used in both white and black box tests.
  - Especially useful against software that implements safeguards at a trust boundary, e.g., to target file formats and network protocols.
- *Static Application testing*—done during code development to show root causes of security vulnerabilities in source code and ensure code is trustworthy (or show what to fix); tools include WhiteHat, Fortify, Veracode, Rational AppScan, Coverity, and FindBugs.

- *Autogenerated Code testing*—the reality is that source code produced automatically from a model can be hard to read and analyze; to deal with this:
  - Before Testing—confirm model correctness (validation in the modeling tool), verify design patterns, confirm states are correct and will not cause loop back, verify architecture (again, validation), implement design in one or more models, and confirm modeling will NOT allow security mechanisms to be bypassed.
  - During autocoding—run static code analysis and respond to identified problems.
  - Include fuzz testing of interfaces.
- *Dynamic Application testing*—essentially penetration test of object code:
  - The goal is to determine how well code resists real-world attacks, if/how code can be compromised, and how vulnerabilities can be mitigated.
  - Should be run in a lab/test environment, not the production environment.
  - Test from the viewpoint of an attacker; may use active exploitation of code security vulnerabilities.
  - Include external dependences to ensure valid results.
  - Use tools that identify type and location of vulnerabilities and suggest fixes.
  - Rescan after vulnerabilities are fixed.
  - Tools include:

    Open Source—Burp Suite, Zed Attack Proxy, Netsparker, Nessus, Valgrind
    Licensed—WhiteHat, Rational AppScan, WebInspect, Nessus, Netsparker, Burp Suite
- *Component Testing*:
  - A component is a software entity or product that is packaged as an autonomous element to be deployed and used as part of a system; may be a business/mission application or part of a computing platform's software infrastructure; components may be:

    Open Source—often "freeware" that benefits from the efforts of an entire interested community to develop, maintain, and enhance capabilities.
    Commercial—purchased/licensed products from vendors.
    Reused—existing software from the same or another organization, intended to provide the same capabilities as those for which it was originally developed.
  - Component security testing is essential because a high percentage of products contain severe or critical security flaws and applications are the primary attack vector.
  - Component testing involves determining the open source and other types of components used in a system; performing analysis to identify security issues, license issues, policy issues, etc.; and performing penetration, and perhaps other types, of testing to define vulnerabilities and assess risks.

Appendix J: Open System Interconnect (OSI) Layers and Protocols

Layers	Functions	Standards	Details
Layer 1 Physical	Send/receive bits across network Repeater—amplify signals Hub/Concentrator (active/ passive)—connect LAN devices	DTE-DCE (User-Network) Interfaces: EIA/TIA-232, EIA/TIA-449, V.24, V.35, X.21bis, HSSI	Topologies: star, mesh, ring, bus Cabling: coax, (thicknet RG8, RG11, thinnet RG58), UTP/STP (Cat 5e, 6, 6a, 7; 10Base-T, 100Base-TX, 1000 Base-T 10GbE), fiber (100Base-F) DTEs include NICs, modems, and CSUs/DSUs
Layer 2 Data Link	Deliver msgs to proper device across physical net Switch—connect devices, send packets to correct device	LAN Protocols: ARCnet, Ethernet (CSMA/CD), Token Ring, FDDI, ARP (MAC addr → IP addr), RARP (IP addr → MAC addr) WLAN Standards: 802.11a, b, g, n WAN Point-to-Point Protocols: L2F, L2TP, PPP, PPTP, SLIP Circuit-Switched Protocols: xDSL (ADSL, SDSL, HDSL, VDSL), DOCSIS, ISDN (BRI, PRI) Packet-Switched Protocols: ATM, Frame Relay, MPLS, SONET, SMDS X.25	Sublayers: • LLC—interface SSAP to DSAP; control, sequence, and acknowledge frames; timing and flow control • MAC—error control; identify MAC addresses; control media access (contention, token passing or polling)

J. M. Borky, T. H. Bradley, *Effective Model-Based Systems Engineering*,
https://doi.org/10.1007/978-3-319-95669-5

Layers	Functions	Standards	Details
Layer 3 Network	Routing and related functions for data transport between systems on the same or interconnected networks Router—link dissimilar networks; forward packets only to destination network Gateway—link dissimilar programs and protocols	Routing Protocols: RIP (distance vector), OSPF (IGP within an AS), IS-IS (IGP), BGP (path vector, EGP between ASs), IP, IPX, ICMP, SKIP, AMQP	Routing protocols can be static, dynamic, distance vector, hop count, link state, or path vector IP does connectionless, best-effort datagram delivery, and datagram fragmentation/reassembly
Layer 4 Transport	Transparent, reliable data transport and end-to-end transmission control • Flow control • Multiplexing • Virtual circuit management • Error checking/recovery	Transport Protocols: TCP, UDP, SPX, SSL/TLS	TCP is full-duplex, connection-oriented, reliable, and slow; connection established by SYN-SYN ACK-SYN ACK ACK handshake UDP is connectionless, best-effort, and fast; used by DNS, SNMP, and streaming SSL/TLS provides session-based encryption/authentication
Layer 5 Session	Establish, coordinate, and terminate comm sessions (service requests/responses)	Session Protocols: NetBIOS, NFS, RPC, SSH/SSH-2, SIP	Session consists of: • Connection establishment (simplex, half-duplex, full-duplex) • Data transfer • Connection release
Layer 6 Presentation	Coding/conversion of data for Layer 7	Presentation Protocols: ASCII, EBCDIC, GIF, JPEG, MPEG	Tasks include: • Data representation/formatting • Character conversion • Data compression • Data encryption
Layer 7 Application	Supports network access for applications; provides user interface	Application Layer Applications: FTP, HTTP, HTTPS, IMAP, POP3, PEM, SET, S-HTTP, S/MIME, S-RPC, SMTP, SNMP, Telnet, TFTP	Tasks include: • Identify/establish availability of communication partners • Determine resource availability • Synchronize communications

Networking Acronyms

ACK	Acknowledge
ADSL/ASL2	Asymmetric Digital Subscriber Line
AMQP	Advanced Message Queuing Protocol
ARP	Address Resolution Protocol
AS	Autonomous System
ASCII	American Standard Code for Info Exchange
ATM	Asynchronous Transfer Mode
BGP	Boundary Gateway Protocol
BRI	Basic Rate Interface (ISDN)
CSMA/CA	Carrier Sense Multiple Access/Collision Avoid
CSMA/CD	Carrier Sense Multiple Access/Collision Detect
CSU	Channel Service Unit
DCE	Data Communications Equipment
DLCI	Data Link Connection Identifier (Frame Relay)
DOCSIS	Data Over Cable Services Interface Spec
DSAP	Destination Service Access Point
DTE	Data Terminal Equipment
DVC	Data Service Unit
EBCDIC	Extended Binary-Coded Decimal Interchange Code
EGP	Exterior Gateway Protocol
FDDI	Fiber Distributed Data Interface
FTP	File Transfer Protocol
GIF	Graphics Interchange Format
HDLC	High-level Data Link Control
HDSL	High-rate Digital Subscriber Line
HSSI	High-Speed Serial Interface
HTTP	HyperText Transfer Protocol (TCP Port 80)
HTTPS	HTTP Secure (HTTP + SSL/TLS)
ICMP	Internet Control Message Protocol
IGP	Interior Gateway Protocol
IMAP	Internet Message Access Protocol (TCP Port 143) (can use SSL or TLS)
IP	Internet Protocol
IPX	Internet Packet Exchange (with SPX)
IS-IS	Intermediate System to Intermediate System
JPEG	Joint Photographic Experts Group
L2F	Layer 2 Forwarding Protocol
L2TP	Layer 2 Tunneling Protocol
LAPB	Link Access Procedure, Balanced (X.25)
LLC	Logical Link Control
MAC	Media Access Control
MPEG	Motion Picture Experts Group
MPLS	Multi-Protocol Label Switching (VPNs)

NAT	Network Address Translation
NBT	NetBios over TCP/IP
NetBIOS	Network Basic Input/Output System
NFS	Network File System (UNIX/TCP/IP)
NIC	Network Interface Card
OSI	Open System Interconnection
OSPF	Open Shortest Path First
PEM	Privacy Enhanced Mail
PLP	Packet Level Protocol
POP3	Post Office Protocol Ver 3 (can use TLS or SSL)
PPP	Point-to-Point Protocol
PPTP	Point-to-Point Tunneling Protocol
PRI	Primary Rate Interface (ISDN)
PVC	Permanent Virtual Circuit
RARP	Reverse Address Resolution Protocol
RIP	Routing Information Protocol
RPC	Remote Procedure Call
SDH	Synchronous Digital Hierarchy
SDLC	Synchronous Data Link Control
SDSL	Single-line Digital Subscriber Line
SET	Secure Electronic Transaction
S-HTTP	Secure HTTP (different from HTTPS)
SIP	Session Initiation Protocol
SKIP	Simple Key Management for IP
SLIP	Serial Line IP
SMDS	Switched Multimegabit Data Service
S/MIME	Secure Multipurpose Internet Mail Extensions
SMTP	Simple Mail Transfer Prot (TCP/UDP Port 25)
SNMP	Simple Network Mgmt Prot (TCP/UDP Port 161)
SONET	Synchronous Optical Network
SPX	Sequenced Packet Exchange (with IPX)
S-RPC	Secure RPC
SSAP	Source Service Access Point
SSH/SSH-2	Secure Shell
SSL	Secure Sockets Layer
SVC	Switched Virtual Circuit
SYN	Synchronize
TCP	Transmission Control Protocol (usually with IP)
Telnet	TErminaL NETwork (TCP/UDP Port 23)
TFTP	Trivial File Transfer Protocol (UDP Port 69)
TLS	Transport Layer Security
UDP	User Datagram Protocol
VDSL	Very-high-rate Digital Subscriber Line

Appendix K: Service-Oriented Architecture (SOA) Policy Requirements for Department of Defense (DoD) Programs

The growing importance of SOA as the enabler of globally networked mission capabilities has given rise to a number of architecture guidance and mandates in Defense programs. The following paragraphs address a few of these that the architects of Defense systems and enterprises will often encounter. Other Government agencies have or can be expected to have similar guidance and required deliverables.

- Open, layered SOA is mandated by Net-Centric Data and Services Strategies, Global Information Grid (GIG) architecture, Net-Centric Enterprise Services for Interoperability (NESI), and other policies and reference architectures. Most of these documents are updated or superseded on a regular basis, but the general SOA requirement is unchanged.
- Programs must develop and maintain Integrated Architecture products in accordance with the DoD Architecture Framework and document them in an Information Support Plan (ISP). Under DoDAF Version 1, products were not well suited to representing services and SOA, but DoDAF 2 shifts the overall emphasis of architecture definition to data and expands the original Operational, System, and Technical Viewpoints to include, among others, a Services Viewpoint.
- The GIG Enterprise Services (GES) initiative of the Defense Information Services Agency (DISA) implements enterprise services to which nodes in the enterprise must conform.
- DoD directives require that programs define data with structural and discovery metadata; the latter must implement the DoD Discovery Metadata Specification (DDMS); services must be published for discovery and sharing, with the default approach being a UDDI registry.
- Under a policy that is not always enforced, data definitions are supposed to employ models developed by Communities of Interest (COIs) that are chartered by the Assistant Secretary of Defense, Networks, and Information Integration (ASD/NII) to progress toward a common vocabulary of data and services definitions and formats.

J. M. Borky, T. H. Bradley, *Effective Model-Based Systems Engineering*,
https://doi.org/10.1007/978-3-319-95669-5

- The preferred service model is Web Services, implemented on a converged, secure IP network-/Internet-carrying data, digitized voice, imagery, video, and other traffic.

Bibliography[3]

1. ISO/IEC/IEEE (2011) Standard 42010:2011, Systems and software engineering—architecture description. http://www.iso-architecture.org/ieee-1471/ANSI-IEEE 1471-2000
2. Bar-Yam Y (1997) Dynamics of complex systems. Addison-Wesley, Reading, MA
3. Berganthal J (2011) Final report model based engineering (MBE). NDIA Systems
4. Bjorkman EA, Sarkani S, Mazzuchi TA (2013) Using model-based systems engineering as a framework for improving test and evaluation activities. Syst Eng 16(3):346–362
5. Booch G (1997) Object-oriented analysis and design with applications, 2nd edn. Addison-Wesley, Reading, MA
6. Booch G, Rumbaugh J, Jacobson I (1999) The unified modeling language user guide. Addison-Wesley, Boston, MA
7. Bowen P et al (2006) Information security handbook: a guide for managers, NIST SP 800-100, June 2006
8. Brown B (2011) Model-based systems engineering: revolution or evolution? IBM Software, Thought Leadership White Paper, Dec 2011
9. Broy M, Feilkas M, Herrmannsdoerfer M, Merenda S, Ratiu D (2010) Seamless model-based development: from isolated tools to integrated model engineering environments. Proc IEEE 98(4):526–545
10. Buede DM, Miller WD (2016) The engineering design of systems: models and methods, 3rd edn. Wiley, New York
11. Chew E et al (2008) Performance measurement guide for information security, NIST SP 800-55, July 2008
12. Chonoles M, Schardt J (2003) UML2 for dummies. Wiley, New York
13. Clements P, Bachmann F, Bass L, Garlan D, Ivers J, Little R, Nord R, Stafford J (2002) Documenting software architectures: views and beyond. Addison-Wesley, Boston, MA
14. Clements P, Kazman R, Klein M (2001) Evaluating software architectures: methods and case studies. Addison-Wesley, Boston, MA
15. Office of Management and Budget (2013) Federal Enterprise Architecture Framework Ver 2. https://obamawhitehouse.archives.gov/sites/default/files/omb/assets/egov_docs/fea_v2.pdf
16. Conrad E, Misenar S, Feldman J (2016) CISSP study guide, 3rd edn. Syngress, New York
17. Constant JN (1972) Introduction to defense radar systems engineering. Spartan Books, New York
18. Date CJ (1981) An introduction to database systems, 3rd edn. Addison-Wesley, Reading, MA
19. Dickerson C, Mavris D (2010) Architecture and principles of systems engineering. CRC Press, Boca Raton, FL

[3] [Primary sources highlighted as **.]

J. M. Borky, T. H. Bradley, *Effective Model-Based Systems Engineering*,
https://doi.org/10.1007/978-3-319-95669-5

20. DoDAF Version 2, Volumes 1–3, available from the Office of the Assistant Secretary of Defense for Networks and Information Integration
21. Dori D (2016) Model-based systems engineering with OPM and SysML. Springer, New York
22. Douglass BP (2004) Real-time UML, Advances in the UML for real-time systems, 3rd edn. Addison-Wesley, New York
23. Douglass BP (2016) Agile systems engineering. Elsevier, New York
24. Erl T (2005) Service-oriented architecture: concepts, technologies, and design. Prentice Hall, New York
25. Estefan JA (2007) Survey of model-based systems engineering (MBSE) methodologies. INCOSE, Seattle, WA
26. Fowler M (2004) UML distilled, A brief guide to the standard object modeling language, 3rd edn. Addison-Wesley, Boston, MA
27. **Friedenthal S, Moore A, Steiner F (2015) A practical guide to SysML: the systems modeling language, 3rd edn. Morgan Kaufmann, Burlington, MA
28. **Friedenthal S, Moore A, Steiner F (2009) OMG Systems Modeling Language (OMG SysML™) Tutorial, Sept 2009. http://www.omgsysml.org/INCOSE-OMGSysML-Tutorial-Final-090901.pdf
29. Gamma E et al (1995) Design patterns: elements of reusable object-oriented software. Addison-Wesley, Boston, MA
30. Guide to the System Engineering Body of Knowledge (SEBoK). http://www.sebokwiki.org/1.0/index.php/Main_Page
31. Gurney K (1997) An introduction to neural networks. Routledge, London
32. Holt J, Perry S (2008) SysML for systems engineering. Institute for Engineering and Technology, London
33. INCOSE (2007) Model-based systems engineering (MBSE) initiative, 29 June 2007. http://www.incose.org/
34. International Council on System Engineering Handbook (free with INCOSE membership). http://www.incose.org/ProductsPubs/products/sehandbook.aspx
35. Jacobson I et al (1992) Object-oriented software engineering: a use case driven approach. Addison-Wesley, Reading, MA
36. Jamshidi M (ed) (2009) System of systems engineering: innovations for the 21st century. Wiley, Hoboken, NJ
37. Jenney J, Gangl M, Kwolek R, Melton D, Ridenour N, Coe M (2010) Modern Methods of Systems Engineering: With an Introduction to Pattern and Model Based Methods, privately published
38. Johnson SB (2006) The secret of Apollo: systems management in American and European space programs. Johns Hopkins University Press, Baltimore, MD
39. Lattanze AJ (2009) Architecting software intensive systems: a practitioner's guide. CRC Press, New York
40. Long D, Scott Z (2011) A primer for model-based system engineering, 2nd edn. Vitech Corp, Blacksburg, VA
41. Maier MW, Rechtin E (2009) The art of systems architecting. CRC Press, Boca Raton, FL
42. Maier MW (2006) System and software architecture reconciliation. Syst Eng 9(2):146–159
43. Marks E, Bell T (2006) Service-oriented architecture: a planning and implementation guide for business and technology. Wiley, Hoboken, NJ
44. Mellor S, Balcer M (2002) Executable UML: a foundation for model-driven architecture. Addison-Wesley, New York
45. Muller G (2005) System architecting. http://www.gaudisite.nl
46. Muller P-A (1997) Instant UML. Wrox Press, Paris
47. Murch R, Johnson T (1998) Intelligent software agents. Prentice Hall, Upper Saddle River, NJ
48. NASA Systems Engineering Handbook, NASA/SP-2007-6105, 2007
49. Object Management Group. Standards and specifications. http://sysml.org/docs/specs/OMGSysML-v1.4-15-06-03.pdf

50. Object Management Group. Model driven architecture. www.omg.org/mda/mda_files/Model-Driven_Architecture.pdf
51. Object Management Group. The UML profile for MARTE: modeling and analysis of real-time and embedded systems. http://www.omgmarte.org
52. Open Group. TOGAF version 9 enterprise edition. www.opengroup.org/togaf/
53. Raistrick C et al (2004) Model driven architecture with executable UML. Cambridge University Press, Cambridge
54. Rhapsody Users Guide. https://www-947.ibm.com/support/entry/portal/documentation_expanded_list/rational/rational_rhapsody_family?productContext=-1713148450
55. Rozanski N, Woods E (2005) Software systems architecture: working with stakeholders using viewpoints and perspectives. Addison-Wesley, New York
56. Senge P (1990) The fifth discipline. Doubleday, New York
57. Shalloway A, Trott J (2001) Design patterns explained: a new perspective on object-oriented design. Addison-Wesley, Boston, MA
58. Stimson GW (1998) Introduction to airborne radar, 2nd edn. Scitech, Mendham, NJ
59. Systems Engineering Tutorial for Rational Rhapsody, IBM, 2009. ftp://public.dhe.ibm.com/software/rationalsdp/documentation/product_doc/Rhapsody/version_7-5/tutorial_Systems_Eng.pdf
60. Thompson RE, Colombi JM, Black J, Ayres BJ (2015) Disaggregated space system concept optimization: model-based conceptual design methods. Syst Eng 18(6):549–567
61. Utting M, Legeard B (2007) Practical model-based testing: a tools approach. Morgan Kaufmann, San Francisco, CA
62. **Weilkiens T (2013) SysML reference card. http://model-based-systems-engineering.com/wp-content/uploads/2012/03/sysmod-sysml-1.3-reference-card-weilkiens.pdf

Index

J. M. Borky, T. H. Bradley, *Effective Model-Based Systems Engineering*,
https://doi.org/10.1007/978-3-319-95669-5

C

H

I

O

P

S